Dick Aufmann

Joanne Lockwood

We have taught math for many years. During that time, we have had students ask us a number of questions about mathematics and this course. Here you find some of the questions we have been asked most often, starting with the big one.

Why do I have to take this cour___ _____ ____ _math is everywhere."_ That is probably a slight exaggeration but math does find its way into many disciplines. There are obvious places like engineering, science, and medicine. There are other disciplines such as business, social science, and political science where math may be less obvious but still essential. If you are going to be an artist, writer, or musician, the direct connection to math may be even less obvious. Even so, as art historians who have studied the Mona Lisa have shown, there is a connection to math. But, suppose you find these reasons not all that compelling. **There is still a reason to learn basic math skills: You will be a better consumer and able to make better financial choices for you and your family.** For instance, is it better to buy a car or lease a car? Math can provide an answer.

I find math difficult. Why is that?   It is true that some people, even very smart people, find math difficult. Some of this can be traced to previous math experiences. If your basic skills are lacking, it is more difficult to understand the math in a new math course. Some of the difficulty can be attributed to the ideas and concepts in math. They can be quite challenging to learn. Nonetheless, most of us can learn and understand the ideas in the math courses that are required for graduation. **If you want math to be less difficult, practice. When you have finished practicing, practice some more.** Ask an athlete, actor, singer, dancer, artist, doctor, skateboarder, or (name a profession) what it takes to become successful and the one common characteristic they all share is that they practiced—a lot.

Why is math important?   As we mentioned earlier, math is found in many fields of study. There are, however, other reasons to take a math course. Primary among these reasons is to become a better problem solver. Math can help you learn critical thinking skills. It can help you develop a logical plan to solve a problem. Math can help you see relationships between ideas and to identify patterns. **When employers are asked what they look for in a new employee, being a problem solver is one of the highest ranked criteria.**

What do I need to do to pass this course?   The most important thing you must do is to know and understand the requirements outlined by your instructor. These requirements are usually given to you in a syllabus. Once you know what is required, you can chart a course of action. Set time aside to study and do homework. If possible, choose your classes so that you have a free hour after your math class. Use this time to review your lecture notes, rework examples given by the instructor, and to begin your homework. All of us eventually need help, so know where you can get assistance with this class. This means knowing your instructor's office hours, know the hours of the math help center, and how to access available online resources. And finally, do not get behind. **Try to do some math EVERY day, even if it is for only 20 minutes.**

Basic College Mathematics
An Applied Approach

NINTH EDITION

Instructor's Annotated Edition

Richard N. Aufmann
Palomar College

Joanne S. Lockwood
Nashua Community College

BROOKS/COLE
CENGAGE Learning™

Australia • Brazil • Japan • Korea • Mexico • Singapore • Spain • United Kingdom • United States

BROOKS/COLE
CENGAGE Learning™

Basic College Mathematics: An Applied Approach,
   Ninth Edition
Richard N. Aufmann and Joanne S. Lockwood

Acquisitions Editor: Marc Bove

Developmental Editor: Erin Brown

Assistant Editor: Shaun Williams

Editorial Assistant: Kyle O'Loughlin

Media Editor: Heleny Wong

Marketing Manager: Gordon Lee

Marketing Assistant: Erica O'Connell

Marketing Communications Manager: Katy Malatesta

Content Project Manager: Cheryll Linthicum

Creative Director: Rob Hugel

Art Director: Vernon Boes

Print Buyer: Becky Cross

Rights Acquisitions Account Manager, Text: Roberta Broyer

Rights Acquisitions Account Manager, Image: Don Schlotman

Production Service: Graphic World Inc.

Text Designer: The Davis Group

Photo Researcher: Jennifer Lim

Illustrator: Graphic World Inc.

Cover Designer: Irene Morris

Cover Image: Panoramic Images/Panoramic Images Collection/Getty Images

Compositor: Graphic World Inc.

For product information and technology assistance, contact us at
Cengage Learning Customer & Sales Support, 1-800-354-9706
For permission to use material from this text or product, submit all requests online at **www.cengage.com/permissions**
Further permissions questions can be e-mailed to
permissionrequest@cengage.com

Library of Congress Control Number: 2009932924

ISBN-13: 978-1-337-70923-1

ISBN-10: 1-337-70923-9

Brooks/Cole
20 Davis Drive
Belmont, CA 94002-3098
USA

Cengage Learning is a leading provider of customized learning solutions with office locations around the globe, including Singapore, the United Kingdom, Australia, Mexico, Brazil, and Japan. Locate your local office at **www.cengage.com/global.**

Cengage Learning products are represented in Canada by Nelson Education, Ltd.

To learn more about Brooks/Cole, visit **www.cengage.com/brookscole**
Purchase any of our products at your local college store or at our preferred online store **www.ichapters.com**

Printed in the United States of America

Contents

CHAPTER 4

Ratio and Proportion

173

CHAPTER 5

Percents

201

CHAPTER 6   Applications for Business and Consumers   233

CHAPTER 11 Introduction to Algebra 459

CHAPTER 12 Geometry 517

Preface

The goal in any textbook revision is to improve upon the previous edition, taking advantage of new information and new technologies, where applicable, in order to make the book more current and appealing to students and instructors. While change goes hand-in-hand with revision, a revision must be handled carefully, without compromise to valued features and pedagogy. In the ninth edition of *Basic College Mathematics: An Applied Approach,* we endeavored to meet these goals.

As in previous editions, the focus remains on the **Aufmann Interactive Method (AIM).** Students are encouraged to be active participants in the classroom and in their own studies as they work through the How To examples and the paired Examples and You Try It problems. The role of "active participant" is crucial to success. Providing students with worked examples, and then affording them the opportunity to immediately work similar problems, helps them build their confidence and eventually master the concepts.

To this point, simplicity plays a key factor in the organization of this edition, as in all other editions. All lessons, exercise sets, tests, and supplements are organized around a carefully constructed hierarchy of objectives. This "objective-based" approach not only serves the needs of students, in terms of helping them to clearly organize their thoughts around the content, but instructors as well, as they work to design syllabi, lesson plans, and other administrative documents.

In order to enhance the AIM and the organization of the text around objectives, we have introduced a new design. We believe students and instructors will find the page even easier to follow. Along with this change, we have introduced several new features and modifications that we believe will increase student interest and renew the appeal of presenting the content to students in the classroom, be it live or virtual.

Changes to the Ninth Edition

With the ninth edition, previous users will recognize many of the features that they have come to trust. Yet, they will notice some new additions and changes:

- Enhanced WebAssign® now accompanies the text
- Revised exercise sets with new applications
- New **In the News** applications
- New **Think About It** exercises
- Revised Chapter Review Exercises and Chapter Tests
- End-of-chapter materials now include Concept Reviews
- Revised Chapter Openers, now with Prep Tests

Take AIM and Succeed!

Basic College Mathematics: An Applied Approach is organized around a carefully constructed hierarchy of **OBJECTIVES**. This "objective-based" approach provides an integrated learning environment that allows students and professors to find resources such as assessment (both within the text and online), videos, tutorials, and additional exercises.

Each Chapter Opener outlines the **OBJECTIVES** that appear in each section. The list of objectives serves as a resource to guide you in your study and review of the topics.

ARE YOU READY? outlines what you need to know to be successful in the coming chapter.

Complete each **PREP TEST** to determine which topics you may need to study more carefully, versus those you may only need to skim over to review.

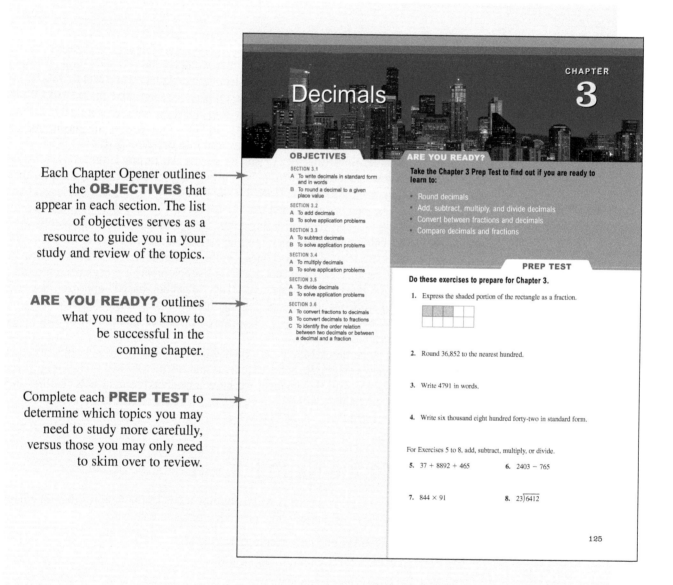

CHAPTER

Decimals

3

OBJECTIVES

SECTION 3.1
A To write decimals in standard form and in words
B To round a decimal to a given place value

SECTION 3.2
A To add decimals
B To solve application problems

SECTION 3.3
A To subtract decimals
B To solve application problems

SECTION 3.4
A To multiply decimals
B To solve application problems

SECTION 3.5
A To divide decimals
B To solve application problems

SECTION 3.6
A To convert fractions to decimals
B To convert decimals to fractions
C To identify the order relation between two decimals or between a decimal and a fraction

ARE YOU READY?

Take the Chapter 3 Prep Test to find out if you are ready to learn to:

- Round decimals
- Add, subtract, multiply, and divide decimals
- Convert between fractions and decimals
- Compare decimals and fractions

PREP TEST

Do these exercises to prepare for Chapter 3.

1. Express the shaded portion of the rectangle as a fraction.

2. Round 36,852 to the nearest hundred.

3. Write 4791 in words.

4. Write six thousand eight hundred forty-two in standard form.

For Exercises 5 to 8, add, subtract, multiply, or divide.

5. $37 + 8892 + 465$

6. $2403 - 765$

7. 844×91

8. $23\overline{)6412}$

125

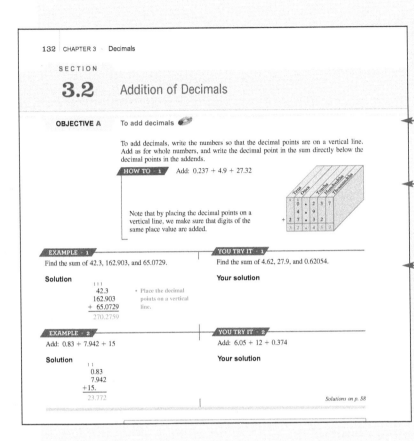

SECTION

3.2 Addition of Decimals

OBJECTIVE A To add decimals

To add decimals, write the numbers so that the decimal points are on a vertical line. Add as for whole numbers, and write the decimal point in the sum directly below the decimal points in the addends.

HOW TO · 1 Add: $0.237 + 4.9 + 27.32$

Note that by placing the decimal points on a vertical line, we make sure that digits of the same place value are added.

EXAMPLE · 1

Find the sum of 42.3, 162.903, and 65.0729.

Solution

$$\begin{array}{r} 1\ 1\ 1 \\ 42.3 \\ 162.903 \\ +\ 65.0729 \\ \hline 270.2759 \end{array}$$

• Place the decimal points on a vertical line.

YOU TRY IT · 1

Find the sum of 4.62, 27.9, and 0.62054.

Your solution

EXAMPLE · 2

Add: $0.83 + 7.942 + 15$

Solution

$$\begin{array}{r} 1\ 1 \\ 0.83 \\ 7.942 \\ +15. \\ \hline 23.772 \end{array}$$

YOU TRY IT · 2

Add: $6.05 + 12 + 0.374$

Your solution

Solutions on p. S8

In each section, **OBJECTIVE STATEMENTS** introduce each new topic of discussion.

In each section, the **HOW TO'S** provide detailed explanations of problems related to the corresponding objectives.

The **EXAMPLE/YOU TRY IT** matched pairs are designed to actively involve you in learning the techniques presented. The You Try Its are based on the Examples. They appear side-by-side so you can easily refer to the steps in the Examples as you work through the You Try Its.

Complete, **WORKED-OUT SOLUTIONS** to the You Try It problems are found in an appendix at the back of the text. Compare your solutions to the solutions in the appendix to obtain immediate feedback and reinforcement of the concept(s) you are studying.

SOLUTIONS TO CHAPTER 3 "YOU TRY IT"

SECTION 3.1

You Try It 1 The digit 4 is in the thousandths place.

You Try It 2 $\dfrac{501}{1000} = 0.501$
(five hundred one thousandths)

You Try It 3 $0.67 = \dfrac{67}{100}$ (sixty-seven hundredths)

You Try It 4 Fifty-five and six thousand eighty-three ten-thousandths

You Try It 5 806.00491 • 1 is in the hundred-thousandths place.

You Try It 6 ┌─ Given place value
3.675849
└─ 4 < 5
3.675849 rounded to the nearest ten-thousandth is 3.6758.

You Try It 7 ┌─ Given place value
48.907
└─ 0 < 5
48.907 rounded to the nearest tenth is 48.9.

65 to 74, and 75 and over.

Solution
$$\begin{array}{r} 4.48 \\ 4.31 \\ 5.41 \\ +3.80 \\ \hline 18.00 \end{array}$$

18 million Americans ages 45 and older are hearing-impaired.

You Try It 4

Strategy To find the total income, add the four commissions (985.80, 791.46, 829.75, and 635.42) to the salary (875).

Solution $875 + 985.80 + 791.46 + 829.75 + 635.42 = 4117.43$

Anita's total income was $4117.43.

SECTION 3.3

You Try It 1
$$\begin{array}{r} 11\ 9 \\ 6\ \cancel{8}\ \cancel{8}13 \\ 72.039 \\ -\ 8.47 \\ \hline 63.569 \end{array}$$

Check:
$$\begin{array}{r} 1\ 11 \\ 8.47 \\ +\ 63.569 \\ \hline 72.039 \end{array}$$

Basic College Mathematics: An Applied Approach contains **A WIDE VARIETY OF EXERCISES** that promote skill building, skill maintenance, concept development, critical thinking, and problem solving.

THINK ABOUT IT exercises promote conceptual understanding. Completing these exercises will deepen your understanding of the concepts being addressed.

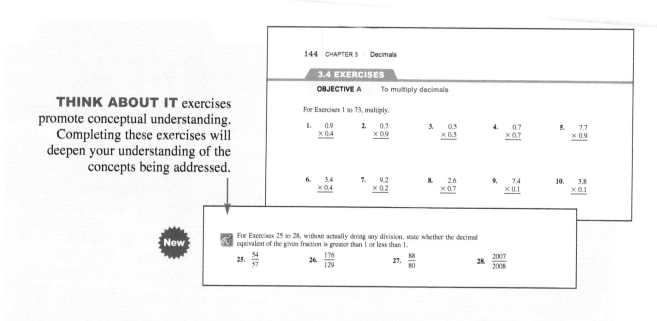

144   CHAPTER 3   ·   Decimals

3.4 EXERCISES

OBJECTIVE A   To multiply decimals

For Exercises 1 to 73, multiply.

| 1. | 0.9 × 0.4 | 2. | 0.7 × 0.9 | 3. | 0.5 × 0.5 | 4. | 0.7 × 0.7 | 5. | 7.7 × 0.9 |

| 6. | 3.4 × 0.4 | 7. | 9.2 × 0.2 | 8. | 2.6 × 0.7 | 9. | 7.4 × 0.1 | 10. | 3.8 × 0.1 |

New

For Exercises 25 to 28, without actually doing any division, state whether the decimal equivalent of the given fraction is greater than 1 or less than 1.

25. $\frac{54}{57}$ 26. $\frac{176}{129}$ 27. $\frac{88}{80}$ 28. $\frac{2007}{2008}$

Working through the application exercises that contain **REAL DATA** will help prepare you to answer questions and/or solve problems based on your own experiences, using facts or information you gather.

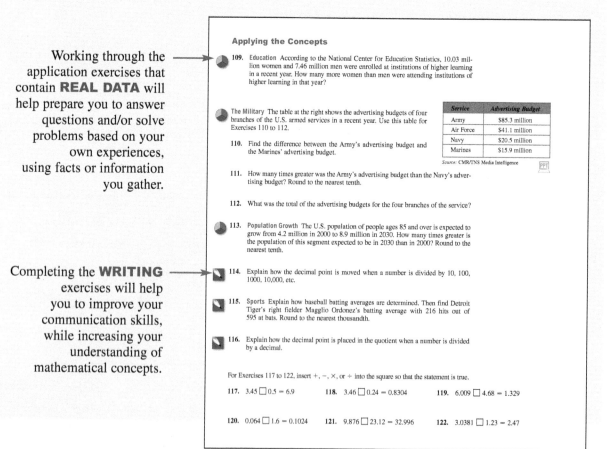

Applying the Concepts

109. **Education** According to the National Center for Education Statistics, 10.03 million women and 7.46 million men were enrolled at institutions of higher learning in a recent year. How many more women than men were attending institutions of higher learning in that year?

The Military The table at the right shows the advertising budgets of four branches of the U.S. armed services in a recent year. Use this table for Exercises 110 to 112.

Service	Advertising Budget
Army	$85.3 million
Air Force	$41.1 million
Navy	$20.5 million
Marines	$15.9 million

Source: CMR/TNS Media Intelligence

110. Find the difference between the Army's advertising budget and the Marines' advertising budget.

111. How many times greater was the Army's advertising budget than the Navy's advertising budget? Round to the nearest tenth.

112. What was the total of the advertising budgets for the four branches of the service?

113. **Population Growth** The U.S. population of people ages 85 and over is expected to grow from 4.2 million in 2000 to 8.9 million in 2030. How many times greater is the population of this segment expected to be in 2030 than in 2000? Round to the nearest tenth.

114. Explain how the decimal point is moved when a number is divided by 10, 100, 1000, 10,000, etc.

115. **Sports** Explain how baseball batting averages are determined. Then find Detroit Tiger's right fielder Magglio Ordonez's batting average with 216 hits out of 595 at bats. Round to the nearest thousandth.

116. Explain how the decimal point is placed in the quotient when a number is divided by a decimal.

For Exercises 117 to 122, insert +, −, ×, or ÷ into the square so that the statement is true.

117. 3.45 ☐ 0.5 = 6.9   118. 3.46 ☐ 0.24 = 0.8304   119. 6.009 ☐ 4.68 = 1.329

120. 0.064 ☐ 1.6 = 0.1024   121. 9.876 ☐ 23.12 = 32.996   122. 3.0381 ☐ 1.23 = 2.47

Completing the **WRITING** exercises will help you to improve your communication skills, while increasing your understanding of mathematical concepts.

23. Girl Scout Cookies Using the information in the news clipping at the right, calculate the cash generated annually **a.** from sales of Thin Mints and **b.** from sales of Trefoil shortbread cookies.

New

In the News
Thin Mints Biggest Seller
Every year, sales from all the Girl Scout cookies sold by about 2.7 million girls total $700 million. The most popular cookie is Thin Mints, which earn 25% of total sales, while sales of the Trefoil shortbread cookies represent only 9% of total sales.
Source: Southwest Airlines Spirit Magazine 2007

24. Charities The American Red Cross spent $185,048,179 for administrative expenses. This amount was 3.16% of its total revenue. Find the American Red Cross's total revenue. Round to the nearest hundred million.

25. Poultry In a recent year, North Carolina produced 1,300,000,000 pounds of turkey. This was 18.6% of the U.S. total in that year. Calculate the U.S. total turkey production for that year. Round to the nearest billion.

26. Mining During 1 year, approximately 2,240,000 ounces of gold went into the manufacturing of electronic equipment in the United States. This is 16% of all the gold mined in the United States that year. How many ounces of gold were mined in the United States that year?

In the News
Over Half of Baby Boomers Have College Experience
Of the 78 million baby boomers living in the United States, 45 million have some college experience but no college degree. Twenty million baby boomers have one

27. Education See the news clipping at the right. What percent of the baby boomers living in the United States have some college experience but have not earned a college degree? Round to the nearest tenth of a percent.

IN THE NEWS application exercises help you master the utility of mathematics in our everyday world. They are based on information found in popular media sources, including newspapers and magazines, and the Web.

APPLYING THE CONCEPTS exercises may involve further exploration of topics, or they may involve analysis. They may also integrate concepts introduced earlier in the text. **Optional** scientific calculator exercises are included, denoted by .

Applying the Concepts

77. Air Pollution An emissions test for cars requires that of the total engine exhaust, less than 1 part per thousand $\left(\frac{1}{1000} = 0.001\right)$ be hydrocarbon emissions. Using this figure, determine which of the cars in the table at the right would fail the emissions test.

Car	Total Engine Exhaust	Hydrocarbon Emission
1	367,921	360
2	401,346	420
3	298,773	210
4	330,045	320
5	432,989	450

PPT

---

164 CHAPTER 3 Decimals

PROJECTS AND GROUP ACTIVITIES

Fractions as Terminating or Repeating Decimals

The fraction $\frac{3}{4}$ is equivalent to 0.75. The decimal 0.75 is a **terminating decimal** because there is a remainder of zero when 3 is divided by 4. The fraction $\frac{1}{3}$ is equivalent to 0.333 . . . . The three dots mean the pattern continues on and on. 0.333 . . . is a **repeating decimal**. To determine whether a fraction can be written as a terminating decimal, first write the fraction in simplest form. Then look at the denominator of the fraction. If it contains prime factors of only 2s and/or 5s, then it can be expressed as a terminating decimal. If it contains prime factors other than 2s or 5s, it represents a repeating decimal.

✓ Take Note
If the denominator of a fraction in simplest form is 20, then it can be written as a terminating decimal because 20 = 2 · 2 · 5 (only prime factors of 2 and 5). If the denominator of a fraction in simplest form is 6, it represents a repeating decimal because it contains the prime factor 3 (a number other than 2 or 5).

1. Assume that each of the following numbers is the denominator of a fraction written in simplest form. Does the fraction represent a terminating or repeating decimal?
 a. 4 **b.** 5 **c.** 7 **d.** 9 **e.** 10 **f.** 12 **g.** 15
 h. 16 **i.** 18 **j.** 21 **k.** 24 **l.** 25 **m.** 28 **n.** 40

2. Write two other numbers that, as denominators of fractions in simplest form, represent terminating decimals, and write two other numbers that, as denominators of fractions in simplest form, represent repeating decimals.

PROJECTS AND GROUP ACTIVITIES appear at the end of each chapter. Your instructor may assign these to you individually, or you may be asked to work through the activity in groups.

Basic College Mathematics: An Applied Approach addresses
students' broad range of study styles
by offering **A WIDE VARIETY OF TOOLS FOR REVIEW**.

At the end of each chapter you will find a **SUMMARY** with **KEY WORDS** and **ESSENTIAL RULES AND PROCEDURES**. Each entry includes an example of the summarized concept, an objective reference, and a page reference to show where each concept was introduced.

CHAPTER 3

SUMMARY

KEY WORDS	EXAMPLES
A number written in *decimal notation* has three parts: a *whole-number part*, a *decimal point*, and a *decimal part*. The decimal part of a number represents a number less than 1. A number written in decimal notation is often simply called a *decimal*. [3.1A, p. 126]	For the decimal 31.25, 31 is the whole-number part and 25 is the decimal part.

ESSENTIAL RULES AND PROCEDURES	EXAMPLES
To write a decimal in words, write the decimal part as if it were a whole number. Then name the place value of the last digit. The decimal point is read as "and." [3.1A, p. 126]	The decimal 12.875 is written in words as twelve and eight hundred seventy-five thousandths.
To write a decimal in standard form when it is written in words, write the whole-number part, replace the word *and* with a decimal point, and write the decimal part so that the last digit is in the given place-value position. [3.1A, p. 127]	The decimal forty-nine and sixty-three thousandths is written in standard form as 49.063.

166   CHAPTER 3 · Decimals

CHAPTER 3

CONCEPT REVIEW

Test your knowledge of the concepts presented in this chapter. Answer each question.
Then check your answers against the ones provided in the Answer Section.

1. How do you round a decimal to the nearest tenth?

2. How do you write the decimal 0.37 as a fraction?

3. How do you write the fraction $\frac{173}{10,000}$ as a decimal?

CONCEPT REVIEWS actively engage you as you study and review the contents of a chapter. The **ANSWERS** to the questions are found in an appendix at the back of the text. After each answer, look for an objective reference that indicates where the concept was introduced.

By completing the chapter **REVIEW EXERCISES**, you can practice working problems that appear in an order that is different from the order they were presented in the chapter. The **ANSWERS** to these exercises include references to the section objectives upon which they are based. This will help you to quickly identify where to go to review the concepts if needed.

CHAPTER 3

REVIEW EXERCISES

1. Find the quotient of 3.6515 and 0.067.

2. Find the sum of 369.41, 88.3, 9.774, and 366.474.

3. Place the correct symbol, $<$ or $>$, between the two numbers.
    0.055   0.1

4. Write 22.0092 in words.

5. Round 0.05678235 to the nearest hundred-thousandth.

6. Convert $2\frac{1}{3}$ to a decimal. Round to the nearest hundredth.

7. Convert 0.375 to a fraction.

8. Add: $3.42 + 0.794 + 32.5$

Each chapter **TEST** is designed to simulate a possible test of the concepts covered in the chapter. The **ANSWERS** include references to section objectives. References to How Tos, worked Examples, and You Try Its, that provide solutions to similar problems, are also included.

CHAPTER 3

TEST

1. Place the correct symbol, $<$ or $>$, between the two numbers.
   0.66   0.666

2. Subtract: 13.027
               $-\ \ 8.94$

3. Write 45.0302 in words.

4. Convert $\frac{9}{13}$ to a decimal. Round to the nearest thousandth.

5. Convert 0.825 to a fraction.

6. Round 0.07395 to the nearest ten-thousandth.

CUMULATIVE REVIEW EXERCISES

1. Divide: $89\overline{)20{,}932}$

2. Simplify: $2^3 \cdot 4^2$

3. Simplify: $2^2 - (7 - 3) \div 2 + 1$

4. Find the LCM of 9, 12, and 24.

5. Write $\frac{22}{5}$ as a mixed number.

6. Write $4\frac{5}{8}$ as an improper fraction.

7. Write an equivalent fraction with the given denominator.
   $\frac{5}{12} = \frac{}{60}$

8. Add: $\frac{3}{8} + \frac{5}{12} + \frac{9}{16}$

CUMULATIVE REVIEW EXERCISES, which appear at the end of each chapter (beginning with Chapter 2), help you maintain skills you previously learned. The **ANSWERS** include references to the section objectives upon which the exercises are based.

FINAL EXAM

1. Subtract: $100{,}914 - 97{,}655$

2. Find 34,821 divided by 657.

3. Find 90,001 decreased by 29,796.

4. Simplify: $3^2 \cdot (5 - 3)^2 \div 3 + 4$

5. Find the LCM of 9, 12, and 16.

6. Add: $\frac{3}{8} + \frac{5}{6} + \frac{1}{5}$

7. Subtract: $7\frac{5}{12} - 3\frac{13}{16}$

8. Find the product of $3\frac{5}{8}$ and $1\frac{5}{7}$.

9. Divide: $1\frac{2}{3} \div 3\frac{3}{4}$

10. Simplify: $\left(\frac{2}{3}\right)^3 \cdot \left(\frac{3}{4}\right)^2$

11. Simplify: $\left(\frac{2}{3}\right)^2 \div \left(\frac{3}{4} + \frac{1}{3}\right) - \frac{1}{3}$

12. Add: 4.972
           28.6

A **FINAL EXAM** appears after the last chapter in the text. It is designed to simulate a possible examination of all the concepts covered in the text. The **ANSWERS** to the exam questions are provided in the answer appendix at the back of the text and include references to the section objectives upon which the questions are based.

Other Key Features

MARGINS Within the margins, students can find the following.

 Take Note boxes alert students to concepts that require special attention.

 Point of Interest boxes, which may be historical in nature or be of general interest, relate to topics under discussion.

 Integrated Technology boxes, which are offered as optional instruction in the proper use of the scientific calculator, appear for selected topics under discussion.

 Tips for Success boxes outline good study habits.

ESTIMATION

Estimating the Sum of Two or More Decimals

Calculate 23.037 + 16.7892. Then use estimation to determine whether the sum is reasonable.

Add to find the exact sum.   23.037 ➕ 16.7892 🟰 39.8262

To estimate the sum, round each number to the same place value. Here we have rounded to the nearest whole number. Then add. The estimated answer is 40, which is very close to the exact sum, 39.8262.

$$\begin{aligned} 23.037 &\approx 23 \\ +16.7892 &\approx +17 \\ \hline &40 \end{aligned}$$

ESTIMATION Throughout the textbook, Estimation boxes appear, where appropriate. Tied to relevant content, the Estimation boxes demonstrate how estimation may be used to check answers for reasonableness.

PROBLEM-SOLVING STRATEGIES The text features a carefully developed approach to problem solving that encourages students to develop a Strategy for a problem and then to create a Solution based on the Strategy.

EXAMPLE · 3

Determine the number of Americans under the age of 45 who are hearing-impaired.

Strategy
To determine the number, add the numbers of hearing impaired ages 0 to 17, 18 to 34, and 35 to 44.

Solution

$$\begin{aligned} &1.37 \\ &2.77 \\ +&4.07 \\ \hline &8.21 \end{aligned}$$

8.21 million Americans under the age of 45 are hearing-impaired.

YOU TRY IT · 3

Determine the number of Americans ages 45 and older who are hearing-impaired.

Your strategy

Your solution

EXAMPLE · 4

Dan Burhoe earned a salary of $210.48 for working 3 days this week as a food server. He also received $82.75, $75.80, and $99.25 in tips during the 3 days. Find his total income for the 3 days of work.

YOU TRY IT · 4

Anita Khavari, an insurance executive, earns a salary of $875 every 4 weeks. During the past 4-week period, she received commissions of $985.80, $791.46, $829.75, and $635.42. Find her total income for the past 4-week period.

FOCUS ON PROBLEM SOLVING At the end of each chapter, the Focus on Problem Solving fosters further discovery of new problem-solving strategies, such as applying solutions to other problems, working backwards, inductive reasoning, and trial and error.

FOCUS ON PROBLEM SOLVING

Relevant Information   Problems in mathematics or real life involve a question or a need and information or circumstances related to that question or need. Solving problems in the sciences usually involves a question, an observation, and measurements of some kind.

One of the challenges of problem solving in the sciences is to separate the information that is relevant to the problem from other information. Following is an example from the physical sciences in which some relevant information was omitted.

Hooke's Law states that the distance that a weight will stretch a spring is directly proportional to the weight on the spring. That is, $d = kF$, where d is the distance the spring is stretched and F is the force. In an experiment to verify this law, some physics students were continually getting inconsistent results. Finally, the instructor discovered that the heat produced when the lights were turned on was affecting the experiment. In this case, relevant information was omitted—namely, that the temperature of the spring can affect the distance it will stretch.

A lawyer drove 8 miles to the train station. After a 35-minute ride of 18 miles, the lawyer walked 10 minutes to the office. Find the total time it took the lawyer to get to work.

From this situation, answer the following before reading on.

a. What is asked for?

b. Is there enough information to answer the question?

c. Is information given that is not needed?

General Revisions

- Chapter Openers now include Prep Tests for students to test their knowledge of prerequisite skills for the new chapter.
- Each exercise set has been thoroughly reviewed to ensure that the pace and scope of the exercises adequately cover the concepts introduced in the section.
- The variety of word problems has increased. This will appeal to instructors who teach to a range of student abilities and want to address different learning styles.
- Think About It exercises, which are conceptual in nature, have been added. They are meant to assess and strengthen a student's understanding of the material presented in an objective.
- In the News exercises have been added and are based on a media source such as a newspaper, a magazine, or the Web. The exercises demonstrate the pervasiveness and utility of mathematics in a contemporary setting.
- Concept Reviews now appear in the end-of-chapter materials to help students more actively study and review the contents of the chapter.
- The Chapter Review Exercises and Chapter Tests have been adjusted to ensure that there are questions that assess the key ideas in the chapter.
- The design has been significantly modified to make the text even easier for students to follow.

Acknowledgments

The authors would like to thank the people who have reviewed this manuscript and provided many valuable suggestions.

Dorothy Fujimura, *CSU East Bay*
Rinav Mehta, *Seattle Central Community College*
Joseph Phillips, *Warren County Community College*
Yan Tian, *Palomar College*

The authors would also like to thank the people who reviewed the eighth edition.

Dorothy A. Brown, *Camden County College, NJ*
Kim Doyle, *Monroe Community College, NY*
Said Fariabi, *San Antonio College, TX*
Kimberly A. Gregor, *Delaware Technical and Community College, DE*
Allen Grommet, *East Arkansas Community College, AR*
Anne Haney
Rose M. Kaniper, *Burlington County College, NJ*
Mary Ann Klicka, *Bucks County Community College, PA*
Helen Medley, *Kent State University, OH*
Steve Meidinger, *Merced College, CA*
James R. Perry, *Owens Community College, OH*
Gowribalan Vamadeva, *University of Cincinnati, OH*
Susan Wessner, *Tallahassee Community College, FL*

Special thanks go to Jean Bermingham for copyediting the manuscript and proofreading pages, to Carrie Green for preparing the solutions manuals, and to Lauri Semarne for her work in ensuring the accuracy of the text. We would also like to thank the many people at Cengage Learning who worked to guide the manuscript from development through production.

Instructor Resources

Print Ancillaries

Complete Solutions Manual (0-538-49394-1)
Carrie Green

The Complete Solutions Manual provides worked-out solutions to all of the problems in the text.

Instructor's Resource Binder (0-538-49773-4)
Maria H. Andersen, *Muskegon Community College*

 The Instructor's Resource Binder contains uniquely designed Teaching Guides, which include instruction tips, examples, activities, worksheets, overheads, and assessments, with answers to accompany them.

Appendix to accompany Instructor's Resource Binder (0-538-49773-4)
Richard N. Aufmann, *Palomar College*
Joanne S. Lockwood, *Nashua Community College*

New! The Appendix to accompany the Instructor's Resource Binder contains teacher resources that are tied directly to *Basic College Mathematics: An Applied Approach,* 9e. Organized by objective, the Appendix contains additional questions and short, in-class activities. The Appendix also includes answers to Writing Exercises, Focus on Problem Solving, and Projects and Group Activities found in the text.

Electronic Ancillaries

Enhanced WebAssign WebAssign

Used by over one million students at more than 1,100 institutions, WebAssign allows you to assign, collect, grade, and record homework assignments via the Web. This proven and reliable homework system includes thousands of algorithmically generated homework problems, links to relevant textbook sections, video examples, problem-specific tutorials, and more.

Solution Builder (0-538-49238-4)

This online solutions manual allows instructors to create customizable solutions that they can print out to distribute or post as needed. This is a convenient and expedient way to deliver solutions to specific homework sets.

PowerLecture with Diploma® (0-538-49405-0)

This CD-ROM provides the instructor with dynamic media tools for teaching. Create, deliver, and customize tests (both print and online) in minutes with Diploma's Computerized Testing featuring algorithmic equations. Easily build solution sets for homework or exams using Solution Builder's online solutions manual. Quickly and easily update your syllabus with the new Syllabus Creator, which was created by the authors and contains the new edition's table of contents. Practice Sheets, First Day of Class PowerPoint® lecture slides, art and figures from the book, and a test bank in electronic format are also included on this CD-ROM.

Text Specific DVDs (0-538-73632-1)

Hosted by Dana Mosely and captioned for the hearing-impaired, these DVDs cover all sections in the text. Ideal for promoting individual study and review, these comprehensive DVDs also support students in online courses or those who may have missed a lecture.

Student Resources

Print Ancillaries

Student Solutions Manual (0-538-49386-0)
Carrie Green

The Student Solutions Manual provides worked-out solutions to the odd-numbered problems in the textbook.

Student Workbook (0-538-49399-2)
Maria H. Andersen, *Muskegon Community College*

Get a head-start! The Student Workbook contains assessments, activities, and worksheets from the Instructor's Resource Binder. Use them for additional practice to help you master the content.

Electronic Ancillaries

Enhanced WebAssign WebAssign

If you are looking for extra practice or additional support, Enhanced WebAssign offers practice problems, videos, and tutorials that are tied directly to the problems found in the textbook.

Text Specific DVDs (0-538-73632-1)

Hosted by Dana Mosley, an experienced mathematics instructor, the DVDs will help you to get a better handle on topics found in the textbook. A comprehensive set of DVDs for the entire course is available to order.

AIM for Success: Getting Started

Welcome to *Basic College Mathematics: An Applied Approach!* Students come to this course with varied backgrounds and different experiences in learning math. We are committed to your success in learning mathematics and have developed many tools and resources to support you along the way. Want to excel in this course? Read on to learn the skills you'll need and how best to use this book to get the results you want.

Motivate Yourself

You'll find many real-life problems in this book, relating to sports, money, cars, music, and more. We hope that these topics will help you understand how you will use mathematics in your real life. However, to learn all of the necessary skills and how you can apply them to your life outside this course, you need to stay motivated.

✓ Take Note

Motivation alone won't lead to success. For example, suppose a person who cannot swim is rowed out to the middle of a lake and thrown overboard. That person has a lot of motivation to swim, but will most likely drown without some help. You'll need motivation and learning in order to succeed.

> **THINK ABOUT WHY YOU WANT TO SUCCEED IN THIS COURSE. LIST THE REASONS HERE (NOT IN YOUR HEAD . . . ON THE PAPER!):**
>
> _____
>
> _____

We also know that this course may be a requirement for you to graduate or complete your major. That's OK. If you have a goal for the future, such as becoming a nurse or a teacher, you will need to succeed in mathematics first. Picture yourself where you want to be, and use this image to stay on track.

Make the Commitment

Stay committed to success! With practice, you will improve your math skills. Skeptical? Think about when you first learned to ride a bike or drive a car. You probably felt self-conscious and worried that you might fail. But with time and practice, it became second nature to you.

You will also need to put in the time and practice to do well in mathematics. Think of us as your "driving" instructors. We'll lead you along the path to success, but we need you to stay focused and energized along the way.

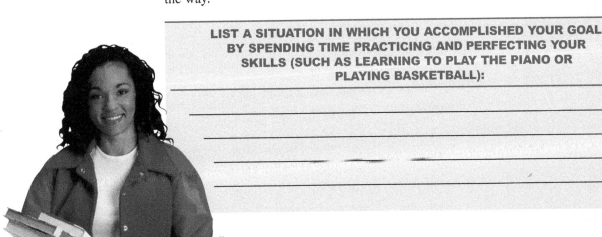

> **LIST A SITUATION IN WHICH YOU ACCOMPLISHED YOUR GOAL BY SPENDING TIME PRACTICING AND PERFECTING YOUR SKILLS (SUCH AS LEARNING TO PLAY THE PIANO OR PLAYING BASKETBALL):**
>
> _____
>
> _____
>
> _____
>
> _____

If you spend time learning and practicing the skills in this book, you will also succeed in math.

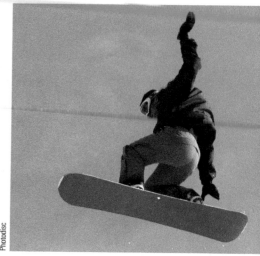

Think You Can't Do Math? Think Again!

You can do math! When you first learned the skills you just listed, you may have not done them well. With practice, you got better, With practice, you will be better at math. Stay focused, motivated, and committed to success.

It is difficult for us to emphasize how important it is to overcome the "I Can't Do Math Syndrome." If you listen to interviews of very successful athletes after a particularly bad performance, you will note that they focus on the positive aspect of what they did, not the negative. Sports psychologists encourage athletes to always be positive—to have a "Can Do" attitude. Develop this attitude toward math and you will succeed.

Skills for Success

GET THE BIG PICTURE If this were an English class, we wouldn't encourage you to look ahead in the book. But this is mathematics—go right ahead! Take a few minutes to read the table of contents. Then, look through the entire book. Move quickly: scan titles, look at pictures, notice diagrams.

Getting this big picture view will help you see where this course is going. To reach your goal, it's important to get an idea of the steps you will need to take along the way.

As you look through the book, find topics that interest you. What's your preference? Horse racing? Sailing? TV? Amusement parks? Find the Index of Applications at the back of the book and pull out three subjects that interest you. Then, flip to the pages in the book where the topics are featured and read the exercises or problems where they appear.

WRITE THE TOPIC HERE:	WRITE THE CORRESPONDING EXERCISE/PROBLEM HERE:
_____	_____
_____	_____
_____	_____

You'll find it's easier to work at learning the material if you are interested in how it can be used in your everyday life.

Use the following activities to think about more ways you might use mathematics in your daily life. Flip open your book to the following exercises to answer the questions.

- (see p. 83, #82) I just started a new job and will be paid hourly, but my hours change every week. I need to use mathematics to . . .

_____

_____

_____

_____

- (see p. 228, #24) I'd like to buy a new video camera, but it's very expensive. I need math to . . .

  _____

  _____

- (see p. 546, #33) I want to rent a car, but I have to find the company that offers the best overall price. I need mathematics to . . .

  _____

  _____

  _____

You know that the activities you just completed are from daily life, but do you notice anything else they have in common? That's right—they are **word problems.** Try not to be intimidated by word problems. You just need a strategy. It's true that word problems can be challenging because we need to use multiple steps to solve them:

- ■ Read the problem.
- ■ Determine the quantity we must find.
- ■ Think of a method to find it.
- ■ Solve the problem.
- ■ Check the answer.

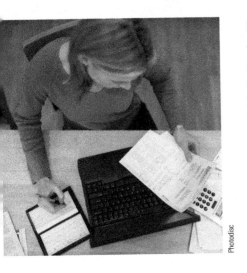

Photodisc

In short, we must come up with a **strategy** and then use that strategy to find the **solution.**

We'll teach you about strategies for tackling word problems that will make you feel more confident in branching out to these problems from daily life. After all, even though no one will ever come up to you on the street and ask you to solve a multiplication problem, you will need to use math every day to balance your checkbook, evaluate credit card offers, etc.

Take a look at the following example. You'll see that solving a word problem includes finding a *strategy* and using that strategy to find a *solution.* If you find yourself struggling with a word problem, try writing down the information you know about the problem. Be as specific as you can. Write out a phrase or a sentence that states what you are trying to find. Ask yourself whether there is a formula that expresses the known and unknown quantities. Then, try again!

EXAMPLE 7	YOU TRY IT 7
It costs $.036 an hour to operate an electric motor. How much does it cost to operate the motor for 120 hours?	The cost of electricity to run a freezer for 1 hour is $.035. This month the freezer has run for 210 hours. Find the total cost of running the freezer this month.
Strategy To find the cost of running the motor for 120 hours, multiply the hourly cost (0.036) by the number of hours the motor is run (120).	**Your strategy**
Solution $$\begin{array}{r} 0.036 \\ \times \quad 120 \\ \hline 720 \\ 36 \quad \\ \hline 4.320 \end{array}$$ The cost of running the motor for 120 hours is $4.32.	**Your solution**

Take a look at your syllabus to see if your instructor has an **attendance policy** that is part of your overall grade in the course.

The attendance policy will tell you:

- How many classes you can miss without a penalty
- What to do if you miss an exam or quiz
- If you can get the lecture notes from the professor if you miss a class

✓ **Take Note**

When planning your schedule, give some thought to how much time you realistically have available each week. For example, if you work 40 hours a week, take 15 units, spend the recommended study time given at the right, and sleep 8 hours a day, you will use over 80% of the available hours in a week. That leaves less than 20% of the hours in a week for family, friends, eating, recreation, and other activities.

Visit http://college. cengage.com/masterstudent/ shared/content/time_chart/ chart.html and use the Interactive Time Chart to see how you're spending your time—you may be surprised.

GET THE BASICS   On the first day of class, your instructor will hand out a **syllabus** listing the requirements of your course. Think of this syllabus as your personal roadmap to success. It shows you the destinations (topics you need to learn) and the dates you need to arrive at those destinations (by when you need to learn the topics). Learning mathematics is a journey. But, to get the most out of this course, you'll need to know what the important stops are and what skills you'll need to learn for your arrival at those stops.

You've quickly scanned the table of contents, but now we want you to take a closer look. Flip open to the table of contents and look at it next to your syllabus. Identify when your major exams are and what material you'll need to learn by those dates. For example, if you know you have an exam in the second month of the semester, how many chapters of this text will you need to learn by then? What homework do you have to do during this time? Managing this important information will help keep you on track for success.

MANAGE YOUR TIME   We know how busy you are outside of school. Do you have a full-time or a part-time job? Do you have children? Visit your family often? Play basketball or write for the school newspaper? It can be stressful to balance all of the important activities and responsibilities in your life. Making a **time management plan** will help you create a schedule that gives you enough time for everything you need to do.

Let's get started! Create a weekly schedule.

First, list all of your responsibilities that take up certain set hours during the week. Be sure to include:

- each class you are taking
- time you spend at work
- any other commitments (child care, tutoring, volunteering, etc.)

Then, list all of your responsibilities that are more flexible. Remember to make time for:

- **STUDYING** You'll need to study to succeed, but luckily you get to choose what times work best for you. Keep in mind:
  - Most instructors ask students to spend twice as much time studying as they do in class (3 hours of class = 6 hours of study).
  - Try studying in chunks. We've found it works better to study an hour each day, rather than studying for 6 hours on one day.
  - Studying can be even more helpful if you're able to do it right after your class meets, when the material is fresh in your mind.
- **MEALS** Eating well gives you energy and stamina for attending classes and studying.
- **ENTERTAINMENT** It's impossible to stay focused on your responsibilities 100% of the time. Giving yourself a break for entertainment will reduce your stress and help keep you on track.
- **EXERCISE** Exercise contributes to overall health. You'll find you're at your most productive when you have both a healthy mind and a healthy body.

Here is a sample of what part of your schedule might look like:

	8–9	9–10	10–11	11–12	12–1	1–2	2–3	3–4	4–5	5–6
Monday	History class Jenkins Hall 8–9:15	Eat 9:15–10	Study/Homework for History 10–12		Lunch and Napl 12–1:30		Work 2–6			
Tuesday	Breakfast	Math Class Douglas Hall 9–9:45	Study/Homework for Math 10–12		Eat 12–1	English Class Scott Hall 1–1:45	Study/Homework for English 2–4		Hang out with Alli and Mike 4–6	

Features for Success in This Text

ORGANIZATION   Let's look again at the Table of Contents. There are 12 chapters in this book. You'll see that every chapter is divided into **sections,** and each section contains a number of **learning objectives.** Each learning objective is labeled with a letter from A to D. Knowing how this book is organized will help you locate important topics and concepts as you're studying.

PREPARATION   Ready to start a new chapter? Take a few minutes to be sure you're ready, using some of the tools in this book.

- ■ **CUMULATIVE REVIEW EXERCISES:** You'll find these exercises after every chapter, starting with Chapter 2. The questions in the Cumulative Review Exercises are taken from the previous chapters. For example, the Cumulative Review for Chapter 3 will test all of the skills you have learned in Chapters 1, 2, and 3. Use this to refresh yourself before moving on to the next chapter, or to test what you know before a big exam.

Here's an example of how to use the Cumulative Review:

- Turn to page 171 and look at the questions for the Chapter 3 Cumulative Review, which are taken from the current chapter and the previous chapters.
- We have the answers to all of the Cumulative Review Exercises in the back of the book. Flip to page A10 to see the answers for this chapter.
- Got the answer wrong? We can tell you where to go in the book for help! For example, scroll down page A10 to find the answer for the first exercise, which is 235 r17. You'll see that after this answer, there is an **objective reference** [1.5C]. This means that the question was taken from Chapter 1, Section 5, Objective C. Go here to restudy the objective.
- ■ **PREP TESTS:** These tests are found at the beginning of every chapter and will help you see if you've mastered all of the skills needed for the new chapter.

Here's an example of how to use the Prep Test:

- Turn to page 173 and look at the Prep Test for Chapter 4.
- All of the answers to the Prep Tests are in the back of the book. You'll find them in the first set of answers in each answer section for a chapter. Turn to page A10 to see the answers for this Prep Test.
- Restudy the objectives if you need some extra help.

Photodisc

■ Before you start a new section, take a few minutes to read the **Objective Statement** for that section. Then, browse through the objective material. Especially note the words or phrases in bold type—these are important concepts that you'll need as you're moving along in the course.

■ As you start moving through the chapter, pay special attention to the **rule boxes.** These rules give you the reasons certain types of problems are solved the way they are. When you see a rule, try to rewrite the rule in your own words.

> **Rule for Adding Two Numbers**
>
> To add numbers with the same sign, add the absolute values of the numbers. Then attach the sign of the addends.
>
> To add numbers with different signs, find the difference between the absolute values of the numbers. Then attach the sign of the addend with the greater absolute value.

Page 411

Knowing what to pay attention to as you move through a chapter will help you study and prepare.

INTERACTION We want you to be actively involved in learning mathematics and have given you many ways to get hands-on with this book.

■ **HOW TO EXAMPLES** Take a look at page 150 shown here. See the HOW TO example? This contains an explanation by each step of the solution to a sample problem.

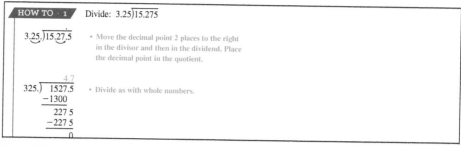

Page 150

Grab a paper and pencil and work along as you're reading through each example. When you're done, get a clean sheet of paper. Write down the problem and try to complete the solution without looking at your notes or at the book. When you're done, check your answer. If you got it right, you're ready to move on.

■ **EXAMPLE/YOU TRY IT PAIRS** You'll need hands-on practice to succeed in mathematics. When we show you an example, work it out beside our solution. Use the Example/You Try It pairs to get the practice you need.

Take a look at page 69, Example 5 and You Try It 5 shown here:

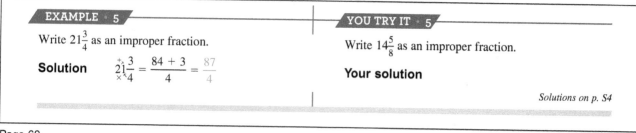

Page 69

You'll see that each Example is fully worked-out. Study this Example carefully by working through each step. Then, try your hand at it by completing the You Try It. If you get stuck, the solutions to the You Try Its are provided in the back of the book. There is a page number following the You Try It, which shows you where you can find the completely worked-out solution. Use the solution to get a hint for the step on which you are stuck. Then, try again!

When you've finished the solution, check your work against the solution in the back of the book. Turn to page S4 to see the solution for You Try It 5.

Remember that sometimes there can be more than one way to solve a problem. But, your answer should always match the answers we've given in the back of the book. If you have any questions about whether your method will always work, check with your instructor.

REVIEW We have provided many opportunities for you to practice and review the skills you have learned in each chapter.

- **SECTION EXERCISES** After you're done studying a section, flip to the end of the section and complete the exercises. If you immediately practice what you've learned, you'll find it easier to master the core skills. Want to know if you answered the questions correctly? The answers to the odd-numbered exercises are given in the back of the book.

- **CHAPTER SUMMARY** Once you've completed a chapter, look at the Chapter Summary. This is divided into two sections: *Key Words* and *Essential Rules and Procedures*. Flip to page 395 to see the Chapter Summary for Chapter 9. This summary shows all of the important topics covered in the chapter. See the reference following each topic? This shows you the objective reference and the page in the text where you can find more information on the concept.

- **CONCEPT REVIEW** Following the Chapter Summary for each chapter is the Concept Review. Flip to page 396 to see the Concept Review for Chapter 9. When you read each question, jot down a reminder note on the right about whatever you feel will be most helpful to remember if you need to apply that concept during an exam. You can also use the space on the right to mark what concepts your instructor expects you to know for the next test. If you are unsure of the answer to a concept review question, flip to the answers appendix at the back of the book.

- **CHAPTER REVIEW EXERCISES** You'll find the Chapter Review Exercises after the Concept Review. Flip to page 333 to see the Chapter Review Exercises for Chapter 7. When you do the review exercises, you're giving yourself an important opportunity to test your understanding of the chapter. The answer to each review exercise is given at the back of the book, along with the objective the question relates to. When you're done with the Chapter Review Exercises, check your answers. If you had trouble with any of the questions, you can restudy the objectives and retry some of the exercises in those objectives for extra help.

Photodisc

■ **CHAPTER TESTS** The Chapter Tests can be found after the Chapter Review Exercises and can be used to prepare for your exams. The answer to each test question is given at the back of the book, along with a reference to a How To, Example, or You Try It that the question relates to. Think of these tests as "practice runs" for your in-class tests. Take the test in a quiet place and try to work through it in the same amount of time you will be allowed for your exam.

Here are some strategies for success when you're taking your exams:

- Scan the entire test to get a feel for the questions (get the big picture).

- Read the directions carefully.

- Work the problems that are easiest for you first.

- Stay calm, and remember that you will have lots of opportunities for success in this class!

EXCEL Visit **www.cengage.com/math/aufmann** to learn about additional study tools!

■ *Enhanced WebAssign*® online practice exercises and homework problems match the textbook exercises.

■ **DVDs** Hosted by Dana Mosley, an experienced mathematics instructor, the DVDs will help you to get a better handle on topics that may be giving you trouble. A comprehensive set of DVDs for the entire course is available to order.

Get Involved Have a question? Ask! Your professor and your classmates are there to help. Here are some tips to help you jump in to the action:

Photodisc

■ Raise your hand in class.

■ If your instructor prefers, email or call your instructor with your question. If your professor has a website where you can post your question, also look there for answers to previous questions from other students. Take advantage of these ways to get your questions answered.

■ Visit a **math center.** Ask your instructor for more information about the math center services available on your campus.

Photodisc

■ Your instructor will have **office hours** where he or she will be available to help you. Take note of where and when your instructor holds office hours. Use this time for one-on-one help, if you need it.

■ Form a **study group** with students from your class. This is a great way to prepare for tests, catch up on topics you may have missed, or get extra help on problems you're struggling with. Here are a few suggestions to make the most of your study group:

- **Test each other by asking questions.** Have each person bring a few sample questions when you get together.

- **Practice teaching each other.** We've found that you can learn a lot about what you know when you have to explain it to someone else.

- **Compare class notes.** Couldn't understand the last five minutes of class? Missed class because you were sick? Chances are someone in your group has the notes for the topics you missed.

- **Brainstorm test questions.**

- **Make a plan for your meeting.** Agree on what topics you'll talk about and how long you'll be meeting. When you make a plan, you'll be sure that you make the most of your meeting.

Photodisc

Ready, Set, Succeed!   It takes hard work and commitment to succeed, but we know you can do it! Doing well in mathematics is just one step you'll take along the path to success.

I succeeded in Basic College Mathematics!

We are confident that if you follow our suggestions, you will succeed. Good luck!

Rubberball

Whole Numbers

OBJECTIVES

SECTION 1.1
A To identify the order relation between two numbers
B To write whole numbers in words and in standard form
C To write whole numbers in expanded form
D To round a whole number to a given place value

SECTION 1.2
A To add whole numbers
B To solve application problems

SECTION 1.3
A To subtract whole numbers without borrowing
B To subtract whole numbers with borrowing
C To solve application problems

SECTION 1.4
A To multiply a number by a single digit
B To multiply larger whole numbers
C To solve application problems

SECTION 1.5
A To divide by a single digit with no remainder in the quotient
B To divide by a single digit with a remainder in the quotient
C To divide by larger whole numbers
D To solve application problems

SECTION 1.6
A To simplify expressions that contain exponents
B To use the Order of Operations Agreement to simplify expressions

SECTION 1.7
A To factor numbers
B To find the prime factorization of a number

ARE YOU READY?

Take the Chapter 1 Prep Test to find out if you are ready to learn to:

- Order whole numbers
- Round whole numbers
- Add, subtract, multiply, and divide whole numbers
- Simplify numerical expressions
- Factor numbers and find their prime factorization

PREP TEST

Do these exercises to prepare for Chapter 1.

1. Name the number of ♦s shown below.

   ♦ ♦ ♦ ♦ ♦ ♦ ♦ ♦

2. Write the numbers from 1 to 10.

   1 ___ ___ ___ ___ ___ ___ ___ ___ 10

3. Match the number with its word form.
   a. 4   A. five
   b. 2   B. one
   c. 5   C. zero
   d. 1   D. four
   e. 3   E. two
   f. 0   F. three

SECTION

1.1 Introduction to Whole Numbers

OBJECTIVE A To identify the order relation between two numbers

The **whole numbers** are 0, 1, 2, 3, 4, 5, 6, 7, 8, 9, 10, 11, 12, 13, 14, ...

The three dots mean that the list continues on and on and that there is no largest whole number.

Just as distances are associated with the markings on the edge of a ruler, the whole numbers can be associated with points on a line. This line is called the **number line.** The arrow on the number line below indicates that there is no largest whole number.

The **graph of a whole number** is shown by placing a heavy dot directly above that number on the number line. Here is the graph of 7 on the number line:

The number line can be used to show the order of whole numbers. A number that appears to the left of a given number **is less than (<)** the given number. A number that appears to the right of a given number **is greater than (>)** the given number.

Four is less than seven.
$4 < 7$

Twelve is greater than seven.
$12 > 7$

EXAMPLE 1

Graph 11 on the number line.

Solution

0 1 2 3 4 5 6 7 8 9 10 11 12 13 14

YOU TRY IT 1

Graph 6 on the number line.

Your solution

0 1 2 3 4 5 6 7 8 9 10 11 12 13 14

EXAMPLE 2

Place the correct symbol, < or >, between the two numbers.

a. 39   24

b. 0   51

Solution

a. 39   24

b. 0   51

YOU TRY IT 2

Place the correct symbol, < or >, between the two numbers.

a. 45   29

b. 27   0

Your solution

a.

b.

Solutions on p. S1

OBJECTIVE B   To write whole numbers in words and in standard form

Point of Interest

The Babylonians had a place-value system based on 60. Its influence is still with us in angle measurement and time: 60 seconds in 1 minute, 60 minutes in 1 hour. It appears that the earliest record of a base-10 place-value system for natural numbers dates from the 8th century.

When a whole number is written using the digits 0, 1, 2, 3, 4, 5, 6, 7, 8, and 9, it is said to be in **standard form.** The position of each digit in the number determines the digit's **place value.** The diagram below shows a **place-value chart** naming the first 12 place values. The number 37,462 is in standard form and has been entered in the chart.

In the number 37,462, the position of the digit 3 determines that its place value is ten-thousands.

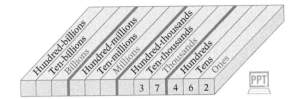

When a number is written in standard form, each group of digits separated from the other digits by a comma (or commas) is called a **period.** The number 3,786,451,294 has four periods. The period names are shown in red in the place-value chart above.

To write a number in words, start from the left. Name the number in each period. Then write the period name in place of the comma.

3,786,451,294 is read "three billion seven hundred eighty-six million four hundred fifty-one thousand two hundred ninety-four."

To write a whole number in standard form, write the number named in each period, and replace each period name with a comma.

Four million sixty-two thousand five hundred eighty-four is written 4,062,584. The zero is used as a place holder for the hundred-thousands place.

EXAMPLE 3

Write 25,478,083 in words.

Solution

Twenty-five million four hundred seventy-eight thousand eighty-three

YOU TRY IT 3

Write 36,462,075 in words.

Your solution

EXAMPLE 4

Write three hundred three thousand three in standard form.

Solution

303,003

YOU TRY IT 4

Write four hundred fifty-two thousand seven in standard form.

Your solution

Solutions on p. S1

OBJECTIVE C   To write whole numbers in expanded form

The whole number 26,429 can be written in **expanded form** as

20,000 + 6000 + 400 + 20 + 9.

The place-value chart can be used to find the expanded form of a number.

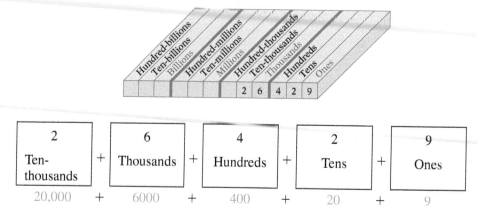

2 Ten-thousands	+	6 Thousands	+	4 Hundreds	+	2 Tens	+	9 Ones
20,000	+	6000	+	400	+	20	+	9

The number 420,806 is written in expanded form below. Note the effect of having zeros in the number.

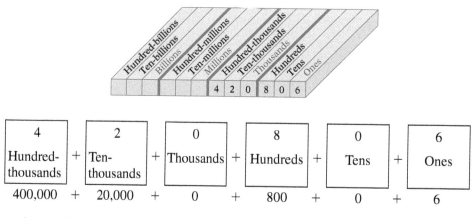

4 Hundred-thousands	+	2 Ten-thousands	+	0 Thousands	+	8 Hundreds	+	0 Tens	+	6 Ones
400,000	+	20,000	+	0	+	800	+	0	+	6

or simply 400,000 + 20,000 + 800 + 6.

EXAMPLE · 5

Write 23,859 in expanded form.

Solution

20,000 + 3000 + 800 + 50 + 9

YOU TRY IT · 5

Write 68,281 in expanded form.

Your solution

EXAMPLE · 6

Write 709,542 in expanded form.

Solution

700,000 + 9000 + 500 + 40 + 2

YOU TRY IT · 6

Write 109,207 in expanded form.

Your solution

Solutions on p. S1

OBJECTIVE D **To round a whole number to a given place value**

When the distance to the moon is given as 240,000 miles, the number represents an approximation to the true distance. Taking an approximate value for an exact number is called **rounding.** A rounded number is always rounded to a given place value.

37 is closer to 40 than it is to 30. 37 rounded to the nearest ten is 40.

673 rounded to the nearest ten is 670. 673 rounded to the nearest hundred is 700.

A whole number is rounded to a given place value without using the number line by looking at the first digit to the right of the given place value.

HOW TO 1 Round 13,834 to the nearest hundred.

Given place value

13,834

3 < 5

• If the digit to the right of the given place value is less than 5, that digit and all digits to the right are replaced by zeros.

13,834 rounded to the nearest hundred is 13,800.

HOW TO 2 Round 386,217 to the nearest ten-thousand.

Given place value

386,217

6 > 5

• If the digit to the right of the given place value is greater than or equal to 5, increase the digit in the given place value by 1, and replace all other digits to the right by zeros.

386,217 rounded to the nearest ten-thousand is 390,000.

EXAMPLE • 7

Round 525,453 to the nearest ten-thousand.

Solution

Given place value

525,453

5 = 5

525,453 rounded to the nearest ten-thousand is 530,000.

YOU TRY IT • 7

Round 368,492 to the nearest ten-thousand.

Your solution

EXAMPLE • 8

Round 1972 to the nearest hundred.

Solution

Given place value

1972

7 > 5

1972 rounded to the nearest hundred is 2000.

YOU TRY IT • 8

Round 3962 to the nearest hundred.

Your solution

Solutions on p. S1

1.1 EXERCISES

OBJECTIVE A   To identify the order relation between two numbers

For Exercises 1 to 4, graph the number on the number line.

1. 3 [number line 0–12]   **2.** 5 [number line 0–12]

3. 9 [number line 0–12]   **4.** 0 [number line 0–12]

For Exercises 5 to 12, place the correct symbol, < or >, between the two numbers.

5. 37   49   **6.** 58   21   **7.** 101   87   **8.** 245   158

9. 2701   2071   **10.** 0   45   **11.** 107   0   **12.** 815   928

 13. Do the inequalities 21 < 30 and 30 > 21 express the same order relation?

OBJECTIVE B   To write whole numbers in words and in standard form

For Exercises 14 to 17, name the place value of the digit 3.

14. 83,479   **15.** 3,491,507   **16.** 2,634,958   **17.** 76,319,204

For Exercises 18 to 25, write the number in words.

18. 2675   **19.** 3790   **20.** 42,928   **21.** 58,473

22. 356,943   **23.** 498,512   **24.** 3,697,483   **25.** 6,842,715

For Exercises 26 to 31, write the number in standard form.

26. Eighty-five   **27.** Three hundred fifty-seven

28. Three thousand four hundred fifty-six   **29.** Sixty-three thousand seven hundred eighty

30. Six hundred nine thousand nine hundred forty-eight

31. Seven million twenty-four thousand seven hundred nine

32. What is the place value of the first number on the left in a seven-digit whole number?

OBJECTIVE C   **To write whole numbers in expanded form**

For Exercises 33 to 40, write the number in expanded form.

33. 5287   **34.** 6295   **35.** 58,943   **36.** 453,921

37. 200,583   **38.** 301,809   **39.** 403,705   **40.** 3,000,642

41. The expanded form of a number consists of four numbers added together. Must the number be a four-digit number?

OBJECTIVE D   **To round a whole number to a given place value**

For Exercises 42 to 53, round the number to the given place value.

42. 926   Tens   **43.** 845   Tens   **44.** 1439   Hundreds

45. 3973   Hundreds   **46.** 43,607   Thousands   **47.** 52,715   Thousands

48. 389,702   Thousands   **49.** 629,513   Thousands   **50.** 647,989   Ten-thousands

51. 253,678   Ten-thousands   **52.** 36,702,599   Millions   **53.** 71,834,250   Millions

54. True or false? If a number rounded to the nearest ten is less than the original number, then the ones digit of the original number is greater than 5.

Applying the Concepts

55. If 3846 is rounded to the nearest ten and then that number is rounded to the nearest hundred, is the result the same as what you get when you round 3846 to the nearest hundred? If not, which of the two methods is correct for rounding to the nearest hundred?

SECTION

1.2 Addition of Whole Numbers

OBJECTIVE A　　**To add whole numbers**

Addition is the process of finding the total of two or more numbers.

✓ **Take Note**

The numbers being added are called **addends**. The result is the **sum**.

By counting, we see that the total of $3 and $4 is $7.

$$\underset{\textbf{Addend}}{\$3} \quad + \quad \underset{\textbf{Addend}}{\$4} \quad = \quad \underset{\textbf{Sum}}{\$7}$$

Addition can be illustrated on the number line by using arrows to represent the addends. The size, or magnitude, of a number can be represented on the number line by an arrow.

The number 3 can be represented anywhere on the number line by an arrow that is 3 units in length.

💿 **Point of Interest**

The first use of the plus sign appeared in 1489 in *Mercantile Arithmetic*. It was used to indicate a surplus, not as the symbol for addition. That use did not appear until about 1515.

To add on the number line, place the arrows representing the addends head to tail, with the first arrow starting at zero. The sum is represented by an arrow starting at zero and stopping at the tip of the last arrow.

$$3 + 4 = 7$$

More than two numbers can be added on the number line.

$$3 + 2 + 4 = 9$$

Some special properties of addition that are used frequently are given below.

Addition Property of Zero

Zero added to a number does not change the number.

$$4 + 0 = 4$$
$$0 + 7 = 7$$

Commutative Property of Addition

Two numbers can be added in either order; the sum will be the same.

$$4 + 8 = 8 + 4$$
$$12 = 12$$

✓ **Take Note**

This is the same addition problem shown on the number line above.

Associative Property of Addition

Grouping the addition in any order gives the same result. The parentheses are grouping symbols and have the meaning "Do the operations inside the parentheses first."

$$(3 + 2) + 4 = 3 + (2 + 4)$$
$$5 + 4 = 3 + 6$$
$$9 = 9$$

The number line is not useful for adding large numbers. The basic addition facts for adding one digit to one digit should be memorized. Addition of larger numbers requires the repeated use of the basic addition facts.

To add large numbers, begin by arranging the numbers vertically, keeping the digits of the same place value in the same column.

HOW TO 1 Add: 321 + 6472

• Add the digits in each column.

There are several words or phrases in English that indicate the operation of addition. Here are some examples:

added to	3 added to 5	5 + 3
more than	7 more than 5	5 + 7
the sum of	the sum of 3 and 9	3 + 9
increased by	4 increased by 6	4 + 6
the total of	the total of 8 and 3	8 + 3
plus	5 plus 10	5 + 10

Integrating Technology

Most scientific calculators use *algebraic logic:* the add (**+**), subtract (**−**), multiply (**x**), and divide (**÷**) keys perform the indicated operation on the number in the display and the next number keyed in. For instance, for the example at the right, enter 24 **+** 71 **=** . The display reads 95.

HOW TO 2 What is the sum of 24 and 71?

$$\begin{array}{r} 24 \\ + 71 \\ \hline 95 \end{array}$$

• The phrase *the sum of* means to add.

The sum of 24 and 71 is 95.

When the sum of the digits in a column exceeds 9, the addition will involve **carrying.**

HOW TO 3 Add: 487 + 369

$$\begin{array}{r} \overset{1}{}487 \\ + 369 \\ \hline 6 \end{array}$$

• Add the ones column.
  7 + 9 = 16 (1 ten + 6 ones).
  Write the 6 in the ones column and carry the 1 ten to the tens column.

$$\begin{array}{r} \overset{1}{4}\overset{1}{8}7 \\ + 369 \\ \hline 56 \end{array}$$

• Add the tens column.
  1 + 8 + 6 = 15 (1 hundred + 5 tens).
  Write the 5 in the tens column and carry the 1 hundred to the hundreds column.

$$\begin{array}{r} \overset{1}{4}\overset{1}{8}7 \\ + 369 \\ \hline 856 \end{array}$$

• Add the hundreds column.
  1 + 4 + 3 = 8 (8 hundreds).
  Write the 8 in the hundreds column.

EXAMPLE • 1

Find the total of 17, 103, and 8.

Solution

$$
\begin{array}{r}
\overset{1}{17} \\
103 \\
+\quad 8 \\
\hline
128
\end{array}
$$

• 7 + 3 + 8 = 18
Write the 8 in the ones column. Carry the 1 to the tens column.

YOU TRY IT • 1

What is 347 increased by 12,453?

Your solution

EXAMPLE • 2

Add: 89 + 36 + 98

Solution

$$
\begin{array}{r}
\overset{2}{89} \\
36 \\
+\ 98 \\
\hline
223
\end{array}
$$

• 9 + 6 + 8 = 23
Write the 3 in the ones column. Carry the 2 to the tens column.

YOU TRY IT • 2

Add: 95 + 88 + 67

Your solution

EXAMPLE • 3

Add: 41,395
          4,327
        497,625
     +   32,991

Solution

$$
\begin{array}{r}
\overset{1\,1\,2\ \ 2\,1}{41,395} \\
4,327 \\
497,625 \\
+\ \ 32,991 \\
\hline
576,338
\end{array}
$$

YOU TRY IT • 3

Add: 392
          4,079
         89,035
      +   4,992

Your solution

Solutions on p. S1

Integrating Technology

This example illustrates that estimation is important when one is using a calculator.

ESTIMATION

Estimation and Calculators

At some places in the text, you will be asked to use your calculator. Effective use of a calculator requires that you estimate the answer to the problem. This helps ensure that you have entered the numbers correctly and pressed the correct keys.

For example, if you use your calculator to find 22,347 + 5896 and the answer in the calculator's display is 131,757,912, you should realize that you have entered some part of the calculation incorrectly. In this case, you pressed x instead of + . By estimating the answer to a problem, you can help ensure the accuracy of your calculations. We have a special symbol for **approximately equal to** (≈).

For example, to estimate the answer to 22,347 + 5896, round each number to the same place value. In this case, we will round to the nearest thousand. Then add.

$$
\begin{array}{r}
22,347 \approx \quad 22,000 \\
+\quad 5,896 \approx +\quad 6,000 \\
\hline
28,000
\end{array}
$$

The sum 22,347 + 5896 is approximately 28,000. Knowing this, you would know that 131,757,912 is much too large and is therefore incorrect.

To estimate the sum of two numbers, first round each whole number to the same place value and then add. Compare this answer with the calculator's answer.

OBJECTIVE B   To solve application problems

To solve an application problem, first read the problem carefully. The **strategy** involves identifying the quantity to be found and planning the steps that are necessary to find that quantity. The **solution of an application problem** involves performing each operation stated in the strategy and writing the answer.

HOW TO 4

 The table below displays the Wal-Mart store count and square footage in the United States as reported in the Wal-Mart 2008 Annual Report.

	Discount Stores	Supercenters	Sam's Clubs	Neighborhood Markets
Number of Units	941	2523	593	134
Square footage (in millions)	105	457	78	5

Find the total number of Wal-Mart discount stores and Supercenters in the United States.

Strategy To find the total number of Wal-Mart discount stores and Supercenters in the United States, read the table to find the number of each type of store in the United States. Then add the numbers.

Solution
```
   941
+ 2523
------
  3464
```

Wal-Mart has a total of 3464 discount stores and Supercenters in the United States.

EXAMPLE 4

Use the table above to find the total number of Sam's Clubs and neighborhood markets that Wal-Mart has in the United States.

Strategy
To determine the total number of Sam's Clubs and neighborhood markets, read the table to find the number of Sam's Clubs and the number of neighborhood markets. Then add the two numbers.

Solution
```
   593
+ 134
------
  727
```

Wal-Mart has a total of 727 Sam's Clubs and neighborhood markets.

YOU TRY IT 4

Use the table above to determine the total square footage of Wal-Mart stores in the United States.

Your strategy

Your solution

Solution on p. S1

1.2 EXERCISES

OBJECTIVE A   To add whole numbers

For Exercises 1 to 32, add.

1. $\begin{array}{r} 17 \\ +\ 11 \\ \hline \end{array}$ 2. $\begin{array}{r} 25 \\ +\ 63 \\ \hline \end{array}$ 3. $\begin{array}{r} 83 \\ +\ 42 \\ \hline \end{array}$ 4. $\begin{array}{r} 63 \\ +\ 94 \\ \hline \end{array}$

5. $\begin{array}{r} 77 \\ +\ 25 \\ \hline \end{array}$ 6. $\begin{array}{r} 63 \\ +\ 49 \\ \hline \end{array}$ 7. $\begin{array}{r} 56 \\ +\ 98 \\ \hline \end{array}$ 8. $\begin{array}{r} 86 \\ +\ 68 \\ \hline \end{array}$

9. $\begin{array}{r} 658 \\ +\ 831 \\ \hline \end{array}$ 10. $\begin{array}{r} 842 \\ +\ 936 \\ \hline \end{array}$ 11. $\begin{array}{r} 735 \\ +\ 93 \\ \hline \end{array}$ 12. $\begin{array}{r} 189 \\ +\ 50 \\ \hline \end{array}$

13. $\begin{array}{r} 859 \\ +\ 725 \\ \hline \end{array}$ 14. $\begin{array}{r} 637 \\ +\ 829 \\ \hline \end{array}$ 15. $\begin{array}{r} 470 \\ +\ 749 \\ \hline \end{array}$ 16. $\begin{array}{r} 427 \\ +\ 690 \\ \hline \end{array}$

17. $\begin{array}{r} 36{,}925 \\ +\ 65{,}392 \\ \hline \end{array}$ 18. $\begin{array}{r} 56{,}772 \\ +\ 51{,}239 \\ \hline \end{array}$ 19. $\begin{array}{r} 50{,}873 \\ +\ 28{,}453 \\ \hline \end{array}$ 20. $\begin{array}{r} 34{,}872 \\ +\ 46{,}079 \\ \hline \end{array}$

21. $\begin{array}{r} 878 \\ 737 \\ +\ 189 \\ \hline \end{array}$ 22. $\begin{array}{r} 768 \\ 461 \\ +\ 669 \\ \hline \end{array}$ 23. $\begin{array}{r} 319 \\ 348 \\ +\ 912 \\ \hline \end{array}$ 24. $\begin{array}{r} 292 \\ 579 \\ +\ 315 \\ \hline \end{array}$

25. $\begin{array}{r} 9409 \\ 3253 \\ +\ 7078 \\ \hline \end{array}$ 26. $\begin{array}{r} 8188 \\ 8020 \\ +\ 7104 \\ \hline \end{array}$ 27. $\begin{array}{r} 2038 \\ 2243 \\ +\ 3139 \\ \hline \end{array}$ 28. $\begin{array}{r} 4252 \\ 6882 \\ +\ 5235 \\ \hline \end{array}$

29. $\begin{array}{r} 67{,}428 \\ 32{,}171 \\ +\ 20{,}971 \\ \hline \end{array}$ 30. $\begin{array}{r} 52{,}801 \\ 11{,}664 \\ +\ 89{,}638 \\ \hline \end{array}$ 31. $\begin{array}{r} 76{,}290 \\ 43{,}761 \\ +\ 87{,}402 \\ \hline \end{array}$ 32. $\begin{array}{r} 43{,}901 \\ 98{,}301 \\ +\ 67{,}943 \\ \hline \end{array}$

For Exercises 33 to 40, add.

33. 20,958 + 3218 + 42

34. 80,973 + 5168 + 29

35. 392 + 37 + 10,924 + 621

36. 694 + 62 + 70,129 + 217

37. 294 + 1029 + 7935 + 65

38. 692 + 2107 + 3196 + 92

39. 97 + 7234 + 69,532 + 276

40. 87 + 1698 + 27,317 + 727

41. What is 9874 plus 4509?

42. What is 7988 plus 5678?

43. What is 3487 increased by 5986?

44. What is 99,567 increased by 126,863?

45. What is 23,569 more than 9678?

46. What is 7894 more than 45,872?

47. What is 479 added to 4579?

48. What is 23,902 added to 23,885?

49. Find the total of 659, 55, and 1278.

50. Find the total of 4561, 56, and 2309.

51. Find the sum of 34, 329, 8, and 67,892.

52. Find the sum of 45, 1289, 7, and 32,876.

For Exercises 53 to 56, use a calculator to add. Then round the numbers to the nearest hundred, and use estimation to determine whether the sum is reasonable.

53. 1234 + 9780 + 6740

54. 919 + 3642 + 8796

55. 241 + 569 + 390 + 1672

56. 107 + 984 + 1035 + 2904

For Exercises 57 to 60, use a calculator to add. Then round the numbers to the nearest thousand, and use estimation to determine whether the sum is reasonable.

57.	**58.**	**59.**	**60.**
32,461	29,036	25,432	66,541
9,844	22,904	62,941	29,365
+ 59,407	+ 7,903	+ 70,390	+ 98,742

For Exercises 61 to 64, use a calculator to add. Then round the numbers to the nearest ten-thousand, and use estimation to determine whether the sum is reasonable.

61.
```
   67,421
   82,984
   66,361
   10,792
+  34,037
```

62.
```
   21,896
    4,235
   62,544
   21,892
+   1,334
```

63.
```
  281,421
    9,874
   34,394
  526,398
+  94,631
```

64.
```
  542,698
   97,327
    7,235
   73,667
+ 173,201
```

 65. Which property of addition (see page 8) allows you to use either arrangement shown at the right to find the sum of 691 and 452?

```
  691      452
+ 452    + 691
```

OBJECTIVE B **To solve application problems**

 66. Use the table of Wal-Mart data on page 11. What does the sum 105 + 457 represent?

67. **Demographics** In a recent year, according to the U.S. Department of Health and Human Services, there were 110,670 twin births in this country, 6919 triplet births, 627 quadruplet deliveries, and 79 quintuplet and other higher-order multiple births. Find the total number of multiple births during the year.

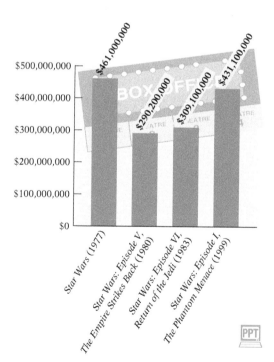

68. **Demographics** The Census Bureau estimates that the U.S. population will grow by 296 million people from 2000 to 2100. Given that the U.S. population in 2000 was 281 million, find the Census Bureau's estimate of the U.S. population in 2100.

 The Film Industry The graph at the right shows the domestic box-office income from the first four *Star Wars* movies. Use this information for Exercises 69 to 71.

69. Estimate the total income from the first four *Star Wars* movies.

70. Find the total income from the first four *Star Wars* movies.

71. **a.** Find the total income from the two movies with the lowest box-office incomes.
b. Does the total income from the two movies with the lowest box-office incomes exceed the income from the 1977 *Star Wars* production?

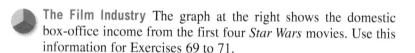

Source: www.worldwideboxoffice.com

72. **Geometry** The perimeter of a triangle is the sum of the lengths of the three sides of the triangle. Find the perimeter of a triangle that has sides that measure 12 inches, 14 inches, and 17 inches.

12 in.   14 in.   17 in.

73. **Travel** The odometer on a moving van reads 68,692. The driver plans to drive 515 miles the first day, 492 miles the second day, and 278 miles the third day.
    a. How many miles will be driven during the three days?
    b. What will the odometer reading be at the end of the trip?

74. **Internet** Thirty-one million U.S. households do not have Internet access. Eighty-three million U.S. households do have Internet access. How many households are there in the United States? (*Source:* U.S. Bureau of the Census)

75. **Trail** Although 685 miles of the Northern Forest Canoe Trail can be paddled, there are another 55 miles of land over which a canoe must be carried. Find the total length of the Northern Forest Canoe Trail. (*Source: Yankee,* May/June 2007)

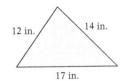

*Northern Forest
Canoe Trail*

Image courtesy of Northern Forest Canoe Trail/www.northernforestcanoetrail.com

76. **Energy** In a recent year, the United States produced 5,102,000 barrels of crude oil per day and imported 10,118,000 barrels of crude oil per day. Find the total number of barrels of crude oil produced and imported per day in the United States. (*Source:* Energy Information Administration)

Applying the Concepts

77. If you roll two ordinary six-sided dice and add the two numbers that appear on top, how many different sums are possible?

78. If you add two *different* whole numbers, is the sum always greater than either one of the numbers? If not, give an example.

79. If you add two whole numbers, is the sum always greater than either one of the numbers? If not, give an example. (Compare this with the previous exercise.)

80. Make up a word problem for which the answer is the sum of 34 and 28

81. Call a number "lucky" if it ends in a 7. How many lucky numbers are less than 100?

SECTION

1.3 Subtraction of Whole Numbers

OBJECTIVE A To subtract whole numbers without borrowing

Subtraction is the process of finding the difference between two numbers.

✓ **Take Note**

The **minuend** is the number from which another number is subtracted. The **subtrahend** is the number that is subtracted from another number. The result is the **difference**.

By counting, we see that the difference between $8 and $5 is $3.

$$\$8 \quad - \quad \$5 \quad = \quad \$3$$

Minuend Subtrahead Difference

The difference 8 − 5 can be shown on the number line.

Note from the number line that addition and subtraction are related.

Subtrahend	5
+ Difference	+ 3
= Minuend	8

🎯 **Point of Interest**

The use of the minus sign dates from the same period as the plus sign, around 1515.

The fact that the sum of the subtrahend and the difference equals the minuend can be used to check subtraction.

To subtract large numbers, begin by arranging the numbers vertically, keeping the digits that have the same place value in the same column. Then subtract the digits in each column.

HOW TO 1 Subtract 8955 − 2432 and check.

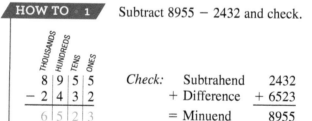

Check:	Subtrahend	2432
	+ Difference	+ 6523
	= Minuend	8955

EXAMPLE 1

Subtract 6594 − 3271 and check.

Solution

	6594	*Check:*	3271
	− 3271		+ 3323
	3323		6594

YOU TRY IT 1

Subtract 8925 − 6413 and check.

Your solution

EXAMPLE 2

Subtract 15,762 − 7541 and check.

Solution

	15,762	*Check:*	7,541
	− 7,541		+ 8,221
	8,221		15,762

YOU TRY IT 2

Subtract 17,504 − 9302 and check.

Your solution

Solutions on p. S1

OBJECTIVE B   To subtract whole numbers with borrowing

In all the subtraction problems in the previous objective, for each place value the lower digit was not larger than the upper digit. When the lower digit is larger than the upper digit, subtraction will involve **borrowing.**

HOW TO 2   Subtract: 692 − 378

HUNDREDS	TENS	ONES
6	$\overset{8+1}{\cancel{9}}$	2
− 3	7	8

Because 8 > 2, borrowing is necessary.
9 tens = 8 tens + 1 ten.

HUNDREDS	TENS	ONES
6	$\overset{8}{\cancel{9}}$	$\overset{10}{2}$
− 3	7	8

Borrow 1 ten from the tens column and write 10 in the ones column.

HUNDREDS	TENS	ONES
6	$\overset{8}{\cancel{9}}$	$\overset{12}{\cancel{2}}$
− 3	7	8

Add the borrowed 10 to 2.

HUNDREDS	TENS	ONES
6	$\overset{8}{\cancel{9}}$	$\overset{12}{\cancel{2}}$
− 3	7	8
3	1	4

Subtract the digits in each column.

The phrases below are used to indicate the operation of subtraction. An example is shown at the right of each phrase.

minus	8 minus 5	8 − 5
less	9 less 3	9 − 3
less than	2 less than 7	7 − 2
the difference between	the difference between 8 and 2	8 − 2
decreased by	5 decreased by 1	5 − 1

HOW TO 3   Find the difference between 1234 and 485, and check.

"The difference between 1234 and 485" means 1234 − 485.

$$1\ 2\ \overset{2}{\cancel{3}}\ \overset{14}{\cancel{4}}$$
$$-\ \ \ 4\ 8\ 5$$
$$9$$

$$\overset{}{\cancel{1}}\ \overset{12}{\cancel{2}}\ \overset{}{\cancel{3}}\ \overset{14}{\cancel{4}}$$
$$-\ \ \ 4\ 8\ 5$$
$$4\ 9$$

$$\overset{0}{1}\ \overset{11}{\cancel{2}}\ \overset{12}{\cancel{3}}\ \overset{14}{\cancel{4}}$$
$$-\ \ \ 4\ 8\ 5$$
$$7\ 4\ 9$$

Check:
$$\overset{1\ 1}{485}$$
$$+\ 749$$
$$\overline{1234}$$

Subtraction with a zero in the minuend involves repeated borrowing.

HOW TO 4   Subtract: 3904 − 1775

$$3\ \overset{8}{\cancel{9}}\ \overset{10}{\cancel{0}}\ 4$$
$$-\ 1\ 7\ 7\ 5$$

5 > 4
There is a 0 in the tens column. Borrow 1 hundred (= 10 tens) from the hundreds column and write 10 in the tens column.

$$3\ \overset{8}{\cancel{9}}\ \overset{\overset{9}{\cancel{10}}}{\cancel{0}}\ \overset{14}{\cancel{4}}$$
$$-\ 1\ 7\ 7\ 5$$

Borrow 1 ten from the tens column and add 10 to the 4 in the ones column.

$$3\ \overset{8}{\cancel{9}}\ \overset{\overset{9}{\cancel{10}}}{\cancel{0}}\ \overset{14}{\cancel{4}}$$
$$-\ 1\ 7\ 7\ 5$$
$$2\ 1\ 2\ 9$$

Subtract the digits in each column.

🏆 Tips for Success
The HOW TO feature indicates an example with explanatory remarks. Using paper and pencil, you should work through the example. See *AIM for Success* at the front of the book.

EXAMPLE · 3

Subtract 4392 − 678 and check.

Solution

$$\begin{array}{r} \overset{3}{\cancel{4}}\ \overset{13}{\cancel{3}}\ \overset{8}{\cancel{9}}\ \overset{12}{\cancel{2}} \\ -\ \ 6\ 7\ 8 \\ \hline 3\ 7\ 1\ 4 \end{array}$$

Check:
$$\begin{array}{r} 678 \\ +\ 3714 \\ \hline 4392 \end{array}$$

YOU TRY IT · 3

Subtract 3481 − 865 and check.

Your solution

EXAMPLE · 4

Find 23,954 less than 63,221 and check.

Solution

$$\begin{array}{r} \overset{5}{\cancel{6}}\ \overset{12}{\cancel{3}},\overset{11}{\cancel{2}}\ \overset{11}{\cancel{2}}\ \overset{11}{\cancel{1}} \\ -\ 2\ 3\ ,\ 9\ 5\ 4 \\ \hline 3\ 9\ ,\ 2\ 6\ 7 \end{array}$$

Check:
$$\begin{array}{r} 23,954 \\ +\ 39,267 \\ \hline 63,221 \end{array}$$

YOU TRY IT · 4

Find 54,562 decreased by 14,485 and check.

Your solution

EXAMPLE · 5

Subtract 46,005 − 32,167 and check.

Solution

$$\begin{array}{r} \overset{5}{\cancel{4}}\ 6,\overset{10}{\cancel{0}}\ 0\ 5 \\ -3\ 2,\ 1\ 6\ 7 \end{array}$$
• There are two zeros in the minuend. Borrow 1 thousand from the thousands column and write 10 in the hundreds column.

$$\begin{array}{r} \overset{5}{\cancel{4}}\ 6,\overset{9}{\cancel{0}}\ \overset{10}{\cancel{0}}\ 5 \\ -3\ 2,\ 1\ 6\ 7 \end{array}$$
• Borrow 1 hundred from the hundreds column and write 10 in the tens column.

$$\begin{array}{r} \overset{5}{\cancel{4}}\ 6,\overset{9}{\cancel{0}}\ \overset{9}{\cancel{0}}\ \overset{15}{\cancel{5}} \\ -3\ 2,\ 1\ 6\ 7 \\ \hline 1\ 3,\ 8\ 3\ 8 \end{array}$$
• Borrow 1 ten from the tens column and add 10 to the 5 in the ones column.

Check:
$$\begin{array}{r} 32,167 \\ +\ 13,838 \\ \hline 46,005 \end{array}$$

YOU TRY IT · 5

Subtract 64,003 − 54,936 and check.

Your solution

Solutions on pp. S1–S2

ESTIMATION

Estimating the Difference Between Two Whole Numbers

Calculate 323,502 − 28,912. Then use estimation to determine whether the difference is reasonable.

Subtract to find the exact difference. To estimate the difference, round each number to the same place value. Here we have rounded to the nearest ten-thousand. Then subtract. The estimated answer is 290,000, which is very close to the exact difference 294,590.

$$\begin{array}{r} 323,502 \approx\ \ \ 320,000 \\ -\ 28,912 \approx -\ 30,000 \\ \hline 294,590\ \ \ \ \ \ 290,000 \end{array}$$

OBJECTIVE C   To solve application problems

© Hutton-Deutsch Collection/Corbis

The table at the right shows the number of personnel on active duty in the branches of the U.S. military in 1940 and 1945. Use this table for Example 6 and You Try It 6.

Branch	1940	1945
U.S. Army	267,767	8,266,373
U.S. Navy	160,997	3,380,817
U.S. Air Force	51,165	2,282,259
U.S. Marine Corps	28,345	474,680

Source: Dept. of the Army, Dept. of the Navy, Air Force Dept., Dept. of the Marines, U.S. Dept. of Defense

---

EXAMPLE • 6

Find the difference between the number of U.S. Army personnel on active duty in 1945 and the number in 1940.

Strategy
To find the difference, subtract the number of U.S. Army personnel on active duty in 1940 (267,767) from the number on active duty in 1945 (8,266,373).

Solution

$$\begin{array}{r} 8{,}266{,}373 \\ -\ 267{,}767 \\ \hline 7{,}998{,}606 \end{array}$$

There were 7,998,606 more personnel on active duty in the U.S. Army in 1945 than in 1940.

YOU TRY IT • 6

Find the difference between the number of personnel on active duty in the Navy and the number in the Air Force in 1945.

Your strategy

Your solution

EXAMPLE • 7

You had a balance of $415 on your student debit card. You then used the card, deducting $197 for books, $48 for art supplies, and $24 for theater tickets. What is your new student debit card balance?

Strategy
To find your new debit card balance:

• Add to find the total of the three deductions (197 + 48 + 24).
• Subtract the total of the three deductions from the old balance (415).

Solution

$$\begin{array}{r} 197 \\ 48 \\ +\ 24 \\ \hline 269 \end{array} \qquad \begin{array}{r} 415 \\ -\ 269 \\ \hline 146 \end{array}$$

269 total deductions

Your new debit card balance is $146.

YOU TRY IT • 7

Your total weekly salary is $638. Deductions of $127 for taxes, $18 for insurance, and $35 for savings are taken from your pay. Find your weekly take-home pay.

Your strategy

Your solution

Solutions on p. S2

1.3 EXERCISES

OBJECTIVE A   To subtract whole numbers without borrowing

For Exercises 1 to 35, subtract.

1. $\begin{array}{r} 9 \\ -5 \end{array}$	**2.** $\begin{array}{r} 8 \\ -7 \end{array}$	**3.** $\begin{array}{r} 8 \\ -4 \end{array}$	**4.** $\begin{array}{r} 7 \\ -3 \end{array}$	**5.** $\begin{array}{r} 10 \\ -0 \end{array}$
6. $\begin{array}{r} 11 \\ -4 \end{array}$	**7.** $\begin{array}{r} 12 \\ -8 \end{array}$	**8.** $\begin{array}{r} 19 \\ -8 \end{array}$	**9.** $\begin{array}{r} 15 \\ -6 \end{array}$	**10.** $\begin{array}{r} 16 \\ -7 \end{array}$
11. $\begin{array}{r} 25 \\ -3 \end{array}$	**12.** $\begin{array}{r} 55 \\ -4 \end{array}$	**13.** $\begin{array}{r} 68 \\ -8 \end{array}$	**14.** $\begin{array}{r} 77 \\ -3 \end{array}$	**15.** $\begin{array}{r} 89 \\ -23 \end{array}$
16. $\begin{array}{r} 54 \\ -21 \end{array}$	**17.** $\begin{array}{r} 88 \\ -57 \end{array}$	**18.** $\begin{array}{r} 1202 \\ -701 \end{array}$	**19.** $\begin{array}{r} 1305 \\ -404 \end{array}$	**20.** $\begin{array}{r} 1763 \\ -801 \end{array}$
21. $\begin{array}{r} 1497 \\ -706 \end{array}$	**22.** $\begin{array}{r} 8974 \\ -3972 \end{array}$	**23.** $\begin{array}{r} 2836 \\ -1711 \end{array}$	**24.** $\begin{array}{r} 8976 \\ -7463 \end{array}$	**25.** $\begin{array}{r} 9273 \\ -6142 \end{array}$

26. $77 - 36$ **27.** $129 - 82$ **28.** $132 - 61$ **29.** $969 - 44$ **30.** $1347 - 103$

31. $4865 - 304$ **32.** $1525 - 702$ **33.** $9999 - 6794$ **34.** $7806 - 3405$ **35.** $8843 - 7621$

36. Suppose three whole numbers, called *minuend, subtrahend,* and *difference,* are related by the subtraction statement *minuend − subtrahend = difference*. State whether the given relationship *must be true, might be true,* or *cannot be true.*
  a. minuend > difference   **b.** subtrahend < difference

OBJECTIVE B   To subtract whole numbers with borrowing

For Exercises 37 to 80, subtract.

37. $\begin{array}{r} 71 \\ -18 \end{array}$	**38.** $\begin{array}{r} 93 \\ -28 \end{array}$	**39.** $\begin{array}{r} 47 \\ -18 \end{array}$	**40.** $\begin{array}{r} 44 \\ -27 \end{array}$
41. $\begin{array}{r} 37 \\ -29 \end{array}$	**42.** $\begin{array}{r} 50 \\ -27 \end{array}$	**43.** $\begin{array}{r} 70 \\ -33 \end{array}$	**44.** $\begin{array}{r} 993 \\ -537 \end{array}$

45.   250
        − 192

46.   840
        − 783

47.   768
        − 194

48.   770
        − 395

49. 674 − 337

50. 3526 − 387

51. 1712 − 289

52. 4350 − 729

53. 1702 − 948

54. 1607 − 869

55. 5933 − 3754

56. 7293 − 3748

57. 9407 − 2918

58. 3706 − 2957

59. 8605 − 7716

60. 8052 − 2709

61. 80,305 − 9176

62. 70,702 − 4239

63. 10,004 − 9306

64. 80,009 − 63,419

65. 70,618 − 41,213

66. 80,053 − 27,649

67. 70,700 − 21,076

68. 80,800 − 42,023

69.   2600
        − 1972

70.   8400
        − 3762

71.   9003
        − 2471

72.   6004
        − 2392

73.   8202
        − 3916

74.   7050
        − 4137

75.   7015
        − 2973

76.   4207
        − 1624

77.   7005
        − 1796

78.   8003
        − 2735

79.   20,005
        − 9,627

80.   80,004
        − 8,237

81. Which of the following phrases represent the subtraction 673 − 571?
(i) 571 less 673   **(ii)** 571 less than 673   **(iii)** 673 decreased by 571

82. Find 10,051 less 9027.

83. Find 17,031 less 5792.

84. Find the difference between 1003 and 447.

85. What is 29,874 minus 21,392?

86. What is 29,797 less than 68,005?

87. What is 69,379 less than 70,004?

88. What is 25,432 decreased by 7994?

89. What is 86,701 decreased by 9976?

For Exercises 90 to 93, use the relationship between addition and subtraction to complete the statement.

90. ___ + 39 = 104   **91.** 67 + ___ = 90   **92.** ___ + 497 = 862   **93.** 253 + ___ = 4901

For Exercises 94 to 99, use a calculator to subtract. Then round the numbers to the nearest ten-thousand and use estimation to determine whether the difference is reasonable.

94.
  80,032
− 19,605

95.
  90,765
− 60,928

96.
  32,574
− 10,961

97.
  96,430
− 59,762

98.
  567,423
− 208,444

99.
  300,712
− 198,714

OBJECTIVE C   To solve application problems

100. **Banking** You have $304 in your checking account. If you write a check for $139, how much is left in your checking account?

101. **Insects** The table at the right shows the number of taste genes and the number of smell genes in the mosquito, fruit fly, and honey bee.

   a. How many more smell genes does the honey bee have than the mosquito?

   b. How many more taste genes does the mosquito have than the fruit fly?

   c. Which of these insects has the best sense of smell?

   d. Which of these insects has the worst sense of taste?

	Mosquito	Fruit Fly	Honey Bee
Taste genes	76	68	10
Smell genes	79	62	170

Source: www.sciencedaily.com

102. **Car Sales** The graph at the right shows the number of cars sold in India for each year from 2003 to 2007.

   a. Has the number of cars sold increased each year from 2003 to 2007?

   b. How many more cars were sold in India in 2007 than in 2003?

   c. Between which two years shown did car sales increase the most?

Tata Motors' One Lakh Car

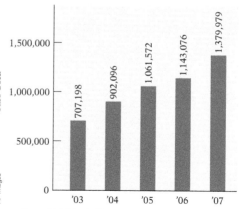

Cars Sold in India

Source: Society of Indian Automobile Manufacturers

103. **Earth Science** Use the graph at the right to find the difference between the maximum height to which Great Fountain
geyser erupts and the maximum height to which Valentine
erupts.

 104. **Earth Science** According to the graph at the right, how much
higher is the eruption of the Giant than that of Old Faithful?

 105. **Education** In a recent year, 775,424 women and 573,079
men earned a bachelor's degree. How many more women
than men earned a bachelor's degree in that year? (*Source:
The National Center for Education Statistics*)

The Maximum Heights of the Eruptions of
Six Geysers at Yellowstone National Park

Demographics The graph at the right shows the expected U.S.
population aged 100 and over for every 2 years from 2010 to 2020.
Use this information for Exercises 106 to 108.

106. What is the expected growth in the population aged 100 and
over during the 10-year period?

107. **a.** Which 2-year period has the smallest expected increase in
the number of people aged 100 and over?
b. Which 2-year period has the greatest expected
increase?

108. What does the difference 208,000 − 166,000 represent?

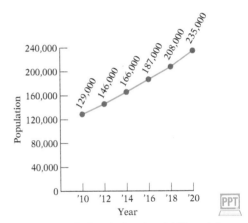

Expected U.S. Population Aged 100 and Over
Source: Census Bureau

109. **Finances** You had a credit card balance of $409 before you
used the card to purchase books for $168, CDs for $36, and
a pair of shoes for $97. You then made a payment to the cred-
it card company of $350. Find your new credit card balance.

Applying the Concepts

110. Answer true or false.
   a. The phrases "the difference between 9 and 5" and "5 less
   than 9" mean the same thing.
   b. $9 − (5 − 3) = (9 − 5) − 3$
   c. Subtraction is an associative operation. *Hint:* See part (b) of this exercise.

 111. Make up a word problem for which the difference between 15 and 8 is the answer.

SECTION

1.4 Multiplication of Whole Numbers

OBJECTIVE A To multiply a number by a single digit

© iStockphoto.com/Ivan Bajic

Six boxes of CD players are ordered. Each box contains eight CD players. How many CD players are ordered?

This problem can be worked by adding 6 eights.

$$8 + 8 + 8 + 8 + 8 + 8 = 48$$

This problem involves repeated addition of the same number and can be worked by a shorter process called **multiplication.** Multiplication is the repeated addition of the same number.

$$8 + 8 + 8 + 8 + 8 + 8 = 48$$

The numbers that are multiplied are called **factors.** The result is called the **product.**

or

6	×	8	=	48
Factor		**Factor**		**Product**

The product of 6 × 8 can be represented on the number line. The arrow representing the whole number 8 is repeated 6 times. The result is the arrow representing 48.

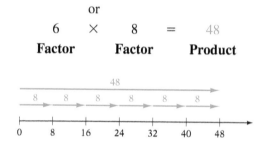

The times sign "×" is only one symbol that is used to indicate multiplication. Each of the expressions that follow represents multiplication.

$$7 \times 8 \qquad 7 \cdot 8 \qquad 7(8) \qquad (7)(8) \qquad (7)8$$

As with addition, there are some useful properties of multiplication.

Multiplication Property of Zero

The product of a number and zero is zero.

$0 \times 4 = 0$
$7 \times 0 = 0$

Tips for Success

Some students think that they can "coast" at the beginning of this course because the topic of Chapter 1 is whole numbers. However, this chapter lays the foundation for the entire course. Be sure you know and understand all the concepts presented. For example, study the properties of multiplication presented in this lesson.

Multiplication Property of One

The product of a number and one is the number.

$1 \times 6 = 6$
$8 \times 1 = 8$

Commutative Property of Multiplication

Two numbers can be multiplied in either order. The product will be the same.

$4 \times 3 = 3 \times 4$
$12 = 12$

Associative Property of Multiplication

Grouping the numbers to be multiplied in any order gives the same result. Do the multiplication inside the parentheses first.

$(4 \times 2) \times 3 = 4 \times (2 \times 3)$
$8 \quad \times 3 = 4 \times \quad 6$
$24 = 24$

The basic facts for multiplying one-digit numbers should be memorized. Multiplication of larger numbers requires the repeated use of the basic multiplication facts.

HOW TO · 1 Multiply: 37 × 4

$$\overset{2}{3}\,7$$
$$\times\quad 4$$
$$\overline{\qquad 8}$$

• 4 × 7 = 28 (2 tens + 8 ones).
  Write the 8 in the ones column and
  carry the 2 to the tens column.

$$\overset{2}{3}\,7$$
$$\times\quad 4$$
$$\overline{14\ \ 8}$$

• The 3 in 37 is 3 tens.

  $4 \times 3 \text{ tens} = 12 \text{ tens}$

  Add the carry digit. $\underline{+\ 2 \text{ tens}}$

  14 tens

• Write the 14. The product is 148.

The phrases below are used to indicate the operation of multiplication. An example is shown at the right of each phrase.

times	7 times 3	7 · 3
the product of	the product of 6 and 9	6 · 9
multiplied by	8 multiplied by 2	2 · 8

EXAMPLE · 1

Multiply: 735 × 9

Solution

$$\overset{3\ 4}{735}$$
$$\times\quad 9$$
$$\overline{6615}$$

• 9 × 5 = 45
  Write the 5 in the ones
  column. Carry the 4 to the
  tens column.
  9 × 3 = 27, 27 + 4 = 31
  9 × 7 = 63, 63 + 3 = 66

YOU TRY IT · 1

Multiply: 648 × 7

Your solution

Solution on p. S2

OBJECTIVE B **To multiply larger whole numbers**

Note the pattern when the following numbers are multiplied.

Multiply the nonzero part of the factors.

Now attach the same number of zeros to the product as the total number of zeros in the factors.

$$4 \times 2$$
$$4 \times 2 = 8$$
zero

$$4 \times 2$$
$$4 \times 2 = 8$$
zeros

$$4 \times 2$$
$$4 \times 2 = 8$$
zeros

$$12 \times 5$$
$$12 \times 5 = 60,$$
zeros

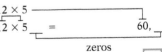

HOW TO 2   Find the product of 47 and 23.

Multiply by the ones digit.	Multiply by the tens digit.	Add.

$$\begin{array}{r} 47 \\ \times\ 23 \\ \hline 141 \end{array} \quad (= 47 \times 3)$$

$$\begin{array}{r} 47 \\ \times\ 23 \\ \hline 141 \\ 940 \end{array} \quad (= 47 \times 20)$$

$$\begin{array}{r} 47 \\ \times\ 23 \\ \hline 141 \\ 940 \\ \hline 1081 \end{array}$$

Writing the 0
is optional.

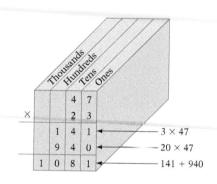

$\times$

Thousands	Hundreds	Tens	Ones	
		4	7	
		2	3	
	1	4	1	← 3 × 47
	9	4	0	← 20 × 47
1	0	8	1	← 141 + 940

The place-value chart on the right above illustrates the placement of the products.

Note the placement of the products when we are multiplying by a factor that contains a zero.

HOW TO 3   Multiply: 439×206

$$\begin{array}{r} 439 \\ \times\ 206 \\ \hline 2634 \\ 000 \quad 0 \times 439 \\ 878 \\ \hline 90{,}434 \end{array}$$

When working the problem, we usually write only one zero. Writing this zero ensures the proper placement of the products.

$$\begin{array}{r} 439 \\ \times\ 206 \\ \hline 2634 \\ 8780 \\ \hline 90{,}434 \end{array}$$

EXAMPLE 2

Find 829 multiplied by 603.

Solution

$$\begin{array}{r} 829 \\ \times\ 603 \\ \hline 2487 \\ 49740 \\ \hline 499{,}887 \end{array}$$

• 3 × 829 = 2487
• Write a zero in the tens column for 0 × 829.
• 6 × 829 = 4974

YOU TRY IT 2

Multiply: 756×305

Your solution

Solution on p. S2

ESTIMATION

Estimating the Product of Two Whole Numbers

Calculate 3267×389. Then use estimation to determine whether the product is reasonable.

Multiply to find the exact product.

3267 **x** 389 **=** $1{,}270{,}863$

To estimate the product, round each number so that it has only one nonzero digit. Then multiply. The estimated answer is 1,200,000, which is very close to the exact product 1,270,863.

$$\begin{array}{r} 3267 \approx \quad 3000 \\ \times\ 389 \approx \quad \times\ 400 \\ \hline 1{,}200{,}000 \end{array}$$

OBJECTIVE C   To solve application problems

EXAMPLE · 3	YOU TRY IT · 3
An auto mechanic receives a salary of $1050 each week. How much does the auto mechanic earn in 4 weeks?	A new-car dealer receives a shipment of 37 cars each month. Find the number of cars the dealer will receive in 12 months.

Strategy	**Your strategy**
To find the mechanic's earnings for 4 weeks, multiply the weekly salary (1050) by the number of weeks (4).	

Solution

```
  1050
×    4
──────
  4200
```

The mechanic earns $4200 in 4 weeks.

Your solution

EXAMPLE · 4	YOU TRY IT · 4
A press operator earns $640 for working a 40-hour week. This week the press operator also worked 7 hours of overtime at $26 an hour. Find the press operator's total pay for the week.	The buyer for Ross Department Store can buy 80 men's suits for $4800. Each sports jacket will cost the store $23. The manager orders 80 men's suits and 25 sports jackets. What is the total cost of the order?

Strategy

To find the press operator's total pay for the week:
- Find the overtime pay by multiplying the hours of overtime (7) by the overtime rate of pay (26).
- Add the weekly salary (640) to the overtime pay.

Your strategy

Solution

```
   26              640
×   7            + 182
─────            ─────
  182              822
  overtime pay
```

The press operator earned $822 this week.

Your solution

Solutions on p. S2

1.4 EXERCISES

OBJECTIVE A   To multiply a number by a single digit

For Exercises 1 to 4, write the expression as a product.

1. $2 + 2 + 2 + 2 + 2 + 2$ **2.** $4 + 4 + 4 + 4 + 4$ **3.** $7 + 7 + 7 + 7$ **4.** $18 + 18 + 18$

For Exercises 5 to 39, multiply.

5.   3
    × 4

6.   2
    × 8

7.   5
    × 7

8.   6
    × 4

9.   5
    × 5

10.   7
     × 7

11.   0
     × 7

12.   8
     × 0

13.   8
     × 9

14.   7
     × 6

15.   66
     × 3

16.   70
     × 4

17.   67
     × 5

18.   127
     × 9

19.   623
     × 4

20.   802
     × 5

21.   607
     × 9

22.   300
     × 5

23.   600
     × 7

24.   906
     × 8

25.   703
     × 9

26.   127
     × 5

27.   632
     × 3

28.   559
     × 4

29.   632
     × 8

30.   524
     × 4

31.   337
     × 5

32.   841
     × 6

33.   6709
     × 7

34.   3608
     × 5

35.   8568
     × 7

36.   5495
     × 4

37.   4780
     × 4

38.   3690
     × 5

39.   9895
     × 2

 40. True or false? The product of two one-digit whole numbers must be a two-digit whole number.

41. Find the product of 5, 7, and 4.

42. Find the product of 6, 2, and 9.

43. What is 3208 multiplied by 7?

44. What is 5009 multiplied by 4?

45. What is 3105 times 6?

46. What is 8957 times 8?

OBJECTIVE B To multiply larger whole numbers

For Exercises 47 to 78, multiply.

47.
$$\begin{array}{r} 16 \\ \times\ 21 \\ \hline \end{array}$$

48.
$$\begin{array}{r} 18 \\ \times\ 24 \\ \hline \end{array}$$

49.
$$\begin{array}{r} 35 \\ \times\ 26 \\ \hline \end{array}$$

50.
$$\begin{array}{r} 27 \\ \times\ 72 \\ \hline \end{array}$$

51.
$$\begin{array}{r} 693 \\ \times\ 91 \\ \hline \end{array}$$

52.
$$\begin{array}{r} 581 \\ \times\ 72 \\ \hline \end{array}$$

53.
$$\begin{array}{r} 419 \\ \times\ 80 \\ \hline \end{array}$$

54.
$$\begin{array}{r} 727 \\ \times\ 60 \\ \hline \end{array}$$

55.
$$\begin{array}{r} 8279 \\ \times\ 46 \\ \hline \end{array}$$

56.
$$\begin{array}{r} 9577 \\ \times\ 35 \\ \hline \end{array}$$

57.
$$\begin{array}{r} 6938 \\ \times\ 78 \\ \hline \end{array}$$

58.
$$\begin{array}{r} 8875 \\ \times\ 67 \\ \hline \end{array}$$

59.
$$\begin{array}{r} 7035 \\ \times\ 57 \\ \hline \end{array}$$

60.
$$\begin{array}{r} 6702 \\ \times\ 48 \\ \hline \end{array}$$

61.
$$\begin{array}{r} 3009 \\ \times\ 35 \\ \hline \end{array}$$

62.
$$\begin{array}{r} 6003 \\ \times\ 57 \\ \hline \end{array}$$

63.
$$\begin{array}{r} 809 \\ \times\ 530 \\ \hline \end{array}$$

64.
$$\begin{array}{r} 607 \\ \times\ 460 \\ \hline \end{array}$$

65.
$$\begin{array}{r} 800 \\ \times\ 325 \\ \hline \end{array}$$

66.
$$\begin{array}{r} 700 \\ \times\ 274 \\ \hline \end{array}$$

67.
$$\begin{array}{r} 987 \\ \times\ 349 \\ \hline \end{array}$$

68.
$$\begin{array}{r} 688 \\ \times\ 674 \\ \hline \end{array}$$

69.
$$\begin{array}{r} 312 \\ \times\ 134 \\ \hline \end{array}$$

70.
$$\begin{array}{r} 423 \\ \times\ 427 \\ \hline \end{array}$$

71.
$$\begin{array}{r} 379 \\ \times\ 500 \\ \hline \end{array}$$

72.
$$\begin{array}{r} 684 \\ \times\ 700 \\ \hline \end{array}$$

73.
$$\begin{array}{r} 985 \\ \times\ 408 \\ \hline \end{array}$$

74.
$$\begin{array}{r} 758 \\ \times\ 209 \\ \hline \end{array}$$

75.
$$\begin{array}{r} 3407 \\ \times\ 309 \\ \hline \end{array}$$

76.
$$\begin{array}{r} 5207 \\ \times\ 902 \\ \hline \end{array}$$

77.
$$\begin{array}{r} 4258 \\ \times\ 986 \\ \hline \end{array}$$

78.
$$\begin{array}{r} 6327 \\ \times\ 876 \\ \hline \end{array}$$

79. Find a one-digit number and a two-digit number whose product is a number that ends in two zeros.

80. What is 5763 times 45?

81. What is 7349 times 27?

82. Find the product of 2, 19, and 34.

83. Find the product of 6, 73, and 43.

84. What is 376 multiplied by 402?

85. What is 842 multiplied by 309?

 For Exercises 86 to 93, use a calculator to multiply. Then use estimation to determine whether the product is reasonable.

86.
$$\begin{array}{r} 8745 \\ \times\ \ \ 63 \\ \hline \end{array}$$

87.
$$\begin{array}{r} 4732 \\ \times\ \ \ 93 \\ \hline \end{array}$$

88.
$$\begin{array}{r} 2937 \\ \times\ 206 \\ \hline \end{array}$$

89.
$$\begin{array}{r} 8941 \\ \times\ 726 \\ \hline \end{array}$$

90.
$$\begin{array}{r} 3097 \\ \times\ 1025 \\ \hline \end{array}$$

91.
$$\begin{array}{r} 6379 \\ \times\ 2936 \\ \hline \end{array}$$

92.
$$\begin{array}{r} 32{,}508 \\ \times\ \ \ 591 \\ \hline \end{array}$$

93.
$$\begin{array}{r} 62{,}504 \\ \times\ \ \ 923 \\ \hline \end{array}$$

OBJECTIVE C   To solve application problems

 94. The price of Braeburn apples is $1.29 per pound, and the price of Cameo apples is $1.79 per pound. Which of the following represents the price of 3 pounds of Braeburn apples and 2 pounds of Cameo apples?
(i) $(3 \times 1.29) + (3 \times 1.79)$ **(ii)** $(2 \times 1.29) + (3 \times 1.79)$
(iii) $5 \times (1.29 + 1.79)$ **(iv)** $(3 \times 1.29) + (2 \times 1.79)$

95. Fuel Efficiency Rob Hill owns a compact car that averages 43 miles on 1 gallon of gas. How many miles could the car travel on 12 gallons of gas?

16 mi

96. Fuel Efficiency A plane flying from Los Angeles to Boston uses 865 gallons of jet fuel each hour. How many gallons of jet fuel were used on a 6-hour flight?

15 m

24 m

97. Geometry The perimeter of a square is equal to four times the length of a side of the square. Find the perimeter of a square whose side measures 16 miles.

98. Geometry The area of a rectangle is equal to the product of the length of the rectangle times its width. Find the area of a rectangle that has a length of 24 meters and a width of 15 meters. The area will be in square meters.

99. Matchmaking Services See the news clipping at the right. **a.** How many marriages occur between eHarmony members each week? **b.** How many marriages occur each year? Use a 365-day year.

> **In the News**
>
> **Find Your Match Online**
>
> eHarmony, the online matchmaking service, boasts marriages among its members at the rate of 90 a day.
>
> *Source: Time*, January 17, 2008

100. College Education See the news clipping at the right. **a.** Find the average cost of tuition, room, and board for 4 years at a public college. **b.** Find the average cost of tuition, room, and board for 4 years at a private college. **c.** Find the difference in cost for tuition, room, and board between 4 years at a private college and 4 years at a public college.

> **In the News**
>
> **Comparing Tuition Costs**
>
> The average annual cost of tuition, room, and board at a four-year public college is $12,796. At a four-year private college, the average cost is $30,367.
>
> *Source:* Kiplinger.com, January 24, 2007

Construction The table at the right shows the hourly wages of four different job classifications at a small construction company. Use this table for Exercises 101 to 103.

101. The owner of this company wants to provide the electrical installation for a new house. On the basis of the architectural plans for the house, it is estimated that it will require 3 electricians, each working 50 hours, to complete the job. What is the estimated cost for the electricians' labor?

Type of Work	Wage per Hour
Electrician	$34
Plumber	$30
Clerk	$16
Bookkeeper	$20

102. Carlos Vasquez, a plumbing contractor, hires 4 plumbers from this company at the hourly wage given in the table. If each plumber works 23 hours, what are the total wages paid by Carlos?

103. The owner of this company estimates that remodeling a kitchen will require 1 electrician working 30 hours and 1 plumber working 33 hours. This project also requires 3 hours of clerical work and 4 hours of bookkeeping. What is the total cost for these four components of this remodeling?

Applying the Concepts

104. Determine whether each of the following statements is always true, sometimes true, or never true.
a. A whole number times zero is zero.
b. A whole number times one is the whole number.
c. The product of two whole numbers is greater than either one of the whole numbers.

105. Safety According to the National Safety Council, in a recent year a death resulting from an accident occurred at the rate of 1 every 5 minutes. At this rate, how many accidental deaths occurred each hour? Each day? Throughout the year? Explain how you arrived at your answers.

106. Demographics According to the Population Reference Bureau, in the world today, 261 people are born every minute and 101 people die every minute. Using this statistic, what is the increase in the world's population every hour? Every day? Every week? Every year? Use a 365-day year. Explain how you arrived at your answers.

SECTION

1.5   Division of Whole Numbers

OBJECTIVE A   **To divide by a single digit with no remainder in the quotient**

Division is used to separate objects into equal groups.

A store manager wants to display 24 new objects equally on 4 shelves. From the diagram, we see that the manager would place 6 objects on each shelf.

The manager's division problem can be written as follows:

✓ **Take Note**

The **divisor** is the number that is divided into another number. The **dividend** is the number into which the divisor is divided. The result is the **quotient**.

Note that the quotient multiplied by the divisor equals the dividend.

$$4\overline{)24}^{\,6} \quad \text{because} \quad \boxed{\overset{6}{\text{Quotient}}} \times \boxed{\overset{4}{\text{Divisor}}} = \boxed{\overset{24}{\text{Dividend}}}$$

$$9\overline{)54}^{\,6} \quad \text{because} \quad 6 \quad \times \quad 9 \quad = \quad 54$$

$$8\overline{)40}^{\,5} \quad \text{because} \quad 5 \quad \times \quad 8 \quad = \quad 40$$

Here are some important quotients and the properties of zero in division:

Properties of One in Division

Any whole number, except zero, divided by itself is 1.

$$8\overline{)8}^{\,1} \qquad 14\overline{)14}^{\,1} \qquad 10\overline{)10}^{\,1}$$

📟 **Integrating Technology**

Enter 8 ÷ 0 = on your calculator. An error message is displayed because division by zero is not allowed.

Any whole number divided by 1 is the whole number.

$$1\overline{)9}^{\,9} \qquad 1\overline{)27}^{\,27} \qquad 1\overline{)10}^{\,10}$$

Properties of Zero in Division

Zero divided by any other whole number is zero.

$$7\overline{)0}^{\,0} \qquad 13\overline{)0}^{\,0} \qquad 10\overline{)0}^{\,0}$$

Division by zero is not allowed.

$$0\overline{)8}^{\,\boxed{?}} \qquad \text{There is no number whose product with 0 is 8.}$$

When the dividend is a larger whole number, the digits in the quotient are found in steps.

HOW TO 1 Divide $4\overline{)3192}$ and check.

$$
\begin{array}{r}
7 \\
4\overline{)3192} \\
-28 \\
\hline
39
\end{array}
$$

• Think $4\overline{)31}$.
• Subtract 7×4.
• Bring down the 9.

$$
\begin{array}{r}
79 \\
4\overline{)3192} \\
-28 \\
\hline
39 \\
-36 \\
\hline
32
\end{array}
$$

• Think $4\overline{)39}$.
• Subtract 9×4.
• Bring down the 2.

$$
\begin{array}{r}
798 \\
4\overline{)3192} \\
-28 \\
\hline
39 \\
-36 \\
\hline
32 \\
-32 \\
\hline
0
\end{array}
$$

• Think $\overline{}$.
• Subtract 8×4.

Check:
$$
\begin{array}{r}
798 \\
\times\quad 4 \\
\hline
3192
\end{array}
$$

The place-value chart can be used to show why this method works.

$$
\begin{array}{r}
\text{HUNDREDS} \ \text{TENS} \ \text{ONES} \\
7\ 9\ 8 \\
4\overline{)3\ 1\ 9\ 2} \\
-\ 2\ 8\ 0\ 0 \\
\hline
3\ 9\ 2 \\
-\ 3\ 6\ 0 \\
\hline
3\ 2 \\
-\ 3\ 2 \\
\hline
0
\end{array}
$$

7 hundreds $\times$ 4

9 tens $\times$ 4

8 ones $\times$ 4

There are other ways of expressing division.

54 divided by 9 equals 6.

54 $\div$ 9 equals 6.

$\dfrac{54}{9}$ equals 6.

EXAMPLE • 1

Divide 7)56 and check.

Solution

$$\begin{array}{r} 8 \\ 7\overline{)56} \end{array}$$

Check: 8 × 7 = 56

YOU TRY IT • 1

Divide 9)63 and check.

Your solution

EXAMPLE • 2

Divide 2808 ÷ 8 and check.

Solution

$$\begin{array}{r} 351 \\ 8\overline{)2808} \\ -24 \\ \hline 40 \\ -40 \\ \hline 08 \\ -8 \\ \hline 0 \end{array}$$

Check: 351 × 8 = 2808

YOU TRY IT • 2

Divide 4077 ÷ 9 and check.

Your solution

EXAMPLE • 3

Divide 7)2856 and check.

Solution

$$\begin{array}{r} 408 \\ 7\overline{)2856} \\ -28 \\ \hline 05 \\ -0 \\ \hline 56 \\ -56 \\ \hline 0 \end{array}$$

• Think 7)5. Place 0 in quotient.
• Subtract 0 × 7.
• Bring down the 6.

Check: 408 × 7 = 2856

YOU TRY IT • 3

Divide 9)6345 and check.

Your solution

Solutions on pp. S2–S3

OBJECTIVE B   **To divide by a single digit with a remainder in the quotient**

Sometimes it is not possible to separate objects into a whole number of equal groups.

A baker has 14 muffins to pack into 3 boxes. Each box holds 4 muffins. From the diagram, we see that after the baker places 4 muffins in each box, there are 2 left over. The 2 is called the **remainder.**

The baker's division problem could be written

Quotient
(Number in each box)

Divisor
(Number of boxes)

$$3\overline{)14}$$
−12

Dividend
(Total number of objects)

2 ← **Remainder**
(Number left over)

The answer to a division problem with a remainder is frequently written

$$\frac{4 \ r2}{3\overline{)14}}$$

Note that

| 4 3 Quotient × Divisor | + | 2 Remainder | = | 14 Dividend | .

EXAMPLE • 4

Divide $4\overline{)2522}$ and check.

Solution

$$\frac{630 \ r2}{4\overline{)2522}}$$
−24
___
12
−12
___
02   • Think $4\overline{)2}$. Place 0 in
−0   quotient.
___
2   • Subtract 0 × 4.

Check: (630 × 4) + 2 =
            2520   + 2 = 2522

YOU TRY IT • 4

Divide $6\overline{)5225}$ and check.

Your solution

EXAMPLE • 5

Divide $9\overline{)27,438}$ and check.

Solution

$$\frac{3,048 \ r6}{9\overline{)27,438}}$$
−27
___
0 4   • Think $9\overline{)4}$.
−0   • Subtract 0 × 9.
___
43
−36
___
78
−72
___
6

Check: (3048 × 9) + 6 =
            27,432   + 6 − 27,438

YOU TRY IT • 5

Divide $7\overline{)21,409}$ and check.

Your solution

Solutions on p. S3

OBJECTIVE C   To divide by larger whole numbers

When the divisor has more than one digit, estimate at each step by using the first digit of the divisor. If that product is too large, lower the guess by 1 and try again.

HOW TO ?   Divide 34)1598 and check.

$$
\begin{array}{r}
5 \\
34)\overline{1598} \\
-170 \\
\end{array}
$$
• Think
• Subtract 5 × 34.

$$
\begin{array}{r}
4 \\
34)\overline{1598} \\
-136 \\
\hline
238 \\
\end{array}
$$
• Subtract 4 × 34.

170 is too large. Lower the guess by 1 and try again.

$$
\begin{array}{r}
47 \\
34)\overline{1598} \\
-136 \\
\hline
238 \\
-238 \\
\hline
0 \\
\end{array}
$$
• Think 3)23.
• Subtract 7 × 34.

Check:
$$
\begin{array}{r}
47 \\
\times 34 \\
\hline
188 \\
141 \\
\hline
1598 \\
\end{array}
$$

Tips for Success
One of the key instructional features of this text is the Example/You Try It pairs. Each Example is completely worked. You are to solve the You Try It problems. When you are ready, check your solution against the one in the Solutions section. The solution for You Try It 6 below is on page S3 (see the reference at the bottom right of the You Try It). See *AIM for Success* at the front of the book.

The phrases below are used to indicate the operation of division. An example is shown at the right of each phrase.

the quotient of	the quotient of 9 and 3	9 ÷ 3
divided by	6 divided by 2	6 ÷ 2

EXAMPLE · 6

Find 7077 divided by 34 and check.

Solution

$$
\begin{array}{r}
208 \text{ r}5 \\
34)\overline{7077} \\
-68 \\
\hline
27 \\
-0 \\
\hline
277 \\
-272 \\
\hline
5 \\
\end{array}
$$
• Think 34)27.
• Place 0 in quotient.
• Subtract 0 × 34.

Check: (208 × 34) + 5 =
          7072   + 5 = 7077

YOU TRY IT · 6

Divide 4578 ÷ 42 and check.

Your solution

Solution on p. S3

EXAMPLE · 7

Find the quotient of 21,312 and 56 and check.

Solution

```
     380 r32
56)21,312        • Think 5)21.
  −168             4 × 56 is too large.
   4 51            Try 3.
  −4 48
     32
    −0
     32
```

Check: $(380 \times 56) + 32 =$
$\qquad 21,280 \ + 32 = 21,312$

YOU TRY IT · 7

Divide $18,359 \div 39$ and check.

Your solution

EXAMPLE · 8

Divide $427)\overline{24,782}$ and check.

Solution

```
        58 r16
427)24,782
   −21 35
     3 432
    −3 416
        16
```

Check: $(58 \times 427) + 16 =$
$\qquad 24,766 \ + 16 = 24,782$

YOU TRY IT · 8

Divide $534)\overline{33,219}$ and check.

Your solution

EXAMPLE · 9

Divide $386)\overline{206,149}$ and check.

Solution

```
        534 r25
386)206,149
   −193 0
     13 14
    −11 58
     1 569
    −1 544
        25
```

Check: $(534 \times 386) + 25 =$
$\qquad 206,124 \ + 25 = 206,149$

YOU TRY IT · 9

Divide $515)\overline{216,848}$ and check.

Your solution

Solutions on p. S3

ESTIMATION

Estimating the Quotient of Two Whole Numbers

Calculate 36,936 ÷ 54. Then use estimation to determine whether the quotient is reasonable.

Divide to find the exact quotient.	36,936 ÷ 54 = 684
To estimate the quotient, round each number so that it contains one nonzero digit. Then divide. The estimated answer is 800, which is close to the exact quotient 684.	36,936 ÷ 54 ≈ 40,000 ÷ 50 = 800

OBJECTIVE D   To solve application problems

The **average** of several numbers is the sum of all the numbers divided by the number of those numbers.

$$\text{Average test score} = \frac{81 + 87 + 80 + 85 + 79 + 86}{6} = \frac{498}{6} = 83$$

HOW TO 3

The table at the right shows what an upper-income family can expect to spend to raise a child to the age of 17 years. Find the average amount spent each year. Round to the nearest dollar.

Expenses to Raise a Child	
Housing	$89,580
Food	$35,670
Transportation	$32,760
Child care/education	$26,520
Clothing	$13,770
Health care	$13,380
Other	$30,090

Source: Department of Agriculture, *Expenditures on Children by Families*

Strategy
To find the average amount spent each year:

• Add all the numbers in the table to find the total amount spent during the 17 years.
• Divide the sum by 17.

Solution

```
   89,580                    14,221
   35,670            17) 241,770
   32,760                 -17
   26,520                  71
   13,770                 -68
   13,380                  37
 + 30,090                 -34
  -------                  37
  241,770  Sum of all     -34
            the costs      30
                          -17
                          13
```

• When rounding to the nearest whole number, compare twice the remainder to the divisor. If twice the remainder is less than the divisor, drop the remainder. If twice the remainder is greater than or equal to the divisor, add 1 to the units digit of the quotient.

• Twice the remainder is 2 × 13 = 26. Because 26 > 17, add 1 to the units digit of the quotient.

The average amount spent each year to raise a child to the age of 17 is $14,222.

EXAMPLE • 10

Ngan Hui, a freight supervisor, shipped 192,600 bushels of wheat in 9 railroad cars. Find the amount of wheat shipped in each car.

Strategy

To find the amount of wheat shipped in each car, divide the number of bushels (192,600) by the number of cars (9).

Solution

$$
\begin{array}{r}
21,400 \\
9{\overline{\smash{\big)}\,192,600}} \\
\underline{-18} \\
12 \\
\underline{-9} \\
36 \\
\underline{-36} \\
0
\end{array}
$$

Each car carried 21,400 bushels of wheat.

YOU TRY IT • 10

Suppose a Michelin retail outlet can store 270 tires on 15 shelves. How many tires can be stored on each shelf?

Your strategy

Your solution

EXAMPLE • 11

The used car you are buying costs $11,216. A down payment of $2000 is required. The remaining balance is paid in 48 equal monthly payments. What is the monthly payment?

Strategy

To find the monthly payment:

- Find the remaining balance by subtracting the down payment (2000) from the total cost of the car (11,216).
- Divide the remaining balance by the number of equal monthly payments (48).

Solution

$$
\begin{array}{r}
11,216 \\
-\ 2,000 \\
\hline
9,216
\end{array}
$$

Remaining balance

$$
\begin{array}{r}
192 \\
48{\overline{\smash{\big)}\,9216}} \\
\underline{-48} \\
441 \\
\underline{-432} \\
96 \\
\underline{-96} \\
0
\end{array}
$$

The monthly payment is $192.

YOU TRY IT • 11

A soft-drink manufacturer produces 12,600 cans of soft drink each hour. Cans are packed 24 to a case. How many cases of soft drink are produced in 8 hours?

Your strategy

Your solution

Solutions on p. S3

1.5 EXERCISES

OBJECTIVE A   To divide by a single digit with no remainder in the quotient

For Exercises 1 to 20, divide.

1. $4\overline{)8}$ **2.** $3\overline{)9}$ **3.** $6\overline{)36}$ **4.** $9\overline{)81}$

5. $7\overline{)49}$ **6.** $5\overline{)80}$ **7.** $6\overline{)96}$ **8.** $6\overline{)480}$

9. $4\overline{)840}$ **10.** $3\overline{)690}$ **11.** $7\overline{)308}$ **12.** $7\overline{)203}$

13. $9\overline{)6327}$ **14.** $4\overline{)2120}$ **15.** $8\overline{)7280}$ **16.** $9\overline{)8118}$

17. $3\overline{)64,680}$ **18.** $4\overline{)50,760}$ **19.** $6\overline{)21,480}$ **20.** $5\overline{)18,050}$

21. What is 7525 divided by 7?   **22.** What is 32,364 divided by 4?

23. If the dividend and the divisor in a division problem are the same number, what is the quotient?

For Exercises 24 to 27, use the relationship between multiplication and division to complete the multiplication problem.

24. ___ × 7 = 364   **25.** 8 × ___ = 376   **26.** 5 × ___ = 170   **27.** ___ × 4 = 92

OBJECTIVE B   To divide by a single digit with a remainder in the quotient

For Exercises 28 to 50, divide.

28. $4\overline{)9}$ **29.** $2\overline{)7}$ **30.** $5\overline{)27}$ **31.** $9\overline{)88}$ **32.** $3\overline{)40}$

33. $6\overline{)97}$ **34.** $8\overline{)83}$ **35.** $5\overline{)54}$ **36.** $7\overline{)632}$ **37.** $4\overline{)363}$

38. $4\overline{)921}$ **39.** $7\overline{)845}$ **40.** $8\overline{)1635}$ **41.** $5\overline{)1548}$ **42.** $7\overline{)9432}$

43. $7\overline{)8124}$ **44.** $3\overline{)5162}$ **45.** $5\overline{)3542}$ **46.** $8\overline{)3274}$

47. $4\overline{)15,301}$ **48.** $7\overline{)43,500}$ **49.** $8\overline{)72,354}$ **50.** $5\overline{)43,542}$

51. What is 45,738 divided by 4? Round to the nearest ten.

52. What is 37,896 divided by 9? Round to the nearest hundred.

53. What is 3572 divided by 7? Round to the nearest ten.

54. What is 78,345 divided by 4? Round to the nearest hundred.

 55. True or false? When a three-digit number is divided by a one-digit number, the quotient can be a one-digit number.

OBJECTIVE C   To divide by larger whole numbers

For Exercises 56 to 83, divide.

56. $27\overline{)96}$ **57.** $44\overline{)82}$ **58.** $42\overline{)87}$ **59.** $67\overline{)93}$

60. $41\overline{)897}$ **61.** $32\overline{)693}$ **62.** $23\overline{)784}$ **63.** $25\overline{)772}$

64. $74\overline{)600}$ **65.** $92\overline{)500}$ **66.** $70\overline{)329}$ **67.** $50\overline{)467}$

68. $36\overline{)7225}$ **69.** $44\overline{)8821}$ **70.** $19\overline{)3859}$ **71.** $32\overline{)9697}$

72. $88\overline{)3127}$ **73.** $92\overline{)6177}$ **74.** $33\overline{)8943}$ **75.** $27\overline{)4765}$

76. $22\overline{)98,654}$ **77.** $77\overline{)83,629}$ **78.** $64\overline{)38,912}$ **79.** $78\overline{)31,434}$

80. $206\overline{)3097}$

81. $504\overline{)6504}$

82. $654\overline{)1217}$

83. $546\overline{)2344}$

84. Find the quotient of 5432 and 21.

85. Find the quotient of 8507 and 53.

86. What is 37,294 divided by 72?

87. What is 76,788 divided by 46?

88. Find 23,457 divided by 43. Round to the nearest hundred.

89. Find 341,781 divided by 43. Round to the nearest ten.

 90. True or false? If the remainder of a division problem is 210, then the divisor was less than 210.

 For Exercises 91 to 102, use a calculator to divide. Then use estimation to determine whether the quotient is reasonable.

91. $76\overline{)389,804}$

92. $53\overline{)117,925}$

93. $29\overline{)637,072}$

94. $67\overline{)738,072}$

95. $38\overline{)934,648}$

96. $34\overline{)906,304}$

97. $309\overline{)876,324}$

98. $642\overline{)323,568}$

99. $209\overline{)632,016}$

100. $614\overline{)332,174}$

101. $179\overline{)5,734,444}$

102. $374\overline{)7,712,254}$

OBJECTIVE D　　To solve application problems

 Insurance The table at the right shows the sources of insurance claims for losses of laptop computers in a recent year. Claims have been rounded to the nearest ten thousand dollars. Use this information for Exercises 103 and 104.

103. What was the average monthly claim for theft?

104. For all sources combined, find the average claims per month.

Source	Claims
Accidents	$560,000
Theft	$300,000
Power surge	$80,000
Lightning	$50,000
Transit	$20,000
Water/flood	$20,000
Other	$110,000

Source: Safeware, The Insurance Company

Work Hours The table at the right shows, for different countries, the average number of hours per year that employees work. Use this information for Exercises 105 and 106. Use a 50-week year. Round answers to the nearest whole number.

Country	Annual Number of Hours Worked
Britian	1731
France	1656
Japan	1889
Norway	1399
United States	1966

Source: International Labor Organization

105. What is the average number of hours worked per week by employees in Britain?

106. On average, how many more hours per week do employees in the United States work than employees in France?

107. **Coins** The U.S. Mint estimates that about 114,000,000,000 of the 312,000,000,000 pennies it has minted over the last 30 years are in active circulation. That works out to how many pennies in circulation for each of the 300,000,000 people living in the United States?

© blickwinkel/Alamy

108. **Toy Sales** Every hour, 25,200 sets of Legos® are sold by retailers worldwide. (*Source: Time,* February 11, 2008) How many sets of Legos are sold each second by retailers worldwide?

In the News

Holiday Mail Delivery

The U.S. Postal Service expects to deliver 20 billion pieces of mail between Thanksgiving and Christmas this year.

Source: www.usps.com

109. **U.S. Postal Service** There are 114 households in the United States. Use the information in the news clipping at the right to determine, on average, how many pieces of mail each household will receive between Thanksgiving and Christmas this year. Round to the nearest whole number.

110. **Arlington National Cemetery** There are approximately 10,200 funerals each year at Arlington National Cemetery. (*Source:* www.arlingtoncemetery.org) Calculate the average number of funerals each day at Arlington National Cemetery. Round to the nearest whole number.

Arlington National Cemetery

© 2009 Jupiterimages

111. Which problems below require division to solve?
   (i) Four friends want to share a restaurant bill of $45.65 equally. Find the amount that each friend should pay.
   (ii) On average, Sam spends $30 a week on gas. Find Sam's average yearly expenditure for gas.
   (iii) Emma's 12 phone bills for last year totaled $660. Find Emma's average monthly phone bill.

Applying the Concepts

112. **Wages** A sales associate earns $374 for working a 40-hour week. Last week the associate worked an additional 9 hours at $13 an hour. Find the sales associate's total pay for last week's work.

113. Payroll Deductions Your paycheck shows deductions of $225 for savings, $98 for taxes, and $27 for insurance. Find the total of the three deductions.

 Dairy Products The topic of the graph at the right is the eggs produced in the United States in a recent year. It shows where the eggs that were produced went or how they were used. Use this table for Exercises 114 and 115.

114. Use the graph to determine the total number of cases of eggs produced during the year.

115. How many more cases of eggs were sold by retail stores than were used for non-shell products?

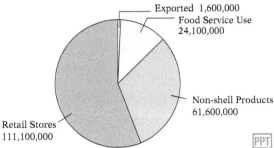

Eggs Produced in the United States (in cases)
Source: American Egg Board

 Finance The graph at the right shows the annual expenditures, in a recent year, of the average household in the United States. Use this information for Exercises 116 to 118. Round answers to the nearest whole number.

116. What is the total amount spent annually by the average household in the United States?

117. What is the average monthly expense for housing?

118. What is the difference between the average monthly expense for food and the average monthly expense for health care?

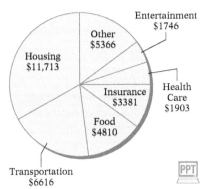

Average Annual Household Expenses
Source: Bureau of Labor Statistics Consumer Expenditure Survey

The Military The graph at the right shows the basic monthly pay for Army officers with over 20 years of service. Use this graph for Exercises 119 and 120.

119. What is a major's annual pay?

120. What is the difference between a colonel's annual pay and a lieutenant colonel's annual pay?

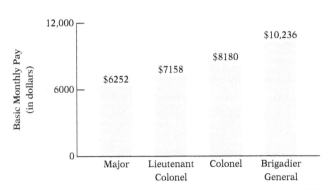

Basic Montly Pay for Army Officers
Source: Department of Defense

121. Finances You purchase a used car with a down payment of $2500 and monthly payments of $195 for 48 months. Find the total amount paid for the car.

SECTION

1.6

Exponential Notation and the Order of Operations Agreement

OBJECTIVE A

To simplify expressions that contain exponents

Repeated multiplication of the same factor can be written in two ways:

$$3 \cdot 3 \cdot 3 \cdot 3 \cdot 3 \quad \text{or} \quad 3^5 \leftarrow \textbf{Exponent}$$

The **exponent** indicates how many times the factor occurs in the multiplication. The expression 3^5 is in **exponential notation.**

It is important to be able to read numbers written in exponential notation.

$6 = 6^1$ is read "six to the first **power**" or just "six." Usually the exponent 1 is not written.

$6 \cdot 6 = 6^2$ is read "six squared" or "six to the second power."

$6 \cdot 6 \cdot 6 = 6^3$ is read "six cubed" or "six to the third power."

$6 \cdot 6 \cdot 6 \cdot 6 = 6^4$ is read "six to the fourth power."

$6 \cdot 6 \cdot 6 \cdot 6 \cdot 6 = 6^5$ is read "six to the fifth power."

Each place value in the place-value chart can be expressed as a power of 10.

Ten =	10	=	10	$= 10^1$
Hundred =	100	=	$10 \cdot 10$	$= 10^2$
Thousand =	1000	=	$10 \cdot 10 \cdot 10$	$= 10^3$
Ten-thousand =	10,000	=	$10 \cdot 10 \cdot 10 \cdot 10$	$= 10^4$
Hundred-thousand =	100,000	=	$10 \cdot 10 \cdot 10 \cdot 10 \cdot 10$	$= 10^5$
Million =	1,000,000	=	$10 \cdot 10 \cdot 10 \cdot 10 \cdot 10 \cdot 10$	$= 10^6$

Integrating Technology

A calculator can be used to evaluate an exponential expression. The $\boxed{y^x}$ key (or, on some calculators, an $\boxed{x^y}$ or $\boxed{\wedge}$ key) is used to enter the exponent. For instance, for the example at the right, enter 4 $\boxed{y^x}$ 3 $\boxed{=}$. The display reads 64.

To simplify a numerical expression containing exponents, write each factor as many times as indicated by the exponent and carry out the indicated multiplication.

$$4^3 = 4 \cdot 4 \cdot 4 = 64$$

$$2^2 \cdot 3^4 = (2 \cdot 2) \cdot (3 \cdot 3 \cdot 3 \cdot 3) = 4 \cdot 81 = 324$$

EXAMPLE 1

Write $3 \cdot 3 \cdot 3 \cdot 5 \cdot 5$ in exponential notation.

Solution $3 \cdot 3 \cdot 3 \cdot 5 \cdot 5 = 3^3 \cdot 5^2$

YOU TRY IT 1

Write $2 \cdot 2 \cdot 2 \cdot 2 \cdot 3 \cdot 3 \cdot 3$ in exponential notation.

Your solution

EXAMPLE 2

Write as a power of 10: $10 \cdot 10 \cdot 10 \cdot 10$

Solution $10 \cdot 10 \cdot 10 \cdot 10 = 10^4$

YOU TRY IT 2

Write as a power of 10: $10 \cdot 10 \cdot 10 \cdot 10 \cdot 10 \cdot 10 \cdot 10$

Your solution

EXAMPLE 3

Simplify $3^2 \cdot 5^3$.

Solution $3^2 \cdot 5^3 = (3 \cdot 3) \cdot (5 \cdot 5 \cdot 5)$
$= 9 \cdot 125 = 1125$

YOU TRY IT 3

Simplify $2^3 \cdot 5^2$.

Your solution

Solutions on p. S4

OBJECTIVE B   To use the Order of Operations Agreement to simplify expressions

More than one operation may occur in a numerical expression. The answer may be different, depending on the order in which the operations are performed. For example, consider $3 + 4 \times 5$.

Multiply first, then add.

$$\underbrace{3 + 4 \times 5}$$
$$\underbrace{3 + 20}$$
$$23$$

Add first, then multiply.

$$\underbrace{3 + 4} \times 5$$
$$\underbrace{7 \times 5}$$
$$35$$

An Order of Operations Agreement is used so that only one answer is possible.

The Order of Operations Agreement

Step 1. Do all the operations inside parentheses.

Step 2. Simplify any number expressions containing exponents.

Step 3. Do multiplication and division as they occur from left to right.

Step 4. Do addition and subtraction as they occur from left to right.

Integrating Technology

Many scientific calculators have an **x²** key. This key is used to square the displayed number. For example, after the user presses 2 **x²** **=**, the display reads 4.

HOW TO 1 $3 \times (2 + 1) - 2^2 + 4 \div 2$ by using the Order of Operations Agreement.

$3 \times \underbrace{(2 + 1)} - 2^2 + 4 \div 2$
$3 \times 3 - \underbrace{2^2} + 4 \div 2$
$\underbrace{3 \times 3} - 4 + \underbrace{4 \div 2}$
$\underbrace{9 - 4} + 2$
$\underbrace{5 + 2}$
7

1. Perform operations in parentheses.
2. Simplify expressions with exponents.
3. Do multiplication and division as they occur from left to right.
4. Do addition and subtraction as they occur from left to right.

One or more of these steps may not be needed to simplify an expression. In that case, proceed to the next step in the Order of Operations Agreement.

HOW TO 2 Simplify $5 + 8 \div 2$.   There are no parentheses or exponents. Proceed to Step 3 of the agreement.

$5 + \underbrace{8 \div 2}$
$\underbrace{5 + 4}$
9

3. Do multiplication or division.
4. Do addition or subtraction.

EXAMPLE 4

Simplify: $64 \div (8 - 4)^2 \cdot 9 - 5^2$

Solution

$64 \div (8 - 4)^2 \cdot 9 - 5^2$
$= 64 \div 4^2 \cdot 9 - 5^2$ • Parentheses
$= 64 \div 16 \cdot 9 - 25$ • Exponents
$= 4 \cdot 9 - 25$ • Division and
    multiplication
$= 36 - 25$ • Subtraction
$= 11$

YOU TRY IT 4

Simplify: $5 \cdot (8 - 4)^2 \div 4 - 2$

Your solution

Solution on p. S4

1.6 EXERCISES

OBJECTIVE A   To simplify expressions that contain exponents

For Exercises 1 to 12, write the number in exponential notation.

1. $2 \cdot 2 \cdot 2$

2. $7 \cdot 7 \cdot 7 \cdot 7 \cdot 7$

3. $6 \cdot 6 \cdot 6 \cdot 7 \cdot 7 \cdot 7 \cdot 7$

4. $6 \cdot 6 \cdot 9 \cdot 9 \cdot 9 \cdot 9$

5. $2 \cdot 2 \cdot 2 \cdot 3 \cdot 3 \cdot 3$

6. $3 \cdot 3 \cdot 10 \cdot 10$

7. $5 \cdot 7 \cdot 7 \cdot 7 \cdot 7 \cdot 7$

8. $4 \cdot 4 \cdot 4 \cdot 5 \cdot 5 \cdot 5$

9. $3 \cdot 3 \cdot 3 \cdot 6 \cdot 6 \cdot 6 \cdot 6$

10. $2 \cdot 2 \cdot 5 \cdot 5 \cdot 5 \cdot 8$

11. $3 \cdot 3 \cdot 3 \cdot 5 \cdot 9 \cdot 9 \cdot 9$

12. $2 \cdot 2 \cdot 2 \cdot 4 \cdot 7 \cdot 7 \cdot 7$

For Exercises 13 to 37, simplify.

13. 2^3

14. 2^6

15. $2^4 \cdot 5^2$

16. $2^6 \cdot 3^2$

17. $3^2 \cdot 10^2$

18. $2^3 \cdot 10^4$

19. $6^2 \cdot 3^3$

20. $4^3 \cdot 5^2$

21. $5 \cdot 2^3 \cdot 3$

22. $6 \cdot 3^2 \cdot 4$

23. $2^2 \cdot 3^2 \cdot 10$

24. $3^2 \cdot 5^2 \cdot 10$

25. $0^2 \cdot 4^3$

26. $6^2 \cdot 0^3$

27. $3^2 \cdot 10^4$

28. $5^3 \cdot 10^3$

29. $2^2 \cdot 3^3 \cdot 5$

30. $5^2 \cdot 7^3 \cdot 2$

31. $2 \cdot 3^4 \cdot 5^2$

32. $6 \cdot 2^6 \cdot 7^2$

33. $5^2 \cdot 3^2 \cdot 7^2$

34. $4^2 \cdot 9^2 \cdot 6^2$

35. $3^4 \cdot 2^6 \cdot 5$

36. $4^3 \cdot 6^3 \cdot 7$

37. $4^2 \cdot 3^3 \cdot 10^4$

 38. Rewrite the expression using the numbers 3 and 5 exactly once. Then simplify the expression.
   a. $3 + 3 + 3 + 3 + 3$
   b. $3 \cdot 3 \cdot 3 \cdot 3 \cdot 3$

OBJECTIVE B   To use the Order of Operations Agreement to simplify expressions

For Exercises 39 to 77, simplify by using the Order of Operations Agreement.

39. $4 - 2 + 3$

40. $6 - 3 + 2$

41. $6 \div 3 + 2$

42. $8 \div 4 + 8$

43. $6 \cdot 3 + 5$　　　　**44.** $5 \cdot 9 + 2$　　　　**45.** $3^2 - 4$　　　　**46.** $5^2 - 17$

47. $4 \cdot (5 - 3) + 2$　　**48.** $3 + (4 + 2) \div 3$　　**49.** $5 + (8 + 4) \div 6$　　**50.** $8 - 2^2 + 4$

51. $16 \cdot (3 + 2) \div 10$　**52.** $12 \cdot (1 + 5) \div 12$　**53.** $10 - 2^3 + 4$　　**54.** $5 \cdot 3^2 + 8$

55. $16 + 4 \cdot 3^2$　　**56.** $12 + 4 \cdot 2^3$　　**57.** $16 + (8 - 3) \cdot 2$　　**58.** $7 + (9 - 5) \cdot 3$

59. $2^2 + 3 \cdot (6 - 2)^2$　**60.** $3^3 + 5 \cdot (8 - 6)^3$　**61.** $2^2 \cdot 3^2 + 2 \cdot 3$　**62.** $4 \cdot 6 + 3^2 \cdot 4^2$

63. $16 - 2 \cdot 4$　　　**64.** $12 + 3 \cdot 5$　　　**65.** $3 \cdot (6 - 2) + 4$

66. $5 \cdot (8 - 4) - 6$　　**67.** $8 - (8 - 2) \div 3$　　**68.** $12 - (12 - 4) \div 4$

69. $8 + 2 - 3 \cdot 2 \div 3$　　**70.** $10 + 1 - 5 \cdot 2 \div 5$　　**71.** $3 \cdot (4 + 2) \div 6$

72. $(7 - 3)^2 \div 2 - 4 + 8$　**73.** $20 - 4 \div 2 \cdot (3 - 1)^3$　**74.** $12 \div 3 \cdot 2^2 + (7 - 3)^2$

75. $(4 - 2) \cdot 6 \div 3 + (5 - 2)^2$　**76.** $18 - 2 \cdot 3 + (4 - 1)^3$　**77.** $100 \div (2 + 3)^2 - 8 \div 2$

 For Exercises 78 to 80, insert parentheses as needed in the expression $8 - 2 \cdot 3 + 1$ in order to make the statement true.

78. $8 - 2 \cdot 3 + 1 = 3$　　　**79.** $8 - 2 \cdot 3 + 1 = 0$　　　**80.** $8 - 2 \cdot 3 + 1 = 24$

Applying the Concepts

 81. Explain the difference that the order of operations makes between **a.** $(14 - 2) \div 2 \cdot 3$ and **b.** $(14 - 2) \div (2 \cdot 3)$. Work the two problems. What is the difference between the larger answer and the smaller answer?

SECTION

1.7   Prime Numbers and Factoring

OBJECTIVE A   To factor numbers

Whole-number **factors of a number** divide that number evenly (there is no remainder).

1, 2, 3, and 6 are whole-number factors of 6 because they divide 6 evenly.

$$\frac{6}{1\overline{)6}} \quad \frac{3}{2\overline{)6}} \quad \frac{2}{3\overline{)6}} \quad \frac{1}{6\overline{)6}}$$

Note that both the divisor and the quotient are factors of the dividend.

To find the factors of a number, try dividing the number by 1, 2, 3, 4, 5, ... Those numbers that divide the number evenly are its factors. Continue this process until the factors start to repeat.

HOW TO 1   Find all the factors of 42.

$42 \div 1 = 42$	1 and 42 are factors.
$42 \div 2 = 21$	2 and 21 are factors.
$42 \div 3 = 14$	3 and 14 are factors.
$42 \div 4$	Will not divide evenly
$42 \div 5$	Will not divide evenly
$42 \div 6 = 7$	6 and 7 are factors. ⎫ Factors are repeating; all the
$42 \div 7 = 6$	7 and 6 are factors. ⎭ factors of 42 have been found.

1, 2, 3, 6, 7, 14, 21, and 42 are factors of 42.

The following rules are helpful in finding the factors of a number.

2 is a factor of a number if the last digit of the number is 0, 2, 4, 6, or 8.

436 ends in 6; therefore, 2 is a factor of 436. ($436 \div 2 = 218$)

3 is a factor of a number if the sum of the digits of the number is divisible by 3.

The sum of the digits of 489 is $4 + 8 + 9 = 21$. 21 is divisible by 3. Therefore, 3 is a factor of 489. ($489 \div 3 = 163$)

5 is a factor of a number if the last digit of the number is 0 or 5.

520 ends in 0; therefore, 5 is a factor of 520. ($520 \div 5 = 104$)

EXAMPLE 1

Find all the factors of 30.

Solution
$30 \div 1 = 30$
$30 \div 2 = 15$
$30 \div 3 = 10$
$30 \div 4$ Will not divide evenly
$30 \div 5 = 6$
$30 \div 6 = 5$ Factors repeating

1 2, 3, 5, 6, 10, 15, and 30 are factors of 30.

YOU TRY IT 1

Find all the factors of 40.

Your solution

Solution on p. S4

OBJECTIVE B   To find the prime factorization of a number

Point of Interest

Prime numbers are an important part of cryptology, the study of secret codes. To make it less likely that codes can be broken, cryptologists use prime numbers that have hundreds of digits.

A number is a **prime number** if its only whole-number factors are 1 and itself. 7 is prime because its only factors are 1 and 7. If a number is not prime, it is called a **composite number.** Because 6 has factors of 2 and 3, 6 is a composite number. The number 1 is not considered a prime number; therefore, it is not included in the following list of prime numbers less than 50.

$$2, 3, 5, 7, 11, 13, 17, 19, 23, 29, 31, 37, 41, 43, 47$$

The **prime factorization** of a number is the expression of the number as a product of its prime factors. We use a "T-diagram" to find the prime factors of 60. Begin with the smallest prime number as a trial divisor, and continue with prime numbers as trial divisors until the final quotient is 1.

$$
\begin{array}{c|c}
\multicolumn{2}{c}{60} \\
\hline
2 & 30 \\
2 & 15 \\
3 & 5 \\
5 & 1 \\
\end{array}
\qquad
\begin{aligned}
60 \div 2 &= 30 \\
30 \div 2 &= 15 \\
15 \div 3 &= 5 \\
5 \div 5 &= 1
\end{aligned}
$$

The prime factorization of 60 is $2 \cdot 2 \cdot 3 \cdot 5$.

Finding the prime factorization of larger numbers can be more difficult. Try each prime number as a trial divisor. Stop when the square of the trial divisor is greater than the number being factored.

HOW TO · 2   Find the prime factorization of 106.

$$
\begin{array}{c|c}
\multicolumn{2}{c}{106} \\
\hline
2 & 53 \\
53 & 1 \\
\end{array}
$$

• 53 cannot be divided evenly by 2, 3, 5, 7, or 11. Prime numbers greater than 11 need not be tested because 11^2 is greater than 53.

The prime factorization of 106 is $2 \cdot 53$.

EXAMPLE · 2

Find the prime factorization of 315.

Solution

$$
\begin{array}{c|c}
\multicolumn{2}{c}{315} \\
\hline
3 & 105 \\
3 & 35 \\
5 & 7 \\
7 & 1 \\
\end{array}
$$

• $315 \div 3 = 105$
• $105 \div 3 = 35$
• $35 \div 5 = 7$
• $7 \div 7 = 1$

$315 = 3 \cdot 3 \cdot 5 \cdot 7$

YOU TRY IT · 2

Find the prime factorization of 44.

Your solution

EXAMPLE · 3

Find the prime factorization of 201.

Solution

$$
\begin{array}{c|c}
\multicolumn{2}{c}{201} \\
\hline
3 & 67 \\
67 & 1 \\
\end{array}
$$

• Try only 2, 3, 5, 7, and 11 because $11^2 > 67$.

$201 = 3 \cdot 67$

YOU TRY IT · 3

Find the prime factorization of 177.

Your solution

Solutions on p. S4

1.7 EXERCISES

OBJECTIVE A   To factor numbers

For Exercises 1 to 40, find all the factors of the number.

1. 4   **2.** 6   **3.** 10   **4.** 20

5. 7   **6.** 12   **7.** 9   **8.** 8

9. 13   **10.** 17   **11.** 18   **12.** 24

13. 56   **14.** 36   **15.** 45   **16.** 28

17. 29   **18.** 33   **19.** 22   **20.** 26

21. 52   **22.** 49   **23.** 82   **24.** 37

25. 57   **26.** 69   **27.** 48   **28.** 64

29. 95   **30.** 46   **31.** 54   **32.** 50

33. 66   **34.** 77   **35.** 80   **36.** 100

37. 96   **38.** 85   **39.** 90   **40.** 101

 41. True or false? A number can have an odd number of factors.

 42. True or false? If a number has exactly four factors, then the product of those four factors must be the number.

OBJECTIVE B   To find the prime factorization of a number

For Exercises 43 to 86, find the prime factorization.

43. 6   **44.** 14   **45.** 17   **46.** 83

47. 24 **48.** 12 **49.** 27 **50.** 9

51. 36 **52.** 40 **53.** 19 **54.** 37

55. 90 **56.** 65 **57.** 115 **58.** 80

59. 18 **60.** 26 **61.** 28 **62.** 49

63. 31 **64.** 42 **65.** 62 **66.** 81

67. 22 **68.** 39 **69.** 101 **70.** 89

71. 66 **72.** 86 **73.** 74 **74.** 95

75. 67 **76.** 78 **77.** 55 **78.** 46

79. 120 **80.** 144 **81.** 160 **82.** 175

83. 216 **84.** 400 **85.** 625 **86.** 225

87. True or false? The prime factorization of 102 is $2 \cdot 51$.

Applying the Concepts

88. In 1742, Christian Goldbach conjectured that every even number greater than 2 could be expressed as the sum of two prime numbers. Show that this conjecture is true for 8, 24, and 72. (*Note:* Mathematicians have not yet been able to determine whether Goldbach's conjecture is true or false.)

89. Explain why 2 is the only even prime number.

FOCUS ON PROBLEM SOLVING

Questions to Ask

© Brownie Harris/Corbis

You encounter problem-solving situations every day. Some problems are easy to solve, and you may mentally solve these problems without considering the steps you are taking in order to draw a conclusion. Others may be more challenging and may require more thought and consideration.

Suppose a friend suggests that you both take a trip over spring break. You'd like to go. What questions go through your mind? You might ask yourself some of the following questions:

How much will the trip cost? What will be the cost for travel, hotel rooms, meals, and so on?

Are some costs going to be shared by both me and my friend?

Can I afford it?

How much money do I have in the bank?

How much more money than I have now do I need?

How much time is there to earn that much money?

How much can I earn in that amount of time?

How much money must I keep in the bank in order to pay the next tuition bill (or some other expense)?

These questions require different mathematical skills. Determining the cost of the trip requires **estimation;** for example, you must use your knowledge of air fares or the cost of gasoline to arrive at an estimate of these costs. If some of the costs are going to be shared, you need to **divide** those costs by 2 in order to determine your share of the expense. The question regarding how much more money you need requires **subtraction:** the amount needed minus the amount currently in the bank. To determine how much money you can earn in the given amount of time requires **multiplication**—for example, the amount you earn per week times the number of weeks to be worked. To determine if the amount you can earn in the given amount of time is sufficient, you need to use your knowledge of **order relations** to compare the amount you can earn with the amount needed.

Facing the problem-solving situation described above may not seem difficult to you. The reason may be that you have faced similar situations before and, therefore, know how to work through this one. You may feel better prepared to deal with a circumstance such as this one because you know what questions to ask. An important aspect of learning to solve problems is learning what questions to ask. As you work through application problems in this text, try to become more conscious of the mental process you are going through. You might begin the process by asking yourself the following questions whenever you are solving an application problem.

1. Have I read the problem enough times to be able to understand the situation being described?

2. Will restating the problem in different words help me to understand the problem situation better?

3. What facts are given? (You might make a list of the information contained in the problem.)

4. What information is being asked for?

5. What relationship exists among the given facts? What relationship exists between the given facts and the solution?

6. What mathematical operations are needed in order to solve the problem?

Try to focus on the problem-solving situation, not on the computation or on getting the answer quickly. And remember, the more problems you solve, the better able you will be to solve other problems in the future, partly because you are learning what questions to ask.

PROJECTS AND GROUP ACTIVITIES

Order of Operations

Does your calculator use the Order of Operations Agreement? To find out, try this problem:

$$2 + 4 \cdot 7$$

If your answer is 30, then the calculator uses the Order of Operations Agreement. If your answer is 42, it does not use that agreement.

Even if your calculator does not use the Order of Operations Agreement, you can still correctly evaluate numerical expressions. The parentheses keys, **(** and **)** , are used for this purpose.

Remember that $2 + 4 \cdot 7$ means $2 + (4 \cdot 7)$ because the multiplication must be completed before the addition. To evaluate this expression, enter the following:

Enter:	2	**+**	**(**	4	**×**	7	**)**	**=**
Display:	2	2	**(**	4	4	7	28	30

When using your calculator to evaluate numerical expressions, insert parentheses around multiplications and around divisions. This has the effect of forcing the calculator to do the operations in the order you want.

For Exercises 1 to 10, evaluate.

1. $3 \cdot 8 - 5$

2. $6 + 8 \div 2$

3. $3 \cdot (8 - 2)^2$

4. $24 - (4 - 2)^2 \div 4$

5. $3 + (6 \div 2 + 4)^2 - 2$

6. $16 \div 2 + 4 \cdot (8 - 12 \div 4)^2 - 50$

7. $3 \cdot (15 - 2 \cdot 3) - 36 \div 3$

8. $4 \cdot 2^2 - (12 + 24 \div 6) + 5$

9. $16 \div 4 \cdot 3 + (3 \cdot 4 - 5) + 2$

10. $15 \cdot 3 \div 9 + (2 \cdot 6 - 3) + 4$

Patterns in Mathematics   For the circle at the left, use a straight line to connect each dot on the circle with every other dot on the circle. How many different straight lines are there?

Follow the same procedure for each of the circles shown below. How many different straight lines are there in each?

Find a pattern to describe the number of dots on a circle and the corresponding number of different lines drawn. Use the pattern to determine the number of different lines that would be drawn in a circle with 7 dots and in a circle with 8 dots.

Now use the pattern to answer the following question. You are arranging a tennis tournament with 9 players. How many singles matches will be played among the 9 players if each player plays each of the other players only once?

Search the World Wide Web   Go to www.census.gov on the Internet.

1. Find a projection for the total U.S. population 10 years from now and a projection for the total population 20 years from now. Record the two numbers.

2. Use the data from Exercise 1 to determine the expected growth in the population over the next 10 years.

3. Use the answer from Exercise 2 to find the average increase in the U.S. population per year over the next 10 years. Round to the nearest million.

4. Use data in the population table you found to write two word problems. Then state whether addition, subtraction, multiplication, or division is required to solve each of the problems.

CHAPTER 1

SUMMARY

KEY WORDS	EXAMPLES
The *whole numbers* are 0, 1, 2, 3, 4, 5, 6, 7, 8, 9, 10, ... [1.1A, p. 2]	
The *graph of a whole number* is shown by placing a heavy dot directly above that number on the number line. [1.1A, p. 2]	This is the graph of 4 on the number line.
The symbol for *is less than* is $<$. The symbol for *is greater than* is $>$. These symbols are used to show the order relation between two numbers. [1.1A, p. 2]	$3 < 7$ $9 > 2$

When a whole number is written using the digits 0, 1, 2, 3, 4, 5, 6, 7, 8, and 9, it is said to be in *standard form*. The position of each digit in the number determines the digit's *place value*. The place values are used to write the expanded form of a number. [1.1B, p. 3]

The number 598,317 is in standard form. The digit 8 is in the thousands place. The number 598,317 is written in expanded form as 500,000 + 90,000 + 8000 + 300 + 10 + 7.

Addition is the process of finding the total of two or more numbers. The numbers being added are called *addends*. The result is the *sum*. [1.2A, p. 8]

$$\begin{array}{r} \overset{1\ \ 11}{8{,}762} \\ +\ 1{,}359 \\ \hline 10{,}121 \end{array}$$

Subtraction is the process of finding the difference between two numbers. The *minuend* minus the *subtrahend* equals the *difference*. [1.3A, p. 16]

$$\begin{array}{r} \overset{4\ \ 11\ 11\ \ 6\ \ 13}{8\ 2{,}\cancel{8}\ \cancel{7}\ \cancel{3}} \\ -\ 3\ 4{,}9\ 6\ 8 \\ \hline 1\ 7{,}2\ 0\ 5 \end{array}$$

Multiplication is the repeated addition of the same number. The numbers that are multiplied are called *factors*. The result is the *product*. [1.4A, p. 24]

$$\begin{array}{r} \overset{4\ 5}{358} \\ \times\ \ 7 \\ \hline 2506 \end{array}$$

Division is used to separate objects into equal groups. The *dividend* divided by the *divisor* equals the *quotient*. [1.5A, p. 32]

For any division problem,

(*quotient · divisor*) + remainder = *dividend*. [1.5B, p. 35]

$$\begin{array}{r} 93\ r3 \\ 7)\overline{654} \\ -63 \\ \hline 24 \\ -21 \\ \hline 3 \end{array}$$

Check: $(7 \cdot 93) + 3 = 651 + 3 = 654$

The expression 4^3 is in *exponential notation*. The *exponent*, 3, indicates how many times 4 occurs as a factor in the multiplication. [1.6A, p. 45]

$5^4 = 5 \cdot 5 \cdot 5 \cdot 5 = 625$

Whole-number *factors of a number* divide that number evenly (there is no remainder). [1.7A, p. 49]

$18 \div 1 = 18$
$18 \div 2 = 9$
$18 \div 3 = 6$
$18 \div 4$ 4 does not divide 18 evenly.
$18 \div 5$ 5 does not divide 18 evenly.
$18 \div 6 = 3$ The factors are repeating.
The factors of 18 are 1, 2, 3, 6, 9, and 18.

A number greater than 1 is a *prime number* if its only whole-number factors are 1 and itself. If a number is not prime, it is a *composite number*. [1.7B, p. 50]

The prime numbers less than 20 are 2, 3, 5, 7, 11, 13, 17, and 19.
The composite numbers less than 20 are 4, 6, 8, 9, 10, 12, 14, 15, 16, and 18.

The *prime factorization* of a number is the expression of the number as a product of its prime factors. [1.7B, p. 50]

$$\begin{array}{r|l} & 42 \\ \hline 2 & 21 \\ 3 & 7 \\ 7 & 1 \end{array}$$

The prime factorization of 42 is $2 \cdot 3 \cdot 7$.

ESSENTIAL RULES AND PROCEDURES	EXAMPLES
To round a number to a given place value: If the digit to the right of the given place value is less than 5, replace that digit and all digits to the right by zeros. If the digit to the right of the given place value is greater than or equal to 5, increase the digit in the given place value by 1, and replace all other digits to the right by zeros. [1.1D, p. 5]	36,178 rounded to the nearest thousand is 36,000. 4592 rounded to the nearest thousand is 5000.

Properties of Addition [1.2A, p. 8]

Addition Property of Zero
Zero added to a number does not change the number.

$7 + 0 = 7$

Commutative Property of Addition
Two numbers can be added in either order; the sum will be the same.

$8 + 3 = 3 + 8$

Associative Property of Addition
Numbers to be added can be grouped in any order; the sum will be the same.

$(2 + 4) + 6 = 2 + (4 + 6)$

To estimate the answer to an addition calculation: Round each number to the same place value. Perform the calculation using the rounded numbers. [1.2A, p. 10]

$$\begin{array}{rr} 39{,}471 & 40{,}000 \\ 12{,}586 & +\ 10{,}000 \\ \hline & 50{,}000 \end{array}$$

50,000 is an estimate of the sum of 39,471 and 12,586.

Properties of Multiplication [1.4A, p. 24]

Multiplication Property of Zero
The product of a number and zero is zero.

$3 \cdot 0 = 0$

Multiplication Property of One
The product of a number and one is the number.

$6 \cdot 1 = 6$

Commutative Property of Multiplication
Two numbers can be multiplied in either order; the product will be the same.

$2 \cdot 8 = 8 \cdot 2$

Associative Property of Multiplication
Grouping numbers to be multiplied in any order gives the same result.

$(2 \cdot 4) \cdot 6 = 2 \cdot (4 \cdot 6)$

Division Properties of Zero and One [1.5A, p. 32]
Any whole number, except zero, divided by itself is 1.
Any whole number divided by 1 is the whole number.
Zero divided by any other whole number is zero.
Division by zero is not allowed.

$3 \div 3 = 1$
$3 \div 1 = 3$
$0 \div 3 = 0$
$3 \div 0$ is not allowed.

Order of Operations Agreement [1.6B, p. 46]

Step 1 Do all the operations inside parentheses.

Step 2 Simplify any number expressions containing exponents.

Step 3 Do multiplications and divisions as they occur from left to right.

Step 4 Do addition and subtraction as they occur from left to right.

$$\begin{aligned} 5^2 - 3(2 + 4) &= 5^2 - 3(6) \\ &= 25 - 3(6) \\ &= 25 - 18 \\ &= 7 \end{aligned}$$

CHAPTER 1

CONCEPT REVIEW

Test your knowledge of the concepts presented in this chapter. Answer each question. Then check your answers against the ones provided in the Answer Section.

1. What is the difference between the symbols $<$ and $>$?

2. How do you round a four-digit whole number to the nearest hundred?

3. What is the difference between the Commutative Property of Addition and the Associative Property of Addition?

4. How do you estimate the sum of two numbers?

5. When is it necessary to borrow when performing subtraction?

6. What is the difference between the Multiplication Property of Zero and the Multiplication Property of One?

7. How do you multiply a whole number by 100?

8. How do you estimate the product of two numbers?

9. What is the difference between $0 \div 9$ and $9 \div 0$?

10. How do you check the answer to a division problem that has a remainder?

11. What are the steps in the Order of Operations Agreement?

12. How do you know if a number is a factor of another number?

13. What is a quick way to determine if 3 is a factor of a number?

CHAPTER 1

REVIEW EXERCISES

1. Simplify: $3 \cdot 2^3 \cdot 5^2$

2. Write 10,327 in expanded form.

3. Find all the factors of 18.

4. Find the sum of 5894, 6301, and 298.

5. Subtract: $\begin{array}{r} 4926 \\ -\ 3177 \\ \hline \end{array}$

6. Divide: $7\overline{)14{,}945}$

7. Place the correct symbol, $<$ or $>$, between the two numbers: 101   87

8. Write $5 \cdot 5 \cdot 7 \cdot 7 \cdot 7 \cdot 7 \cdot 7$ in exponential notation.

9. What is 2019 multiplied by 307?

10. What is 10,134 decreased by 4725?

11. Add: $\begin{array}{r} 298 \\ 461 \\ +\ 322 \\ \hline \end{array}$

12. Simplify: $2^3 - 3 \cdot 2$

13. Round 45,672 to the nearest hundred.

14. Write 276,057 in words.

15. Find the quotient of 109,763 and 84.

16. Write two million eleven thousand forty-four in standard form.

17. What is 3906 divided by 8?

18. Simplify: $3^2 + 2^2 \cdot (5 - 3)$

19. Simplify: $8 \cdot (6 - 2)^2 \div 4$

20. Find the prime factorization of 72.

21. What is 3895 minus 1762?

22. Multiply: 843
                   × 27

23. Wages Vincent Meyers, a sales assistant, earns $480 for working a 40-hour week. Last week Vincent worked an additional 12 hours at $24 an hour. Find Vincent's total pay for last week's work.

24. Fuel Efficiency Louis Reyes, a sales executive, drove a car 351 miles on 13 gallons of gas. Find the number of miles driven per gallon of gasoline.

25. Consumerism A car is purchased for $29,880, with a down payment of $3000. The balance is paid in 48 equal monthly payments. Find the monthly car payment.

26. Compensation An insurance account executive received commissions of $723, $544, $812, and $488 during a 4-week period. Find the total income from commissions for the 4 weeks.

27. Banking You had a balance of $516 in your checking account before making deposits of $88 and $213. Find the total amount deposited, and determine your new account balance.

28. Compensation You have a car payment of $246 per month. What is the total of the car payments over a 12-month period?

Athletics The table at the right shows the athletic participation by males and females at U.S. colleges in 1972 and 2005. Use this information for Exercises 29 to 32.

29. In which year, 1972 or 2005, were there more males involved in sports at U.S. colleges?

Year	Male Athletes	Female Athletes
1972	170,384	29,977
2005	291,797	205,492

Source: U.S. Department of Education commission report PPT

30. What is the difference between the number of males involved in sports and the number of females involved in sports at U.S. colleges in 1972?

31. Find the increase in the number of females involved in sports in U.S. colleges from 1972 to 2005.

© Pete Saloutos/Corbis

32. How many more U.S. college students were involved in athletics in 2005 than in 1972?

CHAPTER 1

TEST

1. Simplify: $3^3 \cdot 4^2$

2. Write 207,068 in words.

3. Subtract: $\begin{array}{r} 17,495 \\ -\ 8,162 \end{array}$

4. Find all the factors of 20.

5. Multiply: $\begin{array}{r} 9736 \\ \times\ 704 \end{array}$

6. Simplify: $4^2 \cdot (4 - 2) \div 8 + 5$

7. Write 906,378 in expanded form.

8. Round 74,965 to the nearest hundred.

9. Divide: $97\overline{)108,764}$

10. Write $3 \cdot 3 \cdot 3 \cdot 7 \cdot 7$ in exponential form.

11. Find the sum of 8756, 9094, and 37,065.

12. Find the prime factorization of 84.

13. Simplify: $16 \div 4 \cdot 2 - (7 - 5)^2$

14. Find the product of 8 and 90,763.

15. Write one million two hundred four thousand six in standard form.

16. Divide: $7\overline{)60,972}$

17. Place the correct symbol, < or >, between the two numbers: 21   19

18. Find the quotient of 5624 and 8.

19. Add: 25,492
             +71,306

20. Find the difference between 29,736 and 9814.

Education The table at the right shows the projected enrollment in public and private elementary and secondary schools in the fall of 2013 and the fall of 2016. Use this information for Exercises 21 and 22.

21. Find the difference between the total enrollment in 2016 and that in 2013.

Year	Pre-Kindergarten through Grade 8	Grades 9 through 12
2013	41,873,000	16,000,000
2016	43,097,000	16,684,000

Source: The National Center for Education Statistics

PPT

22. Find the average enrollment in each of grades 9 through 12 in 2016.

23. **Farming** A farmer harvested 48,290 pounds of lemons from one grove and 23,710 pounds of lemons from another grove. The lemons were packed in boxes with 24 pounds of lemons in each box. How many boxes were needed to pack the lemons?

© Ed Young/Corbis

24. **Investments** An investor receives $237 each month from a corporate bond fund. How much will the investor receive over a 12-month period?

25. **Travel** A family drives 425 miles the first day, 187 miles the second day, and 243 miles the third day of their vacation. The odometer read 47,626 miles at the start of the vacation.
a. How many miles were driven during the 3 days?
b. What is the odometer reading at the end of the 3 days?

Fractions

OBJECTIVES

SECTION 2.1
A To find the least common multiple (LCM)
B To find the greatest common factor (GCF)

SECTION 2.2
A To write a fraction that represents part of a whole
B To write an improper fraction as a mixed number or a whole number, and a mixed number as an improper fraction

SECTION 2.3
A To find equivalent fractions by raising to higher terms
B To write a fraction in simplest form

SECTION 2.4
A To add fractions with the same denominator
B To add fractions with different denominators
C To add whole numbers, mixed numbers, and fractions
D To solve application problems

SECTION 2.5
A To subtract fractions with the same denominator
B To subtract fractions with different denominators
C To subtract whole numbers, mixed numbers, and fractions
D To solve application problems

SECTION 2.6
A To multiply fractions
B To multiply whole numbers, mixed numbers, and fractions
C To solve application problems

SECTION 2.7
A To divide fractions
B To divide whole numbers, mixed numbers, and fractions
C To solve application problems

SECTION 2.8
A To identify the order relation between two fractions
B To simplify expressions containing exponents
C To use the Order of Operations Agreement to simplify expressions

ARE YOU READY?

Take the Chapter 2 Prep Test to find out if you are ready to learn to:

- Write equivalent fractions
- Write fractions in simplest form
- Add, subtract, multiply, and divide fractions
- Compare fractions

PREP TEST

Do these exercises to prepare for Chapter 2.

For Exercises 1 to 6, add, subtract, multiply, or divide.

1. 4×5

2. $2 \cdot 2 \cdot 2 \cdot 3 \cdot 5$

3. 9×1

4. $6 + 4$

5. $10 - 3$

6. $63 \div 30$

7. Which of the following numbers divide evenly into 12?
1 2 3 4 5 6 7 8 9 10 11 12

8. Simplify: $8 \times 7 + 3$

9. Complete: $8 = ? + 1$

10. Place the correct symbol, $<$ or $>$, between the two numbers.
44 48

SECTION

2.1   The Least Common Multiple and Greatest Common Factor

OBJECTIVE A   To find the least common multiple (LCM)

Tips for Success

Before you begin a new chapter, you should take some time to review previously learned skills. One way to do this is to complete the Prep Test. See page 63. This test focuses on the particular skills that will be required for the new chapter.

The **multiples of a number** are the products of that number and the numbers 1, 2, 3, 4, 5, ...

$$3 \times 1 = 3$$
$$3 \times 2 = 6$$
$$3 \times 3 = 9$$
$$3 \times 4 = 12$$ The multiples of 3 are 3, 6, 9, 12, 15, ...
$$3 \times 5 = 15$$

A number that is a multiple of two or more numbers is a **common multiple** of those numbers.

The multiples of 4 are 4, 8, 12, 16, 20, 24, 28, 32, 36, ...
The multiples of 6 are 6, 12, 18, 24, 30, 36, 42, ...
Some common multiples of 4 and 6 are 12, 24, and 36.

The **least common multiple (LCM)** is the smallest common multiple of two or more numbers.

The least common multiple of 4 and 6 is 12.

Listing the multiples of each number is one way to find the LCM. Another way to find the LCM uses the prime factorization of each number.

To find the LCM of 450 and 600, find the prime factorization of each number and write the factorization of each number in a table. Circle the greatest product in each column. The LCM is the product of the circled numbers.

	2	3	5
450 =	2	(3 · 3)	(5 · 5)
600 =	(2 · 2 · 2)	3	5 · 5

• In the column headed by 5, the products are equal. Circle just one product.

The LCM is the product of the circled numbers.
The LCM = 2 · 2 · 2 · 3 · 3 · 5 · 5 = 1800.

EXAMPLE 1

Find the LCM of 24, 36, and 50.

Solution

	2	3	5
24 =	(2 · 2 · 2)	3	
36 =	2 · 2	(3 · 3)	
50 =	2		(5 · 5)

The LCM = 2 · 2 · 2 · 3 · 3 · 5 · 5 = 1800.

YOU TRY IT 1

Find the LCM of 12, 27, and 50.

Your solution

Solution on p. S4

OBJECTIVE D To find the greatest common factor (GCF)

Recall that a number that divides another number evenly is a factor of that number. The number 64 can be evenly divided by 1, 2, 4, 8, 16, 32, and 64, so the numbers 1, 2, 4, 8, 16, 32, and 64 are factors of 64.

A number that is a factor of two or more numbers is a **common factor** of those numbers.

The factors of 30 are 1, 2, 3, 5, 6, 10, 15, and 30.
The factors of 105 are 1, 3, 5, 7, 15, 21, 35, and 105.
The common factors of 30 and 105 are 1, 3, 5, and 15.

The **greatest common factor (GCF)** is the largest *common factor* of two or more numbers.

The greatest common factor of 30 and 105 is 15.

Listing the factors of each number is one way of finding the GCF. Another way to find the GCF is to use the prime factorization of each number.

To find the GCF of 126 and 180, find the prime factorization of each number and write the factorization of each number in a table. Circle the least product in each column that does not have a blank. The GCF is the product of the circled numbers.

	2	3	5	7
126 =	②	③ · 3		7
180 =	2 · 2	3 · 3	5	

• In the column headed by 3, the products are equal. Circle just one product. Columns 5 and 7 have a blank, so 5 and 7 are not common factors of 126 and 180. Do not circle any number in these columns.

The GCF is the product of the circled numbers.
The GCF = 2 · 3 · 3 = 18.

EXAMPLE • 2

Find the GCF of 90, 168, and 420.

Solution

	2	3	5	7
90 =	②	3 · 3	5	
168 =	2 · 2 · 2	③		7
420 =	2 · 2	3	5	7

The GCF = 2 · 3 = 6.

YOU TRY IT • 2

Find the GCF of 36, 60, and 72.

Your solution

EXAMPLE • 3

Find the GCF of 7, 12, and 20.

Solution

	2	3	5	7
7 =				7
12 =	2 · 2	3		
20 =	2 · 2		5	

Because no numbers are circled, the GCF = 1.

YOU TRY IT • 3

Find the GCF of 11, 24, and 30.

Your solution

Solutions on p. S4

2.1 EXERCISES

OBJECTIVE A   To find the least common multiple (LCM)

For Exercises 1 to 34, find the LCM.

1. 5, 8   **2.** 3, 6   **3.** 3, 8   **4.** 2, 5   **5.** 5, 6

6. 5, 7   **7.** 4, 6   **8.** 6, 8   **9.** 8, 12   **10.** 12, 16

11. 5, 12   **12.** 3, 16   **13.** 8, 14   **14.** 6, 18   **15.** 3, 9

16. 4, 10   **17.** 8, 32   **18.** 7, 21   **19.** 9, 36   **20.** 14, 42

21. 44, 60   **22.** 120, 160   **23.** 102, 184   **24.** 123, 234   **25.** 4, 8, 12

26. 5, 10, 15   **27.** 3, 5, 10   **28.** 2, 5, 8   **29.** 3, 8, 12   **30.** 5, 12, 18

31. 9, 36, 64   **32.** 18, 54, 63   **33.** 16, 30, 84   **34.** 9, 12, 15

35. True or false? If two numbers have no common factors, then the LCM of the two numbers is their product.

36. True or false? If one number is a multiple of a second number, then the LCM of the two numbers is the second number.

OBJECTIVE B   To find the greatest common factor (GCF)

For Exercises 37 to 70, find the GCF.

37. 3, 5   **38.** 5, 7   **39.** 6, 9   **40.** 18, 24   **41.** 15, 25

42. 14, 49   **43.** 25, 100   **44.** 16, 80   **45.** 32, 51   **46.** 21, 44

47. 12, 80   **48.** 8, 36   **49.** 16, 140   **50.** 12, 76

51. 24, 30   **52.** 48, 144   **53.** 44, 96   **54.** 18, 32

55. 3, 5, 11   **56.** 6, 8, 10   **57.** 7, 14, 49   **58.** 6, 15, 36

59. 10, 15, 20   **60.** 12, 18, 20   **61.** 24, 40, 72   **62.** 3, 17, 51

63. 17, 31, 81   **64.** 14, 42, 84   **65.** 25, 125, 625   **66.** 12, 68, 92

67. 28, 35, 70   **68.** 1, 49, 153   **69.** 32, 56, 72   **70.** 24, 36, 48

71. True or false? If two numbers have a GCF of 1, then the LCM of the two numbers is their product.

72. True or false? If the LCM of two numbers is one of the two numbers, then the GCF of the numbers is the other of the two numbers.

Applying the Concepts

73. **Work Schedules** Joe Salvo, a lifeguard, works 3 days and then has a day off. Joe's friend works 5 days and then has a day off. How many days after Joe and his friend have a day off together will they have another day off together?

74. Find the LCM of each of the following pairs of numbers: 2 and 3, 5 and 7, and 11 and 19. Can you draw a conclusion about the LCM of two prime numbers? Suggest a way of finding the LCM of three distinct prime numbers.

75. Find the GCF of each of the following pairs of numbers: 3 and 5, 7 and 11, and 29 and 43. Can you draw a conclusion about the GCF of two prime numbers? What is the GCF of three distinct prime numbers?

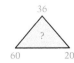

76. Using the pattern for the first two triangles at the right, determine the center number of the last triangle.

SECTION

2.2 Introduction to Fractions

OBJECTIVE A To write a fraction that represents part of a whole

 Take Note

The **fraction bar** separates the numerator from the denominator. The **numerator** is the part of the fraction that appears above the fraction bar. The **denominator** is the part of the fraction that appears below the fraction bar.

Point of Interest

The fraction bar was first used in 1050 by al-Hassar. It is also called a vinculum.

A **fraction** can represent the number of equal parts of a whole.

The shaded portion of the circle is represented by the fraction $\frac{4}{7}$. Four of the seven equal parts of the circle (that is, four-sevenths of it) are shaded.

Each part of a fraction has a name.

Fraction bar $\rightarrow \dfrac{4 \leftarrow \textbf{Numerator}}{7 \leftarrow \textbf{Denominator}}$

A **proper fraction** is a fraction less than 1. The numerator of a proper fraction is smaller than the denominator. The shaded portion of the circle can be represented by the proper fraction $\frac{3}{4}$.

A **mixed number** is a number greater than 1 with a whole-number part and a fractional part. The shaded portion of the circles can be represented by the mixed number $2\frac{1}{4}$.

An **improper fraction** is a fraction greater than or equal to 1. The numerator of an improper fraction is greater than or equal to the denominator. The shaded portion of the circles can be represented by the improperfraction $\frac{9}{4}$. The shaded portion of the square can be represented by $\frac{4}{4}$.

EXAMPLE • 1

Express the shaded portion of the circles as a mixed number.

Solution $3\frac{2}{5}$

YOU TRY IT • 1

Express the shaded portion of the circles as a mixed number.

Your solution

EXAMPLE • 2

Express the shaded portion of the circles as an improper fraction.

Solution $\frac{17}{5}$

YOU TRY IT • 2

Express the shaded portion of the circles as an improper fraction.

Your solution

Solutions on p. S4

OBJECTIVE B   To write an improper fraction as a mixed number or a whole number, and a mixed number as an improper fraction

Note from the diagram that the mixed number $2\frac{3}{5}$ and the improper fraction $\frac{13}{5}$ both represent the shaded portion of the circles.

$$2\frac{3}{5} = \frac{13}{5}$$

An improper fraction can be written as a mixed number or a whole number.

Point of Interest

Archimedes (c. 287–212 B.C.) is the person who calculated that $\pi \approx 3\frac{1}{7}$. He actually showed that $3\frac{10}{71} < \pi < 3\frac{1}{7}$. The approximation $3\frac{10}{71}$ is more accurate but more difficult to use.

HOW TO 1   Write $\frac{13}{5}$ as a mixed number.

Divide the numerator by the denominator.	To write the fractional part of the mixed number, write the remainder over the divisor.	Write the answer.
$\begin{array}{r} 2 \\ 5\overline{)\,13} \\ -10 \\ \hline 3 \end{array}$	$\begin{array}{r} 2\frac{3}{5} \\ 5\overline{)\,13} \\ -10 \\ \hline 3 \end{array}$	$\frac{13}{5} = 2\frac{3}{5}$

To write a mixed number as an improper fraction, multiply the denominator of the fractional part by the whole-number part. The sum of this product and the numerator of the fractional part is the numerator of the improper fraction. The denominator remains the same.

HOW TO 2   Write $7\frac{3}{8}$ as an improper fraction.

$$7\frac{3}{8} = \frac{(8 \times 7) + 3}{8} = \frac{56 + 3}{8} = \frac{59}{8} \qquad 7\frac{3}{8} = \frac{59}{8}$$

EXAMPLE 3

Write $\frac{21}{4}$ as a mixed number.

Solution

$\begin{array}{r} 5 \\ 4\overline{)\,21} \\ -20 \\ \hline 1 \end{array}$ $\frac{21}{4} = 5\frac{1}{4}$

YOU TRY IT 3

Write $\frac{22}{5}$ as a mixed number.

Your solution

EXAMPLE 4

Write $\frac{18}{6}$ as a whole number.

Solution $\frac{18}{6} = 18 \div 6 = 3$

YOU TRY IT 4

Write $\frac{28}{7}$ as a whole number.

Your solution

EXAMPLE 5

Write $21\frac{3}{4}$ as an improper fraction.

Solution $21\frac{3}{4} = \frac{84 + 3}{4} = \frac{87}{4}$

YOU TRY IT 5

Write $14\frac{5}{8}$ as an improper fraction.

Your solution

Solutions on p. S4

2.2 EXERCISES

OBJECTIVE A   To write a fraction that represents part of a whole

For Exercises 1 to 4, identify the fraction as a proper fraction, an improper fraction, or a mixed number.

1. $\dfrac{12}{7}$

2. $5\dfrac{2}{11}$

3. $\dfrac{29}{40}$

4. $\dfrac{19}{13}$

For Exercises 5 to 8, express the shaded portion of the circle as a fraction.

5.

6.

7.

8.

For Exercises 9 to 14, express the shaded portion of the circles as a mixed number.

9.

10.

11.

12.

13.

14.

For Exercises 15 to 20, express the shaded portion of the circles as an improper fraction.

15.

16.

17.

18.

19.

20.

21. Shade $1\dfrac{2}{5}$ of

22. Shade $1\dfrac{3}{4}$ of

23. Shade $\dfrac{6}{5}$ of

24. Shade $\dfrac{7}{3}$ of

25. True or false? The fractional part of a mixed number is an improper fraction.

OBJECTIVE B To write an improper fraction as a mixed number or a whole number, and a mixed number as an improper fraction

For Exercises 26 to 49, write the improper fraction as a mixed number or a whole number.

26. $\dfrac{11}{4}$ **27.** $\dfrac{16}{3}$ **28.** $\dfrac{20}{4}$ **29.** $\dfrac{18}{9}$ **30.** $\dfrac{9}{8}$ **31.** $\dfrac{13}{4}$

32. $\dfrac{23}{10}$ **33.** $\dfrac{29}{2}$ **34.** $\dfrac{48}{16}$ **35.** $\dfrac{51}{3}$ **36.** $\dfrac{8}{7}$ **37.** $\dfrac{16}{9}$

38. $\dfrac{7}{3}$ **39.** $\dfrac{9}{5}$ **40.** $\dfrac{16}{1}$ **41.** $\dfrac{23}{1}$ **42.** $\dfrac{17}{8}$ **43.** $\dfrac{31}{16}$

44. $\dfrac{12}{5}$ **45.** $\dfrac{19}{3}$ **46.** $\dfrac{9}{9}$ **47.** $\dfrac{40}{8}$ **48.** $\dfrac{72}{8}$ **49.** $\dfrac{3}{3}$

For Exercises 50 to 73, write the mixed number as an improper fraction.

50. $2\dfrac{1}{3}$ **51.** $4\dfrac{2}{3}$ **52.** $6\dfrac{1}{2}$ **53.** $8\dfrac{2}{3}$ **54.** $6\dfrac{5}{6}$ **55.** $7\dfrac{3}{8}$

56. $9\dfrac{1}{4}$ **57.** $6\dfrac{1}{4}$ **58.** $10\dfrac{1}{2}$ **59.** $15\dfrac{1}{8}$ **60.** $8\dfrac{1}{9}$ **61.** $3\dfrac{5}{12}$

62. $5\dfrac{3}{11}$ **63.** $3\dfrac{7}{9}$ **64.** $2\dfrac{5}{8}$ **65.** $12\dfrac{2}{3}$ **66.** $1\dfrac{5}{8}$ **67.** $5\dfrac{3}{7}$

68. $11\dfrac{1}{9}$ **69.** $12\dfrac{3}{5}$ **70.** $3\dfrac{3}{8}$ **71.** $4\dfrac{5}{9}$ **72.** $6\dfrac{7}{13}$ **73.** $8\dfrac{5}{14}$

 74. True or false? If an improper fraction is equivalent to 1, then the numerator and the denominator are the same number.

Applying the Concepts

 75. Name three situations in which fractions are used. Provide an example of a fraction that is used in each situation.

SECTION

2.3 Writing Equivalent Fractions

OBJECTIVE A To find equivalent fractions by raising to higher terms

Equal fractions with different denominators are called **equivalent fractions.**

$\frac{4}{6}$ is equivalent to $\frac{2}{3}$.

Remember that the Multiplication Property of One states that the product of a number and one is the number. This is true for fractions as well as whole numbers. This property can be used to write equivalent fractions.

$$\frac{2}{3} \times 1 = \frac{2}{3} \times \frac{1}{1} = \frac{2 \cdot 1}{3 \cdot 1} = \frac{2}{3}$$

$$\frac{2}{3} \times 1 = \frac{2}{3} \times \boxed{\frac{2}{2}} = \frac{2 \cdot 2}{3 \cdot 2} = \frac{4}{6} \qquad \frac{4}{6} \text{ is equivalent to } \frac{2}{3}.$$

$$\frac{2}{3} \times 1 = \frac{2}{3} \times \boxed{\frac{4}{4}} = \frac{2 \cdot 4}{3 \cdot 4} = \frac{8}{12} \qquad \frac{8}{12} \text{ is equivalent to } \frac{2}{3}.$$

$\frac{2}{3}$ was rewritten as the equivalent fractions $\frac{4}{6}$ and $\frac{8}{12}$.

HOW TO 1 Write a fraction that is equivalent to $\frac{5}{8}$ and has a denominator of 32.

$32 \div 8 = 4$ • Divide the larger denominator by the smaller.

$\dfrac{5}{8} = \dfrac{5 \cdot 4}{8 \cdot 4} = \dfrac{20}{32}$ • Multiply the numerator and denominator of the given fraction by the quotient (4).

$\frac{20}{32}$ is equivalent to $\frac{5}{8}$.

EXAMPLE 1

Write $\frac{2}{3}$ as an equivalent fraction that has a denominator of 42.

Solution $42 \div 3 = 14$ $\dfrac{2}{3} = \dfrac{2 \cdot 14}{3 \cdot 14} = \dfrac{28}{42}$

$\frac{28}{42}$ is equivalent to $\frac{2}{3}$.

YOU TRY IT 1

Write $\frac{3}{5}$ as an equivalent fraction that has a denominator of 45.

Your solution

EXAMPLE 2

Write 4 as a fraction that has a denominator of 12.

Solution   Write 4 as $\dfrac{4}{1}$.

$12 \div 1 = 12$ $4 = \dfrac{4 \cdot 12}{1 \cdot 12} = \dfrac{48}{12}$

$\frac{48}{12}$ is equivalent to 4.

YOU TRY IT 2

Write 6 as a fraction that has a denominator of 18.

Your solution

Solutions on p. S4

OBJECTIVE B   To write a fraction in simplest form

Writing the **simplest form of a fraction** means writing it so that the numerator and denominator have no common factors other than 1.

The fractions $\frac{4}{6}$ and $\frac{2}{3}$ are equivalent fractions.

$\frac{4}{6}$ has been written in simplest form as $\frac{2}{3}$.

The Multiplication Property of One can be used to write fractions in simplest form. Write the numerator and denominator of the given fraction as a product of factors. Write factors common to both the numerator and denominator as an improper fraction equivalent to 1.

$$\frac{4}{6} = \frac{2 \cdot 2}{2 \cdot 3} = \boxed{\frac{2}{2}} \cdot \frac{2}{3} = 1 \cdot \frac{2}{3} = \frac{2}{3}$$

The process of eliminating common factors is displayed with slashes through the common factors as shown at the right.

To write a fraction in simplest form, eliminate the common factors.

An improper fraction can be changed to a mixed number.

$$\frac{4}{6} = \frac{\overset{1}{\cancel{2}} \cdot 2}{\underset{1}{\cancel{2}} \cdot 3} = \frac{2}{3}$$

$$\frac{18}{30} = \frac{\overset{1}{\cancel{2}} \cdot \overset{1}{\cancel{3}} \cdot 3}{\underset{1}{\cancel{2}} \cdot \underset{1}{\cancel{3}} \cdot 5} = \frac{3}{5}$$

$$\frac{22}{6} = \frac{\overset{1}{\cancel{2}} \cdot 11}{\underset{1}{\cancel{2}} \cdot 3} = \frac{11}{3} = 3\frac{2}{3}$$

EXAMPLE 3

Write $\frac{15}{40}$ in simplest form.

Solution

$$\frac{15}{40} = \frac{3 \cdot \overset{1}{\cancel{5}}}{2 \cdot 2 \cdot 2 \cdot \underset{1}{\cancel{5}}} = \frac{3}{8}$$

YOU TRY IT

Write $\frac{16}{24}$ in simplest form.

Your solution

EXAMPLE 4

Write $\frac{6}{42}$ in simplest form.

Solution

$$\frac{6}{42} = \frac{\overset{1}{\cancel{2}} \cdot \overset{1}{\cancel{3}}}{\underset{1}{\cancel{2}} \cdot \underset{1}{\cancel{3}} \cdot 7} = \frac{1}{7}$$

YOU TRY IT 4

Write $\frac{8}{56}$ in simplest form.

Your solution

EXAMPLE 5

Write $\frac{8}{9}$ in simplest form.

Solution

$$\frac{8}{9} = \frac{2 \cdot 2 \cdot 2}{3 \cdot 3} = \frac{8}{9}$$

$\frac{8}{9}$ is already in simplest form because there are no common factors in the numerator and denominator.

YOU TRY IT 5

Write $\frac{15}{32}$ in simplest form.

Your solution

EXAMPLE 6

Write $\frac{30}{12}$ in simplest form.

Solution

$$\frac{30}{12} = \frac{\overset{1}{\cancel{2}} \cdot \overset{1}{\cancel{3}} \cdot 5}{\underset{1}{\cancel{2}} \cdot 2 \cdot \underset{1}{\cancel{3}}} = \frac{5}{2} = 2\frac{1}{2}$$

YOU TRY IT 6

Write $\frac{48}{36}$ in simplest form.

Your solution

Solutions on p. S4

2.3 EXERCISES

OBJECTIVE A   To find equivalent fractions by raising to higher terms

For Exercises 1 to 35, write an equivalent fraction with the given denominator.

1. $\dfrac{1}{2} = \dfrac{}{10}$ **2.** $\dfrac{1}{4} = \dfrac{}{16}$ **3.** $\dfrac{3}{16} = \dfrac{}{48}$ **4.** $\dfrac{5}{9} = \dfrac{}{81}$ **5.** $\dfrac{3}{8} = \dfrac{}{32}$

6. $\dfrac{7}{11} = \dfrac{}{33}$ **7.** $\dfrac{3}{17} = \dfrac{}{51}$ **8.** $\dfrac{7}{10} = \dfrac{}{90}$ **9.** $\dfrac{3}{4} = \dfrac{}{16}$ **10.** $\dfrac{5}{8} = \dfrac{}{32}$

11. $3 = \dfrac{}{9}$ **12.** $5 = \dfrac{}{25}$ **13.** $\dfrac{1}{3} = \dfrac{}{60}$ **14.** $\dfrac{1}{16} = \dfrac{}{48}$ **15.** $\dfrac{11}{15} = \dfrac{}{60}$

16. $\dfrac{3}{50} = \dfrac{}{300}$ **17.** $\dfrac{2}{3} = \dfrac{}{18}$ **18.** $\dfrac{5}{9} = \dfrac{}{36}$ **19.** $\dfrac{5}{7} = \dfrac{}{49}$ **20.** $\dfrac{7}{8} = \dfrac{}{32}$

21. $\dfrac{5}{9} = \dfrac{}{18}$ **22.** $\dfrac{11}{12} = \dfrac{}{36}$ **23.** $7 = \dfrac{}{3}$ **24.** $9 = \dfrac{}{4}$ **25.** $\dfrac{7}{9} = \dfrac{}{45}$

26. $\dfrac{5}{6} = \dfrac{}{42}$ **27.** $\dfrac{15}{16} = \dfrac{}{64}$ **28.** $\dfrac{11}{18} = \dfrac{}{54}$ **29.** $\dfrac{3}{14} = \dfrac{}{98}$ **30.** $\dfrac{5}{6} = \dfrac{}{144}$

31. $\dfrac{5}{8} = \dfrac{}{48}$ **32.** $\dfrac{7}{12} = \dfrac{}{96}$ **33.** $\dfrac{5}{14} = \dfrac{}{42}$ **34.** $\dfrac{2}{3} = \dfrac{}{42}$ **35.** $\dfrac{17}{24} = \dfrac{}{144}$

36. When you multiply the numerator and denominator of a fraction by the same number, you are actually multiplying the fraction by the number _____.

OBJECTIVE B   To write a fraction in simplest form

For Exercises 37 to 71, write the fraction in simplest form.

37. $\dfrac{4}{12}$ **38.** $\dfrac{8}{22}$ **39.** $\dfrac{22}{44}$ **40.** $\dfrac{2}{14}$ **41.** $\dfrac{2}{12}$

42. $\dfrac{50}{75}$ **43.** $\dfrac{40}{36}$ **44.** $\dfrac{12}{8}$ **45.** $\dfrac{0}{30}$ **46.** $\dfrac{10}{10}$

47. $\dfrac{9}{22}$ **48.** $\dfrac{14}{35}$ **49.** $\dfrac{75}{25}$ **50.** $\dfrac{8}{60}$ **51.** $\dfrac{16}{84}$

52. $\dfrac{20}{44}$ **53.** $\dfrac{12}{35}$ **54.** $\dfrac{8}{36}$ **55.** $\dfrac{28}{44}$ **56.** $\dfrac{12}{16}$

57. $\dfrac{16}{12}$ **58.** $\dfrac{24}{18}$ **59.** $\dfrac{24}{40}$ **60.** $\dfrac{44}{60}$ **61.** $\dfrac{8}{88}$

62. $\dfrac{9}{90}$ **63.** $\dfrac{144}{36}$ **64.** $\dfrac{140}{297}$ **65.** $\dfrac{48}{144}$ **66.** $\dfrac{32}{120}$

67. $\dfrac{60}{100}$ **68.** $\dfrac{33}{110}$ **69.** $\dfrac{36}{16}$ **70.** $\dfrac{80}{45}$ **71.** $\dfrac{32}{160}$

 72. Suppose the denominator of a fraction is a multiple of the numerator. When the fraction is written in simplest form, what number is its numerator?

Applying the Concepts

73. Make a list of five different fractions that are equivalent to $\dfrac{2}{3}$.

74. Show that $\dfrac{15}{24} = \dfrac{5}{8}$ by using a diagram.

75. **a. Geography** What fraction of the states in the United States of America have names that begin with the letter M?
    b. What fraction of the states have names that begin and end with a vowel?

SECTION

2.4   Addition of Fractions and Mixed Numbers

OBJECTIVE A   **To add fractions with the same denominator**

Fractions with the same denominator are added by adding the numerators and placing the sum over the common denominator. After adding, write the sum in simplest form.

HOW TO · 1   Add: $\frac{2}{7} + \frac{4}{7}$

$$
\begin{array}{r}
\frac{2}{7} \\
+\ \frac{4}{7} \\
\hline
\frac{6}{7}
\end{array}
$$

• Add the numerators and place the sum over the common denominator.

$$\frac{2}{7} + \frac{4}{7} = \frac{2+4}{7} = \frac{6}{7}$$

EXAMPLE · 1

Add: $\frac{5}{12} + \frac{11}{12}$

Solution

$$
\begin{array}{r}
\frac{5}{12} \\
+\ \frac{11}{12} \\
\hline
\frac{16}{12} = \frac{4}{3} = 1\frac{1}{3}
\end{array}
$$

• The denominators are the same. Add the numerators. Place the sum over the common denominator.

YOU TRY IT · 1

Add: $\frac{3}{8} + \frac{7}{8}$

Your solution

Solution on p. S5

OBJECTIVE B   **To add fractions with different denominators**

Integrating Technology

Some scientific calculators have a fraction key, $\boxed{a^{b}/_{c}}$. It is used to perform operations on fractions. To use this key to simplify the expression at the right, enter

$1\ \boxed{a^{b}/_{c}}\ 2 + 1\ \boxed{a^{b}/_{c}}\ 3\ \boxed{=}$

$\underbrace{}_{\frac{1}{2}}\quad \underbrace{}_{\frac{1}{3}}$

To add fractions with different denominators, first rewrite the fractions as equivalent fractions with a common denominator. The common denominator is the LCM of the denominators of the fractions.

HOW TO · 2   Find the total of $\frac{1}{2}$ and $\frac{1}{3}$.

The common denominator is the LCM of 2 and 3. The LCM = 6. The LCM of denominators is sometimes called the **least common denominator (LCD).**

Write equivalent fractions using the LCM.

$$
\begin{array}{r}
\frac{1}{2} = \frac{3}{6} \\
+\ \frac{1}{3} = \frac{2}{6} \\
\hline
\end{array}
$$

Add the fractions.

$$
\begin{array}{r}
\frac{1}{2} = \frac{3}{6} \\
+\ \frac{1}{3} = \frac{2}{6} \\
\hline
\frac{5}{6}
\end{array}
$$

EXAMPLE 2

Find $\frac{7}{12}$ more than $\frac{3}{8}$.

Solution

$$\frac{3}{8} = \frac{9}{24}$$

• The LCM of 8 and 12 is 24.

$$+\frac{7}{12} = \frac{14}{24}$$

$$\frac{23}{24}$$

YOU TRY IT 2

Find the sum of $\frac{5}{12}$ and $\frac{9}{16}$.

Your solution

EXAMPLE 3

Add: $\frac{5}{8} + \frac{7}{9}$

Solution

$$\frac{5}{8} = \frac{45}{72}$$

$$+\frac{7}{9} = \frac{56}{72}$$

$$\frac{101}{72} = 1\frac{29}{72}$$

YOU TRY IT 3

Add: $\frac{7}{8} + \frac{11}{15}$

Your solution

EXAMPLE 4

Add: $\frac{2}{3} + \frac{3}{5} + \frac{5}{6}$

Solution

$$\frac{2}{3} = \frac{20}{30}$$

• The LCM of 3, 5, and 6 is 30.

$$\frac{3}{5} = \frac{18}{30}$$

$$+\frac{5}{6} = \frac{25}{30}$$

$$\frac{63}{30} = 2\frac{3}{30} = 2\frac{1}{10}$$

YOU TRY IT 4

Add: $\frac{3}{4} + \frac{4}{5} + \frac{5}{8}$

Your solution

Solutions on p. S5

OBJECTIVE C   To add whole numbers, mixed numbers, and fractions

✓ **Take Note**

The procedure at the right illustrates why $2 + \frac{2}{3} = 2\frac{2}{3}$. You do not need to show these steps when adding a whole number and a fraction. Here are two more examples:

$$7 + \frac{1}{5} = 7\frac{1}{5}$$

$$6 + \frac{3}{4} = 6\frac{3}{4}$$

The sum of a whole number and a fraction is a mixed number.

HOW TO 3   Add: $2 + \frac{2}{3}$

$$\boxed{2} + \frac{2}{3} = \boxed{\frac{6}{3}} + \frac{2}{3} = \frac{8}{3} = 2\frac{2}{3}$$

To add a whole number and a mixed number, write the fraction and then add the whole numbers.

HOW TO 4   Add: $7\frac{2}{5} + 4$

Write the fraction.
$$7\frac{2}{5}$$
$$+\ 4$$
$$\frac{2}{5}$$

Add the whole numbers.
$$7\frac{2}{5}$$
$$+\ 4$$
$$11\frac{2}{5}$$

Integrating Technology

Use the fraction key on a calculator to enter mixed numbers. For the example at the right, enter

$5\frac{4}{9}$

$6\frac{14}{15}$

To add two mixed numbers, add the fractional parts and then add the whole numbers. Remember to reduce the sum to simplest form.

HOW TO 5 What is $6\frac{14}{15}$ added to $5\frac{4}{9}$?

The LCM of 9 and 15 is 45.

Add the fractional parts.

$$5\frac{4}{9} = 5\frac{20}{45}$$
$$+\ 6\frac{14}{15} = 6\frac{42}{45}$$
$$\frac{62}{45}$$

Add the whole numbers.

$$5\frac{4}{9} = 5\frac{20}{45}$$
$$+\ 6\frac{14}{15} = 6\frac{42}{45}$$
$$11\frac{62}{45} = 11 + 1\frac{17}{45} = 12\frac{17}{45}$$

EXAMPLE 5

Add: $5 + \frac{3}{8}$

Solution $5 + \frac{3}{8} = 5\frac{3}{8}$

YOU TRY IT 5

What is 7 added to $\frac{6}{11}$?

Your solution

EXAMPLE 6

Find 17 increased by $3\frac{3}{8}$.

Solution $17 + 3\frac{3}{8} = 20\frac{3}{8}$

YOU TRY IT 6

Find the sum of 29 and $17\frac{5}{12}$.

Your solution

EXAMPLE 7

Add: $5\frac{2}{3} + 11\frac{5}{6} + 12\frac{7}{9}$

Solution

$$5\frac{2}{3} = 5\frac{12}{18}$$ • LCM = 18
$$11\frac{5}{6} = 11\frac{15}{18}$$
$$+\ 12\frac{7}{9} = 12\frac{14}{18}$$
$$28\frac{41}{18} = 30\frac{5}{18}$$

YOU TRY IT 7

Add: $7\frac{4}{5} + 6\frac{7}{10} + 13\frac{11}{15}$

Your solution

EXAMPLE 8

Add: $11\frac{5}{8} + 7\frac{5}{9} + 8\frac{7}{15}$

Solution

$$11\frac{5}{8} = 11\frac{225}{360}$$ • LCM = 360
$$7\frac{5}{9} = 7\frac{200}{360}$$
$$+\ 8\frac{7}{15} = 8\frac{168}{360}$$
$$26\frac{593}{360} = 27\frac{233}{360}$$

YOU TRY IT 8

Add: $9\frac{3}{8} + 17\frac{7}{12} + 10\frac{14}{15}$

Your solution

Solutions on p. S5

OBJECTIVE D   To solve application problems

EXAMPLE · 9

A rain gauge collected $2\frac{1}{3}$ inches of rain in October, $5\frac{1}{2}$ inches in November, and $3\frac{3}{8}$ inches in December. Find the total rainfall for the 3 months.

Strategy

To find the total rainfall for the 3 months, add the three amounts of rainfall $\left(2\frac{1}{3},\ 5\frac{1}{2},\ \text{and } 3\frac{3}{8}\right)$.

Solution

$$2\frac{1}{3} = 2\frac{8}{24}$$

$$5\frac{1}{2} = 5\frac{12}{24}$$

$$+\ 3\frac{3}{8} = 3\frac{9}{24}$$

$$10\frac{29}{24} = 11\frac{5}{24}$$

The total rainfall for the 3 months was $11\frac{5}{24}$ inches.

YOU TRY IT · 9

On Monday, you spent $4\frac{1}{2}$ hours in class, $3\frac{3}{4}$ hours studying, and $1\frac{1}{3}$ hours driving. Find the total number of hours spent on these three activities.

Your strategy

Your solution

EXAMPLE · 10

Barbara Walsh worked 4 hours, $2\frac{1}{3}$ hours, and $5\frac{2}{3}$ hours this week at a part-time job. Barbara is paid $9 an hour. How much did she earn this week?

Strategy

To find how much Barbara earned:
• Find the total number of hours worked.
• Multiply the total number of hours worked by the hourly wage (9).

Solution

$$
\begin{array}{rr}
4 & 12 \\
2\frac{1}{3} & \times\ \ 9 \\
+\ 5\frac{2}{3} & \overline{108} \\
\hline
11\frac{3}{3} = 12 \text{ hours worked}
\end{array}
$$

Barbara earned $108 this week.

YOU TRY IT · 10

Jeff Sapone, a carpenter, worked $1\frac{2}{3}$ hours of overtime on Monday, $3\frac{1}{3}$ hours of overtime on Tuesday, and 2 hours of overtime on Wednesday. At an overtime hourly rate of $36, find Jeff's overtime pay for these 3 days.

Your strategy

Your solution

Solutions on p. S5

2.4 EXERCISES

OBJECTIVE A   To add fractions with the same denominator

For Exercises 1 to 16, add.

1. $\dfrac{2}{7} + \dfrac{1}{7}$

2. $\dfrac{3}{11} + \dfrac{5}{11}$

3. $\dfrac{1}{2} + \dfrac{1}{2}$

4. $\dfrac{1}{3} + \dfrac{2}{3}$

5. $\dfrac{8}{11} + \dfrac{7}{11}$

6. $\dfrac{9}{13} + \dfrac{7}{13}$

7. $\dfrac{8}{5} + \dfrac{9}{5}$

8. $\dfrac{5}{3} + \dfrac{7}{3}$

9. $\dfrac{3}{5} + \dfrac{8}{5} + \dfrac{3}{5}$

10. $\dfrac{3}{8} + \dfrac{5}{8} + \dfrac{7}{8}$

11. $\dfrac{3}{4} + \dfrac{1}{4} + \dfrac{5}{4}$

12. $\dfrac{2}{7} + \dfrac{4}{7} + \dfrac{5}{7}$

13. $\dfrac{3}{8} + \dfrac{7}{8} + \dfrac{1}{8}$

14. $\dfrac{5}{12} + \dfrac{7}{12} + \dfrac{1}{12}$

15. $\dfrac{4}{15} + \dfrac{7}{15} + \dfrac{11}{15}$

16. $\dfrac{5}{7} + \dfrac{4}{7} + \dfrac{5}{7}$

17. Find the sum of $\dfrac{5}{12}, \dfrac{1}{12},$ and $\dfrac{11}{12}$.

18. Find the total of $\dfrac{5}{8}, \dfrac{3}{8},$ and $\dfrac{7}{8}$.

 For Exercises 19 to 22, each statement concerns a pair of fractions that have the same denominator. State whether the sum of the fractions is a proper fraction, the number 1, a mixed number, or a whole number other than 1.

19. The sum of the numerators is a multiple of the denominator.

20. The sum of the numerators is one more than the denominator.

21. The sum of the numerators is the denominator.

22. The sum of the numerators is smaller than the denominator.

OBJECTIVE B   To add fractions with different denominators

For Exercises 23 to 42, add.

23. $\dfrac{1}{2} + \dfrac{2}{3}$

24. $\dfrac{2}{3} + \dfrac{1}{4}$

25. $\dfrac{3}{14} + \dfrac{5}{7}$

26. $\dfrac{3}{5} + \dfrac{7}{10}$

27. $\dfrac{8}{15} + \dfrac{7}{20}$

28. $\dfrac{1}{6} + \dfrac{7}{9}$

29. $\dfrac{3}{8} + \dfrac{9}{14}$

30. $\dfrac{5}{12} + \dfrac{5}{16}$

31. $\dfrac{3}{20} + \dfrac{1}{30}$ **32.** $\dfrac{5}{12} + \dfrac{7}{30}$ **33.** $\dfrac{1}{3} + \dfrac{5}{6} + \dfrac{7}{9}$ **34.** $\dfrac{2}{3} + \dfrac{5}{6} + \dfrac{7}{12}$

35. $\dfrac{5}{6} + \dfrac{1}{12} + \dfrac{5}{16}$ **36.** $\dfrac{2}{9} + \dfrac{7}{15} + \dfrac{4}{21}$ **37.** $\dfrac{2}{3} + \dfrac{1}{5} + \dfrac{7}{12}$ **38.** $\dfrac{3}{4} + \dfrac{4}{5} + \dfrac{7}{12}$

39. $\dfrac{2}{3} + \dfrac{3}{5} + \dfrac{7}{8}$ **40.** $\dfrac{3}{10} + \dfrac{14}{15} + \dfrac{9}{25}$ **41.** $\dfrac{2}{3} + \dfrac{5}{8} + \dfrac{7}{9}$ **42.** $\dfrac{1}{3} + \dfrac{2}{9} + \dfrac{7}{8}$

43. What is $\dfrac{3}{8}$ added to $\dfrac{3}{5}$?   **44.** What is $\dfrac{5}{9}$ added to $\dfrac{7}{12}$?

45. Find the sum of $\dfrac{3}{8}$, $\dfrac{5}{6}$, and $\dfrac{7}{12}$.   **46.** Find the total of $\dfrac{1}{2}$, $\dfrac{5}{8}$, and $\dfrac{7}{9}$.

47. Which statement describes a pair of fractions for which the least common denominator is the product of the denominators?
  (i) The denominator of one fraction is a multiple of the denominator of the second fraction.
  (ii) The denominators of the two fractions have no common factors.

OBJECTIVE C   **To add whole numbers, mixed numbers, and fractions**

For Exercises 48 to 69, add.

48. $2\dfrac{2}{5}$ **49.** $4\dfrac{1}{2}$ **50.** $3\dfrac{3}{8}$ **51.** 4 **52.** $6\dfrac{8}{9}$
$+ 3\dfrac{3}{10}$ $+ 5\dfrac{7}{12}$ $+ 2\dfrac{5}{16}$ $+ 5\dfrac{2}{7}$ $+ 12$

53. $7\dfrac{5}{12} + 2\dfrac{9}{16}$ **54.** $9\dfrac{1}{2} + 3\dfrac{3}{11}$ **55.** $6 + 2\dfrac{3}{13}$ **56.** $8\dfrac{21}{40} + 6$

57. $8\dfrac{29}{30} + 7\dfrac{11}{40}$ **58.** $17\dfrac{5}{16} + 3\dfrac{11}{24}$ **59.** $17\dfrac{3}{8} + 7\dfrac{7}{20}$ **60.** $14\dfrac{7}{12} + 29\dfrac{13}{21}$

61. $5\dfrac{7}{8} + 27\dfrac{5}{12}$

62. $7\dfrac{5}{6} + 3\dfrac{5}{9}$

63. $7\dfrac{5}{9} + 2\dfrac{7}{12}$

64. $3\dfrac{1}{2} + 2\dfrac{3}{4} + 1\dfrac{5}{6}$

65. $2\dfrac{1}{2} + 3\dfrac{2}{3} + 4\dfrac{1}{4}$

66. $3\dfrac{1}{3} + 7\dfrac{1}{5} + 2\dfrac{1}{7}$

67. $3\dfrac{1}{2} + 3\dfrac{1}{5} + 8\dfrac{1}{9}$

68. $6\dfrac{5}{9} + 6\dfrac{5}{12} + 2\dfrac{5}{18}$

69. $2\dfrac{3}{8} + 4\dfrac{7}{12} + 3\dfrac{5}{16}$

70. Find the sum of $2\dfrac{4}{9}$ and $5\dfrac{7}{12}$.

71. Find $5\dfrac{5}{6}$ more than $3\dfrac{3}{8}$.

72. What is $4\dfrac{3}{4}$ added to $9\dfrac{1}{3}$?

73. What is $4\dfrac{8}{9}$ added to $9\dfrac{1}{6}$?

74. Find the total of 2, $4\dfrac{5}{8}$, and $2\dfrac{2}{9}$.

75. Find the total of $1\dfrac{5}{8}$, 3, and $7\dfrac{7}{24}$.

 For Exercises 76 and 77, state whether the given sum can be a whole number. Answer *yes* or *no*.

76. The sum of two mixed numbers

77. The sum of a mixed number and a whole number

OBJECTIVE D   To solve application problems

78. **Mechanics** Find the length of the shaft.

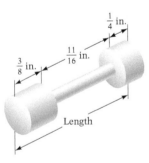

79. **Mechanics** Find the length of the shaft.

80. **Carpentry** A table 30 inches high has a top that is $1\dfrac{1}{8}$ inches thick. Find the total thickness of the table top after a $\dfrac{3}{16}$-inch veneer is applied.

81. For the table pictured at the right, what does the sum $30 + 1\dfrac{1}{8} + \dfrac{3}{16}$ represent?

82. Wages You are working a part-time job that pays $11 an hour. You worked 5, $3\frac{3}{4}$, $2\frac{1}{3}$, $1\frac{1}{4}$, and $7\frac{2}{3}$ hours during the last five days.
   a. Find the total number of hours you worked during the last five days.
   b. Find your total wages for the five days.

83. Sports The course of a yachting race is in the shape of a triangle with sides that measure $4\frac{3}{10}$ miles, $3\frac{7}{10}$ miles, and $2\frac{1}{2}$ miles. Find the total length of the course.

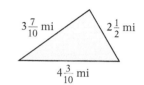

$3\frac{7}{10}$ mi $2\frac{1}{2}$ mi

$4\frac{3}{10}$ mi

Construction The size of an interior door frame is determined by the width of the wall into which it is installed. The width of the wall is determined by the width of the stud in the wall and the thickness of the sheets of dry wall installed on each side of the wall. A 2 × 4 stud is $3\frac{5}{8}$ inches thick. A 2 × 6 stud is $5\frac{5}{8}$ inches thick. Use this information for Exercises 84 to 86.

84. Find the thickness of a wall constructed with 2 × 4 studs and dry wall that is $\frac{1}{2}$ inch thick.

85. Find the thickness of a wall constructed with 2 × 6 studs and dry wall that is $\frac{1}{2}$ inch thick.

86. A fire wall is a physical barrier in a building designed to limit the spread of fire. Suppose a fire wall is built between the garage and the kitchen of a house. Find the width of the fire wall if it is constructed using 2 × 4 studs and dry wall that is $\frac{5}{8}$ inch thick.

87. Construction Two pieces of wood must be bolted together. One piece of wood is $\frac{1}{2}$ inch thick. The second piece is $\frac{5}{8}$ inch thick. A washer will be placed on each of the outer sides of the two pieces of wood. Each washer is $\frac{1}{16}$ inch thick. The nut is $\frac{3}{16}$ inch thick. Find the minimum length of bolt needed to bolt the two pieces of wood together.

Applying the Concepts

88. What is a unit fraction? Find the sum of the three largest unit fractions. Is there a smallest unit fraction? If so, write it down. If not, explain why.

89. A survey was conducted to determine people's favorite color from among blue, green, red, purple, and other. The surveyor claims that $\frac{1}{3}$ of the people responded blue, $\frac{1}{6}$ responded green, $\frac{1}{8}$ responded red, $\frac{1}{12}$ responded purple, and $\frac{2}{5}$ responded some other color. Is this possible? Explain your answer.

SECTION

2.5   Subtraction of Fractions and Mixed Numbers

OBJECTIVE A   To subtract fractions with the same denominator

Fractions with the same denominator are subtracted by subtracting the numerators and placing the difference over the common denominator. After subtracting, write the fraction in simplest form.

> **HOW TO 1**   Subtract: $\dfrac{5}{7} - \dfrac{3}{7}$

$$\begin{array}{r} \dfrac{5}{7} \\ -\dfrac{3}{7} \\ \hline \dfrac{2}{7} \end{array}$$

- Subtract the numerators and place the difference over the common denominator.

$$\frac{5}{7} - \frac{3}{7} = \frac{5-3}{7} = \frac{2}{7}$$

EXAMPLE • 1

Find $\dfrac{17}{30}$ less $\dfrac{11}{30}$.

Solution

$$\begin{array}{r} \dfrac{17}{30} \\ -\dfrac{11}{30} \\ \hline \dfrac{6}{30} = \dfrac{1}{5} \end{array}$$

- The denominators are the same. Subtract the numerators. Place the difference over the common denominator.

YOU TRY IT • 1

Subtract: $\dfrac{16}{27} - \dfrac{7}{27}$

Your solution

Solution on p. S5

OBJECTIVE B   To subtract fractions with different denominators

To subtract fractions with different denominators, first rewrite the fractions as equivalent fractions with a common denominator. As with adding fractions, the common denominator is the LCM of the denominators of the fractions.

> **HOW TO 2**   Subtract: $\dfrac{5}{6} - \dfrac{1}{4}$

The common denominator is the LCM of 6 and 4. The LCM = 12.

Write equivalent fractions using the LCM.

$$\frac{5}{6} = \frac{10}{12}$$
$$-\frac{1}{4} = \frac{3}{12}$$

Subtract the fractions.

$$\begin{array}{r} \dfrac{5}{6} = \dfrac{10}{12} \\ -\dfrac{1}{4} = \dfrac{3}{12} \\ \hline \dfrac{7}{12} \end{array}$$

EXAMPLE • 2

Subtract: $\dfrac{11}{16} - \dfrac{5}{12}$

Solution

$\dfrac{11}{16} = \dfrac{33}{48}$ • LCM = 48

$-\dfrac{5}{12} = \dfrac{20}{48}$

$\dfrac{13}{48}$

YOU TRY IT • 2

Subtract: $\dfrac{13}{18} - \dfrac{7}{24}$

Your solution

Solution on p. S5

OBJECTIVE C To subtract whole numbers, mixed numbers, and fractions

To subtract mixed numbers without borrowing, subtract the fractional parts and then subtract the whole numbers.

HOW TO 3 Subtract: $5\dfrac{5}{6} - 2\dfrac{3}{4}$

Subtract the fractional parts.

$5\dfrac{5}{6} = 5\dfrac{10}{12}$ • The LCM of
                                    6 and 4 is 12.
$-2\dfrac{3}{4} = 2\dfrac{9}{12}$

Subtract the whole numbers.

$5\dfrac{5}{6} = 5\dfrac{10}{12}$

$-2\dfrac{3}{4} = 2\dfrac{9}{12}$

$\dfrac{1}{12}$

Subtraction of mixed numbers sometimes involves borrowing.

HOW TO 4 Subtract: $5 - 2\dfrac{5}{8}$

Borrow 1 from 5.

$5 = \overset{4}{\cancel{5}}\,1$

$-2\dfrac{5}{8} = 2\dfrac{5}{8}$

Write 1 as a fraction so
that the fractions have
the same denominators.

$5 = 4\dfrac{8}{8}$

$-2\dfrac{5}{8} = 2\dfrac{5}{8}$

Subtract the mixed
numbers.

$5 = 4\dfrac{8}{8}$

$-2\dfrac{5}{8} = 2\dfrac{5}{8}$

$2\dfrac{3}{8}$

HOW TO 5 Subtract: $7\dfrac{1}{6} - 2\dfrac{5}{8}$

Write equivalent
fractions using the LCM.

$7\dfrac{1}{6} = 7\dfrac{4}{24}$

$-2\dfrac{5}{8} = 2\dfrac{15}{24}$

Borrow 1 from 7. Add the
1 to $\dfrac{4}{24}$. Write $1\dfrac{4}{24}$ as $\dfrac{28}{24}$.

$7\dfrac{1}{6} = \overset{6}{\cancel{7}}1\dfrac{4}{24} = 6\dfrac{28}{24}$

$-2\dfrac{5}{8} = 2\dfrac{15}{24} = 2\dfrac{15}{24}$

Subtract the mixed
numbers.

$7\dfrac{1}{6} = 6\dfrac{28}{24}$

$-2\dfrac{5}{8} = 2\dfrac{15}{24}$

$4\dfrac{13}{24}$

EXAMPLE · 3

Subtract: $15\frac{7}{8} - 12\frac{2}{3}$

Solution
$$15\frac{7}{8} = 15\frac{21}{24}$$ • LCM = 24
$$- 12\frac{2}{3} = 12\frac{16}{24}$$
$$\overline{\phantom{-12\frac{2}{3} = 1}3\frac{5}{24}}$$

YOU TRY IT · 3

Subtract: $17\frac{5}{9} - 11\frac{5}{12}$

Your solution

EXAMPLE · 4

Subtract: $9 - 4\frac{3}{11}$

Solution
$$9\phantom{\frac{3}{1}} = 8\frac{11}{11}$$ • LCM = 11
$$- 4\frac{3}{11} = 4\frac{3}{11}$$
$$\overline{\phantom{-4\frac{3}{11} = }4\frac{8}{11}}$$

YOU TRY IT · 4

Subtract: $8 - 2\frac{4}{13}$

Your solution

EXAMPLE · 5

Find $11\frac{5}{12}$ decreased by $2\frac{11}{16}$.

Solution
$$11\frac{5}{12} = 11\frac{20}{48} = 10\frac{68}{48}$$ • LCM = 48
$$- 2\frac{11}{16} = 2\frac{33}{48} = 2\frac{33}{48}$$
$$\overline{\phantom{- 2\frac{11}{16} = 2\frac{33}{48} = }8\frac{35}{48}}$$

YOU TRY IT · 5

What is $21\frac{7}{9}$ minus $7\frac{11}{12}$?

Your solution

Solutions on p. S6

OBJECTIVE D　　To solve application problems

HOW TO · 6　The outside diameter of a bushing is $3\frac{3}{8}$ inches and the wall thickness is $\frac{1}{4}$ inch. Find the inside diameter of the bushing.

Outside Diameter

Inside Diameter

$$\frac{1}{4} + \frac{1}{4} = \frac{2}{4} = \frac{1}{2}$$ • Add $\frac{1}{4}$ and $\frac{1}{4}$ to find the total thickness of the two walls.

$$3\frac{3}{8} = 3\frac{3}{8} = 2\frac{11}{8}$$ • Subtract the total thickness of the two walls from the outside diameter to find the inside diameter.
$$- \frac{1}{2} = \frac{4}{8} = \frac{4}{8}$$
$$\overline{\phantom{- \frac{1}{2} = \frac{4}{8} = }2\frac{7}{8}}$$

The inside diameter of the bushing is $2\frac{7}{8}$ inches.

EXAMPLE · 6

A $2\frac{2}{3}$-inch piece is cut from a $6\frac{5}{8}$-inch board. How much of the board is left?

Strategy

To find the length remaining, subtract the length of the piece cut from the total length of the board.

Solution

$$6\frac{5}{8} = 6\frac{15}{24} = 5\frac{39}{24}$$
$$-2\frac{2}{3} = 2\frac{16}{24} = 2\frac{16}{24}$$
$$\overline{ 3\frac{23}{24}}$$

$3\frac{23}{24}$ inches of the board are left.

YOU TRY IT · 6

A flight from New York to Los Angeles takes $5\frac{1}{2}$ hours. After the plane has been in the air for $2\frac{3}{4}$ hours, how much flight time remains?

Your strategy

Your solution

EXAMPLE · 7

Two painters are staining a house. In 1 day one painter stained $\frac{1}{3}$ of the house, and the other stained $\frac{1}{4}$ of the house. How much of the job remains to be done?

Strategy

To find how much of the job remains:

• Find the total amount of the house already stained $\left(\frac{1}{3} + \frac{1}{4}\right)$.

• Subtract the amount already stained from 1, which represents the complete job.

Solution

$$\frac{1}{3} = \frac{4}{12} \qquad\qquad 1 = \frac{12}{12}$$
$$+\frac{1}{4} = \frac{3}{12} \qquad\qquad -\frac{7}{12} = \frac{7}{12}$$
$$\overline{\frac{7}{12}} \qquad\qquad \overline{\frac{5}{12}}$$

$\frac{5}{12}$ of the house remains to be stained.

YOU TRY IT · 7

A patient is put on a diet to lose 24 pounds in 3 months. The patient lost $7\frac{1}{2}$ pounds the first month and $5\frac{3}{4}$ pounds the second month. How much weight must be lost the third month to achieve the goal?

Your strategy

Your solution

Solutions on p. S6

2.5 EXERCISES

OBJECTIVE A   To subtract fractions with the same denominator

For Exercises 1 to 10, subtract.

1. $\dfrac{9}{17}$
$-\dfrac{7}{17}$

2. $\dfrac{11}{15}$
$-\dfrac{3}{15}$

3. $\dfrac{11}{12}$
$-\dfrac{7}{12}$

4. $\dfrac{13}{15}$
$-\dfrac{4}{15}$

5. $\dfrac{9}{20}$
$-\dfrac{7}{20}$

6. $\dfrac{48}{55}$
$-\dfrac{13}{55}$

7. $\dfrac{42}{65}$
$-\dfrac{17}{65}$

8. $\dfrac{11}{24}$
$-\dfrac{5}{24}$

9. $\dfrac{23}{30}$
$-\dfrac{13}{30}$

10. $\dfrac{17}{42}$
$-\dfrac{5}{42}$

11. What is $\dfrac{5}{14}$ less than $\dfrac{13}{14}$?

12. Find the difference between $\dfrac{7}{8}$ and $\dfrac{5}{8}$.

13. Find $\dfrac{17}{24}$ decreased by $\dfrac{11}{24}$.

14. What is $\dfrac{19}{30}$ minus $\dfrac{11}{30}$?

For Exercises 15 and 16, each statement describes the difference between a pair of fractions that have the same denominator. State whether the difference of the fractions will need to be rewritten in order to be in simplest form. Answer *yes* or *no*.

15. The difference between the numerators is a factor of the denominator.

16. The difference between the numerators is 1.

OBJECTIVE B   To subtract fractions with different denominators

For Exercises 17 to 26, subtract.

17. $\dfrac{2}{3}$
$-\dfrac{1}{6}$

18. $\dfrac{7}{8}$
$-\dfrac{5}{16}$

19. $\dfrac{5}{8}$
$-\dfrac{2}{7}$

20. $\dfrac{5}{6}$
$-\dfrac{3}{7}$

21. $\dfrac{5}{7}$
$-\dfrac{3}{14}$

22. $\dfrac{5}{9}$
$-\dfrac{7}{15}$

23. $\dfrac{8}{15}$
$-\dfrac{7}{20}$

24. $\dfrac{7}{9}$
$-\dfrac{1}{6}$

25. $\dfrac{9}{16}$
$-\dfrac{17}{32}$

26. $\dfrac{29}{60}$
$-\dfrac{3}{40}$

27. What is $\frac{3}{5}$ less than $\frac{11}{12}$?

28. What is $\frac{5}{9}$ less than $\frac{11}{15}$?

29. Find the difference between $\frac{11}{24}$ and $\frac{7}{18}$.

30. Find the difference between $\frac{9}{14}$ and $\frac{5}{42}$.

31. Find $\frac{11}{12}$ decreased by $\frac{11}{15}$.

32. Find $\frac{17}{20}$ decreased by $\frac{7}{15}$.

33. What is $\frac{13}{20}$ minus $\frac{1}{6}$?

34. What is $\frac{5}{6}$ minus $\frac{7}{9}$?

35. Which statement describes a pair of fractions for which the least common denominator is one of the denominators?

(i) The denominator of one fraction is a factor of the denominator of the second fraction.

(ii) The denominators of the two fractions have no common factors.

OBJECTIVE C   To subtract whole numbers, mixed numbers, and fractions

For Exercises 36 to 50, subtract.

36. $5\frac{7}{12}$
$-2\frac{5}{12}$

37. $16\frac{11}{15}$
$-11\frac{8}{15}$

38. $6\frac{1}{3}$
-2

39. $5\frac{7}{8}$
-1

40. 10
$-6\frac{1}{3}$

41. 3
$-2\frac{5}{21}$

42. $6\frac{2}{5}$
$-4\frac{4}{5}$

43. $16\frac{3}{8}$
$-10\frac{7}{8}$

44. $25\frac{4}{9}$
$-16\frac{7}{9}$

45. $8\frac{3}{7}$
$-2\frac{6}{7}$

46. $16\frac{2}{5}$
$-8\frac{4}{9}$

47. $23\frac{7}{8}$
$-16\frac{2}{3}$

48. $82\frac{4}{33}$
$-16\frac{5}{22}$

49. 6
$-4\frac{3}{5}$

50. 17
$-7\frac{8}{13}$

51. What is $7\frac{3}{5}$ less than $23\frac{3}{20}$?

52. Find the difference between $12\frac{3}{8}$ and $7\frac{5}{12}$.

53. What is $10\frac{5}{9}$ minus $5\frac{11}{15}$?

54. Find $6\frac{1}{3}$ decreased by $3\frac{3}{5}$.

55. Can the difference between a whole number and a mixed number ever be a whole number?

OBJECTIVE D   To solve application problems

56. Mechanics Find the missing dimension.

57. Mechanics Find the missing dimension.

58. Sports In the Kentucky Derby the horses run $1\frac{1}{4}$ miles. In the Belmont Stakes they run $1\frac{1}{2}$ miles, and in the Preakness Stakes they run $1\frac{3}{16}$ miles. How much farther do the horses run in the Kentucky Derby than in the Preakness Stakes? How much farther do they run in the Belmont Stakes than in the Preakness Stakes?

© Reuters/Corbis

59. Sports In the running high jump in the 1948 Summer Olympic Games, Alice Coachman's distance was $66\frac{1}{8}$ inches. In the same event in the 1972 Summer Olympics, Urika Meyfarth jumped $75\frac{1}{2}$ inches, and in the 1996 Olympic Games, Stefka Kostadinova jumped $80\frac{3}{4}$ inches. Find the difference between Meyfarth's distance and Coachman's distance. Find the difference between Kostadinova's distance and Meyfarth's distance.

60. Fundraising A 12-mile walkathon has three checkpoints. The first checkpoint is $3\frac{3}{8}$ miles from the starting point. The second checkpoint is $4\frac{1}{3}$ miles from the first.
a. How many miles is it from the starting point to the second checkpoint?
b. How many miles is it from the second checkpoint to the finish line?

61. **Hiking** Two hikers plan a 3-day, $27\frac{1}{2}$-mile backpack trip carrying a total of 80 pounds. The hikers plan to travel $7\frac{3}{8}$ miles the first day and $10\frac{1}{3}$ miles the second day.

a. How many total miles do the hikers plan to travel the first two days?
b. How many miles will be left to travel on the third day?

 For Exercises 62 and 63, refer to Exercise 61. Describe what each difference represents.

62. $27\frac{1}{2} - 7\frac{3}{8}$ **63.** $10\frac{1}{3} - 7\frac{3}{8}$

64. **Health** A patient with high blood pressure who weighs 225 pounds is put on a diet to lose 25 pounds in 3 months. The patient loses $8\frac{3}{4}$ pounds the first month and $11\frac{5}{8}$ pounds the second month. How much weight must be lost the third month for the goal to be achieved?

65. **Sports** A wrestler is entered in the 172-pound weight class in the conference finals coming up in 3 weeks. The wrestler needs to lose $12\frac{3}{4}$ pounds. The wrestler loses $5\frac{1}{4}$ pounds the first week and $4\frac{1}{4}$ pounds the second week.

a. Without doing the calculations, determine whether the wrestler can reach his weight class by losing less in the third week than was lost in the second week.

b. How many pounds must be lost in the third week for the desired weight to be reached?

66. **Construction** Find the difference in thickness between a fire wall constructed with 2×6 studs and dry wall that is $\frac{1}{2}$ inch thick and a fire wall constructed with 2×4 studs and dry wall that is $\frac{5}{8}$ inch thick. See Exercises 84 to 86 on page 83.

67. **Finances** If $\frac{4}{15}$ of an electrician's income is spent for housing, what fraction of the electrician's income is not spent for housing?

Applying the Concepts

68. Fill in the square to produce a true statement: $5\frac{1}{3} - \boxed{} = 2\frac{1}{2}$

69. Fill in the square to produce a true statement: $\boxed{} - 4\frac{1}{2} = 1\frac{5}{8}$

70. Fill in the blank squares at the right so that the sum of the numbers is the same along any row, column, or diagonal. The resulting square is called a magic square.

		$\frac{3}{4}$
1	$\frac{5}{8}$	
$\frac{1}{2}$		$\frac{7}{8}$

SECTION

2.6 Multiplication of Fractions and Mixed Numbers

OBJECTIVE A To multiply fractions

The product of two fractions is the product of the numerators over the product of the denominators.

HOW TO 1 Multiply: $\frac{2}{3} \times \frac{4}{5}$

$$\frac{2}{3} \times \frac{4}{5} = \frac{2 \cdot 4}{3 \cdot 5} = \frac{8}{15}$$

• Multiply the numerators.
• Multiply the denominators.

The product $\frac{2}{3} \times \frac{4}{5}$ can be read "$\frac{2}{3}$ times $\frac{4}{5}$" or "$\frac{2}{3}$ of $\frac{4}{5}$."

Reading the times sign as "of" is useful in application problems.

$\frac{4}{5}$ of the bar is shaded.

Shade $\frac{2}{3}$ of the $\frac{4}{5}$ already shaded.

$\frac{8}{15}$ of the bar is then shaded light yellow.

$$\frac{2}{3} \text{ of } \frac{4}{5} = \frac{2}{3} \times \frac{4}{5} = \frac{8}{15}$$

After multiplying two fractions, write the product in simplest form.

HOW TO 2 Multiply: $\frac{3}{4} \times \frac{14}{15}$

$$\frac{3}{4} \times \frac{14}{15} = \frac{3 \cdot 14}{4 \cdot 15}$$

• Multiply the numerators.
• Multiply the denominators.

$$= \frac{3 \cdot 2 \cdot 7}{2 \cdot 2 \cdot 3 \cdot 5}$$

• Write the prime factorization of each number.

$$= \frac{\overset{1}{\cancel{3}} \cdot \overset{1}{\cancel{2}} \cdot 7}{2 \cdot 2 \cdot \underset{1}{\cancel{3}} \cdot 5} = \frac{7}{10}$$

• Eliminate the common factors. Then multiply the remaining factors in the numerator and denominator.

This example could also be worked by using the GCF.

$$\frac{3}{4} \times \frac{14}{15} = \frac{42}{60}$$

• Multiply the numerators.
• Multiply the denominators.

$$= \frac{6 \cdot 7}{6 \cdot 10}$$

• The GCF of 42 and 60 is 6. Factor 6 from 42 and 60.

$$= \frac{\overset{1}{\cancel{6}} \cdot 7}{\underset{1}{\cancel{6}} \cdot 10} = \frac{7}{10}$$

• Eliminate the GCF.

EXAMPLE · 1

Multiply $\frac{4}{15}$ and $\frac{5}{28}$.

Solution

$$\frac{4}{15} \times \frac{5}{28} = \frac{4 \cdot 5}{15 \cdot 28} = \frac{\overset{1}{2} \cdot \overset{1}{2} \cdot \overset{1}{5}}{3 \cdot \underset{1}{5} \cdot \underset{1}{2} \cdot \underset{1}{2} \cdot 7} = \frac{1}{21}$$

YOU TRY IT · 1

Multiply $\frac{4}{21}$ and $\frac{7}{44}$.

Your solution

EXAMPLE · 2

Find the product of $\frac{9}{20}$ and $\frac{33}{35}$.

Solution

$$\frac{9}{20} \times \frac{33}{35} = \frac{9 \cdot 33}{20 \cdot 35} = \frac{3 \cdot 3 \cdot 3 \cdot 11}{2 \cdot 2 \cdot 5 \cdot 5 \cdot 7} = \frac{297}{700}$$

YOU TRY IT · 2

Find the product of $\frac{2}{21}$ and $\frac{10}{33}$.

Your solution

EXAMPLE · 3

What is $\frac{14}{9}$ times $\frac{12}{7}$?

Solution

$$\frac{14}{9} \times \frac{12}{7} = \frac{14 \cdot 12}{9 \cdot 7} = \frac{2 \cdot \overset{1}{7} \cdot 2 \cdot 2 \cdot \overset{1}{3}}{3 \cdot \underset{1}{3} \cdot \underset{1}{7}} = \frac{8}{3} = 2\frac{2}{3}$$

YOU TRY IT · 3

What is $\frac{16}{5}$ times $\frac{15}{24}$?

Your solution

Solutions on p. S6

OBJECTIVE B   To multiply whole numbers, mixed numbers, and fractions

To multiply a whole number by a fraction or a mixed number, first write the whole number as a fraction with a denominator of 1.

HOW TO · 3   Multiply: $4 \times \frac{3}{7}$

$$4 \times \frac{3}{7} = \frac{4}{1} \times \frac{3}{7} = \frac{4 \cdot 3}{1 \cdot 7} = \frac{2 \cdot 2 \cdot 3}{7} = \frac{12}{7} = 1\frac{5}{7}$$

• Write 4 with a denominator of 1; then multiply the fractions.

When one or more of the factors in a product is a mixed number, write the mixed number as an improper fraction before multiplying.

HOW TO · 4   Multiply: $2\frac{1}{3} \times \frac{3}{14}$

$$2\frac{1}{3} \times \frac{3}{14} = \frac{7}{3} \times \frac{3}{14} = \frac{7 \cdot 3}{3 \cdot 14} = \frac{\overset{1}{7} \cdot \overset{1}{3}}{\underset{1}{3} \cdot 2 \cdot \underset{1}{7}} = \frac{1}{2}$$

• Write $2\frac{1}{3}$ as an improper fraction; then multiply the fractions.

EXAMPLE • 4

Multiply: $4\frac{5}{6} \times \frac{12}{13}$

Solution

$$4\frac{5}{6} \times \frac{12}{13} = \frac{29}{6} \times \frac{12}{13} = \frac{29 \cdot 12}{6 \cdot 13}$$

$$= \frac{29 \cdot \overset{1}{\cancel{2}} \cdot 2 \cdot \overset{1}{\cancel{3}}}{\underset{1}{\cancel{2}} \cdot \underset{1}{\cancel{3}} \cdot 13} = \frac{58}{13} = 4\frac{6}{13}$$

YOU TRY IT • 4

Multiply: $5\frac{2}{5} \times \frac{5}{9}$

Your solution

EXAMPLE • 5

Find $5\frac{2}{3}$ times $4\frac{1}{2}$.

Solution

$$5\frac{2}{3} \times 4\frac{1}{2} = \frac{17}{3} \times \frac{9}{2} = \frac{17 \cdot 9}{3 \cdot 2}$$

$$= \frac{17 \cdot \overset{1}{\cancel{3}} \cdot 3}{\underset{1}{\cancel{3}} \cdot 2} = \frac{51}{2} = 25\frac{1}{2}$$

YOU TRY IT • 5

Multiply: $3\frac{2}{5} \times 6\frac{1}{4}$

Your solution

EXAMPLE • 6

Multiply: $4\frac{2}{5} \times 7$

Solution

$$4\frac{2}{5} \times 7 = \frac{22}{5} \times \frac{7}{1} = \frac{22 \cdot 7}{5 \cdot 1}$$

$$= \frac{2 \cdot 11 \cdot 7}{5} = \frac{154}{5} = 30\frac{4}{5}$$

YOU TRY IT • 6

Multiply: $3\frac{2}{7} \times 6$

Your solution

Solutions on p. S6

OBJECTIVE C   To solve application problems

Length (ft)	Weight (lb/ft)
$6\frac{1}{2}$	$\frac{3}{8}$
$8\frac{5}{8}$	$1\frac{1}{4}$
$10\frac{3}{4}$	$2\frac{1}{2}$
$12\frac{7}{12}$	$4\frac{1}{3}$

The table at the left lists the lengths of steel rods and their corresponding weight per foot. The weight per foot is measured in pounds for each foot of rod and is abbreviated as lb/ft.

HOW TO • 5   Find the weight of the steel bar that is $10\frac{3}{4}$ feet long.

Strategy

To find the weight of the steel bar, multiply its length by the weight per foot.

Solution

$$10\frac{3}{4} \times 2\frac{1}{2} = \frac{43}{4} \times \frac{5}{2} = \frac{43 \cdot 5}{4 \cdot 2} = \frac{215}{8} = 26\frac{7}{8}$$

The weight of the $10\frac{3}{4}$-foot rod is $26\frac{7}{8}$ pounds.

EXAMPLE · 7

An electrician earns $206 for each day worked. What are the electrician's earnings for working $4\frac{1}{2}$ days?

Strategy

To find the electrician's total earnings, multiply the daily earnings (206) by the number of days worked $\left(4\frac{1}{2}\right)$.

Solution

$$206 \times 4\frac{1}{2} = \frac{206}{1} \times \frac{9}{2}$$
$$= \frac{206 \cdot 9}{1 \cdot 2}$$
$$= 927$$

The electrician's earnings are $927.

YOU TRY IT · 7

Over the last 10 years, a house increased in value by $2\frac{1}{2}$ times. The price of the house 10 years ago was $170,000. What is the value of the house today?

Your strategy

Your solution

EXAMPLE · 8

The value of a small office building and the land on which it is built is $290,000. The value of the land is $\frac{1}{4}$ the total value. What is the dollar value of the building?

Strategy

To find the value of the building:

- Find the value of the land $\left(\frac{1}{4} \times 290,000\right)$.
- Subtract the value of the land from the total value (290,000).

Solution

$$\frac{1}{4} \times 290,000 = \frac{290,000}{4}$$
$$= 72,500 \qquad \bullet \text{ Value of the land}$$
$$290,000 - 72,500 = 217,500$$

The value of the building is $217,500.

YOU TRY IT · 8

A paint company bought a drying chamber and an air compressor for spray painting. The total cost of the two items was $160,000. The drying chamber's cost was $\frac{4}{5}$ of the total cost. What was the cost of the air compressor?

Your strategy

Your solution

Solutions on pp. S6–S7

2.6 EXERCISES

OBJECTIVE A   To multiply fractions

For Exercises 1 to 32, multiply.

1. $\dfrac{2}{3} \times \dfrac{7}{8}$

2. $\dfrac{1}{2} \times \dfrac{2}{3}$

3. $\dfrac{5}{16} \times \dfrac{7}{15}$

4. $\dfrac{3}{8} \times \dfrac{6}{7}$

5. $\dfrac{1}{6} \times \dfrac{1}{8}$

6. $\dfrac{2}{5} \times \dfrac{5}{6}$

7. $\dfrac{11}{12} \times \dfrac{6}{7}$

8. $\dfrac{11}{12} \times \dfrac{3}{5}$

9. $\dfrac{8}{9} \times \dfrac{27}{4}$

10. $\dfrac{3}{5} \times \dfrac{3}{10}$

11. $\dfrac{5}{6} \times \dfrac{1}{2}$

12. $\dfrac{3}{8} \times \dfrac{5}{12}$

13. $\dfrac{16}{9} \times \dfrac{27}{8}$

14. $\dfrac{5}{8} \times \dfrac{16}{15}$

15. $\dfrac{3}{2} \times \dfrac{4}{9}$

16. $\dfrac{5}{3} \times \dfrac{3}{7}$

17. $\dfrac{7}{8} \times \dfrac{3}{14}$

18. $\dfrac{2}{9} \times \dfrac{1}{5}$

19. $\dfrac{1}{10} \times \dfrac{3}{8}$

20. $\dfrac{5}{12} \times \dfrac{6}{7}$

21. $\dfrac{15}{8} \times \dfrac{16}{3}$

22. $\dfrac{5}{6} \times \dfrac{4}{15}$

23. $\dfrac{1}{2} \times \dfrac{2}{15}$

24. $\dfrac{3}{8} \times \dfrac{5}{16}$

25. $\dfrac{5}{7} \times \dfrac{14}{15}$

26. $\dfrac{3}{8} \times \dfrac{15}{41}$

27. $\dfrac{5}{12} \times \dfrac{42}{65}$

28. $\dfrac{16}{33} \times \dfrac{55}{72}$

29. $\dfrac{12}{5} \times \dfrac{5}{3}$

30. $\dfrac{17}{9} \times \dfrac{81}{17}$

31. $\dfrac{16}{85} \times \dfrac{125}{84}$

32. $\dfrac{19}{64} \times \dfrac{48}{95}$

33. Give an example of a proper and an improper fraction whose product is 1.

34. Multiply $\frac{7}{12}$ and $\frac{15}{42}$.

35. Multiply $\frac{32}{9}$ and $\frac{3}{8}$.

36. Find the product of $\frac{5}{9}$ and $\frac{3}{20}$.

37. Find the product of $\frac{7}{3}$ and $\frac{15}{14}$.

38. What is $\frac{1}{2}$ times $\frac{8}{15}$?

39. What is $\frac{3}{8}$ times $\frac{12}{17}$?

OBJECTIVE B   To multiply whole numbers, mixed numbers, and fractions

For Exercises 40 to 71, multiply.

40. $4 \times \frac{3}{8}$

41. $14 \times \frac{5}{7}$

42. $\frac{2}{3} \times 6$

43. $\frac{5}{12} \times 40$

44. $\frac{1}{3} \times 1\frac{1}{3}$

45. $\frac{2}{5} \times 2\frac{1}{2}$

46. $1\frac{7}{8} \times \frac{4}{15}$

47. $2\frac{1}{5} \times \frac{5}{22}$

48. $4 \times 2\frac{1}{2}$

49. $9 \times 3\frac{1}{3}$

50. $2\frac{1}{7} \times 3$

51. $5\frac{1}{4} \times 8$

52. $3\frac{2}{3} \times 5$

53. $4\frac{2}{9} \times 3$

54. $\frac{1}{2} \times 3\frac{3}{7}$

55. $\frac{3}{8} \times 4\frac{4}{5}$

56. $6\frac{1}{8} \times \frac{4}{7}$

57. $5\frac{1}{3} \times \frac{5}{16}$

58. $\frac{3}{8} \times 4\frac{1}{2}$

59. $\frac{5}{7} \times 2\frac{1}{3}$

60. $0 \times 2\frac{2}{3}$

61. $6\frac{1}{8} \times 0$

62. $2\frac{5}{8} \times 3\frac{2}{5}$

63. $5\frac{3}{16} \times 5\frac{1}{3}$

64. $3\frac{1}{7} \times 2\frac{1}{8}$

65. $16\frac{5}{8} \times 1\frac{1}{16}$

66. $2\frac{2}{5} \times 3\frac{1}{12}$

67. $2\frac{2}{3} \times \frac{3}{20}$

68. $5\frac{1}{5} \times 3\frac{1}{13}$

69. $3\frac{3}{4} \times 2\frac{3}{20}$

70. $12\frac{3}{5} \times 1\frac{3}{7}$

71. $6\frac{1}{2} \times 1\frac{3}{13}$

72. True or false? If the product of a whole number and a fraction is a whole number, then the denominator of the fraction is a factor of the original whole number.

73. Multiply $2\frac{1}{2}$ and $3\frac{3}{5}$.

74. Multiply $4\frac{3}{8}$ and $3\frac{3}{5}$.

75. Find the product of $2\frac{1}{8}$ and $\frac{5}{17}$.

76. Find the product of $12\frac{2}{5}$ and $3\frac{7}{31}$.

77. What is $1\frac{3}{8}$ times $2\frac{1}{5}$?

78. What is $3\frac{1}{8}$ times $2\frac{4}{7}$?

OBJECTIVE C   To solve application problems

For Exercises 79 and 80, give your answer without actually doing a calculation.

79. Read Exercise 81. Will the requested cost be greater than or less than $12?

80. Read Exercise 83. Will the requested length be greater than or less than 4 feet?

81. **Consumerism** Salmon costs $4 per pound. Find the cost of $2\frac{3}{4}$ pounds of salmon.

82. **Exercise** Maria Rivera can walk $3\frac{1}{2}$ miles in 1 hour. At this rate, how far can Maria walk in $\frac{1}{3}$ hour?

$3\frac{1}{2}$ mi
1 h
?
$\frac{1}{3}$ h

83. **Carpentry** A board that costs $6 is $9\frac{1}{4}$ feet long. One-third of the board is cut off. What is the length of the piece cut off?

84. **Geometry** The perimeter of a square is equal to four times the length of a side of the square. Find the perimeter of a square whose side measures $16\frac{3}{4}$ inches.

$16\frac{3}{4}$ in.

85. **Geometry** To find the area of a square, multiply the length of one side of the square times itself. What is the area of a square whose side measures $5\frac{1}{4}$ feet? The area of the square will be in square feet.

86. **Geometry** The area of a rectangle is equal to the product of the length of the rectangle times its width. Find the area of a rectangle that has a length of $4\frac{2}{5}$ miles and a width of $3\frac{3}{10}$ miles. The area will be in square miles.

$4\frac{2}{5}$ mi
$3\frac{3}{10}$ mi

87. **Biofuels** See the news clipping at the right. How many bushels of corn produced each year are turned into ethanol?

Measurement The table at the right below shows the lengths of steel rods and their corresponding weights per foot. Use this table for Exercises 88 to 90.

88. Find the weight of the $6\frac{1}{2}$-foot steel rod.

89. Find the weight of the $12\frac{7}{12}$-foot steel rod.

90. Find the total weight of the $8\frac{5}{8}$-foot and the $10\frac{3}{4}$-foot steel rods.

Length (ft)	Weight (lb/ft)
$6\frac{1}{2}$	$\frac{3}{8}$
$8\frac{5}{8}$	$1\frac{1}{4}$
$10\frac{3}{4}$	$2\frac{1}{2}$
$12\frac{7}{12}$	$4\frac{1}{3}$

91. **Sewing** The Booster Club is making 22 capes for the members of the high school marching band. Each cape is made from $1\frac{3}{8}$ yards of material at a cost of $12 per yard. Find the total cost of the material.

92. **Construction** On an architectural drawing of a kitchen, the front face of the cabinet below the sink is $23\frac{1}{2}$ inches from the back wall. Before the cabinet is installed, a plumber must install a drain in the floor halfway between the wall and the front face of the cabinet. Find the required distance from the wall to the center of the drain.

Applying the Concepts

93. The product of 1 and a number is $\frac{1}{2}$. Find the number.

94. **Time** Our calendar is based on the solar year, which is $365\frac{1}{4}$ days. Use this fact to explain leap years.

95. Which of the labeled points on the number line at the right could be the graph of the product of *B* and *C*?

```
 |--•--•--•--|--•---|----•--|
 0  A  B  C  1  D   2   E   3
```

96. Fill in the circles on the square at the right with the fractions $\frac{1}{6}$, $\frac{5}{18}$, $\frac{4}{9}$, $\frac{5}{9}$, $\frac{2}{3}$, $\frac{3}{4}$, $1\frac{1}{9}$, $1\frac{1}{2}$, and $2\frac{1}{4}$ so that the product of any row is equal to $\frac{5}{18}$. (*Note:* There is more than one possible answer.)

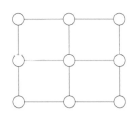

SECTION

2.7   Division of Fractions and Mixed Numbers

OBJECTIVE A   To divide fractions

The **reciprocal of a fraction** is the fraction with the numerator and denominator interchanged.

The reciprocal of $\frac{2}{3}$ is $\frac{3}{2}$.

The process of interchanging the numerator and denominator is called **inverting a fraction.**

To find the reciprocal of a whole number, first write the whole number as a fraction with a denominator of 1. Then find the reciprocal of that fraction.

The reciprocal of 5 is $\frac{1}{5}$. $\left(\text{Think } 5 = \frac{5}{1}.\right)$

Reciprocals are used to rewrite division problems as related multiplication problems. Look at the following two problems:

$$8 \div 2 = 4 \qquad\qquad 8 \times \frac{1}{2} = 4$$

8 divided by 2 **is 4.**   8 times the reciprocal of 2 **is 4.**

"Divided by" means the same as "times the reciprocal of." Thus "÷ 2" can be replaced with "× $\frac{1}{2}$," and the answer will be the same. Fractions are divided by making this replacement.

HOW TO   1   Divide: $\frac{2}{3} \div \frac{3}{4}$

$$\frac{2}{3} \div \frac{3}{4} = \frac{2}{3} \times \frac{4}{3} = \frac{2 \cdot 4}{3 \cdot 3} = \frac{2 \cdot 2 \cdot 2}{3 \cdot 3} = \frac{8}{9}$$

• Multiply the first fraction by the reciprocal of the second fraction.

EXAMPLE • 1

Divide: $\frac{5}{8} \div \frac{4}{9}$

Solution $\frac{5}{8} \div \frac{4}{9} = \frac{5}{8} \times \frac{9}{4} = \frac{5 \cdot 9}{8 \cdot 4}$

$$= \frac{5 \cdot 3 \cdot 3}{2 \cdot 2 \cdot 2 \cdot 2 \cdot 2} = \frac{45}{32} = 1\frac{13}{32}$$

YOU TRY IT • 1

Divide: $\frac{3}{7} \div \frac{2}{3}$

Your solution

EXAMPLE • 2

Divide: $\frac{3}{5} \div \frac{12}{25}$

Solution $\frac{3}{5} \div \frac{12}{25} = \frac{3}{5} \times \frac{25}{12} = \frac{3 \cdot 25}{5 \cdot 12}$

$$= \frac{\overset{1}{\cancel{3}} \cdot \overset{1}{\cancel{5}} \cdot 5}{\underset{1}{\cancel{5}} \cdot 2 \cdot 2 \cdot \underset{1}{\cancel{3}}} = \frac{5}{4} = 1\frac{1}{4}$$

YOU TRY IT • 2

Divide: $\frac{3}{4} \div \frac{9}{10}$

Your solution

Solutions on p. S7

OBJECTIVE B   To divide whole numbers, mixed numbers, and fractions

To divide a fraction and a whole number, first write the whole number as a fraction with a denominator of 1.

> **HOW TO 2**   Divide: $\frac{3}{7} \div 5$
>
> $\frac{3}{7} \div \boxed{5} = \frac{3}{7} \div \boxed{\frac{5}{1}} = \frac{3}{7} \times \frac{1}{5} = \frac{3 \cdot 1}{7 \cdot 5} = \frac{3}{35}$ • Write 5 with a denominator of 1. Then divide the fractions.

When a number in a quotient is a mixed number, write the mixed number as an improper fraction before dividing.

> **HOW TO 3**   Divide: $1\frac{13}{15} \div 4\frac{4}{5}$
>
> Write the mixed numbers as improper fractions. Then divide the fractions.
>
> $1\frac{13}{15} \div 4\frac{4}{5} = \frac{28}{15} \div \frac{24}{5} = \frac{28}{15} \times \frac{5}{24} = \frac{28 \cdot 5}{15 \cdot 24} = \frac{\overset{1}{2} \cdot \overset{1}{2} \cdot 7 \cdot \overset{1}{5}}{3 \cdot \underset{1}{5} \cdot \underset{1}{2} \cdot \underset{1}{2} \cdot 2 \cdot 3} = \frac{7}{18}$

EXAMPLE 3

Divide $\frac{4}{9}$ by 5.

Solution

$\frac{4}{9} \div 5 = \frac{4}{9} \div \frac{5}{1} = \frac{4}{9} \times \frac{1}{5}$ • $5 = \frac{5}{1}$. The reciprocal of $\frac{5}{1}$ is $\frac{1}{5}$.

$= \frac{4 \cdot 1}{9 \cdot 5} = \frac{2 \cdot 2}{3 \cdot 3 \cdot 5} = \frac{4}{45}$

YOU TRY IT 3

Divide $\frac{5}{7}$ by 6.

Your solution

EXAMPLE 4

Find the quotient of $\frac{3}{8}$ and $2\frac{1}{10}$.

Solution

$\frac{3}{8} \div 2\frac{1}{10} = \frac{3}{8} \div \frac{21}{10} = \frac{3}{8} \times \frac{10}{21}$

$= \frac{3 \cdot 10}{8 \cdot 21} = \frac{\overset{1}{3} \cdot \overset{1}{2} \cdot 5}{\underset{}{2} \cdot 2 \cdot 2 \cdot \underset{1}{3} \cdot 7} = \frac{5}{28}$

YOU TRY IT 4

Find the quotient of $12\frac{3}{5}$ and 7.

Your solution

EXAMPLE 5

Divide: $2\frac{3}{4} \div 1\frac{5}{7}$

Solution

$2\frac{3}{4} \div 1\frac{5}{7} = \frac{11}{4} \div \frac{12}{7} = \frac{11}{4} \times \frac{7}{12} = \frac{11 \cdot 7}{4 \cdot 12}$

$= \frac{11 \cdot 7}{2 \cdot 2 \cdot 2 \cdot 2 \cdot 3} = \frac{77}{48} = 1\frac{29}{48}$

YOU TRY IT 5

Divide: $3\frac{2}{3} \div 2\frac{2}{5}$

Your solution

Solutions on p. S7

EXAMPLE • 6

Divide: $1\frac{13}{15} \div 4\frac{1}{5}$

Solution

$$1\frac{13}{15} \div 4\frac{1}{5} = \frac{28}{15} \div \frac{21}{5} = \frac{28}{15} \times \frac{5}{21} = \frac{28 \cdot 5}{15 \cdot 21}$$

$$= \frac{2 \cdot 2 \cdot \overset{1}{\cancel{7}} \cdot \overset{1}{\cancel{5}}}{3 \cdot \underset{1}{\cancel{5}} \cdot 3 \cdot \underset{1}{\cancel{7}}} = \frac{4}{9}$$

YOU TRY IT • 6

Divide: $2\frac{5}{6} \div 8\frac{1}{2}$

Your solution

EXAMPLE • 7

Divide: $4\frac{3}{8} \div 7$

Solution

$$4\frac{3}{8} \div 7 = \frac{35}{8} \div \frac{7}{1} = \frac{35}{8} \times \frac{1}{7}$$

$$= \frac{35 \cdot 1}{8 \cdot 7} = \frac{5 \cdot \overset{1}{\cancel{7}}}{2 \cdot 2 \cdot 2 \cdot \underset{1}{\cancel{7}}} = \frac{5}{8}$$

YOU TRY IT • 7

Divide: $6\frac{2}{5} \div 4$

Your solution

Solutions on p. S7

OBJECTIVE C   To solve application problems

EXAMPLE • 8

A car used $15\frac{1}{2}$ gallons of gasoline on a 310-mile trip. How many miles can this car travel on 1 gallon of gasoline?

Strategy
To find the number of miles, divide the number of miles traveled by the number of gallons of gasoline used.

Solution

$$310 \div 15\frac{1}{2} = \frac{310}{1} \div \frac{31}{2}$$

$$= \frac{310}{1} \times \frac{2}{31} = \frac{310 \cdot 2}{1 \cdot 31}$$

$$= \frac{2 \cdot 5 \cdot \overset{1}{\cancel{31}} \cdot 2}{1 \cdot \underset{1}{\cancel{31}}} = \frac{20}{1} = 20$$

The car travels 20 miles on 1 gallon of gasoline.

YOU TRY IT • 8

A factory worker can assemble a product in $7\frac{1}{2}$ minutes. How many products can the worker assemble in 1 hour?

Your strategy

Your solution

Solutions on p. S7

EXAMPLE • 9

A 12-foot board is cut into pieces $2\frac{1}{4}$ feet long for use as bookshelves. What is the length of the remaining piece after as many shelves as possible have been cut?

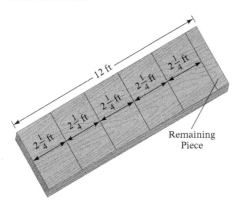

YOU TRY IT • 9

A 16-foot board is cut into pieces $3\frac{1}{3}$ feet long for shelves for a bookcase. What is the length of the remaining piece after as many shelves as possible have been cut?

Strategy

To find the length of the remaining piece:

• Divide the total length of the board (12) by the length of each shelf $\left(2\frac{1}{4}\right)$. This will give you the number of shelves cut, with a certain fraction of a shelf left over.

• Multiply the fractional part of the result in step 1 by the length of one shelf to determine the length of the remaining piece.

Your strategy

Solution

$$12 \div 2\frac{1}{4} = \frac{12}{1} \div \frac{9}{4} = \frac{12}{1} \times \frac{4}{9}$$

$$= \frac{12 \cdot 4}{1 \cdot 9} = \frac{16}{3} = 5\frac{1}{3}$$

There are 5 pieces that are each $2\frac{1}{4}$ feet long.

There is 1 piece that is $\frac{1}{3}$ of $2\frac{1}{4}$ feet long.

$$\frac{1}{3} \times 2\frac{1}{4} = \frac{1}{3} \times \frac{9}{4} = \frac{1 \cdot 9}{3 \cdot 4} = \frac{3}{4}$$

The length of the piece remaining is $\frac{3}{4}$ foot.

Your solution

Solution on p. S7

2.7 EXERCISES

OBJECTIVE A   To divide fractions

For Exercises 1 to 28, divide.

1. $\dfrac{1}{3} \div \dfrac{2}{5}$

2. $\dfrac{3}{7} \div \dfrac{3}{2}$

3. $\dfrac{3}{7} \div \dfrac{3}{7}$

4. $0 \div \dfrac{1}{2}$

5. $0 \div \dfrac{3}{4}$

6. $\dfrac{16}{33} \div \dfrac{4}{11}$

7. $\dfrac{5}{24} \div \dfrac{15}{36}$

8. $\dfrac{11}{15} \div \dfrac{1}{12}$

9. $\dfrac{1}{9} \div \dfrac{2}{3}$

10. $\dfrac{10}{21} \div \dfrac{5}{7}$

11. $\dfrac{2}{5} \div \dfrac{4}{7}$

12. $\dfrac{3}{8} \div \dfrac{5}{12}$

13. $\dfrac{1}{2} \div \dfrac{1}{4}$

14. $\dfrac{1}{3} \div \dfrac{1}{9}$

15. $\dfrac{1}{5} \div \dfrac{1}{10}$

16. $\dfrac{4}{15} \div \dfrac{2}{5}$

17. $\dfrac{7}{15} \div \dfrac{14}{5}$

18. $\dfrac{5}{8} \div \dfrac{15}{2}$

19. $\dfrac{14}{3} \div \dfrac{7}{9}$

20. $\dfrac{7}{4} \div \dfrac{9}{2}$

21. $\dfrac{5}{9} \div \dfrac{25}{3}$

22. $\dfrac{5}{16} \div \dfrac{3}{8}$

23. $\dfrac{2}{3} \div \dfrac{1}{3}$

24. $\dfrac{4}{9} \div \dfrac{1}{9}$

25. $\dfrac{5}{7} \div \dfrac{2}{7}$

26. $\dfrac{5}{6} \div \dfrac{1}{9}$

27. $\dfrac{2}{3} \div \dfrac{2}{9}$

28. $\dfrac{5}{12} \div \dfrac{5}{6}$

29. Divide $\dfrac{7}{8}$ by $\dfrac{3}{4}$.

30. Divide $\dfrac{7}{12}$ by $\dfrac{3}{4}$.

31. Find the quotient of $\dfrac{5}{7}$ and $\dfrac{3}{14}$.

32. Find the quotient of $\dfrac{6}{11}$ and $\dfrac{9}{32}$.

 33. True or false? If a fraction has a numerator of 1, then the reciprocal of the fraction is a whole number.

 34. True or false? The reciprocal of an improper fraction that is not equal to 1 is a proper fraction.

OBJECTIVE B   To divide whole numbers, mixed numbers, and fractions

For Exercises 35 to 73, divide.

35. $4 \div \dfrac{2}{3}$

36. $\dfrac{2}{3} \div 4$

37. $\dfrac{3}{2} \div 3$

38. $3 \div \dfrac{3}{2}$

39. $\dfrac{5}{6} \div 25$

40. $22 \div \dfrac{3}{11}$

41. $6 \div 3\dfrac{1}{3}$

42. $5\dfrac{1}{2} \div 11$

43. $6\dfrac{1}{2} \div \dfrac{1}{2}$

44. $\dfrac{3}{8} \div 2\dfrac{1}{4}$

45. $8\dfrac{1}{4} \div 2\dfrac{3}{4}$

46. $3\dfrac{5}{9} \div 32$

47. $4\dfrac{1}{5} \div 21$

48. $6\dfrac{8}{9} \div \dfrac{31}{36}$

49. $\dfrac{11}{12} \div 2\dfrac{1}{3}$

50. $\dfrac{7}{8} \div 3\dfrac{1}{4}$

51. $35 \div \dfrac{7}{24}$

52. $\dfrac{3}{8} \div 2\dfrac{3}{4}$

53. $\dfrac{11}{18} \div 2\dfrac{2}{9}$

54. $\dfrac{21}{40} \div 3\dfrac{3}{10}$

55. $2\dfrac{1}{16} \div 2\dfrac{1}{2}$

56. $7\dfrac{3}{5} \div 1\dfrac{7}{12}$

57. $1\dfrac{2}{3} \div \dfrac{3}{8}$

58. $16 \div \dfrac{2}{3}$

59. $1\dfrac{5}{8} \div 4$

60. $13\dfrac{3}{8} \div \dfrac{1}{4}$

61. $16 \div 1\dfrac{1}{2}$

62. $9 \div \dfrac{7}{8}$

63. $1\dfrac{1}{3} \div 5\dfrac{8}{9}$

64. $13\dfrac{2}{3} \div 0$

65. $82\dfrac{3}{5} \div 19\dfrac{1}{10}$

66. $45\dfrac{3}{5} \div 15$

67. $102 \div 1\dfrac{1}{2}$

68. $0 \div 3\dfrac{1}{2}$

69. $8\dfrac{2}{7} \div 1$

70. $6\dfrac{9}{16} \div 1\dfrac{3}{32}$

71. $8\dfrac{8}{9} \div 2\dfrac{13}{18}$

72. $10\dfrac{1}{5} \div 1\dfrac{7}{10}$

73. $7\dfrac{3}{8} \div 1\dfrac{27}{32}$

74. Divide $7\dfrac{7}{9}$ by $5\dfrac{5}{6}$.

75. Divide $2\dfrac{3}{4}$ by $1\dfrac{23}{32}$.

76. Find the quotient of $8\dfrac{1}{4}$ and $1\dfrac{5}{11}$.

77. Find the quotient of $\dfrac{14}{17}$ and $3\dfrac{1}{9}$.

 78. True or false? The reciprocal of a mixed number is an improper fraction.

 79. True or false? A fraction divided by its reciprocal is 1.

OBJECTIVE C   **To solve application problems**

 For Exercises 80 and 81, give your answer without actually doing a calculation.

80. Read Exercise 82. Will the requested number of boxes be greater than or less than 600?

81. Read Exercise 83. Will the requested number of servings be greater than or less than 16?

82. **Consumerism** Individual cereal boxes contain $\dfrac{3}{4}$ ounce of cereal. How many boxes can be filled with 600 ounces of cereal?

83. **Consumerism** A box of Post's Great Grains cereal costing $4 contains 16 ounces of cereal. How many $1\dfrac{1}{3}$-ounce servings are in this box?

84. **Gemology** A $\dfrac{5}{8}$-karat diamond was purchased for $1200. What would a similar diamond weighing 1 karat cost?

85. **Real Estate** The Inverness Investor Group bought $8\dfrac{1}{3}$ acres of land for $200,000. What was the cost of each acre?

86. **Fuel Efficiency** A car used $12\dfrac{1}{2}$ gallons of gasoline on a 275-mile trip. How many miles can the car travel on 1 gallon of gasoline?

87. **Mechanics** A nut moves $\dfrac{5}{32}$ inch for each turn. Find the number of turns it will take for the nut to move $1\dfrac{7}{8}$ inches.

88. **Real Estate** The Hammond Company purchased $9\frac{3}{4}$ acres of land for a housing project. One and one-half acres were set aside for a park.

    a. How many acres are available for housing?

    b. How many $\frac{1}{4}$-acre parcels of land can be sold after the land for the park is set aside?

89. **The Food Industry** A chef purchased a roast that weighed $10\frac{3}{4}$ pounds. After the fat was trimmed and the bone removed, the roast weighed $9\frac{1}{3}$ pounds.

    a. What was the total weight of the fat and bone?

    b. How many $\frac{1}{3}$-pound servings can be cut from the trimmed roast?

90. **Carpentry** A 15-foot board is cut into pieces $3\frac{1}{2}$ feet long for a bookcase. What is the length of the piece remaining after as many shelves as possible have been cut?

91. **Construction** The railing of a stairway extends onto a landing. The distance between the end posts of the railing on the landing is $22\frac{3}{4}$ inches. Five posts are to be inserted, evenly spaced, between the end posts. Each post has a square base that measures $1\frac{1}{4}$ inches. Find the distance between each pair of posts.

92. **Construction** The railing of a stairway extends onto a landing. The distance between the end posts of the railing on the landing is $42\frac{1}{2}$ inches. Ten posts are to be inserted, evenly spaced, between the end posts. Each post has a square base that measures $1\frac{1}{2}$ inches. Find the distance between each pair of posts.

Applying the Concepts

Loans The figure at the right shows how the money borrowed on home equity loans is spent. Use this graph for Exercises 93 and 94.

93. What fractional part of the money borrowed on home equity loans is spent on debt consolidation and home improvement?

94. What fractional part of the money borrowed on home equity loans is spent on home improvement, cars, and tuition?

How Money Borrowed on Home Equity Loans Is Spent
Source: Consumer Bankers Association

95. **Puzzles** You completed $\frac{1}{3}$ of a jigsaw puzzle yesterday and $\frac{1}{2}$ of the puzzle today. What fraction of the puzzle is left to complete?

96. Finances A bank recommends that the maximum monthly payment for a home be $\frac{1}{3}$ of your total monthly income. Your monthly income is \$4500. What would the bank recommend as your maximum monthly house payment?

97. Sports During the second half of the 1900s, greenskeepers mowed the grass on golf putting surfaces progressively lower. The table at the right shows the average grass height by decade. What was the difference between the average height of the grass in the 1980s and its average height in the 1950s?

Average Height of Grass on Golf Putting Surfaces	
Decade	Height (in inches)
1950s	$\frac{1}{4}$
1960s	$\frac{7}{32}$
1970s	$\frac{3}{16}$
1980s	$\frac{5}{32}$
1990s	$\frac{1}{8}$

Source: Golf Course Superintendents Association

98. Wages You have a part-time job that pays \$9 an hour. You worked 5 hours, $3\frac{3}{4}$ hours, $1\frac{1}{4}$ hours, and $2\frac{1}{3}$ hours during the four days you worked last week. Find your total earnings for last week's work.

99. Board Games A wooden travel game board has hinges that allow the board to be folded in half. If the dimensions of the open board are 14 inches by 14 inches by $\frac{7}{8}$ inch, what are the dimensions of the board when it is closed?

Nutrition According to the Center for Science in the Public Interest, the average teenage boy drinks $3\frac{1}{3}$ cans of soda per day. The average teenage girl drinks $2\frac{1}{3}$ cans of soda per day. Use this information for Exercises 100 and 101.

100. If a can of soda contains 150 calories, how many calories does the average teenage boy consume each week in soda?

101. How many more cans of soda per week does the average teenage boy drink than the average teenage girl?

Bill Aron/PhotoEdit, Inc.

102. Maps On a map, two cities are $4\frac{5}{8}$ inches apart. If $\frac{3}{8}$ inch on the map represents 60 miles, what is the number of miles between the two cities?

103. Fill in the box to make a true statement.

　a. $\frac{3}{4} \cdot \square = \frac{1}{2}$　　**b.** $\frac{2}{3} \cdot \square = 1\frac{3}{4}$

$\longleftarrow 7\frac{1}{2}$ in. $\longrightarrow$

104. Publishing A page of type in a certain textbook is $7\frac{1}{2}$ inches wide. If the page is divided into three equal columns, with $\frac{3}{8}$ inch between columns, how wide is each column?

$\frac{3}{8}$　　$\frac{3}{8}$

SECTION

2.8 Order, Exponents, and the Order of Operations Agreement

OBJECTIVE A To identify the order relation between two fractions

Point of Interest

Leonardo of Pisa, who was also called Fibonacci (c. 1175–1250), is credited with bringing the Hindu-Arabic number system to the Western world and promoting its use in place of the cumbersome Roman numeral system.

He was also influential in promoting the idea of the fraction bar. His notation, however, was very different from what we use today.

For instance, he wrote $\frac{3}{4}\frac{5}{7}$ to mean $\frac{5}{7} + \frac{3}{7 \cdot 4}$, which equals $\frac{23}{28}$.

Recall that whole numbers can be graphed as points on the number line. Fractions can also be graphed as points on the number line.

The graph of $\frac{3}{4}$ on the number line

The number line can be used to determine the order relation between two fractions. A fraction that appears to the left of a given fraction is less than the given fraction. A fraction that appears to the right of a given fraction is greater than the given fraction.

$\frac{1}{8} < \frac{3}{8} \qquad \frac{6}{8} > \frac{3}{8}$

To find the order relation between two fractions with the same denominator, compare the numerators. The fraction that has the smaller numerator is the smaller fraction. When the denominators are different, begin by writing equivalent fractions with a common denominator; then compare the numerators.

HOW TO 1 Find the order relation between $\frac{11}{18}$ and $\frac{5}{8}$.

The LCM of 18 and 8 is 72.

$\frac{11}{18} = \frac{44}{72}$ ← Smaller numerator $\qquad \frac{11}{18} < \frac{5}{8}$ or $\frac{5}{8} > \frac{11}{18}$

$\frac{5}{8} = \frac{45}{72}$ ← Larger numerator

EXAMPLE · 1

Place the correct symbol, < or >, between the two numbers.

$\frac{5}{12} \qquad \frac{7}{18}$

Solution $\frac{5}{12} = \frac{15}{36} \qquad \frac{7}{18} = \frac{14}{36}$

$\frac{5}{12} > \frac{7}{18}$

YOU TRY IT · 1

Place the correct symbol, < or >, between the two numbers.

$\frac{9}{14} \qquad \frac{13}{21}$

Your solution

Solution on p. S8

OBJECTIVE B To simplify expressions containing exponents

Repeated multiplication of the same fraction can be written in two ways:

$\frac{1}{2} \cdot \frac{1}{2} \cdot \frac{1}{2} \cdot \frac{1}{2}$ or $\left(\frac{1}{2}\right)^4$ ← Exponent

The exponent indicates how many times the fraction occurs as a factor in the multiplication. The expression $\left(\frac{1}{2}\right)^4$ is in exponential notation.

EXAMPLE · 2

Simplify: $\left(\frac{5}{6}\right)^3 \cdot \left(\frac{3}{5}\right)^2$

Solution

$$\left(\frac{5}{6}\right)^3 \cdot \left(\frac{3}{5}\right)^2 = \left(\frac{5}{6} \cdot \frac{5}{6} \cdot \frac{5}{6}\right) \cdot \left(\frac{3}{5} \cdot \frac{3}{5}\right)$$

$$= \frac{\overset{1}{\cancel{5}} \cdot \overset{1}{\cancel{5}} \cdot 5 \cdot \overset{1}{\cancel{3}} \cdot \overset{1}{\cancel{3}}}{2 \cdot \underset{1}{\cancel{3}} \cdot 2 \cdot \underset{1}{\cancel{3}} \cdot 2 \cdot 3 \cdot \underset{1}{\cancel{5}} \cdot \underset{1}{\cancel{5}}} = \frac{5}{24}$$

YOU TRY IT · 2

Simplify: $\left(\frac{7}{11}\right)^2 \cdot \left(\frac{2}{7}\right)$

Your solution

Solution on p. S8

OBJECTIVE C   To use the Order of Operations Agreement to simplify expressions

The Order of Operations Agreement is used for fractions as well as whole numbers.

> **The Order of Operations Agreement**
> **Step 1.** Do all the operations inside parentheses.
> **Step 2.** Simplify any number expressions containing exponents.
> **Step 3.** Do multiplications and divisions as they occur from left to right.
> **Step 4.** Do additions and subtractions as they occur from left to right.

HOW TO · 2   Simplify $\frac{14}{15} - \left(\frac{1}{2}\right)^2 \times \left(\frac{2}{3} + \frac{4}{5}\right)$.

$$\frac{14}{15} - \left(\frac{1}{2}\right)^2 \times \left(\frac{2}{3} + \frac{4}{5}\right)$$

1. Perform operations in parentheses.

$$\frac{14}{15} - \left(\frac{1}{2}\right)^2 \times \frac{22}{15}$$

2. Simplify expressions with exponents.

$$\frac{14}{15} - \frac{1}{4} \times \frac{22}{15}$$

3. Do multiplication and division as they occur from left to right.

$$\frac{14}{15} - \frac{11}{30}$$

4. Do addition and subtraction as they occur from left to right.

$$\frac{17}{30}$$

One or more of the above steps may not be needed to simplify an expression. In that case, proceed to the next step in the Order of Operations Agreement.

EXAMPLE · 3

Simplify: $\left(\frac{3}{4}\right)^2 \div \left(\frac{3}{8} - \frac{1}{12}\right)$

Solution

$$\left(\frac{3}{4}\right)^2 \div \left(\frac{3}{8} - \frac{1}{12}\right)$$

$$= \left(\frac{3}{4}\right)^2 \div \left(\frac{7}{24}\right) = \frac{9}{16} \div \frac{7}{24}$$

$$= \frac{9}{16} \cdot \frac{24}{7} = \frac{27}{14} = 1\frac{13}{14}$$

YOU TRY IT · 3

Simplify: $\left(\frac{1}{13}\right)^2 \cdot \left(\frac{1}{4} + \frac{1}{6}\right) \div \frac{5}{13}$

Your solution

Solution on p. S8

2.8 EXERCISES

OBJECTIVE A   To identify the order relation between two fractions

For Exercises 1 to 12, place the correct symbol, $<$ or $>$, between the two numbers.

1. $\dfrac{11}{40}$ $\dfrac{19}{40}$

2. $\dfrac{92}{103}$ $\dfrac{19}{103}$

3. $\dfrac{2}{3}$ $\dfrac{5}{7}$

4. $\dfrac{2}{5}$ $\dfrac{3}{8}$

5. $\dfrac{5}{8}$ $\dfrac{7}{12}$

6. $\dfrac{11}{16}$ $\dfrac{17}{24}$

7. $\dfrac{7}{9}$ $\dfrac{11}{12}$

8. $\dfrac{5}{12}$ $\dfrac{7}{15}$

9. $\dfrac{13}{14}$ $\dfrac{19}{21}$

10. $\dfrac{13}{18}$ $\dfrac{7}{12}$

11. $\dfrac{7}{24}$ $\dfrac{11}{30}$

12. $\dfrac{13}{36}$ $\dfrac{19}{48}$

 13. Without writing the fractions $\dfrac{4}{5}$ and $\dfrac{1}{7}$ with a common denominator, decide which fraction is larger.

OBJECTIVE B   To simplify expressions containing exponents

For Exercises 14 to 29, simplify.

14. $\left(\dfrac{3}{8}\right)^2$

15. $\left(\dfrac{5}{12}\right)^2$

16. $\left(\dfrac{2}{9}\right)^3$

17. $\left(\dfrac{1}{2}\right)\cdot\left(\dfrac{2}{3}\right)^2$

18. $\left(\dfrac{2}{3}\right)\cdot\left(\dfrac{1}{2}\right)^4$

19. $\left(\dfrac{1}{3}\right)^2\cdot\left(\dfrac{3}{5}\right)^3$

20. $\left(\dfrac{2}{5}\right)^3\cdot\left(\dfrac{5}{7}\right)^2$

21. $\left(\dfrac{5}{9}\right)^3\cdot\left(\dfrac{18}{25}\right)^2$

22. $\left(\dfrac{1}{3}\right)^4\cdot\left(\dfrac{9}{11}\right)^2$

23. $\left(\dfrac{1}{2}\right)^6\cdot\left(\dfrac{32}{35}\right)^2$

24. $\left(\dfrac{2}{3}\right)^4\cdot\left(\dfrac{81}{100}\right)^2$

25. $\left(\dfrac{1}{6}\right)\cdot\left(\dfrac{6}{7}\right)^2\cdot\left(\dfrac{2}{3}\right)$

26. $\left(\dfrac{2}{7}\right)\cdot\left(\dfrac{7}{8}\right)^2\cdot\left(\dfrac{8}{9}\right)$

27. $3\cdot\left(\dfrac{3}{5}\right)^3\cdot\left(\dfrac{1}{3}\right)^2$

28. $4\cdot\left(\dfrac{3}{4}\right)^3\cdot\left(\dfrac{4}{7}\right)^2$

29. $11\cdot\left(\dfrac{3}{8}\right)^3\cdot\left(\dfrac{8}{11}\right)^2$

 30. True or false? When simplified, the expression $\left(\dfrac{1}{2}\right)^{24}\cdot\left(\dfrac{1}{3}\right)^{35}$ is a fraction with a numerator of 1.

OBJECTIVE C To use the Order of Operations Agreement to simplify expressions

For Exercises 31 to 49, simplify.

31. $\dfrac{1}{2} - \dfrac{1}{3} + \dfrac{2}{3}$

32. $\dfrac{2}{5} + \dfrac{3}{10} - \dfrac{2}{3}$

33. $\dfrac{1}{3} \div \dfrac{1}{2} + \dfrac{3}{4}$

34. $\dfrac{4}{5} + \dfrac{3}{7} \cdot \dfrac{14}{15}$

35. $\left(\dfrac{3}{4}\right)^2 - \dfrac{5}{12}$

36. $\left(\dfrac{3}{5}\right)^3 - \dfrac{3}{25}$

37. $\dfrac{5}{6} \cdot \left(\dfrac{2}{3} - \dfrac{1}{6}\right) + \dfrac{7}{18}$

38. $\dfrac{3}{4} \cdot \left(\dfrac{11}{12} - \dfrac{7}{8}\right) + \dfrac{5}{16}$

39. $\dfrac{7}{12} - \left(\dfrac{2}{3}\right)^2 + \dfrac{5}{8}$

40. $\dfrac{11}{16} - \left(\dfrac{3}{4}\right)^2 + \dfrac{7}{12}$

41. $\dfrac{3}{4} \cdot \left(\dfrac{4}{9}\right)^2 + \dfrac{1}{2}$

42. $\dfrac{9}{10} \cdot \left(\dfrac{2}{3}\right)^3 + \dfrac{2}{3}$

43. $\left(\dfrac{1}{2} + \dfrac{3}{4}\right) \div \dfrac{5}{8}$

44. $\left(\dfrac{2}{3} + \dfrac{5}{6}\right) \div \dfrac{5}{9}$

45. $\dfrac{3}{8} \div \left(\dfrac{5}{12} + \dfrac{3}{8}\right)$

46. $\dfrac{7}{12} \div \left(\dfrac{2}{3} + \dfrac{5}{9}\right)$

47. $\left(\dfrac{3}{8}\right)^2 \div \left(\dfrac{3}{7} + \dfrac{3}{14}\right)$

48. $\left(\dfrac{5}{6}\right)^2 \div \left(\dfrac{5}{12} + \dfrac{2}{3}\right)$

49. $\dfrac{2}{5} \div \dfrac{3}{8} \cdot \dfrac{4}{5}$

 50. Insert parentheses into the expression $\dfrac{2}{9} \cdot \dfrac{5}{6} + \dfrac{3}{4} \div \dfrac{3}{5}$ so that **a.** the first operation to be performed is addition and **b.** the first operation to be performed is division.

Fast-Food Patrons' Top Criteria for Fast-Food Restaurants	
Food quality	$\dfrac{1}{4}$
Location	$\dfrac{13}{50}$
Menu	$\dfrac{4}{25}$
Price	$\dfrac{2}{25}$
Speed	$\dfrac{3}{25}$
Other	$\dfrac{13}{100}$

Source: Maritz Marketing Research, Inc.

Applying the Concepts

51. **The Food Industry** The table at the right shows the results of a survey that asked fast-food patrons their criteria for choosing where to go for fast food. For example, 3 out of every 25 people surveyed said that the speed of the service was most important.

   a. According to the survey, do more people choose a fast-food restaurant on the basis of its location or the quality of the food?

   b. Which criterion was cited by the most people?

FOCUS ON PROBLEM SOLVING

Common Knowledge   An application problem may not provide all the information that is needed to solve the problem. Sometimes, however, the necessary information is common knowledge.

HOW TO 1   You are traveling by bus from Boston to New York. The trip is 4 hours long. If the bus leaves Boston at 10 A.M., what time should you arrive in New York?

What other information do you need to solve this problem?

You need to know that, using a 12-hour clock, the hours run

10 A.M.
11 A.M.
12 P.M.
1 P.M.
2 P.M.

Four hours after 10 A.M. is 2 P.M.

You should arrive in New York at 2 P.M.

HOW TO 2   You purchase a 44¢ stamp at the Post Office and hand the clerk a one-dollar bill. How much change do you receive?

What information do you need to solve this problem?

You need to know that there are 100¢ in one dollar.

Your change is 100¢ − 44¢.

$100 - 44 = 56$

You receive 56¢ in change.

What information do you need to know to solve each of the following problems?

1. You sell a dozen tickets to a fundraiser. Each ticket costs $10. How much money do you collect?

2. The weekly lab period for your science course is 1 hour and 20 minutes long. Find the length of the science lab period in minutes.

3. An employee's monthly salary is $3750. Find the employee's annual salary.

4. A survey revealed that eighth graders spend an average of 3 hours each day watching television. Find the total time an eighth grader spends watching TV each week.

5. You want to buy a carpet for a room that is 15 feet wide and 18 feet long. Find the amount of carpet that you need.

PROJECTS AND GROUP ACTIVITIES

Music   In musical notation, notes are printed on a **staff,** which is a set of five horizontal lines and the spaces between them. The notes of a musical composition are grouped into **measures,** or **bars.** Vertical lines separate measures on a staff. The shape of a note indicates how long it should be held. The whole note has the longest time value of any note. Each time value is divided by 2 in order to find the next smallest time value.

The **time signature** is a fraction that appears at the beginning of a piece of music. The numerator of the fraction indicates the number of beats in a measure. The denominator indicates what kind of note receives 1 beat. For example, music written in $\frac{2}{4}$ time has 2 beats to a measure, and a quarter note receives 1 beat. One measure in $\frac{2}{4}$ time may have 1 half note, 2 quarter notes, 4 eighth notes, or any other combination of notes totaling 2 beats. Other common time signatures are $\frac{4}{4}$, $\frac{3}{4}$, and $\frac{6}{8}$.

1. Explain the meaning of the 6 and the 8 in the time signature $\frac{6}{8}$.

2. Give some possible combinations of notes in one measure of a piece written in $\frac{4}{4}$ time.

3. What does a dot at the right of a note indicate? What is the effect of a dot at the right of a half note? At the right of a quarter note? At the right of an eighth note?

4. Symbols called rests are used to indicate periods of silence in a piece of music. What symbols are used to indicate the different time values of rests?

5. Find some examples of musical compositions written in different time signatures. Use a few measures from each to show that the sum of the time values of the notes and rests in each measure equals the numerator of the time signature.

Construction   Suppose you are involved in building your own home. Design a stairway from the first floor of the house to the second floor. Some of the questions you will need to answer follow.

What is the distance from the floor of the first story to the floor of the second story?

Typically, what is the number of steps in a stairway?

What is a reasonable length for the run of each step?

What is the width of the wood being used to build the staircase?

In designing the stairway, remember that each riser should be the same height, that each run should be the same length, and that the width of the wood used for the steps will have to be incorporated into the calculation.

Fractions of Diagrams   The diagram that follows has been broken up into nine areas separated by heavy lines. Eight of the areas have been labeled *A* through *H*. The ninth area is shaded. Determine which lettered areas would have to be shaded so that half of the entire diagram is shaded and half is not shaded. Write down the strategy that you or your group used to arrive at the solution. Compare your strategy with that of other individual students or groups.

Tips for Success

Three important features of this text that can be used to prepare for a test are the
• Chapter Summary
• Chapter Review Exercises
• Chapter Test
See *AIM for Success* at the front of the book.

CHAPTER 2

SUMMARY

KEY WORDS	EXAMPLES
A number that is a multiple of two or more numbers is a *common multiple* of those numbers. The *least common multiple (LCM)* is the smallest common multiple of two or more numbers. [2.1A, p. 64]	12, 24, 36, 48, . . . are common multiples of 4 and 6. The LCM of 4 and 6 is 12.
A number that is a factor of two or more numbers is a *common factor* of those numbers. The *greatest common factor (GCF)* is the largest common factor of two or more numbers. [2.1B, p. 65]	The common factors of 12 and 16 are 1, 2, and 4. The GCF of 12 and 16 is 4.
A *fraction* can represent the number of equal parts of a whole. In a fraction, the *fraction bar* separates the *numerator* and the *denominator*. [2.2A, p. 68]	In the fraction $\frac{3}{4}$, the numerator is 3 and the denominator is 4.

In a *proper fraction*, the numerator is smaller than the denominator; a proper fraction is a number less than 1. In an *improper fraction*, the numerator is greater than or equal to the denominator; an improper fraction is a number greater than or equal to 1. A *mixed number* is a number greater than 1 with a whole-number part and a fractional part. [2.2A, p. 68]

$\frac{2}{5}$ is proper fraction.

$\frac{7}{6}$ is an improper fraction.

$4\frac{1}{10}$ is a mixed number; 4 is the whole-number part and $\frac{1}{10}$ is the fractional part.

Equal fractions with different denominators are called *equivalent fractions*. [2.3A, p. 72]

$\frac{3}{4}$ and $\frac{6}{8}$ are equivalent fractions.

A fraction is in *simplest form* when the numerator and denominator have no common factors other than 1. [2.3B, p. 73]

The fraction $\frac{11}{12}$ is in simplest form.

The *reciprocal* of a fraction is the fraction with the numerator and denominator interchanged. [2.7A, p. 100]

The reciprocal of $\frac{3}{8}$ is $\frac{8}{3}$.

The reciprocal of 5 is $\frac{1}{5}$.

ESSENTIAL RULES AND PROCEDURES

EXAMPLES

To find the LCM of two or more numbers, find the prime factorization of each number and write the factorization of each number in a table. Circle the greatest product in each column. The LCM is the product of the circled numbers. [2.1A, p. 64]

	2	3
12 =	②·②	3
18 =	2	③·③

The LCM of 12 and 18 is
$2 \cdot 2 \cdot 3 \cdot 3 = 36$.

To find the GCF of two or more numbers, find the prime factorization of each number and write the factorization of each number in a table. Circle the least product in each column that does not have a blank. The GCF is the product of the circled numbers. [2.1B, p. 65]

	2	3
12 =	2·2	③
18 =	②	3·3

The GCF of 12 and 18 is $2 \cdot 3 = 6$.

To write an improper fraction as a mixed number or a whole number, divide the numerator by the denominator. [2.2B, p. 69]

$\frac{29}{6} = 29 \div 6 = 4\frac{5}{6}$

To write a mixed number as an improper fraction, multiply the denominator of the fractional part of the mixed number by the whole-number part. Add this product and the numerator of the fractional part. The sum is the numerator of the improper fraction. The denominator remains the same. [2.2B, p. 69]

$3\frac{2}{5} = \frac{5 \times 3 + 2}{5} = \frac{17}{5}$

To find equivalent fractions by raising to higher terms, multiply the numerator and denominator of the fraction by the same number. [2.3A, p. 72]

$\frac{3}{4} = \frac{3 \cdot 5}{4 \cdot 5} = \frac{15}{20}$

$\frac{3}{4}$ and $\frac{15}{20}$ are equivalent fractions.

To write a fraction in simplest form, factor the numerator and denominator of the fraction; then eliminate the common factors. [2.3B, p. 73]

$\frac{30}{45} = \frac{2 \cdot \overset{1}{\cancel{3}} \cdot \overset{1}{\cancel{5}}}{\underset{1}{\cancel{3}} \cdot 3 \cdot \underset{1}{\cancel{5}}} = \frac{2}{3}$

To add fractions with the same denominator, add the numerators and place the sum over the common denominator. [2.4A, p. 76]	$\dfrac{5}{12} + \dfrac{11}{12} = \dfrac{16}{12} = 1\dfrac{4}{12} = 1\dfrac{1}{3}$
To add fractions with different denominators, first rewrite the fractions as equivalent fractions with a common denominator. (The common denominator is the LCM of the denominators of the fractions.) Then add the fractions. [2.4B, p. 76]	$\dfrac{1}{4} + \dfrac{2}{5} = \dfrac{5}{20} + \dfrac{8}{20} = \dfrac{13}{20}$
To subtract fractions with the same denominator, subtract the numerators and place the difference over the common denominator. [2.5A, p. 84]	$\dfrac{9}{16} - \dfrac{5}{16} = \dfrac{4}{16} = \dfrac{1}{4}$
To subtract fractions with different denominators, first rewrite the fractions as equivalent fractions with a common denominator. (The common denominator is the LCM of the denominators of the fractions.) Then subtract the fractions. [2.5B, p. 84]	$\dfrac{2}{3} - \dfrac{7}{16} = \dfrac{32}{48} - \dfrac{21}{48} = \dfrac{11}{48}$
To multiply two fractions, multiply the numerators; this is the numerator of the product. Multiply the denominators; this is the denominator of the product. [2.6A, p. 92]	$\dfrac{3}{4} \cdot \dfrac{2}{9} = \dfrac{3 \cdot 2}{4 \cdot 9} = \dfrac{\overset{1}{\cancel{3}} \cdot \overset{1}{\cancel{2}}}{2 \cdot 2 \cdot \underset{1}{\cancel{3}} \cdot 3} = \dfrac{1}{6}$
To divide two fractions, multiply the first fraction by the reciprocal of the second fraction. [2.7A, p. 100]	$\dfrac{8}{15} \div \dfrac{4}{5} = \dfrac{8}{15} \cdot \dfrac{5}{4} = \dfrac{8 \cdot 5}{15 \cdot 4}$ $= \dfrac{\overset{1}{\cancel{2}} \cdot \overset{1}{\cancel{2}} \cdot 2 \cdot \overset{1}{\cancel{5}}}{3 \cdot \underset{1}{\cancel{5}} \cdot \underset{1}{\cancel{2}} \cdot \underset{1}{\cancel{2}}} = \dfrac{2}{3}$
The find the order relation between two fractions with the same denominator, compare the numerators. The fraction that has the smaller numerator is the smaller fraction. [2.8A, p. 109]	$\dfrac{17}{25} \leftarrow$ Smaller numerator $\dfrac{19}{25} \leftarrow$ Larger numerator $\dfrac{17}{25} < \dfrac{19}{25}$
To find the order relation between two fractions with different denominators, first rewrite the fractions with a common denominator. The fraction that has the smaller numerator is the smaller fraction. [2.8A, p. 109]	$\dfrac{3}{5} = \dfrac{24}{40} \qquad \dfrac{5}{8} = \dfrac{25}{40}$ $\dfrac{24}{40} < \dfrac{25}{40}$ $\dfrac{3}{5} < \dfrac{5}{8}$

Order of Operations Agreement [2.8C, p. 110]

Step 1 Do all the operations inside parentheses.

Step 2 Simplify any numerical expressions containing exponents.

Step 3 Do multiplication and division as they occur from left to right.

Step 4 Do addition and subtraction as they occur from left to right.

$$\left(\dfrac{1}{3}\right)^2 + \left(\dfrac{5}{6} - \dfrac{7}{12}\right) \cdot (4)$$

$$= \left(\dfrac{1}{3}\right)^2 + \left(\dfrac{1}{4}\right) \cdot (4)$$

$$= \dfrac{1}{9} + \left(\dfrac{1}{4}\right) \cdot (4)$$

$$= \dfrac{1}{9} + 1 = 1\dfrac{1}{9}$$

CHAPTER 2

CONCEPT REVIEW

Test your knowledge of the concepts presented in this chapter. Answer each question. Then check your answers against the ones provided in the Answer Section.

1. How do you find the LCM of 75, 30, and 50?

2. How do you find the GCF of 42, 14, and 21?

3. How do you write an improper fraction as a mixed number?

4. When is a fraction in simplest form?

5. When adding fractions, why do you have to convert to equivalent fractions with a common denominator?

6. How do you add mixed numbers?

7. If you are subtracting a mixed number from a whole number, why do you need to borrow?

8. When multiplying two fractions, why is it better to eliminate the common factors before multiplying the remaining factors in the numerator and denominator?

9. When multiplying two fractions that are less than 1, will the product be greater than 1, less than the smaller number, or between the smaller number and the bigger number?

10. How are reciprocals used when dividing fractions?

11. When a fraction is divided by a whole number, why do we write the whole number as a fraction before dividing?

12. When comparing two fractions, why is it important to look at both the numerators and denominators to determine which is larger?

13. In the expression $\left(\frac{5}{6}\right)^2 - \left(\frac{3}{4} - \frac{2}{3}\right) \div \frac{1}{2}$, in what order should the operations be performed?

CHAPTER 2

REVIEW EXERCISES

1. Write $\frac{30}{45}$ in simplest form.

2. Simplify: $\left(\frac{3}{4}\right)^3 \cdot \frac{20}{27}$

3. Express the shaded portion of the circles as an improper fraction.

4. Find the total of $\frac{2}{3}$, $\frac{5}{6}$, and $\frac{2}{9}$.

5. Place the correct symbol, $<$ or $>$, between the two numbers.
$$\frac{11}{18} \quad \frac{17}{24}$$

6. Subtract:
$$18\frac{1}{6}$$
$$-3\frac{5}{7}$$

7. Simplify: $\frac{2}{7}\left(\frac{5}{8} - \frac{1}{3}\right) \div \frac{3}{5}$

8. Multiply: $2\frac{1}{3} \times 3\frac{7}{8}$

9. Divide: $1\frac{1}{3} \div \frac{2}{3}$

10. Find $\frac{17}{24}$ decreased by $\frac{3}{16}$.

11. Divide: $8\frac{2}{3} \div 2\frac{3}{5}$

12. Find the GCF of 20 and 48.

13. Write an equivalent fraction with the given denominator.
$$\frac{2}{3} = \frac{}{36}$$

14. What is $\frac{15}{28}$ divided by $\frac{5}{7}$?

15. Write an equivalent fraction with the given denominator.
$$\frac{8}{11} = \frac{}{44}$$

16. Multiply: $2\frac{1}{4} \times 7\frac{1}{3}$

17. Find the LCM of 18 and 12.

18. Write $\frac{16}{44}$ in simplest form.

19. Add: $\frac{3}{8} + \frac{5}{8} + \frac{1}{8}$

20. Subtract: $\begin{array}{r} 16 \\ -\ 5\frac{7}{8} \\ \hline \end{array}$

21. Add: $4\frac{4}{9} + 2\frac{1}{6} + 11\frac{17}{27}$

22. Find the GCF of 15 and 25.

23. Write $\frac{17}{5}$ as a mixed number.

24. Simplify: $\left(\frac{4}{5} - \frac{2}{3}\right)^2 \div \frac{4}{15}$

25. Add: $\frac{3}{8} + 1\frac{2}{3} + 3\frac{5}{6}$

26. Find the LCM of 18 and 27.

27. Subtract: $\frac{11}{18} - \frac{5}{18}$

28. Write $2\frac{5}{7}$ as an improper fraction.

29. Divide: $\frac{5}{6} \div \frac{5}{12}$

30. Multiply: $\frac{5}{12} \times \frac{4}{25}$

31. What is $\frac{11}{50}$ multiplied by $\frac{25}{44}$?

32. Express the shaded portion of the circles as a mixed number.

33. **Meteorology** During 3 months of the rainy season, $5\frac{7}{8}$, $6\frac{2}{3}$, and $8\frac{3}{4}$ inches of rain fell. Find the total rainfall for the 3 months.

34. **Real Estate** A home building contractor bought $4\frac{2}{3}$ acres of land for $168,000. What was the cost of each acre?

35. **Sports** A 15-mile race has three checkpoints. The first checkpoint is $4\frac{1}{2}$ miles from the starting point. The second checkpoint is $5\frac{3}{4}$ miles from the first checkpoint. How many miles is the second checkpoint from the finish line?

36. **Fuel Efficiency** A compact car gets 36 miles on each gallon of gasoline. How many miles can the car travel on $6\frac{3}{4}$ gallons of gasoline?

CHAPTER 2

TEST

1. Multiply: $\frac{9}{11} \times \frac{44}{81}$

2. Find the GCF of 24 and 80.

3. Divide: $\frac{5}{9} \div \frac{7}{18}$

4. Simplify: $\left(\frac{3}{4}\right)^2 \div \left(\frac{2}{3} + \frac{5}{6}\right) - \frac{1}{12}$

5. Write $9\frac{4}{5}$ as an improper fraction.

6. What is $5\frac{2}{3}$ multiplied by $1\frac{7}{17}$?

7. Write $\frac{40}{64}$ in simplest form.

8. Place the correct symbol, $<$ or $>$, between the two numbers.

$\frac{3}{8} \quad \frac{5}{12}$

9. Simplify: $\left(\frac{1}{4}\right)^3 \div \left(\frac{1}{8}\right)^2 - \frac{1}{6}$

10. Find the LCM of 24 and 40.

11. Subtract: $\frac{17}{24} - \frac{11}{24}$

12. Write $\frac{18}{5}$ as a mixed number.

13. Find the quotient of $6\frac{2}{3}$ and $3\frac{1}{6}$.

14. Write an equivalent fraction with the given denominator.

15. Add:
$$\frac{5}{6}$$
$$\frac{7}{9}$$
$$+\frac{1}{15}$$

16. Subtract:
$$23\frac{1}{8}$$
$$-9\frac{9}{44}$$

17. What is $\frac{9}{16}$ minus $\frac{5}{12}$?

18. Simplify: $\left(\frac{2}{3}\right)^4 \cdot \frac{27}{32}$

19. Add: $\frac{7}{12} + \frac{11}{12} + \frac{5}{12}$

20. What is $12\frac{5}{12}$ more than $9\frac{17}{20}$?

21. Express the shaded portion of the circles as an improper fraction.

22. **Compensation** An electrician earns $240 for each day worked. What is the total of the electrician's earnings for working $3\frac{1}{2}$ days?

23. **Real Estate** Grant Miura bought $7\frac{1}{4}$ acres of land for a housing project. One and three-fourths acres were set aside for a park, and the remaining land was developed into $\frac{1}{2}$-acre lots. How many lots were available for sale?

24. **Architecture** A scale of $\frac{1}{2}$ inch to 1 foot is used to draw the plans for a house. The scale measurements for three walls are given in the table at the right. Complete the table to determine the actual wall lengths for the three walls a, b, and c.

Wall	Scale	Actual Wall Length
a	$6\frac{1}{4}$ in.	?
b	9 in.	?
c	$7\frac{7}{8}$ in.	?

25. **Meteorology** In 3 successive months, the rainfall measured $11\frac{1}{2}$ inches, $7\frac{5}{8}$ inches, and $2\frac{1}{3}$ inches. Find the total rainfall for the 3 months.

CUMULATIVE REVIEW EXERCISES

1. Round 290,496 to the nearest thousand.

2. Subtract: $\begin{array}{r} 390,047 \\ -\ 98,769 \\ \hline \end{array}$

3. Find the product of 926 and 79.

4. Divide: $57\overline{)30{,}792}$

5. Simplify: $4 \cdot (6 - 3) \div 6 - 1$

6. Find the prime factorization of 44.

7. Find the LCM of 30 and 42.

8. Find the GCF of 60 and 80.

9. Write $7\frac{2}{3}$ as an improper fraction.

10. Write $\frac{25}{4}$ as a mixed number.

11. Write an equivalent fraction with the given denominator.

$$\frac{5}{16} = \frac{}{48}$$

12. Write $\frac{24}{60}$ in simplest form.

13. What is $\frac{9}{16}$ more than $\frac{7}{12}$?

14. Add: $\begin{array}{r} 3\frac{7}{8} \\ 7\frac{5}{12} \\ +\ 2\frac{15}{16} \\ \hline \end{array}$

15. Find $\frac{3}{8}$ less than $\frac{11}{12}$.

16. Subtract: $\begin{array}{r} 5\frac{1}{6} \\ -\ 3\frac{7}{18} \\ \hline \end{array}$

17. Multiply: $\frac{3}{8} \times \frac{14}{15}$

18. Multiply: $3\frac{1}{8} \times 2\frac{2}{5}$

19. Divide: $\frac{7}{16} \div \frac{5}{12}$

20. Find the quotient of $6\frac{1}{8}$ and $2\frac{1}{3}$.

21. Simplify: $\left(\frac{1}{2}\right)^3 \cdot \frac{8}{9}$

22. Simplify: $\left(\frac{1}{2} + \frac{1}{3}\right) \div \left(\frac{2}{5}\right)^2$

23. **Banking** Molly O'Brien had $1359 in a checking account. During the week, Molly wrote checks for $128, $54, and $315. Find the amount in the checking account at the end of the week.

24. **Entertainment** The tickets for a movie were $10 for an adult and $4 for a student. Find the total income from the sale of 87 adult tickets and 135 student tickets.

Kevin Lee/Getty Images

25. **Measurement** Find the total weight of three packages that weigh $1\frac{1}{2}$ pounds, $7\frac{7}{8}$ pounds, and $2\frac{2}{3}$ pounds.

26. **Carpentry** A board $2\frac{5}{8}$ feet long is cut from a board $7\frac{1}{3}$ feet long. What is the length of the remaining piece?

27. **Fuel Efficiency** A car travels 27 miles on each gallon of gasoline. How many miles can the car travel on $8\frac{1}{3}$ gallons of gasoline?

28. **Real Estate** Jimmy Santos purchased $10\frac{1}{3}$ acres of land to build a housing development. Jimmy donated 2 acres for a park. How many $\frac{1}{3}$-acre parcels can be sold from the remaining land?

Decimals

OBJECTIVES

Panoramic Images/Getty Images

ARE YOU READY?

Take the Chapter 3 Prep Test to find out if you are ready to learn to:

- Round decimals
- Add, subtract, multiply, and divide decimals
- Convert between fractions and decimals
- Compare decimals and fractions

PREP TEST

Do these exercises to prepare for Chapter 3.

1. Express the shaded portion of the rectangle as a fraction.

2. Round 36,852 to the nearest hundred.

3. Write 4791 in words.

4. Write six thousand eight hundred forty-two in standard form.

For Exercises 5 to 8, add, subtract, multiply, or divide.

5. $37 + 8892 + 465$ **6.** $2403 - 765$

7. 844×91 **8.** $23\overline{)6412}$

3.1　Introduction to Decimals

| OBJECTIVE A | To write decimals in standard form and in words |

✓ Take Note

In decimal notation, the part of the number that appears to the left of the decimal point is the **whole-number part**. The part of the number that appears to the right of the decimal point is the **decimal part**. The **decimal point** separates the whole-number part from the decimal part.

The price tag on a sweater reads $61.88. The number 61.88 is in **decimal notation.** A number written in decimal notation is often called simply a **decimal.**

A number written in decimal notation has three parts.

61	.	88
Whole-number part	Decimal point	Decimal part

The decimal part of the number represents a number less than 1. For example, $.88 is less than $1. The decimal point (.) separates the whole-number part from the decimal part.

The position of a digit in a decimal determines the digit's place value. The place-value chart is extended to the right to show the place value of digits to the right of a decimal point.

◎ Point of Interest

The idea that all fractions should be represented in tenths, hundredths, and thousandths was presented in 1585 in Simon Stevin's publication *De Thiende* and its French translation, *La Disme,* which was widely read and accepted by the French. This may help to explain why the French accepted the metric system so easily two hundred years later.

In *De Thiende,* Stevin argued in favor of his notation by including examples for astronomers, tapestry makers, surveyors, tailors, and the like. He stated that using decimals would enable calculations to be "performed . . . with as much ease as counterreckoning."

In the decimal 458.302719, the position of the digit 7 determines that its place value is ten-thousandths.

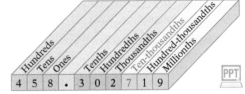

Note the relationship between fractions and numbers written in decimal notation.

Seven tenths	Seven hundredths	Seven thousandths
$\dfrac{7}{10} = 0.7$	$\dfrac{7}{100} = 0.07$	$\dfrac{7}{1000} = 0.007$
1 zero in 10	2 zeros in 100	3 zeros in 1000
1 decimal place in 0.7	2 decimal places in 0.07	3 decimal places in 0.007

To write a decimal in words, write the decimal part of the number as though it were a whole number, and then name the place value of the last digit.

0.9684　Nine thousand six hundred eighty-four ten-thousandths

The decimal point in a decimal is read as "and."

372.516　Three hundred seventy-two and five hundred sixteen thousandths

Point of Interest

The decimal point did not make its appearance until the early 1600s. Stevin's notation used subscripts with circles around them after each digit: 0 for ones, 1 for tenths (which he called "primes"), 2 for hundredths (called "seconds"), 3 for thousandths ("thirds"), and so on. For example, 1.375 would have been written

1 3 7 5
⓪ ① ② ③

To write a decimal in standard form when it is written in words, write the whole-number part, replace the word *and* with a decimal point, and write the decimal part so that the last digit is in the given place-value position.

Four and twenty-three <u>hundredths</u>
3 is in the hundredths place.

When writing a decimal in standard form, you may need to insert zeros after the decimal point so that the last digit is in the given place-value position.

Ninety-one and eight <u>thousandths</u>
8 is in the thousandths place.
Insert two zeros so that the 8 is in the thousandths place.

Sixty-five <u>ten-thousandths</u>
5 is in the ten-thousandths place.
Insert two zeros so that the 5 is in the ten-thousandths place.

EXAMPLE • 1

Name the place value of the digit 8 in the number 45.687.

Solution
The digit 8 is in the hundredths place.

YOU TRY IT • 1

Name the place value of the digit 4 in the number 907.1342.

Your solution

EXAMPLE • 2

Write $\dfrac{43}{100}$ as a decimal.

Solution
$\dfrac{43}{100} = 0.43$ • Forty-three hundredths

YOU TRY IT • 2

Write $\dfrac{501}{1000}$ as a decimal.

Your solution

EXAMPLE • 3

Write 0.289 as a fraction.

Solution
$0.289 = \dfrac{289}{1000}$ • 289 thousandths

YOU TRY IT • 3

Write 0.67 as a fraction.

Your solution

EXAMPLE • 4

Write 293.50816 in words.

Solution
Two hundred ninety-three and fifty thousand eight hundred sixteen hundred-thousandths

YOU TRY IT • 4

Write 55.6083 in words.

Your solution

Solutions on p. S8

EXAMPLE • 5

Write twenty-three and two hundred forty-seven millionths in standard form.

Solution

23.000247   • 7 is in the millionths place

YOU TRY IT • 5

Write eight hundred six and four hundred ninety-one hundred-thousandths in standard form.

Your solution

Solution on p. S8

OBJECTIVE B   To round a decimal to a given place value

Tips for Success

Have you considered joining a study group? Getting together regularly with other students in the class to go over material and quiz each other can be very beneficial. See *AIM for Success* at the front of the book.

In general, rounding decimals is similar to rounding whole numbers except that the digits to the right of the given place value are dropped instead of being replaced by zeros.

If the digit to the right of the given place value is less than 5, that digit and all digits to the right are dropped.

Round 6.9237 to the nearest hundredth.

┌── Given place value (hundredths)
6.9237
       └── $3 < 5$ Drop the digits 3 and 7.

6.9237 rounded to the nearest hundredth is 6.92.

If the digit to the right of the given place value is greater than or equal to 5, increase the digit in the given place value by 1, and drop all digits to its right.

Round 12.385 to the nearest tenth.

┌── Given place value (tenths)
12.385
      └── $8 > 5$ Increase 3 by 1 and drop all digits to the right of 3.

12.385 rounded to the nearest tenth is 12.4.

✓ Take Note

In the example at the right, the zero in the given place value is not dropped. This indicates that the number is rounded to the nearest thousandth. If we dropped the zero and wrote 0.47, it would indicate that the number was rounded to the nearest hundredth.

HOW TO • 1   Round 0.46972 to the nearest thousandth.

┌── Given place value (thousandths)
0.46972
      └── $7 > 5$ Round up by adding 1 to the 9 ($9 + 1 = 10$). Carry the 1 to the hundredths place ($6 + 1 = 7$).

0.46972 rounded to the nearest thousandth is 0.470.

EXAMPLE • 6

Round 0.9375 to the nearest thousandth.

Solution

Given place value

0.9375

5 = 5

0.9375 rounded to the nearest thousandth is 0.938.

YOU TRY IT • 6

Round 3.675849 to the nearest ten-thousandth.

Your solution

EXAMPLE • 7

Round 2.5963 to the nearest hundredth.

Solution

Given place value

2.5963

6 > 5

2.5963 rounded to the nearest hundredth is 2.60.

YOU TRY IT • 7

Round 48.907 to the nearest tenth.

Your solution

EXAMPLE • 8

Round 72.416 to the nearest whole number.

Solution

Given place value

72.416

4 < 5

72.416 rounded to the nearest whole number is 72.

YOU TRY IT • 8

Round 31.8652 to the nearest whole number.

Your solution

EXAMPLE • 9

On average, an American goes to the movies 4.56 times per year. To the nearest whole number, how many times per year does an American go to the movies?

Solution

4.56 rounded to the nearest whole number is 5.
An American goes to the movies about 5 times per year.

YOU TRY IT • 9

One of the driest cities in the Southwest is Yuma, Arizona, with an average annual precipitation of 2.65 inches. To the nearest inch, what is the average annual precipitation in Yuma?

Your solution

Solutions on p. S8

3.1 EXERCISES

OBJECTIVE A   To write decimals in standard form and in words

For Exercises 1 to 6, name the place value of the digit 5.

1. 76.31587   **2.** 291.508   **3.** 432.09157

4. 0.0006512   **5.** 38.2591   **6.** 0.0000853

For Exercises 7 to 12, write the fraction as a decimal.

7. $\dfrac{3}{10}$ **8.** $\dfrac{9}{10}$ **9.** $\dfrac{21}{100}$ **10.** $\dfrac{87}{100}$ **11.** $\dfrac{461}{1000}$ **12.** $\dfrac{853}{1000}$

For Exercises 13 to 18, write the decimal as a fraction.

13. 0.1   **14.** 0.3   **15.** 0.47   **16.** 0.59   **17.** 0.289   **18.** 0.601

For Exercises 19 to 27, write the number in words.

19. 0.37   **20.** 25.6   **21.** 9.4

22. 1.004   **23.** 0.0053   **24.** 41.108

25. 0.045   **26.** 3.157   **27.** 26.04

For Exercises 28 to 35, write the number in standard form.

28. Six hundred seventy-two thousandths

29. Three and eight hundred six ten-thousandths

30. Nine and four hundred seven ten-thousandths

31. Four hundred seven and three hundredths

32. Six hundred twelve and seven hundred four thousandths

33. Two hundred forty-six and twenty-four thousandths

34. Two thousand sixty-seven and nine thousand two ten-thousandths

35. Seventy-three and two thousand six hundred eighty-four hundred-thousandths

36. Suppose the first nonzero digit to the right of the decimal point in a decimal number is in the hundredths place. If the number has three consecutive nonzero digits to the right of the decimal point, and all other digits are zero, what place value names the number?

OBJECTIVE B   To round a decimal to a given place value

For Exercises 37 to 51, round the number to the given place value.

37. 6.249 Tenths

38. 5.398 Tenths

39. 21.007 Tenths

40. 30.0092 Tenths

41. 18.40937 Hundredths

42. 413.5972 Hundredths

43. 72.4983 Hundredths

44. 6.061745 Thousandths

45. 936.2905 Thousandths

46. 96.8027 Whole number

47. 47.3192 Whole number

48. 5439.83 Whole number

49. 7014.96 Whole number

50. 0.023591 Ten-thousandths

51. 2.975268 Hundred-thousandths

52. Measurement A nickel weighs about 0.1763668 ounce. Find the weight of a nickel to the nearest hundredth of an ounce.

53. Sports Runners in the Boston Marathon run a distance of 26.21875 miles. To the nearest tenth of a mile, find the distance that an entrant who completes the Boston Marathon runs.

For Exercises 54 and 55, give an example of a decimal number that satisfies the given condition.

54. The number rounded to the nearest tenth is greater than the number rounded to the nearest hundredth.

55. The number rounded to the nearest hundredth is equal to the number rounded to the nearest thousandth.

Applying the Concepts

56. Indicate which digits of the number, if any, need not be entered on a calculator.
   a. 1.500   **b.** 0.908   **c.** 60.07   **d.** 0.0032

57. **a.** Find a number between 0.1 and 0.2.   **b.** Find a number between 1 and 1.1.
   c. Find a number between 0 and 0.005.

SECTION

3.2　Addition of Decimals

OBJECTIVE A　　To add decimals

To add decimals, write the numbers so that the decimal points are on a vertical line. Add as for whole numbers, and write the decimal point in the sum directly below the decimal points in the addends.

| HOW TO • 1 | Add: $0.237 + 4.9 + 27.32$ |

Note that by placing the decimal points on a vertical line, we make sure that digits of the same place value are added.

EXAMPLE • 1

Find the sum of 42.3, 162.903, and 65.0729.

Solution

```
    111
    42.3        • Place the decimal
   162.903        points on a vertical
 +  65.0729       line.
 ──────────
   270.2759
```

EXAMPLE • 2

Add: $0.83 + 7.942 + 15$

Solution

```
   1 1
    0.83
    7.942
  +15.
  ──────
   23.772
```

YOU TRY IT • 1

Find the sum of 4.62, 27.9, and 0.62054.

Your solution

YOU TRY IT • 2

Add: $6.05 + 12 + 0.374$

Your solution

Solutions on p. S8

ESTIMATION

Estimating the Sum of Two or More Decimals

Calculate $23.037 + 16.7892$. Then use estimation to determine whether the sum is reasonable.

Add to find the exact sum.

To estimate the sum, round each number to the same place value. Here we have rounded to the nearest whole number. Then add. The estimated answer is 40, which is very close to the exact sum, 39.8262.

$$23.037 \;\boxed{+}\; 16.7892 \;\boxed{=}\; 39.8262$$

```
  23.037  ≈   23
+16.7892  ≈ +17
          ──────
            40
```

OBJECTIVE B   **To solve application problems**

The graph at the right shows the breakdown by age group of Americans who are hearing-impaired. Use this graph for Example 3 and You Try It 3.

Breakdown by Age Group of Americans Who Are Hearing-Impaired
Source: American Speech-Language-Hearing Association

EXAMPLE • 3

Determine the number of Americans under the age of 45 who are hearing-impaired.

Strategy
To determine the number, add the numbers of hearing impaired ages 0 to 17, 18 to 34, and 35 to 44.

Solution
```
  1.37
  2.77
+4.07
─────
  8.21
```

8.21 million Americans under the age of 45 are hearing-impaired.

YOU TRY IT • 3

Determine the number of Americans ages 45 and older who are hearing-impaired.

Your strategy

Your solution

EXAMPLE • 4

Dan Burhoe earned a salary of $210.48 for working 3 days this week as a food server. He also received $82.75, $75.80, and $99.25 in tips during the 3 days. Find his total income for the 3 days of work.

Strategy
To find the total income, add the tips (82.75, 75.80, and 99.25) to the salary (210.48).

Solution
210.48 + 82.75 + 75.80 + 99.25 = 468.28

Dan's total income for the 3 days of work was $468.28.

YOU TRY IT • 4

Anita Khavari, an insurance executive, earns a salary of $875 every 4 weeks. During the past 4-week period, she received commissions of $985.80, $791.46, $829.75, and $635.42. Find her total income for the past 4-week period.

Your strategy

Your solution

Solutions on p. S8

3.2 EXERCISES

OBJECTIVE A   To add decimals

For Exercises 1 to 17, add.

1. 16.008 + 2.0385 + 132.06

2. 17.32 + 1.0579 + 16.5

3. 1.792 + 67 + 27.0526

4. 8.772 + 1.09 + 26.5027

5. 3.02 + 62.7 + 3.924

6. 9.06 + 4.976 + 59.6

7. 82.006 + 9.95 + 0.927

8. 0.826 + 8.76 + 79.005

9. 4.307 + 99.82 + 9.078

10.   0.3
      + 0.07

11.   0.29
      + 0.4

12.   1.007
      + 2.1

13.   7.3
      + 9.005

14.   4.9257
      27.05
      + 9.0063

15.   8.72
      99.073
      + 2.9736

16.   62.4
      9.827
      + 692.44

17.   8
      89.43
      + 7.0659

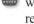 For Exercises 18 to 21, use a calculator to add. Then round the numbers to the nearest whole number and use estimation to determine whether the sum you calculated is reasonable.

18.   342.42
      89.625
      + 176.2

19.   219.9
      0.872
      + 13.42

20.   823.9
      82.65
      + 46.923

21.   678.92
      97.6
      + 5.423

 22. For a certain decimal addition problem, each addend rounded to the nearest whole number is greater than the addend itself. Must the sum of the rounded numbers be greater than the exact sum?

 23. If none of the addends of a decimal addition problem is a whole number, is it possible for the sum to be a whole number?

OBJECTIVE B   To solve application problems

24. Mechanics Find the length of the shaft.

2.15 in.

1.87 in.

1.63 in.

Length

25. Mechanics Find the length of the shaft.

1.52 ft

0.53 ft

2.3 ft

Length

26. **Banking** You have $2143.57 in your checking account. You make deposits of $210.98, $45.32, $1236.34, and $27.99. Find the amount in your checking account after you have made the deposits if no money has been withdrawn.

27. **Geometry** The perimeter of a triangle is the sum of the lengths of the three sides of the triangle. Find the perimeter of a triangle that has sides that measure 4.9 meters, 6.1 meters, and 7.5 meters.

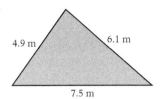

28. **Demography** The world's population in 2050 is expected to be 8.9 billion people. It is projected that in that year, Asia's population will be 5.3 billion and Africa's population will be 1.8 billion. What are the combined populations of Asia and Africa expected to be in 2050? (*Source:* United Nations Population Division, World Population Prospects)

29. **TV Viewership** The table at the right shows the numbers of viewers, in millions, of three network evening news programs for the week of January 28 to February 1, 2008. Calculate the total number of people who watched these three news programs that week.

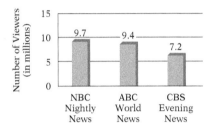

30. **The Stock Market** On May 1, 2008, the Dow Jones Industrial Average climbed 189.87 points after starting the day at 12,820.13. The Nasdaq Composite started the day at 2412.80 and rose 67.91 points during the day. The Standard & Poor 500 Index began the day at 1385.59 and ended the day 23.75 points higher. Find the values of **a.** the Dow Jones Industrial Average, **b.** the Nasdaq Composite, and **c.** the Standard & Poor 500 Index at the end of the trading day on May 1, 2008.

31. **Measurement** Can a piece of rope 4 feet long be wrapped around the box shown at the right?

Applying the Concepts

Consumerism The table at the right gives the prices for selected products in a grocery store. Use this table for Exercises 32 and 33.

32. Does a customer with $10 have enough money to purchase raisin bran, bread, milk, and butter?

33. Name three items that would cost more than $8 but less than $9. (There is more than one answer.)

Product	Cost
Raisin bran	$3.29
Butter	$2.79
Bread	$1.99
Popcorn	$2.19
Potatoes	$3.49
Cola (6-pack)	$2.99
Mayonnaise	$3.99
Lunch meat	$3.39
Milk	$2.59
Toothpaste	$2.69

SECTION

3.3   Subtraction of Decimals

OBJECTIVE A   To subtract decimals

To subtract decimals, write the numbers so that the decimal points are on a vertical line. Subtract as for whole numbers, and write the decimal point in the difference directly below the decimal point in the subtrahend.

HOW TO 1   Subtract 21.532 − 9.875 and check.

Placing the decimal points on a vertical line ensures that digits of the same place value are subtracted.

Check:

Subtrahend		9.875
+ Difference		+ 11.657
= Minuend		21.532

HOW TO 2   Subtract 4.3 − 1.7942 and check.

$$\begin{array}{r} 3 \quad 12\ 9\ 9\ 10 \\ 4.3\,0\,0\,0 \\ -\ 1.7\,9\,4\,2 \\ \hline 2.5\,0\,5\,8 \end{array}$$

If necessary, insert zeros in the minuend before subtracting.

Check:
$$\begin{array}{r} 1\ 1\ 1\ 1 \\ 1.7942 \\ +\ 2.5058 \\ \hline 4.3000 \end{array}$$

EXAMPLE 1

Subtract 39.047 − 7.96 and check.

Solution

$$\begin{array}{r} 8 \quad 9\ 14 \\ 3\,9.\,0\,4\,7 \\ -\ \ \ 7.9\,6 \\ \hline 3\,1.0\,8\,7 \end{array}$$

Check:

$$\begin{array}{r} 1\ 1 \\ 7.96 \\ +\ 31.087 \\ \hline 39.047 \end{array}$$

YOU TRY IT 1

Subtract 72.039 − 8.47 and check.

Your solution

EXAMPLE 2

Find 9.23 less than 29 and check.

Solution

$$\begin{array}{r} 1\ 18 \quad 9\ 10 \\ 2\,9.\,0\,0 \\ -\ \ 9.2\,3 \\ \hline 1\,9.7\,7 \end{array}$$

Check:

$$\begin{array}{r} 1\ 1\ 1 \\ 9.23 \\ +\ 19.77 \\ \hline 29.00 \end{array}$$

YOU TRY IT 2

Subtract 35 − 9.67 and check.

Your solution

EXAMPLE 3

Subtract 1.2 − 0.8235 and check.

Solution

$$\begin{array}{r} 0 \quad 11\ 9\ 9\ 10 \\ 1.\,2\,0\,0\,0 \\ -\ 0.8\,2\,3\,5 \\ \hline 0.3\,7\,6\,5 \end{array}$$

Check:

$$\begin{array}{r} 1\ 1\ 1\ 1 \\ 0.8235 \\ +\ 0.3765 \\ \hline 1.2000 \end{array}$$

YOU TRY IT 3

Subtract 3.7 − 1.9715 and check.

Your solution

Solutions on pp. S8–S9

ESTIMATION

Estimating the Difference Between Two Decimals

Calculate $820.23 - 475.748$. Then use estimation to determine whether the difference is reasonable.

Subtract to find the exact difference.

$$820.23 \boxed{-} 475.748 \boxed{=} 344.482$$

To estimate the difference, round each number to the same place value. Here we have rounded to the nearest ten. Then subtract. The estimated answer is 340, which is very close to the exact difference, 344.482.

$$\begin{array}{r} 820.23 \approx 820 \\ -475.748 \approx -480 \\ \hline 340 \end{array}$$

OBJECTIVE B   To solve application problems

EXAMPLE • 4

You bought a book for $15.87. How much change did you receive from a $20.00 bill?

Strategy
To find the amount of change, subtract the cost of the book (15.87) from $20.00.

Solution

$$\begin{array}{r} 20.00 \\ -15.87 \\ \hline 4.13 \end{array}$$

You received $4.13 in change.

YOU TRY IT • 4

Your breakfast cost $6.85. How much change did you receive from a $10.00 bill?

Your strategy

Your solution

EXAMPLE • 5

You had a balance of $87.93 on your student debit card. You then used the card, deducting $15.99 for a CD, $6.85 for lunch, and $28.50 for a ticket to the football game. What is your new student debit card balance?

Strategy
To find your new debit card balance:
- Add to find the total of the three deductions (15.99 + 6.85 + 28.50).
- Subtract the total of the three deductions from the old balance (87.93).

Solution

$$\begin{array}{r} 15.99 \\ 6.85 \\ +28.50 \\ \hline 51.34 \end{array} \text{ total of deductions}$$

$$\begin{array}{r} 87.93 \\ -51.34 \\ \hline 36.59 \end{array}$$

Your new debit card balance is $36.59.

YOU TRY IT • 5

You had a balance of $2472.69 in your checking account. You then wrote checks for $1025.60, $79.85, and $162.47. Find the new balance in your checking account.

Your strategy

Your solution

Solutions on p. S9

3.3 EXERCISES

OBJECTIVE A   To subtract decimals

For Exercises 1 to 24, subtract and check.

1. $24.037 - 18.41$ **2.** $26.029 - 19.31$ **3.** $123.07 - 9.4273$ **4.** $214 - 7.143$

5. $16.5 - 9.7902$ **6.** $13.2 - 8.6205$ **7.** $235.79 - 20.093$ **8.** $463.27 - 40.095$

9. $63.005 - 9.1274$ **10.** $23.004 - 7.2175$ **11.** $92 - 19.2909$ **12.** $41.2405 - 25.2709$

13. $\begin{array}{r} 0.32 \\ -\,0.0058 \\ \hline \end{array}$ **14.** $\begin{array}{r} 0.78 \\ -\,0.0073 \\ \hline \end{array}$ **15.** $\begin{array}{r} 3.005 \\ -\,1.982 \\ \hline \end{array}$ **16.** $\begin{array}{r} 6.007 \\ -\,2.734 \\ \hline \end{array}$

17. $\begin{array}{r} 352.16 \\ -\,90.994 \\ \hline \end{array}$ **18.** $\begin{array}{r} 872 \\ -\,80.753 \\ \hline \end{array}$ **19.** $\begin{array}{r} 724.32 \\ -\,69 \\ \hline \end{array}$ **20.** $\begin{array}{r} 625.46 \\ -\,77.509 \\ \hline \end{array}$

21. $\begin{array}{r} 362.394 \\ -\,19.4672 \\ \hline \end{array}$ **22.** $\begin{array}{r} 421.385 \\ -\,17.5293 \\ \hline \end{array}$ **23.** $\begin{array}{r} 19 \\ -\,10.372 \\ \hline \end{array}$ **24.** $\begin{array}{r} 23.4 \\ -\,0.921 \\ \hline \end{array}$

For Exercises 25 to 27, use the relationship between addition and subtraction to write the subtraction problem you would use to find the missing addend.

25. _____ $+ 2.325 = 7.01$ **26.** $5.392 +$ _____ $= 8.07$ **27.** _____ $+ 8.967 = 19.35$

For Exercises 28 to 31, use a calculator to subtract. Then round the numbers to the nearest whole number and use estimation to determine whether the difference you calculated is reasonable.

28. $\begin{array}{r} 93.079256 \\ -\,66.09249 \\ \hline \end{array}$ **29.** $\begin{array}{r} 3.7529 \\ -\,1.00784 \\ \hline \end{array}$ **30.** $\begin{array}{r} 76.53902 \\ -\,45.73005 \\ \hline \end{array}$ **31.** $\begin{array}{r} 9.07325 \\ -\,1.924 \\ \hline \end{array}$

OBJECTIVE B   To solve application problems

32. Mechanics Find the missing dimension.

33. Mechanics Find the missing dimension.

34. Business The manager of the Edgewater Cafe takes a reading of the cash register tape each hour. At 1:00 P.M. the tape read $967.54. At 2:00 P.M. the tape read $1437.15. Find the amount of sales between 1:00 P.M. and 2:00 P.M.

35. Moviegoing The graph at the right shows the average annual numbers of theater tickets sold each decade. Find the difference between the average annual number of theater tickets sold in the 1990s and in the 1970s.

36. Coal In a recent year, 1.163 billion tons of coal were produced in the United States. In the same year, U.S. consumption of coal was 1.112 billion tons. (*Source:* Department of Energy) How many more million tons of coal were produced than were consumed that year?

Average Annual Number of Theater Tickets Sold Each Decade

Source: National Association of Theater Owners

37. Super Bowl Super Bowl XLII was watched on the Fox network by 97.4 million people. On the same network, 63.9 million people watched the Super Bowl post-game show. (*Source:* Nielsen Network Research) How many more people watched Super Bowl XLII than watched the Super Bowl post-game show?

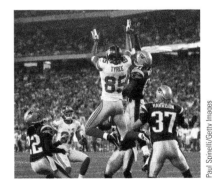

38. You have $30 to spend, and you make purchases that cost $6.74 and $13.68. Which expressions correctly represent the amount of money you have left?
(i) 30 − 6.74 + 13.68   (ii) (6.74 + 13.68) − 30
(iii) 30 − (6.74 + 13.68)   (iv) 30 − 6.74 − 13.68

Applying the Concepts

39. Find the largest amount by which the estimate of the sum of two decimals rounded to the given place value could differ from the exact sum.
a. Tenths   **b.** Hundredths   **c.** Thousandths

SECTION

3.4 Multiplication of Decimals

OBJECTIVE A   To multiply decimals

Point of Interest

Benjamin Banneker (1731–1806) was the first African American to earn distinction as a mathematician and scientist. He was on the survey team that determined the boundaries of Washington, D.C. The mathematics of surveying requires extensive use of decimals.

Decimals are multiplied as though they were whole numbers. Then the decimal point is placed in the product. Writing the decimals as fractions shows where to write the decimal point in the product.

$$0.\underline{3} \times 5 = \frac{3}{10} \times \frac{5}{1} = \frac{15}{10} = 1.\underline{5}$$

1 decimal place   1 decimal place

$$0.\underline{3} \times 0.\underline{5} = \frac{3}{10} \times \frac{5}{10} = \frac{15}{100} = 0.\underline{15}$$

1 decimal place   1 decimal place   2 decimal places

$$0.\underline{3} \times 0.\underline{05} = \frac{3}{10} \times \frac{5}{100} = \frac{15}{1000} = 0.\underline{015}$$

1 decimal place   2 decimal places   3 decimal places

To multiply decimals, multiply the numbers as with whole numbers. Write the decimal point in the product so that the number of decimal places in the product is the sum of the decimal places in the factors.

Integrating Technology

Scientific calculators have a floating decimal point. This means that the decimal point is automatically placed in the answer. For example, for the product at the right, enter

21 · 4 **x** 0 · 36 **=**

The display reads 7.704, with the decimal point in the correct position.

HOW TO 1   Multiply: 21.4×0.36

$$
\begin{array}{r}
21.4 \\
\times\ 0.36 \\
\hline
1284 \\
642 \\
\hline
7.704
\end{array}
$$

21.4   1 decimal place
× 0.36   2 decimal places

7.704   3 decimal places

HOW TO 2   Multiply: 0.037×0.08

$$
\begin{array}{r}
0.037 \\
\times\ 0.08 \\
\hline
0.00296
\end{array}
$$

0.037   3 decimal places   • Two zeros must be inserted between the
× 0.08   2 decimal places   2 and the decimal point so that there are
0.00296   5 decimal places   5 decimal places in the product.

To multiply a decimal by a power of 10 (10, 100, 1000, . . .), move the decimal point to the right the same number of places as there are zeros in the power of 10.

$3.8925 \times 1\underline{0} = 38.925$

1 zero   1 decimal place

$3.8925 \times 1\underline{00} = 389.25$

2 zeros   2 decimal places

$3.8925 \times 1\underline{000} = 3892.5$

3 zeros   3 decimal places

$3.8925 \times 1\underline{0,000} = 38,925.$

4 zeros   4 decimal places

$3.8925 \times 1\underline{00,000} = 389,250.$

5 zeros   5 decimal places

• Note that a zero must be inserted before the decimal point.

Note that if the power of 10 is written in exponential notation, the exponent indicates how many places to move the decimal point.

$$3.8925 \times 10^1 = 38.925$$
1 decimal place

$$3.8925 \times 10^2 = 389.25$$
2 decimal places

$$3.8925 \times 10^3 = 3892.5$$
3 decimal places

$$3.8925 \times 10^4 = 38,925.$$
4 decimal places

$$3.8925 \times 10^5 = 389,250.$$
5 decimal places

EXAMPLE 1

Multiply: 920×3.7

Solution

$$
\begin{array}{r}
920 \\
\times \quad 3.7 \\
\hline
644\,0 \\
2760 \\
\hline
3404.0
\end{array}
$$

• 1 decimal place

• 1 decimal place

YOU TRY IT 1

Multiply: 870×4.6

Your solution

EXAMPLE 2

Find 0.00079 multiplied by 0.025.

Solution

$$
\begin{array}{r}
0.00079 \\
\times \quad 0.025 \\
\hline
395 \\
158 \\
\hline
0.00001975
\end{array}
$$

• 5 decimal places
• 3 decimal places

• 8 decimal places

YOU TRY IT 2

Find 0.000086 multiplied by 0.057.

Your solution

EXAMPLE 3

Find the product of 3.69 and 2.07.

Solution

$$
\begin{array}{r}
3.69 \\
\times \quad 2.07 \\
\hline
2583 \\
7380 \\
\hline
7.6383
\end{array}
$$

• 2 decimal places
• 2 decimal places

• 4 decimal places

YOU TRY IT 3

Find the product of 4.68 and 6.03.

Your solution

EXAMPLE 4

Multiply: $42.07 \times 10,000$

Solution

$42.07 \times 10,000 = 420,700$

YOU TRY IT 4

Multiply: 6.9×1000

Your solution

EXAMPLE 5

Find 3.01 times 10^3.

Solution

$3.01 \times 10^3 = 3010$

YOU TRY IT 5

Find 4.0273 times 10^2.

Your solution

Solutions on p. S9

ESTIMATION

Estimating the Product of Two Decimals

Calculate 28.259 × 0.029. Then use estimation to determine whether the product is reasonable.

Multiply to find the exact product.   28.259 **x** 0.029 **=** 0.819511

To estimate the product, round each   28.259 ≈   30
number so that it contains one nonzero   ×0.029 ≈   ×0.03
digit. Then multiply. The estimated   0.90
answer is 0.90, which is very close
to the exact product, 0.819511.

OBJECTIVE B   To solve application problems

The tables that follow list water rates and meter fees for a city. These tables are used for Example 6 and You Try It 6.

Water Charges	
Commercial	$1.39/1000 gal
Comm Restaurant	$1.39/1000 gal
Industrial	$1.39/1000 gal
Institutional	$1.39/1000 gal
Res—No Sewer	
Residential—SF	
>0 <200 gal per day	$1.15/1000 gal
>200 <1500 gal per day	$1.39/1000 gal
>1500 gal per day	$1.54/1000 gal

Meter Charges	
Meter	*Meter Fee*
5/8" & 3/4"	$13.50
1"	$21.80
1-1/2"	$42.50
2"	$67.20
3"	$133.70
4"	$208.20
6"	$415.10
8"	$663.70

EXAMPLE • 6

Find the total bill for an industrial water user with a 6-inch meter that used 152,000 gallons of water for July and August.

Strategy

To find the total cost of water:

• Find the cost of water by multiplying the cost per 1000 gallons (1.39) by the number of 1000-gallon units used.
• Add the cost of the water to the meter fee (415.10).

Solution

Cost of water = $\dfrac{152,000}{1000}$ · 1.39 = 211.28

Total cost = 211.28 + 415.10 = 626.38

The total cost is $626.38.

YOU TRY IT • 6

Find the total bill for a commercial user that used 5000 gallons of water per day for July and August. The user has a 3-inch meter.

Your strategy

Your solution

Solution on p. S9

EXAMPLE 7

It costs $.036 an hour to operate an electric motor. How much does it cost to operate the motor for 120 hours?

Strategy
To find the cost of running the motor for 120 hours, multiply the hourly cost (0.036) by the number of hours the motor is run (120).

Solution

```
   0.036
×    120
    720
   36
  4.320
```

The cost of running the motor for 120 hours is $4.32.

EXAMPLE 8

Jason Ng earns a salary of $440 for a 40-hour workweek. This week he worked 12 hours of overtime at a rate of $16.50 for each hour of overtime worked. Find his total income for the week.

Strategy
To find Jason's total income for the week:
• Find the overtime pay by multiplying the hourly overtime rate (16.50) by the number of hours of overtime worked (12).
• Add the overtime pay to the weekly salary (440).

Solution

```
   16.50          440.00
×    12         + 198.00
   33 00          638.00
  165 0
  198.00  Overtime pay
```

Jason's total income for the week is $638.00.

YOU TRY IT 7

The cost of electricity to run a freezer for 1 hour is $.035. This month the freezer has run for 210 hours. Find the total cost of running the freezer this month.

Your strategy

Your solution

YOU TRY IT 8

You make a down payment of $175 on an electronic drum kit and agree to make payments of $37.18 a month for the next 18 months to repay the remaining balance. Find the total cost of the electronic drum kit.

Your strategy

Your solution

Solutions on p. S9

3.4 EXERCISES

OBJECTIVE A To multiply decimals

For Exercises 1 to 73, multiply.

1. 0.9
    $\times$ 0.4

2. 0.7
    $\times$ 0.9

3. 0.5
    $\times$ 0.5

4. 0.7
    $\times$ 0.7

5. 7.7
    $\times$ 0.9

6. 3.4
    $\times$ 0.4

7. 9.2
    $\times$ 0.2

8. 2.6
    $\times$ 0.7

9. 7.4
    $\times$ 0.1

10. 3.8
    $\times$ 0.1

11. 7.9
    $\times$ 5

12. 9.3
    $\times$ 7

13. 0.68
    $\times$ 4

14. 0.83
    $\times$ 9

15. 0.67
    $\times$ 0.9

16. 0.84
    $\times$ 0.3

17. 2.5
    $\times$ 5.4

18. 3.9
    $\times$ 1.9

19. 0.83
    $\times$ 5.2

20. 0.24
    $\times$ 2.7

21. 1.47
    $\times$ 0.09

22. 6.37
    $\times$ 0.05

23. 8.92
    $\times$ 0.004

24. 6.75
    $\times$ 0.007

25. 0.49
    $\times$ 0.16

26. 0.38
    $\times$ 0.21

27. 7.6
    $\times$ 0.01

28. 5.1
    $\times$ 0.01

29. 8.62
    $\times$ 4

30. 5.83
    $\times$ 7

31. 64.5
    $\times$ 9

32. 37.8
    $\times$ 8

33. 2.19
    $\times$ 9.2

34. 1.25
    $\times$ 5.6

35. 1.85
    $\times$ 0.023

36. 37.8
    $\times$ 0.052

37. 0.478
    $\times$ 0.37

38. 0.526
    $\times$ 0.22

39. 48.3
    $\times$ 0.0041

40. 67.2
    $\times$ 0.0086

41. 0.413
$\times$ 0.0016

42. 0.517
$\times$ 0.0029

43. 8.005
$\times$ 0.067

44. 9.032
$\times$ 0.019

45. 4.29×0.1

46. 6.78×0.1

47. 5.29×0.4

48. 6.78×0.5

49. 0.68×0.7

50. 0.56×0.9

51. 1.4×0.73

52. 6.3×0.37

53. 5.2×7.3

54. 7.4×2.9

55. 3.8×0.61

56. 7.2×0.72

57. 0.32×10

58. 6.93×10

59. 0.065×100

60. 0.039×100

61. 6.2856×1000

62. 3.2954×1000

63. 3.2×1000

64. $0.006 \times 10,000$

65. $3.57 \times 10,000$

66. 8.52×10^1

67. 0.63×10^1

68. 82.9×10^2

69. 0.039×10^2

70. 6.8×10^3

71. 4.9×10^4

72. 6.83×10^4

73. 0.067×10^2

74. A number is rounded to the nearest thousandth. What is the smallest power of 10 the number must be multiplied by to give a product that is a whole number?

75. Find the product of 0.0035 and 3.45.

76. Find the product of 237 and 0.34.

77. Multiply 3.005 by 0.00392.

78. Multiply 20.34 by 1.008.

79. Multiply 1.348 by 0.23.

80. Multiply 0.000358 by 3.56.

81. Find the product of 23.67 and 0.0035.

82. Find the product of 0.00346 and 23.1.

83. Find the product of 5, 0.45, and 2.3.

84. Find the product of 0.03, 23, and 9.45.

 For Exercises 85 to 96, use a calculator to multiply. Then use estimation to determine whether the product you calculated is reasonable.

85.   28.5
        × 3.2

86.   86.3
        × 4.4

87.   2.38
        × 0.44

88.   9.82
        × 0.77

89.   0.866
        ×   4.5

90.   0.239
        ×   8.2

91.   4.34
        × 2.59

92.   6.87
        × 9.98

93.   8.434
        × 0.044

94.   7.037
        × 0.094

95.   28.44
        ×   1.12

96.   86.57
        ×   7.33

OBJECTIVE B   To solve application problems

97. **Electricity** In the United States, a homeowner's average monthly bill for electricity is $95.66. (*Source:* Department of Energy) What is a U.S. homeowner's average annual cost of electricity?

98. **Transportation** A long-haul truck driver earns $.43 for each mile driven. How much will a truck driver earn for driving 1507 miles from Boston to Miami?

99. **Recycling** Four hundred empty soft drink cans weigh 18.75 pounds. A recycling center pays $.75 per pound for the cans. Find the amount received for the 400 cans. Round to the nearest cent.

100. **Recycling** A recycling center pays $.045 per pound for newspapers.
a. Estimate the payment for recycling 520 pounds of newspapers.
b. Find the actual amount received from recycling the newspapers.

 101. **Taxes** The tax per gallon of gasoline in California is $.192. (*Source:* Tax Foundation) If you fill your gasoline tank with 12.5 gallons of gasoline in California, how much will you pay in taxes?

102. **Geometry** The perimeter of a square is equal to four times the length of a side of the square. Find the perimeter of a square whose side measures 2.8 meters.

2.8 m

103. **Geometry** The area of a rectangle is equal to the product of the length of the rectangle times its width. Find the area of a rectangle that has a length of 6.75 feet and a width of 3.5 feet. The area will be in square feet.

3.5 ft
6.75 ft

104. **Finance** You bought a car for $5000 down and made payments of $499.50 each month for 36 months.
    a. Find the amount of the payments over the 36 months.
    b. Find the total cost of the car.

105. **Compensation** A nurse earns a salary of $1396 for a 40-hour work week. This week the nurse worked 15 hours of overtime at a rate of $52.35 for each hour of overtime worked.
    a. Find the nurse's overtime pay.
    b. Find the nurse's total income for the week.

106. **Consumerism** Bay Area Rental Cars charges $25 a day and $.25 per mile for renting a car. You rented a car for 3 days and drove 235 miles. Find the total cost of renting the car.

107. **Shipping** The graph at the right shows United States Postal Service rates for express mail. How much would it cost a company to mail 25 express mail packages, each weighing $\frac{1}{4}$ pound, post office to addressee?

U.S. Postal Service Rates for Express Mail

For Exercises 108 and 109, use the graph of Postal Service rates from Exercise 107. Write a verbal description of what each expression represents.

108. 17.10×15

109. $(16.25 \times 4) + (19.50 \times 9)$

110. **Taxes** For tax purposes, the standard deduction on tax returns for the business use of a car in 2007 was 48.5¢ per mile. In 2008, the standard deduction was 50.5¢ per mile. Find the amount deductible on a 2008 tax return for driving a business car 11,842 miles during the year.

111. Electronic Checks See the news clipping at the right. Find the added cost to the government for issuing paper checks to the 4 million Social Security recipients who request a paper check because they do not have a bank account.

In the News

Paper Checks Cost Government Millions

The federal government's cost to issue a paper check is $.89, while the cost for an electronic check is $.09.

Source: finance.yahoo.com

112. Transportation A taxi costs $2.50 and $.20 for each $\frac{1}{8}$ mile driven. Find the cost of hiring a taxi to get from the airport to the hotel, a distance of $5\frac{1}{2}$ miles.

113. Business The table at the right lists three pieces of steel required for a repair project.
   a. Find the total cost of grade 1.
   b. Find the total cost of grade 2.
   c. Find the total cost of grade 3.
   d. Find the total cost of the three pieces of steel.

Grade of Steel	Weight (pounds per foot)	Required Number of Feet	Cost per Pound
1	2.2	8	$3.20
2	3.4	6.5	$3.35
3	6.75	15.4	$3.94

Code	Description	Price
112	Almonds 16 oz	$6.75
116	Cashews 8 oz	$5.90
117	Cashews 16 oz	$8.50
130	Macadamias 7 oz	$7.25
131	Macadamias 16 oz	$11.95
149	Pecan halves 8 oz	$8.25
155	Mixed nuts 8 oz	$6.80
160	Cashew brittle 8 oz	$5.95
182	Pecan roll 8 oz	$6.70
199	Chocolate peanuts 8 oz	$5.90

114. Business A confectioner ships holiday packs of candy and nuts anywhere in the United States. At the right is a price list for nuts and candy, and below is a table of shipping charges to zones in the United States. For any fraction of a pound, use the next higher weight. Sixteen ounces (16 oz) is equal to 1 pound.

Pounds	Zone 1	Zone 2	Zone 3	Zone 4
1–3	$7.55	$7.85	$8.25	$8.75
4–6	$8.10	$8.40	$8.80	$9.30
7–9	$8.50	$8.80	$9.20	$9.70
10–12	$8.90	$9.20	$9.60	$10.10

Find the cost of sending the following orders to the given mail zone.

a. Code	Quantity	b. Code	Quantity	c. Code	Quantity
116	2	112	1	117	3
130	1	117	4	131	1
149	3	131	2	155	2
182	4	160	3	160	4
Mail to zone 4.		182	5	182	1
		Mail to zone 3.		199	3
				Mail to zone 2.	

Applying the Concepts

115. Show how the decimal is placed in the product of 1.3 × 2.31 by first writing each number as a fraction and then multiplying. Then change the product back to decimal notation.

116. Automotive Repair Chris works at B & W Garage as an auto mechanic and has just completed an engine overhaul for a customer. To determine the cost of the repair job, Chris keeps a list of times worked and parts used. A parts list and a list of the times worked are shown below.

Parts Used		Time Spent	
Item	*Quantity*	*Day*	*Hours*
Gasket set	1	Monday	7.0
Ring set	1	Tuesday	7.5
Valves	8	Wednesday	6.5
Wrist pins	8	Thursday	8.5
Valve springs	16	Friday	9.0
Rod bearings	8		
Main bearings	5		
Valve seals	16		
Timing chain	1		

Price List		
Item Number	*Description*	*Unit Price*
27345	Valve spring	$9.25
41257	Main bearing	$17.49
54678	Valve	$16.99
29753	Ring set	$169.99
45837	Gasket set	$174.90
23751	Timing chain	$50.49
23765	Fuel pump	$229.99
28632	Wrist pin	$23.55
34922	Rod bearing	$13.69
2871	Valve seal	$1.69

a. Organize a table of data showing the parts used, the unit price for each part, and the price of the quantity used. *Hint:* Use the following headings for the table.

Quantity   Item Number   Description   Unit Price   Total

b. Add up the numbers in the "Total" column to find the total cost of the parts.

c. If the charge for labor is $46.75 per hour, compute the cost of labor.

d. What is the total cost for parts and labor?

117. Explain how the decimal point is placed when a number is multiplied by 10, 100, 1000, 10,000, etc.

118. Explain how the decimal point is placed in the product of two decimals.

SECTION

3.5 Division of Decimals

OBJECTIVE A   To divide decimals

To divide decimals, move the decimal point in the divisor to the right to make the divisor a whole number. Move the decimal point in the dividend the same number of places to the right. Place the decimal point in the quotient directly over the decimal point in the dividend, and then divide as with whole numbers.

HOW TO 1   Divide: $3.25\overline{)15.275}$

$$3.\underset{\curvearrowright}{25.})15.\underset{\curvearrowright}{27.}5$$

• Move the decimal point 2 places to the right in the divisor and then in the dividend. Place the decimal point in the quotient.

$$\begin{array}{r} 4.7 \\ 325.)\overline{1527.5} \\ -1300 \\ \hline 227\ 5 \\ -227\ 5 \\ \hline 0 \end{array}$$

• Divide as with whole numbers.

 Tips for Success
To learn mathematics, you must be an active participant. Listening and watching your professor do mathematics are not enough. Take notes in class, mentally think through every question your instructor asks, and try to answer it even if you are not called on to answer it verbally. Ask questions when you have them. See *AIM for Success* at the front of the book for other ways to be an active learner.

Moving the decimal point the same number of decimal places in the divisor and dividend does not change the value of the quotient, because this process is the same as multiplying the numerator and denominator of a fraction by the same number. In the example above,

$$3.25\overline{)15.275} = \frac{15.275}{3.25} = \frac{15.275 \times 100}{3.25 \times 100} = \frac{1527.5}{325} = 325\overline{)1527.5}$$

When dividing decimals, we usually round the quotient off to a specified place value, rather than writing the quotient with a remainder.

HOW TO 2   Divide: $0.3\overline{)0.56}$

Round to the nearest hundredth.

$$\begin{array}{r} 1.866 \approx 1.87 \\ 0.\underset{\curvearrowright}{3.})0.\underset{\curvearrowright}{5.}600 \\ -\ 3 \\ \hline 2\ 6 \\ -2\ 4 \\ \hline 20 \\ -18 \\ \hline 20 \\ -18 \end{array}$$

We must carry the division to the thousandths place to round the quotient to the nearest hundredth. Therefore, zeros must be inserted in the dividend so that the quotient has a digit in the thousandths place.

Integrating Technology

A calculator displays the quotient to the limit of the calculator's display. Enter

57 . 93 ÷ 3 . 24

=

to determine the number of places your calculator displays.

HOW TO 3 Divide 57.93 by 3.24. Round to the nearest thousandth.

$$17.8796 \approx 17.880$$

$$3.24\overline{)57.93.0000}$$

Zeros must be inserted in the dividend so that the quotient has a digit in the ten-thousandths place.

$$-32\ 4$$
$$\overline{25\ 53}$$
$$-22\ 68$$
$$\overline{2\ 85\ 0}$$
$$-2\ 59\ 2$$
$$\overline{25\ 80}$$
$$-22\ 68$$
$$\overline{3\ 120}$$
$$-2\ 916$$
$$\overline{2040}$$
$$-1944$$

To divide a decimal by a power of 10 (10, 100, 1000, . . .), move the decimal point to the left the same number of places as there are zeros in the power of 10.

$34.65 \div 1\underline{0} = 3.465$
    1 zero   1 decimal place

$34.65 \div 1\underline{00} = 0.3465$
    2 zeros   2 decimal places

$34.65 \div 1\underline{000} = 0.03465$
    3 zeros   3 decimal places

• Note that a zero must be inserted between the 3 and the decimal point.

$34.65 \div 1\underline{0,000} = 0.003465$
    4 zeros   4 decimal places

• Note that two zeros must be inserted between the 3 and the decimal point.

If the power of 10 is written in exponential notation, the exponent indicates how many places to move the decimal point.

$34.65 \div 10^1 = 3.465$ 1 decimal place

$34.65 \div 10^2 = 0.3465$ 2 decimal places

$34.65 \div 10^3 = 0.03465$ 3 decimal places

$34.65 \div 10^4 = 0.003465$ 4 decimal places

EXAMPLE 1

Divide: $0.1344 \div 0.032$

Solution

$$0.032\overline{)0.134.4}$$
    4.2
$$-128$$
$$\overline{6\ 4}$$
$$-6\ 4$$
$$\overline{0}$$

• Move the decimal point 3 places to the right in the divisor and the dividend.

YOU TRY IT 1

Divide: $0.1404 \div 0.052$

Your solution

Solution on p. S9

EXAMPLE • 2

Divide: 58.092 ÷ 82
Round to the nearest thousandth.

Solution

$$
\begin{array}{r}
0.7084 \approx 0.708 \\
82\overline{)\ 58.0920} \\
-57\ 4 \\
\hline
69 \\
-\ 0 \\
\hline
692 \\
-656 \\
\hline
360 \\
-328 \\
\end{array}
$$

YOU TRY IT • 2

Divide: 37.042 ÷ 76
Round to the nearest thousandth.

Your solution

EXAMPLE • 3

Divide: 420.9 ÷ 7.06
Round to the nearest tenth.

Solution

$$
\begin{array}{r}
59.61 \approx 59.6 \\
7.06\overline{)\ 420.90.00} \\
-353\ 0 \\
\hline
67\ 90 \\
-63\ 54 \\
\hline
4\ 36\ 0 \\
-4\ 23\ 6 \\
\hline
12\ 40 \\
-\ 7\ 06 \\
\end{array}
$$

YOU TRY IT • 3

Divide: 370.2 ÷ 5.09
Round to the nearest tenth.

Your solution

EXAMPLE • 4

Divide: 402.75 ÷ 1000

Solution
402.75 ÷ 1000 = 0.40275

YOU TRY IT • 4

Divide: 309.21 ÷ 10,000

Your solution

EXAMPLE • 5

What is 0.625 divided by 10^2?

Solution
$0.625 \div 10^2 = 0.00625$

YOU TRY IT • 5

What is 42.93 divided by 10^4?

Your solution

Solutions on p. S10

ESTIMATION

Estimating the Quotient of Two Decimals

Calculate 282.18 ÷ 0.48. Then use estimation to determine whether the quotient is reasonable.

Divide to find the exact quotient.

282.18 ÷ 0.48 = 587.875

To estimate the quotient, round each number so that it contains one nonzero digit. Then divide. The estimated answer is 600, which is very close to the exact quotient, 587.875.

$282.18 \div 0.48 \approx 300 \div 0.5$
$= 600$

OBJECTIVE B   To solve application problems

 The graph at the right shows average hourly earnings in the United States for selected years. Use this table for Example 6 and You Try It 6.

Average Hourly Earnings in the United States *Source:* Bureau of Labor Statistics

---

EXAMPLE • 6

How many times greater were the average hourly earnings in 2008 than in 1978? Round to the nearest whole number.

Strategy
To find how many times greater the average hourly earnings were, divide the 2008 average hourly earnings (17.85) by the 1978 average hourly earnings (5.70).

Solution
$17.85 \div 5.70 \approx 3$

The average hourly earnings in 2008 were about 3 times greater than in 1978.

YOU TRY IT • 6

How many times greater were the average hourly earnings in 1998 than in 1978? Round to the nearest tenth.

Your strategy

Your solution

EXAMPLE • 7

A 1-year subscription to a monthly magazine costs $90. The price of each issue at the newsstand is $9.80. How much would you save per issue by buying a year's subscription rather than buying each issue at the newsstand?

Strategy
To find the amount saved:
• Find the subscription price per issue by dividing the cost of the subscription (90) by the number of issues (12).
• Subtract the subscription price per issue from the newsstand price (9.80).

Solution
$90 \div 12 = 7.50$
$9.80 - 7.50 = 2.30$

The savings would be $2.30 per issue.

YOU TRY IT • 7

A Nielsen survey of the number of people (in millions) who watch television each day of the week is given in the table below.

Mon.	Tues.	Wed.	Thu.	Fri.	Sat.	Sun.
91.9	89.8	90.6	93.9	78.0	77.1	87.7

Find the average number of people who watch television per day.

Your strategy

Your solution

Solutions on p. S10

3.5 EXERCISES

OBJECTIVE A   To divide decimals

For Exercises 1 to 20, divide.

1. $3\overline{)2.46}$

2. $7\overline{)3.71}$

3. $0.8\overline{)3.84}$

4. $0.9\overline{)6.93}$

5. $0.7\overline{)62.3}$

6. $0.4\overline{)52.8}$

7. $0.4\overline{)24}$

8. $0.5\overline{)65}$

9. $0.7\overline{)59.01}$

10. $0.9\overline{)8.721}$

11. $0.5\overline{)16.15}$

12. $0.8\overline{)77.6}$

13. $0.7\overline{)3.542}$

14. $0.6\overline{)2.436}$

15. $6.3\overline{)8.19}$

16. $3.2\overline{)7.04}$

17. $3.6\overline{)0.396}$

18. $2.7\overline{)0.648}$

19. $6.9\overline{)26.22}$

20. $1.7\overline{)84.66}$

For Exercises 21 to 29, divide. Round to the nearest tenth.

21. $55.62 \div 8.8$

22. $25.43 \div 5.4$

23. $5.427 \div 9.5$

24. $1.837 \div 1.4$

25. $18.4 \div 7.3$

26. $52.9 \div 8.1$

27. $0.183 \div 0.17$

28. $0.381 \div 0.47$

29. $6.924 \div 0.053$

For Exercises 30 to 38, divide. Round to the nearest hundredth.

30. $4.817 \div 16$

31. $6.467 \div 8$

32. $0.0418 \div 0.53$

33. $0.0647 \div 0.72$

34. $7 \div 0.55$

35. $38.665 \div 0.95$

36. $13.97 \div 25.4$

37. $27.738 \div 60.8$

38. $3.171 \div 45.6$

For Exercises 39 to 47, divide. Round to the nearest thousandth.

39. $1.028 \div 54$

40. $6.729 \div 27$

41. $0.0437 \div 0.5$

42. $75.469 \div 77.8$

43. $34.31 \div 95.3$

44. $0.2695 \div 2.67$

45. $0.4871 \div 4.72$

46. $0.1142 \div 17.2$

47. $0.2307 \div 26.7$

For Exercises 48 to 56, divide. Round to the nearest whole number.

48. $16.5 \div 4$

49. $89.76 \div 90$

50. $1.94 \div 0.3$

51. $1.0478 \div 0.413$

52. $2.148 \div 0.519$

53. $0.79 \div 0.778$

54. $3.092 \div 0.075$

55. $392 \div 6.9$

56. $8.729 \div 0.075$

For Exercises 57 to 74, divide.

57. $4.07 \div 10$

58. $0.039 \div 10$

59. $42.67 \div 10$

60. $389.7 \div 100$

61. $1.037 \div 100$

62. $237.835 \div 100$

63. $8.295 \div 1000$

64. $82{,}547 \div 1000$

65. $825.37 \div 1000$

66. $8.35 \div 10^1$

67. $0.32 \div 10^1$

68. $87.65 \div 10^1$

69. $23.627 \div 10^2$

70. $2.954 \div 10^2$

71. $0.0053 \div 10^2$

72. $289.32 \div 10^3$

73. $1.8932 \div 10^3$

74. $0.139 \div 10^3$

75. Divide 44.208 by 2.4.

76. Divide 0.04664 by 0.44.

77. Find the quotient of 723.15 and 45.

78. Find the quotient of 3.3463 and 3.07.

79. Divide 13.5 by 10^3.

80. Divide 0.045 by 10^5.

81. Find the quotient of 23.678 and 1000.

82. Find the quotient of 7.005 and 10,000.

83. What is 0.0056 divided by 0.05?

84. What is 123.8 divided by 0.02?

 For Exercises 85 to 93, use a calculator to divide. Round to the nearest ten-thousandth. Then use estimation to determine whether the quotient you calculated is reasonable.

85. $42.42 \div 3.8$

86. $69.8 \div 7.2$

87. $389 \div 0.44$

88. $642 \div 0.83$

89. $6.394 \div 3.5$

90. $8.429 \div 4.2$

91. $1.235 \div 0.021$

92. $7.456 \div 0.072$

93. $95.443 \div 1.32$

94. A four-digit whole number is divided by 1000. Is the quotient less than 1 or greater than 1?

OBJECTIVE B   To solve application problems

 95. A 12-pack of bottled spring water sells for $3.85. State whether to use *multiplication* or *division* to find the specified amount.
a. The cost of one bottle of spring water
b. The cost of four 12-packs of spring water

 96. A city school district spends $8754 per student, and the school district in a nearby suburb spends $10,296 per student. Which expression represents how many times greater the amount spent per student by the suburban schools is than the amount spent per student by the city schools?
(i) $10,296 - 8754$ (ii) $10,296 \div 8754$
(iii) $8754 \div 10,296$ (iv) $10,296 \times 8754$

97. **Sports** Ramon, a high school football player, gained 162 yards on 26 carries in a high school football game. Find the average number of yards gained per carry. Round to the nearest hundredth.

98. **Consumerism** A case of diet cola costs $6.79. If there are 24 cans in a case, find the cost per can. Round to the nearest cent.

99. Transportation A truck driver is paid by the number of miles driven. If a truck driver earns $.46 per mile, how many miles must the trucker drive in 1 hour to earn $16.00 per mile? Round to the nearest mile.

100. Tourism See the news clipping at the right. Find the average amount spent by each visitor to the United States. Round to the nearest cent.

101. Carpentry Anne is building bookcases that are 3.4 feet long. How many complete shelves can be cut from a 12-foot board?

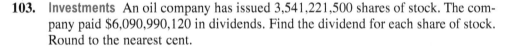

102. Travel When the Massachusetts Turnpike opened, the toll for a passenger car that traveled the entire 136 miles of it was $5.60. Calculate the cost per mile. Round to the nearest cent.

103. Investments An oil company has issued 3,541,221,500 shares of stock. The company paid $6,090,990,120 in dividends. Find the dividend for each share of stock. Round to the nearest cent.

104. Insurance Earl is 52 years old and is buying $70,000 of life insurance for an annual premium of $703.80. If he pays the annual premium in 12 equal installments, how much is each monthly payment?

105. Fuel Efficiency A car with an odometer reading of 17,814.2 is filled with 9.4 gallons of gas. At an odometer reading of 18,130.4, the tank is empty and the car is filled with 12.4 gallons of gas. How many miles does the car travel on 1 gallon of gasoline?

106. Carbon Footprint Depending on the efficiency of a power plant, 1 ton of coal can produce about 3000 kilowatt-hours of electricity. Suppose your family uses 25 kilowatt-hours of electricity per month. How many tons of coal will your family use in 1 year?

107. Carbon Footprint One barrel of oil produces approximately 800 kilowatt-hours of electricity. Suppose you use 27 kilowatt-hours of electricity per month. How many barrels of oil will you use in 1 year?

108. Going Green See the news clipping at the right.
a. Find the reduction in solid waste per month if every U.S. household viewed and paid its bills online.
b. Find the reduction in greenhouse gas emissions per month if every household viewed and paid its bills online.
Write your answers in standard form, rounded to the nearest whole number.

In the News

Tourists Boost the Economy

This past summer, 14.3 million visitors came to the United States, spending a record $30.7 billion.

Source: Commerce Department

In the News

"Green" Banking Has Far-Reaching Effects

Banking and paying bills online not only saves trees; it cuts down on the amount of fuel used by vehicles that transport paper checks. According to Javelin Strategy and Research, if every household in the United States paid its bills online, solid waste would be reduced by 1.6 billion tons a year and greenhouse-gas emissions would be cut by 2.1 million tons a year.

Source: Time, April 9, 2008

Applying the Concepts

109. **Education** According to the National Center for Education Statistics, 10.03 million women and 7.46 million men were enrolled at institutions of higher learning in a recent year. How many more women than men were attending institutions of higher learning in that year?

The Military The table at the right shows the advertising budgets of four branches of the U.S. armed services in a recent year. Use this table for Exercises 110 to 112.

Service	Advertising Budget
Army	$85.3 million
Air Force	$41.1 million
Navy	$20.5 million
Marines	$15.9 million

Source: CMR/TNS Media Intelligence

110. Find the difference between the Army's advertising budget and the Marines' advertising budget.

111. How many times greater was the Army's advertising budget than the Navy's advertising budget? Round to the nearest tenth.

112. What was the total of the advertising budgets for the four branches of the service?

113. **Population Growth** The U.S. population of people ages 85 and over is expected to grow from 4.2 million in 2000 to 8.9 million in 2030. How many times greater is the population of this segment expected to be in 2030 than in 2000? Round to the nearest tenth.

114. Explain how the decimal point is moved when a number is divided by 10, 100, 1000, 10,000, etc.

115. **Sports** Explain how baseball batting averages are determined. Then find Detroit Tiger's right fielder Magglio Ordonez's batting average with 216 hits out of 595 at bats. Round to the nearest thousandth.

116. Explain how the decimal point is placed in the quotient when a number is divided by a decimal.

For Exercises 117 to 122, insert $+$, $-$, $\times$, or $\div$ into the square so that the statement is true.

117. $3.45 \; \square \; 0.5 = 6.9$ **118.** $3.46 \; \square \; 0.24 = 0.8304$ **119.** $6.009 \; \square \; 4.68 = 1.329$

120. $0.064 \; \square \; 1.6 = 0.1024$ **121.** $9.876 \; \square \; 23.12 = 32.996$ **122.** $3.0381 \; \square \; 1.23 = 2.47$

For Exercises 123 to 125, fill in the square to make a true statement.

123. $6.47 - \square = 1.253$ **124.** $6.47 + \square = 9$ **125.** $0.009 \div \square = 0.36$

SECTION

3.6 Comparing and Converting Fractions and Decimals

OBJECTIVE A To convert fractions to decimals

Every fraction can be written as a decimal. To write a fraction as a decimal, divide the numerator of the fraction by the denominator. The quotient can be rounded to the desired place value.

HOW TO 1 Convert $\frac{3}{7}$ to a decimal.

$$7)\overline{3.00000} \atop 0.42857$$

$\frac{3}{7}$ rounded to the nearest hundredth is 0.43.

$\frac{3}{7}$ rounded to the nearest thousandth is 0.429.

$\frac{3}{7}$ rounded to the nearest ten-thousandth is 0.4286.

HOW TO 2 Convert $3\frac{2}{9}$ to a decimal. Round to the nearest thousandth.

$$3\frac{2}{9} = \frac{29}{9} \qquad 9)\overline{29.0000} \atop 3.2222$$

$3\frac{2}{9}$ rounded to the nearest thousandth is 3.222.

EXAMPLE 1

Convert $\frac{3}{8}$ to a decimal.
Round to the nearest hundredth.

Solution
$$\frac{0.375}{8)\overline{3.000}} \approx 0.38$$

YOU TRY IT 1

Convert $\frac{9}{16}$ to a decimal.
Round to the nearest tenth.

Your solution

EXAMPLE 2

Convert $2\frac{3}{4}$ to a decimal.
Round to the nearest tenth.

Solution
$$2\frac{3}{4} = \frac{11}{4} \qquad \frac{2.75}{4)\overline{11.00}} \approx 2.8$$

YOU TRY IT 2

Convert $4\frac{1}{6}$ to a decimal.
Round to the nearest hundredth.

Your solution

Solutions on p. S10

OBJECTIVE B To convert decimals to fractions

To convert a decimal to a fraction, remove the decimal point and place the decimal part over a denominator equal to the place value of the last digit in the decimal.

hundredths
$$0.47 = \frac{47}{100}$$

hundredths
$$7.45 = 7\frac{45}{100} = 7\frac{9}{20}$$

thousandths
$$0.275 = \frac{275}{1000} = \frac{11}{40}$$

hundredths
$$0.16\frac{2}{3} = \frac{16\frac{2}{3}}{100} = 16\frac{2}{3} \div 100 = \frac{50}{3} \times \frac{1}{100} = \frac{1}{6}$$

EXAMPLE • 3

Convert 0.82 and 4.75 to fractions.

Solution $0.82 = \dfrac{82}{100} = \dfrac{41}{50}$

$4.75 = 4\dfrac{75}{100} = 4\dfrac{3}{4}$

YOU TRY IT • 3

Convert 0.56 and 5.35 to fractions.

Your solution

EXAMPLE • 4

Convert $0.15\frac{2}{3}$ to a fraction.

Solution $0.15\dfrac{2}{3} = \dfrac{15\frac{2}{3}}{100} = 15\dfrac{2}{3} \div 100$

$= \dfrac{47}{3} \times \dfrac{1}{100} = \dfrac{47}{300}$

YOU TRY IT • 4

Convert $0.12\frac{7}{8}$ to a fraction.

Your solution

Solutions on p. S10

OBJECTIVE C   To identify the order relation between two decimals or between a decimal and a fraction

Decimals, like whole numbers and fractions, can be graphed as points on the number line. The number line can be used to show the order of decimals. A decimal that appears to the right of a given number is greater than the given number. A decimal that appears to the left of a given number is less than the given number.

```
◄──┼────┼────┼────┼────┼────┼────┼────┼────┼──►
  3.00  3.05  3.10  3.15  3.20  3.25  3.30  3.35  3.40
```

Note that 3, 3.0, and 3.00 represent the same number.

HOW TO • 3   Find the order relation between $\dfrac{3}{8}$ and 0.38.

$\dfrac{3}{8} = 0.375$ $0.38 = 0.380$ • Convert the fraction $\dfrac{3}{8}$ to a decimal.

$0.375 < 0.380$ • Compare the two decimals.

$\dfrac{3}{8} < 0.38$ • Convert 0.375 back to a fraction.

EXAMPLE • 5

Place the correct symbol, < or >, between the numbers.

$\dfrac{5}{16}$ 0.32

Solution $\dfrac{5}{16} \approx 0.313$ • Convert $\dfrac{5}{16}$ to a decimal.

$0.313 < 0.32$ • Compare the two decimals.

$\dfrac{5}{16} < 0.32$ • Convert 0.313 back to a fraction.

YOU TRY IT • 5

Place the correct symbol, < or >, between the numbers.

0.63 $\dfrac{5}{8}$

Your solution

Solution on p. S10

3.6 EXERCISES

OBJECTIVE A To convert fractions to decimals

For Exercises 1 to 24, convert the fraction to a decimal.
Round to the nearest thousandth.

1. $\dfrac{5}{8}$ **2.** $\dfrac{7}{12}$ **3.** $\dfrac{2}{3}$ **4.** $\dfrac{5}{6}$ **5.** $\dfrac{1}{6}$ **6.** $\dfrac{7}{8}$

7. $\dfrac{5}{12}$ **8.** $\dfrac{9}{16}$ **9.** $\dfrac{7}{4}$ **10.** $\dfrac{5}{3}$ **11.** $1\dfrac{1}{2}$ **12.** $2\dfrac{1}{3}$

13. $\dfrac{16}{4}$ **14.** $\dfrac{36}{9}$ **15.** $\dfrac{3}{1000}$ **16.** $\dfrac{5}{10}$ **17.** $7\dfrac{2}{25}$ **18.** $16\dfrac{7}{9}$

19. $37\dfrac{1}{2}$ **20.** $\dfrac{5}{24}$ **21.** $\dfrac{4}{25}$ **22.** $3\dfrac{1}{3}$ **23.** $8\dfrac{2}{5}$ **24.** $5\dfrac{4}{9}$

For Exercises 25 to 28, without actually doing any division, state whether the decimal equivalent of the given fraction is greater than 1 or less than 1.

25. $\dfrac{54}{57}$ **26.** $\dfrac{176}{129}$ **27.** $\dfrac{88}{80}$ **28.** $\dfrac{2007}{2008}$

OBJECTIVE B To convert decimals to fractions

For Exercises 29 to 53, convert the decimal to a fraction.

29. 0.8   **30.** 0.4   **31.** 0.32   **32.** 0.48   **33.** 0.125

34. 0.485   **35.** 1.25   **36.** 3.75   **37.** 16.9   **38.** 17.5

39. 8.4   **40.** 10.7   **41.** 8.437   **42.** 9.279   **43.** 2.25

44. 7.75   **45.** $0.15\frac{1}{3}$ **46.** $0.17\frac{2}{3}$ **47.** $0.87\frac{7}{8}$ **48.** $0.12\frac{5}{9}$

49. 7.38   **50.** 0.33   **51.** 0.57   **52.** $0.33\frac{1}{3}$ **53.** $0.66\frac{2}{3}$

 54. Is $0.27\frac{4}{9}$ greater than 0.27 or less than 0.27?

 OBJECTIVE C   To identify the order relation between two decimals or between a decimal and a fraction

For Exercises 55 to 74, place the correct symbol, < or >, between the numbers.

55. 0.15 0.5   **56.** 0.6 0.45   **57.** 6.65 6.56   **58.** 3.89 3.98

59. 2.504 2.054   **60.** 0.025 0.105   **61.** $\frac{3}{8}$ 0.365   **62.** $\frac{4}{5}$ 0.802

63. $\frac{2}{3}$ 0.65   **64.** 0.85 $\frac{7}{8}$ **65.** $\frac{5}{9}$ 0.55   **66.** $\frac{7}{12}$ 0.58

67. 0.62 $\frac{7}{15}$ **68.** $\frac{11}{12}$ 0.92   **69.** 0.161 $\frac{1}{7}$ **70.** 0.623 0.6023

71. 0.86 0.855   **72.** 0.87 0.087   **73.** 1.005 0.5   **74.** 0.033 0.3

 75. Use the inequality symbol < to rewrite the order relation expressed by the inequality 17.2 > 0.172.

76. Use the inequality symbol > to rewrite the order relation expressed by the inequality 0.0098 < 0.98.

Applying the Concepts

 77. **Air Pollution** An emissions test for cars requires that of the total engine exhaust, less than 1 part per thousand $\left(\frac{1}{1000} = 0.001\right)$ be hydrocarbon emissions. Using this figure, determine which of the cars in the table at the right would fail the emissions test.

Car	Total Engine Exhaust	Hydrocarbon Emission
1	367,921	360
2	401,346	420
3	298,773	210
4	330,045	320
5	432,989	450

78. Explain how terminating, repeating, and nonrepeating decimals differ. Give an example of each kind of decimal.

FOCUS ON PROBLEM SOLVING

Relevant Information
Problems in mathematics or real life involve a question or a need and information or circumstances related to that question or need. Solving problems in the sciences usually involves a question, an observation, and measurements of some kind.

One of the challenges of problem solving in the sciences is to separate the information that is relevant to the problem from other information. Following is an example from the physical sciences in which some relevant information was omitted.

Hooke's Law states that the distance that a weight will stretch a spring is directly proportional to the weight on the spring. That is, $d = kF$, where d is the distance the spring is stretched and F is the force. In an experiment to verify this law, some physics students were continually getting inconsistent results. Finally, the instructor discovered that the heat produced when the lights were turned on was affecting the experiment. In this case, relevant information was omitted—namely, that the temperature of the spring can affect the distance it will stretch.

A lawyer drove 8 miles to the train station. After a 35-minute ride of 18 miles, the lawyer walked 10 minutes to the office. Find the total time it took the lawyer to get to work.

From this situation, answer the following before reading on.

a. What is asked for?

b. Is there enough information to answer the question?

c. Is information given that is not needed?

Here are the answers.

a. We want the total time for the lawyer to get to work.

b. No. We do not know the time it takes the lawyer to get to the train station.

c. Yes. Neither the distance to the train station nor the distance of the train ride is necessary to answer the question.

For each of the following problems, answer the questions printed in red above.

1. A customer bought 6 boxes of strawberries and paid with a $20 bill. What was the change?

2. A board is cut into two pieces. One piece is 3 feet longer than the other piece. What is the length of the original board?

3. A family rented a car for their vacation and drove 680 miles. The cost of the rental car was $21 per day with 150 free miles per day and $.15 for each mile driven above the number of free miles allowed. How many miles did the family drive per day?

4. An investor bought 8 acres of land for $80,000. One and one-half acres were set aside for a park, and the remaining land was developed into one-half-acre lots. How many lots were available for sale?

5. You wrote checks of $43.67, $122.88, and $432.22 after making a deposit of $768.55. How much do you have left in your checking account?

PROJECTS AND GROUP ACTIVITIES

Fractions as Terminating or Repeating Decimals

The fraction $\frac{3}{4}$ is equivalent to 0.75. The decimal 0.75 is a **terminating decimal** because there is a remainder of zero when 3 is divided by 4. The fraction $\frac{1}{3}$ is equivalent to 0.333 . . . . The three dots mean the pattern continues on and on. 0.333 . . . is a **repeating decimal.** To determine whether a fraction can be written as a terminating decimal, first write the fraction in simplest form. Then look at the denominator of the fraction. If it contains prime factors of only 2s and/or 5s, then it can be expressed as a terminating decimal. If it contains prime factors other than 2s or 5s, it represents a repeating decimal.

✓ **Take Note**

If the denominator of a fraction in simplest form is 20, then it can be written as a terminating decimal because $20 = 2 \cdot 2 \cdot 5$ (only prime factors of 2 and 5). If the denominator of a fraction in simplest form is 6, it represents a repeating decimal because it contains the prime factor 3 (a number other than 2 or 5).

1. Assume that each of the following numbers is the denominator of a fraction written in simplest form. Does the fraction represent a terminating or repeating decimal?

   a. 4 **b.** 5 **c.** 7 **d.** 9 **e.** 10 **f.** 12 **g.** 15
   h. 16 **i.** 18 **j.** 21 **k.** 24 **l.** 25 **m.** 28 **n.** 40

2. Write two other numbers that, as denominators of fractions in simplest form, represent terminating decimals, and write two other numbers that, as denominators of fractions in simplest form, represent repeating decimals.

CHAPTER 3

SUMMARY

KEY WORDS	EXAMPLES
A number written in *decimal notation* has three parts: a *whole-number part*, a *decimal point*, and a *decimal part*. The decimal part of a number represents a number less than 1. A number written in decimal notation is often simply called a *decimal*. [3.1A, p. 126]	For the decimal 31.25, 31 is the whole-number part and 25 is the decimal part.

ESSENTIAL RULES AND PROCEDURES	EXAMPLES
To write a decimal in words, write the decimal part as if it were a whole number. Then name the place value of the last digit. The decimal point is read as "and." [3.1A, p. 126]	The decimal 12.875 is written in words as twelve and eight hundred seventy-five thousandths.
To write a decimal in standard form when it is written in words, write the whole-number part, replace the word *and* with a decimal point, and write the decimal part so that the last digit is in the given place-value position. [3.1A, p. 127]	The decimal forty-nine and sixty-three thousandths is written in standard form as 49.063.
To round a decimal to a given place value, use the same rules used with whole numbers, except drop the digits to the right of the given place value instead of replacing them with zeros. [3.1B, p. 128]	2.7134 rounded to the nearest tenth is 2.7. 0.4687 rounded to the nearest hundredth is 0.47.

To add decimals, write the decimals so that the decimal points are on a vertical line. Add as you would with whole numbers. Then write the decimal point in the sum directly below the decimal points in the addends. [3.2A, p. 132]	$\begin{array}{r} {\scriptstyle 1\ \ 1}\\ 1.35\\ 20.8\\ +\ 0.76\\ \hline 22.91 \end{array}$
To subtract decimals, write the decimals so that the decimal points are on a vertical line. Subtract as you would with whole numbers. Then write the decimal point in the difference directly below the decimal point in the subtrahend. [3.3A, p. 136]	$\begin{array}{r} {\scriptstyle 2\ 15\ \ \ \ \ 6\ 10}\\ 3\cancel{3}.8\cancel{7}\cancel{0}\\ -\ 9.641\\ \hline 26.229 \end{array}$
To multiply decimals, multiply the numbers as you would whole numbers. Then write the decimal point in the product so that the number of decimal places in the product is the sum of the decimal places in the factors. [3.4A, p. 140]	$\begin{array}{r} 26.83 \quad \text{2 decimal places}\\ \times\ \ \ 0.45 \quad \text{2 decimal places}\\ \hline 13415\\ 10732\ \\ \hline 12.0735 \quad \text{4 decimal places} \end{array}$
To multiply a decimal by a power of 10, move the decimal point to the right the same number of places as there are zeros in the power of 10. If the power of 10 is written in exponential notation, the exponent indicates how many places to move the decimal point. [3.4A, pp. 140, 141]	$3.97 \cdot 10,000 = 39,700$ $0.641 \cdot 10^5 = 64,100$
To divide decimals, move the decimal point in the divisor to the right so that it is a whole number. Move the decimal point in the dividend the same number of places to the right. Place the decimal point in the quotient directly above the decimal point in the dividend. Then divide as you would with whole numbers. [3.5A, p. 150]	$\begin{array}{r} 6.2\ \ \\ 0.39.\overline{)2.41.8}\\ -2\,34\ \ \ \\ \hline 7\,8\\ -7\,8\\ \hline 0 \end{array}$
To divide a decimal by a power of 10, move the decimal point to the left the same number of places as there are zeros in the power of 10. If the power of 10 is written in exponential notation, the exponent indicates how many places to move the decimal point. [3.5A, p. 151]	$972.8 \div 1000 = 0.9728$ $61.305 \div 10^4 = 0.0061305$
To convert a fraction to a decimal, divide the numerator of the fraction by the denominator. [3.6A, p. 159]	$\dfrac{7}{8} = 7 \div 8 = 0.875$
To convert a decimal to a fraction, remove the decimal point and place the decimal part over a denominator equal to the place value of the last digit in the decimal. [3.6B, p. 159]	0.85 is eighty-five <u>hundredths</u>. $0.85 = \dfrac{85}{100} = \dfrac{17}{20}$
To find the order relation between a decimal and a fraction, first rewrite the fraction as a decimal. Then compare the two decimals. [3.6C, p. 160]	Because $\dfrac{3}{11} \approx 0.273$, and $0.273 > 0.26$, $\dfrac{3}{11} > 0.26$.

CONCEPT REVIEW

Test your knowledge of the concepts presented in this chapter. Answer each question. Then check your answers against the ones provided in the Answer Section.

1. How do you round a decimal to the nearest tenth?

2. How do you write the decimal 0.37 as a fraction?

3. How do you write the fraction $\frac{173}{10,000}$ as a decimal?

4. When adding decimals of different place values, what do you do with the decimal points?

5. Where do you put the decimal point in the product of two decimals?

6. How do you estimate the product of two decimals?

7. What do you do with the decimal point when dividing decimals?

8. Which is greater, the decimal 0.63 or the fraction $\frac{5}{8}$?

9. How many zeros must be inserted when dividing 0.763 by 0.6 and rounding to the nearest hundredth?

10. How do you subtract a decimal from a whole number that has no decimal point?

CHAPTER 3

REVIEW EXERCISES

1. Find the quotient of 3.6515 and 0.067.

2. Find the sum of 369.41, 88.3, 9.774, and 366.474.

3. Place the correct symbol, $<$ or $>$, between the two numbers.
   0.055   0.1

4. Write 22.0092 in words.

5. Round 0.05678235 to the nearest hundred-thousandth.

6. Convert $2\frac{1}{3}$ to a decimal. Round to the nearest hundredth.

7. Convert 0.375 to a fraction.

8. Add: $3.42 + 0.794 + 32.5$

9. Write thirty-four and twenty-five thousandths in standard form.

10. Place the correct symbol, $<$ or $>$, between the two numbers.
    $\frac{5}{8}$ 0.62

11. Convert $\frac{7}{9}$ to a decimal. Round to the nearest thousandth.

12. Convert 0.66 to a fraction.

13. Subtract: $27.31 - 4.4465$

14. Round 7.93704 to the nearest hundredth.

15. Find the product of 3.08 and 2.9.

16. Write 342.37 in words.

17. Write three and six thousand seven hundred fifty-three hundred-thousandths in standard form.

18. Multiply: $\begin{array}{r} 34.79 \\ \times\ \ 0.74 \end{array}$

19. Divide: $0.053\overline{)0.349482}$

20. What is 7.796 decreased by 2.9175?

21. **Banking** You had a balance of $895.68 in your checking account. You then wrote checks for $145.72 and $88.45. Find the new balance in your checking account.

For Exercises 22 and 23, use the news clipping at the right.

22. **Fuel Consumption** Find the difference between the amount United expects to pay per gallon of fuel and the amount Southwest expects to pay per gallon of fuel.

23. **Fuel Consumption** What is Northwest's cost per gallon of fuel? Round to the nearest cent. Is Northwest's cost per gallon of fuel greater than or less than United's cost per gallon?

24. **Travel** In a recent year, 30.6 million Americans drove to their destinations over Thanksgiving, and 4.8 million Americans traveled by plane. (*Source:* AAA) How many times greater is the number who drove than the number who flew? Round to the nearest tenth.

25. **Nutrition** According to the American School Food Service Association, 1.9 million gallons of milk are served in school cafeterias every day. How many gallons of milk are served in school cafeterias during a 5-day school week?

CHAPTER 3

TEST

1. Place the correct symbol, < or >, between the two numbers.

0.66   0.666

2. Subtract: 13.027
  − 8.94

3. Write 45.0302 in words.

4. Convert $\frac{9}{13}$ to a decimal. Round to the nearest thousandth.

5. Convert 0.825 to a fraction.

6. Round 0.07395 to the nearest ten-thousandth.

7. Find 0.0569 divided by 0.037. Round to the nearest thousandth.

8. Find 9.23674 less than 37.003.

9. Round 7.0954625 to the nearest thousandth.

10. Divide: $0.006\overline{)1.392}$

11. Add: 270.93
     97.
      1.976
  + 88.675

12. **Mechanics** Find the missing dimension.

4.86 in.

?

6.23 in.

13. Multiply: 1.37
                   × 0.004

14. What is the total of 62.3, 4.007, and 189.65?

15. Write two hundred nine and seven thousand eighty-six hundred-thousandths in standard form.

16. Finances A car was bought for $16,734.40, with a down payment of $2500. The balance was paid in 36 monthly payments. Find the amount of each monthly payment.

17. Compensation You received a salary of $727.50, a commission of $1909.64, and a bonus of $450. Find your total income.

18. Consumerism A long-distance telephone call costs $.85 for the first 3 minutes and $.42 for each additional minute. Find the cost of a 12-minute long-distance telephone call.

Computers The table at the right shows the average number of hours per week that students use a computer. Use this table for Exercises 19 and 20.

19. On average, how many hours per year does a 10th-grade student use a computer? Use a 52-week year.

Grade Level	Average Number of Hours of Computer Use per Week
Prekindergarten–kindergarten	3.9
1st – 3rd	4.9
4th – 6th	4.2
7th – 8th	6.9
9th – 12th	6.7

Source: Find/SVP American Learning Household Survey

20. On average, how many more hours per year does a 2nd-grade student use a computer than a 5th-grade student? Use a 52-week year.

CUMULATIVE REVIEW EXERCISES

1. Divide: $89\overline{)20{,}932}$

2. Simplify: $2^3 \cdot 4^2$

3. Simplify: $2^2 - (7 - 3) \div 2 + 1$

4. Find the LCM of 9, 12, and 24.

5. Write $\frac{22}{5}$ as a mixed number.

6. Write $4\frac{5}{8}$ as an improper fraction.

7. Write an equivalent fraction with the given denominator.

$$\frac{5}{12} = \frac{}{60}$$

8. Add: $\frac{3}{8} + \frac{5}{12} + \frac{9}{16}$

9. What is $5\frac{7}{12}$ increased by $3\frac{7}{18}$?

10. Subtract: $9\frac{5}{9} - 3\frac{11}{12}$

11. Multiply: $\frac{9}{16} \times \frac{4}{27}$

12. Find the product of $2\frac{1}{8}$ and $4\frac{5}{17}$.

13. Divide: $\frac{11}{12} \div \frac{3}{4}$

14. What is $2\frac{3}{8}$ divided by $2\frac{1}{2}$?

15. Simplify: $\left(\frac{2}{3}\right)^2 \cdot \left(\frac{3}{4}\right)^3$

16. Simplify: $\left(\frac{2}{3}\right)^2 - \left(\frac{2}{3} - \frac{1}{2}\right) + 2$

17. Write 65.0309 in words.

18. Add: 379.006
            27.523
             9.8707
         + 88.2994

19. What is 29.005 decreased by 7.9286?

20. Multiply: 9.074
                       $\times$ 6.09

21. Divide: $8.09\overline{)17.42963}$
Round to the nearest thousandth

22. Convert $\frac{11}{15}$ to a decimal. Round to the nearest thousandth.

23. Convert $0.16\frac{2}{3}$ to a fraction.

24. Place the correct symbol, $<$ or $>$, between the two numbers.

$\frac{8}{9}$ 0.98

25. **Vacation** The graph at the right shows the number of vacation days per year that are legally mandated in several countries. How many more vacation days does Sweden mandate than Germany?

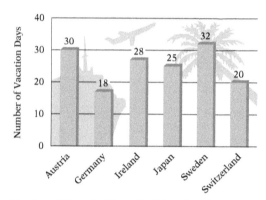

Number of Legally Mandated Vacation Days
Sources: Economic Policy Institute; *World Almanac*

26. **Health** A patient is put on a diet to lose 24 pounds in 3 months. The patient loses $9\frac{1}{2}$ pounds the first month and $6\frac{3}{4}$ pounds the second month. How much weight must this patient lose the third month to achieve the goal?

27. **Banking** You have a checking account balance of $814.35. You then write checks for $42.98, $16.43, and $137.56. Find your checking account balance after you write the checks.

28. **Mechanics** A machine lathe takes 0.017 inch from a brass bushing that is 1.412 inches thick. Find the resulting thickness of the bushing.

29. **Taxes** The state income tax on your business is $820 plus 0.08 times your profit. You made a profit of $64,860 last year. Find the amount of income tax you paid last year.

30. **Finances** You bought a camera costing $410.96. The down payment was $40, and the balance is to be paid in 8 equal monthly payments. Find the monthly payment.

Ratio and Proportion

OBJECTIVES

SECTION 4.1
A To write the ratio of two quantities in simplest form
B To solve application problems

SECTION 4.2
A To write rates
B To write unit rates
C To solve application problems

SECTION 4.3
A To determine whether a proportion is true
B To solve proportions
C To solve application problems

ARE YOU READY?

Take the Chapter 4 Prep Test to find out if you are ready to learn to:

- Write ratios, rates, and unit rates
- Solve proportions

PREP TEST

Do these exercises to prepare for Chapter 4.

1. Simplify: $\dfrac{8}{10}$

2. Simplify: $\dfrac{450}{650 + 250}$

3. Write as a decimal: $\dfrac{372}{15}$

4. Which is greater, 4×33 or 62×2?

5. Complete: $? \times 5 = 20$

SECTION

4.1 Ratio

OBJECTIVE A **To write the ratio of two quantities in simplest form** 💿

Point of Interest

In the 1990s, the major-league pitchers with the best strikeout-to-walk ratios (having pitched a minimum of 100 innings) were

Dennis Eckersley	6.46:1
Shane Reynolds	4.13:1
Greg Maddux	4:1
Bret Saberhagen	3.92:1
Rod Beck	3.81:1

The best single-season strikeout-to-walk ratio for starting pitchers in the same period was that of Bret Saberhagen, 11:1. (*Source:* Elias Sports Bureau)

Quantities such as 3 feet, 12 cents, and 9 cars are number quantities written with units.

3 feet
12 cents   These are some examples of units. Shirts, dollars, trees, miles,
9 cars   and gallons are further examples.
↑
units

A **ratio** is a comparison of two quantities that have the *same* units. This comparison can be written three different ways:

1. As a fraction
2. As two numbers separated by a colon (:)
3. As two numbers separated by the word *to*

The ratio of the lengths of two boards, one 8 feet long and the other 10 feet long, can be written as

1. $\dfrac{8 \text{ feet}}{10 \text{ feet}} = \dfrac{8}{10} = \dfrac{4}{5}$

2. 8 feet:10 feet = 8:10 = 4:5

3. 8 feet to 10 feet = 8 to 10 = 4 to 5

Writing the **simplest form of a ratio** means writing it so that the two numbers have no common factor other than 1.

This ratio means that the smaller board is $\dfrac{4}{5}$ the length of the longer board.

EXAMPLE 1

Write the comparison $6 to $8 as a ratio in simplest form using a fraction, a colon, and the word *to*.

Solution $\dfrac{\$6}{\$8} = \dfrac{6}{8} = \dfrac{3}{4}$

$6:$8 = 6:8 = 3:4
$6 to $8 = 6 to 8 = 3 to 4

YOU TRY IT 1

Write the comparison 20 pounds to 24 pounds as a ratio in simplest form using a fraction, a colon, and the word *to*.

Your solution

EXAMPLE 2

Write the comparison 18 quarts to 6 quarts as a ratio in simplest form using a fraction, a colon, and the word *to*.

Solution $\dfrac{18 \text{ quarts}}{6 \text{ quarts}} = \dfrac{18}{6} = \dfrac{3}{1}$

18 quarts:6 quarts = 18:6 = 3:1
18 quarts to 6 quarts =
    18 to 6 = 3 to 1

YOU TRY IT 2

Write the comparison 64 miles to 8 miles as a ratio in simplest form using a fraction, a colon, and the word *to*.

Your solution

Solutions on p. S10

OBJECTIVE B To solve application problems

© Charles O'Rear/Corbis

Use the table below for Example 3 and You Try It 3.

Board Feet of Wood at a Lumber Store			
Pine	Ash	Oak	Cedar
20,000	18,000	10,000	12,000

EXAMPLE 3

Find, as a fraction in simplest form, the ratio of the number of board feet of pine to the number of board feet of oak.

Strategy
To find the ratio, write the ratio of board feet of pine (20,000) to board feet of oak (10,000) in simplest form.

Solution

$$\frac{20,000}{10,000} = \frac{2}{1}$$

The ratio is $\frac{2}{1}$.

YOU TRY IT 3

Find, as a fraction in simplest form, the ratio of the number of board feet of cedar to the number of board feet of ash.

Your strategy

Your solution

EXAMPLE 4

The cost of building a patio cover was $500 for labor and $700 for materials. What, as a fraction in simplest form, is the ratio of the cost of materials to the total cost for labor and materials?

Strategy
To find the ratio, write the ratio of the cost of materials ($700) to the total cost ($500 + $700) in simplest form.

Solution

$$\frac{\$700}{\$500 + \$700} = \frac{700}{1200} = \frac{7}{12}$$

The ratio is $\frac{7}{12}$.

YOU TRY IT 4

A company spends $600,000 a month for television advertising and $450,000 a month for radio advertising. What, as a fraction in simplest form, is the ratio of the cost of radio advertising to the total cost of radio and television advertising?

Your strategy

Your solution

Solutions on p. S10

4.1 EXERCISES

OBJECTIVE A To write the ratio of two quantities in simplest form

For Exercises 1 to 18, write the comparison as a ratio in simplest form using a fraction, a colon (:), and the word *to*.

1. 3 pints to 15 pints

2. 6 pounds to 8 pounds

3. $40 to $20

4. 10 feet to 2 feet

5. 3 miles to 8 miles

6. 2 hours to 3 hours

7. 6 minutes to 6 minutes

8. 8 days to 12 days

9. 35 cents to 50 cents

10. 28 inches to 36 inches

11. 30 minutes to 60 minutes

12. 25 cents to 100 cents

13. 32 ounces to 16 ounces

14. 12 quarts to 4 quarts

15. 30 yards to 12 yards

16. 12 quarts to 18 quarts

17. 20 gallons to 28 gallons

18. 14 days to 7 days

19. To write a ratio that compares 3 days to 3 weeks, change 3 weeks into an equivalent number of _____.

20. Is the ratio 3 : 4 the same as the ratio 4 : 3?

OBJECTIVE B To solve application problems

For Exercises 21 to 23, write ratios in simplest form using a fraction.

Family Budget						
Housing	Food	Transportation	Taxes	Utilities	Miscellaneous	Total
$1600	$800	$600	$700	$300	$800	$4800

21. **Budgets** Use the table to find the ratio of housing costs to total expenses.

22. **Budgets** Use the table to find the ratio of food costs to total expenses.

23. **Budgets** Use the table to find the ratio of utilities costs to food costs.

24. Refer to the table above. Write a verbal description of the ratio represented by 1 : 2. (*Hint:* There is more than one answer.)

25. Facial Hair Using the data in the news clipping at the right and the figure 50 million for the number of adult males in the United States, write the ratio of the number of men who participated in Movember to the number of adult males in the U.S. (*Source: Time*, February 18, 2008) Write the ratio as a fraction in simplest form.

26. Real Estate A house with an original value of $180,000 increased in value to $220,000 in 5 years. What is the ratio of the increase in value to the original value of the house?

27. Energy Prices The price of gasoline jumped from $2.70 per gallon to $3.24 per gallon in 1 year. What is the ratio of the increase in price to the original price?

In the News

Grow a Mustache, Save a Life

Last fall, in an effort to raise money for the Prostate Cancer Foundation, approximately 2000 men participated in a month-long mustache-growing competition. The event was dubbed Movember.

Source: Time, February 18, 2008

28. Sports National Collegiate Athletic Association (NCAA) statistics show that for every 154,000 high school seniors playing basketball, only 4000 will play college basketball as first-year students. Write the ratio of the number of first-year students playing college basketball to the number of high school seniors playing basketball.

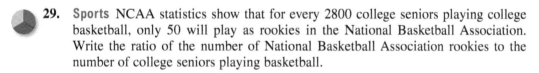

29. Sports NCAA statistics show that for every 2800 college seniors playing college basketball, only 50 will play as rookies in the National Basketball Association. Write the ratio of the number of National Basketball Association rookies to the number of college seniors playing basketball.

Female Vocalists The table at the right shows the concert earnings for Madonna, Barbra Streisand, and Celine Dion for performances between June 2006 and June 2007.

30. Find the ratio of the amount earned by Celine Dion to the amount earned by Barbra Stresand. Write the ratio in simplest form using the word *to*.

31. Find the ratio of the amount earned by Madonna to the total amount earned by the three women. Write the ratio in simplest form using the word *to*.

32. Consumerism In a recent year, women spent $2 million on swimwear and purchased 92,000 swimsuits. During the same year, men spent $500,000 on swimwear and purchased 37,000 swimsuits. (*Source:* NPD Group) **a.** Find the ratio of the amount men spent on swimwear to the amount women spent on swimwear. **b.** Find the ratio of the amout men spent on swimwear to the total amount men and women spent on swimwear. Write the ratios as fractions in simplest form.

Earnings from Concerts, June 2006 to June 2007

Source: Time, Feburary 18, 2008

Applying the Concepts

33. Is the value of a ratio always less than 1? Explain.

SECTION

4.2 Rates

OBJECTIVE A   To write rates

 Point of Interest

Listed below are rates at which various crimes are committed in our nation.

Crime	Every
Larceny	4 seconds
Burglary	14 seconds
Robbery	60 seconds
Rape	6 minutes
Murder	31 minutes

A **rate** is a comparison of two quantities that have *different* units. A rate is written as a fraction.

A distance runner ran 26 miles in 4 hours. The distance-to-time rate is written

$$\frac{26 \text{ miles}}{4 \text{ hours}} = \frac{13 \text{ miles}}{2 \text{ hours}}$$

Writing the **simplest form of a rate** means writing it so that the two numbers that form the rate have no common factor other than 1.

EXAMPLE • 1

Write "6 roof supports for every 9 feet" as a rate in simplest form.

Solution

$$\frac{6 \text{ supports}}{9 \text{ feet}} = \frac{2 \text{ supports}}{3 \text{ feet}}$$

YOU TRY IT • 1

Write "15 pounds of fertilizer for 12 trees" as a rate in simplest form.

Your solution

Solution on p. S11

OBJECTIVE B   To write unit rates

 Point of Interest

According to a Gallup Poll, women see doctors more often than men do. On average, men visit the doctor 3.8 times per year, whereas women go to the doctor 5.8 times per year.

A **unit rate** is a rate in which the number in the denominator is 1.

$$\frac{\$3.25}{1 \text{ pound}}$$ or $3.25/pound is read "$3.25 per pound."

To find a unit rate, divide the number in the numerator of the rate by the number in the denominator of the rate.

A car traveled 344 miles on 16 gallons of gasoline. To find the miles per gallon (unit rate), divide the numerator of the rate by the denominator of the rate.

$$\frac{344 \text{ miles}}{16 \text{ gallons}} \text{ is the rate.} \qquad 16\overline{)344.0} \quad \begin{array}{c} 21.5 \end{array} \qquad 21.5 \text{ miles/gallon is the unit rate.}$$

EXAMPLE • 2

Write "300 feet in 8 seconds" as a unit rate.

Solution

$$\frac{300 \text{ feet}}{8 \text{ seconds}} \qquad 8\overline{)300.0} \quad 37.5$$

37.5 feet/second

YOU TRY IT • 2

Write "260 miles in 8 hours" as a unit rate.

Your solution

Solution on p. S11

OBJECTIVE C   To solve application problems

Denver Airport

HOW TO 1   The table at the right shows air fares for some routes in the continental United States. Find the cost per mile for the four routes in order to determine the most expensive route and the least expensive route on the basis of mileage flown.

Long Routes	Miles	Fare
New York–Los Angeles	2475	$683
San Francisco–Dallas	1464	$536
Denver–Pittsburgh	1302	$525
Minneapolis–Hartford	1050	$483

Integrating Technology

To calculate the costs per mile using a calculator, perform four divisions:

683 ÷ 2475 =

536 ÷ 1464 =

525 ÷ 1302 =

483 ÷ 1050 =

In each case, round the number in the display to the nearest hundredth.

Strategy

To find the cost per mile, divide the fare by the miles flown for each route. Compare the costs per mile to determine the most expensive and least expensive routes per mile.

Solution

New York–Los Angeles $\dfrac{683}{2475} \approx 0.28$

San Francisco–Dallas $\dfrac{536}{1464} \approx 0.37$

Denver–Pittsburgh $\dfrac{525}{1302} \approx 0.40$

Minneapolis–Hartford $\dfrac{483}{1050} = 0.46$

$0.28 < 0.37 < 0.40 < 0.46$

The Minneapolis–Hartford route is the most expensive per mile, and the New York–Los Angeles route is the least expensive per mile.

EXAMPLE 3

As an investor, Jung Ho purchased 100 shares of stock for $1500. One year later, Jung sold the 100 shares for $1800. What was his profit per share?

Strategy

To find Jung's profit per share:

• Find the total profit by subtracting the original cost ($1500) from the selling price ($1800).

• Find the profit per share (unit rate) by dividing the total profit by the number of shares of stock (100).

Solution

$1800 - 1500 = 300$

$300 \div 100 = 3$

Jung Ho's profit was $3/share.

YOU TRY IT 3

Erik Peltier, a jeweler, purchased 5 ounces of a gold alloy for $1625. Later, he sold the 5 ounces for $1720. What was Erik's profit per ounce?

Your strategy

Your solution

Solution on p. S11

4.2 EXERCISES

OBJECTIVE A   To write rates

For Exercises 1 to 8, write each phrase as a rate in simplest form.

1. 3 pounds of meat for 4 people

2. 30 ounces in 24 glasses

3. $80 for 12 boards

4. 84 cents for 3 candy bars

5. 300 miles on 15 gallons

6. 88 feet in 8 seconds

7. 16 gallons in 2 hours

8. 25 ounces in 5 minutes

9. For television advertising rates, what units are **a.** in the numerator and **b.** in the denominator?

OBJECTIVE B   To write unit rates

For Exercises 10 to 12, complete the unit rate.

10. 5 miles in ___ hour

11. 15 feet in ___ second

12. 5 grams of fat in ___ serving

For Exercises 13 to 22, write each phrase as a unit rate.

13. 10 feet in 4 seconds

14. 816 miles in 6 days

15. $3900 earned in 4 weeks

16. $51,000 earned in 12 months

17. 1100 trees planted on 10 acres

18. 3750 words on 15 pages

19. $131.88 earned in 7 hours

20. 628.8 miles in 12 hours

21. 409.4 miles on 11.5 gallons of gasoline

22. $11.05 for 3.4 pounds

OBJECTIVE C   To solve application problems

Miles per Dollar One measure of how expensive it is to drive your car is calculated as miles per dollar, which is the number of miles you drive on 1 dollar's worth of gasoline.

23. Suppose you get 26 miles per gallon of gasoline and gasoline costs $3.49 per gallon. Calculate your miles per dollar. Round to the nearest tenth.

24. Suppose you get 23 miles per gallon of gasoline and gasoline costs $3.15 per gallon. It costs you $44.10 to fill the tank. Calculate your miles per dollar. Round to the nearest tenth.

25. **Corn Production** See the news clipping at the right. Find the average number of bushels harvested from each acre of corn grown in Iowa. Round to the nearest hundredth.

26. **Consumerism** The Pierre family purchased a 250-pound side of beef for $365.75 and had it packaged. During the packaging, 75 pounds of beef were discarded as waste. What was the cost per pound for the packaged beef?

27. **Manufacturing** Regency Computer produced 5000 thumb drives for $13,268.16. Of the disks made, 122 did not meet company standards. What was the cost per disk for those disks that met company standards?

<table>
<tr><td colspan="2">In the News</td></tr>
</table>

Iowa Grows Record Amounts of Corn

In 2007, Iowa corn farmers grew 2.5 billion bushels of corn on 13.9 million acres.

Source: Iowa Corn Promotion Board/Iowa Corn Growers Association

28. **Advertising** The advertising fee for a 30-second spot on the TV show *Deal or No Deal* is $165,000. The show averages 16.1 million viewers. (*Source: USA Today,* December 18, 2006) What is the advertiser's cost per viewer for a 30-second ad? Round to the nearest cent.

29. **Demography** The table at the right shows the population and area of three countries. The population density of a country is the number of people per square mile.

   a. Which country has the least population density?

   b. How many more people per square mile are there in India than in the United States? Round to the nearest whole number.

Country	Population	Area (in square miles)
Australia	20,264,000	2,968,000
India	1,129,866,000	1,269,000
United States	301,140,000	3,619,000

Exchange Rates Another application of rates is in the area of international trade. Suppose a company in Canada purchases a shipment of sneakers from an American company. The Canadian company must exchange Canadian dollars for U.S. dollars in order to pay for the order. The number of Canadian dollars that are equivalent to 1 U.S. dollar is called the **exchange rate.**

30. The table at the right shows the exchange rates per U.S. dollar for three foreign countries and for the euro at the time of this writing.
   a. How many euros would be paid for an order of American computer hardware costing $120,000?
   b. Calculate the cost, in Japanese yen, of an American car costing $34,000.

Exchange Rates per U.S. Dollar	
Australian Dollar	1.0694
Canadian Dollar	1.0179
Japanese Yen	105.3300
The Euro	0.6483

31. Use the table in Exercise 30. What does the quantity 1.0179 × 2500 represent?

Applying the Concepts

32. **Compensation** You have a choice of receiving a wage of $34,000 per year, $2840 per month, $650 per week, or $18 per hour. Which pay choice would you take? Assume a 40-hour work week with 52 weeks per year.

33. The price–earnings ratio of a company's stock is one measure used by stock market analysts to assess the financial well-being of the company. Explain the meaning of the price–earnings ratio.

SECTION

4.3 Proportions

OBJECTIVE A To determine whether a proportion is true

Point of Interest

Proportions were studied by the earliest mathematicians. Clay tablets uncovered by archaeologists show evidence of proportions in Egyptian and Babylonian cultures dating from 1800 B.C.

A **proportion** is an expression of the equality of two ratios or rates.

$$\frac{50 \text{ miles}}{4 \text{ gallons}} = \frac{25 \text{ miles}}{2 \text{ gallons}}$$

Note that the units of the numerators are the same and the units of the denominators are the same.

$$\frac{3}{6} = \frac{1}{2}$$

This is the equality of two ratios.

A proportion is **true** if the fractions are equal when written in lowest terms.

In any true proportion, the **cross products** are equal.

HOW TO 1 Is $\frac{2}{3} = \frac{8}{12}$ a true proportion?

$$\frac{2}{3} \diagup\!\!\!\diagdown \frac{8}{12} \rightarrow \begin{array}{l} 3 \times 8 = 24 \\ 2 \times 12 = 24 \end{array}$$

The cross products *are* equal.

$\frac{2}{3} = \frac{8}{12}$ is a true proportion.

A proportion is **not true** if the fractions are not equal when reduced to lowest terms.

If the cross products are not equal, then the proportion is not true.

HOW TO 2 Is $\frac{4}{5} = \frac{8}{9}$ a true proportion?

$$\frac{4}{5} \diagup\!\!\!\diagdown \frac{8}{9} \rightarrow \begin{array}{l} 5 \times 8 = 40 \\ 4 \times 9 = 36 \end{array}$$

The cross products *are not* equal.

$\frac{4}{5} = \frac{8}{9}$ is not a true proportion.

EXAMPLE 1

Is $\frac{5}{8} = \frac{10}{16}$ a true proportion?

Solution

$$\frac{5}{8} \diagup\!\!\!\diagdown \frac{10}{16} \rightarrow \begin{array}{l} 8 \times 10 = 80 \\ 5 \times 16 = 80 \end{array}$$

The cross products are equal.
The proportion is true.

YOU TRY IT 1

Is $\frac{6}{10} = \frac{9}{15}$ a true proportion?

Your solution

EXAMPLE 2

Is $\frac{62 \text{ miles}}{4 \text{ gallons}} = \frac{33 \text{ miles}}{2 \text{ gallons}}$ a true proportion?

Solution

$$\frac{62}{4} \diagup\!\!\!\diagdown \frac{33}{2} \rightarrow \begin{array}{l} 4 \times 33 = 132 \\ 62 \times 2 = 124 \end{array}$$

The cross products are not equal.
The proportion is not true.

YOU TRY IT 2

Is $\frac{\$32}{6 \text{ hours}} = \frac{\$90}{8 \text{ hours}}$ a true proportion?

Your solution

Solutions on p. S11

OBJECTIVE B   To solve proportions

Tips for Success
An important element of success is practice. We cannot do anything well if we do not practice it repeatedly. Practice is crucial to success in mathematics. In this objective you are learning a new skill: how to solve a proportion. You will need to practice this skill over and over again in order to be successful at it.

Sometimes one of the numbers in a proportion is unknown. In this case, it is necessary to *solve* the proportion.

To **solve a proportion,** find a number to replace the unknown so that the proportion is true.

> **HOW TO 3**   Solve: $\frac{9}{6} = \frac{3}{n}$
>
> $$\frac{9}{6} = \frac{3}{n}$$
> $9 \times n = 6 \times 3$ • Find the cross products.
> $9 \times n = 18$
> $\quad\quad n = 18 \div 9$ • Think of $9 \times n = 18$ as $\begin{smallmatrix}\times n\\ =\end{smallmatrix}$.
> $\quad\quad n = 2$
>
> *Check:*
> $$\frac{9}{6} \diagdown \frac{3}{2} \rightarrow \begin{matrix}6 \times 3 = 18\\ 9 \times 2 = 18\end{matrix}$$

EXAMPLE 3

Solve $\frac{n}{12} = \frac{25}{60}$ and check.

Solution
$n \times 60 = 12 \times 25$ • Find the cross
$n \times 60 = 300$ products. Then
$\quad\quad n = 300 \div 60$ solve for n.
$\quad\quad n = 5$

Check:
$$\frac{5}{12} \diagdown \frac{25}{60} \rightarrow \begin{matrix}12 \times 25 = 300\\ 5 \times 60 = 300\end{matrix}$$

YOU TRY IT 3

Solve $\frac{n}{14} = \frac{3}{7}$ and check.

Your solution

EXAMPLE 4

Solve $\frac{4}{9} = \frac{n}{16}$. Round to the nearest tenth.

Solution
$4 \times 16 = 9 \times n$ • Find the cross
$\quad\quad 64 = 9 \times n$ products. Then
$64 \div 9 = n$ solve for n.
$\quad 7.1 \approx n$

Note: A rounded answer is an approximation. Therefore, the answer to a check will not be exact.

YOU TRY IT 4

Solve $\frac{5}{7} = \frac{n}{20}$. Round to the nearest tenth.

Your solution

Solutions on p. S11

EXAMPLE • 5

Solve $\frac{28}{52} = \frac{7}{n}$ and check.

Solution

$28 \times n = 52 \times 7$ • Find the cross
$28 \times n = 364$ products. Then
    $n = 364 \div 28$ solve for n.
    $n = 13$

Check:

$\frac{28}{52} \diagdown \frac{7}{13} \rightarrow \begin{matrix} 52 \times 7 = 364 \\ 28 \times 13 = 364 \end{matrix}$

YOU TRY IT • 5

Solve $\frac{15}{20} = \frac{12}{n}$ and check.

Your solution

EXAMPLE • 6

Solve $\frac{15}{n} = \frac{8}{3}$. Round to the nearest hundredth.

Solution

$15 \times 3 = n \times 8$
$\quad\quad 45 = n \times 8$
$45 \div 8 = n$
$\quad 5.63 \approx n$

YOU TRY IT • 6

Solve $\frac{12}{n} = \frac{7}{4}$. Round to the nearest hundredth.

Your solution

EXAMPLE • 7

Solve $\frac{n}{9} = \frac{3}{1}$ and check.

Solution

$n \times 1 = 9 \times 3$
$n \times 1 = 27$
$\quad\quad n = 27 \div 1$
$\quad\quad n = 27$

Check:

$\frac{27}{9} \diagdown \frac{3}{1} \rightarrow \begin{matrix} 9 \times 3 = 27 \\ 27 \times 1 = 27 \end{matrix}$

YOU TRY IT • 7

Solve $\frac{n}{12} = \frac{4}{1}$ and check.

Your solution

Solutions on p. S11

OBJECTIVE C   To solve application problems

The application problems in this objective require you to write and solve a proportion.
When setting up a proportion, remember to keep the same units in the numerator and the
same units in the denominator.

EXAMPLE • 8

The dosage of a certain medication is 2 ounces for every 50 pounds of body weight. How many ounces of this medication are required for a person who weighs 175 pounds?

Strategy

To find the number of ounces of medication for a person weighing 175 pounds, write and solve a proportion using n to represent the number of ounces of medication for a 175-pound person.

Solution

$$\frac{2 \text{ ounces}}{50 \text{ pounds}} = \frac{n \text{ ounces}}{175 \text{ pounds}}$$

- The unit "ounces" is in the numerator. The unit "pounds" is in the denominator.

$$2 \times 175 = 50 \times n$$
$$350 = 50 \times n$$
$$350 \div 50 = n$$
$$7 = n$$

A 175-pound person requires 7 ounces of medication.

YOU TRY IT • 8

Three tablespoons of a liquid plant fertilizer are to be added to every 4 gallons of water. How many tablespoons of fertilizer are required for 10 gallons of water?

Your strategy

Your solution

EXAMPLE • 9

A mason determines that 9 cement blocks are required for a retaining wall 2 feet long. At this rate, how many cement blocks are required for a retaining wall that is 24 feet long?

Strategy

To find the number of cement blocks for a retaining wall 24 feet long, write and solve a proportion using n to represent the number of blocks required.

Solution

$$\frac{9 \text{ cement blocks}}{2 \text{ feet}} = \frac{n \text{ cement blocks}}{24 \text{ feet}}$$

$$9 \times 24 = 2 \times n$$
$$216 = 2 \times n$$
$$216 \div 2 = n$$
$$108 = n$$

A 24-foot retaining wall requires 108 cement blocks.

YOU TRY IT • 9

Twenty-four jars can be packed in 6 identical boxes. At this rate, how many jars can be packed in 15 boxes?

Your strategy

Your solution

Solutions on p. S11

4.3 EXERCISES

OBJECTIVE A   **To determine whether a proportion is true**

For Exercises 1 to 18, determine whether the proportion is true or not true.

1. $\dfrac{4}{8} = \dfrac{10}{20}$ **2.** $\dfrac{39}{48} = \dfrac{13}{16}$ **3.** $\dfrac{7}{8} = \dfrac{11}{12}$ **4.** $\dfrac{15}{7} = \dfrac{17}{8}$

5. $\dfrac{27}{8} = \dfrac{9}{4}$ **6.** $\dfrac{3}{18} = \dfrac{4}{19}$ **7.** $\dfrac{45}{135} = \dfrac{3}{9}$ **8.** $\dfrac{3}{4} = \dfrac{54}{72}$

9. $\dfrac{50 \text{ miles}}{2 \text{ gallons}} = \dfrac{25 \text{ miles}}{1 \text{ gallon}}$ **10.** $\dfrac{16 \text{ feet}}{10 \text{ seconds}} = \dfrac{24 \text{ feet}}{15 \text{ seconds}}$

11. $\dfrac{6 \text{ minutes}}{5 \text{ cents}} = \dfrac{30 \text{ minutes}}{25 \text{ cents}}$ **12.** $\dfrac{16 \text{ pounds}}{12 \text{ days}} = \dfrac{20 \text{ pounds}}{14 \text{ days}}$

13. $\dfrac{\$15}{4 \text{ pounds}} = \dfrac{\$45}{12 \text{ pounds}}$ **14.** $\dfrac{270 \text{ trees}}{6 \text{ acres}} = \dfrac{90 \text{ trees}}{2 \text{ acres}}$

15. $\dfrac{300 \text{ feet}}{4 \text{ rolls}} = \dfrac{450 \text{ feet}}{7 \text{ rolls}}$ **16.** $\dfrac{1 \text{ gallon}}{4 \text{ quarts}} = \dfrac{7 \text{ gallons}}{28 \text{ quarts}}$

17. $\dfrac{\$65}{5 \text{ days}} = \dfrac{\$26}{2 \text{ days}}$ **18.** $\dfrac{80 \text{ miles}}{2 \text{ hours}} = \dfrac{110 \text{ miles}}{3 \text{ hours}}$

 19. Suppose that in a true proportion you switch the numerator of the first fraction with the denominator of the second fraction. Must the result be another true proportion?

 20. Write a true proportion in which the cross products are equal to 36.

OBJECTIVE B **To solve proportions**

21. Consider the proportion $\frac{n}{7} = \frac{9}{21}$ in Exercise 23. In lowest terms, $\frac{9}{21} = \frac{3}{7}$. Will solving the proportion $\frac{n}{7} = \frac{3}{7}$ give the same result for n as found in Exercise 23?

For Exercises 22 to 41, solve. Round to the nearest hundredth, if necessary.

22. $\frac{n}{4} = \frac{6}{8}$

23. $\frac{n}{7} = \frac{9}{21}$

24. $\frac{12}{18} = \frac{n}{9}$

25. $\frac{7}{21} = \frac{35}{n}$

26. $\frac{6}{n} = \frac{24}{36}$

27. $\frac{3}{n} = \frac{15}{10}$

28. $\frac{n}{6} = \frac{2}{3}$

29. $\frac{5}{12} = \frac{n}{144}$

30. $\frac{n}{5} = \frac{7}{8}$

31. $\frac{4}{n} = \frac{9}{5}$

32. $\frac{5}{12} = \frac{n}{8}$

33. $\frac{36}{20} = \frac{12}{n}$

34. $\frac{n}{15} = \frac{21}{12}$

35. $\frac{40}{n} = \frac{15}{8}$

36. $\frac{28}{8} = \frac{12}{n}$

37. $\frac{n}{30} = \frac{65}{120}$

38. $\frac{0.3}{5.6} = \frac{n}{25}$

39. $\frac{1.3}{16} = \frac{n}{30}$

40. $\frac{0.7}{9.8} = \frac{3.6}{n}$

41. $\frac{1.9}{7} = \frac{13}{n}$

OBJECTIVE C **To solve application problems**

42. Jesse walked 3 miles in 40 minutes. Let n be the number of miles Jesse can walk in 60 minutes at the same rate. To determine how many miles Jesse can walk in 60 minutes, a student used the proportion $\frac{40}{3} = \frac{60}{n}$. Is this a valid proportion to use in solving this problem?

For Exercises 43 to 61, solve. Round to the nearest hundredth.

43. **Nutrition** A 6-ounce package of Puffed Wheat contains 600 calories. How many calories are in a 0.5-ounce serving of the cereal?

44. **Health** Using the data at the right and a figure of 300 million for the number of Americans, determine the number of morbidly obese Americans.

45. **Fuel Efficiency** A car travels 70.5 miles on 3 gallons of gas. Find the distance the car can travel on 14 gallons of gas.

46. **Landscaping** Ron Stokes uses 2 pounds of fertilizer for every 100 square feet of lawn for landscape maintenance. At this rate, how many pounds of fertilizer did he use on a lawn that measures 3500 square feet?

47. **Gardening** A nursery prepares a liquid plant food by adding 1 gallon of water for each 2 ounces of plant food. At this rate, how many gallons of water are required for 25 ounces of plant food?

48. **Masonry** A brick wall 20 feet in length contains 1040 bricks. At the same rate, how many bricks would it take to build a wall 48 feet in length?

49. **Cartography** The scale on the map at the right is "1.25 inches equals 10 miles." Find the distance between Carlsbad and Del Mar, which are 2 inches apart on the map.

50. **Architecture** The scale on the plans for a new house is "1 inch equals 3 feet." Find the width and the length of a room that measures 5 inches by 8 inches on the drawing.

51. **Medicine** The dosage for a medication is $\frac{1}{3}$ ounce for every 40 pounds of body weight. At this rate, how many ounces of medication should a physician prescribe for a patient who weighs 150 pounds? Write the answer as a decimal.

52. **Banking** A bank requires a monthly payment of $33.45 on a $2500 loan. At the same rate, find the monthly payment on a $10,000 loan.

53. **Elections** A pre-election survey showed that 2 out of every 3 eligible voters would cast ballots in the county election. At this rate, how many people in a county of 240,000 eligible voters would vote in the election?

54. **Interior Design** A paint manufacturer suggests using 1 gallon of paint for every 400 square feet of wall. At this rate, how many gallons of paint would be required for a room that has 1400 square feet of wall?

55. **Insurance** A 60-year-old male can obtain $10,000 of life insurance for $35.35 per month. At this rate, what is the monthly cost for $50,000 of life insurance?

In the News

Number of Obese Americans Increasing

In the past 20 years, the number of obese Americans (those at least 30 pounds overweight) has doubled. The number of morbidly obese (those at least 100 pounds overweight) has quadrupled to 1 in 50.

Source: Time, July 9, 2006

Carlsbad
Encinitas
Solana Beach
Del Mar

0   5   10
Miles

56. Food Waste At the rate given in the news clipping, find the cost of food wasted yearly by **a.** the average family of three and **b.** the average family of five.

In the News

How Much Food Do You Waste?

In the United States, the estimated cost of food wasted each year by the average family of four is $590.

Source: University of Arizona

57. Manufacturing Suppose a computer chip manufacturer knows from experience that in an average production run of 2000 circuit boards, 60 will be defective. How many defective circuit boards can be expected in a run of 25,000 circuit boards?

58. Investments You own 240 shares of stock in a computer company. The company declares a stock split of 5 shares for every 3 owned. How many shares of stock will you own after the stock split?

 59. Physics The ratio of weight on the moon to weight on Earth is 1:6. If a bowling ball weighs 16 pounds on Earth, what would it weigh on the moon?

60. Automobiles When engineers designed a new car, they first built a model of the car. The ratio of the size of a part on the model to the actual size of the part is 2:5. If a door is 1.3 feet long on the model, what is the length of the door on the car?

61. Investments Carlos Capasso owns 50 shares of Texas Utilities that pay dividends of $153. At this rate, what dividend would Carlos receive after buying 300 additional shares of Texas Utilities?

Applying the Concepts

 62. Publishing In the first quarter of 2008, *USA Today* reported that Eckhart Tolle's *A New Earth* outsold John Grisham's *The Appeal* by 3.7 copies to 1. Explain how a proportion can be used to determine the number of copies of *A New Earth* sold given the number of copies of *The Appeal* sold.

  63. Social Security According to the Social Security Administration, the numbers of workers per retiree in the future are expected to be as given in the table below.

Year	2020	2030	2040
Number of workers per retiree	2.5	2.1	2.0

Why is the shrinking number of workers per retiree of importance to the Social Security Administration?

 64. Elections A survey of voters in a city claimed that 2 people of every 5 who voted cast a ballot in favor of city amendment A and that 3 people of every 4 who voted cast a ballot against amendment A. Is this possible? Explain your answer.

 65. Write a word problem that requires solving a proportion to find the answer.

FOCUS ON PROBLEM SOLVING

Looking for a Pattern   A very useful problem-solving strategy is looking for a pattern.

Problem   A legend says that a peasant invented the game of chess and gave it to a very rich king as a present. The king so enjoyed the game that he gave the peasant the choice of anything in the kingdom. The peasant's request was simple: "Place one grain of wheat on the first square, 2 grains on the second square, 4 grains on the third square, 8 on the fourth square, and continue doubling the number of grains until the last square of the chessboard is reached." How many grains of wheat must the king give the peasant?

Solution   A chessboard consists of 64 squares. To find the total number of grains of wheat on the 64 squares, we begin by looking at the amount of wheat on the first few squares.

Square 1	Square 2	Square 3	Square 4	Square 5	Square 6	Square 7	Square 8
1	2	4	8	16	32	64	128
1	3	7	15	31	63	127	255

The bottom row of numbers represents the sum of the number of grains of wheat up to and including that square. For instance, the number of grains of wheat on the first 7 squares is $1 + 2 + 4 + 8 + 16 + 32 + 64 = 127$.

Notice that the number of grains of wheat on a square can be expressed as a power of 2.

The number of grains on square $n = 2^{n-1}$.

For example, the number of grains on square $7 = 2^{7-1} = 2^6 = 64$.

A second pattern of interest is that **the number *below* a square** (the total number of grains up to and including that square) **is 1 less than the number of grains of wheat *on the next square*.** For example, the number *below* square 7 is 1 less than the number on square 8 ($128 - 1 = 127$). From this observation, the number of grains of wheat on the first 8 squares is the number on square 8 (128) plus 1 less than the number on square 8 (127): The total number of grains of wheat on the first 8 squares is $128 + 127 = 255$.

From this observation,

$$\text{Number of grains of wheat on the chessboard} = \text{number of grains on square 64} + \text{1 less than the number of grains on square 64}$$

$$= 2^{64-1} + (2^{64-1} - 1)$$

$$= 2^{63} + 2^{63} - 1 \approx 18{,}000{,}000{,}000{,}000{,}000{,}000$$

To give you an idea of the magnitude of this number, this is more wheat than has been produced in the world since chess was invented.

The same king decided to have a banquet in the long banquet room of the palace to celebrate the invention of chess. The king had 50 square tables, and each table could seat only one person on each side. The king pushed the tables together to form one long banquet table. How many people could sit at this table? *Hint:* Try constructing a pattern by using 2 tables, 3 tables, and 4 tables.

PROJECTS AND GROUP ACTIVITIES

The Golden Ratio There are certain designs that have been repeated over and over in both art and architecture. One of these involves the **golden rectangle.**

A golden rectangle is drawn at the right. Begin with a square that measures, say, 2 inches on a side. Let *A* be the midpoint of a side (halfway between two corners). Now measure the distance from *A* to *B*. Place this length along the bottom of the square, starting at *A*. The resulting rectangle is a golden rectangle.

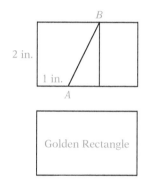

The **golden ratio** is the ratio of the length of the golden rectangle to its width. If you have drawn the rectangle following the procedure above, you will find that the golden ratio is approximately 1.6 to 1.

The golden ratio appears in many different situations. Some historians claim that some of the great pyramids of Egypt are based on the golden ratio. The drawing at the right shows the Pyramid of Giza, which dates from approximately 2600 B.C. The ratio of the height to a side of the base is approximately 1.6 to 1.

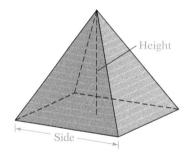

1. There are instances of the golden rectangle in the Mona Lisa painted by Leonardo da Vinci. Do some research on this painting and write a few paragraphs summarizing your findings.

2. What do 3 × 5 and 5 × 8 index cards have to do with the golden rectangle?

3. What does the United Nations Building in New York City have to do with the golden rectangle?

4. When was the Parthenon in Athens, Greece, built? What does the front of that building have to do with the golden rectangle?

Dallas & John Heaton/Corbis

Drawing the Floor Plans for a Building

The drawing at the left is a sketch of the floor plan for a cabin at a resort in the mountains of Utah. The measurements are missing. Assume that you are the architect and will finish the drawing. You will have to decide the size of the rooms and put in the measurements to scale.

Design a cabin that you would like to own. Select a scale and draw all the rooms to scale.

If you are interested in architecture, visit an architect who is using CAD (computer-aided design) software to create a floor plan. Computer technology has revolutionized the field of architectural design.

The U.S. House of Representatives

The framers of the Constitution decided to use a ratio to determine the number of representatives from each state. It was determined that each state would have one representative for every 30,000 citizens, with a minimum of one representative. Congress has changed this ratio over the years, and we now have 435 representatives.

Find the number of representatives from your state. Determine the ratio of citizens to representatives. Also do this for the most populous state and for the least populous state.

You might consider getting information on the number of representatives for each state and the populations of different states via the Internet.

CHAPTER 4

SUMMARY

KEY WORDS	EXAMPLES
A *ratio* is the comparison of two quantities with the same units. A ratio can be written in three ways: as a fraction, as two numbers separated by a colon (:), or as two numbers separated by the word *to*. A ratio is in *simplest form* when the two numbers do not have a common factor. [4.1A, p. 174]	The comparison 16 to 24 ounces can be written as a ratio in simplest form as $\frac{2}{3}$, 2:3, or 2 to 3.
A *rate* is the comparison of two quantities with different units. A rate is written as a fraction. A rate is in *simplest form* when the numbers that form the rate do not have a common factor. [4.2A, p. 178]	You earned \$63 for working 6 hours. The rate is written in simplest form as $\frac{\$21}{2 \text{ hours}}$.
A *unit rate* is a rate in which the number in the denominator is 1. [4.2B, p. 178]	You traveled 144 miles in 3 hours. The unit rate is 48 miles per hour.
A *proportion* is an expression of the equality of two ratios or rates. A proportion is true if the fractions are equal when written in lowest terms; in any true proportion, the *cross products* are equal. A proportion is not true if the fractions are not equal when written in lowest terms; if the cross products are not equal, the proportion is not true. [4.3A, p. 182]	The proportion $\frac{3}{5} = \frac{12}{20}$ is true because the cross products are equal: $3 \times 20 = 5 \times 12$. The proportion $\frac{3}{4} = \frac{12}{20}$ is not true because the cross products are not equal: $3 \times 20 \neq 4 \times 12$.

ESSENTIAL RULES AND PROCEDURES	EXAMPLES
To find a unit rate, divide the number in the numerator of the rate by the number in the denominator of the rate. [4.2B, p. 178]	You earned \$41 for working 4 hours. $41 \div 4 = 10.25$ The unit rate is \$10.25/hour.
To solve a proportion, find a number to replace the unknown so that the proportion is true. [4.3B, p. 183]	$\frac{6}{24} = \frac{9}{n}$ $6 \times n = 24 \times 9$ • Find the cross products. $6 \times n = 216$ $n = 216 \div 6$ $n = 36$
To set up a proportion, keep the same units in the numerator and the same units in the denominator. [4.3C, p. 184]	Three machines fill 5 cereal boxes per minute. How many boxes can 8 machines fill per minute? $\frac{3 \text{ machines}}{5 \text{ cereal boxes}} = \frac{8 \text{ machines}}{n \text{ cereal boxes}}$

CONCEPT REVIEW

Test your knowledge of the concepts presented in this chapter. Answer each question.
Then check your answers against the ones provided in the Answer Section.

1. If the units in a comparison are different, is it a ratio or a rate?

2. How do you find a unit rate?

3. How do you write the ratio $\frac{6}{7}$ using a colon?

4. How do you write the ratio 12 : 15 in simplest form?

5. How do you write the rate $\frac{342 \text{ miles}}{9.5 \text{ gallons}}$ as a unit rate?

6. When is a proportion true?

7. How do you solve a proportion?

8. How do the units help you to set up a proportion?

9. How do you check the solution of a proportion?

10. How do you write the ratio 19 : 6 as a fraction?

CHAPTER 4

REVIEW EXERCISES

1. Determine whether the proportion is true or not true.
$$\frac{2}{9} = \frac{10}{45}$$

2. Write the comparison 32 dollars to 80 dollars as a ratio in simplest form using a fraction, a colon (:), and the word *to*.

3. Write "250 miles in 4 hours" as a unit rate.

4. Determine whether the proportion is true or not true.
$$\frac{8}{15} = \frac{32}{60}$$

5. Solve the proportion.
$$\frac{16}{n} = \frac{4}{17}$$

6. Write "$500 earned in 40 hours" as a unit rate.

7. Write "$8.75 for 5 pounds" as a unit rate.

8. Write the comparison 8 feet to 28 feet as a ratio in simplest form using a fraction, a colon (:), and the word *to*.

9. Solve the proportion.
$$\frac{n}{8} = \frac{9}{2}$$

10. Solve the proportion. Round to the nearest hundredth.
$$\frac{18}{35} = \frac{10}{n}$$

11. Write the comparison 6 inches to 15 inches as a ratio in simplest form using a fraction, a colon (:), and the word *to*.

12. Determine whether the proportion is true or not true.
$$\frac{3}{8} = \frac{10}{24}$$

13. Write "$35 in 4 hours" as a rate in simplest form.

14. Write "326.4 miles on 12 gallons" as a unit rate.

15. Write the comparison 12 days to 12 days as a ratio in simplest form using a fraction, a colon (:), and the word *to*.

16. Determine whether the proportion is true or not true.
$$\frac{5}{7} = \frac{25}{35}$$

17. Solve the proportion. Round to the nearest hundredth.

$$\frac{24}{11} = \frac{n}{30}$$

18. Write "100 miles in 3 hours" as a rate in simplest form.

19. **Business** In 5 years, the price of a calculator went from $80 to $48. What is the ratio, as a fraction in simplest form, of the decrease in price to the original price?

20. **Taxes** The property tax on a $245,000 home is $4900. At the same rate, what is the property tax on a home valued at $320,000?

21. **Consumerism** Rita Sterling bought a computer system for $2400. Five years later, she sold the computer for $900. Find the ratio of the amount she received for the computer to the cost of the computer.

22. **Manufacturing** The total cost of manufacturing 1000 camera phones was $36,600. Of the phones made, 24 did not pass inspection. What is the cost per phone of the phones that *did* pass inspection?

23. **Masonry** A brick wall 40 feet in length contains 448 concrete blocks. At the same rate, how many blocks would it take to build a wall that is 120 feet in length?

24. **Advertising** A retail computer store spends $30,000 a year on radio advertising and $12,000 on newspaper advertising. Find the ratio, as a fraction in simplest form, of radio advertising to newspaper advertising.

25. **Consumerism** A 15-pound turkey costs $13.95. What is the cost per pound?

26. **Travel** Mahesh drove 198.8 miles in 3.5 hours. Find the average number of miles he drove per hour.

27. **Insurance** An insurance policy costs $9.87 for every $1000 of insurance. At this rate, what is the cost of $50,000 of insurance?

28. **Investments** Pascal Hollis purchased 80 shares of stock for $3580. What was the cost per share?

29. **Landscaping** Monique uses 1.5 pounds of fertilizer for every 200 square feet of lawn. How many pounds of fertilizer will she have to use on a lawn that measures 3000 square feet?

30. **Real Estate** A house had an original value of $160,000, but its value increased to $240,000 in 2 years. Find the ratio, as a fraction in simplest form, of the increase to the original value.

CHAPTER 4

TEST

1. Write "$46,036.80 earned in 12 months" as a unit rate.

2. Write the comparison 40 miles to 240 miles as a ratio in simplest form using a fraction, a colon (:), and the word *to*.

3. Write "18 supports for every 8 feet" as a rate in simplest form.

4. Determine whether the proportion is true or not true.
$$\frac{40}{125} = \frac{5}{25}$$

5. Write the comparison 12 days to 8 days as a ratio in simplest form using a fraction, a colon (:), and the word *to*.

6. Solve the proportion.
$$\frac{5}{12} = \frac{60}{n}$$

7. Write "256.2 miles on 8.4 gallons of gas" as a unit rate.

8. Write the comparison 27 dollars to 81 dollars as a ratio in simplest form using a fraction, a colon (:), and the word *to*.

9. Determine whether the proportion is true or not true.
$$\frac{5}{14} = \frac{25}{70}$$

10. Solve the proportion.
$$\frac{n}{18} = \frac{9}{4}$$

11. Write "$81 for 6 boards" as a rate in simplest form.

12. Write the comparison 18 feet to 30 feet as a ratio in simplest form using a fraction, a colon (:), and the word *to*.

13. **Investments** Fifty shares of a utility stock pay a dividend of $62.50. At the same rate, what is the dividend paid on 500 shares of the utility stock?

Primary coil   Secondary coil

14. **Electricity** A transformer has 40 turns in the primary coil and 480 turns in the secondary coil. State the ratio of the number of turns in the primary coil to the number of turns in the secondary coil.

15. **Travel** A plane travels 2421 miles in 4.5 hours. Find the plane's speed in miles per hour.

16. **Physiology** A research scientist estimates that the human body contains 88 pounds of water for every 100 pounds of body weight. At this rate, estimate the number of pounds of water in a college student who weighs 150 pounds.

17. **Business** If 40 feet of lumber costs $69.20, what is the per-foot cost of the lumber?

18. **Medicine** The dosage of a certain medication is $\frac{1}{4}$ ounce for every 50 pounds of body weight. How many ounces of this medication are required for a person who weighs 175 pounds? Write the answer as a decimal.

19. **Sports** A basketball team won 20 games and lost 5 games during the season. Write, as a fraction in simplest form, the ratio of the number of games won to the total number of games played.

20. **Manufacturing** A computer manufacturer discovers through experience that an average of 3 defective hard drives are found in every 100 hard drives manufactured. How many defective hard drives are expected to be found in the production of 1200 hard drives?

CUMULATIVE REVIEW EXERCISES

1. Subtract: $\begin{array}{r} 20{,}095 \\ -\ 10{,}937 \\ \hline \end{array}$

2. Write $2 \cdot 2 \cdot 2 \cdot 2 \cdot 3 \cdot 3 \cdot 3$ in exponential notation.

3. Simplify: $4 - (5 - 2)^2 \div 3 + 2$

4. Find the prime factorization of 160.

5. Find the LCM of 9, 12, and 18.

6. Find the GCF of 28 and 42.

7. Write $\frac{40}{64}$ in simplest form.

8. Find $4\frac{7}{15}$ more than $3\frac{5}{6}$.

9. What is $4\frac{5}{9}$ less than $10\frac{1}{6}$?

10. Multiply: $\frac{11}{12} \times 3\frac{1}{11}$

11. Find the quotient of $3\frac{1}{3}$ and $\frac{5}{7}$.

12. Simplify: $\left(\frac{2}{5} + \frac{3}{4} \right) \div \frac{3}{2}$

13. Write 4.0709 in words.

14. Round 2.09762 to the nearest hundredth.

15. Divide: $8.09\overline{)16.0976}$
    Round to the nearest thousandth.

16. Convert $0.06\frac{2}{3}$ to a fraction.

17. Write the comparison 25 miles to 200 miles as a ratio in simplest form using a fraction.

18. Write "87 cents for 6 pencils" as a rate in simplest form.

19. Write "250.5 miles on 7.5 gallons of gas" as a unit rate.

20. Solve $\dfrac{40}{n} = \dfrac{160}{17}$.

21. **Travel** A car traveled 457.6 miles in 8 hours. Find the car's speed in miles per hour.

22. Solve the proportion.

$$\frac{12}{5} = \frac{n}{15}$$

23. **Banking** You had $1024 in your checking account. You then wrote checks for $192 and $88. What is your new checking account balance?

24. **Finance** Malek Khatri buys a tractor for $32,360. A down payment of $5000 is required. The balance remaining is paid in 48 equal monthly installments. What is the monthly payment?

25. **Homework Assignments** Yuko is assigned to read a book containing 175 pages. She reads $\frac{2}{5}$ of the book during Thanksgiving vacation. How many pages of the assignment remain to be read?

26. **Real Estate** A building contractor bought $2\frac{1}{3}$ acres of land for $84,000. What was the cost of each acre?

27. **Consumerism** Benjamin Eli bought a shirt for $45.58 and a tie for $19.18. He used a $100 bill to pay for the purchases. Find the amount of change.

28. **Compensation** If you earn an annual salary of $41,619, what is your monthly salary?

29. **Erosion** A soil conservationist estimates that a river bank is eroding at the rate of 3 inches every 6 months. At this rate, how many inches will be eroded in 50 months?

30. **Medicine** The dosage of a certain medication is $\frac{1}{2}$ ounce for every 50 pounds of body weight. How many ounces of this medication are required for a person who weighs 160 pounds? Write the answer as a decimal.

Percents

OBJECTIVES

SECTION 5.1
A To write a percent as a fraction or a decimal
B To write a fraction or a decimal as a percent

SECTION 5.2
A To find the amount when the percent and the base are given
B To solve application problems

SECTION 5.3
A To find the percent when the base and amount are given
B To solve application problems

SECTION 5.4
A To find the base when the percent and amount are given
B To solve application problems

SECTION 5.5
A To solve percent problems using proportions
B To solve application problems

ARE YOU READY?

Take the Chapter 5 Prep Test to find out if you are ready to learn to:

- Convert fractions, decimals, and percents
- Solve percent problems using the basic percent equation
- Solve percent problems using proportions

PREP TEST

Do these exercises to prepare for Chapter 5.

For Exercises 1 to 6, multiply or divide.

1. $19 \times \dfrac{1}{100}$

2. 23×0.01

3. 0.47×100

4. $0.06 \times 47{,}500$

5. $60 \div 0.015$

6. $8 \div \dfrac{1}{4}$

7. Multiply $\dfrac{5}{8} \times 100$. Write the answer as a decimal.

8. Write $\dfrac{200}{3}$ as a mixed number.

9. Divide $28 \div 16$. Write the answer as a decimal.

SECTION

5.1 Introduction to Percents

OBJECTIVE A   **To write a percent as a fraction or a decimal**

Percent means "parts of 100." In the figure at the right, there are 100 parts. Because 13 of the 100 parts are shaded, 13% of the figure is shaded. The symbol % is the **percent sign.**

In most applied problems involving percents, it is necessary either to rewrite a percent as a fraction or a decimal or to rewrite a fraction or a decimal as a percent.

To write a percent as a fraction, remove the percent sign and multiply by $\frac{1}{100}$.

$$13\% = 13 \times \frac{1}{100} = \frac{13}{100}$$

To write a percent as a decimal, remove the percent sign and multiply by 0.01.

$$13\% \quad = \quad 13 \times 0.01 \quad = \quad 0.13$$

Move the decimal point two places to the left. Then remove the percent sign.

> ✓ **Take Note**
> Recall that division is defined as multiplication by the reciprocal. Therefore, multiplying by $\frac{1}{100}$ is equivalent to dividing by 100.

EXAMPLE • 1

a. Write 120% as a fraction.
b. Write 120% as a decimal.

Solution   **a.** $120\% = 120 \times \dfrac{1}{100} = \dfrac{120}{100}$

$$= 1\frac{1}{5}$$

b. $120\% = 120 \times 0.01 = 1.2$

Note that percents larger than 100 are greater than 1.

YOU TRY IT • 1

a. Write 125% as a fraction.
b. Write 125% as a decimal.

Your solution

EXAMPLE • 2

Write $16\frac{2}{3}\%$ as a fraction.

Solution $16\dfrac{2}{3}\% = 16\dfrac{2}{3} \times \dfrac{1}{100}$

$$= \frac{50}{3} \times \frac{1}{100} = \frac{50}{300} = \frac{1}{6}$$

YOU TRY IT • 2

Write $33\frac{1}{3}\%$ as a fraction.

Your solution

EXAMPLE • 3

Write 0.5% as a decimal.

Solution $0.5\% = 0.5 \times 0.01 = 0.005$

YOU TRY IT • 3

Write 0.25% as a decimal.

Your solution

Solutions on pp. S11–S12

OBJECTIVE B   **To write a fraction or a decimal as a percent**

A fraction or a decimal can be written as a percent by multiplying by 100%.

> **HOW TO 1**   Write $\frac{3}{8}$ as a percent.
>
> $$\frac{3}{8} = \frac{3}{8} \times 100\% = \frac{3}{8} \times \frac{100}{1}\% = \frac{300}{8}\% = 37\frac{1}{2}\% \text{ or } 37.5\%$$

> **HOW TO 2**   Write 0.37 as a percent.
>
> $$0.37 \quad = \quad 0.37 \times 100\% \quad = \quad 37\%$$
>
> Move the decimal point two places to the right. Then write the percent sign.

EXAMPLE 4

Write 0.015, 2.15, and $0.33\frac{1}{3}$ as percents.

Solution

$$0.015 = 0.015 \times 100\%$$
$$= 1.5\%$$

$$2.15 = 2.15 \times 100\% = 215\%$$

$$0.33\frac{1}{3} = 0.33\frac{1}{3} \times 100\%$$

$$= 33\frac{1}{3}\%$$

YOU TRY IT 4

Write 0.048, 3.67, and $0.62\frac{1}{2}$ as percents.

Your solution

EXAMPLE 5

Write $\frac{2}{3}$ as a percent.
Write the remainder in fractional form.

Solution $\frac{2}{3} = \frac{2}{3} \times 100\% = \frac{200}{3}\%$

$$= 66\frac{2}{3}\%$$

YOU TRY IT 5

Write $\frac{5}{6}$ as a percent.
Write the remainder in fractional form.

Your solution

EXAMPLE 6

Write $2\frac{2}{7}$ as a percent.
Round to the nearest tenth.

Solution $2\frac{2}{7} = \frac{16}{7} = \frac{16}{7} \times 100\%$

$$= \frac{1600}{7}\% \approx 228.6\%$$

YOU TRY IT 6

Write $1\frac{4}{9}$ as a percent.
Round to the nearest tenth.

Your solution

Solutions on p. S12

5.1 EXERCISES

OBJECTIVE A   To write a percent as a fraction or a decimal

For Exercises 1 to 16, write as a fraction and as a decimal.

1. 25%   **2.** 40%   **3.** 130%   **4.** 150%

5. 100%   **6.** 87%   **7.** 73%   **8.** 45%

9. 383%   **10.** 425%   **11.** 70%   **12.** 55%

13. 88%   **14.** 64%   **15.** 32%   **16.** 18%

For Exercises 17 to 28, write as a fraction.

17. $66\frac{2}{3}\%$ **18.** $12\frac{1}{2}\%$ **19.** $83\frac{1}{3}\%$ **20.** $3\frac{1}{8}\%$ **21.** $11\frac{1}{9}\%$ **22.** $\frac{3}{8}\%$

23. $45\frac{5}{11}\%$ **24.** $15\frac{3}{8}\%$ **25.** $4\frac{2}{7}\%$ **26.** $5\frac{3}{4}\%$ **27.** $6\frac{2}{3}\%$ **28.** $8\frac{2}{3}\%$

For Exercises 29 to 40, write as a decimal.

29. 6.5%   **30.** 9.4%   **31.** 12.3%   **32.** 16.7%   **33.** 0.55%   **34.** 0.45%

35. 8.25%   **36.** 6.75%   **37.** 5.05%   **38.** 3.08%   **39.** 2%   **40.** 7%

 41. When a certain percent is written as a fraction, the result is an improper fraction. Is the percent less than, equal to, or greater than 100%?

OBJECTIVE B   To write a fraction or a decimal as a percent

For Exercises 42 to 53, write as a percent.

42. 0.16   **43.** 0.73   **44.** 0.05   **45.** 0.01   **46.** 1.07   **47.** 2.94

48. 0.004   **49.** 0.006   **50.** 1.012   **51.** 3.106   **52.** 0.8   **53.** 0.7

For Exercises 54 to 65, write as a percent. If necessary, round to the nearest tenth of a percent.

54. $\dfrac{27}{50}$ **55.** $\dfrac{37}{100}$ **56.** $\dfrac{1}{3}$ **57.** $\dfrac{2}{5}$ **58.** $\dfrac{5}{8}$ **59.** $\dfrac{1}{8}$

60. $\dfrac{1}{6}$ **61.** $1\dfrac{1}{2}$ **62.** $\dfrac{7}{40}$ **63.** $1\dfrac{2}{3}$ **64.** $1\dfrac{7}{9}$ **65.** $\dfrac{7}{8}$

For Exercises 66 to 73, write as a percent. Write the remainder in fractional form.

66. $\dfrac{15}{50}$ **67.** $\dfrac{12}{25}$ **68.** $\dfrac{7}{30}$ **69.** $\dfrac{1}{3}$

70. $2\dfrac{3}{8}$ **71.** $1\dfrac{2}{3}$ **72.** $2\dfrac{1}{6}$ **73.** $\dfrac{7}{8}$

74. Does a mixed number represent a percent greater than 100% or less than 100%?

75. A decimal number less than 0 has zeros in the tenths and hundredths places. Does the decimal represent a percent greater than 1% or less than 1%?

76. Write the part of the square that is shaded as a fraction, as a decimal, and as a percent. Write the part of the square that is not shaded as a fraction, as a decimal, and as a percent.

Applying the Concepts

77. **The Food Industry** In a survey conducted by Opinion Research Corp. for Lloyd's Barbeque Co., people were asked to name their favorite barbeque side dishes. 38% named corn on the cob, 35% named cole slaw, 11% named corn bread, and 10% named fries. What percent of those surveyed named something other than corn on the cob, cole slaw, corn bread, or fries?

David Chasey/Photodisc/Getty Images

78. **Consumerism** A sale on computers advertised $\dfrac{1}{3}$ off the regular price. What percent of the regular price does this represent?

79. **Consumerism** A suit was priced at 50% off the regular price. What fraction of the regular price does this represent?

80. **Elections** If $\dfrac{2}{5}$ of the population voted in an election, what percent of the population did not vote?

SECTION

5.2 Percent Equations: Part 1

OBJECTIVE A   **To find the amount when the percent and the base are given**

A real estate broker receives a payment that is 4% of a $285,000 sale. To find the amount the broker receives requires answering the question "4% of $285,000 is what?"

This sentence can be written using mathematical symbols and then solved for the unknown number.

4%	of	$285,000	is	what?
↓	↓	↓	↓	↓

of   is written as × (times)
is   is written as = (equals)
what is written as *n* (the unknown number)

Percent 4%	×	base 285,000	=	amount *n*

$$0.04 \times 285,000 = n$$
$$11,400 = n$$

Note that the percent is written as a decimal.

The broker receives a payment of $11,400.

The solution was found by solving the **basic percent equation** for amount.

The Basic Percent Equation

| Percent | × | base | = | amount |

In most cases, the percent is written as a decimal before the basic percent equation is solved. However, some percents are more easily written as a fraction than as a decimal. For example,

$$33\frac{1}{3}\% = \frac{1}{3} \qquad 66\frac{2}{3}\% = \frac{2}{3} \qquad 16\frac{2}{3}\% = \frac{1}{6} \qquad 83\frac{1}{3}\% = \frac{5}{6}$$

EXAMPLE · 1

Find 5.7% of 160.

Solution

Percent × base = amount
$$0.057 \times 160 = n$$
$$9.12 = n$$

• The word *Find* is used instead of the words *what is.*

YOU TRY IT · 1

Find 6.3% of 150.

Your solution

EXAMPLE · 2

What is $33\frac{1}{3}\%$ of 90?

Solution

Percent × base = amount
$$\frac{1}{3} \times 90 = n$$
$$30 = n$$

• $33\frac{1}{3}\% = \frac{1}{3}$

YOU TRY IT · 2

What is $16\frac{2}{3}\%$ of 66?

Your solution

Solutions on p. S12

OBJECTIVE B   To solve application problems

Solving percent problems requires identifying the three elements of the basic percent equation. Recall that these three parts are the *percent*, the *base*, and the *amount*. Usually the base follows the phrase "percent of."

 During a recent year, Americans gave $212 billion to charities. The circle graph at the right shows where that money came from. Use these data for Example 3 and You Try It 3.

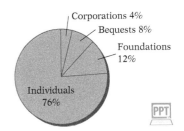

Corporations 4%
Bequests 8%
Foundations 12%
Individuals 76%

Charitable Giving
Sources: American Association of Fundraising Counsel; AP

EXAMPLE · 3	YOU TRY IT · 3

EXAMPLE · 3

How much of the amount given to charities came from individuals?

Strategy
To determine the amount that came from individuals, write and solve the basic percent equation using n to represent the amount. The percent is 76%. The base is $212 billion.

Solution
Percent × base = amount
$$76\% \times 212 = n$$
$$0.76 \times 212 = n$$
$$161.12 = n$$

Individuals gave $161.12 billion to charities.

YOU TRY IT · 3

How much of the amount given to charities was given by corporations?

Your strategy

Your solution

EXAMPLE · 4

A quality control inspector found that 1.2% of 2500 camera phones inspected were defective. How many camera phones inspected were not defective?

Strategy
To find the number of nondefective phones:
• Find the number of defective phones. Write and solve the basic percent equation using n to represent the number of defective phones (amount). The percent is 1.2% and the base is 2500.
• Subtract the number of defective phones from the number of phones inspected (2500).

Solution
$$1.2\% \times 2500 = n$$
$$0.012 \times 2500 = n$$
$$30 = n \text{ defective phones}$$

$$2500 - 30 = 2470$$

2470 camera phones were not defective.

YOU TRY IT · 4

An electrician's hourly wage was $33.50 before an 8% raise. What is the new hourly wage?

Your strategy

Your solution

Solutions on p. S12

5.2 EXERCISES

OBJECTIVE A   To find the amount when the percent and the base are given

1. 8% of 100 is what?

2. 16% of 50 is what?

3. 27% of 40 is what?

4. 52% of 95 is what?

5. 0.05% of 150 is what?

6. 0.075% of 625 is what?

7. 125% of 64 is what?

8. 210% of 12 is what?

9. Find 10.7% of 485.

10. Find 12.8% of 625.

11. What is 0.25% of 3000?

12. What is 0.06% of 250?

13. 80% of 16.25 is what?

14. 26% of 19.5 is what?

15. What is $1\frac{1}{2}$% of 250?

16. What is $5\frac{3}{4}$% of 65?

17. $16\frac{2}{3}$% of 120 is what?

18. What is $66\frac{2}{3}$% of 891?

19. Which is larger: 5% of 95, or 75% of 6?

20. Which is larger: 112% of 5, or 0.45% of 800?

21. Which is smaller: 79% of 16, or 20% of 65?

22. Which is smaller: 15% of 80, or 95% of 15?

 23. Is 15% of a number greater than or less than the number?

 24. Is 150% of a number greater than or less than the number?

OBJECTIVE B   To solve application problems

 25. Read Exercise 26. Without doing any calculations, determine whether the number of people in the United States aged 18 to 24 who do not have health insurance is *less than, equal to,* or *greater than* 44 million.

 26. **Health Insurance** Approximately 30% of the 44 million people in the United States who do not have health insurance are between the ages of 18 and 24. (*Source:* U.S. Census Bureau) About how many people in the United States aged 18 to 24 do not have health insurance?

 27. **Aviation** The Federal Aviation Administration reported that 55,422 new student pilots were flying single-engine planes last year. The number of new student pilots flying single-engine planes this year is 106% of the number flying single-engine planes last year. How many new student pilots are flying single-engine planes this year?

28. **Jewelry** An 18-carat yellow-gold necklace contains 75% gold, 16% silver, and 9% copper. If the necklace weighs 25 grams, how many grams of copper are in the necklace?

29. **Jewelry** Fourteen-carat yellow gold contains 58.5% gold, 17.5% silver, and 24% copper. If a jeweler has a 50-gram piece of 14-carat yellow gold, how many grams of gold, silver, and copper are in the piece?

30. **Lifestyles** There are 114 million households in the United States. Opposite-sex cohabitating couples comprise 4.4% of these households. (*Source:* Families and Living Arrangements) Find the number of opposite-sex cohabitating couples who maintain households in the United States. Round to the nearest million.

> **In the News**
>
> **More Taxpayers Filing Electronically**
>
> The IRS reported that, as of May 4, it has received 128 million returns. Sixty percent of the returns were filed electronically.
>
> *Source:* IRS

31. **e-Filed Tax Returns** See the news clipping at the right. How many of the 128 million returns were filed electronically? Round to the nearest million.

32. **Taxes** A sales tax of 6% of the cost of a car is added to the purchase price of $29,500. What is the total cost of the car, including sales tax?

33. **Email** The number of email messages sent each day has risen to 171 billion, of which 71% are spam. (*Source:* FeedsFarm.com) How many email messages sent per day are not spam?

34. **Prison Population** The prison population in the United States is 1,596,127 people. Male prisoners comprise 91% of this population. (*Source: Time,* March 17, 2008) How many inmates are male? How many are female?

35. **Entertainment** A USA TODAY.com online poll asked 8878 Internet users, "Would you use software to cut out objectionable parts of movies?" 29.8% of the respondents answered yes. How many respondents did not answer yes to the question? Round to the nearest whole number.

Applying the Concepts

36. **Jewelry** Eighteen-carat white gold contains 75% gold, 15% silver, and 10% platinum. A jeweler wants to make a 2-ounce, 18-carat, white gold ring. If gold costs $900 per ounce, silver costs $17.20 per ounce, and platinum costs $1900 per ounce, what is the cost of the metal used to make the ring?

SECTION

5.3   Percent Equations: Part II

OBJECTIVE A   **To find the percent when the base and amount are given**

A recent promotional game at a grocery store listed the probability of winning a prize as "1 chance in 2." A percent can be used to describe the chance of winning. This requires answering the question "What percent of 2 is 1?"

The chance of winning can be found by solving the basic percent equation for *percent*.

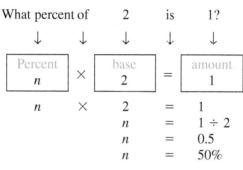

Integrating Technology

The percent key % on a scientific calculator moves the decimal point to the right two places when pressed after a multiplication or division computation. For the example at the right, enter

1 ÷ 2 % =

The display reads 50.

$$n \times 2 = 1$$
$$n = 1 \div 2$$
$$n = 0.5$$
$$n = 50\%$$

There is a 50% chance of winning a prize.

• The solution must be written as a percent in order to answer the question.

EXAMPLE • 1

What percent of 40 is 30?

Solution   Percent × base = amount
$$n \times 40 = 30$$
$$n = 30 \div 40$$
$$n = 0.75$$
$$n = 75\%$$

YOU TRY IT • 1

What percent of 32 is 16?

Your solution

EXAMPLE • 2

What percent of 12 is 27?

Solution   Percent × base = amount
$$n \times 12 = 27$$
$$n = 27 \div 12$$
$$n = 2.25$$
$$n = 225\%$$

YOU TRY IT • 2

What percent of 15 is 48?

Your solution

EXAMPLE • 3

25 is what percent of 75?

Solution   Percent × base = amount
$$n \times 75 = 25$$
$$n = 25 \div 75$$
$$n = \frac{1}{3} = 33\frac{1}{3}\%$$

YOU TRY IT • 3

30 is what percent of 45?

Your solution

Solutions on p. S12

OBJECTIVE B To solve application problems

To solve percent problems, remember that it is necessary to identify the percent, base, and amount. Usually the base follows the phrase "percent of."

EXAMPLE · 4	YOU TRY IT · 4

The monthly house payment for the Kaminski family is $787.50. What percent of the Kaminskis' monthly income of $3750 is the house payment?

Tomo Nagata had an income of $33,500 and paid $5025 in income tax. What percent of the income is the income tax?

Strategy

To find what percent of the income the house payment is, write and solve the basic percent equation using n to represent the percent. The base is $3750 and the amount is $787.50.

Your strategy

Solution

$n \times 3750 = 787.50$
$n = 787.50 \div 3750$
$n = 0.21 = 21\%$

The house payment is 21% of the monthly income.

Your solution

EXAMPLE · 5	YOU TRY IT · 5

On one Monday night, 31.39 million of the approximately 40.76 million households watching television were not watching David Letterman. What percent of these households were watching David Letterman? Round to the nearest percent.

 According to the U.S. Department of Defense, of the 518,921 enlisted personnel in the U.S. Army in 1950, 512,370 people were men. What percent of the enlisted personnel in the U.S. Army in 1950 were women? Round to the nearest tenth of a percent.

Strategy

To find the percent of households watching David Letterman:

* Subtract to find the number of households that were watching David Letterman (40.76 million − 31.39 million).
* Write and solve the basic percent equation using n to represent the percent. The base is 40.76, and the amount is the number of households watching David Letterman.

Your strategy

Solution

40.76 million − 31.39 million = 9.37 million

9.37 million households were watching David Letterman.

$n \times 40.76 = 9.37$
$n = 9.37 \div 40.76$
$n \approx 0.23$

Approximately 23% of the households were watching David Letterman.

Your solution

Solutions on p. S12

5.3 EXERCISES

OBJECTIVE A   To find the percent when the base and amount are given

1. What percent of 75 is 24?

2. What percent of 80 is 20?

3. 15 is what percent of 90?

4. 24 is what percent of 60?

5. What percent of 12 is 24?

6. What percent of 6 is 9?

7. What percent of 16 is 6?

8. What percent of 24 is 18?

9. 18 is what percent of 100?

10. 54 is what percent of 100?

11. 5 is what percent of 2000?

12. 8 is what percent of 2500?

13. What percent of 6 is 1.2?

14. What percent of 2.4 is 0.6?

15. 16.4 is what percent of 4.1?

16. 5.3 is what percent of 50?

17. 1 is what percent of 40?

18. 0.3 is what percent of 20?

19. What percent of 48 is 18?

20. What percent of 11 is 88?

21. What percent of 2800 is 7?

22. What percent of 400 is 12?

23. True or false? If the *base* is larger than the *amount* in the basic percent equation, then the *percent* is larger than 100.

OBJECTIVE B   To solve application problems

24. Read Exercise 26. Without doing any calculations, determine whether the percent of those surveyed who were irked by tailgaters is *less than* or *greater than* 25%.

25. **Sociology** Seven in ten couples disagree about financial issues. (*Source: Yankelovich Partners for Lutheran Brotherhood*) What percent of couples disagree about financial matters?

26. **Sociology** In a survey, 1236 adults nationwide were asked, "What irks you most about the actions of other motorists?" The response "tailgaters" was given by 293 people. (*Source: Reuters/Zogby*) What percent of those surveyed were most irked by tailgaters? Round to the nearest tenth of a percent.

27. Agriculture According to the U.S. Department of Agriculture, of the 63 billion pounds of vegetables produced in the United States in 1 year, 16 billion pounds were wasted. What percent of the vegetables produced were wasted? Round to the nearest tenth of a percent.

 28. Wind Energy In a recent year, wind machines in the United States generated 17.8 billion kilowatt-hours of electricity, enough to serve over 1.6 million households. The nation's total electricity production that year was 4,450 billion kilowatt-hours. (*Source:* Energy Information Administration) What percent of the total energy production was generated by wind machines?

 29. Diabetes Approximately 7% of the American population has diabetes. Within this group, 14.6 million are diagnosed, while 6.2 million are undiagnosed. (*Source:* The National Diabetes Education Program) What percent of Americans with diabetes have not been diagnosed with the disease? Round to the nearest tenth of a percent.

 30. Internal Revenue Service See the news clipping at the right. Given that the number of millionaires in the United States is 9.3 million, what percent of U.S. millionaires were audited by the IRS? Round to the nearest hundredth of a percent.

> **In the News**
>
> **More Millionaires Audited**
>
> The Internal Revenue Service reported that 17,015 millionaires were audited this year. This figure is 33% more than last year.
>
> *Source:* The Internal Revenue Service; TSN Financial Services

31. Construction In a test of the breaking strength of concrete slabs for freeway construction, 3 of the 200 slabs tested did not meet safety requirements. What percent of the slabs did meet safety requirements?

Applying the Concepts

 Pets The graph at the right shows several categories of average lifetime costs of dog ownership. Use this graph for Exercises 32 to 34. Round answers to the nearest tenth of a percent.

32. What percent of the total amount is spent on food?

33. What percent of the total is spent on veterinary care?

34. What percent of the total is spent on all categories except training?

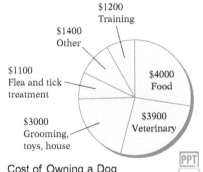

Cost of Owning a Dog
Source: Based on data from the American Kennel Club, *USA Today* research

 35. Sports The Fun in the Sun organization claims to have taken a survey of 350 people, asking them to give their favorite outdoor temperature for hiking. The results are given in the table at the right. Explain why these results are not possible.

Favorite Temperature	Percent
Greater than 90	5%
80–89	28%
70–79	35%
60–69	32%
Below 60	13%

SECTION

5.4 Percent Equations: Part III

OBJECTIVE A To find the base when the percent and amount are given

In 1780, the population of Virginia was 538,000. This was 19% of the total population of the United States at that time. To find the total population at that time, you must answer the question "19% of what number is 538,000?"

Tips for Success

After completing this objective, you will have learned to solve the basic percent equation for each of the three elements: percent, base, and amount. You will need to be able to recognize these three different types of problems. To test yourself, try the Chapter 5 Review Exercises.

19%	of	what	is	538,000?
↓	↓	↓	↓	↓

Percent		base		amount
19%		n		538,000

- The population of the United States in 1780 can be found by solving the basic percent equation for the base.

$$0.19 \times n = 538{,}000$$
$$n = 538{,}000 \div 0.19$$
$$n \approx 2{,}832{,}000$$

The population of the United States in 1780 was approximately 2,832,000.

EXAMPLE • 1

18% of what is 900?

Solution
$$\text{Percent} \times \text{base} = \text{amount}$$
$$0.18 \times n = 900$$
$$n = 900 \div 0.18$$
$$n = 5000$$

YOU TRY IT • 1

86% of what is 215?

Your solution

EXAMPLE • 2

30 is 1.5% of what?

Solution
$$\text{Percent} \times \text{base} = \text{amount}$$
$$0.015 \times n = 30$$
$$n = 30 \div 0.015$$
$$n = 2000$$

YOU TRY IT • 2

15 is 2.5% of what?

Your solution

EXAMPLE • 3

$33\frac{1}{3}\%$ of what is 7?

Solution
$$\text{Percent} \times \text{base} = \text{amount}$$
$$\frac{1}{3} \times n = 7$$
$$n = 7 \div \frac{1}{3}$$
$$n = 21$$

- Note that the percent is written as a fraction.

YOU TRY IT • 3

$16\frac{2}{3}\%$ of what is 5?

Your solution

Solutions on p. S13

OBJECTIVE B To solve application problems

To solve percent problems, it is necessary to identify the percent, the base, and the amount. Usually the base follows the phrase "percent of."

EXAMPLE • 4

A business office bought a used copy machine for $900, which was 75% of the original cost. What was the original cost of the copier?

Strategy

To find the original cost of the copier, write and solve the basic percent equation using n to represent the original cost (base). The percent is 75% and the amount is $900.

Solution

$75\% \times n = 900$
$0.75 \times n = 900$
$\qquad n = 900 \div 0.75$
$\qquad n = 1200$

The original cost of the copier was $1200.

YOU TRY IT • 4

A used car has a value of $10,458, which is 42% of the car's original value. What was the car's original value?

Your strategy

Your solution

EXAMPLE • 5

A carpenter's wage this year is $26.40 per hour, which is 110% of last year's wage. What was the increase in the hourly wage over last year?

Strategy

To find the increase in the hourly wage over last year:

• Find last year's wage. Write and solve the basic percent equation using n to represent last year's wage (base). The percent is 110% and the amount is $26.40.
• Subtract last year's wage from this year's wage (26.40).

Solution

$110\% \times n = 26.40$
$1.10 \times n = 26.40$
$\qquad n = 26.40 \div 1.10$
$\qquad n = 24.00$ • Last year's wage

$26.40 - 24.00 = 2.40$

The increase in the hourly wage was $2.40.

YOU TRY IT • 5

Chang's Sporting Goods has a tennis racket on sale for $89.60, which is 80% of the original price. What is the difference between the original price and the sale price?

Your strategy

Your solution

Solutions on p. S13

5.4 EXERCISES

OBJECTIVE A   To find the base when the percent and amount are given

1. 12% of what is 9?

2. 38% of what is 171?

3. 8 is 16% of what?

4. 54 is 90% of what?

5. 10 is 10% of what?

6. 37 is 37% of what?

7. 30% of what is 25.5?

8. 25% of what is 21.5?

9. 2.5% of what is 30?

10. 10.4% of what is 52?

11. 125% of what is 24?

12. 180% of what is 21.6?

13. 18 is 240% of what?

14. 24 is 320% of what?

15. 4.8 is 15% of what?

16. 87.5 is 50% of what?

17. 25.6 is 12.8% of what?

18. 45.014 is 63.4% of what?

19. 30% of what is 2.7?

20. 78% of what is 3.9?

21. 84 is $16\frac{2}{3}$% of what?

22. 120 is $33\frac{1}{3}$% of what?

 23. Consider the question "*P*% of what number is 50?" If the percent *P* is greater than 100%, is the unknown number greater than 50 or less than 50?

OBJECTIVE B   To solve application problems

 24. Read Exercise 25. Without doing any calculations, determine whether the number of travelers who allowed their children to miss school to go on a trip is *less than, equal to,* or *greater than* 1.738 million.

 25. **Travel** Of the travelers who, during a recent year, allowed their children to miss school to go along on a trip, approximately 1.738 million allowed their children to miss school for more than a week. This represented 11% of the travelers who allowed their children to miss school. (*Source:* Travel Industry Association) About how many travelers allowed their children to miss school to go along on a trip?

26. e-Commerce Using the information in the news clipping at the right, calculate the total retail sales during the fourth quarter of last year. Round to the nearest billion.

27. Marathons In 2008, 98.2% of the runners who started the Boston Marathon, or 21,963 people, crossed the finish line. (*Source:* www.bostonmarathon.org) How many runners started the Boston Marathon in 2008?

28. Education In the United States today, 23.1% of women and 27.5% of men have earned a bachelor's or graduate degree. (*Source:* Census Bureau) How many women in the United States have earned a bachelor's or graduate degree?

29. Wind-Powered Ships Using the information in the news clipping at the right, calculate the cargo ships' daily fuel bill.

Courtesy SkySails

30. Taxes A TurboTax online survey asked people how they planned to use their tax refunds. Seven hundred forty people, or 22% of the respondents, said they would save the money. How many people responded to the survey?

31. Manufacturing During a quality control test, Micronics found that 24 computer boards were defective. This amount was 0.8% of the computer boards tested.
a. How many computer boards were tested?
b. How many of the computer boards tested were not defective?

32. Directory Assistance Of the calls a directory assistance operator received, 441 were requests for telephone numbers listed in the current directory. This accounted for 98% of the calls for assistance that the operator received.
a. How many calls did the operator receive?
b. How many telephone numbers requested were not listed in the current directory?

Applying the Concepts

33. Nutrition The table at the right contains nutrition information about a breakfast cereal. The amount of thiamin in one serving of this cereal with skim milk is 0.45 milligram. Find the recommended daily allowance of thiamin for an adult.

34. Increase a number by 10%. Now decrease the number by 10%. Is the result the original number? Explain.

In the News

eCommerce on the Rise

Retail e-commerce sales for the fourth quarter of last year exceeded e-commerce sales for the first three quarters of the year. E-commerce sales during October, November, and December totaled $35.3 billion, or 3.4% of total retail sales during the quarter.

Source: Service Sector Statistics

In the News

Kite-Powered Cargo Ships

In January 2008, the first cargo ship partially powered by a giant kite set sail from Germany bound for Venezuela. The 1722-square-foot kite helped to propel the ship, which consequently used 20% less fuel, cutting approximately $1600 from the ship's daily fuel bill.

Source: The Internal Revenue Service; TSN Financial Services

NUTRITION INFORMATION

SERVING SIZE: 1.4 OZ WHEAT FLAKES WITH
0.4 OZ. RAISINS: 39.4 g. ABOUT 1/2 CUP
SERVINGS PER PACKAGE: 14

	CEREAL & RAISINS	WITH 1/2 CUP VITAMINS A & D SKIM MILK

PERCENTAGE OF U.S. RECOMMENDED DAILY ALLOWANCES (U.S. RDA)

	CEREAL & RAISINS	WITH 1/2 CUP SKIM MILK
PROTEIN	4	15
VITAMIN A	15	20
VITAMIN C	**	2
THIAMIN	25	30
RIBOFLAVIN	25	35
NIACIN	25	35
CALCIUM	**	15
IRON	100	100
VITAMIN D	10	25
VITAMIN B$_6$	25	25
FOLIC ACID...............	25	25
VITAMIN B$_{12}$	25	30
PHOSPHOROUS.........	10	15
MAGNESIUM	10	20
ZINC	25	30
COPPER...................	2	4

* 2% MILK SUPPLIES AN ADDITIONAL 20 CALORIES.
2 g FAT, AND 10 mg CHOLESTEROL.
** CONTAINS LESS THAN 2% OF THE U.S. RDA OF THIS NUTRIENT

SECTION

5.5 Percent Problems: Proportion Method

OBJECTIVE A To solve percent problems using proportions

Problems that can be solved using the basic percent equation can also be solved using proportions.

The proportion method is based on writing two ratios. One ratio is the percent ratio, written as $\frac{percent}{100}$. The second ratio is the amount-to-base ratio, written as $\frac{amount}{base}$. These two ratios form the proportion

$$\frac{\text{percent}}{100} = \frac{\text{amount}}{\text{base}}$$

To use the proportion method, first identify the percent, the amount, and the base (the base usually follows the phrase "percent of").

Integrating Technology

To use a calculator to solve the proportions at the right for n, enter

23 **X** 45 **÷** 100 **=**

100 **X** 4 **÷** 25 **=**

100 **X** 12 **÷** 60 **=**

What is 23% of 45?

$$\frac{23}{100} = \frac{n}{45}$$

$23 \times 45 = 100 \times n$
$1035 = 100 \times n$
$1035 \div 100 = n$
$10.35 = n$

What percent of 25 is 4?

$$\frac{n}{100} = \frac{4}{25}$$

$n \times 25 = 100 \times 4$
$n \times 25 = 400$
$n = 400 \div 25$
$n = 16$

12 is 60% of what number?

$$\frac{60}{100} = \frac{12}{n}$$

$60 \times n = 100 \times 12$
$60 \times n = 1200$
$n = 1200 \div 60$
$n = 20$

EXAMPLE • 1

15% of what is 7? Round to the nearest hundredth.

Solution

$$\frac{15}{100} = \frac{7}{n}$$

$15 \times n = 100 \times 7$
$15 \times n = 700$
$n = 700 \div 15$
$n \approx 46.67$

YOU TRY IT • 1

26% of what is 22? Round to the nearest hundredth.

Your solution

EXAMPLE • 2

30% of 63 is what?

Solution

$$\frac{30}{100} = \frac{n}{63}$$

$30 \times 63 = 100 \times n$
$1890 = 100 \times n$
$1890 \div 100 = n$
$18.90 = n$

YOU TRY IT • 2

16% of 132 is what?

Your solution

Solutions on p. S13

OBJECTIVE B   To solve application problems

EXAMPLE · 3

An antiques dealer found that 86% of the 250 items that were sold during one month sold for under $1000. How many items sold for under $1000?

Strategy
To find the number of items that sold for under $1000, write and solve a proportion using n to represent the number of items sold for less than $1000 (amount). The percent is 86%, and the base is 250.

Solution

$$\frac{86}{100} = \frac{n}{250}$$
$$86 \times 250 = 100 \times n$$
$$21{,}500 = 100 \times n$$
$$21{,}500 \div 100 = n$$
$$215 = n$$

215 items sold for under $1000.

Last year it snowed 64% of the 150 days of the ski season at a resort. How many days did it snow?

Your strategy

Your solution

EXAMPLE · 4

In a test of the strength of nylon rope, 5 pieces of the 25 pieces tested did not meet the test standards. What percent of the nylon ropes tested did meet the standards?

Strategy
To find the percent of ropes tested that met the standards:
• Find the number of ropes that met the test standards (25 − 5).
• Write and solve a proportion using n to represent the percent of ropes that met the test standards. The base is 25, and the amount is the number of ropes that met the standards.

Solution
$25 - 5 = 20$ ropes met test standards

$$\frac{n}{100} = \frac{20}{25}$$
$$n \times 25 = 100 \times 20$$
$$n \times 25 = 2000$$
$$n = 2000 \div 25$$
$$n = 80$$

80% of the ropes tested did meet the test standards.

The Rincon Fire Department received 24 false alarms out of a total of 200 alarms received. What percent of the alarms received were not false alarms?

Your strategy

Your solution

Solutions on p. S13

5.5 EXERCISES

OBJECTIVE A   To solve percent problems using proportions

1. 26% of 250 is what?

2. What is 18% of 150?

3. 37 is what percent of 148?

4. What percent of 150 is 33?

5. 68% of what is 51?

6. 126 is 84% of what?

7. What percent of 344 is 43?

8. 750 is what percent of 50?

9. 82 is 20.5% of what?

10. 2.4% of what is 21?

11. What is 6.5% of 300?

12. 96% of 75 is what?

13. 7.4 is what percent of 50?

14. What percent of 1500 is 693?

15. 50.5% of 124 is what?

16. What is 87.4% of 255?

17. 33 is 220% of what?

18. 160% of what is 40?

19. **a.** Which equation(s) below can be used to answer the question "What is 12% of 75?"
   b. Which equation(s) below can be used to answer the question "75 is 12% of what?"

   (i) $\dfrac{12}{100} = \dfrac{75}{n}$ (ii) $0.12 \times 75 = n$ (iii) $\dfrac{12}{100} = \dfrac{n}{75}$ (iv) $0.12 \times n = 75$

OBJECTIVE B   To solve application problems

20. Read Exercise 21. Without doing any calculations, determine whether the length of time the drug will be effective is *less than* or *greater than* 6 hours.

21. **Medicine** A manufacturer of an anti-inflammatory drug claims that the drug will be effective for 6 hours. An independent testing service determined that the drug was effective for only 80% of the length of time claimed by the manufacturer. Find the length of time the drug will be effective as determined by the testing service.

22. **Geography** The land area of North America is approximately 9,400,000 square miles. This represents approximately 16% of the total land area of the world. What is the approximate total land area of the world?

23. Girl Scout Cookies Using the information in the news clipping at the right, calculate the cash generated annually **a.** from sales of Thin Mints and **b.** from sales of Trefoil shortbread cookies.

Jeff Greenberg/age fotostock

24. Charities The American Red Cross spent $185,048,179 for administrative expenses. This amount was 3.16% of its total revenue. Find the American Red Cross's total revenue. Round to the nearest hundred million.

25. Poultry In a recent year, North Carolina produced 1,300,000,000 pounds of turkey. This was 18.6% of the U.S. total in that year. Calculate the U.S. total turkey production for that year. Round to the nearest billion.

26. Mining During 1 year, approximately 2,240,000 ounces of gold went into the manufacturing of electronic equipment in the United States. This is 16% of all the gold mined in the United States that year. How many ounces of gold were mined in the United States that year?

27. Education See the news clipping at the right. What percent of the baby boomers living in the United States have some college experience but have not earned a college degree? Round to the nearest tenth of a percent.

28. Demography According to a 25-city survey of the status of hunger and homelessness by the U.S. Conference of Mayors, 41% of the homeless in the United States are single men, 41% are families with children, 13% are single women, and 5% are unaccompanied minors. How many homeless people in the United States are single men?

29. Police Officers The graph at the right shows the causes of death for all police officers killed in the line of duty during a recent year. What percent of the deaths were due to traffic accidents? Round to the nearest tenth of a percent.

Job-related illness
19

Other
6

Violent attacks
58

Traffic accidents
73

Causes of Death for Police Officers Killed in the Line of Duty
Source: International Union of Police Associations

Applying the Concepts

30. The Federal Government In the 110th Senate, there were 49 Republicans, 49 Democrats, and 2 Independents. In the 110th House of Representatives, there were 202 Republicans, 233 Democrats, and 0 Independents. Which had the larger percentage of Republicans, the 110th Senate or the 110th House of Representatives?

FOCUS ON PROBLEM SOLVING

Using a Calculator as a Problem-Solving Tool

A calculator is an important tool for problem solving. Here are a few problems to solve with a calculator. You may need to research some of the questions to find information you do not know.

1. Choose any single-digit positive number. Multiply the number by 1507 and 7373. What is the answer? Choose another positive single-digit number and again multiply by 1507 and 7373. What is the answer? What pattern do you see? Why does this work?

2. The gross domestic product in 2007 was $13,841,300,000. Is this more or less than the amount of money that would be placed on the last square of a standard checkerboard if 1 cent were placed on the first square, 2 cents were placed on the second square, 4 cents were placed on the third square, 8 cents were placed on the fourth square, and so on, until the 64th square was reached?

3. Which of the reciprocals of the first 16 natural numbers have a terminating-decimal representation and which have a repeating-decimal representation?

4. What is the largest natural number n for which $4^n . 1 \cdot 2 \cdot 3 \cdot 4 \cdot 5 \cdots \cdot n$?

5. If $1000 bills are stacked one on top of another, is the height of $1 billion less than or greater than the height of the Washington Monument?

6. What is the value of $1 + \cfrac{1}{1 + \cfrac{1}{1 + \cfrac{1}{1 + \cfrac{1}{1 + 1}}}}$?

7. Calculate 15^2, 35^2, 65^2, and 85^2. Study the results. Make a conjecture about a relationship between a number ending in 5 and its square. Use your conjecture to find 75^2 and 95^2. Does your conjecture work for 125^2?

8. Find the sum of the first 1000 natural numbers. (*Hint:* You could just start adding $1 + 2 + 3 + \cdots$, but even if you performed one operation every 3 seconds, it would take you an hour to find the sum. Instead, try pairing the numbers and then adding the pairs. Pair 1 and 1000, 2 and 999, 3 and 998, and so on. What is the sum of each pair? How many pairs are there? Use this information to answer the original question.)

9. For a borrower to qualify for a home loan, a bank requires that the monthly mortgage payment be less than 25% of the borrower's monthly take-home income. A laboratory technician has deductions for taxes, insurance, and retirement that amount to 25% of the technician's monthly gross income. What minimum monthly income must this technician earn to receive a bank loan that has a mortgage payment of $1200 per month?

Using Estimation as a Problem-Solving Tool

You can use your knowledge of rounding, your understanding of percent, and your experience with the basic percent equation to quickly estimate the answer to a percent problem. Here is an example.

| HOW TO 1 | What is 11.2% of 978? |

Round the given numbers.
$$11.2\% \approx 10\%$$
$$978 \approx 1000$$

Mentally calculate with the rounded numbers.
$$10\% \text{ of } 1000 = \frac{1}{10} \text{ of } 1000 = 100$$

11.2% of 978 is approximately 100.

✓ **Take Note**

The exact answer is
$0.112 \times 978 = 109.536$.
The exact answer 109.536 is close to the approximation of 100.

For Exercises 1 to 8, state which quantity is greater.

1. 49% of 51, or 201% of 15
2. 99% of 19, or 22% of 55
3. 8% of 31, or 78% of 10
4. 24% of 402, or 76% of 205
5. 10.2% of 51, or 20.9% of 41
6. 51.8% of 804, or 25.3% of 1223
7. 26% of 39.217, or 9% of 85.601
8. 66% of 31.807, or 33% of 58.203

For Exercises 9 to 12, use estimation to provide an approximate number.

9. A company found that 24% of its 2096 employees favored a new dental plan. How many employees favored the new dental plan?

10. A local newspaper reported that 52.3% of the 29,875 eligible voters in the town voted in the last election. How many people voted in the last election?

11. 19.8% of the 2135 first-year students at a community college have part-time jobs. How many of the first-year students at the college have part-time jobs?

12. A couple made a down payment of 33% of the $310,000 cost of a home. Find the down payment.

© Ariel Skelly/Corbis

PROJECTS AND GROUP ACTIVITIES

Health

The American College of Sports Medicine (ACSM) recommends that you know how to determine your target heart rate in order to get the full benefit of exercise. Your **target heart rate** is the rate at which your heart should beat during any aerobic exercise such as running, cycling, fast walking, or participating in an aerobics class. According to the ACSM, you should reach your target rate and then maintain it for 20 minutes or more to achieve cardiovascular fitness. The intensity level varies for different individuals. A sedentary person might begin at the 60% level and gradually work up to 70%, whereas athletes and very fit individuals might work at the 85% level. The ACSM suggests that you calculate both 50% and 85% of your maximum heart rate. This will give you the low and high ends of the range within which your heart rate should stay.

To calculate your target heart rate:

	Example
Subtract your age from 220. This is your maximum heart rate.	$220 - 20 = 200$
Multiply your maximum heart rate by 50%. This is the low end of your range.	$200(0.50) = 100$
Divide the low end by 6. This is your low 10-second heart rate.	$100 \div 6 \approx 17$
Multiply your maximum heart rate by 85%. This is the high end of your range.	$200(0.85) = 170$
Divide the high end by 6. This is your high 10-second heart rate.	$170 \div 6 \approx 28$

1. Why are the low end and high end divided by 6 in order to determine the low and high 10-second heart rates?

2. Calculate your target heart rate, both the low and high end of your range.

Consumer Price Index   The consumer price index (CPI) is a percent that is written without the percent sign. For instance, a CPI of 160.1 means 160.1%. This number means that an item that cost $100 between 1982 and 1984 (the base years) would cost $160.10 today. Determining the cost is an application of the basic percent equation.

$$\text{Percent} \times \text{base} = \text{amount}$$
$$\text{CPI} \times \text{cost in base year} = \text{cost today}$$
$$1.601 \times 100 = 160.1 \qquad \bullet \; 160.1\% = 1.601$$

The table below gives the CPI for various products in March of 2008. If you have Internet access, you can obtain current data for the items below, as well as other items not on this list, by visiting the website of the Bureau of Labor Statistics.

Product	*CPI*
All items	213.5
Food and beverages	209.7
Housing	214.4
Clothes	120.9
Transportation	195.2
Medical care	363.0
Entertainment[1]	112.7
Education[1]	121.8

[1]Indexes on December 1997 = 100

1. Of the items listed, are there any items that in 2008 cost more than twice as much as they cost during the base year? If so, which items?

2. Of the items listed, are there any items that in 2008 cost more than one-and-one-half times as much as they cost during the base years but less than twice as much as they cost during the base years? If so, which items?

3. If the cost for textbooks for one semester was $120 in the base years, how much did similar textbooks cost in 2008? Use the "Education" category.

4. If a new car cost $40,000 in 2008, what would a comparable new car have cost during the base years? Use the "Transportation" category.

5. If a movie ticket cost $10 in 2008, what would a comparable movie ticket have cost during the base years? Use the "Entertainment" category.

6. The base year for the CPI was 1967 before the change to 1982–1984. If 1967 were still used as the base year, the CPI for all items in 2008 (not just those listed above) would be 639.6.

    a. Using the base year of 1967, explain the meaning of a CPI of 639.6.

    b. Using the base year of 1967 and a CPI of 639.6, if textbooks cost $75 for one semester in 1967, how much did similar textbooks cost in 2008?

    c. Using the base year of 1967 and a CPI of 639.6, if a family's food budget in 2008 is $1000 per month, what would a comparable family budget have been in 1967?

CHAPTER 5

SUMMARY

KEY WORDS	EXAMPLES
Percent means "parts of 100." [5.1A, p. 202]	23% means 23 of 100 equal parts.

ESSENTIAL RULES AND PROCEDURES	EXAMPLES
To write a percent as a fraction, drop the percent sign and multiply by $\frac{1}{100}$. [5.1A, p. 202]	$56\% = 56\left(\dfrac{1}{100}\right) = \dfrac{56}{100} = \dfrac{14}{25}$
To write a percent as a decimal, drop the percent sign and multiply by 0.01. [5.1A, p. 202]	$87\% = 87(0.01) = 0.87$
To write a fraction as a percent, multiply by 100%. [5.1B, p. 203]	$\dfrac{7}{20} = \dfrac{7}{20}(100\%) = \dfrac{700}{20}\% = 35\%$
To write a decimal as a percent, multiply by 100%. [5.1B, p. 203]	$0.325 = 0.325(100\%) = 32.5\%$
The Basic Percent Equation [5.2A, p. 206] The basic percent equation is $\qquad$ Percent $\times$ base $=$ amount Solving percent problems requires identifying the three elements of this equation. Usually the base follows the phrase "percent of."	8% of 250 is what number? Percent $\times$ base $=$ amount $0.08 \times 250 = n$ $\qquad\quad 20 = n$
Proportion Method of Solving a Percent Problem [5.5A, p. 218] The following proportion can be used to solve percent problems. $\qquad \dfrac{\text{percent}}{100} = \dfrac{\text{amount}}{\text{base}}$ To use the proportion method, first identify the percent, the amount, and the base. The base usually follows the phrase "percent of."	8% of 250 is what number? $\dfrac{\text{percent}}{100} = \dfrac{\text{amount}}{\text{base}}$ $\dfrac{8}{100} = \dfrac{n}{250}$ $8 \times 250 = 100 \times n$ $2000 = 100 \times n$ $2000 \div 100 = n$ $20 = n$

CHAPTER 5

CONCEPT REVIEW

Test your knowledge of the concepts presented in this chapter. Answer each question.
Then check your answers against the ones provided in the Answer Section.

1. How do you write 197% as a fraction?

2. How do you write 6.7% as a decimal?

3. How do you write $\frac{9}{5}$ as a percent?

4. How do you write 56.3 as a percent?

5. What is the basic percent equation?

6. What percent of 40 is 30? Did you multiply or divide?

7. Find 11.7% of 532. Did you multiply or divide?

8. 36 is 240% of what number? Did you multiply or divide?

9. How do you use the proportion method to solve a percent problem?

10. What percent of 1400 is 763? Use the proportion method to solve.

CHAPTER 5

REVIEW EXERCISES

1. What is 30% of 200?

2. 16 is what percent of 80?

3. Write $1\frac{3}{4}$ as a percent.

4. 20% of what is 15?

5. Write 12% as a fraction.

6. Find 22% of 88.

7. What percent of 20 is 30?

8. $16\frac{2}{3}\%$ of what is 84?

9. Write 42% as a decimal.

10. What is 7.5% of 72?

11. $66\frac{2}{3}\%$ of what is 105?

12. Write 7.6% as a decimal.

13. Find 125% of 62.

14. Write $16\frac{2}{3}\%$ as a fraction.

15. Use the proportion method to find what percent of 25 is 40.

16. 20% of what number is 15? Use the proportion method.

17. Write 0.38 as a percent.

18. 78% of what is 8.5? Round to the nearest tenth.

19. What percent of 30 is 2.2? Round to the nearest tenth of a percent.

20. What percent of 15 is 92? Round to the nearest tenth of a percent.

21. Education Trent missed 9 out of 60 questions on a history exam. What percent of the questions did he answer correctly? Use the proportion method.

22. Advertising A company used 7.5% of its $60,000 advertising budget for newspaper advertising. How much of the advertising budget was spent for newspaper advertising?

Where Your Energy Dollar Goes

The average U.S. household spent $2868 on energy use in a recent year. How it was spent:

Motor gasoline $1492

Fuel oil, kerosene $83

Natural gas $383

Electricity $910

23. Energy The graph at the right shows the amounts that the average U.S. household spends for energy use. What percent of these costs is for electricity? Round to the nearest tenth of a percent.

24. Consumerism Joshua purchased a camcorder for $980 and paid a sales tax of 6.25% of the cost. What was the total cost of the camcorder?

Source: Energy Information Administration

25. Health In a survey of 350 women and 420 men, 275 of the women and 300 of the men reported that they wore sunscreen often. To the nearest tenth of a percent, what percent of the women wore sunscreen often?

26. Demography It is estimated that the world's population will be 9,100,000,000 by the year 2050. This is 149% of the population in 2000. (*Source:* U.S. Census Bureau). What was the world's population in 2000? Round to the nearest hundred million.

27. Computers A computer system can be purchased for $1800. This is 60% of what the computer cost 4 years ago. What was the cost of the computer 4 years ago? Use the proportion method.

28. Agriculture In a recent year, Wisconsin growers produced 281.72 million pounds of cranberries. This represented 49.25% of the total cranberry crop in the United States that year. Find the total cranberry crop in the United States that year. Round to the nearest million.

CHAPTER 5

TEST

1. Write 97.3% as a decimal.

2. Write $83\frac{1}{3}\%$ as a fraction.

3. Write 0.3 as a percent.

4. Write 1.63 as a percent.

5. Write $\frac{3}{2}$ as a percent.

6. Write $\frac{2}{3}$ as a percent.

7. What is 77% of 65?

8. 47.2% of 130 is what?

9. Which is larger:
   7% of 120, or 76% of 13?

10. Which is smaller:
    13% of 200, or 212% of 12?

11. **Advertising** A travel agency uses 6% of its $750,000 budget for advertising. What amount of the budget is spent on advertising?

12. **Agriculture** During the packaging process for vegetables, spoiled vegetables are discarded by an inspector. In one day an inspector found that 6.4% of the 1250 pounds of vegetables were spoiled. How many pounds of vegetables were not spoiled?

Nutrition The table at the right contains nutrition information about a breakfast cereal. Solve Exercises 13 and 14 with information taken from this table.

13. The recommended amount of potassium per day for an adult is 3000 milligrams (mg). What percent, to the nearest tenth of a percent, of the daily recommended amount of potassium is provided by one serving of this cereal with skim milk?

14. The daily recommended number of calories for a 190-pound man is 2200 calories. What percent, to the nearest tenth of a percent, of the daily recommended number of calories is provided by one serving of this cereal with 2% milk?

NUTRITION INFORMATION

SERVING SIZE: 1.4 OZ WHEAT FLAKES WITH 0.4 OZ. RAISINS: 39.4 g. ABOUT 1/2 CUP
SERVINGS PER PACKAGE: 14

	CEREAL & RAISINS	WITH 1/2 CUP VITAMINS A & D SKIM MILK
CALORIES	120	180
PROTEIN, g	3	7
CARBOHYDRATE, g ...	28	34
FAT, TOTAL g	1	1*
UNSATURATED, g 1		
SATURATED, g 0		
CHOLESTEROL, mg	0	0*
SODIUM, mg	125	190
POTASSIUM, mg	240	440

* 2% MILK SUPPLIES AN ADDITIONAL 20 CALORIES, 2 g FAT, AND 10 mg CHOLESTEROL.
** CONTAINS LESS THAN 2% OF THE U.S. RDA OF THIS NUTRIENT

15. **Employment** The Urban Center Department Store has 125 permanent employees and must hire an additional 20 temporary employees for the holiday season. What percent of the number of permanent employees is the number hired as temporary employees for the holiday season?

16. **Education** Conchita missed 7 out of 80 questions on a math exam. What percent of the questions did she answer correctly? Round to the nearest tenth of a percent.

17. 12 is 15% of what?

18. 42.5 is 150% of what? Round to the nearest tenth.

19. **Manufacturing** A manufacturer of PDAs found 384 defective PDAs during a quality control study. This amount was 1.2% of the PDAs tested. Find the number of PDAs tested.

20. **Real Estate** A new house was bought for $285,000. Five years later the house sold for $456,000. The increase was what percent of the original price?

21. 123 is 86% of what number? Use the proportion method. Round to the nearest tenth.

22. What percent of 12 is 120? Use the proportion method.

23. **Wages** An administrative assistant receives a wage of $16.24 per hour. This amount is 112% of last year's wage. What is the dollar increase in the hourly wage over last year? Use the proportion method.

24. **Demography** A city has a population of 71,500. Ten years ago the population was 32,500. The population now is what percent of the population 10 years ago? Use the proportion method.

25. **Fees** The annual license fee on a car is 1.4% of the value of the car. If the license fee during a year is $350, what is the value of the car? Use the proportion method.

CUMULATIVE REVIEW EXERCISES

1. Simplify: $18 \div (7 - 4)^2 + 2$

2. Find the LCM of 16, 24, and 30.

3. Find the sum of $2\frac{1}{3}$, $3\frac{1}{2}$, and $4\frac{5}{8}$.

4. Subtract: $27\frac{5}{12} - 14\frac{9}{16}$

5. Multiply: $7\frac{1}{3} \times 1\frac{5}{7}$

6. What is $\frac{14}{27}$ divided by $1\frac{7}{9}$?

7. Simplify: $\left(\frac{3}{4}\right)^3 \cdot \left(\frac{8}{9}\right)^2$

8. Simplify: $\left(\frac{2}{3}\right)^2 - \left(\frac{3}{8} - \frac{1}{3}\right) \div \frac{1}{2}$

9. Round 3.07973 to the nearest hundredth.

10. Subtract: $\begin{array}{r} 3.0902 \\ -1.9706 \\ \hline \end{array}$

11. Divide: $0.032\overline{)1.097}$
Round to the nearest ten-thousandth.

12. Convert $3\frac{5}{8}$ to a decimal.

13. Convert 1.75 to a fraction.

14. Place the correct symbol, $<$ or $>$, between the two numbers.
$\frac{3}{8}$ 0.87

15. Solve the proportion $\frac{3}{8} = \frac{20}{n}$.
Round to the nearest tenth.

16. Write "$153.60 earned in 8 hours" as a unit rate.

17. Write $18\frac{1}{3}\%$ as a fraction.

18. Write $\frac{5}{6}$ as a percent.

19. 16.3% of 120 is what?

20. 24 is what percent of 18?

21. 12.4 is 125% of what?

22. What percent of 35 is 120? Round to the nearest tenth.

23. Taxes Sergio has an income of $740 per week. One-fifth of his income is deducted for income tax payments. Find his take-home pay.

24. Finance Eunice bought a used car for $12,530, with a down payment of $2000. The balance was paid in 36 equal monthly payments. Find the monthly payment.

25. Taxes The gasoline tax is $.41 a gallon. Find the number of gallons of gasoline used during a month in which $172.20 was paid in gasoline taxes.

26. Taxes The real estate tax on a $344,000 home is $6880. At the same rate, find the real estate tax on a home valued at $500,000.

27. Lodging The graph at the right shows the breakdown of the locations of the 53,500 hotels throughout the United States. How many hotels in the United States are located along highways?

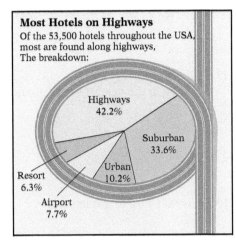

Most Hotels on Highways
Of the 53,500 hotels throughout the USA, most are found along highways. The breakdown:

Highways 42.2%
Suburban 33.6%
Urban 10.2%
Resort 6.3%
Airport 7.7%

Source: American Hotel and Lodging Association

28. Elections A survey of 300 people showed that 165 people favored a certain candidate for mayor. What percent of the people surveyed did not favor this candidate?

29. Television According to the Cabletelevision Advertising Bureau, cable households watch television 36.5% of the time. On average, how many hours per week do cable households spend watching TV? Round to the nearest tenth.

30. Health The Environmental Protection Agency found that 990 out of 5500 children tested had levels of lead in their blood that exceeded federal guidelines. What percent of the children tested had levels of lead in the blood that exceeded federal standards?

Applications for Business and Consumers

CHAPTER 6

OBJECTIVES

SECTION 6.1
A To find unit cost
B To find the most economical purchase
C To find total cost

SECTION 6.2
A To find percent increase
B To apply percent increase to business—markup
C To find percent decrease
D To apply percent decrease to business—discount

SECTION 6.3
A To calculate simple interest
B To calculate finance charges on a credit card bill
C To calculate compound interest

SECTION 6.4
A To calculate the initial expenses of buying a home
B To calculate the ongoing expenses of owning a home

SECTION 6.5
A To calculate the initial expenses of buying a car
B To calculate the ongoing expenses of owning a car

SECTION 6.6
A To calculate commissions, total hourly wages, and salaries

SECTION 6.7
A To calculate checkbook balances
B To balance a checkbook

ARE YOU READY?

Take the Chapter 6 Prep Test to find out if you are ready to learn to:

- Find unit cost, total cost, and the most economical purchase
- Find percent increase and percent decrease and apply them to markup and discount
- Calculate simple interest and compound interest
- Calculate expenses associated with buying and owning a home or a car
- Calculate commissions, wages, and salaries
- Calculate checkbook balances and balance a checkbook

PREP TEST

Do these exercises to prepare for Chapter 6.

For Exercises 1 to 6, add, subtract, multiply, or divide.

1. Divide: $3.75 \div 5$

2. Multiply: 3.47×15

3. Subtract: $874.50 - 369.99$

4. Multiply: $0.065 \times 150,000$

5. Multiply: $1500 \times 0.06 \times 0.5$

6. Add: $1372.47 + 36.91 + 5.00 + 2.86$

7. Divide $10 \div 3$. Round to the nearest hundredth.

8. Divide $345 \div 570$. Round to the nearest thousandth.

9. Place the correct symbol, $<$ or $>$, between the two numbers.
0.379   0.397

233

Vito Palmisano/Getty Images

SECTION

6.1 Applications to Purchasing

OBJECTIVE A To find unit cost

Frequently, stores promote items for purchase by advertising, say, 2 Red Baron Bake to Rise Pizzas for $10.50 or 5 cans of StarKist tuna for $4.25. The **unit cost** is the cost of *one* Red Baron Pizza or of *one* can of StarKist tuna. To find the unit cost, divide the total cost by the number of units.

2 pizzas for $10.50	5 cans for $4.25
$10.50 \div 2 = 5.25$	$4.25 \div 5 = 0.85$
$5.25 is the cost of one pizza.	$.85 is the cost of one can.
Unit cost: $5.25 per pizza	Unit cost: $.85 per can

EXAMPLE 1

Find the unit cost. Round to the nearest tenth of a cent.
a. 3 gallons of mint chip ice cream for $17
b. 4 ounces of Crest toothpaste for $2.29

Strategy
To find the unit cost, divide the total cost by the number of units.

Solution
a. $17 \div 3 \approx 5.667$
$5.667 per gallon
b. $2.29 \div 4 = 0.5725$
$.573 per ounce

YOU TRY IT 1

Find the unit cost. Round to the nearest tenth of a cent.
a. 8 size-AA Energizer batteries for $7.67
b. 15 ounces of Suave shampoo for $2.29

Your strategy

Your solution

Solution on p. S13

OBJECTIVE B To find the most economical purchase

Comparison shoppers often find the most economical buy by comparing unit costs.

One store is selling 6 twelve-ounce cans of ginger ale for $2.99, and a second store is selling 24 twelve-ounce cans of ginger ale for $11.79. To find the better buy, compare the unit costs.

$2.99 \div 6 \approx 0.498$	$11.79 \div 24 \approx 0.491$
Unit cost: $.498 per can	Unit cost: $.491 per can

Because $.491 < $.498, the better buy is 24 cans for $11.79.

EXAMPLE • 2

Find the more economical purchase: 5 pounds of nails for $4.80, or 4 pounds of nails for $3.78.

Strategy
To find the more economical purchase, compare the unit costs.

Solution
4.80 ÷ 5 = 0.96
3.78 ÷ 4 = 0.945
$.945 < $.96

The more economical purchase is 4 pounds for $3.78.

YOU TRY IT • 2

Find the more economical purchase: 6 cans of fruit for $8.70, or 4 cans of fruit for $6.96.

Your strategy

Your solution

Solution on p. S14

OBJECTIVE C To find total cost

An installer of floor tile found the unit cost of identical floor tiles at three stores.

Store 1	Store 2	Store 3
$1.22 per tile	$1.18 per tile	$1.28 per tile

By comparing the unit costs, the installer determined that store 2 would provide the most economical purchase.

The installer also uses the unit cost to find the total cost of purchasing 300 floor tiles at store 2. The **total cost** is found by multiplying the unit cost by the number of units purchased.

Unit cost	×	number of units	=	total cost
1.18	×	300	=	354

The total cost is $354.

EXAMPLE • 3

Clear redwood lumber costs $5.43 per foot. How much would 25 feet of clear redwood cost?

Strategy
To find the total cost, multiply the unit cost (5.43) by the number of units (25).

Solution

Unit cost	×	number of units	=	total cost
5.43	×	25	=	135.75

The total cost is $135.75.

YOU TRY IT • 3

Pine saplings cost $9.96 each. How much would 7 pine saplings cost?

Your strategy

Your solution

Solution on p. S14

6.1 EXERCISES

OBJECTIVE A   To find unit cost

For Exercises 1 to 10, find the unit cost. Round to the nearest tenth of a cent.

1. Heinz B·B·Q sauce, 18 ounces for $.99

2. Birds-eye maple, 6 feet for $18.75

3. Diamond walnuts, $2.99 for 8 ounces

4. A&W root beer, 6 cans for $2.99

5. Ibuprofen, 50 tablets for $3.99

6. Visine eye drops, 0.5 ounce for $3.89

7. Adjustable wood clamps, 2 for $13.95

8. Corn, 6 ears for $2.85

9. Cheerios cereal, 15 ounces for $2.99

10. Doritos Cool Ranch chips, 14.5 ounces for $2.99

11. A store advertises a "buy one, get one free" sale on pint containers of ice cream. How would you find the unit cost of one pint of ice cream?

OBJECTIVE B   To find the most economical purchase

For Exercises 12 to 21, suppose your local supermarket offers the following products at the given prices. Find the more economical purchase.

12. Sutter Home pasta sauce, 25.5 ounces for $3.29, or Muir Glen Organic pasta sauce, 26 ounces for $3.79

13. Kraft mayonnaise, 40 ounces for $3.98, or Springfield mayonnaise, 32 ounces for $3.39

14. Ortega salsa, 20 ounces for $3.29 or 12 ounces for $1.99

15. L'Oréal shampoo, 13 ounces for $4.69, or Cortexx shampoo, 12 ounces for $3.99

16. Golden Sun vitamin E, 200 tablets for $12.99 or 400 tablets for $18.69

17. Ultra Mr. Clean, 20 ounces for $2.67, or Ultra Spic and Span, 14 ounces for $2.19

18. 16 ounces of Kraft cheddar cheese for $4.37, or 9 ounces of Land O'Lakes cheddar cheese for $2.29

19. Bertolli olive oil, 34 ounces for $9.49, or Pompeian olive oil, 8 ounces for $2.39

20. Maxwell House coffee, 4 ounces for $3.99, or Sanka coffee, 2 ounces for $2.39

21. Wagner's vanilla extract, $3.95 for 1.5 ounces, or Durkee vanilla extract, 1 ounce for $2.84

 For Exercises 22 and 23, suppose a box of Tea A contains twice as many tea bags as a box of Tea B. Decide which box of tea is the more economical purchase.

22. The price of a box of Tea A is less than twice the price of a box of Tea B.

23. The price of a box of Tea B is greater than half the price of a box of Tea A.

OBJECTIVE C To find total cost

24. If sliced bacon costs $4.59 per pound, find the total cost of 3 pounds.

25. Used red brick costs $.98 per brick. Find the total cost of 75 bricks.

26. Kiwi fruit cost $.43 each. Find the total cost of 8 kiwi.

27. Boneless chicken filets cost $4.69 per pound. Find the cost of 3.6 pounds. Round to the nearest cent.

28. Herbal tea costs $.98 per ounce. Find the total cost of 6.5 ounces.

29. If Stella Swiss Lorraine cheese costs $5.99 per pound, find the total cost of 0.65 pound. Round to the nearest cent.

30. Red Delicious apples cost $1.29 per pound. Find the total cost of 2.1 pounds. Round to the nearest cent.

31. Choice rib eye steak costs $9.49 per pound. Find the total cost of 2.8 pounds. Round to the nearest cent.

 32. Suppose a store flyer advertises cantaloupes as "buy one, get one free." True or false? The total cost of 6 cantaloupes at the sale price is the same as the total cost of 3 cantaloupes at the regular price.

Applying the Concepts

 33. Explain in your own words the meaning of unit pricing.

 34. What is the UPC (Universal Product Code) and how is it used?

ISBN 0-395-75524-7

SECTION

6.2   Percent Increase and Percent Decrease

OBJECTIVE A   To find percent increase

Percent increase is used to show how much a quantity has increased over its original value. The statements "Food prices increased by 2.3% last year" and "City council members received a 4% pay increase" are examples of percent increase.

Point of Interest

According to the U.S. Census Bureau, the number of persons aged 65 and over in the United States will increase to about 82.0 million by 2050, a 136% increase from 2000.

HOW TO • 1   According to the Energy Information Administration, the number of alternative-fuel vehicles increased from approximately 277,000 to 352,000 in four years. Find the percent increase in alternative-fuel vehicles. Round to the nearest percent.

$$\boxed{\text{New value}} - \boxed{\text{original value}} = \boxed{\text{amount of increase}}$$

$$352,000 - 277,000 = 75,000$$

Now solve the basic percent equation for percent.

$$\text{Percent} \times \text{base} = \text{amount}$$

$$\boxed{\text{Percent increase}} \times \boxed{\text{original value}} = \boxed{\text{amount of increase}}$$

$$n \times 277,000 = 75,000$$
$$n = 75,000 \div 277,000$$
$$n \approx 0.27$$

The number of alternative-fuel vehicles increased by approximately 27%.

EXAMPLE • 1

The average wholesale price of coffee increased from $2 per pound to $3 per pound in one year. What was the percent increase in the price of 1 pound of coffee?

Strategy

To find the percent increase:

• Find the amount of the increase.
• Solve the basic percent equation for *percent*.

Solution

$$\boxed{\text{New value}} - \boxed{\text{original value}} = \boxed{\text{amount of increase}}$$

$$3 - 2 = 1$$

$$\text{Percent} \times \text{base} = \text{amount}$$
$$n \times 2 = 1$$
$$n = 1 \div 2$$
$$n = 0.5 = 50\%$$

The percent increase was 50%.

YOU TRY IT • 1

The average price of gasoline rose from $3.46 to $3.83 in 5 months. What was the percent increase in the price of gasoline? Round to the nearest percent.

Your strategy

Your solution

Solution on p. S14

EXAMPLE 2

Chris Carley was earning $13.50 an hour as a nursing assistant before receiving a 10% increase in pay. What is Chris's new hourly pay?

Strategy

To find the new hourly wage:

• Solve the basic percent equation for *amount*.
• Add the amount of the increase to the original wage.

Solution

Percent × base = amount

$$0.10 \times 13.50 = n$$
$$1.35 = n$$

The amount of the increase was $1.35.

$$13.50 + 1.35 = 14.85$$

The new hourly wage is $14.85.

YOU TRY IT 2

Yolanda Liyama was making a wage of $12.50 an hour as a baker before receiving a 14% increase in hourly pay. What is Yolanda's new hourly wage?

Your strategy

Your solution

Solution on p. S14

OBJECTIVE B   To apply percent increase to business—markup

Some of the expenses involved in operating a business are salaries, rent, equipment, and utilities. To pay these expenses and earn a profit, a business must sell a product at a higher price than it paid for the product.

Cost is the price a business pays for a product, and **selling price** is the price at which a business sells a product to a customer. The difference between selling price and cost is called **markup.**

| Selling price | − | cost | = | markup |

or

| Cost | + | markup | = | selling price |

Markup is frequently expressed as a percent of a product's cost. This percent is called the **markup rate.**

| Markup rate | × | cost | = | markup |

Point of Interest

According to *Managing a Small Business*, from Liraz Publishing Company, goods in a store are often marked up 50% to 100% of the cost. This allows a business to make a profit of 5% to 10%.

HOW TO 2   Suppose Bicycles Galore purchases an AMP Research B-5 bicycle for $2119.20 and sells it for $2649. What markup rate does Bicycles Galore use?

| Selling price | − | cost | = | markup |

$$2649.00 - 2119.20 = 529.80$$ • First find the markup.

| Percent | × | base | = | amount |

| Markup rate | × | cost | = | markup |

$$n \times 2119.20 = 529.80$$

• Then solve the basic percent equation for *percent.*

$$n = 529.80 \div 2119.20 = 0.25$$

The markup rate is 25%.

David Madison/STONE/Getty Images

EXAMPLE • 3

The manager of a sporting goods store determines that a markup rate of 36% is necessary to make a profit. What is the markup on a pair of skis that costs the store $225?

Strategy

To find the markup, solve the basic percent equation for *amount*.

Solution

Percent × base = amount

$$\boxed{\text{Markup rate}} \times \boxed{\text{cost}} = \boxed{\text{markup}}$$

$$0.36 \quad \times \quad 225 \quad = \quad n$$
$$81 = n$$

The markup is $81.

YOU TRY IT • 3

A bookstore manager determines that a markup rate of 20% is necessary to make a profit. What is the markup on a book that costs the bookstore $32?

Your strategy

Your solution

EXAMPLE • 4

A plant nursery bought a yellow twig dogwood for $9.50 and used a markup rate of 46%. What is the selling price?

Strategy

To find the selling price:

• Find the markup by solving the basic percent equation for *amount*.
• Add the markup to the cost.

Solution

Percent × base = amount

$$\boxed{\text{Markup rate}} \times \boxed{\text{cost}} = \boxed{\text{markup}}$$

$$0.46 \quad \times \quad 9.50 \quad = \quad n$$
$$4.37 = n$$

$$\boxed{\text{Cost}} + \boxed{\text{markup}} = \boxed{\text{selling price}}$$

$$9.50 \quad + \quad 4.37 \quad = \quad 13.87$$

The selling price is $13.87.

YOU TRY IT • 4

A clothing store bought a leather jacket for $72 and used a markup rate of 55%. What is the selling price?

Your strategy

Your solution

Solutions on p. S14

OBJECTIVE C To find percent decrease

Percent decrease is used to show how much a quantity has decreased from its original value. The statements "The number of family farms decreased by 2% last year" and "There has been a 50% decrease in the cost of a Pentium chip" are examples of percent decrease.

HOW TO 3 During a 2-year period, the value of U.S. agricultural products exported decreased from approximately $60.6 billion to $52.0 billion. Find the percent decrease in the value of U.S. agricultural exports. Round to the nearest tenth of a percent.

Original value	−	new value	=	amount of decrease
60.6	−	52.0	=	8.6

Now solve the basic percent equation for percent.

Percent	×	base	=	amount
Percent decrease	×	original value	=	amount of decrease
n	×	60.6	=	8.6

$$n = 8.6 \div 60.6$$
$$n \approx 0.142$$

The value of agricultural exports decreased approximately 14.2%.

Tips for Success
Note in the example below that solving a word problem involves stating a strategy and using the strategy to find a solution. If you have difficulty with a word problem, write down the known information. Be very specific. Write out a phrase or sentence that states what you are trying to find. See *AIM for Success* at the front of the book.

EXAMPLE 5

 During an 8-year period, the population of Baltimore, Maryland, decreased from approximately 736,000 to 646,000. Find the percent decrease in Baltimore's population. Round to the nearest tenth of a percent.

Strategy
To find the percent decrease:
• Find the amount of the decrease.
• Solve the basic percent equation for *percent.*

Solution

Original value	−	new value	=	amount of decrease
736,000	−	646,000	=	90,000

Percent × base = amount
$$n \times 736{,}000 = 90{,}000$$
$$n = 90{,}000 \div 736{,}000$$
$$n \approx 0.122$$

Baltimore's population decreased approximately 12.2%.

YOU TRY IT 5

 During an 8-year period, the population of Norfolk, Virginia, decreased from approximately 261,000 to 215,000. Find the percent decrease in Norfolk's population. Round to the nearest tenth of a percent.

Your strategy

Your solution

Solution on p. S14

EXAMPLE · 6

The total sales for December for a stationery store were $96,000. For January, total sales showed an 8% decrease from December's sales. What were the total sales for January?

Strategy

To find the total sales for January:

- Find the amount of decrease by solving the basic percent equation for *amount*.
- Subtract the amount of decrease from the December sales.

Solution

$$\text{Percent} \times \text{base} = \text{amount}$$
$$0.08 \times 96,000 = n$$
$$7680 = n$$

The decrease in sales was $7680.

$$96,000 - 7680 = 88,320$$

The total sales for January were $88,320.

YOU TRY IT · 6

Fog decreased the normal 5-mile visibility at an airport by 40%. What was the visibility in the fog?

Your strategy

Your solution

Solution on p. S14

OBJECTIVE D **To apply percent decrease to business—discount**

To promote sales, a store may reduce the regular price of some of its products temporarily. The reduced price is called the **sale price.** The difference between the regular price and the sale price is called the **discount.**

$$\boxed{\text{Regular price}} - \boxed{\text{sale price}} = \boxed{\text{discount}}$$

or

$$\boxed{\text{Regular price}} - \boxed{\text{discount}} = \boxed{\text{sale price}}$$

Discount is frequently stated as a percent of a product's regular price. This percent is called the **discount rate.**

$$\boxed{\text{Discount rate}} \times \boxed{\text{regular price}} = \boxed{\text{discount}}$$

EXAMPLE • 7

A GE 25-inch stereo television that regularly sells for $299 is on sale for $250. Find the discount rate. Round to the nearest tenth of a percent.

Strategy

To find the discount rate:

• Find the discount.
• Solve the basic percent equation for *percent.*

Solution

$$\boxed{\text{Regular price}} - \boxed{\text{sale price}} = \boxed{\text{discount}}$$

$$299 \quad - \quad 250 \quad = \quad 49$$

$$\text{Percent} \quad \times \quad \text{base} \quad = \quad \text{amount}$$

$$\boxed{\text{Discount rate}} \times \boxed{\text{regular price}} = \boxed{\text{discount}}$$

$$n \quad \times \quad 299 \quad = \quad 49$$

$$n = 49 \div 299$$

$$n \approx 0.164$$

The discount rate is 16.4%.

YOU TRY IT • 7

A white azalea that regularly sells for $12.50 is on sale for $10.99. Find the discount rate. Round to the nearest tenth of a percent.

Your strategy

Your solution

EXAMPLE • 8

A 20-horsepower lawn mower is on sale for 25% off the regular price of $1525. Find the sale price.

Strategy

To find the sale price:

• Find the discount by solving the basic percent equation for *amount.*
• Subtract to find the sale price.

Solution

$$\text{Percent} \quad \times \quad \text{base} \quad = \quad \text{amount}$$

$$\boxed{\text{Discount rate}} \times \boxed{\text{regular price}} = \boxed{\text{discount}}$$

$$0.25 \quad \times \quad 1525 \quad = \quad n$$

$$381.25 = n$$

$$\boxed{\text{Regular price}} - \boxed{\text{discount}} = \boxed{\text{sale price}}$$

$$1525 \quad - \quad 381.25 \quad = \quad 1143.75$$

The sale price is $1143.75.

YOU TRY IT • 8

A hardware store is selling a Newport security door for 15% off the regular price of $225. Find the sale price.

Your strategy

Your solution

Solutions on p. S15

6.2 EXERCISES

OBJECTIVE A   To find percent increase

Solve. If necessary, round percents to the nearest tenth of a percent.

1. **Bison** See the news clipping at the right. Find the percent increase in human consumption of bison from 2005 to the date of this news article.

2. **Fuel Efficiency** An automobile manufacturer increased the average mileage on a car from 17.5 miles per gallon to 18.2 miles per gallon. Find the percent increase in mileage.

3. **Business** In the 1990s, the number of Target stores increased from 420 stores to 914 stores. (*Source:* Target) What was the percent increase in the number of Target stores in the 1990s?

4. **Demography** The graph at the right shows the number of unmarried American couples living together. (*Source:* U.S. Census Bureau) Find the percent increase in the number of unmarried couples living together from 1980 to 2000.

5. **Sports** In 1924, the number of events in the Winter Olympics was 14. The 2006 Winter Olympics in Salt Lake City included 84 medal events. (*Source: David Wallenchinsky's The Complete Book of the Winter Olympics*) Find the percent increase in the number of events in the Winter Olympics from 1924 to 2006.

6. **Television** During 1 year, the number of people subscribing to direct broadcasting satellite systems increased 87%. If the number of subscribers at the beginning of the year was 2.3 million, how many subscribers were there at the end of the year?

7. **Pets** In a recent year, Americans spent $35.9 billion on their pets. This was up from $17 billion a decade earlier. (*Source: Time,* February 4, 2008) Find the percent increase in the amount Americans spent on their pets during the 10-year period.

8. **Demography** From 1970 to 2000, the average age of American mothers giving birth to their first child rose 16.4%. (*Source:* Centers for Disease Control and Prevention) If the average age in 1970 was 21.4 years, what was the average age in 2000? Round to the nearest tenth.

9. **Compensation** A welder earning $12 per hour is given a 10% raise. To find the new wage, we can multiply $12 by 0.10 and add the product to $12. Can the new wage be found by multiplying $12 by 1.10?

In the News

A Taste for Bison

In 2005, the meat of 17,674 bison was consumed in the United States. This year, that number will reach 50,000. However, the consumption of bison is still a small fraction of beef consumption. Every day, the meat of 90,000 cattle is consumed in this country.

Source: Time, March 26, 2007

Unmarried U.S. Couples Living Together

OBJECTIVE B   To apply percent increase to business—markup

The three important markup equations are:
  (1) Selling price − cost = markup
  (2) Cost + markup = selling price
  (3) Markup rate × cost = markup

For Exercises 10 and 11, list, in the order in which they will be used, the equations needed to solve each problem.

10. A book that cost the seller $17 is sold for $23. Find the markup rate.

11. A DVD that cost the seller $12 has a markup rate of 55%. Find the selling price.

12. A window air conditioner cost AirRite Air Conditioning Systems $285. Find the markup on the air conditioner if the markup rate is 25% of the cost.

13. The manager of Brass Antiques has determined that a markup rate of 38% is necessary for a profit to be made. What is the markup on a brass doorknob that costs $45?

14. Computer Inc. uses a markup of $975 on a computer system that costs $3250. What is the markup rate on this system?

15. Saizon Pen & Office Supply uses a markup of $12 on a calculator that costs $20. What markup rate does this amount represent?

16. Giant Photo Service uses a markup rate of 48% on its Model ZA cameras, which cost the shop $162. What is the selling price?

17. The Circle R golf pro shop uses a markup rate of 45% on a set of Tour Pro golf clubs that costs the shop $210. What is the selling price?

18. Resner Builders' Hardware uses a markup rate of 42% for a table saw that costs $225. What is the selling price of the table saw?

19. Brad Burt's Magic Shop uses a markup rate of 48%. What is the selling price of a telescoping sword that costs $50?

OBJECTIVE C   To find percent decrease

Solve. If necessary, round to the nearest tenth of a percent.

20. **Law School** Use the news clipping at the right to find the percent decrease in the number of people who took the LSATs in the last three years.

21. **Travel** A new bridge reduced the normal 45-minute travel time between two cities by 18 minutes. What percent decrease does this represent?

In the News

Fewer Students Take LSATs

This year 137,444 people took the Law School Admission Test (LSATs). Three years ago, the LSATs were administered to 148,014 people.

Source: Law School Admission Council

22. Energy By installing energy-saving equipment, the Pala Rey Youth Camp reduced its normal $800-per-month utility bill by $320. What percent decrease does this amount represent?

	1990 Census	2000 Census	2005 Population Estimate
Chicago	1,783,726	2,896,016	2,842,518
Detroit	1,027,974	951,270	886,671
Phildelphia	1,585,577	1,517,550	1,463,281

Source: Census Bureau

23. Urban Populations The table at the right above shows the populations of three cities in the United States.
   a. Find the percent decrease in the population of Detroit from 1990 to 2005.
   b. Find the percent decrease in the population of Philadelphia from 1990 to 2005.
   c. Find the percent decrease in the population of Chicago from 2000 to 2005.

24. Missing Persons See the news clipping at the right. Find the percent decrease over the last 10 years in the number of people entered into the National Crime Information Center's Missing Person File.

In the News

Missing-Person Cases Decrease

This year, 834,536 missing-person cases were entered into the National Crime Information Center's Missing Person File. Ten years ago, the number was 969,264.

Source: National Crime Information Center

25. Depreciation It is estimated that the value of a new car is reduced 30% after 1 year of ownership. Using this estimate, find how much value a $28,200 new car loses after 1 year.

26. Employment A department store employs 1200 people during the holiday. At the end of the holiday season, the store reduces the number of employees by 45%. What is the decrease in the number of employees?

27. Finance Juanita's average monthly expense for gasoline was $176. After joining a car pool, she was able to reduce the expense by 20%.
   a. What was the amount of the decrease?
   b. What is the average monthly gasoline bill now?

28. Investments An oil company paid a dividend of $1.60 per share. After a reorganization, the company reduced the dividend by 37.5%.
   a. What was the amount of the decrease?
   b. What is the new dividend?

29. The Military In 2000, the Pentagon revised its account of the number of Americans killed in the Korean War from 54,246 to 36,940. (*Source: Time,* June 12, 2000) What is the percent decrease in the reported number of military personnel killed in the Korean War? Round to nearest tenth of a percent.

© Todd A. Gipstein/Corbis

30. In a math class, the average grade on the second test was 5% lower than the average grade on the first test. What should you multiply the first test average by to find the difference between the average grades on the two tests?

OBJECTIVE D   **To apply percent decrease to business—discount**

 The three important discount equations are:
(1) Regular price − sale price = discount
(2) Regular price − discount = sale price
(3) Discount rate × regular price = discount

For Exercises 31 and 32, list, in the order in which they will be used, the equations needed to solve each problem.

31. Shoes that regularly sell for $65 are on sale for 15% off the regular price. Find the sale price.

32. A radio with a regular price of $89 is on sale for $59. Find the discount rate.

33. The Austin College Bookstore is giving a discount of $8 on calculators that normally sell for $24. What is the discount rate?

34. A discount clothing store is selling a $72 sport jacket for $24 off the regular price. What is the discount rate?

35. A disc player that regularly sells for $400 is selling for 20% off the regular price. What is the discount?

36. Dacor Appliances is selling its $450 washing machine for 15% off the regular price. What is the discount?

37. An electric grill that regularly sells for $140 is selling for $42 off the regular price. What is the discount rate?

38. Quick Service Gas Station has its regularly priced $125 tune-up on sale for 16% off the regular price.
 a. What is the discount?
 b. What is the sale price?

39. Tomatoes that regularly sell for $1.25 per pound are on sale for 20% off the regular price.
 a. What is the discount?
 b. What is the sale price?

40. An outdoor supply store has its regularly priced $160 sleeping bags on sale for $120. What is the discount rate?

41. Standard Brands ceiling paint that regularly sells for $20 per gallon is on sale for $16 per gallon. What is the discount rate?

Applying the Concepts

 42. **Business** A promotional sale at a department store offers 25% off the sale price. The sale price itself is 25% off the regular price. Is this the same as a sale that offers 50% off the regular price? If not, which sale gives the better price? Explain your answer.

SECTION

6.3   Interest

OBJECTIVE A   **To calculate simple interest**

When you deposit money in a bank—for example, in a savings account—you are permitting the bank to use your money. The bank may use the deposited money to lend customers the money to buy cars or make renovations on their homes. The bank pays you for the privilege of using your money. The amount paid to you is called **interest.** If you are the one borrowing money from the bank, the amount you pay for the privilege of using that money is also called interest.

✓ **Take Note**

If you deposit $1000 in a savings account paying 5% interest, the $1000 is the principal and 5% is the interest rate.

The original amount deposited or borrowed is called the **principal.** The amount of interest paid is usually given as a percent of the principal. The percent used to determine the amount of interest is the **interest rate.**

Interest paid on the original principal is called **simple interest.** To calculate simple interest, multiply the principal by the interest rate per period by the number of time periods. In this objective, we are working with annual interest rates, so the time periods are years. The simple interest formula for an annual interest rate is given below.

Simple Interest Formula for Annual Interest Rates

Principal $\times$ annual interest rate $\times$ time (in years) = interest

---

Interest rates are generally given as percents. Before performing calculations involving an interest rate, write the interest rate as a decimal.

> **HOW TO 1**   Calculate the simple interest due on a 2-year loan of $1500 that has an annual interest rate of 7.5%.
>
> | Principal | $\times$ | annual interest rate | $\times$ | time (in years) | = | interest |
> |---|---|---|---|---|---|---|
> | 1500 | $\times$ | 0.075 | $\times$ | 2 | = | 225 |
>
> The simple interest due is $225.

When we borrow money, the total amount to be repaid to the lender is the sum of the principal and the interest. This amount is called the **maturity value of a loan.**

Maturity Value Formula for Simple Interest Loans

Principal $+$ interest = maturity value

In the example above, the simple interest due on the loan of $1500 was $225. The maturity value of the loan is therefore $1500 + $225 = $1725.

HOW TO 2 Calculate the maturity value of a simple interest, 8-month loan of $8000 if the annual interest rate is 9.75%.

First find the interest due on the loan.

Principal	×	annual interest rate	×	time (in years)	=	interest
8000	×	0.0975	×	$\frac{8}{12}$	=	520

Find the maturity value.

Principal	+	interest	=	maturity value
8000	+	520	=	8520

The maturity value of the loan is $8520.

✓ **Take Note**
The time of the loan must be in years. Eight months is $\frac{8}{12}$ of a year.

See Example 1. The time of the loan must be in years. 180 days is $\frac{180}{365}$ of a year.

The monthly payment on a loan can be calculated by dividing the maturity value by the length of the loan in months.

> **Monthly Payment on a Simple Interest Loan**
>
> Maturity value ÷ length of the loan in months = monthly payment

In the example above, the maturity value of the loan is $8520. To find the monthly payment on the 8-month loan, divide 8520 by 8.

Maturity value	÷	length of the loan in months	=	monthly payment
8520	÷	8	=	1065

The monthly payment on the loan is $1065.

EXAMPLE 1

Kamal borrowed $500 from a savings and loan association for 180 days at an annual interest rate of 7%. What is the simple interest due on the loan?

Strategy
To find the simple interest due, multiply the principal (500) times the annual interest rate (7% = 0.07) times the time in years (180 days = $\frac{180}{365}$ year).

Solution

Principal	×	annual interest rate	×	time (in years)	=	interest
500	×	0.07	×	$\frac{180}{365}$	≈	17.26

The simple interest due is $17.26.

YOU TRY IT 1

A company borrowed $15,000 from a bank for 18 months at an annual interest rate of 8%. What is the simple interest due on the loan?

Your strategy

Your solution

Solution on p. S15

EXAMPLE • 2

Calculate the maturity value of a simple interest, 9-month loan of $4000 if the annual interest rate is 8.75%.

Strategy

To find the maturity value:

• Use the simple interest formula to find the simple interest due.
• Find the maturity value by adding the principal and the interest.

Solution

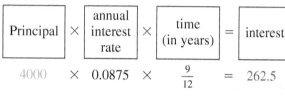

$$\boxed{\text{Principal}} \times \boxed{\begin{array}{c}\text{annual}\\\text{interest}\\\text{rate}\end{array}} \times \boxed{\begin{array}{c}\text{time}\\\text{(in years)}\end{array}} = \boxed{\text{interest}}$$

$$4000 \quad \times \quad 0.0875 \quad \times \quad \frac{9}{12} \quad = \quad 262.5$$

$$\boxed{\text{Principal}} + \boxed{\text{interest}} = \boxed{\begin{array}{c}\text{maturity}\\\text{value}\end{array}}$$

$$4000 \quad + \quad 262.50 \quad = \quad 4262.50$$

The maturity value is $4262.50.

EXAMPLE • 3

The simple interest due on a 3-month loan of $1400 is $26.25. Find the monthly payment on the loan.

Strategy

To find the monthly payment:

• Find the maturity value by adding the principal and the interest.
• Divide the maturity value by the length of the loan in months (3).

Solution

Principal + interest = maturity value
   1400 + 26.25 =   1426.25

Maturity value ÷ length of the loan = payment
   1426.25   ÷   3   ≈ 475.42

The monthly payment is $475.42.

YOU TRY IT • 2

Calculate the maturity value of a simple interest, 90-day loan of $3800. The annual interest rate is 6%.

Your strategy

Your solution

YOU TRY IT • 3

The simple interest due on a 1-year loan of $1900 is $152. Find the monthly payment on the loan.

Your strategy

Your solution

Solutions on p. S15

OBJECTIVE B   **To calculate finance charges on a credit card bill**

When a customer uses a credit card to make a purchase, the customer is actually receiving a loan. Therefore, there is frequently an added cost to the consumer who purchases on credit. This may be in the form of an annual fee and interest charges on purchases. The interest charges on purchases are called **finance charges.**

The finance charge on a credit card bill is calculated using the simple interest formula. In the last objective, the interest rates were annual interest rates. However, credit card companies generally issue *monthly* bills and express interest rates on credit card purchases as *monthly* interest rates. Therefore, when using the simple interest formula to calculate finance charges on credit card purchases, use a monthly interest rate and express the time in months.

Note: In the simple interest formula, the time must be expressed in the same period as the rate. For an *annual* interest rate, the time must be expressed in years. For a *monthly* interest rate, the time must be expressed in months.

EXAMPLE 4

A credit card company charges a customer 1.5% per month on the unpaid balance of charges on the credit card. What is the finance charge in a month in which the customer has an unpaid balance of $254?

Strategy

To find the finance charge, multiply the principal, or unpaid balance (254), times the monthly interest rate (1.5%) times the number of months (1).

Solution

254 × 0.015 × 1 = 3.81

The finance charge is $3.81.

YOU TRY IT 4

The credit card that Francesca uses charges her 1.6% per month on her unpaid balance. Find the finance charge when her unpaid balance for the month is $1250.

Your strategy

Your solution

Solution on p. S15

OBJECTIVE C   To calculate compound interest

Usually, the interest paid on money deposited or borrowed is compound interest. **Compound interest** is computed not only on the original principal but also on interest already earned. Here is an illustration.

Suppose $1000 is invested for 3 years at an annual interest rate of 9% compounded annually. Because this is an *annual* interest rate, we will calculate the interest earned each year.

During the first year, the interest earned is calculated as follows:

Principal	×	annual interest rate	×	time (in years)	=	interest
1000	×	0.09	×	1	=	90

At the end of the first year, the total amount in the account is

$$1000 + 90 = 1090$$

During the second year, the interest earned is calculated on the amount in the account at the end of the first year.

Principal	×	annual interest rate	×	time (in years)	=	interest
1090	×	0.09	×	1	=	98.10

Note that the interest earned during the second year ($98.10) is greater than the interest earned during the first year ($90). This is because the interest earned during the first year was added to the original principal, and the interest for the second year was calculated using this sum. If the account earned simple interest, the interest earned would be the same every year ($90).

At the end of the second year, the total amount in the account is the sum of the amount in the account at the end of the first year and the interest earned during the second year.

$$1090 + 98.10 = 1188.10$$

The interest earned during the third year is calculated using the amount in the account at the end of the second year ($1188.10).

Principal	×	annual interest rate	×	time (in years)	=	interest
1188.10	×	0.09	×	1	≈	106.93

Take Note
The interest earned each year keeps increasing. This is the effect of compound interest.

The amount in the account at the end of the third year is

$$1188.10 + 106.93 = 1295.03$$

To find the interest earned for the three years, subtract the original principal from the new principal.

New principal	−	original principal	=	interest earned
1295.03	−	1000	=	295.03

Note that the compound interest earned is $295.03. The simple interest earned on the investment would have been only $1000 × 0.09 × 3 = $270.

In this example, the interest was compounded annually. However, interest can be compounded

Compounding periods:	annually (once a year)
	semiannually (twice a year)
	quarterly (four times a year)
	monthly (12 times a year)
	daily (365 times a year)

The more frequent the compounding periods, the more interest the account earns. For example, if, in the above example, the interest had been compounded quarterly rather than annually, the interest earned would have been greater.

Calculating compound interest can be very tedious, so there are tables that can be used to simplify these calculations. A portion of a Compound Interest Table is given in the Appendix.

HOW TO 3 What is the value after 5 years of $1000 invested at 7% annual interest, compounded quarterly?

To find the interest earned, multiply the original principal (1000) by the factor found in the Compound Interest Table. To find the factor, first find the table headed "Compounded Quarterly" in the Compound Interest Table in the Appendix. Then look at the number where the 7% column and the 5-year row meet.

	Compounded Quarterly						
	4%	*5%*	*6%*	*7%*	*8%*	*9%*	*10%*
1 year	1.04060	1.05094	1.06136	1.07186	1.08243	1.09308	1.10381
5 years	1.22019	1.28204	1.34686	**1.41478**	1.48595	1.56051	1.63862
10 years	1.48886	1.64362	1.81402	2.00160	2.20804	2.43519	2.68506
15 years	1.81670	2.10718	2.44322	2.83182	3.28103	3.80013	4.39979
20 years	2.21672	2.70148	3.29066	4.00639	4.87544	5.93015	7.20957

The factor is 1.41478.

$$1000 \times 1.41478 = 1414.78$$

The value of the investment after 5 years is $1414.78.

EXAMPLE 5

An investment of $650 pays 8% annual interest, compounded semiannually. What is the interest earned in 5 years?

Strategy

To find the interest earned:

- Find the new principal by multiplying the original principal (650) by the factor found in the Compound Interest Table (1.48024).
- Subtract the original principal from the new principal.

Solution

$650 \times 1.48024 \approx 962.16$

The new principal is $962.16.

$962.16 - 650 = 312.16$

The interest earned is $312.16.

YOU TRY IT 5

An investment of $1000 pays 6% annual interest, compounded quarterly. What is the interest earned in 20 years?

Your strategy

Your solution

Solution on pp. S15–S16

6.3 EXERCISES

OBJECTIVE A   To calculate simple interest

1. A 2-year student loan of $10,000 is made at an annual simple interest rate of 4.25%. The simple interest on the loan is $850. Identify **a.** the principal, **b.** the interest, **c.** the interest rate, and **d.** the time period of the loan.

2. A contractor obtained a 9-month loan for $80,000 at an annual simple interest rate of 9.75%. The simple interest on the loan is $5850. Identify **a.** the principal, **b.** the interest, **c.** the interest rate, and **d.** the time period of the loan.

3. Find the simple interest Jacob Zucker owes on a 2-year student loan of $8000 at an annual interest rate of 6%.

4. Find the simple interest Kara Tanamachi owes on a $1\frac{1}{2}$-year loan of $1500 at an annual interest rate of 7.5%.

5. To finance the purchase of 15 new cars, the Tropical Car Rental Agency borrowed $100,000 for 9 months at an annual interest rate of 4.5%. What is the simple interest due on the loan?

6. A home builder obtained a preconstruction loan of $50,000 for 8 months at an annual interest rate of 9.5%. What is the simple interest due on the loan?

7. A bank lent Gloria Masters $20,000 at an annual interest rate of 8.8%. The period of the loan was 9 months. Find the simple interest due on the loan.

8. Eugene Madison obtained an 8-month loan of $4500 at an annual interest rate of 6.2%. Find the simple interest Eugene owes on the loan.

9. Jorge Elizondo took out a 75-day loan of $7500 at an annual interest rate of 5.5%. Find the simple interest due on the loan.

10. Kristi Yang borrowed $15,000. The term of the loan was 90 days, and the annual simple interest rate was 7.4%. Find the simple interest due on the loan.

11. The simple interest due on a 4-month loan of $4800 is $320. What is the maturity value of the loan?

12. The simple interest due on a 60-day loan of $6500 is $80.14. Find the maturity value of the loan.

13. William Carey borrowed $12,500 for 8 months at an annual simple interest rate of 4.5%. Find the total amount due on the loan.

14. You arrange for a 9-month bank loan of $9000 at an annual simple interest rate of 8.5%. Find the total amount you must repay to the bank.

15. Capital City Bank approves a home-improvement loan application for $14,000 at an annual simple interest rate of 5.25% for 270 days. What is the maturity value of the loan?

16. A credit union lends a member $5000 for college tuition. The loan is made for 18 months at an annual simple interest rate of 6.9%. What is the maturity value of this loan?

17. Action Machining Company purchased a robot-controlled lathe for $225,000 and financed the full amount at 8% annual simple interest for 4 years. The simple interest on the loan is $72,000. Find the monthly payment.

18. For the purchase of an entertainment center, a $1900 loan is obtained for 2 years at an annual simple interest rate of 9.4%. The simple interest due on the loan is $357.20. What is the monthly payment on the loan?

19. To attract new customers, Heller Ford is offering car loans at an annual simple interest rate of 4.5%.
    a. Find the interest charged to a customer who finances a car loan of $12,000 for 2 years.
    b. Find the monthly payment.

20. Cimarron Homes Inc. purchased a snow plow for $57,000 and financed the full amount for 5 years at an annual simple interest rate of 9%.
    a. Find the interest due on the loan.
    b. Find the monthly payment.

21. Dennis Pappas decided to build onto his present home instead of buying a new, larger house. He borrowed $142,000 for $5\frac{1}{2}$ years at an annual simple interest rate of 7.5%. Find the monthly payment.

22. Rosalinda Johnson took out a 6-month, $12,000 loan. The annual simple interest rate on the loan was 8.5%. Find the monthly payment.

23. Student A and Student B borrow the same amount of money at the same annual interest rate. Student A has a 2-year loan and Student B has a 1-year loan. In each case, state whether the first quantity is *less than, equal to,* or *greater than* the second quantity.
    a. Student A's principal; Student B's principal
    b. Student A's maturity value; Student B's maturity value
    c. Student A's monthly payment; Student B's monthly payment

OBJECTIVE B   To calculate finance charges on a credit card bill

24. A credit card company charges a customer 1.25% per month on the unpaid balance of charges on the credit card. What is the finance charge in a month in which the customer has an unpaid balance of $118.72?

25. The credit card that Dee Brown uses charges her 1.75% per month on her unpaid balance. Find the finance charge when her unpaid balance for the month is $391.64.

26. What is the finance charge on an unpaid balance of $12,368.92 on a credit card that charges 1.5% per month on any unpaid balance?

27. Suppose you have an unpaid balance of $995.04 on a credit card that charges 1.2% per month on any unpaid balance. What finance charge do you owe the company?

28. A credit card customer has an unpaid balance of $1438.20. What is the difference between monthly finance charges of 1.15% per month on the unpaid balance and monthly finance charges of 1.85% per month?

29. One credit card company charges 1.25% per month on any unpaid balance, and a second company charges 1.75%. What is the difference between the finance charges that these two companies assess on an unpaid balance of $687.45?

 Your credit card company requires a minimum monthly payment of $10. You plan to pay off the balance on your credit card by paying the minimum amount each month and making no further purchases using this credit card. For Exercises 30 and 31, state whether the finance charge for the second month will be *less than, equal to,* or *greater than* the finance charge for the first month, and state whether you will eventually be able to pay off the balance.

30. The finance charge for the first month was less than $10.

31. The finance charge for the first month was exactly $10.

OBJECTIVE C   To calculate compound interest

32. North Island Federal Credit Union pays 4% annual interest, compounded daily, on time savings deposits. Find the value after 1 year of $750 deposited in this account.

33. Tanya invested $2500 in a tax-sheltered annuity that pays 8% annual interest, compounded daily. Find the value of her investment after 20 years.

34. Sal Travato invested $3000 in a corporate retirement account that pays 6% annual interest, compounded semiannually. Find the value of his investment after 15 years.

35. To replace equipment, a farmer invested $20,000 in an account that pays 7% annual interest, compounded monthly. What is the value of the investment after 5 years?

36. Green River Lodge invests $75,000 in a trust account that pays 8% interest, compounded quarterly.
 a. What will the value of the investment be in 5 years?
 b. How much interest will be earned in the 5 years?

37. To save for retirement, a couple deposited $3000 in an account that pays 7% annual interest, compounded daily.
 a. What will the value of the investment be in 10 years?
 b. How much interest will be earned in the 10 years?

38. To save for a child's education, the Petersens deposited $2500 into an account that pays 6% annual interest, compounded daily. Find the amount of interest earned on this account over a 20-year period.

39. How much interest is earned in 2 years on $4000 deposited in an account that pays 6% interest, compounded quarterly?

 40. The compound interest factor for a 5-year investment at an annual interest rate of 6%, compounded semiannually, is 1.34392. What does the expression $3500 - (3500 \times 1.34392)$ represent?

Applying the Concepts

 41. **Banking** At 4 P.M. on July 31, you open a savings account that pays 5% annual interest and you deposit $500 in the account. Your deposit is credited as of August 1. At the beginning of September, you receive a statement from the bank that shows that during the month of August, you received $2.12 in interest. The interest has been added to your account, bringing the total on deposit to $502.12. At the beginning of October, you receive a statement from the bank that shows that during the month of September, you received $2.06 in interest on the $502.12 on deposit. Explain why you received less interest during the second month when there was more money on deposit.

42. **Banking** Suppose you have a savings account that earns interest at the rate of 6% per year, compounded monthly. On January 1, you open this account with a deposit of $100.
 a. On February 1, you deposit an additional $100 into the account. What is the value of the account after the deposit?
 b. On March 1, you deposit an additional $100 into the account. What is the value of the account after the deposit?
 Note: This type of savings plan, wherein equal amounts ($100) are saved at equal time intervals (every month), is called an annuity.

SECTION

6.4 Real Estate Expenses

OBJECTIVE A To calculate the initial expenses of buying a home

One of the largest investments most people ever make is the purchase of a home. The major initial expense in the purchase is the **down payment,** which is normally a percent of the purchase price. This percent varies among banks, but it usually ranges from 5% to 25%.

The **mortgage** is the amount that is borrowed to buy real estate. The mortgage amount is the difference between the purchase price and the down payment.

> **HOW TO 1** A home is purchased for $140,000, and a down payment of $21,000 is made. Find the mortgage.
>
> | Purchase price | − | down payment | = | mortgage |
> |---|---|---|---|---|
> | 140,000 | − | 21,000 | = | 119,000 |
>
> The mortgage is $119,000.

 Take Note

Because *points* means percent, a loan origination fee of $2\frac{1}{2}$ points = $2\frac{1}{2}\% =$ 2.5% = 0.025.

Another initial expense in buying a home is the **loan origination fee,** which is a fee that the bank charges for processing the mortgage papers. The loan origination fee is usually a percent of the mortgage and is expressed in **points,** which is the term banks use to mean percent. For example, "5 points" means "5 percent."

Points	×	mortgage	=	loan origination fee

EXAMPLE 1

A house is purchased for $250,000, and a down payment, which is 20% of the purchase price, is made. Find the mortgage.

Strategy

To find the mortgage:

- Find the down payment by solving the basic percent equation for *amount.*
- Subtract the down payment from the purchase price.

Solution

Percent	×	base	=	amount

Percent	×	purchase price	=	down payment
0.20	×	250,000	=	n
		50,000	=	n

Purchase price	−	down payment	=	mortgage
250,000	−	50,000	=	200,000

The mortgage is $200,000.

YOU TRY IT 1

An office building is purchased for $1,500,000, and a down payment, which is 25% of the purchase price, is made. Find the mortgage.

Your strategy

Your solution

Solution on p. S16

EXAMPLE · 2

A home is purchased with a mortgage of $165,000. The buyer pays a loan origination fee of $3\frac{1}{2}$ points. How much is the loan origination fee?

Strategy

To find the loan origination fee, solve the basic percent equation for *amount*.

Solution

Percent × base = amount

| Points | × | mortgage | = | fee |

0.035 × 165,000 = n

5775 = n

The loan origination fee is $5775.

YOU TRY IT · 2

The mortgage on a real estate investment is $180,000. The buyer paid a loan origination fee of $4\frac{1}{2}$ points. How much was the loan origination fee?

Your strategy

Your solution

Solution on p. S16

OBJECTIVE B To calculate the ongoing expenses of owning a home

Point of Interest

The number-one response of adults when asked what they would spend money on first if they suddenly became wealthy (for example, by winning the lottery) was a house; 31% gave this response. (*Source:* Yankelovich Partners for Lutheran Brotherhood)

Besides the initial expenses of buying a house, there are continuing monthly expenses involved in owning a home. The **monthly mortgage payment** (one of 12 payments due each year to the lender of money to buy real estate), utilities, insurance, and **property tax** (a tax based on the value of real estate) are some of these ongoing expenses. Of these expenses, the largest one is normally the monthly mortgage payment.

For a **fixed-rate mortgage,** the monthly mortgage payment remains the same throughout the life of the loan. The calculation of the monthly mortgage payment is based on the amount of the loan, the interest rate on the loan, and the number of years required to pay back the loan. Calculating the monthly mortgage payment is fairly difficult, so tables such as the one in the Appendix are used to simplify these calculations.

 Integrating Technology

In general, when a problem requests a monetary payment, the answer is rounded to the nearest cent. For the example at the right, enter

160000 **X** 0.0080462 **=**

The display reads 1287.392. Round this number to the nearest hundredth: 1287.39. The answer is $1287.39.

HOW TO · 2 Find the monthly mortgage payment on a 30-year, $160,000 mortgage at an interest rate of 9%. Use the Monthly Payment Table in the Appendix.

$$160,000 \times \underset{\underset{\text{From the table}}{\uparrow}}{0.0080462} \approx 1287.39$$

The monthly mortgage payment is $1287.39.

The monthly mortgage payment includes the payment of both principal and interest on the mortgage. The interest charged during any one month is charged on the unpaid balance of the loan. Therefore, during the early years of the mortgage, when the unpaid balance is high, most of the monthly mortgage payment is interest charged on the loan. During the last few years of a mortgage, when the unpaid balance is low, most of the monthly mortgage payment goes toward paying off the loan.

Point of Interest

Home buyers rated the following characteristics "extremely important" in their purchase decision.

Natural, open space: 77%
Walking and biking paths: 74%
Gardens with native plants: 56%
Clustered retail stores: 55%
Wilderness area: 52%
Outdoor pool: 52%
Community recreation center: 52%
Interesting little parks: 50%

(*Sources:* American Lives, Inc; Intercommunications, Inc.)

HOW TO 3 Find the interest paid on a mortgage during a month in which the monthly mortgage payment is $886.26 and $358.08 of that amount goes toward paying off the principal.

Monthly mortgage payment	−	principal	=	interest
886.26	−	358.08	=	528.18

The interest paid on the mortgage is $528.18.

Property tax is another ongoing expense of owning a house. Property tax is normally an annual expense that may be paid on a monthly basis. The monthly property tax, which is determined by dividing the annual property tax by 12, is usually added to the monthly mortgage payment.

HOW TO 4 A homeowner must pay $3120 in property tax annually. Find the property tax that must be added each month to the homeowner's monthly mortgage payment.

$3120 \div 12 = 260$

Each month, $260 must be added to the monthly mortgage payment for property tax.

EXAMPLE 3

Serge purchased some land for $120,000 and made a down payment of $25,000. The savings and loan association charges an annual interest rate of 8% on Serge's 25-year mortgage. Find the monthly mortgage payment.

Strategy

To find the monthly mortgage payment:
• Subtract the down payment from the purchase price to find the mortgage.
• Multiply the mortgage by the factor found in the Monthly Payment Table in the Appendix.

Solution

Purchase price	−	down payment	=	mortgage
120,000	−	25,000	=	95,000

$95,000 \times 0.0077182 \approx 733.23$
↑
From the table

The monthly mortgage payment is $733.23.

YOU TRY IT 3

A new condominium project is selling townhouses for $175,000. A down payment of $17,500 is required, and a 20-year mortgage at an annual interest rate of 9% is available. Find the monthly mortgage payment.

Your strategy

Your solution

Solution on p. S16

EXAMPLE • 4

A home has a mortgage of $134,000 for 25 years at an annual interest rate of 7%. During a month in which $375.88 of the monthly mortgage payment is principal, how much of the payment is interest?

Strategy

To find the interest:

- Multiply the mortgage by the factor found in the Monthly Payment Table in the Appendix to find the monthly mortgage payment.
- Subtract the principal from the monthly mortgage payment.

Solution

134,000 × 0.0070678 ≈ 947.09

      ↑   ↑

   From the   Monthly mortgage
    table   payment

Monthly mortgage payment	−	principal	=	interest
947.09	−	375.88	=	571.21

$571.21 of the payment is interest on the mortgage.

YOU TRY IT • 4

An office building has a mortgage of $625,000 for 25 years at an annual interest rate of 7%. During a month in which $2516.08 of the monthly mortgage payment is principal, how much of the payment is interest?

Your strategy

Your solution

EXAMPLE • 5

The monthly mortgage payment for a home is $998.75. The annual property tax is $4020. Find the total monthly payment for the mortgage and property tax.

Strategy

To find the monthly payment:

- Divide the annual property tax by 12 to find the monthly property tax.
- Add the monthly property tax to the monthly mortgage payment.

Solution

4020 ÷ 12 = 335   • Monthly property tax

998.75 + 335 = 1333.75

The total monthly payment is $1333.75.

YOU TRY IT • 5

The monthly mortgage payment for a home is $815.20. The annual property tax is $3000. Find the total monthly payment for the mortgage and property tax.

Your strategy

Your solution

Solutions on p. S16

6.4 EXERCISES

OBJECTIVE A   To calculate the initial expenses of buying a home

1. A condominium at Mt. Baldy Ski Resort was purchased for $197,000, and a down payment of $24,550 was made. Find the mortgage.

2. An insurance business was purchased for $173,000, and a down payment of $34,600 was made. Find the mortgage.

3. Brian Stedman made a down payment of 25% of the $850,000 purchase price of an apartment building. How much was the down payment?

4. A clothing store was purchased for $625,000, and a down payment that was 25% of the purchase price was made. Find the down payment.

5. A loan of $150,000 is obtained to purchase a home. The loan origination fee is $2\frac{1}{2}$ points. Find the amount of the loan origination fee.

6. Security Savings & Loan requires a borrower to pay $3\frac{1}{2}$ points for a loan. Find the amount of the loan origination fee for a loan of $90,000.

7. Baja Construction Inc. is selling homes for $350,000. A down payment of 10% is required. Find the mortgage.

8. A cattle rancher purchased some land for $240,000. The bank requires a down payment of 15% of the purchase price. Find the mortgage.

9. Vivian Tom purchased a home for $210,000. Find the mortgage if the down payment Vivian made is 10% of the purchase price.

10. A mortgage lender requires a down payment of 5% of the $180,000 purchase price of a condominium. How much is the mortgage?

11. A home is purchased for $435,000. The mortgage lender requires a 10% down payment. Which expression below represents the mortgage?
    (i) $0.10 \times 435,000$ (ii) $0.10 \times 435,000 - 435,000$
    (iii) $435,000 - 0.10 \times 435,000$ (iv) $435,000 + 0.10 \times 435,000$

OBJECTIVE B   To calculate the ongoing expenses of owning a home

For Exercises 12 to 23, solve. Use the Monthly Payment Table in the Appendix. Round to the nearest cent.

12. An investor obtained a loan of $850,000 to buy a car wash business. The monthly mortgage payment was based on 25 years at 8%. Find the monthly mortgage payment.

13. A beautician obtained a 20-year mortgage of $90,000 to expand the business. The credit union charges an annual interest rate of 6%. Find the monthly mortgage payment.

14. A couple interested in buying a home determines that they can afford a monthly mortgage payment of $800. Can they afford to buy a home with a 30-year, $110,000 mortgage at 8% interest?

15. A lawyer is considering purchasing a new office building with a 15-year, $400,000 mortgage at 6% interest. The lawyer can afford a monthly mortgage payment of $3500. Can the lawyer afford the monthly mortgage payment on the new office building?

16. The county tax assessor has determined that the annual property tax on a $325,000 house is $3032.40. Find the monthly property tax.

17. The annual property tax on a $155,000 home is $1992. Find the monthly property tax.

18. Abacus Imports Inc. has a warehouse with a 25-year mortgage of $200,000 at an annual interest rate of 9%.
    a. Find the monthly mortgage payment.
    b. During a month in which $941.72 of the monthly mortgage payment is principal, how much of the payment is interest?

19. A vacation home has a mortgage of $135,000 for 30 years at an annual interest rate of 7%.
    a. Find the monthly mortgage payment.
    b. During a month in which $392.47 of the monthly mortgage payment is principal, how much of the payment is interest?

20. The annual mortgage payment on a duplex is $20,844.40. The owner must pay an annual property tax of $1944. Find the total monthly payment for the mortgage and property tax.

21. The monthly mortgage payment on a home is $716.40, and the homeowner pays an annual property tax of $1512. Find the total monthly payment for the mortgage and property tax.

22. Maria Hernandez purchased a home for $210,000 and made a down payment of $15,000. The balance was financed for 15 years at an annual interest rate of 6%. Find the monthly mortgage payment.

23. A customer of a savings and loan purchased a $385,000 home and made a down payment of $40,000. The savings and loan charges its customers an annual interest rate of 7% for 30 years for a home mortgage. Find the monthly mortgage payment.

24. The monthly mortgage payment for a home is $623.57. The annual property tax is $1400. Which expression below represents the total monthly payment for the mortgage and property tax? Which expression represents the total amount of money the owner will spend on the mortgage and property tax in one year?
    (i) $623.57 + 1400$ (ii) $12 \times 623.57 + 1400$
    (iii) $\dfrac{623.57 + 1400}{12}$ (iv) $623.57 + \dfrac{1400}{12}$

Applying the Concepts

25. **Mortgages** A couple considering a mortgage of $100,000 have a choice of loans. One loan is an 8% loan for 20 years, and the other loan is at 8% for 30 years. Find the amount of interest that the couple can save by choosing the 20-year loan.

SECTION

6.5 Car Expenses

OBJECTIVE A **To calculate the initial expenses of buying a car**

The initial expenses in the purchase of a car usually include the down payment, the **license fees** (fees charged for authorization to operate a vehicle), and the **sales tax** (a tax levied by a state or municipality on purchases). The down payment may be very small or as much as 25% or 30% of the purchase price of the car, depending on the lending institution. License fees and sales tax are regulated by each state, so these expenses vary from state to state.

EXAMPLE 1	**YOU TRY IT 1**
A car is purchased for $38,500, and the lender requires a down payment of 15% of the purchase price. Find the amount financed.	A down payment of 20% of the $19,200 purchase price of a new car is made. Find the amount financed.

Strategy
To find the amount financed:

- Find the down payment by solving the basic percent equation for *amount*.
- Subtract the down payment from the purchase price.

Your strategy

Solution

Percent	×	base	=	amount
Percent	×	purchase price	=	down payment
0.15	×	38,500	=	n
			5775	= n

$38,500 - 5775 = $ **32,725**

The amount financed is $32,725.

Your solution

EXAMPLE 2	**YOU TRY IT 2**
A sales clerk purchases a used car for $16,500 and pays a sales tax that is 5% of the purchase price. How much is the sales tax?	A car is purchased for $27,350. The car license fee is 1.5% of the purchase price. How much is the license fee?

Strategy
To find the sales tax, solve the basic percent equation for *amount*.

Your strategy

Solution

Percent	×	base	=	amount
Percent	×	purchase price	=	sales tax
0.05	×	16,500	=	n
			825	= n

The sales tax is $825.

Your solution

Solutions on pp. S16–S17

OBJECTIVE B To calculate the ongoing expenses of owning a car

 Take Note

The same formula that is used to calculate a monthly mortgage payment is used to calculate a monthly car payment.

Besides the initial expenses of buying a car, there are continuing expenses involved in owning a car. These ongoing expenses include car insurance, gas and oil, general maintenance, and the monthly car payment. The monthly car payment is calculated in the same manner as the monthly mortgage payment on a home loan. A monthly payment table, such as the one in the Appendix, is used to simplify the calculation of monthly car payments.

EXAMPLE · 3

At a cost of $.38 per mile, how much does it cost to operate a car during a year in which the car is driven 15,000 miles?

Strategy

To find the cost, multiply the cost per mile (0.38) by the number of miles driven (15,000).

Solution

$15,000 \times 0.38 = 5700$

The cost is $5700.

YOU TRY IT · 3

At a cost of $.41 per mile, how much does it cost to operate a car during a year in which the car is driven 23,000 miles?

Your strategy

Your solution

EXAMPLE · 4

During one month, your total gasoline bill was $252 and the car was driven 1200 miles. What was the cost per mile for gasoline?

Strategy

To find the cost per mile, divide the cost for gasoline (252) by the number of miles driven (1200).

Solution

$252 \div 1200 = 0.21$

The cost per mile was $.21.

YOU TRY IT · 4

In a year in which your total car insurance bill was $360 and the car was driven 15,000 miles, what was the cost per mile for car insurance?

Your strategy

Your solution

EXAMPLE · 5

A car is purchased for $18,500 with a down payment of $3700. The balance is financed for 3 years at an annual interest rate of 6%. Find the monthly car payment.

Strategy

To find the monthly payment:

• Subtract the down payment from the purchase price to find the amount financed.
• Multiply the amount financed by the factor found in the Monthly Payment Table in the Appendix.

Solution

$18,500 - 3700 = 14,800$

$14,800 \times 0.0304219 \approx 450.24$

The monthly payment is $450.24.

YOU TRY IT · 5

A truck is purchased for $25,900 with a down payment of $6475. The balance is financed for 4 years at an annual interest rate of 8%. Find the monthly car payment.

Your strategy

Your solution

Solutions on p. S17

6.5 EXERCISES

OBJECTIVE A   To calculate the initial expenses of buying a car

1. Amanda has saved $780 to make a down payment on a used minivan that costs $7100. The car dealer requires a down payment of 12% of the purchase price. Has she saved enough money to make the down payment?

2. A sedan was purchased for $23,500. A down payment of 15% of the purchase price was required. How much was the down payment?

3. A drapery installer bought a van to carry drapery samples. The purchase price of the van was $26,500, and a 4.5% sales tax was paid. How much was the sales tax?

4. A & L Lumber Company purchased a delivery truck for $28,500. A sales tax of 4% of the purchase price was paid. Find the sales tax.

5. A license fee of 2% of the purchase price is paid on a pickup truck costing $32,500. Find the license fee for the truck.

6. Your state charges a license fee of 1.5% on the purchase price of a car. How much is the license fee for a car that costs $16,998?

7. An electrician bought a $32,000 flatbed truck. A state license fee of $275 and a sales tax of 3.5% of the purchase price are required.
    a. Find the sales tax.
    b. Find the total cost of the sales tax and the license fee.

8. A physical therapist bought a used car for $9375 and made a down payment of $1875. The sales tax is 5% of the purchase price.
    a. Find the sales tax.
    b. Find the total cost of the sales tax and the down payment.

9. Martin bought a motorcycle for $16,200 and made a down payment of 25% of the purchase price. Find the amount financed.

10. A carpenter bought a utility van for $24,900 and made a down payment of 15% of the purchase price. Find the amount financed.

11. An author bought a sports car for $45,000 and made a down payment of 20% of the purchase price. Find the amount financed.

12. Tania purchased a used car for $13,500 and made a down payment of 25% of the cost. Find the amount financed.

 13. The purchase price of a car is $25,700. The car dealer requires a down payment of 15% of the purchase price. There is a license fee of 2.5% of the purchase price and sales tax of 6% of the purchase price. What does the following expression represent?
    $25{,}700 + 0.025 \times 25{,}700 + 0.06 \times 25{,}700$

OBJECTIVE B   To calculate the ongoing expenses of owning a car

 14. A driver had $1100 in car expenses and drove his car 8500 miles. Would you use *multiplication* or *division* to find the cost per mile to operate the car?

 15. A car costs $.36 per mile to operate. Would you use *multiplication* or *division* to find the cost of driving the car 23,000 miles?

For Exercises 16 to 25, solve. Use the Monthly Payment Table in the Appendix. Round to the nearest cent.

16. A rancher financed $24,000 for the purchase of a truck through a credit union at 5% interest for 4 years. Find the monthly truck payment.

17. A car loan of $18,000 is financed for 3 years at an annual interest rate of 4%. Find the monthly car payment.

18. An estimate of the cost of owning a compact car is $.38 per mile. Using this estimate, find how much it costs to operate a car during a year in which the car is driven 16,000 miles.

19. An estimate of the cost of care and maintenance of automobile tires is $.018 per mile. Using this estimate, find how much it costs for care and maintenance of tires during a year in which the car is driven 14,000 miles.

20. A family spent $2600 on gas, oil, and car insurance during a period in which the car was driven 14,000 miles. Find the cost per mile for gas, oil, and car insurance.

21. Last year you spent $2400 for gasoline for your car. The car was driven 15,000 miles. What was your cost per mile for gasoline?

22. The city of Colton purchased a fire truck for $164,000 and made a down payment of $10,800. The balance is financed for 5 years at an annual interest rate of 6%.
   a. Find the amount financed.
   b. Find the monthly truck payment.

23. A used car is purchased for $14,999, and a down payment of $2999 is made. The balance is financed for 3 years at an annual interest rate of 5%.
   a. Find the amount financed.
   b. Find the monthly car payment.

24. An artist purchased a new car costing $27,500 and made a down payment of $5500. The balance is financed for 3 years at an annual interest rate of 4%. Find the monthly car payment.

25. A camper is purchased for $39,500, and a down payment of $5000 is made. The balance is financed for 4 years at an annual interest rate of 6%. Find the monthly payment.

Ulrich Mueller/Flickr/Getty Images

Applying the Concepts

26. **Car Loans** One bank offers a 4-year car loan at an annual interest rate of 7% plus a loan application fee of $45. A second bank offers 4-year car loans at an annual interest rate of 8% but charges no loan application fee. If you need to borrow $5800 to purchase a car, which of the two bank loans has the lesser loan costs? Assume you keep the car for 4 years.

27. **Car Loans** How much interest is paid on a 5-year car loan of $19,000 if the interest rate is 9%? Round to the nearest dollar.

6.6   Wages

OBJECTIVE A   To calculate commissions, total hourly wages, and salaries

Commissions, hourly wage, and salary are three ways to receive payment for doing work.

Commissions are usually paid to salespersons and are calculated as a percent of total sales.

HOW TO 1   As a real estate broker, Emma Smith receives a commission of 4.5% of the selling price of a house. Find the commission she earned for selling a home for $275,000.

To find the commission Emma earned, solve the basic percent equation for *amount*.

Percent	×	base	=	amount
Commission rate	×	total sales	=	commission
0.045	×	275,000	=	12,375

The commission is $12,375.

An employee who receives an **hourly wage** is paid a certain amount for each hour worked.

HOW TO 2   A plumber receives an hourly wage of $28.25. Find the plumber's total wages for working 37 hours.

To find the plumber's total wages, multiply the hourly wage by the number of hours worked.

Hourly wage	×	number of hours worked	=	total wages
28.25	×	37	=	1045.25

The plumber's total wages for working 37 hours are $1045.25.

An employee who is paid a **salary** receives payment based on a weekly, biweekly (every other week), monthly, or annual time schedule. Unlike the employee who receives an hourly wage, the salaried worker does not receive additional pay for working more than the regularly scheduled workday.

HOW TO 3   Ravi Basar is a computer operator who receives a weekly salary of $895. Find his salary for 1 month (4 weeks).

To find Ravi's salary for 1 month, multiply the salary per pay period by the number of pay periods.

Salary per pay period	×	number of pay periods	=	total salary
895	×	4	=	3580

Ravi's total salary for 1 month is $3580.

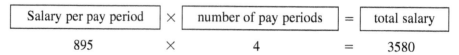

EXAMPLE · 1

A pharmacist's hourly wage is $48. On Saturday, the pharmacist earns time and a half (1.5 times the regular hourly wage). How much does the pharmacist earn for working 6 hours on Saturday?

Strategy

To find the pharmacist's earnings:

- Find the hourly wage for working on Saturday by multiplying the hourly wage by 1.5.
- Multiply the hourly wage by the number of hours worked.

Solution

$48 \times 1.5 = 72 \qquad 72 \times 6 = 432$

The pharmacist earns $432.

YOU TRY IT · 1

A construction worker, whose hourly wage is $28.50, earns double time (2 times the regular hourly wage) for working overtime. Find the worker's wages for working 8 hours of overtime.

Your strategy

Your solution

EXAMPLE · 2

An efficiency expert received a contract for $3000. The expert spent 75 hours on the project. Find the consultant's hourly wage.

Strategy

To find the hourly wage, divide the total earnings by the number of hours worked.

Solution

$3000 \div 75 = 40$

The hourly wage was $40.

YOU TRY IT · 2

A contractor for a bridge project receives an annual salary of $70,980. What is the contractor's salary per month?

Your strategy

Your solution

EXAMPLE · 3

Dani Greene earns $38,500 per year plus a 5.5% commission on sales over $100,000. During one year, Dani sold $150,000 worth of computers. Find Dani's total earnings for the year.

Strategy

To find the total earnings:

- Find the sales over $100,000.
- Multiply the commission rate by sales over $100,000.
- Add the commission to the annual pay.

Solution

$150,000 - 100,000 = 50,000$
$50,000 \times 0.055 = 2750$ • Commission
$38,500 + 2750 = 41,250$

Dani earned $41,250.

YOU TRY IT · 3

An insurance agent earns $37,000 per year plus a 9.5% commission on sales over $50,000. During one year, the agent's sales totaled $175,000. Find the agent's total earnings for the year.

Your strategy

Your solution

Solutions on p. S17

6.6 EXERCISES

OBJECTIVE A   To calculate commissions, total hourly wages, and salaries

1. Lewis works in a clothing store and earns $11.50 per hour. How much does he earn in a 40-hour work week?

2. Sasha pays a gardener an hourly wage of $11. How much does she pay the gardener for working 25 hours?

3. A real estate agent receives a 3% commission for selling a house. Find the commission that the agent earned for selling a house for $131,000.

4. Ron Caruso works as an insurance agent and receives a commission of 40% of the first year's premium. Find Ron's commission for selling a life insurance policy with a first-year premium of $1050.

5. A stockbroker receives a commission of 1.5% of the price of stock that is bought or sold. Find the commission on 100 shares of stock that were bought for $5600.

6. The owner of the Carousel Art Gallery receives a commission of 20% on paintings that are sold on consignment. Find the commission on a painting that sold for $22,500.

7. Keisha Brown receives an annual salary of $38,928 as a teacher of Italian. How much does Keisha receive each month?

8. An apprentice plumber receives an annual salary of $27,900. How much does the plumber receive per month?

9. Carlos receives a commission of 12% of his weekly sales as a sales representative for a medical supply company. Find the commission he earned during a week in which sales were $4500.

10. A golf pro receives a commission of 25% for selling a golf set. Find the commission the pro earned for selling a golf set costing $450.

11. Steven receives $5.75 per square yard to install carpet. How much does he receive for installing 160 square yards of carpet?

12. A typist charges $3.75 per page for typing technical material. How much does the typist earn for typing a 225-page book?

13. A nuclear chemist received $15,000 in consulting fees while working on a nuclear power plant. The chemist worked 120 hours on the project. Find the chemist's hourly wage.

14. Maxine received $3400 for working on a project as a computer consultant for 40 hours. Find her hourly wage.

15. Gil Stratton's hourly wage is $10.78. For working overtime, he receives double time.
a. What is Gil's hourly wage for working overtime?
b. How much does he earn for working 16 hours of overtime?

16. Mark is a lathe operator and receives an hourly wage of $15.90. When working on Saturday, he receives time and a half.
a. What is Mark's hourly wage on Saturday?
b. How much does he earn for working 8 hours on Saturday?

17. A stock clerk at a supermarket earns $8.20 an hour. For working the night shift, the clerk's wage increases by 15%.
a. What is the increase in hourly pay for working the night shift?
b. What is the clerk's hourly wage for working the night shift?

18. A nurse earns $31.50 an hour. For working the night shift, the nurse receives a 10% increase in pay.
a. What is the increase in hourly pay for working the night shift?
b. What is the hourly pay for working the night shift?

19. Nicole Tobin, a salesperson, receives a salary of $250 per week plus a commission of 15% on all sales over $1500. Find her earnings during a week in which sales totaled $3000.

20. A veterinarian's assistant works 35 hours a week at $20 an hour. The assistant is paid time and a half for overtime hours. Which expression represents the assistant's earnings for a week in which the assistant worked 41 hours?
(i)   41 × 20
(ii) (35 × 20) + (41 × 30)
(iii) (35 × 20) + (6 × 30)
(iv) 41 × 30

Applying the Concepts

Compensation The table at the right shows the average starting salaries for recent college graduates. Use this table for Exercises 21 to 24. Round to the nearest dollar.

21. What was the starting salary in the previous year for an accountant?

22. How much did the starting salary for a chemical engineer increase over that of the previous year?

23. What was the starting salary in the previous year for a computer science major?

Average Starting Salaries		
Bachelor's Degree	**Average Starting Salary**	**Change from Previous Year**
Chemical Engineering	$52,169	1.8% increase
Electrical Engineering	$50,566	0.4% increase
Computer Science	$46,536	7.6% decrease
Accounting	$41,360	2.6% increase
Business	$36,515	3.7% increase
Biology	$29,554	1.0% decrease
Political Science	$28,546	12.6% decrease
Psychology	$26,738	10.7% decrease

Source: National Association of Colleges

24. How much did the starting salary for a political science major decrease from that of the previous year?

SECTION

6.7   Bank Statements

OBJECTIVE A   To calculate checkbook balances

A checking account can be opened at most banks and savings and loan associations by depositing an amount of money in the bank. A checkbook contains checks and deposit slips and a checkbook register in which to record checks written and amounts deposited in the checking account. A sample check is shown below.

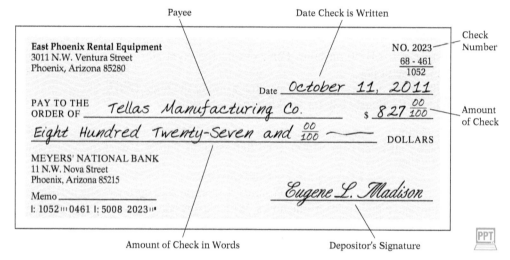

Payee   Date Check is Written   Check Number   Amount of Check   Amount of Check in Words   Depositor's Signature

Each time a check is written, the amount of the check is subtracted from the amount in the account. When a deposit is made, the amount deposited is added to the amount in the account.

A portion of a checkbook register is shown below. The account holder had a balance of $587.93 before writing two checks, one for $286.87 and the other for $202.38, and making one deposit of $345.00.

		RECORD ALL CHARGES OR CREDITS THAT AFFECT YOUR ACCOUNT					BALANCE	
NUMBER	DATE	DESCRIPTION OF TRANSACTION	PAYMENT/DEBIT (−)	√ T	FEE (IF ANY) (−)	DEPOSIT/CREDIT (+)	$ 587	93
108	8/4	Plumber	$286 87	$	$		301	06
109	8/10	Car Payment	202 38				98	68
	8/14	Deposit				345 00	443	68

To find the current checking account balance, subtract the amount of each check from the previous balance. Then add the amount of the deposit.

The current checking account balance is $443.68.

| EXAMPLE · 1 | | YOU TRY IT · 1 |

A mail carrier had a checking account balance of $485.93 before writing two checks, one for $18.98 and another for $35.72, and making a deposit of $250. Find the current checking account balance.

A cement mason had a checking account balance of $302.46 before writing a check for $20.59 and making two deposits, one in the amount of $176.86 and another in the amount of $94.73. Find the current checking account balance.

Strategy

To find the current balance:

- Subtract the amount of each check from the old balance.
- Add the amount of the deposit.

Your strategy

Solution

```
  485.93
−  18.98    first check
  ──────
  466.95
−  35.72    second check
  ──────
  431.23
+ 250.00    deposit
  ──────
  681.23
```

The current checking account balance is $681.23.

Your solution

Solution on p. S17

OBJECTIVE B   To balance a checkbook

Each month a bank statement is sent to the account holder. A **bank statement** is a document showing all the transactions in a bank account during the month. It shows the checks that the bank has paid, the deposits received, and the current bank balance.

A bank statement and checkbook register are shown on the next page.

Balancing a checkbook, or determining whether the checking account balance is accurate, requires a number of steps.

1. In the checkbook register, put a check mark (✓) by each check paid by the bank and by each deposit recorded by the bank.

RECORD ALL CHARGES OR CREDITS THAT AFFECT YOUR ACCOUNT

NUMBER	DATE	DESCRIPTION OF TRANSACTION	PAYMENT/DEBIT (–)		√ T	FEE (IF ANY) (–)	DEPOSIT/CREDIT (+)		BALANCE $ 840	27
263	5/20	Dentist	$ 75	00	√	$	$		765	27
264	5/22	Post Office	33	61	√				731	66
265	5/22	Gas Company	67	14					664	52
	5/29	Deposit			√		192	00	856	52
266	5/29	Pharmacy	38	95	√				817	57
267	5/30	Telephone	63	85					753	72
268	6/2	Groceries	73	19	√				680	53
	6/3	Deposit			√		215	00	895	53
269	6/7	Insurance	103	00	√				792	53
	6/10	Deposit					225	00	1017	53
270	6/15	Photo Shop	16	63	√				1000	90
271	6/18	Newspaper	27	00					973	90

CHECKING ACCOUNT Monthly Statement   Account Number: 924-297-8

Date	Transaction	Amount	Balance
5/20	OPENING BALANCE		840.27
5/21	CHECK	75.00	765.27
5/23	CHECK	33.61	731.66
5/29	DEPOSIT	192.00	923.66
6/1	CHECK	38.95	884.71
6/1	INTEREST	4.47	889.18
6/3	CHECK	73.19	815.99
6/3	DEPOSIT	215.00	1030.99
6/9	CHECK	103.00	927.99
6/16	CHECK	16.63	911.36
6/20	SERVICE CHARGE	3.00	908.36
6/20	CLOSING BALANCE		908.36

---

2. Add to the current checkbook balance all checks that have been written but have not yet been paid by the bank and any interest paid on the account.

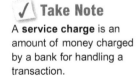

√ Take Note

A **service charge** is an amount of money charged by a bank for handling a transaction.

3. Subtract any service charges and any deposits not yet recorded by the bank. This is the checkbook balance.

4. Compare the balance with the bank balance listed on the bank statement. If the two numbers are equal, the bank statement and the checkbook "balance."

Current checkbook balance:	973.90
Checks: 265	67.14
267	63.85
271	27.00
Interest:	+ 4.47
	1136.36
Service charge:	− 3.00
	1133.36
Deposit:	− 225.00
Checkbook balance:	908.36

Closing bank balance from bank statement $908.36	Checkbook balance = $908.36

The bank statement and checkbook balance.

HOW TO 1

RECORD ALL CHARGES OR CREDITS THAT AFFECT YOUR ACCOUNT									
NUMBER	DATE	DESCRIPTION OF TRANSACTION	PAYMENT/DEBIT (–)		√ T	FEE (IF ANY) (–)	DEPOSIT/CREDIT (+)		BALANCE $ 1620 42
413	3/2	Car Payment	$232	15	√	$	$		1388 27
414	3/2	Utilities	67	14	√				1321 13
415	3/5	Restaurant	78	14					1242 99
	3/8	Deposit			√		1842	66	3085 65
416	3/10	House Payment	672	14	√				2413 51
417	3/14	Insurance	177	10					2236 41

CHECKING ACCOUNT Monthly Statement			Account Number: 924-297-8
Date	Transaction	Amount	Balance
3/1	OPENING BALANCE		1620.42
3/4	CHECK	232.15	1388.27
3/5	CHECK	67.14	1321.13
3/8	DEPOSIT	1842.66	3163.79
3/10	INTEREST	6.77	3170.56
3/12	CHECK	672.14	2498.42
3/25	SERVICE CHARGE	2.00	2496.42
3/30	CLOSING BALANCE		2496.42

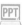

Balance the checkbook shown above.

1. In the checkbook register, put a check mark (✓) by each check paid by the bank and by each deposit recorded by the bank.

2. Add to the current checkbook balance all checks that have been written but have not yet been paid by the bank and any interest paid on the account.

Current checkbook balance:	2236.41
Checks: 415	78.14
417	177.10
Interest:	+ 6.77
	2498.42

3. Subtract any service charges and any deposits not yet recorded by the bank. This is the checkbook balance.

Service charge:	− 2.00
Checkbook balance:	2496.42

4. Compare the balance with the bank balance listed on the bank statement. If the two numbers are equal, the bank statement and the checkbook "balance."

Closing bank balance from bank statement $2496.42

Checkbook balance = $2496.42

The bank statement and checkbook balance.

EXAMPLE · 2

Balance the checkbook shown below.

RECORD ALL CHARGES OR CREDITS THAT AFFECT YOUR ACCOUNT

NUMBER	DATE	DESCRIPTION OF TRANSACTION	PAYMENT/DEBIT (−)	√T	FEE (IF ANY) (−)	DEPOSIT/CREDIT (+)	BALANCE $ 412 64
345	1/14	Phone Bill	$ 54 75	√	$	$	357 89
346	1/19	News Shop	18 98	√			338 91
347	1/23	Theater Tickets	95 00				243 91
	1/31	Deposit		√		947 00	1190 91
348	2/5	Cash	250 00	√			940 91
349	2/12	Rent	840 00				100 91

CHECKING ACCOUNT Monthly Statement		Account Number: 924-297-8	
Date	**Transaction**	**Amount**	**Balance**
1/10	OPENING BALANCE		412.64
1/18	CHECK	54.75	357.89
1/23	CHECK	18.98	338.91
1/31	DEPOSIT	947.00	1285.91
2/1	INTEREST	4.52	1290.43
2/10	CHECK	250.00	1040.43
2/10	CLOSING BALANCE		1040.43

Solution

Current checkbook balance: 100.91

Checks: 347   95.00

       349   840.00

Interest: +   4.52

            1040.43

Service charge: −   0.00

            1040.43

Deposit: −   0.00

Checkbook balance: 1040.43

Closing bank balance from bank statement: $1040.43

Checkbook balance: $1040.43

The bank statement and the checkbook balance.

Balance the checkbook shown below.

RECORD ALL CHARGES OR CREDITS THAT AFFECT YOUR ACCOUNT

NUMBER	DATE	DESCRIPTION OF TRANSACTION	PAYMENT/DEBIT (−)		√ T	FEE (IF ANY) (−)	DEPOSIT/CREDIT (+)		BALANCE $ 903	17
	2/15	Deposit	$			$	$ 523	84	1427	01
234	2/20	Mortgage	773	21					653	80
235	2/27	Cash	200	00					453	80
	3/1	Deposit					523	84	977	64
236	3/12	Insurance	275	50					702	14
237	3/12	Telephone	78	73					623	41

CHECKING ACCOUNT Monthly Statement			Account Number: 314-271-4
Date	Transaction	Amount	Balance
2/14	OPENING BALANCE		903.17
2/15	DEPOSIT	523.84	1427.01
2/21	CHECK	773.21	653.80
2/28	CHECK	200.00	453.80
3/1	INTEREST	2.11	455.91
3/14	CHECK	275.50	180.41
3/14	CLOSING BALANCE		180.41

Your solution

Solution on p. S17

6.7 EXERCISES

OBJECTIVE A   To calculate checkbook balances

1. You had a checking account balance of $342.51 before making a deposit of $143.81. What is your new checking account balance?

2. The business checking account for R and R Tires showed a balance of $1536.97. What is the balance in this account after a deposit of $439.21 has been made?

3. A nutritionist had a checking account balance of $1204.63 before writing one check for $119.27 and another check for $260.09. Find the current checkbook balance.

4. Sam had a checking account balance of $3046.93 before writing a check for $1027.33 and making a deposit of $150.00. Find the current checkbook balance.

5. The business checking account for Rachael's Dry Cleaning had a balance of $3476.85 before a deposit of $1048.53 was made. The store manager then wrote two checks, one for $848.37 and another for $676.19. Find the current checkbook balance.

6. Joel had a checking account balance of $427.38 before a deposit of $127.29 was made. Joel then wrote two checks, one for $43.52 and one for $249.78. Find the current checkbook balance.

7. A carpenter had a checkbook balance of $404.96 before making a deposit of $350 and writing a check for $71.29. Is there enough money in the account to purchase a refrigerator for $675?

8. A taxi driver had a checkbook balance of $149.85 before making a deposit of $245 and writing a check for $387.68. Is there enough money in the account for the bank to pay the check?

9. A sporting goods store has the opportunity to buy downhill skis and cross-country skis at a manufacturer's closeout sale. The downhill skis will cost $3500, and the cross-country skis will cost $2050. There is currently $5625.42 in the sporting goods store's checking account. Is there enough money in the account to make both purchases by check?

10. A lathe operator's current checkbook balance is $1143.42. The operator wants to purchase a utility trailer for $525 and a used piano for $650. Is there enough money in the account to make the two purchases?

 For Exercises 11 and 12, suppose the given transactions take place on an account in one day. State whether the account's ending balance on that day *must be less than, might be less than,* or *cannot be less than* its starting balance on that day.

11. Two deposits and one check written   12. Three checks written

OBJECTIVE B   To balance a checkbook

13. Balance the checkbook.

		RECORD ALL CHARGES OR CREDITS THAT AFFECT YOUR ACCOUNT						BALANCE	
NUMBER	DATE	DESCRIPTION OF TRANSACTION	PAYMENT/DEBIT (−)	√ T	FEE (IF ANY) (−)	DEPOSIT/CREDIT (+)		$ 2466	79
223	3/2	Groceries	$ 167 32		$	$		2299	47
	3/5	Deposit				960 70		3260	17
224	3/5	Rent	860 00					2400	17
225	3/7	Gas & Electric	142 35					2257	82
226	3/7	Cash	300 00					1957	82
227	3/7	Insurance	218 44					1739	38
228	3/7	Credit Card	419 32					1320	06
229	3/12	Dentist	92 00					1228	06
230	3/13	Drug Store	47 03					1181	03
	3/19	Deposit				960 70		2141	73
231	3/22	Car Payment	241 35					1900	38
232	3/25	Cash	300 00					1600	38
233	3/25	Oil Company	166 40					1433	98
234	3/28	Plumber	155 73					1278	25
235	3/29	Department Store	288 39					989	86

CHECKING ACCOUNT Monthly Statement		Account Number: 122-345-1	
Date	Transaction	Amount	Balance
3/1	OPENING BALANCE		2466.79
3/5	DEPOSIT	960.70	3427.49
3/7	CHECK	167.32	3260.17
3/8	CHECK	860.00	2400.17
3/8	CHECK	300.00	2100.17
3/9	CHECK	142.35	1957.82
3/12	CHECK	218.44	1739.38
3/14	CHECK	92.00	1647.38
3/18	CHECK	47.03	1600.35
3/19	DEPOSIT	960.70	2561.05
3/25	CHECK	241.35	2319.70
3/27	CHECK	300.00	2019.70
3/29	CHECK	155.73	1863.07
3/30	INTEREST	13.22	1877.19
4/1	CLOSING BALANCE		1877.19

14. Balance the checkbook.

RECORD ALL CHARGES OR CREDITS THAT AFFECT YOUR ACCOUNT

NUMBER	DATE	DESCRIPTION OF TRANSACTION	PAYMENT/DEBIT (−)		√ T	FEE (IF ANY) (−)	DEPOSIT/CREDIT (+)		BALANCE $ 1219 43	
	5/1	Deposit	$			$	$ 619	14	1838	57
515	5/2	Electric Bill	42	35					1796	22
516	5/2	Groceries	95	14					1701	08
517	5/4	Insurance	122	17					1578	91
518	5/5	Theatre Tickets	84	50					1494	41
	5/8	Deposit					619	14	2113	55
519	5/10	Telephone	37	39					2076	16
520	5/12	Newspaper	22	50					2053	66
	5/15	Deposit					619	14	2672	80
521	5/20	Computer Store	172	90					2499	90
522	5/21	Credit Card	313	44					2186	46
523	5/22	Eye Exam	82	00					2104	46
524	5/24	Groceries	107	14					1997	32
525	5/24	Deposit					619	14	2616	46
526	5/25	Oil Company	144	16					2472	30
527	5/30	Car Payment	288	62					2183	68
528	5/30	Mortgage Payment	877	42					1306	26

CHECKING ACCOUNT Monthly Statement

Account Number: 122-345-1

Date	Transaction	Amount	Balance
5/1	OPENING BALANCE		1219.43
5/1	DEPOSIT	619.14	1838.57
5/3	CHECK	95.14	1743.43
5/4	CHECK	42.35	1701.08
5/6	CHECK	84.50	1616.58
5/8	CHECK	122.17	1494.41
5/8	DEPOSIT	619.14	2113.55
5/15	INTEREST	7.82	2121.37
5/15	CHECK	37.39	2083.98
5/15	DEPOSIT	619.14	2703.12
5/23	CHECK	82.00	2621.12
5/23	CHECK	172.90	2448.22
5/24	CHECK	107.14	2341.08
5/24	DEPOSIT	619.14	2960.22
5/30	CHECK	288.62	2671.60
6/1	CLOSING BALANCE		2671.60

15. Balance the checkbook.

NUMBER	DATE	DESCRIPTION OF TRANSACTION	PAYMENT/DEBIT (−)	√ T	FEE (IF ANY) (−)	DEPOSIT/CREDIT (+)	BALANCE $ 2035 18
218	7/2	Mortgage	$ 984 60		$	$	1050 58
219	7/4	Telephone	63 36				987 22
220	7/7	Cash	200 00				787 22
	7/12	Deposit				792 60	1579 82
221	7/15	Insurance	292 30				1287 52
222	7/18	Investment	500 00				787 52
223	7/20	Credit Card	414 83				372 69
	7/26	Deposit				792 60	1165 29
224	7/27	Department Store	113 37				1051 92

CHECKING ACCOUNT Monthly Statement   Account Number: 122-345-1

Date	Transaction	Amount	Balance
7/1	OPENING BALANCE		2035.18
7/1	INTEREST	5.15	2040.33
7/4	CHECK	984.60	1055.73
7/6	CHECK	63.36	992.37
7/12	DEPOSIT	792.60	1784.97
7/20	CHECK	292.30	1492.67
7/24	CHECK	500.00	992.67
7/26	DEPOSIT	792.60	1785.27
7/28	CHECK	200.00	1585.27
7/30	CLOSING BALANCE		1585.27

16. The ending balance on a monthly bank statement is greater than the beginning balance, and the bank did not include a service charge. Was the total of all deposits recorded *less than* or *greater than* the total of all checks paid?

17. When balancing your checkbook, you find that all the deposits in your checkbook register have been recorded by the bank, four checks in the register have not yet been paid by the bank, and the bank did not include a service charge. Is the ending balance on the monthly bank statement *less than* or *greater than* the ending balance on the check register?

Applying the Concepts

18. Define the words *credit* and *debit* as they apply to checkbooks.

FOCUS ON PROBLEM SOLVING

Counterexamples   An example that is given to show that a statement is not true is called a **counterexample.** For instance, suppose someone makes the statement "All colors are red." A counterexample to that statement would be to show someone the color blue or some other color.

If a statement is *always* true, there are no counterexamples. The statement "All even numbers are divisible by 2" is always true. It is not possible to give an example of an even number that is not divisible by 2.

In mathematics, statements that are always true are called *theorems,* and mathematicians are always searching for theorems. Sometimes a conjecture by a mathematician appears to be a theorem. That is, the statement appears to be always true, but later on someone finds a counterexample.

One example of this occurred when the French mathematician Pierre de Fermat (1601–1665) conjectured that $2^{(2^n)} + 1$ is always a prime number for any natural number n. For instance, when $n = 3$, we have $2^{(2^3)} + 1 = 2^8 + 1 = 257$, and 257 is a prime number. However, in 1732 Leonhard Euler (1707–1783) showed that when $n = 5$, $2^{(2^5)} + 1 = 4{,}294{,}967{,}297$, and that $4{,}294{,}967{,}297 = 641 \cdot 6{,}700{,}417$—without a calculator! Because $4{,}294{,}967{,}297$ is the product of two numbers (other than itself and 1), it is not a prime number. This counterexample showed that Fermat's conjecture is not a theorem.

For Exercises 1 to 5, find at least one counterexample.

1. All composite numbers are divisible by 2.

2. All prime numbers are odd numbers.

3. The square of any number is always bigger than the number.

4. The reciprocal of a number is always less than 1.

5. A number ending in 9 is always larger than a number ending in 3.

When a problem is posed, it may not be known whether the problem statement is true or false. For instance, Christian Goldbach (1690–1764) stated that every even number greater than 2 can be written as the sum of two prime numbers. For example,

$$12 = 5 + 7 \qquad 32 = 3 + 29$$

Although this problem is approximately 250 years old, mathematicians have not been able to prove it is a theorem, nor have they been able to find a counterexample.

For Exercises 6 to 9, answer true if the statement is always true. If there is an instance in which the statement is false, give a counterexample.

6. The sum of two positive numbers is always larger than either of the two numbers.

7. The product of two positive numbers is always larger than either of the two numbers.

8. Percents always represent a number less than or equal to 1.

9. It is never possible to divide by zero.

PROJECTS AND GROUP ACTIVITIES

Buying a Car   Suppose a student has an after-school job to earn money to buy and maintain a car. We will make assumptions about the monthly costs in several categories in order to determine how many hours per week the student must work to support the car. Assume the student earns $10.50 per hour.

1. **Monthly payment**

   Assume that the car cost $18,500 with a down payment of $2220. The remainder is financed for 3 years at an annual simple interest rate of 9%.

   Monthly payment = _____

2. **Insurance**

   Assume that insurance costs $3000 per year.

   Monthly insurance payment = _____

3. **Gasoline**

   Assume that the student travels 750 miles per month, that the car travels 25 miles per gallon of gasoline, and that gasoline costs $3.50 per gallon.

   Number of gallons of gasoline purchased per month = _____

   Monthly cost for gasoline = _____

4. **Miscellaneous**

   Assume $.42 per mile for upkeep.

   Monthly expense for upkeep = _____

5. Total monthly expenses for the monthly payment, insurance, gasoline, and miscellaneous

   = _____

6. To find the number of hours per month that the student must work to finance the car, divide the total monthly expenses by the hourly rate.

   Number of hours per month = _____

7. To find the number of hours per week that the student must work, divide the number of hours per month by 4.

   Number of hours per week = _____

   The student has to work _____ hours per week to pay the monthly car expenses.

If you own a car, make out your own expense record. If you do not own a car, make assumptions about the kind of car that you would like to purchase, and calculate the total monthly expenses that you would have. An insurance company will give you rates on different kinds of insurance. An automobile club can give you approximations of miscellaneous expenses.

CHAPTER 6

SUMMARY

KEY WORDS	EXAMPLES
The *unit cost* is the cost of one item. [6.1A, p. 234]	Three paperback books cost $36. The unit cost is the cost of one paperback book, $12.
Percent increase is used to show how much a quantity has increased over its original value. [6.2A, p. 238]	The city's population increased 5%, from 10,000 people to 10,500 people.
Cost is the price a business pays for a product. *Selling price* is the price at which a business sells a product to a customer. *Markup* is the difference between selling price and cost. *Markup rate* is the markup expressed as a percent of the product's cost. [6.2B, p. 239]	A business pays $90 for a pair of cross trainers; the cost is $90. The business sells the cross trainers for $135; the selling price is $135. The markup is $135 − $90 = $45.
Percent decrease is used to show how much a quantity has decreased from its original value. [6.2C, p. 241]	Sales decreased 10%, from 10,000 units in the third quarter to 9000 units in the fourth quarter.
Sale price is the price after a reduction from the regular price. *Discount* is the difference between the regular price and the sale price. *Discount rate* is the discount as a percent of the product's regular price. [6.2D, p. 242]	A skateboard deck that regularly sells for $50 is on sale for $40. The regular price is $50. The sale price is $40. The discount is $50 − $40 = $10.
Interest is the amount paid for the privilege of using someone else's money. *Principal* is the amount of money originally deposited or borrowed. The percent used to determine the amount of interest is the *interest rate*. Interest computed on the original amount is called *simple interest*. The principal plus the interest owed on a loan is called the *maturity value*. [6.3A, p. 248]	Consider a 1-year loan of $5000 at an annual simple interest rate of 8%. The principal is $5000. The interest rate is 8%. The interest paid on the loan is $5000 × 0.08 = $400. The maturity value is $5000 + $400 = $5400.
The interest charged on purchases made with a credit card is called a *finance charge*. [6.3B, p. 250]	A credit card company charges 1.5% per month on any unpaid balance. The finance charge on an unpaid balance of $1000 is $1000 × 0.015 × 1 = $15.
Compound interest is computed not only on the original principal but also on the interest already earned. [6.3C, p. 251]	$10,000 is invested at 5% annual interest, compounded monthly. The value of the investment after 5 years can be found by multiplying 10,000 by the factor found in the Compound Interest Table in the Appendix. $10,000 × 1.283359 = $12,833.59
A *mortgage* is an amount that is borrowed to buy real estate. The *loan origination fee* is usually a percent of the mortgage and is expressed as *points*. [6.4A, p. 258]	The loan origination fee of 3 points paid on a mortgage of $200,000 is 0.03 × $200,000 = $6000.
A *commission* is usually paid to a salesperson and is calculated as a percent of sales. [6.6A, p. 268]	A commission of 5% on sales of $50,000 is 0.05 × $50,000 = $2500.

An employee who receives an *hourly wage* is paid a certain amount for each hour worked.   [6.6A, p. 268]

An employee is paid an hourly wage of $15. The employee's wages for working 10 hours are $15 \times 10 = $150.

An employee who is paid a *salary* receives payment based on a weekly, biweekly, monthly, or annual time schedule. [6.6A, p. 268]

An employee paid an annual salary of $60,000 is paid $60,000 \div 12 = $5000 per month.

Balancing a checkbook is determining whether the checkbook balance is accurate.   [6.7B, pp. 273–274]

To balance a checkbook:

(1) Put a check mark in the checkbook register by each check paid by the bank and by each deposit recorded by the bank.

(2) Add to the current checkbook balance all checks that have been written but have not yet been paid by the bank and any interest paid on the account.

(3) Subtract any charges and any deposits not yet recorded by the bank. This is the checkbook balance.

(4) Compare the balance with the bank balance listed on the bank statement. If the two numbers are equal, the bank statement and the checkbook "balance."

ESSENTIAL RULES AND PROCEDURES

EXAMPLES

To find unit cost, divide the total cost by the number of units. [6.1A, p. 234]

Three paperback books cost $36. The unit cost is $36 \div 3 = $12 per book.

To find total cost, multiply the unit cost by the number of units purchased.   [6.1C, p. 235]

One melon costs $3. The total cost for 5 melons is $3 \times 5 = $15.

Basic Markup Equations   [6.2B, p. 239]
Selling price − cost = markup
Cost + markup = selling price
Markup rate × cost = markup

A pair of cross trainers that cost a business $90 has a 50% markup rate. The markup is 0.50 \times $90 = $45. The selling price is $90 + $45 = $135.

Basic Discount Equations   [6.2D, p. 242]
Regular price − sale price = discount
Regular price − discount = sale price
Discount rate × regular price = discount

A movie DVD is on sale for 20% off the regular price of $50. The discount is 0.20 \times $50 = $10. The sale price is $50 − $10 = $40.

Simple Interest Formula for Annual Interest Rates   [6.3A, p. 248]
Principal × annual interest rate × time (in years) = interest

The simple interest due on a 2-year loan of $5000 that has an annual interest rate of 5% is $5000 \times 0.05 \times 2 = $500.

Maturity Value Formula for a Simple Interest Loan   [6.3A, p. 248]
Principal + interest = maturity value

The interest to be paid on a 2-year loan of $5000 is $500. The maturity value of the loan is $5000 + $500 = $5500.

Monthly Payment on a Simple Interest Loan   [6.3A, p. 249]
Maturity value ÷ length of the loan in months = monthly payment

The maturity value of a simple interest 8-month loan is $8000. The monthly payment is $8000 \div 8 = $1000.

CHAPTER 6

CONCEPT REVIEW

Test your knowledge of the concepts presented in this chapter. Answer each question.
Then check your answers against the ones provided in the Answer Section.

1. Find the unit cost if 4 cans cost $2.96.

2. Find the total cost of 3.4 pounds of apples if apples cost $.85 per pound.

3. How do you find the selling price if you know the cost and the markup?

4. How do you use the markup rate to find the markup?

5. How do you find the amount of decrease if you know the percent decrease?

6. How do you find the discount if you know the regular price and the sale price?

7. How do you find the discount rate?

8. How do you find simple interest?

9. How do you find the maturity value for a simple interest loan?

10. What is the principal?

11. How do you find the monthly payment for a loan of 18 months if you know the maturity value of the loan?

12. What is compound interest?

13. What is a fixed-rate mortgage?

14. What expenses are involved in owning a car?

15. How do you balance a checkbook?

CHAPTER 6

REVIEW EXERCISES

1. **Consumerism** A 20-ounce box of cereal costs $3.90. Find the unit cost.

2. **Car Expenses** An account executive had car expenses of $1025.58 for insurance, $1805.82 for gas, $37.92 for oil, and $288.27 for maintenance during a year in which 11,320 miles were driven. Find the cost per mile for these four items taken as a group. Round to the nearest tenth of a cent.

3. **Investments** An oil stock was bought for $42.375 per share. Six months later, the stock was selling for $55.25 per share. Find the percent increase in the price of the stock over the 6 months. Round to the nearest tenth of a percent.

4. **Markup** A sporting goods store uses a markup rate of 40%. What is the markup on a ski suit that costs the store $180?

5. **Simple Interest** A contractor borrowed $100,000 from a credit union for 9 months at an annual interest rate of 4%. What is the simple interest due on the loan?

6. **Compound Interest** A computer programmer invested $25,000 in a retirement account that pays 6% interest, compounded daily. What is the value of the investment in 10 years? Use the Compound Interest Table in the Appendix. Round to the nearest cent.

7. **Investments** Last year an oil company had earnings of $4.12 per share. This year the earnings are $4.73 per share. What is the percent increase in earnings per share? Round to the nearest percent.

8. **Real Estate** The monthly mortgage payment for a condominium is $923.67. The owner must pay an annual property tax of $2582.76. Find the total monthly payment for the mortgage and property tax.

9. **Car Expenses** A used pickup truck is purchased for $24,450. A down payment of 8% is made, and the remaining cost is financed for 4 years at an annual interest rate of 5%. Find the monthly payment. Use the Monthly Payment Table in the Appendix. Round to the nearest cent.

10. **Compound Interest** A fast-food restaurant invested $50,000 in an account that pays 7% annual interest, compounded quarterly. What is the value of the investment in 1 year? Use the Compound Interest Table in the Appendix.

11. **Real Estate** Paula Mason purchased a home for $195,000. The lender requires a down payment of 15%. Find the amount of the down payment.

12. **Car Expenses** A plumber bought a truck for $28,500. A state license fee of $315 and a sales tax of 6.25% of the purchase price are required. Find the total cost of the sales tax and the license fee.

13. **Markup** Techno-Center uses a markup rate of 35% on all computer systems. Find the selling price of a computer system that costs the store $1540.

14. **Car Expenses** Mien pays a monthly car payment of $222.78. During a month in which $65.45 is principal, how much of the payment is interest?

15. **Compensation** The manager of the retail store at a ski resort receives a commission of 3% on all sales at the alpine shop. Find the total commission received during a month in which the shop had $108,000 in sales.

16. **Discount** A suit that regularly costs $235 is on sale for 40% off the regular price. Find the sale price.

17. **Banking** Luke had a checking account balance of $1568.45 before writing checks for $123.76, $756.45, and $88.77. He then deposited a check for $344.21. Find Luke's current checkbook balance.

18. **Simple Interest** Pros' Sporting Goods borrowed $30,000 at an annual interest rate of 8% for 6 months. Find the maturity value of the loan.

19. **Real Estate** A credit union requires a borrower to pay $2\frac{1}{2}$ points for a loan. Find the origination fee for a $75,000 loan.

20. **Consumerism** Sixteen ounces of mouthwash cost $3.49. A 33-ounce container of the same brand of mouthwash costs $6.99. Which is the better buy?

21. **Real Estate** The Sweeneys bought a home for $356,000. The family made a 10% down payment and financed the remainder with a 30-year loan at an annual interest rate of 7%. Find the monthly mortgage payment. Use the Monthly Payment Table in the Appendix. Round to the nearest cent.

22. **Compensation** Richard Valdez receives $12.60 per hour for working 40 hours a week and time and a half for working over 40 hours. Find his total income during a week in which he worked 48 hours.

23. **Banking** The business checking account of a donut shop showed a balance of $9567.44 before checks of $1023.55, $345.44, and $23.67 were written and checks of $555.89 and $135.91 were deposited. Find the current checkbook balance.

24. **Simple Interest** The simple interest due on a 4-month loan of $55,000 is $1375. Find the monthly payment on the loan.

25. **Simple Interest** A credit card company charges a customer 1.25% per month on the unpaid balance of charges on the card. What is the finance charge in a month in which the customer has an unpaid balance of $576?

CHAPTER 6

TEST

1. **Consumerism** Twenty feet of lumber cost $138.40. What is the cost per foot?

2. **Consumerism** Which is the more economical purchase: 3 pounds of tomatoes for $7.49 or 5 pounds of tomatoes for $12.59?

3. **Consumerism** Red snapper costs $4.15 per pound. Find the cost of $3\frac{1}{2}$ pounds. Round to the nearest cent.

4. **Business** An exercise bicycle increased in price from $415 to $498. Find the percent increase in the cost of the exercise bicycle.

5. **Markup** A department store uses a 40% markup rate. Find the selling price of a blu-ray disc player that the store purchased for $315.

6. **Investments** The price of gold rose from $790 per ounce to $860 per ounce. What percent increase does this amount represent? Round to the nearest tenth of a percent.

7. **Consumerism** The price of a video camera dropped from $1120 to $896. What percent decrease does this price drop represent?

8. **Discount** A corner hutch with a regular price of $299 is on sale for 30% off the regular price. Find the sale price.

9. **Discount** A box of stationery that regularly sells for $9.50 is on sale for $5.70. Find the discount rate.

10. **Simple Interest** A construction company borrowed $75,000 for 4 months at an annual interest rate of 8%. Find the simple interest due on the loan.

11. **Simple Interest** Craig Allen borrowed $25,000 at an annual interest rate of 9.2% for 9 months. Find the maturity value of the loan.

12. **Simple Interest** A credit card company charges a customer 1.2% per month on the unpaid balance of charges on the card. What is the finance charge in a month in which the customer has an unpaid balance of $374.95?

13. **Compound Interest** Jorge, who is self-employed, placed $30,000 in an account that pays 6% annual interest, compounded quarterly. How much interest was earned in 10 years? Use the Compound Interest Table in the Appendix.

14. **Real Estate** A savings and loan institution is offering mortgage loans that have a loan origination fee of $2\frac{1}{2}$ points. Find the loan origination fee when a home is purchased with a loan of $134,000.

15. **Real Estate** A new housing development offers homes with a mortgage of $222,000 for 25 years at an annual interest rate of 8%. Find the monthly mortgage payment. Use the Monthly Payment Table in the Appendix.

16. **Car Expenses** A Chevrolet was purchased for $23,750, and a 20% down payment was made. Find the amount financed.

17. **Car Expenses** A rancher purchased an SUV for $33,714 and made a down payment of 15% of the cost. The balance was financed for 4 years at an annual interest rate of 7%. Find the monthly truck payment. Use the Monthly Payment Table in the Appendix.

18. **Compensation** Shaney receives an hourly wage of $30.40 an hour as an emergency room nurse. When called in at night, she receives time and a half. How much does Shaney earn in a week when she works 30 hours at normal rates and 15 hours during the night?

19. **Banking** The business checking account for a pottery store had a balance of $7349.44 before checks for $1349.67 and $344.12 were written. The store manager then made a deposit of $956.60. Find the current checkbook balance.

20. **Banking** Balance the checkbook shown.

RECORD ALL CHARGES OR CREDITS THAT AFFECT YOUR ACCOUNT

NUMBER	DATE	DESCRIPTION OF TRANSACTION	PAYMENT/DEBIT (−)	√ T	FEE (IF ANY) (−)	DEPOSIT/CREDIT (+)	BALANCE $ 1422 13	
843	8/1	House Payment	$ 713 72	$	$		708	41
	8/4	Deposit				852 60	1561	01
844	8/5	Loan Payment	162 40				1398	61
845	8/6	Groceries	166 44				1232	17
846	8/10	Car Payment	322 37				909	80
	8/15	Deposit				852 60	1762	40
847	8/16	Credit Card	413 45				1348	95
848	8/18	Pharmacy	92 14				1256	81
849	8/22	Utilities	72 30				1184	51
850	8/28	Telephone	78 20				1106	31

CHECKING ACCOUNT Monthly Statement   Account Number: 122-345-1

Date	Transaction	Amount	Balance
8/1	OPENING BALANCE		1422.13
8/3	CHECK	713.72	708.41
8/4	DEPOSIT	852.60	1561.01
8/8	CHECK	166.44	1394.57
8/8	CHECK	162.40	1232.17
8/15	DEPOSIT	852.60	2084.77
8/23	CHECK	72.30	2012.47
8/24	CHECK	92.14	1920.33
9/1	CLOSING BALANCE		1920.33

CUMULATIVE REVIEW EXERCISES

1. Simplify: $12 - (10 - 8)^2 \div 2 + 3$

2. Add: $3\frac{1}{3} + 4\frac{1}{8} + 1\frac{1}{12}$

3. Find the difference between $12\frac{3}{16}$ and $9\frac{5}{12}$.

4. Find the product of $5\frac{5}{8}$ and $1\frac{9}{15}$.

5. Divide: $3\frac{1}{2} \div 1\frac{3}{4}$

6. Simplify: $\left(\frac{3}{4}\right)^2 \div \left(\frac{3}{8} - \frac{1}{4}\right) + \frac{1}{2}$

7. Divide: $0.059\overline{)3.0792}$
   Round to the nearest tenth.

8. Convert $\frac{17}{12}$ to a decimal. Round to the nearest thousandth.

9. Write "$410 in 8 hours" as a unit rate.

10. Solve the proportion $\frac{5}{n} = \frac{16}{35}$.
    Round to the nearest hundredth.

11. Write $\frac{5}{8}$ as a percent.

12. Find 6.5% of 420.

13. Write 18.2% as a decimal.

14. What percent of 20 is 8.4?

15. 30 is 12% of what?

16. 65 is 42% of what? Round to the nearest hundredth.

17. Meteorology A series of late-summer storms produced rainfall of $3\frac{3}{4}$, $8\frac{1}{2}$, and $1\frac{2}{3}$ inches during a 3-week period. Find the total rainfall during the 3 weeks.

18. Taxes The Homer family pays $\frac{1}{5}$ of its total monthly income for taxes. The family has a total monthly income of $4850. Find the amount of their monthly income that the Homers pay in taxes.

19. Consumerism In 5 years, the cost of a scientific calculator went from $75 to $30. What is the ratio of the decrease in price to the original price?

20. Fuel Efficiencies A compact car was driven 417.5 miles on 12.5 gallons of gasoline. Find the number of miles driven per gallon of gasoline.

21. Consumerism A 14-pound turkey costs $15.40. Find the unit cost. Round to the nearest cent.

22. Investments Eighty shares of a stock paid a dividend of $112. At the same rate, find the dividend on 200 shares of the stock.

23. Discount A laptop computer that regularly sells for $900 is on sale for 20% off the regular price. What is the sale price?

24. Markup A pro skate shop bought a grinding rail for $85 and used a markup rate of 40%. Find the selling price of the grinding rail.

25. Compensation Sook Kim, an elementary school teacher, received an increase in salary from $2800 per month to $3024 per month. Find the percent increase in her salary.

26. Simple Interest A contractor borrowed $120,000 for 6 months at an annual interest rate of 4.5%. How much simple interest is due on the loan?

27. Car Expenses A red Ford Mustang was purchased for $26,900, and a down payment of $2000 was made. The balance is financed for 3 years at an annual interest rate of 9%. Find the monthly payment. Use the Monthly Payment Table in the Appendix. Round to the nearest cent.

28. Banking A family had a checking account balance of $1846.78. A check of $568.30 was deposited into the account, and checks of $123.98 and $47.33 were written. Find the new checking account balance.

29. Car Expenses During 1 year, Anna Gonzalez spent $1840 on gasoline and oil, $820 on insurance, $185 on tires, and $432 on repairs. Find the cost per mile to drive the car 10,000 miles during the year. Round to the nearest cent.

30. Real Estate A house has a mortgage of $172,000 for 20 years at an annual interest rate of 6%. Find the monthly mortgage payment. Use the Monthly Payment Table in the Appendix. Round to the nearest cent.

Statistics and Probability

OBJECTIVES

SECTION 7.1
A To read a pictograph
B To read a circle graph

SECTION 7.2
A To read a bar graph
B To read a broken-line graph

SECTION 7.3
A To read a histogram
B To read a frequency polygon

SECTION 7.4
A To find the mean, median, and mode of a distribution
B To draw a box-and-whiskers plot

SECTION 7.5
A To calculate the probability of simple events

ARE YOU READY?

Take the Chapter 7 Prep Test to find out if you are ready to learn to:

- Read pictographs, circle graphs, bar graphs, and broken-line graphs
- Read histograms and frequency polygons
- Find the mean, median, and mode of data
- Draw a box-and-whiskers plot
- Calculate the probability of an event

PREP TEST

Do these exercises to prepare for Chapter 7.

1. **Mail** Bill-related mail accounted for 49 billion of the 102 billion pieces of first-class mail handled by the U.S. Postal Service during a recent year. (*Source:* U.S. Postal Service) What percent of the pieces of first-class mail handled by the U.S. Postal Service was bill-related mail? Round to the nearest tenth of a percent.

2. **Education** The table at the right shows the estimated costs of funding an education at a public college.
   a. Between which two enrollment years is the increase in cost greatest?

   b. What is the increase between these two years?

Enrollment Year	Cost of Public College
2005	$70,206
2006	$74,418
2007	$78,883
2008	$83,616
2009	$88,633
2010	$93,951

Source: The College Board's Annual Survey of Colleges

3. **Sports** During the 1924 Summer Olympics in Paris, France, the United States won 45 gold medals, 27 silver medals, and 27 bronze medals. (*Source: The Ultimate Book of Sports Lists*) Find the ratio of gold medals won by the United States to silver medals won by the United States during the 1924 Summer Olympics. Write the ratio as a fraction in simplest form.

4. **The Military** Approximately 198,000 women serve in the U.S. military. Six percent of these women serve in the Marine Corps. (*Source:* www.fedstats.gov) What fractional amount of women in the military are in the Marine Corps?

7.1   Pictographs and Circle Graphs

OBJECTIVE A   To read a pictograph

Statistics is the branch of mathematics concerned with **data,** or numerical information. **Graphs** are displays that provide a pictorial representation of data. The advantage of graphs is that they present information in a way that is easily read.

A **pictograph** uses symbols to represent information. The pictograph in Figure 1 represents the net worth of America's richest billionaires. Each symbol represents 10 billion dollars.

Bill Gates

Net Worth (in tens of billions of dollars)

Bill Gates

Warren Buffet

Sheldon Adelson

Larry Ellison

Sergey Brin

Larry Page

FIGURE 1 Net worth of America's richest billionaires
Source: www.Forbes.com

From the pictograph, we can determine that Bill Gates has the greatest net worth. Larry Ellison's net worth is $10 billion more than Sergey Brin's net worth.

HOW TO 1   The pictograph in Figure 2 represents the responses of 600 young Americans when asked what they would like to have with them on a desert island. "Books" was the response of what percent of the respondents?

Strategy
Use the basic percent equation. The base is 600 (the total number of responses), and the amount is 90 (the number responding "Books").

Solution
Percent × base = amount

$$n \ \times 600 = \ 90$$

$$n = 90 \div 600$$

$$n = 0.15$$

15% of the respondents wanted books on a desert island.

Music

Parents

Computer

Books

TV

☘ = 30 responses

FIGURE 2 What 600 young Americans want on a desert island
Source: Time Magazine

The pictograph in Figure 3 shows the number of new cellular phones purchased in a particular city during a 4-month period.

The ratio of the number of cellular phones purchased in March to the number purchased in January is

$$\frac{3000}{4500} = \frac{2}{3}$$

 = 1000 cellular phones

FIGURE 3 Monthly cellular phone purchases

EXAMPLE · 1

Use Figure 3 to find the total number of cellular phones purchased during the 4-month period.

Strategy

To find the total number of cellular phones purchased in the 4-month period:

- Read the pictograph to determine the number of cellular phones purchased each month.
- Add the four numbers.

Solution

Purchases for January: 4500
Purchases for February: 3500
Purchases for March: 3000
Purchases for April: 1500

Total purchases for the 4-month period:

```
    4,500
    3,500
    3,000
+   1,500
   _____
   12,500
```

There were 12,500 cellular phones purchased in the 4-month period.

YOU TRY IT · 1

According to Figure 3, the number of cellular phones purchased in March represents what percent of the total number of cellular phones purchased during the 4-month period?

Your strategy

Your solution

Solution on p. S18

OBJECTIVE B   To read a circle graph

A **circle graph** represents data by the size of the sectors. A **sector of a circle** is one of the "pieces of the pie" into which a circle graph is divided.

The circle graph in Figure 4 shows the consumption of energy sources in the United States during a recent year. The complete circle graph represents the total amount of energy consumed, 96.6 quadrillion Btu. Each sector of the circle represents the consumption of energy from a different source.

✓ Take Note
One quadrillion is 1,000,000,000,000,000.

🎯 Point of Interest
Fossil fuels include coal, natural gas, and petroleum. Renewable energy includes hydroelectric power, solar energy, wood burning, and wind energy.

To find the percent of the total energy consumed that originated from nuclear power, solve the basic percent equation for percent. The base is 96.6 quadrillion Btu, and the amount is 8.2 quadrillion Btu.

Percent × base = amount

$$n \times 96.6 = 8.2$$
$$n = 8.2 \div 96.6$$
$$n \approx 0.085$$

To the nearest tenth of a percent, 8.5% of the energy consumed originated from nuclear power.

FIGURE 4 Annual energy consumption in quadrillion Btu in the United States
Source: The World Almanac and Book of Facts 2003

In a recent year, the top 25 companies in the United States spent a total of $17.8 billion for national advertising. The circle graph in Figure 5 shows what percents of the $17.8 billion went to the various advertising media. The complete circle represents 100% of all the money spent by these companies. Each sector of the graph represents the percent of the total spent for a particular medium.

HOW TO 2   According to Figure 5, how much money was spent for magazine advertising? Round to the nearest hundred million dollars.

Strategy
Use the basic percent equation. The base is $17.8 billion, and the percent is 16%.

Solution
Percent × base = amount

$$0.16 \times 17.8 = n$$
$$2.848 = n$$
$$2.848 \text{ billion} = 2,848,000,000$$

To the nearest hundred million, the amount spent for magazine advertising was $2,800,000,000.

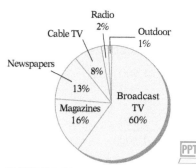

FIGURE 5 Distribution of advertising dollars for 25 companies
Source: Interep research

The circle graph in Figure 6 shows typical annual expenses of owning, operating, and financing a car. Use this figure for Example 2 and You Try It 2.

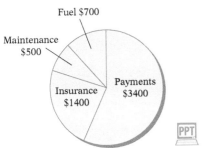

FIGURE 6 Annual expenses of $6000 for owning, operating, and financing a car
Source: Based on data from IntelliChoice

The circle graph in Figure 7 shows the distribution of an employee's gross monthly income. Use this figure for Example 3 and You Try It 3.

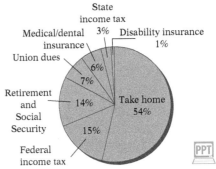

FIGURE 7 Distribution of gross monthly income of $2900

EXAMPLE • 2

Use Figure 6 to find the ratio of the annual insurance expense to the total annual cost of the car.

Strategy
To find the ratio:

- Locate the annual insurance expense in the circle graph.
- Write in simplest form the ratio of the annual insurance expense to the total annual cost of operating the car.

Solution
Annual insurance expense: $1400

$$\frac{1400}{6000} = \frac{7}{30}$$

The ratio is $\frac{7}{30}$.

YOU TRY IT • 2

Use Figure 6 to find the ratio of the annual cost of fuel to the annual cost of maintenance.

Your strategy

Your solution

EXAMPLE • 3

Use Figure 7 to find the employee's take-home pay.

Strategy
To find the take-home pay:
- Locate the percent of the distribution that is take-home pay.
- Solve the basic percent equation for amount.

Solution
Take-home pay: 54%

Percent × base = amount
0.54 × 2900 = n
1566 = n

The employee's take-home pay is $1566.

YOU TRY IT • 3

Use Figure 7 to find the amount paid for medical/dental insurance.

Your strategy

Your solution

Solutions on p. S18

7.1 EXERCISES

OBJECTIVE A   To read a pictograph

The Film Industry The pictograph in Figure 8 shows the approximate gross domestic revenues for four popular action movies. Use this graph for Exercises 1 to 3.

1. Find the total gross revenue from the four movies.

2. Find the ratio of the gross revenue of *Ocean's Thirteen* to the gross revenue of *I Am Legend*.

3. Find the percent of the total gross revenue that was earned by *Spider-Man 3*.

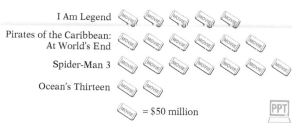

FIGURE 8 Gross revenues of four popular action movies
Source: www.worldwideboxoffice.com

Space Exploration The pictograph in Figure 9 is based on a survey of adults who were asked whether they agreed with each statement. Use this graph for Exercises 4 to 6.

4. Find the ratio of the number of people who agreed that space exploration impacts daily life to the number of people who agreed that space will be colonized in their lifetime.

5. How many more people agreed that humanity should explore planets than agreed that space exploration impacts daily life?

6. Is the number of people who said they would travel in space more than twice the number of people who agreed that space would be colonized in their lifetime?

FIGURE 9 Number of adults who agree with the statement
Source: Opinion Research for Space Day Partners

Children's Behavior The pictograph in Figure 10 is based on a survey of children aged 7 through 12. The percents of children's responses to the survey are shown. Assume that 500 children were surveyed. Use this graph for Exercises 7 to 9.

7. Find the number of children who said they hid vegetables under a napkin.

8. What is the difference between the number of children who fed vegetables to the dog and the number who dropped them on the floor?

9. Were the responses given in the graph the only responses given by the children? Explain your answer.

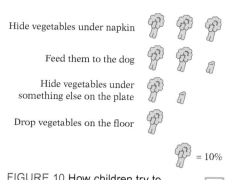

FIGURE 10 How children try to hide vegetables
Source: Strategic Consulting and Research for Del Monte

For Exercises 10 and 11, refer to the pictograph in Figure 1 on page 294.

10. Which statement in blue below the pictograph does not depend upon knowing what each small dollar bill symbol represents?

11. Write down two more facts you can determine from the pictograph without knowing what each small dollar bill symbol represents.

OBJECTIVE B To read a circle graph

Education An accounting major recorded the number of units required in each discipline to graduate with a degree in accounting. The results are shown in the circle graph in Figure 11. Use this graph for Exercises 12 to 14.

12. How many units are required to graduate with a degree in accounting?

 13. Is the number of units required in humanities *less than* or *greater than* twice the number of units required in science?

 14. Is the ratio of the number of units required in accounting to the number of units required in English *less than, equal to,* or *greater than* 5?

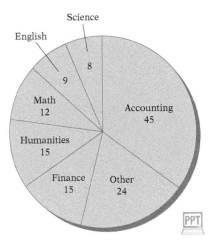

FIGURE 11 Number of units required to graduate with an accounting degree

Theaters The circle graph in Figure 12 shows the result of a survey in which people were asked, "What bothers you most about movie theaters?" Use this graph for Exercises 15 to 18.

15. **a.** What complaint was mentioned the most often?

   b. What complaint was mentioned the least often?

16. How many people were surveyed?

17. What is the ratio of the number of people responding "Dirty floors" to the number responding "High ticket prices"?

18. What percent of the respondents said that people talking bothered them most?

FIGURE 12 Distribution of responses in a survey

Video Games The circle graph in Figure 13 shows the breakdown of the approximately $3,100,000,000 that Americans spent on home video game equipment in one year. Use this graph for Exercises 19 to 22.

19. Find the amount of money spent on TV game machines.

20. Find the amount of money spent on portable game machines.

FIGURE 13 Percents of $3,100,000,000 spent annually on home video games
Source: The NPD Group, Toy Manufacturers of America

21. What fractional amount of the total money spent was spent on accessories?

22. Is the amount spent for TV game machines more than three times the amount spent for portable game machines?

 Demographics The circle graph in Figure 14 shows a breakdown, according to age, of the homeless in America. Use this graph for Exercises 23 to 26.

23. What age group represents the largest segment of the homeless population?

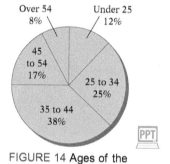

24. Is the number of homeless who are aged 25 to 34 more or less than twice the number who are under the age of 25?

FIGURE 14 Ages of the homeless in America
Source: The Department of Housing and Urban Development

25. What percent of the homeless population is under the age of 35?

26. On average, how many of every 100,000 homeless people in America are over the age of 54?

Geography The circle graph in Figure 15 shows the land area of each of the seven continents in square miles. Use this graph for Exercises 27 to 30.

27. Find the total land area of the seven continents.

28. How much larger is North America than South America?

29. What percent of the total land area is the land area of Asia? Round to the nearest tenth of a percent.

30. What percent of the total land area is the land area of Australia? Round to the nearest tenth of a percent.

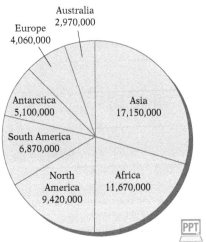

FIGURE 15 Land area of the seven continents (in square miles)

 Demographics There are approximately 300,000,000 people living in the United States. The circle graph in Figure 16 shows the breakdown of the U.S. population by ethnic origin. Use this graph for Exercises 31 to 33.

31. Approximately how many people living in the United States are of Asian ethnic origin?

32. Approximately how many more people of American Indian/Alaska Native ethnic origin live in the United States than people of Hawaiian/Pacific Islander ethnic origin?

33. On average, how many people of black ethnic origin would there be in a random sample of 500,000 people living in the United States?

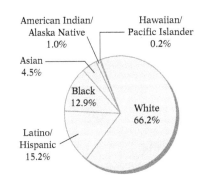

FIGURE 16 U.S. population by ethnic origin
Source: Entertainment Weekly, June 20, 2008

Applying the Concepts

 34. a. What are the advantages of presenting data in the form of a pictograph?
 b. What are the disadvantages?

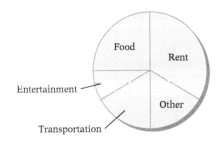

35. The circle graph at the right shows a couple's expenditures last month. Write two observations about this couple's expenses.

SECTION

7.2 Bar Graphs and Broken-Line Graphs

OBJECTIVE A To read a bar graph

Point of Interest

The first bar graph appeared in 1786 in the book *The Commercial and Political Atlas.* The author, William Playfalr (1759–1823), was a pioneer in the use of graphical displays.

A **bar graph** represents data by the height of the bars. The bar graph in Figure 17 shows temperature data recorded for Cincinnati, Ohio, for the months March through November. For each month, the height of the bar indicates the average daily high temperature during that month. The jagged line near the bottom of the graph indicates that the vertical scale is missing the numbers between 0 and 50.

The daily high temperature in September was 78°F. Because the bar for July is the tallest, the daily high temperature was highest in July.

FIGURE 17 Daily high temperatures in Cincinnati, Ohio
Source: U.S. Weather Bureau

Take Note

The bar for athletic females is halfway between the marks for 50 and 60. Therefore, we estimate that the lung capacity is halfway between these two numbers, at 55.

A **double-bar graph** is used to display data for purposes of comparison. The double-bar graph in Figure 18 compares the lung capacities of inactive and athletic 45-year-olds.

The lung capacity of an athletic female is 55 milliliters of oxygen per kilogram of body weight per minute.

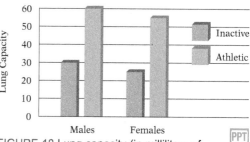

FIGURE 18 Lung capacity (in milliliters of oxygen per kilogram of body weight per minute)

EXAMPLE 1

What is the ratio of the lung capacity of an inactive male to that of an athletic male?

Strategy

To write the ratio:

• Read the graph to find the lung capacity of an inactive male and of an athletic male.
• Write the ratio in simplest form.

Solution

Lung capacity of inactive male: 30
Lung capacity of athletic male: 60

$$\frac{30}{60} = \frac{1}{2}$$

The ratio is $\frac{1}{2}$.

YOU TRY IT 1

What is the ratio of the lung capacity of an inactive female to that of an athletic female?

Your strategy

Your solution

Solution on p. S18

OBJECTIVE B   To read a broken-line graph

A **broken-line graph** represents data by the positions of the lines. It is used to show trends.

The broken-line graph in Figure 19 shows the effect of inflation on the value of a $100,000 life insurance policy. The height of each dot indicates the value of the policy.

After 10 years, the purchasing power of the $100,000 has decreased to approximately $60,000.

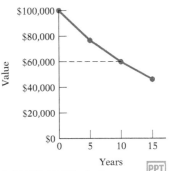

FIGURE 19 Effect of inflation on the value of a $100,000 life insurance policy

Two broken-line graphs are often shown in the same figure for comparison. Figure 20 shows the net incomes of two software companies, Math Associates and MatheMentors, before their merger.

Several things can be determined from the graph:

The net income for Math Associates in 2004 was $12 million.

The net income for MatheMentors declined from 2000 to 2001.

The net income for Math Associates increased for each year shown.

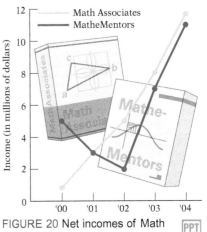

FIGURE 20 Net incomes of Math Associates and MatheMentors

EXAMPLE • 2

Use Figure 20 to approximate the difference between the net income of Math Associates and that of MatheMentors in 2002.

Strategy
To write the difference:

- Read the line graph to determine the net income of Math Associates and that of MatheMentors in 2002.
- Subtract to find the difference.

Solution
Net income for Math Associates: $5 million
Net income for MatheMentors: $2 million

$5 - 2 = 3$

The difference between the net incomes in 2002 was $3 million.

YOU TRY IT • 2

Use Figure 20 to determine between which two years the net income of Math Associates increased the most.

Your strategy

Your solution

Solution on p. S18

7.2 EXERCISES

OBJECTIVE A   To read a bar graph

Ziplining Ziplining is a high-adrenaline sport in which participants sail through the air and enjoy the view below while harnessed to steel cables. The bar graph in Figure 21 shows the maximum heights of various ziplines in the United States. Use this graph for Exercises 1 to 3.

1. What is the maximum height of Zip Idaho?

2. How much higher is the highest point of the Icy Straight Point Ziprider than the highest point of the Moaning Caverns zipline?

3. Find the ratio of the maximum height of Alpine Adventures to the maximum height of Zip Idaho.

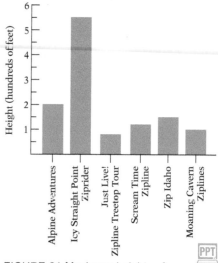

FIGURE 21 Maximum heights of ziplines in the U.S.
Source: USA Today, June 13, 2008

Automobile Production The bar graph in Figure 22 shows the regions in which all passenger cars were produced during a recent year. Use this graph for Exercises 4 to 6.

4. How many passenger cars were produced worldwide?

5. What is the difference between the number of passenger cars produced in Western Europe and the number produced in North America?

6. What percent of the passenger cars were produced in Asia? Round to the nearest percent.

FIGURE 22 Number of passenger cars produced (in a recent year)
Source: Los Angeles Times, 2000.

Compensation The double-bar graph in Figure 23 shows maximum salaries for police officers in selected cities and the corresponding maximum salaries for officers in the suburbs of that city. Use this graph for Exercises 7 to 10.

7. Estimate the difference between the maximum salaries of police officers in the suburbs of New York City and in the city of New York.

8. Is there a city for which the maximum salary of a police officer in the city is greater than the maximum salary in the suburbs?

9. For which city is the difference between the maximum salary in the suburbs and that in the city the greatest?

10. Of the cities shown on the graph, which city has the lowest maximum salary for police officers in the suburbs?

FIGURE 23 Maximum salaries of police officers in the city and the suburbs
Source: USA Today

For Exercises 11 and 12, refer to Figure 18 on page 302. Match the given statement about the double-bar graph to one of the following statements.

(i) The lung capacity of inactive males is less than the lung capacity of athletic males.
(ii) The lung capacity of inactive males is less than the lung capacity of inactive females.
(iii) The lung capacity of inactive males is greater than the lung capacity of inactive females.

11. The brown bar for males is longer than the brown bar for females.

12. The brown bar for males is shorter than the green bar for males.

OBJECTIVE B   To read a broken-line graph

Meteorology The broken-line graph in Figure 24 shows the average monthly snowfall during ski season around Aspen, Colorado. Use this graph for Exercises 13 to 16.

13. What is the average snowfall during January?

14. During which month is the snowfall the greatest?

15. What is the total average snowfall during March and April?

16. Find the ratio of the average snowfall in November to the average snowfall in December.

FIGURE 24 **Average snowfall in** Aspen, Colorado
Source: Weather America, by Alfred Garwood

Wind Power The broken-line graph in Figure 25 shows how the wind power capacity of the United States has changed over recent years. Wind power capacity is measured in megawatts. Use this graph for Exercises 17 to 19.

17. How much wind power capacity was produced in 2002?

18. How much more wind power was available as an energy source in 2006 than in 2000?

19. Between which two consecutive years did the wind power capacity increase the most?

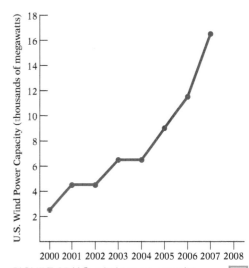

FIGURE 25 U.S. wind power capacity
Source: www.eere.energy.gov

Health The double-broken-line graph in Figure 26 shows the number of Calories per day that should be consumed by women and men in various age groups. Use this graph for Exercises 20 to 22.

20. What is the difference between the number of Calories recommended for men and the number recommended for women 19–22 years of age?

21. People of what age and gender have the lowest recommended number of Calories?

FIGURE 26 Recommended number of Calories per day for women and men PPT

Source: Numbers, by Andrea Sutcliffe (HarperCollins)

22. Find the ratio of the number of Calories recommended for women 15 to 18 years old to the number recommended for women 51 to 74 years old.

For Exercises 23 and 24, each statement refers to a line graph (not shown) that displays the population of a particular state every 10 years between 1950 and 2000. Determine whether the statement is *true* or *false.*

23. If the population decreased between 1990 and 2000, then the segment joining the point for 1990 and the point for 2000 slants down from left to right.

24. If the points for 1960 and 1970 are connected by a horizontal line, the population in 1970 was the same as the population in 1960.

Applying the Concepts

Wind Power The graph in Figure 27 shows how wind power capacity increased from 2000 to 2007 for the states with the largest wind energy capacity, Texas and California. Use this graph for Exercises 25 and 26.

25. Create a table that shows the wind power capacity of each state for each of the years 2000 through 2007.

26. Create a table that shows the difference in the wind power capacities of Texas and California for each year from 2000 to 2007, and indicate which state had the greater wind power capacity. During which years did the wind power capacity of Texas exceed that of California?

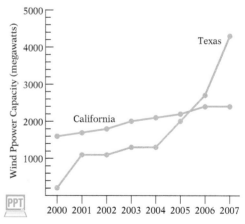

FIGURE 27 Wind power capacity in Texas and California

Source: www.eere.energy.gov

SECTION

7.3 Histograms and Frequency Polygons

OBJECTIVE A To read a histogram

A research group measured the fuel usage of 92 cars. The results are recorded in the histogram in Figure 28. A **histogram** is a special type of bar graph. The width of each bar corresponds to a range of numbers called a **class interval.** The height of each bar corresponds to the number of occurrences of data in each class interval and is called the **class frequency.**

Class Intervals (miles per gallon)	Class Frequencies (number of cars)
18–20	12
20–22	19
22–24	24
24–26	17
26–28	15
28–30	5

FIGURE 28

Twenty-four cars get between 22 and 24 miles per gallon.

A precision tool company has 85 employees. Their hourly wages are recorded in the histogram in Figure 29.

The ratio of the number of employees whose hourly wage is between $14 and $16 to the total number of employees is $\frac{17 \text{ employees}}{85 \text{ employees}} = \frac{1}{5}$.

FIGURE 29

EXAMPLE · 1

Use Figure 29 to find the number of employees whose hourly wage is between $16 and $20.

Strategy

To find the number of employees:

- Read the histogram to find the number of employees whose hourly wage is between $16 and $18 and the number whose hourly wage is between $18 and $20.
- Add the two numbers.

Solution

Number with wages between $16 and $18: 20
Number with wages between $18 and $20; 14

20 + 14 = 34

34 employees have an hourly wage between $16 and $20.

YOU TRY IT · 1

Use Figure 29 to find the number of employees whose hourly wage is between $10 and $14.

Your strategy

Your solution

Solution on p. S19

OBJECTIVE B   To read a frequency polygon

 Take Note

The blue portion of the graph at the right is a histogram. The red portion of the graph is a frequency polygon

The speeds of 70 cars on a highway were measured by radar. The results are recorded in the frequency polygon in Figure 30. A **frequency polygon** is a graph that displays information in a manner similar to a histogram. A dot is placed above the center of each class interval at a height corresponding to that class's frequency. The dots are then connected to form a broken-line graph. The center of a class interval is called the **class midpoint.**

Class Interval (miles per hour)	Class Midpoint	Class Frequency
30–40	35	7
40–50	45	13
50–60	55	25
60–70	65	21
70–80	75	4

FIGURE 30

Twenty-five cars were traveling between 50 and 60 miles per hour.

The per capita incomes in a recent year for the 50 states are recorded in the frequency polygon in Figure 31.

The number of states with a per capita income between $28,000 and $32,000 is 17.

FIGURE 31

Source: Bureau of Economic Analysis

EXAMPLE • 2

According to Figure 31, what percent of the states have a per capita income between $28,000 and $32,000?

Strategy

To find the percent, solve the basic percent equation for percent. The base is 50. The amount is 17.

Solution

Percent × base = amount

n × 50 =   17

$n = 17 \div 50$

$n = 0.34$

34% of the states have a per capita income between $28,000 and $32,000.

YOU TRY IT • 2

Use Figure 31 to find the ratio of the number of states with a per capita income between $36,000 and $40,000 to the number with a per capita income between $44,000 and $48,000.

Your strategy

Your solution

Solution on p. S19

7.3 EXERCISES

OBJECTIVE A   To read a histogram

Customer Credit A total of 50 monthly credit account balances were recorded. Figure 32 is a histogram of these data. Use this figure for Exercises 1 to 4.

1. How many account balances were between $1500 and $2000?

2. How many account balances were less than $2000?

3. What percent of the account balances were between $2000 and $2500?

4. What percent of the account balances were greater than $1500?

FIGURE 32

 Automobiles The histogram in Figure 33 is based on data from the American Automobile Manufacturers Association. It shows the ages of a sample of 1000 cars in a typical city in the United States. Use this figure for Exercises 5 to 9.

5. How many cars are between 6 and 12 years old?

6. Find the ratio of the number of cars between 12 and 15 years old to the total number of cars.

7. Find the number of cars more than 12 years old.

8. Find the percent of cars that are less than 9 years old.

 9. Which two consecutive class intervals have the greatest difference in class frequency?

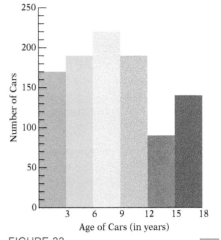

FIGURE 33
Source: American Automobile Manufacturers Association

 Malls According to a Maritz AmeriPoll, the average U.S. adult goes to a shopping mall about two times a month. The histogram in Figure 34 shows the average time 100 adults spend in the mall per trip. Use this figure for Exercises 10 to 12.

10. Find the number of adults who spend between 1 and 2 hours at the mall.

11. Find the number of adults who spend between 3 and 4 hours at the mall.

12. What percent of adults spend less than 1 hour at the mall?

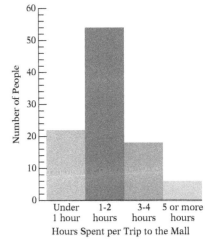

FIGURE 34
Source: Maritz AmeriPoll

OBJECTIVE B To read a frequency polygon

Marathons The frequency polygon in Figure 35 shows the approximate numbers of runners in the 2008 Boston Marathon to finish in each of the given time slots (times are given in hours and minutes). Use this figure for Exercises 13 to 15.

13. Determine the approximate number of runners who finished with times between $2\frac{1}{2}$ hours and 6 hours.

14. Find the approximate number of marathoners who finished with times of more than 4 hours.

15. State whether the frequency polygon can be used to draw the following conclusion:

   More runners had finishing times between 4 and $4\frac{1}{2}$ hours than had finishing times between $4\frac{1}{2}$ and 5 hours.

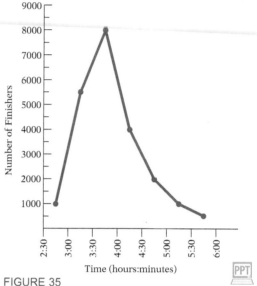

FIGURE 35
Source: www.marathonguide.com

The Lottery The frequency polygon in Figure 36 is based on data from a Gallup poll survey of 74 people who purchased lottery tickets. Use this figure for Exercises 16 to 18.

16. How many people purchased between 0 and 10 tickets?

17. What percent of the people purchased between 20 and 30 tickets each month? Round to the nearest tenth of a percent.

18. Is it possible to determine from the graph how many people purchased 15 lottery tickets? Explain.

FIGURE 36

Education The frequency polygon in Figure 37 shows the distribution of scores of the approximately 1,080,000 students who took an SAT exam. Use this figure for Exercises 19 to 21.

19. How many students scored between 1200 and 1400 on the exam?

20. What percent of the number of students who took the exam scored between 800 and 1000? Round to the nearest tenth of a percent.

21. How many students scored below 1000?

Applying the Concepts

22. Write a paragraph explaining the difference between a histogram and a bar graph.

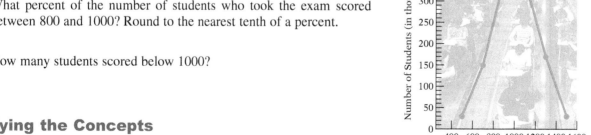

FIGURE 37
Source: Educational Testing Service

SECTION

7.4 Statistical Measures

OBJECTIVE A   To find the mean, median, and mode of a distribution

The average score on the math portion of the SAT was 432. The EPA estimates that a 2007 Toyota Camry Hybrid averages 35 miles per gallon on the highway. The average rainfall for portions of Kauai is 350 inches per year. Each of these statements uses one number to describe an entire collection of numbers. Such a number is called an *average*.

In statistics there are various ways to calculate an average. Three of the most common— *mean, median,* and *mode*—are discussed here.

An automotive engineer tests the miles-per-gallon ratings of 15 cars and records the results as follows:

Miles-per-Gallon Ratings of 15 Cars														
25	22	21	27	25	35	29	31	25	26	21	39	34	32	28

The **mean** of the data is the sum of the measurements divided by the number of measurements. The symbol for the mean is $\bar{x}$.

Formula for the Mean

$$\bar{x} = \frac{\text{sum of the data values}}{\text{number of data values}}$$

To find the mean for the data above, add the numbers and then divide by 15.

$$\bar{x} = \frac{25 + 22 + 21 + 27 + 25 + 35 + 29 + 31 + 25 + 26 + 21 + 39 + 34 + 32 + 28}{15}$$

$$= \frac{420}{15} = 28$$

The mean number of miles per gallon for the 15 cars tested was 28 miles per gallon.

The mean is one of the most frequently computed averages. It is the one that is commonly used to calculate a student's performance in a class.

Integrating Technology

When using a calculator to calculate the mean, use parentheses to group the sum in the numerator.

(78 + 82 + 91
+ 87 + 93) ÷
5 =

HOW TO 1   The test scores for a student taking American history were 78, 82, 91, 87, and 93. What was the mean score for this student?

Strategy

To find the mean, divide the sum of the test scores by 5, the number of scores.

Solution

$$\bar{x} = \frac{78 + 82 + 91 + 87 + 93}{5} = \frac{431}{5} = 86.2$$

The mean score for the history student was 86.2.

The **median** of the data is the number that separates the data into two equal parts when the numbers are arranged from least to greatest (or from greatest to least). There is an equal number of values above the median and below the median.

To find the median of a set of numbers, first arrange the numbers from least to greatest. The median is the number in the middle.

The result of arranging the miles-per-gallon ratings given on the previous page from least to greatest is shown below.

21   21   22   25   25   25   26   27   28   29   31   32   34   35   39

7 values below
the median

Middle number
Median

7 values above
the median

The median is 27 miles per gallon.

If data contain an *even* number of values, the median is the mean of the two middle numbers.

Tips for Success

Word problems are difficult because we must read the problem, determine the quantity we must find, think of a method to find it, actually solve the problem, and then check the answer. In short, we must devise a *strategy* and then use that strategy to find the *solution*. See *AIM for Success* at the front of the book.

| HOW TO   2 | The selling prices of the last six homes sold by a real estate agent were $275,000, $250,000, $350,000, $230,000, $345,000, and $290,000. Find the median selling price of these homes.

Strategy
To find the median, arrange the numbers from least to greatest. Because there is an even number of values, the median is the mean of the two middle numbers.

Solution
230,000   250,000   275,000   290,000   345,000   350,000

Middle 2 numbers

$$\text{Median} = \frac{275,000 + 290,000}{2} = 282,500$$

The median selling price was $282,500.

The **mode** of a set of numbers is the value that occurs most frequently. If a set of numbers has no number occurring more than once, then the data have no mode.

Here again are the data for the gasoline mileage ratings of 15 cars.

Miles-per-Gallon Ratings of 15 Cars														
25	22	21	27	25	35	29	31	25	26	21	39	34	32	28

25 is the number that occurs most frequently.

The mode is 25 miles per gallon.

Note from the miles-per-gallon example that the mean, median, and mode may be different.

EXAMPLE · 1

EXAMPLE · 1

Twenty students were asked the number of units in which they were enrolled. The responses were as follows:

15	12	13	15	17	18	13	20	9	16
14	10	15	12	17	16	6	14	15	12

Find the mean number of units taken by these students.

Strategy

To find the mean number of units:

- Find the sum of the 20 numbers.
- Divide the sum by 20.

Solution

$15 + 12 + 13 + 15 + 17 + 18 + 13 + 20 + 9 +$
$\quad 16 + 14 + 10 + 15 + 12 + 17 + 16 + 6 +$
$\quad 14 + 15 + 12 = 279$

$$\bar{x} = \frac{279}{20} = 13.95$$

The mean is 13.95 units.

YOU TRY IT · 1

The amounts spent by 12 customers at a McDonald's restaurant were as follows:

11.01	10.75	12.09	15.88	13.50	12.29
10.69	9.36	11.66	15.25	10.09	12.72

Find the mean amount spent by these customers. Round to the nearest cent.

Your strategy

Your solution

EXAMPLE · 2

The starting hourly wages for an apprentice electrician for six different work locations are $12.50, $11.25, $10.90, $11.56, $13.75, and $14.55. Find the median starting hourly wage.

Strategy

To find the median starting hourly wage:

- Arrange the numbers from least to greatest.
- Because there is an even number of values, the median is the mean of the two middle numbers.

Solution

10.90, 11.25, 11.56, 12.50, 13.75, 14.55

$$\text{Median} = \frac{11.56 + 12.50}{2} = 12.03$$

The median starting hourly wage is $12.03.

YOU TRY IT · 2

The amounts of weight lost, in pounds, by 10 participants in a 6-month weight-reduction program were 22, 16, 31, 14, 27, 16, 29, 31, 40, and 10. Find the median weight loss for these participants.

Your strategy

Your solution

Solutions on p. S19

OBJECTIVE B   To draw a box-and-whiskers plot

Recall from the last objective that an average is one number that helps to describe all the numbers in a set of data. For example, we know from the following statement that Erie gets a lot of snow each winter.

> The average annual snowfall in Erie, Pennsylvania, is 85 inches.

Now look at these two statements.

> The average annual temperature in San Francisco, California, is 57°F.

> The average annual temperature in St. Louis, Missouri, is 57°F.

San Francisco

The average annual temperature in both cities is the same. However, we do not expect the climate in St. Louis to be like San Francisco's climate. Although both cities have the same average annual temperature, their temperature ranges differ. In fact, the difference between the average monthly high temperatures in July and January in San Francisco is 14°F, whereas the difference between the average monthly high temperatures in July and January in St. Louis is 50°F.

Note that for this example, a single number (the average annual temperature) does not provide us with a very comprehensive picture of the climate of either of these two cities.

One method used to picture an entire set of data is a box-and-whiskers plot. To prepare a box-and-whiskers plot, we begin by separating a set of data into four parts, called **quartiles.** We will illustrate this by using the average monthly high temperatures for St. Louis, in degrees Fahrenheit. These are listed below from January through December.

St. Louis

39	47	58	72	81	88	89	89	85	76	49	47

Source: The Weather Channel

First list the numbers in order from least to greatest and determine the median.

39	47	47	49	58	72	76	81	85	88	89	89

↑
Median = 74

Now find the median of the data values below the median. The median of the data values below the median is called the **first quartile,** symbolized by Q_1. Also find the median of the data values above the median. The median of the data values above the median is called the **third quartile,** symbolized by Q_3.

|← 3 values →|← 3 values →|← 3 values →|← 3 values →|

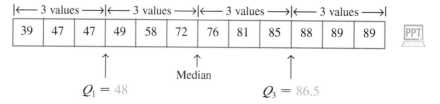

39	47	47	49	58	72	76	81	85	88	89	89

  ↑   ↑   ↑
      Median
$Q_1 = 48$ $Q_3 = 86.5$

The first quartile, Q_1, is the number that one-quarter of the data lie below. This means that 25% of the data lie below the first quartile. The third quartile, Q_3, is the number that one-quarter of the data lie above. This means that 25% of the data lie above the third quartile.

The **range** of a set of numbers is the difference between the greatest number and the least number in the set. The range describes the spread of the data. For the data above,

$$\text{Range} = \text{greatest value} - \text{least value} = 89 - 39 = 50$$

The **interquartile range** is the difference between the third quartile, Q_3, and the first quartile, Q_1. For the data above,

$$\text{Interquartile range} = Q_3 - Q_1 = 86.5 - 48 = 38.5$$

The interquartile range is the distance that spans the "middle" 50% of the data values. Because it excludes the bottom fourth of the data values and the top fourth of the data values, it excludes any extremes in the numbers of the set.

A **box-and-whiskers plot,** or **boxplot,** is a graph that shows five numbers: the least value, the first quartile, the median, the third quartile, and the greatest value. Here are these five values for the data on St. Louis temperatures.

The least number	39
The first quartile, Q_1	48
The median	74
The third quartile, Q_3	86.5
The greatest number	89

Think of a number line that includes the five values listed above. With this in mind, mark off the five values. Draw a box that spans the distance from Q_1 to Q_3. Draw a vertical line the height of the box at the median.

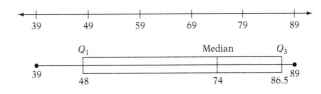

Listed below are the average monthly high temperatures for San Francisco.

57	60	61	64	68	71	71	73	74	73	60	59

Source: The Weather Channel

We can perform the same calculations on these data to determine the five values needed for the box-and-whiskers plot.

The least number	57
The first quartile, Q_1	60
The median	66
The third quartile, Q_3	72
The greatest number	74

The box-and-whiskers plot is shown at the right with the same scale used for the data on the St. Louis temperatures.

Note that by comparing the two boxplots, we can see that the range of temperatures in St. Louis is greater than the range of temperatures in San Francisco. For the St. Louis temperatures, there is a greater spread of the data below the median than above the median, whereas the data above and below the median of the San Francisco boxplot are spread nearly equally.

✓ **Take Note**
50% of the data in a distribution lie in the interquartile range.

✓ **Take Note**
The number line at the right is shown as a reference only. It is not a part of the boxplot. It shows that the numbers labeled on the boxplot are plotted according to the distances on the number line above it.

✓ **Take Note**
It is the "whiskers" on the box-and-whiskers plot that show the range of the data. The "box" on the box-and-whiskers plot shows the interquartile range of the data.

HOW TO 3

The numbers of avalanche deaths in the United States during each of nine consecutive winters were 8, 24, 29, 13, 28, 30, 22, 26, and 32. (*Source:* Colorado Avalanche Information Center) Draw a box-and-whiskers plot of the data, and determine the interquartile range.

Strategy

To draw the box-and-whiskers plot, arrange the data from least to greatest. Then find the median, Q_1, and Q_3. Use the least value, Q_1, the median, Q_3, and the greatest value to draw the box-and-whiskers plot.

To find the interquartile range, find the difference between Q_3 and Q_1.

✓ **Take Note**

Note that the left whisker in this box-and-whiskers plot is quite long, and the length of the box from Q_1 to the median is longer than the length of the box from the median to Q_3. This illustrates a set of data in which the median is closer to the greatest data value. If the two whiskers are approximately the same length, and the distances from Q_1 to the median and from the median to Q_3 are approximately equal, then the least and greatest values are about the same distance from the median. See Example 3 below.

Solution

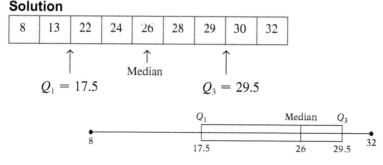

Interquartile range = $Q_3 - Q_1 = 29.5 - 17.5 = 12$

The interquartile range is 12 deaths.

EXAMPLE 3

The average monthly snowfall amounts, in inches, in Buffalo, New York, from October through April are 1, 12, 24, 25, 18, 12, and 3. (*Source:* The Weather Channel) Draw a box-and-whiskers plot of the data.

YOU TRY IT 3

The average monthly snowfall amounts, in inches, in Denver, Colorado, from October through April are 4, 7, 7, 8, 8, 9, and 13. (*Source:* The Weather Channel)
a. Draw a box-and-whiskers plot of the data.
b. How does the spread of the data within the interquartile range compare with that in Example 3?

Strategy

To draw the box-and-whiskers plot:

• Arrange the data from least to greatest.
• Find the median, Q_1, and Q_3.
• Use the least value, Q_1, the median, Q_3, and the greatest value to draw the box-and-whiskers plot.

Your strategy

Solution

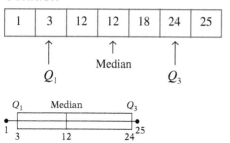

Your solution

Solution on p. S19

7.4 EXERCISES

OBJECTIVE A   **To find the mean, median, and mode of a distribution**

1. State whether the mean, median, or mode is being used.
    a. Half of the houses in the new development are priced under $350,000.

    b. The average bill for lunch at the college union is $11.95.

    c. The college bookstore sells more green college sweatshirts than any other color.

    d. In a recent year, there were as many people age 26 and younger in the world as there were people age 26 and older.

    e. The majority of full-time students carry a load of 12 credit hours per semester.

    f. The average annual return on an investment is 6.5%.

2. **Consumerism** The number of high-definition televisions sold each month for one year was recorded by an electronics store. The results were 15, 12, 20, 20, 19, 17, 22, 24, 17, 20, 15, and 27. Calculate the mean, the median, and the mode of the number of televisions sold per month.

3. **The Airline Industry** The number of seats occupied on a jet for 16 trans–Atlantic flights was recorded. The numbers were 309, 422, 389, 412, 401, 352, 367, 319, 410, 391, 330, 408, 399, 387, 411, and 398. Calculate the mean, the median, and the mode of the number of seats occupied per flight.

4. **Sports** The times, in seconds, for a 100-meter dash at a college track meet were 10.45, 10.23, 10.57, 11.01, 10.26, 10.90, 10.74, 10.64, 10.52, and 10.78.
    a. Calculate the mean time for the 100-meter dash.
    b. Calculate the median time for the 100-meter dash.

5. **Consumerism** A consumer research group purchased identical items in eight grocery stores. The costs for the purchased items were $85.89, $92.12, $81.43, $80.67, $88.73, $82.45, $87.81, and $85.82. Calculate the mean and the median costs of the purchased items.

6. **Computers** One measure of a computer's hard-drive speed is called access time; this is measured in milliseconds (thousandths of a second). Find the mean and median for 11 hard drives whose access times were 5, 4.5, 4, 4.5, 5, 5.5, 6, 5.5, 3, 4.5, and 4.5. Round to the nearest tenth.

7. **Health Plans** Eight health maintenance organizations (HMOs) presented group health insurance plans to a company. The monthly rates per employee were $423, $390, $405, $396, $426, $355, $404, and $430. Calculate the mean and the median monthly rates for these eight companies.

8. **Government** The lengths of the terms, in years, of all the former Supreme Court chief justices are given in the table below. Find the mean and median length of term for the chief justices. Round to the nearest tenth.

5	0	4	34	28	8	14	21
10	8	11	4	7	15	17	19

9. **Life Expectancy** The life expectancies, in years, in ten selected Central and South American countries are given at the right.
   a. Find the mean life expectancy in this group of countries.
   b. Find the median life expectancy in this group of countries.

Country	Life Expectancy
Brazil	72
Chile	77
Costa Rica	77
Ecuador	77
Guatemala	70
Panama	75
Peru	70
Trinidad and Tobago	67
Uruguay	76
Venezuela	73

10. **Education** Your scores on six history tests were 78, 92, 95, 77, 94, and 88. If an "average score" of 90 receives an A for the course, which average, the mean or the median, would you prefer that the instructor use?

11. **Education** One student received scores of 85, 92, 86, and 89. A second student received scores of 90, 97, 91, and 94 (exactly 5 points more on each exam). Are the means of the two students the same? If not, what is the relationship between the means of the two students?

12. **Defense Spending** The table below shows the defense expenditures, in billions of dollars, by the federal government for 1965 through 1973, years during which the United States was actively involved in the Vietnam War.
   a. Calculate the mean annual defense expenditure for these years. Round to the nearest tenth of a billion.

   b. Find the median annual defense expenditure.

   c. If the year 1965 were eliminated from the data, how would that affect the mean? The median?

Year	1965	1966	1967	1968	1969	1970	1971	1972	1973
Expenditures	$49.6	$56.8	$70.1	$80.5	$81.2	$80.3	$77.7	$78.3	$76.0

Source: Statistical Abstract of the United States

OBJECTIVE B   **To draw a box-and-whiskers plot**

13. a. What percent of the data in a set of numbers lie above Q_3?
    b. What percent of the data in a set of numbers lie above Q_1?
    c. What percent of the data in a set of numbers lie below Q_3?
    d. What percent of the data in a set of numbers lie below Q_1?

14. **U.S. Presidents** The box-and-whiskers plot below shows the distribution of the ages of presidents of the United States at the time of their inauguration.
    a. What is the youngest age in the set of data?
    b. What is the oldest age?
    c. What is the first quartile?
    d. What is the third quartile?
    e. What is the median?
    f. Find the range.
    g. Find the interquartile range.

15. Compensation The box-and-whiskers plot below shows the distribution of median incomes for 50 states and the District of Columbia. What is the lowest value in the set of data? The highest value? The first quartile? The third quartile? The median? Find the range and the interquartile range.

46,596   56,067 61,036   66,507   82,879

16. Education An aptitude test was taken by 200 students at the Fairfield Middle School. The box-and-whiskers plot at the right shows the distribution of their scores.

43   54   72   88   98

a. How many students scored over 88?
b. How many students scored below 72?
c. How many scores are represented in each quartile?
d. What percent of the students had scores of at least 54?

17. Health The cholesterol levels for 80 adults were recorded and then displayed in the box-and-whiskers plot shown at the right.

172 198 217   254   345

a. How many adults had a cholesterol level above 217?
b. How many adults had a cholesterol level below 254?
c. How many cholesterol levels are represented in each quartile?

d. What percent of the adults had a cholesterol level of not more than 198?

18. Fuel Efficiency The gasoline consumption of 19 cars was tested, and the results were recorded in the table below.
a. Find the range, the first quartile, the third quartile, and the interquartile range.

b. Draw a box-and-whiskers plot of the data.
c. Is the data value 21 in the interquartile range?

Miles per Gallon for 19 Cars									
33	21	30	32	20	31	25	20	16	24
22	31	30	28	26	19	21	17	26	

19. Environment Carbon dioxide is among the gases that contribute to global warming. The world's biggest emitters of carbon dioxide are listed below. The figures are emissions in millions of metric tons per year.
a. Find the range, the first quartile, the third quartile, and the interquartile range.

b. Draw a box-and-whiskers plot of the data.

c. What data value is responsible for the long whisker at the right?

Carbon Dioxide Emissions (in millions of metric tons per year)			
Canada	0.64	Japan	1.26
China	5.01	Russian Federation	1.52
Germany	0.81	South Korea	0.47
India	1.34	United Kingdom	0.59
Italy	0.45	United States	6.05

Source: U.S. Department of Energy

20. **Meteorology** The average monthly amounts of rainfall, in inches, from January through December for Seattle, Washington, and Houston, Texas, are listed below.

    a. Is the difference between the means greater than 1 inch?

    b. What is the difference between the medians?

    c. Draw a box-and-whiskers plot of each set of data. Use the same scale.

    d. Describe the difference between the distributions of the data for Seattle and Houston.

Seattle	6.0	4.2	3.6	2.4	1.6	1.4	0.7	1.3	2.0	3.4	5.6	6.3
Houston	3.2	3.3	2.7	4.2	4.7	4.1	3.3	3.7	4.9	3.7	3.4	3.7

Source: The Weather Channel

21. **Meteorology** The average monthly amounts of rainfall, in inches, from January through December for Orlando, Florida, and Portland, Oregon, are listed below.

    a. Is the difference between the means greater than 1 inch?

    b. What is the difference between the medians?

    c. Draw a box-and-whiskers plot of each set of data. Use the same scale.

    d. Describe the difference between the distributions of the data for Orlando and Portland.

Orlando	2.1	2.8	3.2	2.2	4.0	7.4	7.8	6.3	5.6	2.8	1.8	1.8
Portland	6.2	3.9	3.6	2.3	2.1	1.5	0.5	1.1	1.6	3.1	5.2	6.4

Source: The Weather Channel

22. Refer to the box-and-whiskers plot in Exercise 15. Which of the following fractions most accurately represents the fraction of states with median incomes less than $66,507?

    (i) $\dfrac{1}{4}$ (ii) $\dfrac{1}{3}$ (iii) $\dfrac{1}{2}$ (iv) $\dfrac{3}{4}$

23. Write a set of data with five data values for which the mean, median, and mode are all 55.

Applying the Concepts

24. A set of data has a mean of 16, a median of 15, and a mode of 14. Which of these numbers must be a value in the data set? Explain your answer.

25. Explain each notation.

    a. Q_1 **b.** Q_3 **c.** $\bar{x}$

26. The box in a box-and-whiskers plot represents 50%, or one-half, of the data in a set. Why is the box in Example 3 of this section not one-half of the entire length of the box-and-whiskers plot?

27. Create a set of data containing 25 numbers that would correspond to the box-and-whiskers plot shown at the right.

SECTION

7.5 Introduction to Probability

OBJECTIVE A

To calculate the probability of simple events

 Point of Interest

It was dice playing that led Antoine Gombaud, Chevalier de Mere, to ask Blaise Pascal, a French mathematician, to figure out the probability of throwing two sixes. Pascal and Pierre Fermat solved the problem, and their explorations led to the birth of probability theory.

A weather forecaster estimates that there is a 75% chance of rain. A state lottery director claims that there is a $\frac{1}{9}$ chance of winning a prize in a new game offered by the lottery. Each of these statements involves uncertainty to some extent. The degree of uncertainty is called **probability.** For the statements above, the probability of rain is 75%, and the probability of winning a prize in the new lottery game is $\frac{1}{9}$.

A probability is determined from an **experiment,** which is any activity that has an observable outcome. Examples of experiments include

> Tossing a coin and observing whether it lands heads up or tails up
>
> Interviewing voters to determine their preference for a political candidate
>
> Drawing a card from a standard deck of 52 cards

All the possible outcomes of an experiment are called the **sample space** of the experiment. The outcomes are listed between braces. For example:

The number cube shown at the left is rolled once. Any of the numbers from 1 to 6 could show on the top of the cube. The sample space is

$$\{1, 2, 3, 4, 5, 6\}$$

A fair coin is tossed once. (A fair coin is one for which heads and tails have an equal chance of landing face up.) If H represents "heads up" and T represents "tails up," then the sample space is

$$\{H, T\}$$

An **event** is one or more outcomes of an experiment. For the experiment of rolling the six-sided cube described above, some possible events are

> The number is even: $\{2, 4, 6\}$
>
> The number is a multiple of 3: $\{3, 6\}$
>
> The number is less than 10: $\{1, 2, 3, 4, 5, 6\}$

Note that in the last case, the event is the entire sample space.

HOW TO 1 The spinner at the left is spun once. Assume that the spinner does not come to rest on a line.

a. What is the sample space?

The arrow could come to rest on any one of the four sectors.
The sample space is $\{1, 2, 3, 4\}$.

b. List the outcomes in the event that the spinner points to an odd number.
$\{1, 3\}$

In discussing experiments and events, it is convenient to refer to the **favorable outcomes** of an experiment. These are the outcomes of an experiment that satisfy the requirements of a particular event.

For instance, consider the experiment of rolling a fair die once. The sample space is

$$\{1, 2, 3, 4, 5, 6\}$$

and one possible event would be rolling a number that is divisible by 3. The outcomes of the experiment that are favorable to the event are 3 and 6:

$$\{3, 6\}$$

The outcomes of the experiment of tossing a fair coin are *equally likely*. Either one of the outcomes is just as likely as the other. If a fair coin is tossed once, the probability of a head is $\frac{1}{2}$, and the probability of a tail is $\frac{1}{2}$. Both events are equally likely. The theoretical probability formula, given below, applies to experiments for which the outcomes are equally likely.

Theoretical Probability Formula

The theoretical probability of an event is a fraction with the number of favorable outcomes of the experiment in the numerator and the total number of possible outcomes in the denominator.

$$\text{Probability of an event} = \frac{\text{number of favorable outcomes}}{\text{number of possible outcomes}}$$

A probability of an event is a number from 0 to 1 that tells us how likely it is that this outcome will happen.

A probability of 0 means that the event is impossible.
The probability of getting a heads when rolling the die shown at the left is 0.

A probability of 1 means that the event must happen.
The probability of getting either heads or tails when tossing a coin is 1.

A probability of $\frac{1}{4}$ means that it is expected that the outcome will happen 1 in every 4 times the experiment is performed.

✓ **Take Note**

The phrase **at random** means that each card has an equal chance of being drawn.

HOW TO 2   Each of the letters of the word *TENNESSEE* is written on a card, and the cards are placed in a hat. If one card is drawn at random from the hat, what is the probability that the card has the letter *E* on it?

Count the possible outcomes of the experiment.

> There are 9 letters in *TENNESSEE*.

> There are 9 possible outcomes of the experiment.

Count the number of outcomes of the experiment that are favorable to the event that a card with the letter *E* on it is drawn.

> There are 4 cards with an *E* on them.

Use the probability formula.

$$\text{Probability of the event} = \frac{\text{number of favorable outcomes}}{\text{number of possible outcomes}} = \frac{4}{9}$$

The probability of drawing an *E* is $\frac{4}{9}$.

As just discussed, calculating the probability of an event requires counting the number of possible outcomes of an experiment and the number of outcomes that are favorable to the event. One way to do this is to list the outcomes of the experiment in a systematic way. Using a table is often helpful.

When two dice are rolled, the sample space for the experiment can be recorded systematically as in the following table.

Point of Interest

Romans called a die that was marked on four faces a *talus*, which meant "anklebone." The anklebone was considered an ideal die because it is roughly a rectangular solid and it has no marrow, so loose anklebones from sheep were more likely than other bones to be lying around after the wolves had left their prey.

Possible Outcomes from Rolling Two Dice

(1, 1)	(2, 1)	(3, 1)	(4, 1)	(5, 1)	(6, 1)
(1, 2)	(2, 2)	(3, 2)	(4, 2)	(5, 2)	(6, 2)
(1, 3)	(2, 3)	(3, 3)	(4, 3)	(5, 3)	(6, 3)
(1, 4)	(2, 4)	(3, 4)	(4, 4)	(5, 4)	(6, 4)
(1, 5)	(2, 5)	(3, 5)	(4, 5)	(5, 5)	(6, 5)
(1, 6)	(2, 6)	(3, 6)	(4, 6)	(5, 6)	(6, 6)

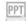

HOW TO 3 Two dice are rolled once. Calculate the probability that the sum of the numbers on the two dice is 7.

Use the table above to count the number of possible outcomes of the experiment.

There are 36 possible outcomes.

Count the number of outcomes of the experiment that are favorable to the event that a sum of 7 is rolled.

There are 6 favorable outcomes: (1, 6), (2, 5), (3, 4), (4, 3), (5, 2), and (6, 1).

Use the probability formula.

$$\text{Probability of the event} = \frac{\text{number of favorable outcomes}}{\text{number of possible outcomes}} = \frac{6}{36} = \frac{1}{6}$$

The probability of a sum of 7 is $\frac{1}{6}$.

The probabilities calculated above are theoretical probabilities. The calculation of a **theoretical probability** is based on theory—for example, that either side of a coin is equally likely to land face up or that each of the six sides of a fair die is equally likely to land face up. Not all probabilities arise from such assumptions.

An **empirical probability** is based on observations of certain events. For instance, a weather forecast of a 75% chance of rain is an empirical probability. From historical records kept by the weather bureau, when a similar weather pattern existed, rain occurred 75% of the time. It is theoretically impossible to predict the weather, and only observations of past weather patterns can be used to predict future weather conditions.

Empirical Probability Formula

The empirical probability of an event is the ratio of the number of observations of the event to the total number of observations.

$$\text{Probability of an event} = \frac{\text{number of observations of the event}}{\text{total number of observations}}$$

For example, suppose the records of an insurance company show that of 2549 claims for theft filed by policy holders, 927 were claims for more than $5000. The empirical probability that the next claim for theft that this company receives will be a claim for more than $5000 is the ratio of the number of claims for over $5000 to the total number of claims.

$$\frac{927}{2549} \approx 0.36$$

The probability is approximately 0.36.

EXAMPLE • 1

There are three choices, *a, b,* or *c,* for each of the two questions on a multiple-choice quiz. If the instructor randomly chooses which questions will have an answer of *a, b,* or *c,* what is the probability that the two correct answers on this quiz will be the same letter?

YOU TRY IT • 1

A professor writes three true/false questions for a quiz. If the professor randomly chooses which questions will have a true answer and which will have a false answer, what is the probability that the test will have 2 true questions and 1 false question?

Strategy
To find the probability:
• List the outcomes of the experiment in a systematic way.
• Count the number of possible outcomes of the experiment.
• Count the number of outcomes of the experiment that are favorable to the event that the two correct answers on the quiz will be the same letter.
• Use the probability formula.

Your strategy

Solution
Possible outcomes: (a, a) (b, a) (c, a)
(a, b) (b, b) (c, b)
(a, c) (b, c) (c, c)

There are 9 possible outcomes.

There are 3 favorable outcomes:
(a, a), (b, b), (c, c)

$$\text{Probability} = \frac{\text{number of favorable outcomes}}{\text{number of possible outcomes}}$$

$$= \frac{3}{9} = \frac{1}{3}$$

The probability that the two correct answers will be the same letter is $\frac{1}{3}$.

Your solution

Solution on pp. S19–S20

7.5 EXERCISES

OBJECTIVE A **To calculate the probability of simple events**

1. A coin is tossed four times. List all the possible outcomes of the experiment as a sample space.

2. Three cards—one red, one green, and one blue—are to be arranged in a stack. Using R for red, G for green, and B for blue, list all the different stacks that can be formed. (Some computer monitors are called RGB monitors for the colors red, green, and blue.)

3. A tetrahedral die is one with four triangular sides. The sides show the numbers from 1 to 4. Say two tetrahedral dice are rolled. List all the possible outcomes of the experiment as a sample space.

4. A coin is tossed and then a die is rolled. List all the possible outcomes of the experiment as a sample space. [To get you started, (H, 1) is one of the possible outcomes.]

Tetrahedral die

5. The spinner at the right is spun once. Assume that the spinner does not come to rest on a line.
   a. What is the sample space?
   b. List the outcomes in the event that the number is less than 4.

6. A coin is tossed four times. Find the probability of the given event.
   a. The outcomes are exactly in the order HHTT. (See Exercise 1.)
   b. The outcomes consist of two heads and two tails.
   c. The outcomes consist of one head and three tails.

7. Two dice are rolled. Find the probability of the given outcome.
   a. The sum of the dots on the upward faces is 5.
   b. The sum of the dots on the upward faces is 15.
   c. The sum of the dots on the upward faces is less than 15.
   d. The sum of the dots on the upward faces is 2.

8. A dodecahedral die has 12 sides numbered from 1 to 12. The die is rolled once. Find the probability of the given outcome.
   a. The upward face shows an 11.
   b. The upward face shows a 5.

9. A dodecahedral die has 12 sides numbered from 1 to 12. The die is rolled once. Find the probability of the given outcome.
   a. The upward face shows a number that is divisible by 4.
   b. The upward face shows a number that is a multiple of 3.

Dodecahedral die

10. Two tetrahedral dice are rolled (see Exercise 3).
    a. What is the probability that the sum on the upward faces is 4?
    b. What is the probability that the sum on the upward faces is 6?

11. Two dice are rolled. Which has the greater probability, throwing a sum of 10 or throwing a sum of 5?

12. Two dice are rolled once. Calculate the probability that the two numbers on the dice are equal.

13. Each of the letters of the word *MISSISSIPPI* is written on a card, and the cards are placed in a hat. One card is drawn at random from the hat.
   a. What is the probability that the card has the letter *I* on it?
   b. Which is greater, the probability of choosing an *S* or that of choosing a *P*?

14. Use the situation described in Exercise 12. Suppose you decide to test your result empirically by rolling a pair of dice 30 times and recording the results. Which number of "doubles" would confirm the result found in Exercise 12?
   (i) 1   (ii) 5   (iii) 6   (iv) 30

15. Use the situation described in Exercise 13. What probability does the fraction $\frac{1}{11}$ represent?

16. Three blue marbles, four green marbles, and five red marbles are placed in a bag. One marble is chosen at random.
   a. What is the probability that the marble chosen is green?
   b. Which is greater, the probability of choosing a blue marble or that of choosing a red marble?

17. Which has the greater probability, drawing a jack, queen, or king from a deck of cards or drawing a spade?

18. In a history class, a set of exams earned the following grades: 4 A's, 8 B's, 22 C's, 10 D's, and 3 F's. If a single student's exam is chosen from this class, what is the probability that it received a B?

19. A survey of 95 people showed that 37 preferred (to using a credit card) a cash discount of 2% if an item was purchased using cash or a check. Judging on the basis of this survey, what is the empirical probability that a person prefers a cash discount? Write the answer as a decimal rounded to the nearest hundredth.

20. A survey of 725 people showed that 587 had a group health insurance plan where they worked. On the basis of this survey, what is the empirical probability that an employee has a group health insurance plan? Write the answer as a decimal rounded to the nearest hundredth.

21. A television cable company surveyed some of its customers and asked them to rate the cable service as excellent, satisfactory, average, unsatisfactory, or poor. The results are recorded in the table at the right. What is the probability that a customer who was surveyed rated the service as satisfactory or excellent?

Quality of Service	Number Who Voted
Excellent	98
Satisfactory	87
Average	129
Unsatisfactory	42
Poor	21

Applying the Concepts

22. If the spinner at the right is spun once, is each of the numbers 1 through 5 equally likely? Why or why not?

23. Why can the probability of an event not be $\frac{5}{3}$?

FOCUS ON PROBLEM SOLVING

Inductive Reasoning   Suppose that, beginning in January, you save $25 each month. The total amount you have saved at the end of each month can be described by a list of numbers.

25	50	75	100	125	150	175	
Jan	Feb	Mar	Apr	May	June	July	...

The list of numbers that indicates your total savings is an *ordered* list of numbers called a **sequence.** Each of the numbers in a sequence is called a **term** of the sequence. The list is ordered because the position of a number in the list indicates the month in which that total amount has been saved. For example, the 7th term of the sequence (indicating July) is 175. This number means that a total of $175 has been saved by the end of the 7th month.

Now consider a person who has a different savings plan. The total amount saved by this person for the first seven months is given by the sequence

$$20, 35, 50, 65, 80, 95, 110, \ldots$$

The process you use to discover the next number in the above sequence is *inductive reasoning.* **Inductive reasoning** involves making generalizations from specific examples; in other words, we reach a conclusion by making observations about particular facts or cases. In the case of the above sequence, the person saved $15 per month after the first month.

Here is another example of inductive reasoning. Find the next two letters of the sequence A, B, E, F, I, J, ...

By trying different patterns, we can determine that a pattern for this sequence is

$$\underline{A}, \underline{B}, C, D, \underline{E}, \underline{F}, G, H, \underline{I}, \underline{J}, \ldots$$

That is, write two letters, skip two letters, write two letters, skip two letters, and so on. The next two letters are M, N.

Use inductive reasoning to solve the following problems.

1.   What is the next term of the sequence, ban, ben, bin, bon, . . . ?

2.   Using a calculator, determine the decimal representation of several proper fractions that have a denominator of 99. For instance, you may use $\frac{8}{99}$, $\frac{23}{99}$, and $\frac{75}{99}$. Now use inductive reasoning to explain the pattern, and use your reasoning to find the decimal representation of $\frac{53}{99}$ without a calculator.

3.   Find the next number in the sequence 1, 1, 2, 3, 5, 8, 13, 21, ...

4.   The decimal representation of a number begins 0.10100100010000100000 . . . . What are the next 10 digits in this number?

5.   The first seven rows of a triangle of numbers called Pascal's triangle are given below. Find the next row.

$$
\begin{array}{ccccccccccccc}
 & & & & & & 1 & & & & & & \\
 & & & & & 1 & & 1 & & & & & \\
 & & & & 1 & & 2 & & 1 & & & & \\
 & & & 1 & & 3 & & 3 & & 1 & & & \\
 & & 1 & & 4 & & 6 & & 4 & & 1 & & \\
 & 1 & & 5 & & 10 & & 10 & & 5 & & 1 & \\
1 & & 6 & & 15 & & 20 & & 15 & & 6 & & 1
\end{array}
$$

PROJECTS AND GROUP ACTIVITIES

Collecting, Organizing, Displaying, and Analyzing Data

Before standardized units of measurement became commonplace, measurements were made in terms of the human body. For example, the cubit was the distance from the end of the elbow to the tips of the fingers. The yard was the distance from the tip of the nose to the tip of the fingers on an outstretched arm.

For each student in the class, find the measure from the tip of the nose to the tip of the fingers on an outstretched arm. Round each measure to the nearest centimeter. Record all the measurements on the board.

1. From the data collected, determine each of the following.

   Mean _____
   Median _____
   Mode _____
   Range _____
   First quartile, Q_1 _____
   Third quartile, Q_3 _____
   Interquartile range _____

2. Prepare a box-and-whiskers plot of the data.

 3. Write a description of the spread of the data.

 4. Explain why we need standardized units of measurement.

CHAPTER 7

SUMMARY

KEY WORDS

Statistics is the branch of mathematics concerned with *data,* or numerical information. A *graph* is a pictorial representation of data. A *pictograph* represents data by using a symbol that is characteristic of the data. [7.1A, p. 294]

EXAMPLES

The pictograph shows the annual per capita turkey consumption in different countries.

Britain	🦃🦃🦃🦃
Canada	🦃🦃🦃🦃🦃
France	🦃🦃🦃🦃🦃🦃
Ireland	🦃🦃🦃🦃
Israel	🦃🦃🦃🦃🦃🦃🦃🦃🦃🦃
Italy	🦃🦃🦃🦃🦃
U.S.	🦃🦃🦃🦃🦃🦃🦃🦃🦃

Each 🦃 represents 2 lb.

Per Capita Turkey Consumption
Source: National Turkey Federation

A *circle graph* represents data by the sizes of the sectors.
[7.1B, p. 296]

The circle graph shows the results of a survey of 300 people who were asked to name their favorite sport.

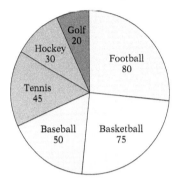

Distribution of Responses in a Survey

A *bar graph* represents data by the heights of the bars.
[7.2A, p. 302]

The bar graph shows the expected U.S. population aged 100 and over.

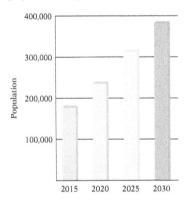

Expected U.S. Population Aged 100 and Over

Source: Census Bureau

A *broken-line graph* represents data by the positions of the lines and shows trends or comparisons.   [7.2B, p. 303]

The line graph shows a recent graduate's cumulative debt in college loans at the end of each of the four years of college.

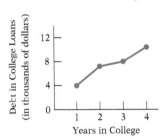

Cumulative Debt in College Loans

A *histogram* is a special kind of bar graph. In a histogram, the width of each bar corresponds to a range of numbers called a *class interval*. The height of each bar corresponds to the number of occurrences of data in each class interval and is called the *class frequency*.   [7.3A, p. 307]

An Internet service provider (ISP) surveyed 1000 of its subscribers to determine the time required for each subscriber to download a particular file. The results of the survey are shown in the histogram below.

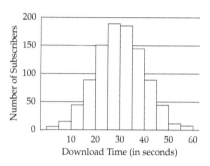

A *frequency polygon* is a graph that displays information in a manner similar to a histogram. A dot is placed above the center of each class interval at a height corresponding to that class's frequency. The dots are connected to form a broken-line graph. The center of a class interval is called the *class midpoint*. [7.3B, p. 308]

Below is a frequency polygon for the data in the histogram above.

The *mean, median,* and *mode* are three types of averages used in statistics. The *mean* of a set of data is the sum of the data values divided by the number of values in the set. The *median* of a set of data is the number that separates the data into two equal parts when the data have been arranged from least to greatest (or greatest to least). There is an equal number of values above the median and below the median. The *mode* of a set of numbers is the value that occurs most frequently.   [7.4A, pp. 311, 312]

Consider the following set of data.
$$24, 28, 33, 45, 45$$
The mean is 35.
The median is 33.
The mode is 45.

A *box-and-whiskers plot,* or *boxplot,* is a graph that shows five numbers: the least value, the first quartile, the median, the third quartile, and the greatest value. The *first quartile, Q_1,* is the number below which one-fourth of the data lie. The *third quartile, Q_3,* is the number above which one-fourth of the data lie. The box is placed around the values between the first quartile and the third quartile. The *range* is the difference between the greatest number and the least number in the set. The range describes the spread of the data. The *interquartile range* is the difference between Q_3 and Q_1. [7.4B, pp. 314–315]

The box-and-whiskers plot for a set of test scores is shown below.

Range $= 96 - 45 = 51$

$Q_1 = 65$

$Q_3 = 86$

Interquartile range
$$= Q_3 - Q_1 = 86 - 65 = 21$$

Probability is a number from 0 to 1 that tells us how likely it is that a certain outcome of an experiment will happen. An *experiment* is an activity with an observable outcome. All the possible outcomes of an experiment are called the *sample space* of the experiment. An *event* is one or more outcomes of an experiment. The *favorable outcomes* of an experiment are the outcomes that satisfy the requirements of a particular event.   [7.5A, p. 321]

Tossing a single die is an example of an experiment. The sample space for this experiment is the set of possible outcomes:

$$\{1, 2, 3, 4, 5, 6\}$$

The event that the number landing face up is an odd number is represented by

$$\{1, 3, 5\}$$

ESSENTIAL RULES AND PROCEDURES

EXAMPLES

To Find the Mean of a Set of Data   [7.4A, p. 311]
Divide the sum of the numbers by the number of values in the set.

$$\bar{x} = \frac{\text{sum of the data values}}{\text{number of data values}}$$

Consider the following set of data.

$$24, 28, 33, 45, 45$$

$$\bar{x} = \frac{24 + 28 + 33 + 45 + 45}{5} = 35$$

To Find the Median   [7.4A, p. 312]

1. Arrange the numbers from least to greatest.

2. If there is an *odd* number of values in the set of data, the median is the middle number. If there is an *even* number of values in the set of data, the median is the mean of the two middle numbers.

Consider the following set of data.

$$24, 28, 33, 35, 45, 45$$

The median is $\dfrac{33 + 35}{2} = 34$.

To Find Q_1   [7.4B, p. 314]

Arrange the numbers from least to greatest and locate the median. Q_1 is the median of the lower half of the data.

Consider the following data.

8	10	12	14	16	19	22
		↑		↑		
		Q_1		Median		

To Find Q_3   [7.4B, p. 314]

Arrange the numbers from least to greatest and locate the median. Q_3 is the median of the upper half of the data.

Consider the following data.

8	10	12	14	16	19	22
			↑		↑	
			Median		Q_3	

Theoretical Probability Formula   [7.5A, p. 322]

$$\text{Probability of an event} = \frac{\text{number of favorable outcomes}}{\text{number of possible outcomes}}$$

A die is rolled. The probability of rolling a 2 or a 4 is $\dfrac{2}{6} = \dfrac{1}{3}$.

Empirical Probability Formula   [7.5A, p. 324]

$$\text{Probability of an event} = \frac{\text{number of observations of the event}}{\text{total number of observations}}$$

A thumbtack is tossed 100 times. It lands point up 15 times and lands on its side 85 times. From this experiment, the empirical probability of "point up" is $\dfrac{15}{100} = \dfrac{3}{20}$.

CHAPTER 7

CONCEPT REVIEW

Test your knowledge of the concepts presented in this chapter. Answer each question. Then check your answers against the ones provided in the Answer Section.

1. What is a sector of a circle?

2. How does a pictograph give numerical information?

3. Why is a portion of the vertical axis jagged on some bar graphs?

4. How does a broken-line graph show changes over time?

5. What is class frequency in a histogram?

6. What is a class interval in a histogram?

7. What is a class midpoint?

8. What is the formula for the mean?

9. To find the median, why must the data be arranged in order from least to greatest?

10. When does a set of data have no mode?

11. What five values are shown in a box-and-whiskers plot?

12. How do you find the first quartile for a set of data values?

13. What is the empirical probability formula?

14. What is the theoretical probability formula?

CHAPTER 7

REVIEW EXERCISES

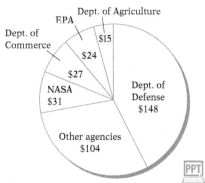

Internet The circle graph in Figure 38 shows the approximate amounts of money that government agencies spent on maintaining Internet websites for a 3-year period. Use this graph for Exercises 1 to 3.

1. Find the total amount of money that these agencies spent on maintaining websites.

2. What is the ratio of the amount spent by the Department of Commerce to the amount spent by the EPA?

3. What percent of the total money spent did NASA spend? Round to the nearest tenth of a percent.

FIGURE 38 Millions of dollars that federal agencies spent on websites
Source: General Accounting Office

Demographics The double-line graph in Figure 39 shows the populations of California and Texas for selected years. Use this graph for Exercises 4 to 6.

4. In 1900, which state had the larger population?

5. In 2000, approximately how much greater was the population of California than the population of Texas?

FIGURE 39 Populations of California and Texas

6. During which 25-year period did the population of Texas increase the least?

Sports The frequency polygon in Figure 40 shows the range of scores for the first 80 games of a season for the New York Knicks basketball team. Use this figure for Exercises 7 to 9.

7. Find the number of games in which fewer than 100 points were scored by the Knicks.

8. What is the ratio of the number of games in which 90 to 100 points were scored to the number of games in which 110 to 120 points were scored?

9. In what percent of the games were 110 points or more scored? Round to the nearest tenth of a percent.

FIGURE 40
Source: Sports Illustrated website at http://CNNSI.com

Airports The pictograph in Figure 41 shows the numbers of passengers that boarded planes in the five busiest U.S. airports in a recent year. Use this graph for Exercises 10 and 11.

10. How many more passengers boarded planes in O'Hare each year than boarded planes in the Denver airport each year?

11. What is the ratio of the number of passengers boarding planes in the Hartsfield airport to the number of passengers boarding planes in the Dallas/Ft. Worth airport each year? Write your answer using a colon.

 = 10 million passengers per year

FIGURE 41 **The busiest U.S. airports**
Source: Federal Aviation Administration

Sports The double-bar graph in Figure 42 shows the total days open and the days of full operation of ski resorts in different regions of the country. Use this graph for Exercises 12 to 14.

12. Find the difference between the total days open and the days of full operation for Midwest ski areas.

13. What percent of the total days open were the days of full operation for the Rocky Mountain ski areas?

14. Which region had the lowest number of days of full operation? How many days of full operation did this region have?

15. A coin is tossed four times. What is the probability that the outcomes of the tosses consist of one tail and three heads?

FIGURE 42
Source: Economic Analysis of United States Ski Areas

Health Based on a Gallup poll, the numbers of hours that the 46 people surveyed slept during a typical weekday night are shown in the histogram in Figure 43. Use this figure for Exercises 16 and 17.

16. How many people slept 8 hours or more?

17. What percent of the people surveyed slept 7 hours? Round to the nearest tenth of a percent.

18. **Sports** The heart rates of 24 women tennis players were measured after each of them had run one-quarter of a mile. The results are listed in the table below.

80	82	99	91	93	87	103	94	73	96	86	80
97	94	108	81	100	109	91	84	78	96	96	100

Hours of Sleep per Night

FIGURE 43

a. Find the mean, median, and mode for the data. Round to the nearest tenth.
b. Find the range and the interquartile range for the data.

CHAPTER 7

TEST

Consumerism Forty college students were surveyed to see how much money they spent each week on dining out in restaurants. The results are recorded in the frequency polygon shown in Figure 44. Use this figure for Exercises 1 to 3.

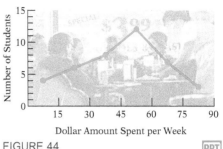

FIGURE 44

1. How many students spent between $45 and $75 per week?

2. Find the ratio of the number of students who spent between $30 and $45 to the number who spent between $45 and $60.

3. What percent of the students surveyed spent less than $45 per week?

Marriage The pictograph in Figure 45 is based on the results of a Gallup poll survey of married couples. Each individual was asked to give a letter grade to the marriage. Use this graph for Exercises 4 to 6.

 = 2 responses

FIGURE 45 Survey of married couples rating their marriage

4. Find the total number of people who were surveyed.

5. Find the ratio of the number of people who gave their marriage a B to the number who gave it a C.

6. What percent of the total number of people surveyed gave their marriage an A? Round to the nearest tenth of a percent.

Amusement Rides The bar graph in Figure 46 shows the number of fatalities that occurred during accidents on amusement rides in the 1990s in the United States. Use this graph for Exercises 7 to 9.

FIGURE 46 Number of fatal accidents on amusement rides
Source: USA Today, April 7, 2000

7. During which two consecutive years were the numbers of fatalities the same?

8. Find the total number of fatalities on amusement rides from 1991 through 1999.

9. How many more fatalities occurred during the years 1995 through 1998 than occurred during the years 1991 through 1994?

The Film Industry The circle graph in Figure 47 categorizes the 655 films released during a recent year by their ratings. Use this graph for Exercises 10 to 12.

10. How many more R-rated films were released than PG films?

11. How many times more PG-13 films were released than NC-17?

12. What percent of the films released were rated G? Round to the nearest tenth of a percent.

FIGURE 47 **Ratings of films released**
Source: MPA Worldwide Market Research

Compensation The histogram in Figure 48 gives information about median incomes, by state, in the United States. Use this figure for Exercises 13 to 15.

13. How many states have median incomes between $40,000 and $60,000?

14. What percent of the states have a median income that is between $50,000 and $70,000?

15. What percent of the states have a median income that is $70,000 or more?

FIGURE 48
Source: U.S. Census Bureau

16. **Probability** A box contains 50 balls, of which 15 are red. If 1 ball is randomly selected from the box, what is the probability of the ball's being red?

Education The broken-line graph in Figure 49 shows the numbers of students enrolled in colleges for selected years. Use this figure for Exercises 17 and 18.

17. During which decade did the student population increase the least?

18. Approximate the increase in college enrollment from 1960 to 2000.

19. **Quality Control** The lengths of time (in days) that various batteries operated a portable CD player continuously are given in the table below.

2.9	2.4	3.1	2.5	2.6	2.0	3.0	2.3	2.4	2.7
2.0	2.4	2.6	2.7	2.1	2.9	2.8	2.4	2.0	2.8

a. Find the mean for the data.
b. Find the median for the data.
c. Draw a box-and-whiskers plot for the data.

FIGURE 49 **Student enrollment in public and private colleges**
Source: National Center for Educational Statistics

CUMULATIVE REVIEW EXERCISES

1. Simplify: $2^2 \cdot 3^3 \cdot 5$

2. Simplify: $3^2 \cdot (5 - 2) \div 3 + 5$

3. Find the LCM of 24 and 40.

4. Write $\frac{60}{144}$ in simplest form.

5. Find the total of $4\frac{1}{2}$, $2\frac{3}{8}$, and $5\frac{1}{5}$.

6. Subtract: $12\frac{5}{8} - 7\frac{11}{12}$

7. Multiply: $\frac{5}{8} \times 3\frac{1}{5}$

8. Find the quotient of $3\frac{1}{5}$ and $4\frac{1}{4}$.

9. Simplify: $\frac{5}{8} \div \left(\frac{3}{4} - \frac{2}{3}\right) + \frac{3}{4}$

10. Write two hundred nine and three hundred five thousandths in standard form.

11. Find the product of 4.092 and 0.69.

12. Convert $16\frac{2}{3}$ to a decimal. Round to the nearest hundredth.

13. Write "330 miles on 12.5 gallons of gas" as a unit rate.

14. Solve the proportion: $\frac{n}{5} = \frac{16}{25}$

15. Write $\frac{4}{5}$ as a percent.

16. 8 is 10% of what?

17. What is 38% of 43?

18. What percent of 75 is 30?

19. Compensation Tanim Kamal, a salesperson at a department store, receives $100 per week plus 2% commission on sales. Find the income for a week in which Tanim had $27,500 in sales.

20. Insurance A life insurance policy costs $8.15 for every $1000 of insurance. At this rate, what is the cost for $50,000 of life insurance?

21. Simple Interest A contractor borrowed $125,000 for 6 months at an annual simple interest rate of 6%. Find the interest due on the loan.

22. Markup A compact disc player with a cost of $180 is sold for $279. Find the markup rate.

23. Finance The circle graph in Figure 50 shows how a family's monthly income of $4500 is budgeted. How much is budgeted for food?

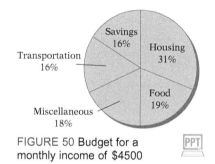

FIGURE 50 Budget for a monthly income of $4500

24. Education The double-broken-line graph in Figure 51 shows two students' scores on 5 math tests of 30 problems each. Find the difference between the numbers of problems that the two students answered correctly on Test 1.

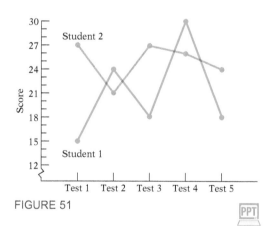

FIGURE 51

25. Meteorology The average daily high temperatures, in degrees Fahrenheit, for a week in Newtown were 56°, 72°, 80°, 75°, 68°, 62°, and 74°. Find the mean high temperature for the week. Round to the nearest tenth of a degree.

26. Probability Two dice are rolled. What is the probability that the sum of the dots on the upward faces is 8?

CHAPTER

8

U.S. Customary Units of Measurement

OBJECTIVES

SECTION 8.1
A To convert measurements of length in the U.S. Customary System
B To perform arithmetic operations with measurements of length
C To solve application problems

SECTION 8.2
A To convert measurements of weight in the U.S. Customary System
B To perform arithmetic operations with measurements of weight
C To solve application problems

SECTION 8.3
A To convert measurements of capacity in the U.S. Customary System
B To perform arithmetic operations with measurements of capacity
C To solve application problems

SECTION 8.4
A To convert units of time

SECTION 8.5
A To use units of energy in the U.S. Customary System
B To use units of power in the U.S. Customary System

ARE YOU READY?

Take the Chapter 8 Prep Test to find out if you are ready to learn to:

- Convert units of length, weight, and capacity in the U.S. Customary System
- Perform arithmetic operations with measurements of length, weight, and capacity
- Convert units of time
- Use units of energy and power in the U.S. Customary System

PREP TEST

Do these exercises to prepare for Chapter 8.

For Exercises 1 to 8, add, subtract, multiply, or divide.

1. $\begin{array}{r} 485 \\ + 217 \end{array}$

2. $\begin{array}{r} 145 \\ - 87 \end{array}$

3. $36 \times \dfrac{1}{9}$

4. $\dfrac{5}{3} \times 6$

5. $400 \times \dfrac{1}{8} \times \dfrac{1}{2}$

6. $5\dfrac{3}{4} \times 8$

7. $3\overline{)714}$

8. $12\overline{)18}$

SECTION

8.1 Length

OBJECTIVE A

To convert measurements of length in the U.S. Customary System

 Point of Interest

The ancient Greeks devised the foot measurement, which they usually divided into 16 fingers. It was the Romans who subdivided the foot into 12 units called inches. The word *inch* is derived from the Latin word *uncia,* which means "a twelfth part."

The Romans also used a unit called *pace,* which equaled two steps. One thousand paces equaled 1 mile. The word *mile* is derived from the Latin word *mille,* which means "1000."

A **measurement** includes a number and a unit.

$$\underbrace{\begin{matrix} 3 \\ 7 \\ 12 \end{matrix}}_{\text{Number}} \quad \underbrace{\begin{matrix} \text{feet} \\ \text{miles} \\ \text{yards} \end{matrix}}_{\text{Unit}}$$

Standard units of measurement have been established to simplify trade and commerce.

The unit of **length,** or distance, that is called the yard was originally defined as the length of a specified bronze bar located in London.

The standard U.S. Customary System units of length are **inch, foot, yard,** and **mile.**

Equivalences Between Units of Length in the U.S. Customary System

12 inches (in.) = 1 foot (ft)

3 ft = 1 yard (yd)

36 in. = 1 yard (yd)

5280 ft = 1 mile (mi)

[PPT]

These equivalences can be used to form conversion rates; a **conversion rate** is a relationship used to change one unit of measurement to another. For example, because $3 \text{ ft} = 1 \text{ yd}$, the conversion rates $\dfrac{3 \text{ ft}}{1 \text{ yd}}$ and $\dfrac{1 \text{ yd}}{3 \text{ ft}}$ are both equivalent to 1.

HOW TO 1 Convert 27 ft to yards.

$$27 \text{ ft} = 27 \text{ ft} \times \boxed{\dfrac{1 \text{ yd}}{3 \text{ ft}}}$$

$$= 27 \text{ ft} \times \dfrac{1 \text{ yd}}{3 \text{ ft}}$$

$$= \dfrac{27 \text{ yd}}{3}$$

$$= 9 \text{ yd}$$

HOW TO 2 Convert 5 yd to feet.

$$5 \text{ yd} = 5 \text{ yd} \times \boxed{\dfrac{3 \text{ ft}}{1 \text{ yd}}}$$

$$= 5 \text{ yd} \times \dfrac{3 \text{ ft}}{1 \text{ yd}}$$

$$= \dfrac{15 \text{ ft}}{1}$$

$$= 15 \text{ ft}$$

Note that in the conversion rate chosen, the unit in the numerator is the same as the unit desired in the answer. The unit in the denominator is the same as the unit in the given measurement.

EXAMPLE · 1

Convert 40 in. to feet.

Solution

$$40 \text{ in.} = 40 \text{ in.} \times \frac{1 \text{ ft}}{12 \text{ in.}}$$

$$= \frac{40 \text{ ft}}{12} = 3\frac{1}{3} \text{ ft}$$

YOU TRY IT · 1

Convert 14 ft to yards.

Your solution

EXAMPLE · 2

Convert $3\frac{1}{4}$ yd to feet.

Solution

$$3\frac{1}{4} \text{ yd} = \frac{13}{4} \text{ yd} = \frac{13}{4} \text{ yd} \times \frac{3 \text{ ft}}{1 \text{ yd}}$$

$$= \frac{39 \text{ ft}}{4} = 9\frac{3}{4} \text{ ft}$$

YOU TRY IT · 2

Convert 9240 ft to miles.

Your solution

Solutions on p. S20

OBJECTIVE B To perform arithmetic operations with measurements of length

When performing arithmetic operations with measurements of length, write the answer in simplest form. For example, 1 ft 14 in. should be written as 2 ft 2 in.

HOW TO · 3 Convert: 50 in. = ____ ft ____ in.

$$\begin{array}{r} 4 \text{ ft } 2 \text{ in.} \\ 12\overline{)\,50} \\ \underline{-48} \\ 2 \end{array}$$

50 in. = 4 ft 2 in.

• Because 12 in. = 1 ft, divide 50 in. by 12. The whole-number part of the quotient is the number of feet. The remainder is the number of inches.

EXAMPLE · 3

Convert: 17 in. = ____ ft ____ in.

Solution

$$\begin{array}{r} 1 \text{ ft } 5 \text{ in.} \\ 12\overline{)\,17} \\ \underline{-12} \\ 5 \end{array}$$

17 in. = 1 ft 5 in.

• 12 in. = 1 ft

YOU TRY IT · 3

Convert: 42 in. = ____ ft ____ in.

Your solution

EXAMPLE · 4

Convert: 31 ft = ____ yd ____ ft

Solution

$$\begin{array}{r} 10 \text{ yd } 1 \text{ ft} \\ 3\overline{)\,31} \\ \underline{-30} \\ 1 \end{array}$$

31 ft = 10 yd 1 ft

• 3 ft = 1 yd

YOU TRY IT · 4

Convert: 14 ft = ____ yd ____ ft

Your solution

Solutions on p. S20

EXAMPLE • 5

Find the sum of 4 ft 4 in. and 1 ft 11 in.

Solution

$$
\begin{array}{r}
4\text{ ft }\;4\text{ in.} \\
+\;1\text{ ft }11\text{ in.} \\
\hline
5\text{ ft }15\text{ in.}
\end{array}
$$
• 15 in. = 1 ft 3 in.

5 ft 15 in. = 6 ft 3 in.

YOU TRY IT • 5

Find the sum of 3 ft 5 in. and 4 ft 9 in.

Your solution

EXAMPLE • 6

Subtract: 9 ft 6 in. − 3 ft 8 in.

Solution

$$
\begin{array}{r}
\overset{8\text{ ft}}{\cancel{9}}\text{ ft }\;\overset{18\text{ in.}}{\cancel{6}}\text{ in.} \\
-\;3\text{ ft }\;\;8\text{ in.} \\
\hline
5\text{ ft }10\text{ in.}
\end{array}
$$
• Borrow 1 ft (12 in.) from 9 ft and add to 6 in.

YOU TRY IT • 6

Subtract: 4 ft 2 in. − 1 ft 8 in.

Your solution

EXAMPLE • 7

Multiply: 3 yd 2 ft × 4

Solution

$$
\begin{array}{r}
3\text{ yd }2\text{ ft} \\
\times\qquad 4 \\
\hline
12\text{ yd }8\text{ ft}
\end{array}
$$
• 8 ft = 2 yd 2 ft

12 yd 8 ft = 14 yd 2 ft

YOU TRY IT • 7

Multiply: 4 yd 1 ft × 8

Your solution

EXAMPLE • 8

Find the quotient of 4 ft 3 in. and 3.

Solution

$$
\begin{array}{r}
1\text{ ft}\qquad 5\text{ in.} \\
3\overline{)\;4\text{ ft}\qquad 3\text{ in.}} \\
\underline{-\,3\text{ ft}}\qquad\quad \\
1\text{ ft} = \underline{12\text{ in.}} \\
15\text{ in.} \\
\underline{-15\text{ in.}} \\
0
\end{array}
$$

YOU TRY IT • 8

Find the quotient of 7 yd 1 ft and 2.

Your solution

EXAMPLE • 9

Multiply: $2\dfrac{3}{4}$ ft × 3

Solution

$$2\frac{3}{4}\text{ ft} \times 3 = \frac{11}{4}\text{ ft} \times 3$$

$$= \frac{33}{4}\text{ ft}$$

$$= 8\frac{1}{4}\text{ ft}$$

YOU TRY IT • 9

Subtract: $6\dfrac{1}{4}$ ft − $3\dfrac{2}{3}$ ft

Your solution

Solutions on p. S20

OBJECTIVE C   To solve application problems

EXAMPLE · 10

A concrete block is 9 in. high. How many rows of blocks are required for a retaining wall that is 6 ft high?

Strategy

To find the number of rows of blocks, convert 9 in. to feet. Then divide the height of the wall (6 ft) by the height of each block.

Solution

$$9 \text{ in.} = \frac{9 \text{ in.}}{1} \cdot \frac{1 \text{ ft}}{12 \text{ in.}} = \frac{9 \text{ ft}}{12} = 0.75 \text{ ft}$$

$$\frac{6 \text{ ft}}{0.75 \text{ ft}} = 8$$

The wall will have 8 rows of blocks.

YOU TRY IT · 10

The floor of a storage room is being tiled. Eight tiles, each a 9-inch square, fit across the width of the floor. Find the width, in feet, of the storage room.

Your strategy

Your solution

EXAMPLE · 11

A plumber used 3 ft 9 in., 2 ft 6 in., and 11 in. of copper tubing to install a sink. Find the total length of copper tubing used.

Strategy

To find the total length of copper tubing used, add the three lengths of copper tubing (3 ft 9 in., 2 ft 6 in., and 11 in.).

Solution

$$
\begin{array}{r}
3 \text{ ft } 9 \text{ in.} \\
2 \text{ ft } 6 \text{ in.} \\
+ \quad\quad 11 \text{ in.} \\
\hline
5 \text{ ft } 26 \text{ in.}
\end{array}
$$ • 26 in. = 2 ft 2 in.

5 ft 26 in. = 7 ft 2 in.

The plumber used 7 ft 2 in. of copper tubing.

YOU TRY IT · 11

A board 9 ft 8 in. is cut into four pieces of equal length. How long is each piece?

Your strategy

Your solution

Solutions on p. S20

8.1 EXERCISES

OBJECTIVE A To convert measurements of length in the U.S. Customary System

 For Exercises 1 to 3, suppose you convert units of measurement as given. Will the number part of the converted measurement be *less than* or *greater than* the number part of the original measurement?

1. Convert feet to inches

2. Convert inches to miles

3. Convert yards to feet

For Exercises 4 to 15, convert.

4. 6 ft = _____ in.

5. 9 ft = _____ in.

6. 30 in. = _____ ft

7. 64 in. = _____ ft

8. 13 yd = _____ ft

9. $4\frac{1}{2}$ yd = _____ ft

10. 16 ft = _____ yd

11. $4\frac{1}{2}$ ft = _____ yd

12. $2\frac{1}{3}$ yd = _____ in.

13. 5 yd = _____ in.

14. 2 mi = _____ ft

15. $1\frac{1}{2}$ mi = _____ ft

OBJECTIVE B To perform arithmetic operations with measurements of length

 For Exercises 16 and 17, look at the indicated exercise. The number that goes in the second blank must be less than what number?

16. Exercise 18

17. Exercise 19

For Exercises 18 to 29, perform the arithmetic operation.

18. 100 in. =
___ ft ___ in.

19. 6400 ft =
___ mi ___ ft

20. 6 ft 7 in.
+ 3 ft 4 in.

21. 9 ft 11 in.
+ 3 ft 6 in.

22. 1 mi 4200 ft
+ 2 mi 3600 ft

23. $4\frac{2}{3}$ ft + $6\frac{1}{2}$ ft

24. 5 ft 3 in.
− 2 ft 6 in.

25. 9 yd 1 ft
− 3 yd 2 ft

26. 2 ft 5 in.
× 6

27. $3\frac{2}{3}$ ft × 4

28. $2\overline{)5\text{ ft 4 in.}}$

29. $12\frac{1}{2}$ in. ÷ 3

OBJECTIVE C   To solve application problems

30. **Interior Decorating** A kitchen counter is to be covered with tile that is 4 in. square. How many tiles can be placed along one row of a countertop that is 4 ft 8 in. long?

31. **Interior Decorating** Thirty-two yards of material were used for making pleated draperies. How many feet of material were used?

32. **Measurement** Find the missing dimension.

33. **Measurement** Find the total length of the shaft.

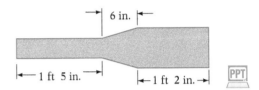

34. **Basketball** The average height of a player in the NBA is 6 ft 6.98 in. (*Source:* National Basketball Association) Find the average height, in inches, of a player in the NBA.

35. **Measurement** Find the missing dimension in the figure.

36. **Carpentry** A board $6\frac{2}{3}$ ft long is cut into four equal pieces. How long is each piece?

37. **Carpentry** How long must a board be if four pieces, each 3 ft 4 in. long, are to be cut from it?

38. **Interior Decorating** A picture is 1 ft 9 in. high and 1 ft 6 in. wide. Find the length of framing needed to frame the picture.

39. **Interior Decorating** You bought 32 ft of baseboard to install in the kitchen of your house. How many inches of baseboard did you purchase?

40. **Masonry** Forty-five bricks, each 9 in. long, are laid end-to-end to make the base for a wall. Find the length of the wall in feet.

41. **Construction** A roof is constructed with nine rafters, each 8 ft 4 in. long. Find the total number of feet of material needed to build the rafters.

For Exercises 42 and 43, use the following information. A ribbon is cut into five equal pieces. The length of each piece is 2 ft and a number of inches (the number of inches is less than 12). Determine whether each of the following statements is *true* or *false*.

42. The ribbon must be longer than 11 ft.   **43.** The ribbon must be shorter than 15 ft.

Applying the Concept

44. **Measurement** There are approximately 200,000,000 adults living in the United States. Assume that the average adult is 19 in. wide from shoulder to shoulder. If all the adults in the United States stood shoulder-to-shoulder, could they reach around Earth at the equator, a distance of approximately 25,000 mi?

SECTION

8.2   Weight

OBJECTIVE A   **To convert measurements of weight in the U.S. Customary System**

Point of Interest

The Romans used two different systems of weights. In both systems, the smallest unit was the *uncia,* abbreviated to "oz," from which the term *ounce* is derived. In one system, there were 16 ounces to 1 pound. In the second system, a pound, which was called a *libra,* equaled 12 unciae. The abbreviation "lb" for pound comes from the word *libra.*

The avoirdupois system of measurement and the troy system of measurement have their heritage in the two Roman systems.

Weight is a measure of how strongly Earth is pulling on an object. The unit of weight called the pound is defined as the weight of a standard solid kept at the Bureau of Standards in Washington, D.C. The U.S. Customary System units of weight are **ounce, pound,** and **ton.**

Equivalences Between Units of Weight in the U.S. Customary System

16 ounces (oz) = 1 pound (lb)

2000 lb = 1 ton

These equivalences can be used to form conversion rates to change one unit of measurement to another. For example, because $16 \text{ oz} = 1 \text{ lb}$, the conversion rates $\dfrac{16 \text{ oz}}{1 \text{ lb}}$ and $\dfrac{1 \text{ lb}}{16 \text{ oz}}$ are both equivalent to 1.

HOW TO · 1   Convert 62 oz to pounds.

$$62 \text{ oz} = 62 \text{ oz} \times \boxed{\frac{1 \text{ lb}}{16 \text{ oz}}}$$

$$= \frac{62 \cancel{\text{ oz}}}{1} \times \frac{1 \text{ lb}}{16 \cancel{\text{ oz}}}$$

$$= \frac{62 \text{ lb}}{16}$$

$$= 3\frac{7}{8} \text{ lb}$$

- The conversion rate must contain lb (the unit desired in the answer) in the numerator and must contain oz (the original unit) in the denominator.

EXAMPLE · 1

Convert $3\frac{1}{2}$ tons to pounds.

Solution

$$3\frac{1}{2} \text{ tons} = \frac{7}{2} \cancel{\text{tons}} \times \frac{2000 \text{ lb}}{1 \cancel{\text{ton}}}$$

$$= \frac{14{,}000 \text{ lb}}{2} = 7000 \text{ lb}$$

YOU TRY IT · 1

Convert 3 lb to ounces.

Your solution

EXAMPLE · 2

Convert 42 oz to pounds.

Solution

$$42 \text{ oz} = 42 \cancel{\text{ oz}} \times \frac{1 \text{ lb}}{16 \cancel{\text{ oz}}}$$

$$= \frac{42 \text{ lb}}{16} = 2\frac{5}{8} \text{ lb}$$

YOU TRY IT · 2

Convert 4200 lb to tons.

Your solution

Solutions on p. S20

OBJECTIVE B **To perform arithmetic operations with measurements of weight**

When performing arithmetic operations with measurements of weight, write the answer in simplest form. For example, 1 lb 22 oz should be written 2 lb 6 oz.

EXAMPLE • 3

Find the difference between 14 lb 5 oz and 8 lb 14 oz.

Solution

$$\begin{array}{cc} 13\ \text{lb} & 21\ \text{oz} \\ \cancel{14\ \text{lb}} & \cancel{5\ \text{oz}} \\ -\quad 8\ \text{lb} & 14\ \text{oz} \\ \hline 5\ \text{lb} & 7\ \text{oz} \end{array}$$

• Borrow 1 lb (16 oz) from 14 lb and add it to 5 oz.

YOU TRY IT • 3

Find the difference between 7 lb 1 oz and 3 lb 4 oz.

Your solution

EXAMPLE • 4

Divide: 7 lb 14 oz ÷ 3

Solution

$$\begin{array}{r} 2\ \text{lb}\quad 10\ \text{oz} \\ 3\overline{)\ 7\ \text{lb}\quad 14\ \text{oz}} \\ -6\ \text{lb} \\ \hline 1\ \text{lb} = 16\ \text{oz} \\ \hline 30\ \text{oz} \\ -30\ \text{oz} \\ \hline 0 \end{array}$$

YOU TRY IT • 4

Multiply: 3 lb 6 oz × 4

Your solution

Solutions on p. S21

OBJECTIVE C **To solve application problems**

EXAMPLE • 5

Sirina Jasper purchased 4 lb 8 oz of oat bran and 2 lb 11 oz of wheat bran. She plans to blend the two brans and then repackage the mixture in 3-ounce packages for a diet supplement. How many 3-ounce packages can she make?

Strategy

To find the number of 3-ounce packages:

• Add the amount of oat bran (4 lb 8 oz) to the amount of wheat bran (2 lb 11 oz).
• Convert the sum to ounces.
• Divide the total ounces by the weight of each package (3 oz).

Solution

$$\begin{array}{r} 4\ \text{lb}\quad 8\ \text{oz} \\ +\ 2\ \text{lb}\ 11\ \text{oz} \\ \hline 6\ \text{lb}\ 19\ \text{oz} = 7\ \text{lb}\ 3\ \text{oz} = 115\ \text{oz} \end{array}$$

$$\frac{115\ \text{oz}}{3\ \text{oz}} \approx 38.3$$

She can make 38 packages.

YOU TRY IT • 5

Find the weight in pounds of 12 bars of soap. Each bar weighs 7 oz.

Your strategy

Your solution

Solution on p. S21

8.2 EXERCISES

OBJECTIVE A To convert measurements of weight in the U.S. Customary System

 For Exercises 1 to 3, suppose you convert units of measurement as given. Will the number part of the converted measurement be *less than* or *greater than* the number part of the original measurement?

1. Convert pounds to tons

2. Convert pounds to ounces

3. Convert tons to pounds

For Exercises 4 to 18, convert.

4. 64 oz = _____ lb

5. 36 oz = _____ lb

6. 8 lb = _____ oz

7. 7 lb = _____ oz

8. 3200 lb = _____ tons

9. 9000 lb = _____ tons

10. 6 tons = _____ lb

11. $1\frac{1}{4}$ tons = _____ lb

12. 66 oz = _____ lb

13. 90 oz = _____ lb

14. $1\frac{1}{2}$ lb = _____ oz

15. $2\frac{5}{8}$ lb = _____ oz

16. $\frac{4}{5}$ ton = _____ lb

17. 5000 lb = _____ tons

18. 180 oz = _____ lb

OBJECTIVE B To perform arithmetic operations with measurements of weight

 For Exercises 19 and 20, look at the indicated exercise. The number that goes in the second blank must be less than what number?

19. Exercise 21

20. Exercise 22

For Exercises 21 to 32, perform the arithmetic operation.

21. 9000 lb =
_____ tons _____ lb

22. 85 oz =
_____ lb _____ oz

23. $\begin{aligned} 4\ \text{lb}\quad 7\ \text{oz} \\ +\ 3\ \text{lb}\ 12\ \text{oz} \end{aligned}$

24. $\begin{aligned} 1\ \text{ton}\quad 800\ \text{lb} \\ +\ 3\ \text{tons}\ 1600\ \text{lb} \end{aligned}$

25. $\begin{aligned} 7\ \text{lb}\ 5\ \text{oz} \\ -\ 3\ \text{lb}\ 8\ \text{oz} \end{aligned}$

26. $\begin{aligned} 3\ \text{tons}\ 500\ \text{lb} \\ -\ 1\ \text{ton}\quad 800\ \text{lb} \end{aligned}$

27. $6\frac{3}{8}$ lb $- 2\frac{5}{6}$ lb

28. $\begin{aligned} 3\ \text{lb}\ 6\ \text{oz} \\ \times\qquad 4 \end{aligned}$

29. $5\frac{1}{2}$ lb × 6

30. $4\frac{2}{3}$ lb × 3

31. $2\overline{)3\ \text{lb}\ 8\ \text{oz}}$

32. 5 lb 12 oz ÷ 4

OBJECTIVE C To solve application problems

33. Read Exercise 35. Without actually finding the total weight of the rods, determine whether the total weight will be *less than* or *greater than* 25 pounds.

34. Read Exercise 37. Without actually finding the total weight of the textbooks, determine whether the total weight will be *less than* or *greater than* 1 ton.

35. **Iron Works** A machinist has 25 iron rods to mill. Each rod weighs 20 oz. Find the total weight of the rods in pounds.

36. **Masonry** A fireplace brick weighs $2\frac{1}{2}$ lb. What is the weight of a load of 800 bricks?

37. **Weights** A college bookstore received 1200 textbooks, each weighing 9 oz. Find the total weight of the 1200 textbooks in pounds.

38. **Ranching** A farmer ordered 20 tons of feed for 100 cattle. After 15 days, the farmer has 5 tons of feed left. On average, how many pounds of feed has each cow eaten per day?

39. **Weights** A case of soft drinks contains 24 cans, each weighing 6 oz. Find the weight, in pounds, of the case of soft drinks.

40. **Child Development** A baby weighed 7 lb 8 oz at birth. At 6 months of age, the baby weighed 15 lb 13 oz. Find the baby's increase in weight during the 6 months.

41. **Packaging** Shampoo weighing 5 lb 4 oz is divided equally and poured into four containers. How much shampoo is in each container?

42. **Weights** A steel rod weighing 16 lb 11 oz is cut into three pieces. Find the weight of each piece of steel rod.

43. **Recycling** Use the news clipping at the right. How many tons of plastic bottles are *not* recycled each year?

44. **Markup** A candy store buys peppermint candy weighing 12 lb for $14.40. The candy is repackaged and sold in 6-ounce packages for $1.15 each. Find the markup on the 12 lb of candy.

45. **Shipping** A manuscript weighing 2 lb 3 oz is mailed at the rate of $.34 per ounce. Find the cost of mailing the manuscript.

In the News

Recycling Efforts Fall Short

Every year, 2 billion pounds of plastic bottles are dumped in landfills because consumers recycle only 25% of them.

Source: Time, August 20, 2007

Applying the Concepts

46. Estimate the weight of a nickel, a textbook, a friend, and a car. Then find the actual weights and compare them with your estimates.

SECTION

8.3　Capacity

OBJECTIVE A　To convert measurements of capacity in the U.S. Customary System

Liquid substances are measured in units of **capacity.** The standard U.S. Customary units of capacity are the **fluid ounce, cup, pint, quart,** and **gallon.**

Equivalences Between Units of Capacity in the U.S. Customary System

8 fluid ounces (fl oz) = 1 cup (c)

2 c = 1 pint (pt)

2 pt = 1 quart (qt)

4 qt = 1 gallon (gal)

 Point of Interest

The word *quart* has its root in the medieval Latin word *quartus,* which means "fourth." Thus a quart is $\frac{1}{4}$ of a gallon.

The same Latin word is the source of such other English words as *quarter, quartile, quadrilateral,* and *quartet.*

These equivalences can be used to form conversion rates to change one unit of measurement to another. For example, because $8 \text{ fl oz} = 1 \text{ c}$, the conversion rates $\frac{8 \text{ fl oz}}{1 \text{ c}}$ and $\frac{1 \text{ c}}{8 \text{ fl oz}}$ are both equivalent to 1.

HOW TO 1　Convert 36 fl oz to cups.

$$36 \text{ fl oz} = 36 \text{ fl oz} \times \frac{1 \text{ c}}{8 \text{ fl oz}}$$

• The conversion rate must contain c in the numerator and fl oz in the denominator.

$$= \frac{36 \text{ fl oz}}{1} \times \frac{1 \text{ c}}{8 \text{ fl oz}}$$

$$= \frac{36 \text{ c}}{8} = 4\frac{1}{2} \text{ c}$$

HOW TO 2　Convert 3 qt to cups.

$$3 \text{ qt} = 3 \text{ qt} \times \frac{2 \text{ pt}}{1 \text{ qt}} \times \frac{2 \text{ c}}{1 \text{ pt}}$$

• The direct equivalence is not given above. Use two conversion rates. First convert quarts to pints, and then convert pints to cups. The unit in the denominator of the second conversion rate and the unit in the numerator of the first conversion rate must be the same in order to cancel.

$$= \frac{3 \text{ qt}}{1} \times \frac{2 \text{ pt}}{1 \text{ qt}} \times \frac{2 \text{ c}}{1 \text{ pt}}$$

$$= \frac{12 \text{ c}}{1} = 12 \text{ c}$$

EXAMPLE 1

Convert 42 c to quarts.

Solution

$$42 \text{ c} = 42 \text{ c} \times \frac{1 \text{ pt}}{2 \text{ c}} \times \frac{1 \text{ qt}}{2 \text{ pt}}$$

$$= \frac{42 \text{ qt}}{4} = 10\frac{1}{2} \text{ qt}$$

YOU TRY IT 1

Convert 18 pt to gallons.

Your solution

Solutions on p. S21

OBJECTIVE B To perform arithmetic operations with measurements of capacity

When performing arithmetic operations with measurements of capacity, write the answer in simplest form. For example, 1 c 12 fl oz should be written as 2 c 4 fl oz.

EXAMPLE · 2

What is 4 gal 1 qt decreased by 2 gal 3 qt?

Solution

$$\begin{array}{r} \overset{3\ \text{gal}}{\cancel{4\ \text{gal}}}\ \overset{5\ \text{qt}}{\cancel{1\ \text{qt}}} \\ -\ 2\ \text{gal}\ 3\ \text{qt} \\ \hline 1\ \text{gal}\ 2\ \text{qt} \end{array}$$

· Borrow 1 gal (4 qt) from 4 gal and add to 1 qt.

YOU TRY IT · 2

Find the quotient of 4 gal 2 qt and 3.

Your solution

Solution on p. S21

OBJECTIVE C To solve application problems

EXAMPLE · 3

A can of apple juice contains 25 fl oz. Find the number of quarts of apple juice in a case of 24 cans.

Strategy

To find the number of quarts of apple juice in one case:

• Multiply the number of cans (24) by the number of fluid ounces per can (25) to find the total number of fluid ounces in the case.

• Convert the number of fluid ounces in the case to quarts.

Solution

$24 \times 25\ \text{fl oz} = 600\ \text{fl oz}$

$$600\ \text{fl oz} = \frac{600\ \cancel{\text{fl oz}}}{1} \cdot \frac{1\ \cancel{c}}{8\ \cancel{\text{fl oz}}} \cdot \frac{1\ \cancel{\text{pt}}}{2\ \cancel{c}} \cdot \frac{1\ \text{qt}}{2\ \cancel{\text{pt}}}$$

$$= \frac{600\ \text{qt}}{32} = 18\frac{3}{4}\ \text{qt}$$

One case of apple juice contains $18\frac{3}{4}$ qt.

YOU TRY IT · 3

Five students are going backpacking in the desert. Each student requires 5 qt of water per day. How many gallons of water should they take for a 3-day trip?

Your strategy

Your solution

Solution on p. S21

8.3 EXERCISES

OBJECTIVE A To convert measurements of capacity in the U.S. Customary System

 For Exercises 1 to 3, suppose you convert units of measurement as given. Will the number part of the converted measurement be *less than* or *greater than* the number part of the original measurement?

1. Convert cups to fluid ounces

2. Convert quarts to gallons

3. Convert fluid ounces to pints

For Exercises 4 to 18, convert.

4. 60 fl oz = _____ c

5. 48 fl oz = _____ c

6. 3 c = _____ fl oz

7. $2\frac{1}{2}$ c = _____ fl oz

8. 8 c = _____ pt

9. 5 c = _____ pt

10. $7\frac{1}{2}$ pt = _____ qt

11. 12 pt = _____ qt

12. 22 qt = _____ gal

13. 10 qt = _____ gal

14. $2\frac{1}{4}$ gal = _____ qt

15. 7 gal = _____ qt

16. $1\frac{1}{2}$ pt = _____ fl oz

17. 17 c = _____ qt

18. $1\frac{1}{2}$ qt = _____ c

OBJECTIVE B To perform arithmetic operations with measurements of capacity

 For Exercises 19 and 20, look at the indicated exercise. The number that goes in the second blank must be less than what number?

19. Exercise 21

20. Exercise 22

For Exercises 21 to 36, perform the arithmetic operation.

21. 14 qt =
___ gal ___ qt

22. 9 pt =
___ qt ___ pt

23. 3 gal 2 qt
+ 4 gal 3 qt

24. 5 c 3 fl oz
+ 3 c 6 fl oz

25. 3 gal 3 qt
+ 1 gal 2 qt

26. $1\frac{1}{2}$ pt + $2\frac{2}{3}$ pt

27. 3 gal 1 qt
− 1 gal 2 qt

28. 3 c 3 fl oz
− 2 c 5 fl oz

29. 4 c 6 fl oz
− 2 c 7 fl oz

30. 3 gal
− 1 gal 2 qt

31. $4\frac{1}{2}$ gal − $1\frac{3}{4}$ gal

32. 2 qt 1 pt
× 5

33. $3\frac{1}{2}$ pt × 5   **34.** $5\overline{)6\text{ gal 1 qt}}$ **35.** $3\frac{1}{2}$ gal ÷ 4   **36.** $2\overline{)3\text{ gal 2 qt}}$

OBJECTIVE C   To solve application problems

37. Catering Sixty adults are expected to attend a book signing. Each adult will drink 2 c of coffee. How many gallons of coffee should be prepared?

38. Catering The Bayside Playhouse serves punch during intermission. Assume that 200 people will each drink 1 c of punch. How many gallons of punch should be ordered?

39. Consumerism One brand of tomato juice costs $1.59 for 1 qt. Another brand costs $1.25 for 24 fl oz. Which is the more economical purchase?

40. Camping Mandy carried 12 qt of water for 3 days of desert camping. Water weighs $8\frac{1}{3}$ lb per gallon. Find the weight of water that she carried.

41. Bottled Water Use the news clipping at the right. On average, how many cups of bottled water does an American drink per month? Round to the nearest tenth.

42. Food Service A cafeteria sold 124 cartons of milk in 1 day. Each carton contained 1 c of milk. How many quarts of milk were sold that day?

43. Vehicle Maintenance A farmer changed the oil in a tractor seven times during the year. Each oil change required 5 qt of oil. How many gallons of oil did the farmer use in the seven oil changes?

44. Business A spa owner bought hand lotion in a 5-quart container and then repackaged the lotion in 8-fluid-ounce bottles. The lotion and bottles cost $81.50, and each 8-fluid-ounce bottle was sold for $8.25. How much profit was made on the 5-quart package of lotion?

45. Business Orlando bought oil in 50-gallon containers for changing the oil in his customers' cars. He paid $960 for the 50 gal of oil and charged customers $9.25 per quart. Find the profit Orlando made on one 50-gallon container of oil.

For Exercises 46 to 48, use the following information. A punch is made from 3 qt of lemonade and 5 qt of sparkling water. What does each product represent?

46. $\dfrac{8\text{ qt}}{1} \cdot \dfrac{1\text{ gal}}{4\text{ qt}}$ **47.** $\dfrac{3\text{ qt}}{1} \cdot \dfrac{4\text{ c}}{1\text{ qt}}$ **48.** $\dfrac{8\text{ qt}}{1} \cdot \dfrac{4\text{ c}}{1\text{ qt}} \cdot \dfrac{8\text{ fl oz}}{1\text{ c}}$

Michael Newman/PhotoEdit, Inc.

Gary Burchell/Taxi/Getty Images

In the News

Bottled Water Consumption Increasing

Americans drink 28.3 gal of bottled water per year, up from 18.8 gal per year five years ago.

Source: Beverage Marketing Corp.

Applying the Concepts

49. Assume that you want to invent a new measuring system. Discuss some of the features that would have to be incorporated into the system.

SECTION

8.4　Time

OBJECTIVE A　To convert units of time

The units in which time is generally measured are the **second, minute, hour, day,** and **week**.

Equivalences Between Units of Time

60 seconds (s) = 1 minute (min)

60 min — 1 hour (h)

24 h = 1 day

7 days = 1 week

These equivalences can be used to form conversion rates to change one unit of time to another. For example, because $24\text{ h} = 1\text{ day}$, the conversion rates $\dfrac{24\text{ h}}{1\text{ day}}$ and $\dfrac{1\text{ day}}{24\text{ h}}$ are both equivalent to 1. An example using each of these two rates is shown below.

HOW TO 1　Convert $5\frac{1}{2}$ days to hours.

$$5\frac{1}{2}\text{ days} = 5\frac{1}{2}\text{ days} \times \boxed{\dfrac{24\text{ h}}{1\text{ day}}}$$

$$= \dfrac{11\ \cancel{\text{days}}}{2} \times \dfrac{24\text{ h}}{1\ \cancel{\text{day}}}$$

$$= \dfrac{264\text{ h}}{2} = 132\text{ h}$$

• The conversion rate must contain h (the unit desired in the answer) in the numerator and must contain day (the original unit) in the denominator.

HOW TO 2　Convert 156 h to days.

$$156\text{ h} = 156\text{ h} \times \boxed{\dfrac{1\text{ day}}{24\text{ h}}}$$

$$= \dfrac{156\ \cancel{\text{h}}}{1} \times \dfrac{1\text{ day}}{24\ \cancel{\text{h}}}$$

$$= \dfrac{156\text{ days}}{24} = 6\frac{1}{2}\text{ days}$$

• The conversion rate must contain day (the unit desired in the answer) in the numerator and must contain h (the original unit) in the denominator.

EXAMPLE 1

Convert 2880 min to days.

Solution

$$2880\text{ min} = 2880\ \cancel{\text{min}} \times \dfrac{1\ \cancel{\text{h}}}{60\ \cancel{\text{min}}} \times \dfrac{1\text{ day}}{24\ \cancel{\text{h}}}$$

$$= \dfrac{2880\text{ days}}{1440} = 2\text{ days}$$

YOU TRY IT 1

Convert 18,000 s to hours.

Your solution

Solution on p. S21

8.4 EXERCISES

OBJECTIVE A   To convert units of time

For Exercises 1 to 3, suppose you convert units of measurement as given. Will the number part of the converted measurement be *less than* or *greater than* the number part of the original measurement?

1. Convert days to minutes

2. Convert seconds to hours

3. Convert weeks to minutes

For Exercises 4 to 27, convert.

4. 98 days = _____ weeks

5. 12 weeks = _____ days

6. $6\frac{1}{4}$ days = _____ h

7. 114 h = _____ days

8. 555 min = _____ h

9. $7\frac{3}{4}$ h = _____ min

10. $18\frac{1}{2}$ min = _____ s

11. 750 s = _____ min

12. 12,600 s = _____ h

13. 15,300 s = _____ h

14. $6\frac{1}{2}$ h = _____ s

15. $5\frac{3}{4}$ h = _____ s

16. 5040 min = _____ days

17. 6840 min = _____ days

18. $2\frac{1}{2}$ days = _____ min

19. $6\frac{1}{4}$ days = _____ min

20. 672 h = _____ weeks

21. 588 h = _____ weeks

22. 3 weeks = _____ h

23. $5\frac{1}{2}$ weeks = _____ h

24. 172,800 s = _____ days

25. 20,160 min = _____ weeks

26. 3 days = _____ s

27. 3 weeks = _____ min

Applying the Concepts

Another unit of time is the year. One year is equivalent to $365\frac{1}{4}$ days. However, our calendar does not include quarter days. Instead, we say that a year is 365 days, and every fourth year is a leap year of 366 days. If a year is divisible by 4, it is a leap year, unless it is a year at the beginning of a century not divisible by 400. 1600, 2000, 2004, 2008, and 2012 are leap years. 1700, 1800, and 1900 are not leap years. For Exercises 28 to 30, state whether the given year is a leap year.

28. 1984

29. 1994

30. 2144

SECTION

8.5 Energy and Power

OBJECTIVE A **To use units of energy in the U.S. Customary System**

Energy can be defined as the ability to do work. Energy is stored in coal, in gasoline, in water behind a dam, and in one's own body.

One **foot-pound** (ft · lb) of energy is the amount of energy necessary to lift 1 pound a distance of 1 foot.

To lift 50 lb a distance of 5 ft requires
50 × 5 = 250 ft · lb of energy.

Consumer items that use energy, such as furnaces, stoves, and air conditioners, are rated in terms of the **British thermal unit** (Btu). For example, a furnace might have a rating of 35,000 Btu per hour, which means that it releases 35,000 Btu of energy in 1 hour (1 h).

Because 1 Btu = 778 ft · lb, the following conversion rate, equivalent to 1, can be written:

$$\frac{778 \text{ ft} \cdot \text{lb}}{1 \text{ Btu}} = 1$$

EXAMPLE 1

Convert 250 Btu to foot-pounds.

Solution

$$250 \text{ Btu} = 250 \text{ Btu} \times \frac{778 \text{ ft} \cdot \text{lb}}{1 \text{ Btu}}$$

$$= 194,500 \text{ ft} \cdot \text{lb}$$

YOU TRY IT 1

Convert 4.5 Btu to foot-pounds.

Your solution

EXAMPLE 2

Find the energy required for a 125-pound person to climb a mile-high mountain.

Solution

In climbing the mountain, the person is lifting 125 lb a distance of 5280 ft.

Energy = 125 lb × 5280 ft
        = 660,000 ft · lb

YOU TRY IT 2

Find the energy required for a motor to lift 800 lb a distance of 16 ft.

Your solution

Solutions on p. S21

EXAMPLE • 3

A furnace is rated at 80,000 Btu per hour. How many foot-pounds of energy are released in 1 h?

Solution

$80,000 \text{ Btu} = 80,000 \text{ Btu} \times \dfrac{778 \text{ ft} \cdot \text{lb}}{1 \text{ Btu}}$

$= 62,240,000 \text{ ft} \cdot \text{lb}$

YOU TRY IT • 3

A furnace is rated at 56,000 Btu per hour. How many foot-pounds of energy are released in 1 h?

Your solution

Solutions on p. S21

OBJECTIVE B   To use units of power in the U.S. Customary System

Power is the rate at which work is done or the rate at which energy is released. Power is measured in **foot-pounds per second** $\left(\dfrac{\text{ft} \cdot \text{lb}}{\text{s}}\right)$. In each of the following examples, the amount of energy released is the same, but the time taken to release the energy is different; thus the power is different.

100 lb is lifted 10 ft in 10 s.

$$\text{Power} = \frac{10 \text{ ft} \times 100 \text{ lb}}{10 \text{ s}} = 100 \frac{\text{ft} \cdot \text{lb}}{\text{s}}$$

100 lb is lifted 10 ft in 5 s.

$$\text{Power} = \frac{10 \text{ ft} \times 100 \text{ lb}}{5 \text{ s}} = 200 \frac{\text{ft} \cdot \text{lb}}{\text{s}}$$

The U.S. Customary unit of power is the **horsepower.** A horse doing average work can pull 550 lb a distance of 1 ft in 1 s and can continue this work all day.

$$1 \text{ horsepower (hp)} = 550 \frac{\text{ft} \cdot \text{lb}}{\text{s}}$$

EXAMPLE • 4

Find the power needed to raise 300 lb a distance of 30 ft in 15 s.

Solution $\text{Power} = \dfrac{30 \text{ ft} \times 300 \text{ lb}}{15 \text{ s}}$

$= 600 \dfrac{\text{ft} \cdot \text{lb}}{\text{s}}$

YOU TRY IT • 4

Find the power needed to raise 1200 lb a distance of 90 ft in 24 s.

Your solution

EXAMPLE • 5

A motor has a power of 2750 $\dfrac{\text{ft} \cdot \text{lb}}{\text{s}}$. Find the horsepower of the motor.

Solution $\dfrac{2750}{550} = 5 \text{ hp}$

YOU TRY IT • 5

A motor has a power of 3300 $\dfrac{\text{ft} \cdot \text{lb}}{\text{s}}$. Find the horsepower of the motor.

Your solution

Solution on p. S21

8.5 EXERCISES

OBJECTIVE A   To use units of energy in the U.S. Customary System

1. Convert 25 Btu to foot-pounds.

2. Convert 6000 Btu to foot-pounds.

3. Convert 25,000 Btu to foot-pounds.

4. Convert 40,000 Btu to foot-pounds.

5. Find the energy required to lift 150 lb a distance of 10 ft.

6. Find the energy required to lift 300 lb a distance of 16 ft.

7. Find the energy required to lift a 3300-pound car a distance of 9 ft.

8. Find the energy required to lift a 3680-pound elevator a distance of 325 ft.

9. Three tons are lifted 5 ft. Find the energy required in foot-pounds.

10. Seven tons are lifted 12 ft. Find the energy required in foot-pounds.

11. A construction worker carries 3-pound blocks up a 10-foot flight of stairs. How many foot-pounds of energy are required to carry 850 blocks up the stairs?

12. A crane lifts an 1800-pound steel beam to the roof of a building 36 ft high. Find the amount of energy the crane requires in lifting the beam.

Corbis

13. A furnace is rated at 45,000 Btu per hour. How many foot-pounds of energy are released by the furnace in 1 h?

14. A furnace is rated at 22,500 Btu per hour. How many foot-pounds of energy does the furnace release in 1 h?

15. Find the amount of energy in foot-pounds given off when 1 lb of coal is burned. One pound of coal gives off 12,000 Btu of energy when burned.

16. Find the amount of energy in foot-pounds given off when 1 lb of gasoline is burned. One pound of gasoline gives off 21,000 Btu of energy when burned.

17. Without finding the equivalent number of foot-pounds, determine whether 360 Btu is *less than* or *greater than* 360,000 ft · lb.

OBJECTIVE B   To use units of power in the U.S. Customary System

18. When you convert horsepower to foot-pounds per second, is the number part of the converted measurement *less than* or *greater than* the number part of the original measurement?

19. Convert $1100 \frac{\text{ft} \cdot \text{lb}}{\text{s}}$ to horsepower.

20. Convert $6050 \frac{\text{ft} \cdot \text{lb}}{\text{s}}$ to horsepower.

21. Convert $4400 \frac{\text{ft} \cdot \text{lb}}{\text{s}}$ to horsepower.

22. Convert $1650 \frac{\text{ft} \cdot \text{lb}}{\text{s}}$ to horsepower.

23. Convert 9 hp to foot-pounds per second.

24. Convert 4 hp to foot-pounds per second.

25. Convert 7 hp to foot-pounds per second.

26. Convert 8 hp to foot-pounds per second.

27. Find the power in foot-pounds per second needed to raise 125 lb a distance of 12 ft in 3 s.

28. Find the power in foot-pounds per second needed to raise 500 lb a distance of 60 ft in 8 s.

29. Find the power in foot-pounds per second needed to raise 3000 lb a distance of 40 ft in 25 s.

30. Find the power in foot-pounds per second of an engine that can raise 180 lb to a height of 40 ft in 5 s.

31. Find the power in foot-pounds per second of an engine that can raise 1200 lb to a height of 18 ft in 30 s.

32. A motor has a power of $4950 \frac{\text{ft} \cdot \text{lb}}{\text{s}}$. Find the horsepower of the motor.

33. A motor has a power of $16{,}500 \frac{\text{ft} \cdot \text{lb}}{\text{s}}$. Find the horsepower of the motor.

34. A motor has a power of $6600 \frac{\text{ft} \cdot \text{lb}}{\text{s}}$. Find the horsepower of the motor.

Applying the Concepts

35. Pick a source of energy and write an article about it. Include the source, possible pollution problems, and future prospects associated with this form of energy.

FOCUS ON PROBLEM SOLVING

Applying Solutions to Other Problems

Problem solving in the previous chapters concentrated on solving specific problems. After a problem is solved, however, there is an important question to be asked. "Does the solution to this problem apply to other types of problems?"

To illustrate this extension of problem solving, we will consider *triangular numbers,* which were studied by ancient Greek mathematicians. The numbers 1, 3, 6, 10, 15, and 21 are the first six triangular numbers. What is the next triangular number?

To answer this question, note in the diagram below that a triangle can be formed using the number of dots that correspond to a triangular number.

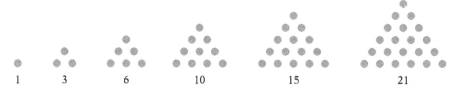

1	3	6	10	15	21

Observe that the number of dots in each row is one more than the number of dots in the row above. The total number of dots can be found by addition.

The pattern suggests that the next triangular number (the seventh one) is the sum of the first seven natural numbers. The seventh triangular number is 28. The diagram at the right shows the seventh triangular number.

Using the pattern for triangular numbers, it is easy to determine that the tenth triangular number is

$$1 + 2 + 3 + 4 + 5 + 6 + 7 + 8 + 9 + 10 = 55$$

Now consider a situation that may seem to be totally unrelated to triangular numbers. Suppose you are in charge of scheduling softball games for a league. There are seven teams in the league, and each team must play every other team once. How many games must be scheduled?

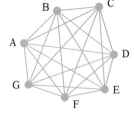

We label the teams A, B, C, D, E, F, and G. (See the figure at the left.) A line between two teams indicates that the two teams play each other. Beginning with A, there are 6 lines for the 6 teams that A must play. There are 6 teams that B must play, but the line between A and B has already been drawn, so there are only 5 remaining games to schedule for B. Now move on to C. The lines between C and A and between C and B have already been drawn, so there are only 4 additional lines to be drawn to represent the teams C will play. Moving on to D, we see that the lines between D and A, D and B, and D and C have already been drawn, so there are 3 more lines to be drawn to represent the teams D will play.

Note that each time we move from team to team, one fewer line needs to be drawn. When we reach F, there is only one line to be drawn, the one between F and G. The total number of lines drawn is $6 + 5 + 4 + 3 + 2 + 1 = 21$, the sixth triangular number. For a league with seven teams, the number of games that must be scheduled so that each team plays every other team once is the sixth triangular number. If there were ten teams in the league, the number of games that must be scheduled would be the ninth triangular number, which is 45.

A college chess team wants to schedule a match so that each of its 15 members plays each other member of the team once. How many matches must be scheduled?

PROJECTS AND GROUP ACTIVITIES

Nomographs

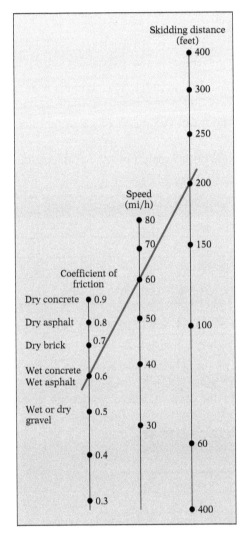

A chart is another tool that is used in problem solving. The chart at the left is a nomograph. A **nomograph** is a chart that represents numerical relationships among variables.

One of the details a traffic accident investigator checks when looking into a car accident is the length of the skid marks made by the car. This length can help the investigator determine the speed of the car when the brakes were applied.

The nomograph at the left can be used to determine the speed of a car under given conditions. It shows the relationship among the speed of the car, the skidding distance, and the *coefficient of friction.*

The coefficient of friction is an experimentally obtained value that reflects how easy or hard it is to drag one object over another. For instance, it is easier to drag a box across ice than to drag it across a carpet. The coefficient of friction is smaller for the box and ice than it is for the box and carpet.

To use the nomograph at the left, an investigator would draw a line from the coefficient of friction to the skidding distance. The point at which the line crosses the speed line shows how fast the car was going when the brakes were applied. The line from 0.6 to 200 intersects the speed line at 60. This indicates that a car that skidded 200 ft on wet asphalt or wet concrete was traveling at 60 mi/h.

1. Use the nomograph to determine the speed of a car when the brakes were applied for a car traveling on gravel and for skid marks of 100 ft.
2. Use the nomograph to determine the speed of a car when the brakes were applied for a car traveling on dry concrete and for skid marks of 150 ft.
3. Suppose a car is traveling 80 mi/h when the brakes are applied. Find the difference in skidding distance if the car is traveling on wet concrete rather than dry concrete.

Averages

If two towns are 150 mi apart and you drive between the two towns in 3 h, then your

$$\text{Average speed} = \frac{\text{total distance}}{\text{total time}} = \frac{150 \text{ mi}}{3 \text{ h}} = 50 \text{ mi/h}$$

It is highly unlikely that your speed was *exactly* 50 mi/h the entire time of the trip. Sometimes you will have traveled faster than 50 mi/h, and other times you will have traveled slower than 50 mi/h. Dividing the total distance you traveled by the total time it took to go that distance is an example of calculating an average.

There are many other averages that may be calculated. For instance, the Environmental Protection Agency calculates an estimated miles per gallon (mpg) for new cars. Miles per gallon is an average calculated from the formula

$$\frac{\text{Miles traveled}}{\text{Gallons of gasoline consumed}}$$

For instance, the miles per gallon for a car that travels 308 mi on 11 gal of gas is

$$\frac{308\ mi}{11\ gal} = 28\ mpg.$$

A pilot would not use miles per gallon as a measure of fuel efficiency. Rather, pilots use gallons per hour. A plane that travels 5 h and uses 400 gal of fuel has an average fuel efficiency of

$$\frac{Gallons\ of\ fuel}{Hours\ flown} = \frac{400\ gal}{5\ h} = 80\ gal/h$$

Using the examples above, calculate the following averages.

1. Determine the average speed of a car that travels 355 mi in 6 h. Round to the nearest tenth.

2. Determine the miles per gallon of a car that can travel 405 mi on 12 gal of gasoline. Round to the nearest tenth.

3. If a plane flew 2000 mi in 5 h and used 1000 gal of fuel, determine the average number of gallons per hour that the plane used.

Another type of average is grade-point average (GPA). It is calculated by multiplying the units for each class by the grade point for that class, adding the results, and dividing by the total number of units taken. Here is an example using the grading scale A = 4, B = 3, C = 2, D = 1, and F = 0.

Class	Units	Grade
Math	4	B (= 3)
English	3	A (= 4)
French	5	C (= 2)
Biology	3	B (= 3)

$$GPA = \frac{4 \cdot 3 + 3 \cdot 4 + 5 \cdot 2 + 3 \cdot 3}{4 + 3 + 5 + 3} = \frac{43}{15} \approx 2.87$$

4. A grading scale that provides for plus or minus grades uses A = 4, A− = 3.7, B+ = 3.3, B = 3, B− = 2.7, C+ = 2.3, C = 2, C− = 1.7, D+ = 1.3, D = 1, D− = 0.7, and F = 0. Calculate the GPA of the student whose grades are given below.

Class	Units	Grade
Math	5	B+
English	3	C+
Spanish	5	A−
Physical science	3	B−

CHAPTER 8

SUMMARY

KEY WORDS	EXAMPLES
A *measurement* includes a number and a unit. [8.1A, p. 340]	9 inches, 6 feet, 3 yards, and 50 miles are measurements.
Equivalent measures are used to form *conversion rates* to change one unit in the U.S. Customary System of measurement to another. In the conversion rate chosen, the unit in the numerator is the same as the unit desired in the answer. The unit in the denominator is the same as the unit in the given measurement. [8.1A, p. 340]	Because 12 in. = 1 ft, the conversion rate $\frac{12\ in.}{1\ ft}$ is used to convert feet to inches. The conversion rate $\frac{1\ ft}{12\ in.}$ is used to convert inches to feet.

Energy is the ability to do work. One *foot-pound* (ft · lb) of energy is the amount of energy necessary to lift 1 pound a distance of 1 foot. Consumer items that use energy are rated in *British thermal units* (Btu). [8.5A, p. 356]

Find the energy required for a 110-pound person to climb a set of stairs 12 ft high.

Energy = 110 lb × 12 ft = 1320 ft · lb

Power is the rate at which work is done or energy is released.

Power is measured in *foot-pounds per second* $\left(\dfrac{\text{ft} \cdot \text{lb}}{\text{s}}\right)$ and *horsepower* (hp). [8.5B, p. 357]

Find the power needed to raise 250 lb a distance of 20 ft in 10 s.

$$\text{Power} = \frac{20 \text{ ft} \times 250 \text{ lb}}{10 \text{ s}} = 500 \frac{\text{ft} \cdot \text{lb}}{\text{s}}$$

ESSENTIAL RULES AND PROCEDURES

EXAMPLES

Equivalences Between Units of Length [8.1A, p. 340]
The U.S. Customary units of length are inch (in.), foot (ft), yard (yd), and mile (mi).
12 in. = 1 ft
3 ft = 1 yd
36 in. = 1 yd
5280 ft = 1 mi

Convert 52 in. to ft.

$$52 \text{ in.} = 52 \text{ in.} \times \frac{1 \text{ ft}}{12 \text{ in.}}$$
$$= \frac{52 \text{ ft}}{12} = 4\frac{1}{3} \text{ ft}$$

Equivalences Between Units of Weight [8.2A, p. 346]
Weight is a measure of how strongly Earth is pulling on an object. The U.S. Customary units of weight are ounce (oz), pound (lb), and ton.
16 oz = 1 lb
2000 lb = 1 ton

Convert 9 lb to ounces.

$$9 \text{ lb} = 9 \text{ lb} \times \frac{16 \text{ oz}}{1 \text{ lb}} = 144 \text{ oz}$$

Equivalences Between Units of Capacity [8.3A, p. 350]
Liquid substances are measured in units of *capacity*. The U.S. Customary units of capacity are fluid ounce (fl oz), cup (c), pint (pt), quart (qt), and gallon (gal).
8 fl oz = 1 c
2 c = 1 pt
2 pt = 1 qt
4 qt = 1 gal

Convert 14 qt to gallons.

$$14 \text{ qt} = 14 \text{ qt} \times \frac{1 \text{ gal}}{4 \text{ qt}}$$
$$= \frac{14 \text{ gal}}{4} = 3\frac{1}{2} \text{ gal}$$

Equivalences Between Units of Time [8.4A, p. 354]
Units of time are seconds (s), minutes (min), hours (h), days, and weeks.
60 s = 1 min
60 min = 1 h
24 h = 1 day
7 days = 1 week

Convert 8 days to hours.

$$8 \text{ days} = 8 \text{ days} \times \frac{24 \text{ h}}{1 \text{ day}} = 192 \text{ h}$$

Equivalences Between Units of Energy [8.5A, p. 356]
1 Btu = 778 ft · lb

Convert 70 Btu to foot-pounds.

$$70 \text{ Btu} = 70 \text{ Btu} \times \frac{778 \text{ ft} \cdot \text{lb}}{1 \text{ Btu}}$$
$$= 54,460 \text{ ft} \cdot \text{lb}$$

Equivalences Between Units of Power [8.5B, p. 357]
The U.S. Customary unit of power is the horsepower (hp).
$$1 \text{ hp} = 550 \frac{\text{ft} \cdot \text{lb}}{\text{s}}$$

Convert 5 hp to foot-pounds per second.

$$5 \times 550 = 2750 \frac{\text{ft} \cdot \text{lb}}{\text{s}}$$

CHAPTER 8

CONCEPT REVIEW

Test your knowledge of the concepts presented in this chapter. Answer each question.
Then check your answers against the ones provided in the Answer Section.

1. What operation is used to convert from feet to inches?

2. How do you convert 5 ft 7 in. to all inches?

3. How do you divide 7 ft 8 in. by 2?

4. What conversion rate is used to convert 5240 lb to tons?

5. How do you multiply 6 lb 9 oz by 3?

6. Name five measures of capacity.

7. What conversion rate is used to convert 7 gal to quarts?

8. What conversion rate is used to convert 374 min to hours?

9. What is a foot-pound of energy?

10. What operations are needed to find the power to raise 200 lb a distance of 24 ft in 12 s?

CHAPTER 8

REVIEW EXERCISES

1. Convert 4 ft to inches.

2. What is 7 ft 6 in. divided by 3?

3. Find the energy needed to lift 200 lb a distance of 8 ft.

4. Convert $2\frac{1}{2}$ pt to fluid ounces.

5. Convert 14 ft to yards.

6. Convert 2400 lb to tons.

7. Find the quotient of 7 lb 5 oz and 3.

8. Convert $3\frac{3}{8}$ lb to ounces.

9. Add: 3 ft 9 in.
         + 5 ft 6 in.

10. Subtract: 3 tons   500 lb
              − 1 ton   1500 lb

11. Add: 4 c 7 fl oz
         + 2 c 3 fl oz

12. Subtract: 5 yd 1 ft
              − 3 yd 2 ft

13. Convert 12 c to quarts.

14. Convert 375 min to hours.

15.   Convert 2.5 hp to foot-pounds per second. $\left(1 \text{ hp} = 550 \dfrac{\text{ft} \cdot \text{lb}}{\text{s}} \right)$

16.   Multiply: 5 lb 8 oz
                   $\underline{\times \qquad 8}$

17.   Convert 50 Btu to foot-pounds. (1 Btu = 778 ft · lb)

18.   Convert 3850 $\dfrac{\text{ft} \cdot \text{lb}}{\text{s}}$ to horsepower. $\left(1 \text{ hp} = 550 \dfrac{\text{ft} \cdot \text{lb}}{\text{s}} \right)$

19.   **Carpentry**   A board 6 ft 11 in. long is cut from a board 10 ft 5 in. long. Find the length of the remaining piece of board.

20.   **Shipping**   A book weighing 2 lb 3 oz is mailed at the rate of $.29 per ounce. Find the cost of mailing the book.

21.   **Capacity**   A can of pineapple juice contains 18 fl oz. Find the number of quarts in a case of 24 cans.

22.   **Food Service**   A cafeteria sold 256 cartons of milk in one school day. Each carton contains 1 c of milk. How many gallons of milk were sold that day?

AP Images

23.   **Energy**   A furnace is rated at 35,000 Btu per hour. How many foot-pounds of energy does the furnace release in 1 h? (1 Btu = 778 ft · lb)

24.   **Power**   Find the power in foot-pounds per second of an engine that can raise 800 lb to a height of 15 ft in 25 s.

CHAPTER 8

TEST

1. Convert $2\frac{1}{2}$ ft to inches.

2. Subtract: 4 ft 2 in. − 1 ft 9 in.

3. **Carpentry**   A board $6\frac{2}{3}$ ft long is cut into five equal pieces. How long is each piece?

4. **Masonry**   Seventy-two bricks, each 8 in. long, are laid end-to-end to make the base for a wall. Find the length of the wall in feet.

5. Convert $2\frac{7}{8}$ lb to ounces.

6. Convert: 40 oz = ___ lb ___ oz

7. Find the sum of 9 lb 6 oz and 7 lb 11 oz.

8. Divide: 6 lb 12 oz ÷ 4

9. **Weights**   A college bookstore received 1000 workbooks, each weighing 12 oz. Find the total weight of the 1000 workbooks in pounds.

10. **Recycling**   An elementary school class gathered 800 aluminum cans for recycling. Four aluminum cans weigh 3 oz. Find the amount the class received if the rate of pay was $.75 per pound for the aluminum cans. Round to the nearest cent.

11. Convert 13 qt to gallons.

12. Convert $3\frac{1}{2}$ gal to pints.

13. What is $1\frac{3}{4}$ gal times 7?

14. Add: 5 gal 2 qt + 2 gal 3 qt

15. Convert 756 h to weeks.

16. Convert $3\frac{1}{4}$ days to minutes.

17. Capacity A can of grapefruit juice contains 20 fl oz. Find the number of cups of grapefruit juice in a case of 24 cans.

18. Business Nick, a mechanic, bought oil in 40-gallon containers for changing the oil in customers' cars. He paid $810 for a 40-gallon container of oil and charged customers $9.35 per quart. Find the profit Nick made on one 40-gallon container of oil.

Tim Boyle/Getty Images

19. Energy Find the energy required to lift 250 lb a distance of 15 ft.

20. Energy A furnace is rated at 40,000 Btu per hour. How many foot-pounds of energy are released by the furnace in 1 h? (1 Btu = 778 ft · lb)

21. Power Find the power needed to lift 200 lb a distance of 20 ft in 25 s.

22. Power A motor has a power of 2200 $\frac{\text{ft} \cdot \text{lb}}{\text{s}}$. Find the motor's horsepower.

$$\left(1 \text{ hp} = 550 \frac{\text{ft} \cdot \text{lb}}{\text{s}}\right)$$

CUMULATIVE REVIEW EXERCISES

1. Find the LCM of 9, 12, and 15.

2. Write $\frac{43}{8}$ as a mixed number.

3. Subtract: $5\frac{7}{8} - 2\frac{7}{12}$

4. What is $5\frac{1}{3}$ divided by $2\frac{2}{3}$?

5. Simplify: $\frac{5}{8} \div \left(\frac{3}{8} - \frac{1}{4}\right) - \frac{5}{8}$

6. Round 2.0972 to the nearest hundredth.

7. Multiply: $\begin{array}{r} 0.0792 \\ \times \quad 0.49 \\ \hline \end{array}$

8. Solve the proportion: $\frac{n}{12} = \frac{44}{60}$

9. Find $2\frac{1}{2}\%$ of 50.

10. 18 is 42% of what? Round to the nearest hundredth.

11. **Consumerism** A 7.2-pound roast costs $37.08. Find the unit cost.

12. Add: $3\frac{2}{5}$ in. $+ 5\frac{1}{3}$ in.

13. Convert: 24 oz = ___ lb ___ oz

14. Multiply: 3 lb 8 oz $\times$ 9

15. Subtract: $4\frac{1}{3}$ qt $- 1\frac{5}{6}$ qt

16. Find 2 lb 10 oz less than 4 lb 6 oz.

17. Investments　An investor receives a dividend of $56 from 40 shares of stock. At the same rate, find the dividend that 200 shares of stock would yield.

18. Banking　Anna had a balance of $578.56 in her checkbook. She wrote checks for $216.98 and $34.12 and made a deposit of $315.33. What is her new checking account balance?

19. Compensation　An account executive receives a salary of $1800 per month plus a commission of 2% on all sales over $25,000. Find the total monthly income of an account executive who has monthly sales of $140,000.

20. Transportation　A truck driver is paid by the number of miles driven. If the truck driver earns $.46 per mile, how many miles must the trucker drive in 1 h to earn $16? Round to the nearest mile.

21. Education　The scores on the final exam of a trigonometry class are recorded in the histogram at the right. What percent of the class received a score between 80% and 90%? Round to the nearest percent.

22. Markup　Hayes Department Store uses a markup rate of 40% on all merchandise. What is the selling price of a DMB television that cost the store $220?

23. Simple Interest　A construction firm received a loan of $200,000 for 8 months at a simple interest rate of 6%. Find the interest paid on the loan.

24. Income　Six college students spent several weeks panning for gold during their summer vacation. The students obtained 1 lb 3 oz of gold, which they sold for $800 per ounce. How much money did each student receive if they shared the money equally? Round to the nearest dollar.

25. Shipping　Four books were mailed at the postal rate of $.28 per ounce. The books weighed 1 lb 3 oz, 13 oz, 1 lb 8 oz, and 1 lb. Find the cost of mailing the books.

26. Consumerism　One brand of yogurt costs $.79 for 8 oz, and 36 oz of another brand can be bought for $2.98. Which purchase is the better buy?

27. Probability　Two dice are rolled. What is the probability that the sum of the dots on the upward faces is 9?

28. Energy　Find the energy required to lift 400 lb a distance of 8 ft.

29. Power　Find the power, in foot-pounds per second, needed to raise 600 lb a distance of 8 ft in 12 s.

The Metric System of Measurement

OBJECTIVES

SECTION 9.1
A   To convert units of length in the metric system of measurement
B   To solve application problems

SECTION 9.2
A   To convert units of mass in the metric system of measurement
B   To solve application problems

SECTION 9.3
A   To convert units of capacity in the metric system of measurement
B   To solve application problems

SECTION 9.4
A   To use units of energy in the metric system of measurement

SECTION 9.5
A   To convert U.S. Customary units to metric units
B   To convert metric units to U.S. Customary units

ARE YOU READY?

Take the Chapter 9 Prep Test to find out if you are ready to learn to:

- Convert units of length, mass, and capacity in the metric system
- Use units of energy in the metric system
- Convert between U.S. Customary units and metric units

PREP TEST

Do these exercises to prepare for Chapter 9.

For Exercises 1 to 10, add, subtract, multiply, or divide.

1. $3.732 \times 10,000$

2. 65.9×10^4

3. $41.07 \div 1000$

4. $28,496 \div 10^3$

5. $6 - 0.875$

6. $5 + 0.96$

7. 3.25×0.04

8. $35 \times \dfrac{1.61}{1}$

9. $1.67 \times \dfrac{1}{3.34}$

10. $4\dfrac{1}{2} \times 150$

SECTION

9.1 Length

OBJECTIVE A   To convert units of length in the metric system of measurement

≈1 meter

In 1789, an attempt was made to standardize units of measurement internationally in order to simplify trade and commerce between nations. A commission in France developed a system of measurement known as the **metric system.**

The basic unit of length in the metric system is the **meter.** One meter is approximately the distance from a doorknob to the floor. All units of length in the metric system are derived from the meter. Prefixes to the basic unit denote the length of each unit. For example, the prefix "centi-" means one-hundredth, so 1 centimeter is 1 one-hundredth of a meter.

North Pole

Equator

Prefixes and Units of Length in the Metric System	
kilo- = 1000	1 kilometer (km) = 1000 meters (m)
hecto- = 100	1 hectometer (hm) = 100 m
deca- = 10	1 decameter (dam) = 10 m
	1 meter (m) = 1 m
deci- = 0.1	1 decimeter (dm) = 0.1 m
centi- = 0.01	1 centimeter (cm) = 0.01 m
milli- = 0.001	1 millimeter (mm) = 0.001 m

PPT

Conversion between units of length in the metric system involves moving the decimal point to the right or to the left. Listing the units in order from largest to smallest will indicate how many places to move the decimal point and in which direction.

To convert 4200 cm to meters, write the units in order from largest to smallest.

km   hm   dam   m   dm   cm   mm

2 positions

4200 cm = 42.00 m

2 places

• Converting cm to m requires moving 2 positions to the left.

• Move the decimal point the same number of places and in the same direction.

A metric measurement that involves two units is customarily written in terms of one unit. Convert the smaller unit to the larger unit and then add.

Tips for Success

The prefixes introduced here are used throughout the chapter. As you study the material in the remaining sections, use the table above for a reference or refer to the Chapter Summary at the end of this chapter.

To convert 8 km 32 m to kilometers, first convert 32 m to kilometers.

km   hm   dam   m   dm   cm   mm

32 m = 0.032 km

8 km 32 m = 8 km + 0.032 km

= 8.032 km

• Converting m to km requires moving 3 positions to the left.

• Move the decimal point the same number of places and in the same direction.

• Add the result to 8 km.

EXAMPLE · 1

Convert 0.38 m to millimeters.

Solution   0.38 m = 380 mm

YOU TRY IT · 1

Convert 3.07 m to centimeters.

Your solution

EXAMPLE · 2

Convert 4 m 62 cm to meters.

Solution   62 cm = 0.62 m

4 m 62 cm = 4 m + 0.62 m
= 4.62 m

YOU TRY IT · 2

Convert 3 km 750 m to kilometers.

Your solution

Solutions on p. S21

OBJECTIVE B   To solve application problems

✓ **Take Note**

Although in this text we will always change units to the larger unit, it is possible to perform the calculation by changing to the smaller unit.

2 m − 85 cm
= 200 cm − 85 cm
= 115 cm

Note that 115 cm = 1.15 m.

In the application problems in this section, we perform arithmetic operations with the measurements of length in the metric system. It is important to remember that before measurements can be added or subtracted, they must be expressed in terms of the same unit. In this textbook, unless otherwise stated, the units should be changed to the larger unit before the arithmetic operation is performed.

To subtract 85 cm from 2 m, convert 85 cm to meters.

$$2 \text{ m} - 85 \text{ cm} = 2 \text{ m} - 0.85 \text{ m}$$
$$= 1.15 \text{ m}$$

EXAMPLE · 3

A piece measuring 142 cm is cut from a board 4.20 m long. Find the length of the remaining piece.

Strategy

To find the length of the remaining piece:
- Convert the length of the piece cut (142 cm) to meters.
- Subtract the length of the piece cut from the original length.

Solution

142 cm = 1.42 m

4.20 m − 142 cm = 4.20 m − 1.42 m
= 2.78 m

The length of the remaining piece is 2.78 m.

YOU TRY IT · 3

A bookcase 175 cm long has four shelves. Find the cost of the shelves when the price of the lumber is $15.75 per meter.

Your strategy

Your solution

Solution on p. S21

9.1 EXERCISES

OBJECTIVE A   To convert units of length in the metric system of measurement

For Exercises 1 to 27, convert.

1. 42 cm = _____ mm

2. 62 cm = _____ mm

3. 81 mm = _____ cm

4. 68.2 mm = _____ cm

5. 6804 m = _____ km

6. 3750 m = _____ km

7. 2.109 km = _____ m

8. 32.5 km = _____ m

9. 432 cm = _____ m

10. 61.7 cm = _____ m

11. 0.88 m = _____ cm

12. 3.21 m = _____ cm

13. 7038 m = _____ km

14. 2589 m = _____ km

15. 3.5 km = _____ m

16. 9.75 km = _____ m

17. 260 cm = _____ m

18. 705 cm = _____ m

19. 1.685 m = _____ cm

20. 0.975 m = _____ cm

21. 14.8 cm = _____ mm

22. 6 m 42 cm = _____ m

23. 62 m 7 cm = _____ m

24. 42 cm 6 mm = _____ cm

25. 31 cm 9 mm = _____ cm

26. 62 km 482 m = _____ km

27. 8 km 75 m = _____ km

For Exercises 28 to 30, fill in the blank with the correct unit of measurement.

28. 5.8 m = 580 _____

29. 0.6 km = 600 _____

30. 54 mm = 0.054 _____

OBJECTIVE B   To solve application problems

31. Carpentry How many shelves, each 240 cm long, can be cut from a board that is 7.20 m in length? Find the length of the board remaining after the shelves are cut.

32. Measurements Find the missing dimension, in centimeters, in the diagram at the right.

27.4 cm

40 mm   ?   15.6 cm

33. **Sports** A walk-a-thon had two checkpoints. One checkpoint was 1400 m from the starting point. The second checkpoint was 1200 m from the first checkpoint. The second checkpoint was 1800 m from the finish line. How long was the walk? Express the answer in kilometers.

34. **Metal Works** Twenty rivets are used to fasten two steel plates together. The plates are 3.4 m long, and the rivets are equally spaced, with a rivet at each end. Find the distance between the rivets. Round to the nearest tenth of a centimeter.

35. **Measurements** Find the total length, in centimeters, of the shaft in the diagram at the right.

36. **Fencing** You purchase a 50-meter roll of fencing, at a cost of $14.95 per meter, in order to build a dog run that is 340 cm wide and 1380 cm long. After you cut the four pieces of fencing from the roll, how much of the fencing is left on the roll?

 37. **Adopt-A-Highway** Use the news clipping at the right. Find the average number of meters adopted by a group in the Missouri Adopt-a-Highway program. Round to the nearest whole number.

 38. A board 3.25 m long is cut into shelves of equal length. For each statement, answer *true* or *false*.
  a. If the length of each shelf is less than 100 cm, then at least three shelves can be cut from the board.
  b. If the length of each shelf is greater than 100 cm, then at most two shelves can be cut from the board.

> **In the News**
>
> **Highway Adoption Proves Popular**
>
> The Missouri Adopt-a-Highway Program, which has been in existence for 20 years, currently has 3772 groups in the program. The groups have adopted 8502 km along the roadways.
>
> *Source:* www.modot.org

 39. **Astronomy** The distance between Earth and the sun is 150,000,000 km. Light travels 300,000,000 m in 1 s. How long does it take for light to reach Earth from the sun?

 40. **Earth Science** The circumference of Earth is 40,000 km. How long would it take to travel the circumference of Earth at a speed of 85 km per hour? Round to the nearest tenth.

 41. **Physics** Light travels 300,000 km in 1 s. How far does light travel in 1 day?

Applying the Concepts

42. Other prefixes in the metric system are becoming more commonly used as a result of technological advances. Find the meaning of the following prefixes: tera-, giga-, mega-, micro-, nano-, and pico.

9.2   Mass

OBJECTIVE A   **To convert units of mass in the metric system of measurement**

1 gram = the mass of water in the box

Mass and weight are closely related. Weight is a measure of how strongly Earth is pulling on an object. Therefore, an object's weight is less in space than on Earth's surface. However, the amount of material in the object, its **mass,** remains the same. On the surface of Earth, mass and weight can be used interchangeably.

The basic unit of mass in the metric system is the **gram.** If a box that is 1 cm long on each side is filled with water, then the mass of that water is 1 gram.

The gram is a very small unit of mass. A paper clip weighs about 1 gram. The kilogram (1000 grams) is a more useful unit of mass in consumer applications. This textbook weighs about 1 kilogram.

 Point of Interest

An average snowflake weighs about $\frac{1}{300}$ g and contains approximately 100,000,000,000,000,000,000 water molecules. You may have heard the expression "No two snowflakes are the same." It was the concept of such a large number of water molecules, and the huge number of their possible arrangements, that led to this expression.

The units of mass in the metric system have the same prefixes as the units of length.

> **Units of Mass in the Metric System**
>
> 1 kilogram (kg) = 1000 grams (g)
> 1 hectogram (hg) = 100 g
> 1 decagram (dag) = 10 g
> 1 gram (g) = 1 g
> 1 decigram (dg) = 0.1 g
> 1 centigram (cg) = 0.01 g
> 1 milligram (mg) = 0.001 g

PPT

Weight ≈ 1 gram

Conversion between units of mass in the metric system involves moving the decimal point to the right or to the left. Listing the units in order from largest to smallest will indicate how many places to move the decimal point and in which direction.

To convert 324 g to kilograms, first write the units in order from largest to smallest.

kg   hg   dag   g   dg   cg   mg
        3 positions

324 g = 0.324 kg
        3 places

• Converting g to kg requires moving 3 positions to the left.

• Move the decimal point the same number of places and in the same direction.

EXAMPLE • 1

Convert 4.23 g to milligrams.

Solution   4.23 g = 4230 mg

YOU TRY IT • 1

Convert 42.3 mg to grams.

Your solution

Solution on p. S22

EXAMPLE · 2

Convert 2 kg 564 g to kilograms.

Solution   564 g = 0.564 kg

                 2 kg 564 g = 2 kg + 0.564 kg
                            = 2.564 kg

YOU TRY IT · 2

Convert 3 g 54 mg to grams.

Your solution

Solution on p. S22

OBJECTIVE B   To solve application problems

✓ **Take Note**

Although in this text we will always change units to the larger unit, it is possible to perform the calculation by changing to the smaller unit.

3 kg − 750 g
 = 3000 g − 750 g
 = 2250 g

Note that

2250 g = 2.250 kg.

In the application problems in this section, we perform arithmetic operations with the measurements of mass in the metric system. Remember that before measurements can be added or subtracted, they must be expressed in terms of the same unit. In this textbook, unless otherwise stated, the units should be changed to the larger unit before the arithmetic operation is performed.

To subtract 750 g from 3 kg, convert 750 g to kilograms.

$$3 \text{ kg} - 750 \text{ g} = 3 \text{ kg} - 0.750 \text{ kg}$$
$$= 2.250 \text{ kg}$$

EXAMPLE · 3

Find the cost of three packages of ground meat weighing 540 g, 670 g, and 890 g if the price per kilogram is $9.89. Round to the nearest cent.

Strategy

To find the cost of the meat:

• Find the total weight of the three packages.
• Convert the total weight to kilograms.
• Multiply the weight by the cost per kilogram ($9.89).

Solution

540 g + 670 g + 890 g = 2100 g

2100 g = 2.1 kg

2.1 × 9.89 = 20.769

The cost of the meat is $20.77.

YOU TRY IT · 3

How many kilograms of fertilizer are required to fertilize 400 trees in an apple orchard if 300 g of fertilizer are used for each tree?

Your strategy

Your solution

Solution on p. S22

9.2 EXERCISES

OBJECTIVE A   To convert units of mass in the metric system of measurement

For Exercises 1 to 24, convert.

1. 420 g = _____ kg

2. 7421 g = _____ kg

3. 127 mg = _____ g

4. 43 mg = _____ g

5. 4.2 kg = _____ g

6. 0.027 kg = _____ g

7. 0.45 g = _____ mg

8. 325 g = _____ mg

9. 1856 g = _____ kg

10. 8900 g = _____ kg

11. 4057 mg = _____ g

12. 1970 mg = _____ g

13. 1.37 kg = _____ g

14. 5.1 kg = _____ g

15. 0.0456 g = _____ mg

16. 0.2 g = _____ mg

17. 18,000 g = _____ kg

18. 0.87 kg = _____ g

19. 3 kg 922 g = _____ kg

20. 1 kg 47 g = _____ kg

21. 7 g 891 mg = _____ g

22. 209 g 42 mg = _____ g

23. 4 kg 63 g = _____ kg

24. 18 g 5 mg = _____ g

For Exercises 25 to 27, fill in the blank with the correct unit of measurement.

25. 0.105 g = 105 _____

26. 70 g = 0.07 _____

27. 31 mg = 0.031 _____

OBJECTIVE B   To solve application problems

28. Consumerism A 1.19-kilogram container of Quaker Oats contains 30 servings. Find the number of grams in one serving of the oatmeal. Round to the nearest whole number.

29. Nutrition A patient is advised to supplement her diet with 2 g of calcium per day. The calcium tablets she purchases contain 500 mg of calcium per tablet. How many tablets per day should the patient take?

30. Gemology A carat is a unit of weight equal to 200 mg. Find the weight in grams of a 10-carat precious stone.

31. Nutrition One egg contains 274 mg of cholesterol. How many grams of cholesterol are in a dozen eggs?

32. Consumerism The nutrition label for a corn bread mix is shown at the right.
a. How many kilograms of mix are in the package?
b. How many grams of sodium are contained in two servings of the corn bread?

Nutrition Facts
Serving Size ⅙ pkg. (31g mix)
Servings Per Container 6

Amount Per Serving	Mix	Prepared
Calories	110	160
Calories from Fat	10	50

	% Daily Value*	
Total Fat 1g	1%	9%
Saturated Fat 0g	0%	7%
Cholesterol 0mg	0%	12%
Sodium 210mg	9%	11%
Total Carbohydrate 24g	8%	8%
Sugars 6g		
Protein 2g		

33. Consumerism Find the cost of three packages of ground meat weighing 470 g, 680 g, and 590 g if the price per kilogram is $8.40.

34. Landscaping Eighty grams of grass seed are used for every 100 m² of lawn. How many kilograms of grass seed are needed to cover 2000 m²?

A New Kind of Power Walking

Max Donelan has invented a high-tech, 1.6-kilogram knee brace that generates enough electricity to power portable gadgets. One minute of walking while wearing one of the braces on each leg generates enough electricity to charge 10 cell phones.

Source: Time, March 17, 2008

35. Energy Use the news clipping at the right. Find the weight, in grams, of two of the power-generating knee braces.

36. Airlines American Airlines charges $50 for a checked bag weighing over 23 kg but not more than 32 kg. The charge for a checked bag weighing over 32 kg but not more than 45 kg is $100. How much must be paid for three pieces of luggage weighing 27.9 kg, 31.5 kg, and 40.2 kg?

37. Business A health food store buys organic pumpkin seeds in 10-kilogram containers and repackages the seeds for resale. The store packages the seeds in 200-gram bags, costing $.04 each, and resells them for $3.89 per bag. Find the profit on a 10-kilogram container of organic pumpkin seeds costing $75.

38. Agriculture During 1 year, the United States exported 37,141 million kg of wheat, 2680 million kg of rice, and 40,365 million kg of corn. What percent of the total of these grain exports was corn? Round to the nearest tenth of a percent.

39. Measurements A trailer is loaded with nine automobiles weighing 1405 kg each. Find the total weight of the automobiles.

40. A serving of orange juice contains 450 mg of potassium. A carton of orange juice contains 8 servings. Without finding the number of grams of potassium in a carton of orange juice, determine whether the following statment is *true* or *false*:

A carton of orange juice contains less than 4 g of potassium.

Applying the Concepts

41. Discuss the advantages and disadvantages of the U.S. Customary System and the metric system of measurement.

SECTION

9.3 Capacity

OBJECTIVE A **To convert units of capacity in the metric system of measurement**

The basic unit of capacity in the metric system is the liter. One **liter** is defined as the capacity of a box that is 10 cm long on each side.

The units of capacity in the metric system have the same prefixes as the units of length.

1-liter bottle

Units of Capacity in the Metric System

1 kiloliter (kl) = 1000 L
1 hectoliter (hl) = 100 L
1 decaliter (dal) = 10 L
1 liter (L) = 1 L
1 deciliter (dl) = 0.1 L
1 centiliter (cl) = 0.01 L
1 milliliter (ml) = 0.001 L

PPT

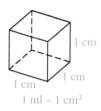

1 ml = 1 cm³

The milliliter is equal to 1 **cubic centimeter** (cm³). See the diagram at the left. In medicine, cubic centimeter is often abbreviated cc.

Conversion between units of capacity in the metric system involves moving the decimal point to the right or to the left. Listing the units in order from largest to smallest will indicate how many places to move the decimal point and in which direction.

To convert 824 ml to liters, first write the units in order from largest to smallest.

kl   hl   dal   **L**   dl   cl   ml

3 positions

* Converting ml to L requires moving 3 positions to the left.

824 ml = 0.824 L

3 places

* Move the decimal point the same number of places and in the same direction.

EXAMPLE · 1

Convert 4 L 32 ml to liters.

Solution   32 ml = 0.032 L

4 L 32 ml = 4 L + 0.032 L
= 4.032 L

YOU TRY IT · 1

Convert 2 kl 167 L to liters.

Your solution

Solution on p. S22

EXAMPLE • 2

Convert 1.23 L to cubic centimeters.

Solution 1.23 L = 1230 ml = 1230 cm³

YOU TRY IT • 2

Convert 325 cm³ to liters.

Your solution

Solution on p. S22

OBJECTIVE B To solve application problems

 Take Note

Although in this text we will always change units to the larger unit, it is possible to perform the calculation by changing to the smaller unit.

2.5 kl + 875 L
= 2500 L + 875 L
= 3375 L

Note that

3375 L = 3.375 kl.

In the application problems in this section, we perform arithmetic operations with the measurements of capacity in the metric system. Remember that before measurements can be added or subtracted, they must be expressed in terms of the same unit. In this textbook, unless otherwise stated, the units should be changed to the larger unit before the arithmetic operation is performed.

To add 2.5 kl and 875 L, convert 875 L to kiloliters.

$$2.5 \text{ kl} + 875 \text{ L} = 2.5 \text{ kl} + 0.875 \text{ kl}$$
$$= 3.375 \text{ kl}$$

EXAMPLE • 3

A laboratory assistant is in charge of ordering acid for three chemistry classes of 30 students each. Each student requires 80 ml of acid. How many liters of acid should be ordered? (The assistant must order by the whole liter.)

Strategy
To find the number of liters to be ordered:

• Find the number of milliliters of acid needed by multiplying the number of classes (3) by the number of students per class (30) by the number of milliliters of acid required by each student (80).
• Convert milliliters to liters.
• Round up to the nearest whole number.

Solution
3(30)(80) = 7200 ml

7200 ml = 7.2 L

7.2 rounded up to the nearest whole number is 8.

The assistant should order 8 L of acid.

YOU TRY IT • 3

For $299.50, a cosmetician buys 5 L of moisturizer and repackages it in 125-milliliter jars. Each jar costs the cosmetician $.85. Each jar of moisturizer is sold for $29.95. Find the profit on the 5 L of moisturizer.

Your strategy

Your solution

Solution on p. S22

9.3 EXERCISES

OBJECTIVE A   To convert units of capacity in the metric system of measurement

For Exercises 1 to 24, convert.

1. 4200 ml = _____ L

2. 7.5 ml = _____ L

3. 3.42 L = _____ ml

4. 0.037 L = _____ ml

5. 423 ml = _____ cm^3

6. 0.32 ml = _____ cm^3

7. 642 cm^3 = _____ ml

8. 0.083 cm^3 = _____ ml

9. 42 cm^3 = _____ L

10. 3075 cm^3 = _____ L

11. 0.435 L = _____ cm^3

12. 2.57 L = _____ cm^3

13. 4.62 kl = _____ L

14. 0.035 kl = _____ L

15. 1423 L = _____ kl

16. 897 L = _____ kl

17. 1.267 L = _____ cm^3

18. 4.105 L = _____ cm^3

19. 3 L 42 ml = _____ L

20. 1 L 127 ml = _____ L

21. 3 kl 4 L = _____ kl

22. 6 kl 32 L = _____ kl

23. 8 L 200 ml = _____ L

24. 9 kl 505 L = _____ kl

For Exercises 25 to 27, fill in the blank with the correct unit of measurement.

25. 620 ml = 0.62 _____

26. 950 cm^3 = 0.95 _____

27. 4.1 kl = 4100 _____

OBJECTIVE B   To solve application problems

28. **Earth Science** The air in Earth's atmosphere is 78% nitrogen and 21% oxygen. Find the amount of oxygen in 50 L of air.

29. **Consumerism** The printed label from a container of milk is shown at the right. How many 230-milliliter servings are in the container? Round to the nearest whole number.

30. **Measurements** An athletic club uses 800 ml of chlorine each day for its swimming pool. How many liters of chlorine are used in a month of 30 days?

Dairy Hill

Skim Milk
Vitamin A & D Added
Pasteurized • Homogenized

INGREDIENTS: PASTEURIZED SKIM
MILK, NONFAT MILK SOLIDS, VITAMIN
A PALMITATE AND VITAMIN D3 ADDED.

0 15400 20209 1

1 GAL (3.78 L)

31. **Medicine** A flu vaccine is being given for the coming winter season. A medical corporation buys 12 L of flu vaccine. How many patients can be immunized if each person receives 3 cm³ of the vaccine?

32. **Consumerism** A can of tomato juice contains 1.36 L. How many 170-milliliter servings are in one can of tomato juice?

 33. **Beverages** Use the news clipping at the right. How many 140-milliliter servings are in one carafe of Pure Premium?

34. **Chemistry** A chemistry experiment requires 12 ml of an acid solution. How many liters of acid should be ordered when 4 classes of 90 students each are going to perform the experiment? (The acid must be ordered by the whole liter.)

35. **Consumerism** A case of 12 one-liter bottles of apple juice costs $19.80. A case of 24 cans, each can containing 340 ml of apple juice, costs $14.50. Which case of apple juice is the better buy?

 36. **Dairies** Use the news clipping at the right. Find the increase in monthly milk production, in kiloliters, from March 1988 to March 2008.

37. **Business** For $195, a pharmacist purchases 5 L of cough syrup and repackages it in 250-milliliter bottles. Each bottle costs the pharmacist $.55. Each bottle of cough syrup is sold for $23.89. Find the profit on the 5 L of cough syrup.

38. **Business** A service station operator bought 85 kl of gasoline for $63,750. The gasoline was sold for $.945 per liter. Find the profit on the 85 kl of gasoline.

39. **Business** A wholesale distributor purchased 32 kl of cooking oil for $64,480. The wholesaler repackaged the cooking oil in 1.25-liter bottles. The bottles cost $.42 each. Each bottle of cooking oil was sold for $5.94. Find the distributor's profit on the 32 kl of cooking oil.

 40. A bottle of spring water holds 0.5 L of water. A case of spring water contains 24 bottles. Without finding the number of milliliters of water in a case of spring water, determine whether the following statement is *true* or *false*:

A case of spring water contains more than 1000 ml of water.

Applying the Concepts

41. After a 280-milliliter serving is taken from a 3-liter bottle of water, how much water remains in the container? Write the answer in three different ways.

In the News

Tropicana Goes Metric

Tropicana has introduced metric-sized packaging. The first was the Pure Premium 1.75-liter carafe. Other sizes of packaging include 200 ml, 400 ml, 450 ml, and 1 L.

Source: www.gometric.us

In the News

U.S. Cows Producing More Milk

On average, a U.S. dairy cow produced 6803 L of milk in the month of March 2008. Compare this with an average of 4684 L of milk during the month of March 1988.

Source: National Agricultural Statistics Service

© iStockphoto.com/shannon drawe

SECTION

9.4 Energy

OBJECTIVE A

To use units of energy in the metric system of measurement

Two commonly used units of energy in the metric system are the calorie and the watt-hour.

Heat is generally measured in units called calories or in larger units called Calories (with a capital C). A **Calorie** is 1000 calories and should be called a kilocalorie, but it is common practice in nutritional references and food labeling to simply call it a Calorie. A Calorie is the amount of heat required to raise the temperature of 1 kg of water 1 degree Celsius. One Calorie is also the energy required to lift 1 kg a distance of 427 m.

Corbis

HOW TO 1 Swimming uses 480 Calories per hour. How many Calories are used by swimming $\frac{1}{2}$ h each day for 30 days?

Strategy To find the number of Calories used:

- Find the number of hours spent swimming.
- Multiply the number of hours spent swimming by the Calories used per hour.

Solution $\frac{1}{2} \times 30 = 15$

$15(480) = 7200$

7200 Calories are used by swimming $\frac{1}{2}$ h each day for 30 days.

The **watt-hour** is used for measuring electrical energy. One watt-hour is the amount of energy required to lift 1 kg a distance of 370 m. A light bulb rated at 100 watts (W) will emit 100 watt-hours (Wh) of energy each hour. A **kilowatt-hour** is 1000 watt-hours.

✓ Take Note

Recall that the prefix kilo-means 1000.

1000 watt-hours (Wh) = 1 kilowatt-hour (kWh)

Integrating Technology

To convert watt-hours to kilowatt-hours, divide by 1000. To use a calculator to determine the number of kilowatt-hours used in the problem at the right, enter the following:

150 **×** 8 **÷** 1000 **=**

The calculator display reads 1.2.

HOW TO 2 A 150-watt bulb is on for 8 h. At 11¢ per kilowatt-hour, find the cost of the energy used.

Strategy To find the cost:

- Find the number of watt-hours used.
- Convert to kilowatt-hours.
- Multiply the number of kilowatt-hours used by the cost per kilowatt-hour.

Solution $150 \times 8 = 1200$

$1200 \text{ Wh} = 1.2 \text{ kWh}$

$1.2 \times 0.11 = 0.132$

The cost of the energy used is $.132.

EXAMPLE · 1

Walking uses 180 Calories per hour. How many Calories will you burn off by walking $5\frac{1}{4}$ h during one week?

Strategy

To find the number of Calories, multiply the number of hours spent walking by the Calories used per hour.

Solution

$$5\frac{1}{4} \times 180 = \frac{21}{4} \times 180 = 945$$

You will burn off 945 Calories.

YOU TRY IT · 1

Housework requires 240 Calories per hour. How many Calories are burned off by doing $4\frac{1}{2}$ h of housework?

Your strategy

Your solution

EXAMPLE · 2

A clothes iron is rated at 1200 W. If the iron is used for 1.5 h, how much energy, in kilowatt-hours, is used?

Strategy

To find the energy used:

• Find the number of watt-hours used.
• Convert watt-hours to kilowatt-hours.

Solution

$1200 \times 1.5 = 1800$

$1800 \text{ Wh} = 1.8 \text{ kWh}$

1.8 kWh of energy are used.

YOU TRY IT · 2

Find the number of kilowatt-hours of energy used when a 150-watt light bulb burns for 200 h.

Your strategy

Your solution

EXAMPLE · 3

A TV set rated at 1800 W is on for an average of 3.5 h per day. At 12.2¢ per kilowatt-hour, find the cost of operating the set for 1 week.

Strategy

To find the cost:

• Multiply 3.5 by the number of days in 1 week to find the total number of hours the set is used per week.
• Multiply the product by the number of watts to find the watt-hours.
• Convert watt-hours to kilowatt-hours.
• Multiply the number of kilowatt-hours by the cost per kilowatt-hour.

Solution

$3.5 \times 7 = 24.5$

$24.5 \times 1800 = 44,100$

$44,100 \text{ Wh} = 44.1 \text{ kWh}$

$44.1 \times 0.122 = 5.3802$

The cost is $5.3802.

YOU TRY IT · 3

A microwave oven rated at 500 W is used an average of 20 min per day. At 13.7¢ per kilowatt-hour, find the cost of operating the oven for 30 days.

Your strategy

Your solution

Solutions on pp. S22–S23

9.4 EXERCISES

OBJECTIVE A   To use units of energy in the metric system of measurement

1. **Health** How many Calories can you eliminate from your diet by omitting 1 slice of bread per day for 30 days? One slice of bread contains 110 Calories.

Nutrition Facts
Serving Size 2 Slices (18g)
Servings Per Container about 15

Amount Per Serving		
Calories 60	Calories from Fat 10	
		% Daily Value*
Total Fat 1g		2%
Saturated Fat 0g		0%
Polyunsaturated Fat 0.5g		
Monounsaturated Fat 0.5g		
Cholesterol 0mg		0%
Sodium 60mg		3%
Total Carbohydrate 10g		3%
Dietary Fiber 3g		10%
Sugars 1g		
Protein 2g		

Vitamin A 0%	•	Vitamin C 0%	
Calcium 0%	•	Iron	4%

* Percent Daily Values are based on a 2,000 calorie diet. Your daily values may be higher or lower depending on your calorie needs.

	Calories:	2,000	2,500
Total Fat	Less than	65g	80g
Saturated Fat	Less than	20g	25g
Cholesterol	Less than	300mg	300mg
Sodium	Less than	2,400mg	2,400mg
Total Carbohydrate		300g	375g
Dietary Fiber		25g	30g

Calories per gram:
Fat 9 • Carbohydrate 4 • Protein 4

2. **Health** How many Calories can you eliminate from your diet in 2 weeks by omitting 400 Calories per day?

3. **Nutrition** A nutrition label from a package of crisp bread is shown at the right.
   a. How many Calories are in $1\frac{1}{2}$ servings?
   b. How many Calories from fat are in 6 slices of the bread?

4. **Health** Moderately active people need 20 Calories per pound of body weight to maintain their weight. How many Calories should a 150-pound, moderately active person consume per day to maintain that weight?

5. **Health** People whose daily activity level would be described as light need 15 Calories per pound of body weight to maintain their weight. How many Calories should a 135-pound, lightly active person consume per day to maintain that weight?

6. **Health** For a healthful diet, it is recommended that 55% of the daily intake of Calories come from carbohydrates. Find the daily intake of Calories from carbohydrates that is appropriate if you want to limit your Calorie intake to 1600 Calories.

7. **Health** Playing singles tennis requires 450 Calories per hour. How many Calories do you burn in 30 days playing 45 min per day?

8. **Health** After playing golf for 3 h, Ruben had a banana split containing 550 Calories. Playing golf uses 320 Calories per hour.
   a. Without doing the calculations, did the banana split contain more or fewer Calories than Ruben burned off playing golf?
   b. Find the number of Calories Ruben gained or lost from these two activities.

9. **Health** Hiking requires approximately 315 Calories per hour. How many hours would you have to hike to burn off the Calories in a 375-Calorie sandwich, a 150-Calorie soda, and a 280-Calorie ice cream cone? Round to the nearest tenth.

10. **Health** Riding a bicycle requires 265 Calories per hour. How many hours would Shawna have to ride a bicycle to burn off the Calories in a 320-Calorie milkshake, a 310-Calorie cheeseburger, and a 150-Calorie apple? Round to the nearest tenth.

11. **Energy** An oven uses 500 W of energy. How many watt-hours of energy are used to cook a 5-kilogram roast for $2\frac{1}{2}$ h?

12. **Energy** A 21-inch color TV set is rated at 90 W. The TV is used an average of $3\frac{1}{2}$ h each day for a week. How many kilowatt-hours of energy are used during the week?

13. **Energy** A fax machine is rated at 9 W when the machine is in standby mode and at 36 W when in operation. How many kilowatt-hours of energy are used during a week in which the fax machine is in standby mode for 39 h and in operation for 6 h?

14. **Energy** A 120-watt CD player is on an average of 2 h a day. Find the cost of listening to the CD player for 2 weeks at a cost of 14.4¢ per kilowatt-hour. Round to the nearest cent.

15. **Energy** How much does it cost to run a 2200-watt air conditioner for 8 h at 12¢ per kilowatt-hour? Round to the nearest cent.

16. **Energy** A space heater is used for 3 h. The heater uses 1400 W per hour. Electricity costs 11.1¢ per kilowatt-hour. Find the cost of using the electric heater. Round to the nearest cent.

17. **Energy** A 60-watt Sylvania Long Life Soft White Bulb has a light output of 835 lumens and an average life of 1250 h. A 34-watt Sylvania Energy Saver Bulb has a light output of 400 lumens and an average life of 1500 h.
    a. Is the light output of the Energy Saver Bulb more or less than half that of the Long Life Soft White Bulb?
    b. If electricity costs 10.8¢ per kilowatt-hour, what is the difference in cost between using the Long Life Soft White Bulb for 150 h and using the Energy Saver Bulb for 150 h? Round to the nearest cent.

18. **Energy** A welder uses 6.5 kWh of energy each hour. Find the cost of using the welder for 6 h a day for 30 days. The cost is 12.4¢ per kilowatt-hour.

19. A package of four 40-watt light bulbs states that each light bulb lasts 2000 hours. Which expression represents the total number of kilowatt-hours of energy available from the package of light bulbs?
    (i) $\dfrac{40 \times 2000}{1000}$ (ii) $4 \times 40 \times 2000$ (iii) $4 \times 40 \times 2$

Applying the Concepts

20. **Human Energy** A moderately active woman who is 5 ft 5 in. tall and weighs 120 lb needs 2100 Calories per day to maintain her weight. Show that this is approximately the power of a 100-watt light bulb. Use the fact that 1 Calorie per day is equal to approximately 0.049 W.

21. Write an essay on how to improve the energy efficiency of a home.

SECTION

SECTION

9.5 Conversion Between the U.S. Customary and the Metric Systems of Measurement

---

OBJECTIVE A To convert U.S. Customary units to metric units

 Point of Interest

The definition of 1 inch has been changed as a consequence of the wide acceptance of the metric system. One inch is now exactly 25.4 mm.

More than 90% of the world's population uses the metric system of measurement. Therefore, converting U.S. Customary units to metric units is essential in trade and commerce—for example, in importing foreign goods and exporting domestic goods. Approximate equivalences between the two systems follow.

Units of Length	Units of Weight	Units of Capacity
1 in. = 2.54 cm	1 oz ≈ 28.35 g	1 L ≈ 1.06 ql
1 m ≈ 3.28 ft	1 lb ≈ 454 g	1 gal ≈ 3.79 L
1 m ≈ 1.09 yd	1 kg ≈ 2.2 lb	
1 mi ≈ 1.61 km		

 Point of Interest

In September of 1999, CNN reported that "NASA lost a $125-million Mars orbiter because a Lockheed Martin engineering team used English units of measurement while the agency's team used the . . . metric system for a key spacecraft operation." One team used miles, the other kilometers. This resulted in the orbiter coming within 60 km of the surface of Mars. This was 100 km closer than the planned approach. The propulsion system of the spacecraft overheated when it hit the Martian atmosphere and broke into pieces.

These equivalences can be used to form conversion rates to change one unit of measurement to another. For example, because $1\text{ mi} \approx 1.61\text{ km}$, the conversion rates $\dfrac{1\text{ mi}}{1.61\text{ km}}$ and $\dfrac{1.61\text{ km}}{1\text{ mi}}$ are both approximately equal to 1.

HOW TO 1 Convert 55 mi to kilometers.

$$55\text{ mi} \approx 55\text{ mi} \times \boxed{\dfrac{1.61\text{ km}}{1\text{ mi}}}$$

$$= \dfrac{55\text{ mi}}{1} \times \dfrac{1.61\text{ km}}{1\text{ mi}}$$

$$= \dfrac{88.55\text{ km}}{1}$$

$$55\text{ mi} \approx 88.55\text{ km}$$

• The conversion rate must contain km in the numerator and mi in the denominator.

EXAMPLE 1

Convert 45 mi/h to kilometers per hour.

Solution

$$\dfrac{45\text{ mi}}{h} \approx \dfrac{45\text{ mi}}{h} \times \dfrac{1.61\text{ km}}{1\text{ mi}}$$

$$= \dfrac{72.45\text{ km}}{1\text{ h}}$$

$$45\text{ mi/h} \approx 72.45\text{ km/h}$$

• The conversion rate is $\dfrac{1.61\text{ km}}{1\text{ mi}}$ with km in the numerator and mi in the denominator.

YOU TRY IT 1

Convert 60 ft/s to meters per second. Round to the nearest hundredth.

Your solution

EXAMPLE · 2	YOU TRY IT · 2

The price of gasoline is \$3.89/gal. Find the cost per liter. Round to the nearest tenth of a cent.

The price of milk is \$3.69/gal. Find the cost per liter. Round to the nearest cent.

Solution

$$\frac{\$3.89}{\text{gal}} \approx \frac{\$3.89}{\text{gal}} \times \frac{1 \text{ gal}}{3.79 \text{ L}} = \frac{\$3.89}{3.79 \text{ L}} \approx \frac{\$1.026}{1 \text{ L}}$$

\$3.89/gal ≈ \$1.026/L

Your solution

Solution on p. S23

OBJECTIVE B   To convert metric units to U.S. Customary units

Metric units are used in the United States. Cereal is sold by the gram, 35-mm film is available, and soda is sold by the liter. The same conversion rates used in Objective A are used for converting metric units to U.S. Customary units.

EXAMPLE · 3	YOU TRY IT · 3

Convert 200 m to feet.

Convert 45 cm to inches. Round to the nearest hundredth.

Solution

$$200 \text{ m} \approx 200 \text{ m} \times \frac{3.28 \text{ ft}}{1 \text{ m}} = \frac{656 \text{ ft}}{1}$$

200 m ≈ 656 ft

Your solution

EXAMPLE · 4	YOU TRY IT · 4

Convert 90 km/h to miles per hour. Round to the nearest hundredth.

Express 75 km/h in miles per hour. Round to the nearest hundredth.

Solution

$$\frac{90 \text{ km}}{\text{h}} \approx \frac{90 \text{ km}}{\text{h}} \times \frac{1 \text{ mi}}{1.61 \text{ km}} = \frac{90 \text{ mi}}{1.61 \text{ h}} \approx \frac{55.90 \text{ mi}}{1 \text{ h}}$$

90 km/h ≈ 55.90 mi/h

Your solution

EXAMPLE · 5	YOU TRY IT · 5

The price of gasoline is \$1.125/L. Find the cost per gallon. Round to the nearest cent.

The price of ice cream is \$1.75/L. Find the cost per gallon. Round to the nearest cent.

Solution

$$\frac{\$1.125}{1 \text{ L}} \approx \frac{\$1.125}{1 \text{ L}} \times \frac{3.79 \text{ L}}{1 \text{ gal}} \approx \frac{\$4.26}{1 \text{ gal}}$$

\$1.125/L ≈ \$4.26/gal

Your solution

Solutions on p. S23

9.5 EXERCISES

OBJECTIVE A   To convert U.S. Customary units to metric units

1. Write a product of conversion rates that you could use to convert cups to liters.

For Exercises 2 to 17, convert. Round to the nearest hundredth if necessary.

2. Convert 100 yd to meters.

3. Find the weight in kilograms of a 145-pound person.

4. Find the height in meters of a person 5 ft 8 in. tall.

5. Find the number of liters in 2 c of soda.

6. How many kilograms does a 15-pound turkey weigh?

7. Find the number of liters in 14.3 gal of gasoline.

8. Find the number of milliliters in 1 c.

9. The winning long jump at a track meet was 29 ft 2 in. Convert this distance to meters.

10. Express 65 mi/h in kilometers per hour.

11. Express 30 mi/h in kilometers per hour.

12. Fat-free hot dogs cost $3.49/lb. Find the cost per kilogram.

13. Seedless watermelon costs $.59/lb. Find the cost per kilogram.

14. The cost of gasoline is $3.87/gal. Find the cost per liter.

15. Deck stain costs $32.99/gal. Find the cost per liter.

16. **Earth Science**   The distance around Earth is 24,887 mi. Convert this distance to kilometers.

17. **Race Car Driving**   The distance driven by drivers in the Grand Prix of Monaco 2008 was 161.887 mi. Convert this distance to kilometers.

OBJECTIVE B   To convert metric units to U.S. Customary units

18. Write a product of conversion rates that you could use to convert kilometers to feet.

For Exercises 19 to 32, convert. Round to the nearest hundredth if necessary.

19. Convert 100 m to feet.

20. Find the weight in pounds of an 86-kilogram person.

21. Find the number of gallons in 6 L of antifreeze.

22. Find the height in inches of a person 1.85 m tall.

23. Find the distance of the 1500-meter race in feet.

24. Find the weight in ounces of 327 g of cereal.

25. How many gallons of water does a 24-liter aquarium hold?

26. Find the width in inches of 35-mm film.

27. Express 80 km/h in miles per hour.

28. Express 30 m/s in feet per second.

29. Gasoline costs $1.015/L. Find the cost per gallon.

30. A 5-kilogram ham costs $10/kg. Find the cost per pound.

31. A backpack tent weighs 2.1 kg. Find its weight in pounds.

32. A 2.5-kilogram bag of grass seed costs $10.99. Find the cost per pound.

33. **Health** Gary is planning a 5-day backpacking trip and decides to hike an average of 5 h each day. Hiking requires an extra 320 Calories per hour. How many pounds will Gary lose during the trip if he consumes an extra 900 Calories each day? (3500 Calories are equivalent to 1 lb.)

34. **Health** Swimming requires 550 Calories per hour. How many pounds could be lost by swimming $1\frac{1}{2}$ h each day for 5 days if no extra calories were consumed? (3500 Calories are equivalent to 1 lb.)

 35. **Motorcycle Speed Racing** Use the news clipping at the right. Convert Cliff Gullet's speed to miles per hour.

Applying the Concepts

36. Determine whether the statement is *true* or *false*.
 a. A liter is more than a gallon.
 b. A meter is less than a yard.
 c. 30 mi/h is less than 60 km/h.
 d. A kilogram is greater than a pound.
 e. An ounce is less than a gram.

In the News

Extreme Racer Killed in Record Attempt

On September 4, 2008, on Utah's Bonneville Salt Flats, Cliff Gullet crashed and was killed after losing control of his motorcycle. He was traveling at 385 km/h.

Source: www.yahoo.sports

FOCUS ON PROBLEM SOLVING

Working Backward   Sometimes the solution to a problem can be found by *working backward*. This problem-solving technique can be used to find a winning strategy for a game called Nim.

There are many variations of this game. For our game, there are two players, Player A and Player B, who alternately place 1, 2, or 3 matchsticks in a pile. The object of the game is to place the 32nd matchstick in the pile. Is there a strategy that Player A can use to guarantee winning the game?

Working backward, if there are 29, 30, or 31 matchsticks in the pile when it is A's turn to play, A can win by placing 3 matchsticks (29 + 3 = 32), 2 matchsticks (30 + 2 = 32), or 1 matchstick (31 + 1 = 32) on the pile. If there are to be 29, 30, or 31 matchsticks in the pile when it is A's turn, there must be 28 matchsticks in the pile when it is B's turn.

Working backward from 28, if there are to be 28 matches in the pile at B's turn, there must be 25, 26, or 27 at A's turn. Player A can then add 3 matchsticks, 2 matchsticks, or 1 matchstick to the pile to bring the number to 28. For there to be 25, 26, or 27 matchsticks in the pile at A's turn, there must be 24 matchsticks at B's turn.

Now working backward from 24, if there are to be 24 matches in the pile at B's turn, there must be 21, 22, or 23 at A's turn. Player A can then add 3 matchsticks, 2 matchsticks, or 1 matchstick to the pile to bring the number to 24. For there to be 21, 22, or 23 matchsticks in the pile at A's turn, there must be 20 matchsticks at B's turn.

So far, we have found that for Player A to win, there must be 28, 24, or 20 matchsticks in the pile when it is B's turn to play. Note that each time, the number is decreasing by 4. Continuing this pattern, Player A will win if there are 16, 12, 8, or 4 matchsticks in the pile when it is B's turn.

Player A can guarantee winning by making sure that the number of matchsticks in the pile is a multiple of 4. To ensure this, Player A allows Player B to go first and then adds exactly enough matchsticks to the pile to bring the total to a multiple of 4.

For example, suppose B places 3 matchsticks in the pile; then A places 1 matchstick (3 + 1 = 4). Now B places 2 matchsticks in the pile. The total is now 6 matchsticks. Player A then places 2 matchsticks in the pile to bring the total to 8, a multiple of 4. If play continues in this way, Player A will win.

Here are some variations of Nim. See whether you can develop a winning strategy for Player A.

1. Suppose the goal is to place the last matchstick in a pile of 30 matches.

2. Suppose the players make two piles of matchsticks, with the final number of matchsticks in each pile to be 20.

3. In this variation of Nim, there are 40 matchsticks in a pile. Each player alternately removes 1, 2, or 3 matches from the pile. The player who removes the last match wins.

PROJECTS AND GROUP ACTIVITIES

Name That Metric Unit

What unit in the metric system would be used to measure each of the following? If you are working in a group, be sure that each member agrees on the unit to be used and understands why that unit is used before going on to the next item.

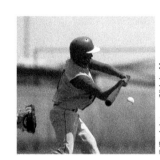

1. The distance from Los Angeles to New York

2. The weight of a truck

3. A person's waist

4. The amount of coffee in a mug

5. The weight of a thumbtack

6. The amount of water in a swimming pool

7. The distance a baseball player hits a baseball

8. A person's hat size

9. The amount of protein needed daily

10. A person's weight

11. The amount of maple syrup served with pancakes

12. The amount of water in a water cooler

13. The amount of medication in an aspirin

14. The distance to the grocery store

15. The width of a hair

16. A person's height

17. The weight of a lawn mower

18. The amount of water a family uses monthly

19. The contents of a bottle of salad dressing

20. The weight of newspapers collected at a recycling center

Metric Measurements for Computers

Other prefixes in the metric system are becoming more commonplace as a result of technological advances in the computer industry and as we learn more and more about objects in our universe that are great distances away.

tera-	= 1,000,000,000,000
giga-	= 1,000,000,000
mega-	= 1,000,000
micro-	= 0.000001
nano-	= 0.000000001
pico-	= 0.000000000001

1. Complete the table.

Metric System Prefix	Symbol	Magnitude	Means Multiply the Basic Unit By:
tera-	T	10^{12}	1,000,000,000,000
giga-	G	___	1,000,000,000
mega-	M	10^6	_____
kilo-	___	___	1000
hecto-	h	___	100
deca-	da	10^1	_____
deci-	d	$\dfrac{1}{10}$	_____
centi-	___	$\dfrac{1}{10^2}$	_____
milli-	___	___	0.001
micro-	μ	$\dfrac{1}{10^6}$	_____
nano-	n	$\dfrac{1}{10^9}$	_____
pico-	p	___	0.000000000001

2. How can the Magnitude column in the table above be used to determine how many places to move the decimal point when converting to the basic unit in the metric system?

A **bit** is the smallest unit of code that computers can read; it is a binary digit, either a 0 or a 1. A bit is abbreviated b. Usually bits are grouped into **bytes** of 8 bits. Each byte stands for a letter, a number, or any other symbol we might use in communicating information. For example, the letter W can be represented 01010111. A byte is abbreviated B.

The amount of memory in a computer hard drive is measured in terabytes (TB), gigabytes (GB), and megabytes (MB). Often a gigabyte is referred to as a gig, and a megabyte is referred to as a meg. Using the definitions of the prefixes given above, a kilobyte is 1000 bytes, a megabyte is 1,000,000 bytes, and a gigabyte is 1,000,000,000 bytes. However, these are not exact equivalences. Bytes are actually computed in powers of 2. Therefore, kilobytes, megabytes, gigabytes, and terabytes are powers of 2. The exact equivalences are shown below.

Apple MacBook 2.66GHz 15" (M9009LL/A) Memory

SALE
Maximum Memory
4 GB

> 1 byte = 2^3 bits
>
> 1 kilobyte = 2^{10} bytes = 1024 bytes
>
> 1 megabyte = 2^{20} bytes = 1,048,576 bytes
>
> 1 gigabyte = 2^{30} bytes = 1,073,741,824 bytes
>
> 1 terabyte = 2^{40} bytes = 1,099,511,627,776 bytes

3. Find an advertisement for a computer system. What is the computer's storage capacity? Convert the capacity to bytes. Use the exact equivalences given above.

CHAPTER 9

SUMMARY

KEY WORDS	EXAMPLES
The *metric system* of measurement is an internationally standardized system of measurement. It is based on the decimal system. The basic unit of length in the metric system is the *meter*. [9.1A, p. 372]	1 km = 1000 m 1 kg = 1000 g 1 kl = 1000 L
The basic unit of mass is the *gram*. [9.2A, p. 376]	
The basic unit of capacity is the *liter*. [9.3A, p. 380]	1 m = 100 cm
Heat is commonly measured in units called *Calories*. [9.4A, p. 384]	1 m = 1000 mm 1 g = 1000 mg 1 L = 1000 ml
The *watt-hour* is used in the metric system for measuring electrical energy. [9.4A, p. 384]	
In the metric system, prefixes to the basic unit denote the magnitude of each unit. [9.1A, p. 372]	

kilo- = 1000   deci- = 0.1
hecto- = 100   centi- = 0.01
deca- = 10   milli- = 0.001

ESSENTIAL RULES AND PROCEDURES	EXAMPLES
Converting between units in the metric system involves moving the decimal point to the right or to the left. Listing the units in order from largest to smallest will indicate how many places to move the decimal point and in which direction. [9.1A, 9.2A, 9.3A, pp. 372, 376, 380] **1.** When converting from a larger unit to a smaller unit, move the decimal point to the *right*. **2.** When converting from a smaller unit to a larger unit, move the decimal point to the *left*.	Convert 3.7 kg to grams. $$3.7 \text{ kg} = 3700 \text{ g}$$ Convert 2387 m to kilometers. $$2387 \text{ m} = 2.387 \text{ km}$$ Convert 9.5 L to milliliters. $$9.5 \text{ L} = 9500 \text{ ml}$$

Approximate equivalences between units in the U.S. Customary and the metric systems of measurement are used to form conversion rates to change one unit of measurement to another. [9.5A, p. 388]

Convert 20 mi/h to kilometers per hour.

$$\frac{20 \text{ mi}}{\text{h}} \approx \frac{20 \text{ mi}}{\text{h}} \times \frac{1.61 \text{ km}}{1 \text{ mi}}$$

$$\approx \frac{32.2 \text{ km}}{1 \text{ h}}$$

$$\approx 32.2 \text{ km/h}$$

Units of Length
1 in. = 2.54 cm
1 m ≈ 3.28 ft
1 m ≈ 1.09 yd
1 mi ≈ 1.61 km

Convert 1000 m to yards.

$$1000 \text{ m} \approx 1000 \text{ m} \times \frac{1.09 \text{ yd}}{1 \text{ m}}$$

$$\approx 1090 \text{ yd}$$

Units of Weight
1 oz ≈ 28.35 g
1 lb ≈ 454 g
1 kg ≈ 2.2 lb

Units of Capacity
1 L ≈ 1.06 qt
1 gal ≈ 3.79 L

CHAPTER 9

CONCEPT REVIEW

Test your knowledge of the concepts presented in this chapter. Answer each question. Then check your answers against the ones provided in the Answer Section.

1. To convert from millimeters to meters, the decimal point is moved in what direction and how many places?

2. How can you convert 51 m 2 cm to meters?

3. On the surface of Earth, how do mass and weight differ?

4. How can you convert 4 kg 7 g to kilograms?

5. How is a liter related to the capacity of a box or cube that is 10 cm long on each side?

6. How can you convert 4 L 27 ml to liters?

7. What operations are needed to find the cost per week to run a microwave oven rated at 650 W that is used for 30 min per day, at 8.8¢ per kilowatt-hour?

8. If 30% of a person's daily intake of Calories is from fat, how do you find the Calories from fat for a daily intake of 1800 Calories?

9. What conversion rate is needed to convert 100 ft to meters?

10. What conversion rate is needed to convert the price of gas from $3.19/gal to a cost per liter?

CHAPTER 9

REVIEW EXERCISES

1. Convert 1.25 km to meters.

2. Convert 0.450 g to milligrams.

3. Convert 0.0056 L to milliliters.

4. Convert the 1000-meter run to yards. (1 m ≈ 1.09 yd)

5. Convert 79 mm to centimeters.

6. Convert 5 m 34 cm to meters.

7. Convert 990 g to kilograms.

8. Convert 2550 ml to liters.

9. Convert 4870 m to kilometers.

10. Convert 0.37 cm to millimeters.

11. Convert 6 g 829 mg to grams.

12. Convert 1.2 L to cubic centimeters.

13. Convert 4.050 kg to grams.

14. Convert 8.7 m to centimeters.

15. Convert 192 ml to cubic centimeters.

16. Convert 356 mg to grams.

17. Convert 372 cm to meters.

18. Convert 8.3 kl to liters.

19. Convert 2 L 89 ml to liters.

20. Convert 5410 cm^3 to liters.

21. Convert 3792 L to kiloliters.

22. Convert 468 cm^3 to milliliters.

23. Measurements   Three pieces of wire are cut from a 50-meter roll. The three pieces measure 240 cm, 560 cm, and 480 cm. How much wire is left on the roll after the three pieces are cut?

24. Consumerism   Find the total cost of three packages of chicken weighing 790 g, 830 g, and 655 g if the cost is $5.59 per kilogram.

25. Consumerism   Cheese costs $4.40 per pound. Find the cost per kilogram. (1 kg ≈ 2.2 lb)

26. Measurements   One hundred twenty-five guests are expected to attend a reception. Assuming that each person drinks 400 ml of coffee, how many liters of coffee should be prepared?

27. Nutrition   A large egg contains approximately 90 Calories. How many Calories can you eliminate from your diet in a 30-day month by eliminating one large egg per day from your usual breakfast?

28. Energy   A TV uses 240 W of energy. The set is on an average of 5 h a day in a 30-day month. At a cost of 9.5¢ per kilowatt-hour, how much does it cost to run the set for 30 days?

29. Measurements   A backpack weighs 1.90 kg. Find the weight in pounds. Round to the nearest hundredth. (1 kg ≈ 2.2 lb)

30. Health   Cycling burns approximately 400 Calories per hour. How many hours of cycling are necessary to lose 1 lb? (3500 Calories are equivalent to 1 lb.)

© Randy M. Ury/Corbis

31. Business   Six liters of liquid soap were bought for $11.40 per liter. The soap was repackaged in 150-milliliter plastic containers. The cost of each container was $.26. Each container of soap sold for $3.29 per bottle. Find the profit on the 6 L of liquid soap.

32. Energy   A color TV is rated at 80 W. The TV is used an average of 2 h each day for a week. How many kilowatt-hours of energy are used during the week?

33. Agriculture   How many kilograms of fertilizer are necessary to fertilize 500 trees in an orchard if 250 g of fertilizer are used for each tree?

CHAPTER 9

TEST

1. Convert 2.96 km to meters.

2. Convert 0.378 g to milligrams.

3. Convert 0.046 L to milliliters.

4. Convert 919 cm^3 to milliliters.

5. Convert 42.6 mm to centimeters.

6. Convert 7 m 96 cm to meters.

7. Convert 847 g to kilograms.

8. Convert 3920 ml to liters.

9. Convert 5885 m to kilometers.

10. Convert 1.5 cm to millimeters.

11. Convert 3 g 89 mg to grams.

12. Convert 1.6 L to cubic centimeters.

13. Convert 3.29 kg to grams.

14. Convert 4.2 m to centimeters.

15. Convert 96 ml to cubic centimeters.

16. Convert 1375 mg to grams.

17. Convert 402 cm to meters.

18. Convert 8.92 kl to liters.

19. **Health**   Sedentary people need 15 Calories per pound of body weight to maintain their weight. How many Calories should a 140-pound, sedentary person consume per day to maintain that weight?

20. **Energy**   A color television is rated at 100 W. The television is used an average of $4\frac{1}{2}$ h each day for a week. Find the number of kilowatt-hours of energy used during the week to operate the television.

21. **Carpentry**   A carpenter needs 30 rafters, each 380 cm long. Find the total length of the rafters in meters.

22. **Measurements**   A tile measuring 20 cm × 20 cm weighs 250 g. Find the weight, in kilograms, of a box of 144 tiles.

23. **Medicine**   A community health clinic is giving flu shots for the coming flu season. Each flu shot contains 2 cm³ of vaccine. How many liters of vaccine are needed to inoculate 2600 people?

24. **Measurements**   Convert 35 mi/h to kilometers per hour. Round to the nearest tenth. (1 mi ≈ 1.61 km)

25. **Metal Works**   Twenty-five rivets are used to fasten two steel plates together. The plates are 4.20 m long, and the rivets are equally spaced, with a rivet at each end. Find the distance, in centimeters, between the rivets.

26. **Agriculture**   Two hundred grams of fertilizer are used for each tree in an orchard containing 1200 trees. At $2.75 per kilogram of fertilizer, how much does it cost to fertilize the orchard?

27. **Energy**   An air conditioner rated at 1600 W is operated an average of 4 h per day. Electrical energy costs 12.5¢ per kilowatt-hour. How much does it cost to operate the air conditioner for 30 days?

28. **Chemistry**   A laboratory assistant is in charge of ordering acid for three chemistry classes of 40 students each. Each student requires 90 ml of acid. How many liters of acid should be ordered? (The assistant must order by the whole liter.)

29. **Sports**   Three ski jumping events are held at the Olympic Games: the individual normal hill, the individual large hill, and the team large hill. The normal hill measures 90 m. The large hill measures 120 m. Convert the measure of the large hill to feet. Round to the nearest tenth. (1 m ≈ 3.28 ft)

GERO BRELOER/dpa/Landov

30. **Sports**   In the archery competition at the Olympic Games, the center ring, or bull's eye, of the target is approximately 4.8 in. in diameter. Convert 4.8 in. to centimeters. Round to the nearest tenth. (1 in. = 2.54 cm)

CUMULATIVE REVIEW EXERCISES

1. Simplify: $12 - 8 \div (6 - 4)^2 \cdot 3$

2. Find the total of $5\frac{3}{4}$, $1\frac{5}{6}$, and $4\frac{7}{9}$.

3. Subtract: $4\frac{2}{9} - 3\frac{5}{12}$

4. Divide: $5\frac{3}{8} \div 1\frac{3}{4}$

5. Simplify: $\left(\frac{2}{3}\right)^4 \cdot \left(\frac{9}{4}\right)^2$

6. Subtract: $12.0072 - 9.937$

7. Solve the proportion $\frac{5}{8} = \frac{n}{50}$. Round to the nearest tenth.

8. Write $1\frac{3}{4}$ as a percent.

9. 6.09 is 4.2% of what number?

10. Convert 18 pt to gallons.

11. Convert 875 cm to meters.

12. Convert 3420 m to kilometers.

13. Convert 5.05 kg to grams.

14. Convert 3 g 672 mg to grams.

15. Convert 6 L to milliliters.

16. Convert 2.4 kl to liters.

17. **Finances**   The Guerrero family has a monthly income of $5244 per month. The family spends one-fourth of its monthly income on rent. How much money is left after the rent is paid?

18. **Taxes**   The state income tax on a business is $620 plus 0.08 times the profit the business makes. The business made a profit of $82,340.00 last year. Find the amount of state income tax the business paid.

19. **Real Estate**   The property tax on a $245,000 home is $4900. At the same rate, what is the property tax on a home worth $275,000?

20. **Consumerism**   A car dealer offers new-car buyers a 12% rebate on some models. What rebate would a new-car buyer receive on a car that cost $23,500?

21. **Investments**   Rob Akullian received a dividend of $533 on an investment of $8200. What percent of the investment is the dividend?

22. **Education**   You received grades of 78, 92, 45, 80, and 85 on five English exams. Find your mean grade.

23. **Compensation**   Karla Perella, a ski instructor, receives a salary of $22,500. Her salary will increase by 12% next year. What will her salary be next year?

24. **Discount**   A sporting goods store has fishing rods that are regularly priced at $180 on sale for $140.40. What is the discount rate?

25. **Masonry**   Forty-eight bricks, each 9 in. long, are laid end-to-end to make the base for a wall. Find the length of the wall in feet.

26. **Fuel Efficiency**   An M1A1 Abrams main battle tank travels 11.2 mi on 20 gal of fuel. How many miles does the tank travel on 1 gal of fuel?

27. **Business**   A garage mechanic bought oil in a 40-gallon container. The mechanic bought the oil for $24.40 per gallon and sold the oil for $9.95 per quart. Find the profit on the 40-gallon container of oil.

28. **Measurements**   A school swimming pool uses 1200 ml of chlorine each school day. How many liters of chlorine are used for 20 school days during the month?

29. **Energy**   A 1200-watt hair dryer is used an average of 30 min a day. At a cost of 13.5¢ per kilowatt-hour, how much does it cost to operate the hair dryer for 30 days?

30. **Measurements**   Convert 60 mi/h to kilometers per hour. Round to the nearest tenth. (1.61 km ≈ 1 mi)

Rational Numbers

OBJECTIVES

SECTION 10.1

A To identify the order relation between two integers

B To evaluate expressions that contain the absolute value symbol

SECTION 10.2

A To add integers

B To subtract integers

C To solve application problems

SECTION 10.3

A To multiply integers

B To divide integers

C To solve application problems

SECTION 10.4

A To add or subtract rational numbers

B To multiply or divide rational numbers

C To solve application problems

SECTION 10.5

A To write a number in scientific notation

B To use the Order of Operations Agreement to simplify expressions

ARE YOU READY?

Take the Chapter 10 Prep Test to find out if you are ready to learn to:

- Order integers
- Evaluate expressions that contain the absolute value symbol
- Add, subtract, multiply, and divide integers and rational numbers
- Write a number in scientific notation
- Simplify numerical expressions

PREP TEST

Do these exercises to prepare for Chapter 10.

1. Place the correct symbol, $<$ or $>$, between the two numbers.
   54   45

2. What is the distance from 4 to 8 on the number line?

For Exercises 3 to 14, add, subtract, multiply, or divide.

3. $7654 + 8193$

4. $6097 - 2318$

5. 472×56

6. $\dfrac{144}{24}$

7. $\dfrac{2}{3} + \dfrac{3}{5}$

8. $\dfrac{3}{4} - \dfrac{5}{16}$

9. $0.75 + 3.9 + 6.408$

10. $5.4 - 1.619$

11. $\dfrac{3}{4} \times \dfrac{8}{15}$

12. $\dfrac{5}{12} \div \dfrac{3}{4}$

13. 23.5×0.4

14. $0.96 \div 2.4$

15. Simplify: $(8 - 6)^2 + 12 \div 4 \cdot 3^2$

SECTION

10.1 Introduction to Integers

OBJECTIVE A To identify the order relation between two integers

Thus far in the text, we have encountered only zero and the numbers greater than zero. The numbers greater than zero are called **positive numbers.** However, the phrases "12 degrees below zero," "$25 in debt," and "15 feet below sea level" refer to numbers less than zero. These numbers are called **negative numbers.**

The **integers** are . . . , -4, -3, -2, -1, 0, 1, 2, 3, 4, ...

> **Point of Interest**
> Among the slang words for zero are *zilch*, *zip*, and *goose egg*. The word *love* for zero in scoring a tennis game comes from the French word *l'oeuf*, which means "the egg."

Each integer can be shown on a number line. The integers to the left of zero on the number line are called **negative integers** and are represented by a negative sign ($-$) placed in front of the number. The integers to the right of zero are called **positive integers.** The positive integers are also called **natural numbers.** Zero is neither a positive nor a negative integer.

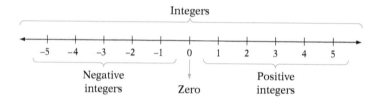

A number line can be used to visualize the order relation between two integers. A number that appears to the left of a given number is less than ($<$) the given number. A number that appears to the right of a given number is greater than ($>$) the given number.

2 is greater than negative 4.
$2 > -4$

Negative 5 is less than negative 3.
$-5 < -3$

EXAMPLE 1

The lowest recorded temperature in Antarctica is 129° below zero. Represent this temperature as an integer.

Solution $-129°$

YOU TRY IT 1

The surface of the Salton Sea is 232 ft below sea level. Represent this depth as an integer.

Your solution

EXAMPLE 2

Graph -2 on the number line.

Solution

YOU TRY IT 2

Graph -4 on the number line.

Your solution

EXAMPLE 3

Place the correct symbol, $<$ or $>$, between the numbers -5 and -7.

Solution $-5 > -7$ • -5 is to the right of -7 on the number line.

YOU TRY IT 3

Place the correct symbol, $<$ or $>$, between the numbers -12 and -8.

Your solution

Solutions on p. S23

OBJECTIVE B   To evaluate expressions that contain the absolute value symbol

Two numbers that are the same distance from zero on the number line but on opposite sides of zero are called **opposites.**

−4 is the opposite of 4
and
4 is the opposite of −4.

Note that a negative sign can be read as "the opposite of."

−(4) = −4   The opposite of positive 4 is negative 4.
−(−4) = 4   The opposite of negative 4 is positive 4.

The **absolute value of a number** is the distance between zero and the number on the number line. Therefore, the absolute value of a number is a positive number or zero. The symbol for absolute value is | |.

The distance from 0 to 4 is 4.
Thus |4| = 4 (the absolute value of 4 is 4).

The distance from 0 to −4 is 4.
Thus |−4| = 4 (the absolute value of −4 is 4).

The absolute value of a positive number is the number itself. The absolute value of a negative number is the opposite of the negative number. The absolute value of zero is zero.

EXAMPLE · 4

Find the absolute value of 2 and −3.

Solution
|2| = 2
|−3| = 3

YOU TRY IT · 4

Find the absolute value of −7 and 21.

Your solution

EXAMPLE · 5

Evaluate |−34| and |0|.

Solution
|−34| = 34
|0| = 0

YOU TRY IT · 5

Evaluate |2| and |−9|.

Your solution

EXAMPLE · 6

Evaluate −|−4|.

Solution
−|−4| = −4

The minus sign *in front of* the absolute value sign is not affected by the absolute value sign.

YOU TRY IT · 6

Evaluate −|−12|.

Your solution

Solutions on p. S23

10.1 EXERCISES

OBJECTIVE A To identify the order relation between two integers

For Exercises 1 to 4, represent the quantity as an integer.

1. A lake 120 ft below sea level

2. A temperature that is 15° below zero

3. A loss of 324 dollars

4. A share of stock up 2 dollars

For Exercises 5 to 8, graph the numbers on the number line.

5. 3 and −3

6. −2 and 0

7. −4 and 1

8. 4 and −1

For Exercises 9 to 14, state which number on the number line is in the location given.

9. 3 units to the right of −2

10. 5 units to the right of −3

11. 4 units to the left of 3

12. 2 units to the left of −1

13. 6 units to the right of −3

14. 4 units to the right of −4

For Exercises 15 to 18, use the following number line.

A B C D E F G H I

15. **a.** If F is 1 and G is 2, what number is A?

b. If F is 1 and G is 2, what number is C?

16. **a.** If G is 1 and H is 2, what number is B?

b. If G is 1 and H is 2, what number is D?

17. **a.** If H is 0 and I is 1, what number is A?

b. If H is 0 and I is 1, what number is D?

18. **a.** If G is 2 and I is 4, what number is B?

b. If G is 2 and I is 4, what number is E?

For Exercises 19 to 42, place the correct symbol, < or >, between the two numbers.

19. −2   −5   **20.** −6   −1   **21.** −16   1   **22.** −2   13

23. 3   −7   **24.** 5   −6   **25.** −11   −8   **26.** −4   −10

27. 35   28   **28.** 42   19   **29.** −42   27   **30.** −36   49

31. 21   −34   **32.** 53   −46   **33.** −27   −39   **34.** −51   −20

35. −87   63   **36.** −75   92   **37.** 86   −79   **38.** 95   −71

39. −62   −84   **40.** −91   −70   **41.** −131   101   **42.** 127   −150

For Exercises 43 to 51, write the given numbers in order from smallest to largest.

43. 3, −7, 0, −2   **44.** −4, 8, 6, −1   **45.** −3, 1, −5, 4

46. −6, 2, −8, 7   **47.** 9, −4, 5, 0   **48.** 6, −9, −12, 8

49. −10, 4, 12, −5, −7   **50.** 11, −8, −1, 7, −6   **51.** 10, −11, −2, 5, −7

 For Exercises 52 to 55, determine whether the statement is *always true, never true,* or *sometimes true.*

52. A number that is to the right of −6 on the number line is a negative number.

53. A number that is to the left of −2 on the number line is a negative number.

54. A number that is to the right of 7 on the number line is a negative number.

55. A number that is to the left of 4 on the number line is a negative number.

OBJECTIVE B   To evaluate expressions that contain the absolute value symbol

For Exercises 56 to 65, find the opposite number.

56. 4   **57.** 16   **58.** −2   **59.** −3   **60.** 22

61. 45　　　　**62.** −31　　　　**63.** −59　　　　**64.** 70　　　　**65.** −88

For Exercises 66 to 73, find the absolute value of the number.

66. 4　　　　**67.** −4　　　　**68.** −7　　　　**69.** 9

70. −1　　　　**71.** −11　　　　**72.** 10　　　　**73.** −12

For Exercises 74 to 103, evaluate.

74. $|2|$　　**75.** $|-2|$　　**76.** $|-6|$　　**77.** $|6|$　　**78.** $|8|$

79. $|5|$　　**80.** $|-9|$　　**81.** $|-1|$　　**82.** $-|-1|$　　**83.** $-|-5|$

84. $-|0|$　　**85.** $|16|$　　**86.** $|19|$　　**87.** $|-12|$　　**88.** $|-22|$

89. $-|29|$　　**90.** $-|20|$　　**91.** $-|-14|$　　**92.** $-|-18|$　　**93.** $|-15|$

94. $|-23|$　　**95.** $-|33|$　　**96.** $-|27|$　　**97.** $|32|$　　**98.** $|25|$

99. $-|-42|$　　**100.** $|-74|$　　**101.** $|-61|$　　**102.** $-|88|$　　**103.** $-|52|$

For Exercises 104 to 111, place the correct symbol, <, =, or >, between the two numbers.

104. $|7|$　$|-9|$　　**105.** $|-12|$　$|8|$　　**106.** $|-5|$　$|-2|$　　**107.** $|6|$　$|13|$

108. $|-8|$　$|3|$　　**109.** $|-1|$　$|-17|$　　**110.** $|-14|$　$|14|$　　**111.** $|17|$　$|-17|$

For Exercises 112 to 117, write the given numbers in order from smallest to largest.

112. $|-8|, -3, |2|, -|-5|$　　　　**113.** $-|6|, -4, |-7|, -9$　　　　**114.** $-1, |-6|, |0|, -|3|$

115. $-|-7|, -9, 5, |4|$ **116.** $-|2|, -8, 6, |1|$ **117.** $-3, -|-8|, |5|, -|10|$

For Exercises 118 to 121, determine whether the statement is true for *positive integers, negative integers,* or *all integers.*

118. The absolute value of an integer is the opposite of the integer.

119. The opposite of an integer is less than the integer.

120. The opposite of an integer is negative.

121. The absolute value of an integer is greater than the integer.

Applying the Concepts

122. **Meteorology** The graph at the right shows the lowest recorded temperatures, in degrees Fahrenheit, for selected states in the United States. Which state has the lowest recorded temperature?

Lowest Recorded Temperatures

Sources: National Climatic Data Center; NESDIS; NOAA; U.S. Dept. of Commerce

123. **a.** Name two numbers that are 5 units from 3 on the number line.
   b. Name two numbers that are 3 units from -1 on the number line.

124. **a.** Find a number that is halfway between -7 and -5.
   b. Find a number that is halfway between -10 and -6.
   c. Find a number that is one-third of the way between -12 and -3.

125. **Rocketry** Which is closer to blastoff, -12 min and counting or -17 min and counting?

126. **Investments** In the stock market, the net change in the price of a share of stock is recorded as a positive or a negative number. If the price rises, the net change is positive. If the price falls, the net change is negative. If the net change for a share of Stock A is -2 and the net change for a share of Stock B is -1, which stock showed the least net change?

127. **Business** Some businesses show a profit as a positive number and a loss as a negative number. During the first quarter of this year, the loss experienced by a company was recorded as $-12,575$. During the second quarter of this year, the loss experienced by the company was $-11,350$. During which quarter was the loss greater?

128. **a.** Find the values of a for which $|a| = 7$.
   b. Find the values of y for which $|y| = 11$.

SECTION

10.2 Addition and Subtraction of Integers

OBJECTIVE A   **To add integers**

An integer can be graphed as a dot on a number line, as shown in the last section. An integer also can be represented anywhere along a number line by an arrow. A positive number is represented by an arrow pointing to the right. A negative number is represented by an arrow pointing to the left. The absolute value of the number is represented by the length of the arrow. The integers 5 and −4 are shown on the number lines below.

The sum of two integers can be shown on a number line. To add two integers, use arrows to represent the addends, with the first arrow starting at zero. The sum is the number directly below the tip of the arrow that represents the second addend.

$4 + 2 = 6$

$-4 + (-2) = -6$

$-4 + 2 = -2$

$4 + (-2) = 2$

The sums of the integers shown above can be categorized by the signs of the addends.

Here the addends have the same sign:

$4 + 2$ *positive* 4 plus *positive* 2
$-4 + (-2)$ *negative* 4 plus *negative* 2

Here the addends have different signs:

$-4 + 2$ *negative* 4 plus *positive* 2
$4 + (-2)$ *positive* 4 plus *negative* 2

The rule for adding two integers depends on whether the signs of the addends are the same or different.

Rule for Adding Two Numbers

To add numbers with the same sign, add the absolute values of the numbers. Then attach the sign of the addends.

To add numbers with different signs, find the difference between the absolute values of the numbers. Then attach the sign of the addend with the greater absolute value.

Point of Interest

Although mathematical symbols are fairly standard in every country, that has not always been true. Italian mathematicians in the 15th century used a "p" to indicate plus. The "p" was from the Italian word *piu*, meaning "more" or "plus."

HOW TO 1   Add: $(-4) + (-9)$

$|-4| = 4, |-9| = 9$
$4 + 9 = 13$
- Because the signs of the addends are the same, add the absolute values of the numbers.

$(-4) + (-9) = -13$
- Then attach the sign of the addends.

HOW TO 2   Add: $6 + (-13)$

$|6| = 6, |-13| = 13$
$13 - 6 = 7$
- Because the signs of the addends are different, subtract the smaller absolute value from the larger absolute value.

$6 + (-13) = -7$
- Then attach the sign of the number with the larger absolute value. Because $|-13| > |6|$, attach the negative sign.

HOW TO 3   Add: $162 + (-247)$

$162 + (-247) = -85$
- Because the signs are different, find the difference between the absolute values of the numbers and attach the sign of the number with the greater absolute value.

Integrating Technology

To add $-14 + (-47)$ on your calculator, enter the following:

14 **+/−** **+** 47 **+/−** **=**

HOW TO 4   Find the sum of -14 and -47.

$-14 + (-47) = -61$
- Because the signs are the same, add the absolute values of the numbers and attach the sign of the addends.

When adding more than two integers, start from the left and add the first two numbers. Then add the sum to the third number. Continue this process until all the numbers have been added.

HOW TO 5   Add: $(-4) + (-6) + (-8) + 9$

$(-4) + (-6) + (-8) + 9 = (-10) + (-8) + 9$
$= (-18) + 9$
$= -9$
- Add the first two numbers.
- Add the sum to the next number.
- Continue adding until all numbers have been added.

EXAMPLE 1

What is -162 added to 98?

Solution

$98 + (-162) = -64$
- The signs of the addends are different.

YOU TRY IT 1

Add: $-154 + (-37)$

Your solution

Solution on p. S23

EXAMPLE · 2

Add: $-2 + (-7) + 4 + (-6)$

Solution

$$-2 + (-7) + 4 + (-6) = -9 + 4 + (-6)$$
$$= -5 + (-6)$$
$$= -11$$

YOU TRY IT · 2

Add: $-5 + (-2) + 9 + (-3)$

Your solution

Solution on p. S23

OBJECTIVE B   To subtract integers

 Tips for Success

Be sure to do all you need to do in order to be successful at adding and subtracting integers: Read through the introductory material, work through the examples indicated by the HOW TO feature, study the paired Examples, do the You Try Its and check your solutions against those in the back of the book, and do the exercises on pages 414 to 418. See *AIM for Success* at the front of the book.

Before the rules for subtracting two integers are explained, look at the translation into words of an expression that is the difference of two integers:

$9 - 3$	positive 9 minus positive 3
$(-9) - 3$	negative 9 minus positive 3
$9 - (-3)$	positive 9 minus negative 3
$(-9) - (-3)$	negative 9 minus negative 3

Note that the sign $-$ is used in two different ways. One way is as a negative sign, as in (-9), *negative* 9. The second way is to indicate the operation of subtraction, as in $9 - 3$, 9 *minus* 3.

Look at the next four subtraction expressions and decide whether the second number in each expression is a positive number or a negative number.

1. $(-10) - 8$ **2.** $(-10) - (-8)$ **3.** $10 - (-8)$ **4.** $10 - 8$

In expressions 1 and 4, the second number is a positive 8. In expressions 2 and 3, the second number is a negative 8.

> **Rule for Subtracting Two Numbers**
>
> To subtract two numbers, add the opposite of the second number to the first number.

This rule states that to subtract two integers, we rewrite the subtraction expression as the sum of the first number and the opposite of the second number.

Here are some examples:

First number	$-$	second number	$=$	first number	$+$	the opposite of the second number	
8	$-$	15	$=$	8	$+$	(-15)	$= -7$
8	$-$	(-15)	$=$	8	$+$	15	$= 23$
(-8)	$-$	15	$=$	(-8)	$+$	(-15)	$= -23$
(-8)	$-$	(-15)	$=$	(-8)	$+$	15	$= 7$

 Integrating Technology

The **+/−** key on your calculator is used to find the opposite of a number. The **−** key is used to perform the operation of subtraction.

HOW TO · 6   Subtract: $(-15) - 75$

$$(-15) - 75 = (-15) + (-75)$$
$$= -90$$

• To subtract, add the opposite of the second number to the first number.

HOW TO 7 Subtract: $27 - (-32)$

$$27 - (-32) = 27 + 32$$
$$= 59$$

• To subtract, add the opposite of the second number to the first number.

When subtraction occurs several times in an expression, rewrite each subtraction as addition of the opposite and then add.

HOW TO 8 Subtract: $-13 - 5 - (-8)$

$$-13 - 5 - (-8) = -13 + (-5) + 8$$
$$= -18 + 8$$
$$= -10$$

• Rewrite each subtraction as the addition of the opposite and then add.

EXAMPLE 3

Find 8 less than -12.

Solution

$$-12 - 8 = -12 + (-8)$$
$$= -20$$

• Rewrite "−" as "+"; the opposite of 8 is −8.

YOU TRY IT 3

Find -8 less 14.

Your solution

EXAMPLE 4

Subtract: $6 - (-20)$

Solution

$$6 - (-20) = 6 + 20$$
$$= 26$$

• Rewrite "−" as "+"; the opposite of −20 is 20.

YOU TRY IT 4

Subtract: $3 - (-15)$

Your solution

EXAMPLE 5

Subtract: $-8 - 30 - (-12) - 7$

Solution

$$-8 - 30 - (-12) - 7$$
$$= -8 + (-30) + 12 + (-7)$$
$$= -38 + 12 + (-7)$$
$$= -26 + (-7)$$
$$= -33$$

YOU TRY IT 5

Subtract: $4 - (-3) - 12 - (-7) - 20$

Your solution

Solutions on p. S23

OBJECTIVE C To solve application problems

EXAMPLE 6

Find the temperature after an increase of 9°C from −6°C.

Strategy

To find the temperature, add the increase (9°C) to the previous temperature (−6°C).

Solution

$$-6 + 9 = 3$$

The temperature is 3°C.

YOU TRY IT 6

Find the temperature after an increase of 12°C from −10°C.

Your strategy

Your solution

Solution on p. S23

10.2 EXERCISES

OBJECTIVE A   **To add integers**

For Exercises 1 and 2, name the negative integers in the list of numbers.

1. $-14, 28, 0, -\dfrac{5}{7}, -364, -9.5$ **2.** $-37, 90, -\dfrac{7}{10}, -88.8, 42, -561$

For Exercises 3 to 30, add.

3. $3 + (-5)$ **4.** $-4 + 2$ **5.** $8 + 12$ **6.** $16 + 23$

7. $-3 + (-8)$ **8.** $-12 + (-1)$ **9.** $-4 + (-5)$ **10.** $-12 + (-12)$

11. $6 + (-9)$ **12.** $4 + (-9)$ **13.** $-6 + 7$ **14.** $-12 + 6$

15. $2 + (-3) + (-4)$ **16.** $7 + (-2) + (-8)$ **17.** $-3 + (-12) + (-15)$

18. $9 + (-6) + (-16)$ **19.** $-17 + (-3) + 29$ **20.** $13 + 62 + (-38)$

21. $-3 + (-8) + 12$ **22.** $-27 + (-42) + (-18)$

23. $13 + (-22) + 4 + (-5)$ **24.** $-14 + (-3) + 7 + (-6)$

25. $-22 + 10 + 2 + (-18)$ **26.** $-6 + (-8) + 13 + (-4)$

27. $-16 + (-17) + (-18) + 10$ **28.** $-25 + (-31) + 24 + 19$

29. $-126 + (-247) + (-358) + 339$ **30.** $-651 + (-239) + 524 + 487$

31. What is -8 more than -12?   **32.** What is -5 more than 3?

33. What is -7 added to -16?   **34.** What is 7 added to -25?

35. What is −4 plus 2?

36. What is −22 plus −17?

37. Find the sum of −2, 8, and −12.

38. Find the sum of 4, −4, and −6.

39. What is the total of 2, −3, 8, and −13?

40. What is the total of −6, −8, 13, and −2?

 For Exercises 41 to 44, determine whether the statement is *always true, never true,* or *sometimes true.*

41. The sum of an integer and its opposite is zero.

42. The sum of two negative integers is a positive integer.

43. The sum of two negative integers and one positive integer is a negative integer.

44. If the absolute value of a negative integer is greater than the absolute value of a positive integer, then the sum of the integers is negative.

OBJECTIVE B   **To subtract integers**

For Exercises 45 to 48, translate the expression into words. Represent each number as positive or negative.

45. −6 − 4

46. −6 − (−4)

47. 6 − (−4)

48. 6 − 4

For Exercises 49 to 52, rewrite the subtraction as the sum of the first number and the opposite of the second number.

49. 9 − (−5)

50. −3 − 7

51. 1 − 8

52. −2 − (−10)

For Exercises 53 to 80, subtract.

53. 16 − 8

54. 12 − 3

55. 7 − 14

56. 6 − 9

57. −7 − 2

58. −9 − 4

59. 7 − (−29)

60. 3 − (−4)

61. −6 − (−3)

62. −4 − (−2)

63. 6 − (−12)

64. −12 − 16

65. −4 − 3 − 2

66. 4 − 5 − 12

67. 12 − (−7) − 8

68. −12 − (−3) − (−15)

69. 4 − 12 − (−8)

70. 13 − 7 − 15

71. $-6 - (-8) - (-9)$

72. $7 - 8 - (-1)$

73. $-30 - (-65) - 29 - 4$

74. $42 - (-82) - 65 - 7$

75. $-16 - 47 - 63 - 12$

76. $42 - (-30) - 65 - (-11)$

77. $47 - (-67) - 13 - 15$

78. $-18 - 49 - (-84) - 27$

79. $167 - 432 - (-287) - 359$

80. $-521 - (-350) - 164 - (-299)$

81. Subtract -8 from -4.

82. Subtract -12 from 3.

83. What is the difference between -8 and 4?

84. What is the difference between 8 and 3?

85. What is -4 decreased by 8?

86. What is -13 decreased by 9?

87. Find -2 less than 1.

88. Find -3 less than -5.

 For Exercises 89 to 92, determine whether the statement is *always true, never true,* or *sometimes true.*

89. The difference between a positive integer and a negative integer is zero.

90. A negative integer subtracted from a positive integer is a positive integer.

91. The difference between two negative integers is a positive integer.

92. The difference between an integer and its absolute value is zero.

OBJECTIVE C   To solve application problems

 93. **Temperature** The news clipping at the right was written on February 11, 2008. The record low temperature for Minnesota is $-51°C$. Find the difference between the low temperature in International Falls on February 11, 2008, and the record low temperature for Minnesota.

> **In the News**
>
> **Minnesota Town Named "Icebox of the Nation"**
>
> In International Falls, Minnesota, the temperature fell to $-40°C$ just days after the citizens received word that the town had won a federal trademark naming it the "Icebox of the Nation."
>
> *Source:* news.yahoo.com

 94. **Temperature** The record high temperature in Illinois is $117°F$. The record low temperature is $-36°F$. Find the difference between the record high and record low temperatures in Illinois.

95. **Temperature** Find the temperature after a rise of 7°C from −8°C.

96. **Temperature** Find the temperature after a rise of 5°C from −19°C.

97. If the temperature begins at −54°C and rises more than 60°C, is the new temperature above or below 0°C?

98. If the temperature begins at −37°C and falls more than 40°C, is the new temperature above or below 0°C?

99. **Games** During a card game of Hearts, Nick had a score of 11 points before his opponent "shot the moon," subtracting a score of 26 from Nick's total. What was Nick's score after his opponent shot the moon?

100. **Games** In a card game of Hearts, Monique had a score of −19 before she "shot the moon," entitling her to add 26 points to her score. What was Monique's score after she shot the moon?

101. **Investments** The price of Byplex Corporation's stock fell each trading day of the first week of June. Use the figure at the right to find the change in the price of Byplex stock over the week's time.

102. **Astronomy** The daytime temperature on the moon can reach 266°F, and the nighttime temperature can go as low as −292°F. Find the difference between these extremes.

Change in Price of Byplex
Corporation Stock (in dollars)

103. **Earth Science** The average temperature throughout Earth's stratosphere is −70°F. The average temperature on Earth's surface is 45°F. Find the difference between these average temperatures.

104. **Golf Scores** In golf, a player's score on a hole is 0 if he completes the hole in *par*. **Par** is the number of strokes in which a golfer should complete a particular hole. In a golf match, scores are given both as the total number of strokes taken on all holes and as a value relative to par, such as −4 ("4 under par") or +2 ("2 over par"). The Masters Tournament is a four-day golf tournament in which players' daily scores are added. In 2008, Tiger Woods' daily scores in the Masters Tournament were 0, −1, −4, and 0. His total of −5 is found by adding the four numbers. Use the table below to find the totals for other players in the same tournament.

Tiger Woods

Player	Day 1	Day 2	Day 3	Day 4	Total
Trevor Immelman	−4	−4	−3	+3	
Brandt Snedeker	−3	−4	−2	+5	
Phil Mickelson	−1	−4	+1	0	

Trevor Immelman

Geography The elevation, or height, of places on Earth is measured in relation to sea level, or the average level of the ocean's surface. The table below shows height above sea level as a positive number and depth below sea level as a negative number. Use the table for Exercises 105 to 107.

Continent	Highest Elevation (in meters)		Lowest Elevation (in meters)	
Africa	Mt. Kilimanjaro	5895	Lake Assal	−156
Asia	Mt. Everest	8850	Dead Sea	−411
North America	Mt. McKinley	5642	Death Valley	−28
South America	Mt. Aconcagua	6960	Valdes Peninsula	−86

Mt. Everest

105. What is the difference in elevation between Mt. Kilimanjaro and Lake Assal?

106. What is the difference in elevation between Mt. Aconcagua and the Valdes Peninsula?

107. For which continent shown is the difference between the highest and lowest elevations greatest?

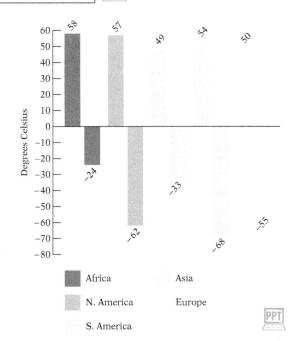

Africa   Asia

N. America   Europe

S. America

Highest and Lowest Temperatures Recorded (in degrees Celsius)
Source: National Climatic Data Center

Meteorology The figure at the right shows the highest and lowest temperatures ever recorded for selected regions of the world. Use this graph for Exercises 108 to 110.

108. What is the difference between the highest and lowest temperatures recorded in Africa?

109. What is the difference between the highest and lowest temperatures recorded in South America?

110. What is the difference between the lowest temperature recorded in Europe and the lowest temperature recorded in Asia?

Applying the Concepts

111. **Number Problems** Consider the numbers 4, −7, −5, 13, and −9. What is the largest difference that can be obtained by subtracting one number in the list from another number in the list? Find the smallest positive difference.

112. **Number Problems** Fill in the blank squares at the right with integers so that the sum of the integers along any row, column, or diagonal is zero.

−3		1
		3

113. **Number Problems** The sum of two negative integers is −8. Find the integers.

114. Explain the difference between the words *negative* and *minus*.

SECTION

10.3 Multiplication and Division of Integers

OBJECTIVE A To multiply integers

Multiplication is the repeated addition of the same number.

Several different symbols are used to indicate multiplication:

$3 \times 2 = 6 \qquad 3 \cdot 2 = 6 \qquad (3)(2) = 6$

When 5 is multiplied by a sequence of decreasing integers, the products decrease by 5.

$5 \times 3 = 15$
$5 \times 2 = 10$
$5 \times 1 = 5$
$5 \times 0 = 0$

The pattern developed can be continued so that 5 is multiplied by a sequence of negative numbers. The resulting products must be negative in order to maintain the pattern of decreasing by 5.

$5 \times (-1) = -5$
$5 \times (-2) = -10$
$5 \times (-3) = -15$
$5 \times (-4) = -20$

This example illustrates that the product of a positive number and a negative number is negative.

When -5 is multiplied by a sequence of decreasing integers, the products increase by 5.

$-5 \times 3 = -15$
$-5 \times 2 = -10$
$-5 \times 1 = -5$
$-5 \times 0 = 0$

The pattern developed can be continued so that -5 is multiplied by a sequence of negative numbers. The resulting products must be positive in order to maintain the pattern of increasing by 5.

$-5 \times (-1) = 5$
$-5 \times (-2) = 10$
$-5 \times (-3) = 15$
$-5 \times (-4) = 20$

This example illustrates that the product of two negative numbers is positive.

The pattern for multiplication shown above is summarized in the following rules for multiplying integers.

Integrating Technology

To multiply $(-4)(-8)$ on your calculator, enter the following:

4 **+/-** **x** 8 **+/-** **=**

Rule for Multiplying Two Numbers

To multiply numbers with the same sign, multiply the absolute values of the factors. The product is positive.

$4 \cdot 8 = 32$

$(-4)(-8) = 32$

To multiply numbers with different signs, multiply the absolute values of the factors. The product is negative.

$-4 \cdot 8 = -32$

$(4)(-8) = -32$

HOW TO • 1   Multiply: $2(-3)(-5)(-7)$

$$2(-3)(-5)(-7) = -6(-5)(-7)$$

• To multiply more than two numbers, first multiply the first two numbers.

$$= 30(-7)$$

• Then multiply the product by the third number.

$$= -210$$

• Continue until all the numbers have been multiplied.

EXAMPLE • 1

Multiply: $(-2)(6)$

Solution

$(-2)(6) = -12$ • The signs are different. The product is negative.

YOU TRY IT • 1

Multiply: $(-3)(-5)$

Your solution

EXAMPLE • 2

Find the product of -42 and 62.

Solution

$-42 \cdot 62 = -2604$ • The signs are different. The product is negative.

YOU TRY IT • 2

Find -38 multiplied by 51.

Your solution

EXAMPLE • 3

Multiply: $-5(-4)(6)(-3)$

Solution

$$-5(-4)(6)(-3) = 20(6)(-3)$$
$$= 120(-3)$$
$$= -360$$

YOU TRY IT • 3

Multiply: $-7(-8)(9)(-2)$

Your solution

Solutions on pp. S23–S24

OBJECTIVE B   To divide integers

For every division problem, there is a related multiplication problem.

Division: $\dfrac{8}{2} = 4$ Related multiplication: $4 \cdot 2 = 8$

This fact can be used to illustrate the rules for dividing signed numbers.

Rule for Dividing Two Numbers

To divide numbers with the same sign, divide the absolute values of the numbers. The quotient is positive.

To divide numbers with different signs, divide the absolute values of the numbers. The quotient is negative.

$\dfrac{8}{2} = 4$ because $4 \cdot 2 = 8$.

$\dfrac{-8}{-2} = 4$ because $4(-2) = -8$.

$\dfrac{8}{-2} = -4$ because $-4(-2) = 8$.

$\dfrac{-8}{2} = -4$ because $-4(2) = -8$.

Note that $\frac{8}{-2}$, $\frac{-8}{2}$, and $-\frac{8}{2}$ are all equal to -4.

If a and b are two integers, then $\frac{a}{-b} = \frac{-a}{b} = -\frac{a}{b}$.

> **Properties of Zero and One in Division**
>
> Zero divided by any number other than zero is zero.
>
> Any number other than zero divided by itself is 1.
>
> Any number divided by 1 is the number.
>
> Division by zero is not defined.

$\frac{0}{a} = 0$ because $0 \cdot a = 0$

$\frac{a}{a} = 1$ because $1 \cdot a = a$

$\frac{a}{1} = a$ because $a \cdot 1 = a$

$\frac{4}{0} = ?$ $? \cdot 0 = 4$

There is no number whose product with zero is 4.

The examples below illustrate these properties of division.

$$\frac{0}{-8} = 0 \qquad \frac{-7}{-7} = 1 \qquad \frac{-9}{1} = -9 \qquad \frac{-3}{0} \text{ is undefined.}$$

EXAMPLE · 4

Divide: $(-120) \div (-8)$

Solution

$(-120) \div (-8) = 15$ • The signs are the same. The quotient is positive.

YOU TRY IT · 4

Divide: $(-135) \div (-9)$

Your solution

EXAMPLE · 5

Divide: $95 \div (-5)$

Solution

$95 \div (-5) = -19$ • The signs are different. The quotient is negative.

YOU TRY IT · 5

Divide: $84 \div (-6)$

Your solution

EXAMPLE · 6

Find the quotient of -81 and 3.

Solution

$-81 \div 3 = -27$

YOU TRY IT · 6

What is -72 divided by 4?

Your solution

EXAMPLE · 7

Divide: $0 \div (-24)$

Solution

$0 \div (-24) = 0$ • Zero divided by a nonzero number is zero.

YOU TRY IT · 7

Divide: $-39 \div 0$

Your solution

Solutions on p. S24

OBJECTIVE C   To solve application problems

EXAMPLE · 8	YOU TRY IT · 8
The combined scores of the top five golfers in a tournament equaled -10 (10 under par). What was the average score of the five golfers?	The melting point of mercury is $-38°C$. The melting point of argon is five times the melting point of mercury. Find the melting point of argon.
Strategy To find the average score, divide the combined scores (-10) by the number of golfers (5).	**Your strategy**
Solution $-10 \div 5 = -2$ The average score was -2.	**Your solution**

EXAMPLE · 9	YOU TRY IT · 9
The daily high temperatures during one week were recorded as follows: $-9°F$, $3°F$, $0°F$, $-8°F$, $2°F$, $1°F$, $4°F$. Find the average daily high temperature for the week.	The daily low temperatures during one week were recorded as follows: $-6°F$, $-7°F$, $1°F$, $0°F$, $-5°F$, $-10°F$, $-1°F$. Find the average daily low temperature for the week.
Strategy To find the average daily high temperature: • Add the seven temperature readings. • Divide by 7.	**Your strategy**
Solution $-9 + 3 + 0 + (-8) + 2 + 1 + 4 = -7$ $-7 \div 7 = -1$ The average daily high temperature was $-1°F$.	**Your solution**

Solutions on p. S24

10.3 EXERCISES

OBJECTIVE A   To multiply integers

For Exercises 1 to 4, state whether the operation in the expression is addition, subtraction, or multiplication.

1. $5 - (-6)$ **2.** $4(-9)$ **3.** $-8(-5)$ **4.** $-3 + (-7)$

For Exercises 5 to 46, multiply.

5. 14×3 **6.** 62×9 **7.** $-4 \cdot 6$

8. $-7 \cdot 3$ **9.** $-2 \cdot (-3)$ **10.** $-5 \cdot (-1)$

11. $(9)(2)$ **12.** $(3)(8)$ **13.** $5(-4)$

14. $4(-7)$ **15.** $-8(2)$ **16.** $-9(3)$

17. $(-5)(-5)$ **18.** $(-3)(-6)$ **19.** $(-7)(0)$

20. -32×4 **21.** -24×3 **22.** $19 \cdot (-7)$

23. $6(-17)$ **24.** $-8(-26)$ **25.** $-4(-35)$

26. $-5 \cdot (23)$ **27.** $-6 \cdot (38)$ **28.** $9(-27)$

29. $8(-40)$ **30.** $-7(-34)$ **31.** $-4(39)$

32. $4 \cdot (-8) \cdot 3$ **33.** $5 \times 7 \times (-2)$ **34.** $8 \cdot (-6) \cdot (-1)$

35. $(-9)(-9)(2)$ **36.** $-8(-7)(-4)$ **37.** $-5(8)(-3)$

38. $(-6)(5)(7)$ **39.** $-1(4)(-9)$ **40.** $6(-3)(-2)$

41. $4(-4) \cdot 6(-2)$

42. $-5 \cdot 9(-7) \cdot 3$

43. $-9(4) \cdot 3(1)$

44. $8(8)(-5)(-4)$

45. $(-6) \cdot 7 \cdot (-10)(-5)$

46. $-9(-6)(11)(-2)$

47. What is -5 multiplied by -4?

48. What is 6 multiplied by -5?

49. What is -8 times 6?

50. What is -8 times -7?

51. Find the product of -4, 7, and -5.

52. Find the product of -2, -4, and -7.

 For Exercises 53 to 56, state whether the given product will be *positive, negative,* or *zero.*

53. The product of three negative integers

54. The product of two negative integers and one positive integer

55. The product of one negative integer, one positive integer, and zero

56. The product of five positive integers and one negative integer

---

OBJECTIVE B   **To divide integers**

For Exercises 57 to 60, write the related multiplication problem.

57. $\dfrac{-36}{-12} = 3$

58. $\dfrac{28}{-7} = -4$

59. $\dfrac{-55}{11} = -5$

60. $\dfrac{-20}{-10} = 2$

For Exercises 61 to 111, divide.

61. $12 \div (-6)$

62. $18 \div (-3)$

63. $(-72) \div (-9)$

64. $(-64) \div (-8)$

65. $0 \div (-6)$

66. $-49 \div 7$

67. $45 \div (-5)$

68. $-24 \div 4$

69. $-36 \div 4$

70. $-56 \div 7$

71. $-81 \div (-9)$

72. $-40 \div (-5)$

73. $72 \div (-3)$

74. $44 \div (-4)$

75. $-60 \div 5$

76. $-66 \div 6$

77. $-93 \div (-3)$

78. $-98 \div (-7)$

79. $(-85) \div (-5)$

80. $(-60) \div (-4)$

81. $120 \div 8$

82. $144 \div 9$

83. $78 \div (-6)$

84. $84 \div (-7)$

85. $-72 \div 4$

86. $-80 \div 5$

87. $-114 \div (-6)$

88. $-91 \div (-7)$

89. $-104 \div (-8)$

90. $-126 \div (-9)$

91. $57 \div (-3)$

92. $162 \div (-9)$

93. $-136 \div (-8)$

94. $-128 \div 4$

95. $-130 \div (-5)$

96. $(-280) \div 8$

97. $(-92) \div (-4)$

98. $-196 \div (-7)$

99. $-150 \div (-6)$

100. $(-261) \div 9$

101. $204 \div (-6)$

102. $165 \div (-5)$

103. $-132 \div (-12)$

104. $-156 \div (-13)$

105. $-182 \div 14$

106. $-144 \div 12$

107. $143 \div 11$

108. $168 \div 14$

109. $-180 \div (-15)$

110. $-169 \div (-13)$

111. $154 \div (-11)$

112. Find the quotient of -132 and -11.

113. Find the quotient of 182 and -13.

114. What is −60 divided by −15?

115. What is 144 divided by −24?

116. Find the quotient of −135 and 15.

117. Find the quotient of −88 and 22.

 For Exercises 118 to 121, determine whether the statement is *always true, never true,* or *sometimes true.*

118. The quotient of a negative integer and its absolute value is −1.

119. The quotient of zero and a positive integer is a positive integer.

120. A negative integer divided by zero is zero.

121. The quotient of two negative numbers is the same as the quotient of the absolute values of the two numbers.

OBJECTIVE C   To solve application problems

122. **Meteorology** The daily low temperatures during one week were recorded as follows: 4°F, −5°F, 8°F, −1°F, −12°F, −14°F, −8°F. Find the average daily low temperature for the week.

123. **Meteorology** The daily high temperatures during one week were recorded as follows: −6°F, −11°F, 1°F, 5°F, −3°F, −9°F, −5°F. Find the average daily high temperature for the week.

124. True or false? If five temperatures are all below 0°C, then the average of the five temperatures is also below 0°C.

125. True or false? If the average of 10 temperatures is below 0°C, then all 10 temperatures are below 0°C.

126. **Chemistry** The graph at the right shows the boiling points of three chemical elements. The boiling point of neon is seven times the highest boiling point shown in the graph.
  a. Without calculating the boiling point, determine whether the boiling point of neon is above 0°C or below 0°C.
  b. What is the boiling point of neon?

127. **Sports** The combined scores of the top 10 golfers in a tournament equaled −20 (20 under par). What was the average score of the 10 golfers?

128. **Sports** The combined scores of the top four golfers in a tournament equaled −12 (12 under par). What was the average score of the four golfers?

129. **Golf** Use the news clipping at the right and the information on golf scores given in Exercise 104 on page 417. The next seven players listed on the leaderboard, along with their scores, are shown below. Find the average score of the top eight players.

Maria Hjorth	−12
Lorena Ochoa	−11
Annika Sorenstam	−11
Laura Diaz	−10
Morgan Pressel	−8
Shi Hyun Ahn	−8
Kelli Kuehne	−8

Andy Lyons/Getty Images

130. **Economics** A nation's balance of trade is the difference between its exports and its imports. If the exports are greater than the imports, the result is a positive number and a *favorable balance of trade*. If the exports are less than the imports, the result is a negative number and an *unfavorable balance of trade*. The table at the right shows the U.S. unfavorable balance of trade, in billions of dollars, for each of the first six months of 2008. Find the average monthly balance of trade for January through June.

U.S. Balance of Trade in 2008 (in billions of dollar)	
January	−57
February	−60
March	−56
April	−60
May	−59
June	−56

Source: U.S. Census Bureau

131. **Meteorology** The wind chill factor when the temperature is −20°F and the wind is blowing at 15 mph is five times the wind chill factor when the temperature is 10°F and the wind is blowing at 20 mph. If the wind chill factor at 10°F with a 20-mph wind is −9°F, what is the wind chill factor at −20°F with a 15-mph wind?

132. **Education** To discourage guessing on a multiple-choice exam, an instructor graded the test by giving 5 points for a correct answer, −2 points for an answer left blank, and −5 points for an incorrect answer. How many points did a student score who answered 20 questions correctly, answered 5 questions incorrectly, and left 2 questions blank?

Applying the Concepts

133. **a. Number Problem** Find the greatest possible product of two negative integers whose sum is −10.

b. Number Problem Find the least possible sum of two negative integers whose product is 16.

134. Use repeated addition to show that the product of two integers with different signs is a negative number.

135. Determine whether the statement is *true* or *false*.
a. The product of a nonzero number and its opposite is negative.
b. The square of a negative number is a positive number.

136. In your own words, describe the rules for multiplying and dividing integers.

SECTION

10.4   Operations with Rational Numbers

OBJECTIVE A   **To add or subtract rational numbers**

In this section, operations with rational numbers are discussed. A **rational number** is the quotient of two integers.

> **Rational Numbers**
>
> A rational number is a number that can be written in the form $\frac{a}{b}$, where a and b are integers and $b \neq 0$.

Each of the three numbers shown at the right is a rational number.

$$\frac{3}{4} \qquad \frac{-2}{9} \qquad \frac{13}{-5}$$

An integer can be written as the quotient of the integer and 1. Therefore, every integer is a rational number.

$$6 = \frac{6}{1} \qquad -8 = \frac{-8}{1}$$

A mixed number can be written as the quotient of two integers. Therefore, every mixed number is a rational number.

$$1\frac{4}{7} = \frac{11}{7} \qquad 3\frac{2}{5} = \frac{17}{5}$$

Recall that every fraction can be written as a decimal by dividing the numerator of the fraction by the denominator. The result is either a terminating decimal or a repeating decimal.

We can write the fraction $\frac{3}{4}$ as the terminating decimal 0.75.

> **✓ Take Note**
>
> The fraction bar can be read "divided by."
>
> $\frac{3}{4} = 3 \div 4$

$$
\begin{array}{r}
0.75 \\
4{\overline{\smash{\big)}\,3.00}} \\
\underline{-2\,8} \\
20 \\
\underline{-20} \\
0
\end{array}
$$

0.75 ⟵ This is a **terminating decimal.**

0 ⟵ The remainder is zero.

We can write the fraction $\frac{2}{3}$ as the repeating decimal $0.\overline{6}$.

> **✓ Take Note**
>
> A **terminating decimal** is a decimal that has a finite number of digits after the decimal point, which means that it comes to an end and does not go on forever. A **repeating decimal** is a decimal that does not end; it has a repeating pattern of digits after the decimal point.

$$
\begin{array}{r}
0.666 = 0.\overline{6} \\
3{\overline{\smash{\big)}\,2.000}} \\
\underline{-1\,8} \\
20 \\
\underline{-18} \\
20 \\
\underline{-18} \\
2
\end{array}
$$

$0.666 = 0.\overline{6}$ ⟵ This is a **repeating decimal.** The bar over the digit 6 in $0.\overline{6}$ is used to show that this digit repeats.

2 ⟵ The remainder is never zero.

All terminating and repeating decimals are rational numbers.

To add or subtract rational numbers in fractional form, first find the least common multiple (LCM) of the denominators.

| HOW TO 1 | Add: $-\dfrac{7}{8} + \dfrac{5}{6}$ |

$8 = 2 \cdot 2 \cdot 2$
$6 = 2 \cdot 3$
$\text{LCM} = 2 \cdot 2 \cdot 2 \cdot 3 = 24$

• Find the LCM of the denominators.

$-\dfrac{7}{8} + \dfrac{5}{6} = -\dfrac{21}{24} + \dfrac{20}{24}$

• Rewrite each fraction using the LCM of the denominators as the common denominator.

$= \dfrac{-21 + 20}{24}$

• Add the numerators.

$= \dfrac{-1}{24} = -\dfrac{1}{24}$

✓ Take Note

In this text, answers that are negative fractions are written with the negative sign in front of the fraction.

| HOW TO 2 | Subtract: $-\dfrac{7}{9} - \dfrac{5}{12}$ |

$9 = 3 \cdot 3$
$12 = 2 \cdot 2 \cdot 3$
$\text{LCM} = 2 \cdot 2 \cdot 3 \cdot 3 = 36$

• Find the LCM of the denominators.

$-\dfrac{7}{9} - \dfrac{5}{12} = -\dfrac{28}{36} - \dfrac{15}{36}$

• Rewrite each fraction using the LCM of the denominators as the common denominator.

$= \dfrac{-28}{36} + \dfrac{-15}{36}$

• Rewrite subtraction as addition of the opposite. Rewrite negative fractions with the negative sign in the numerator.

$= \dfrac{-28 + (-15)}{36}$

• Add the numerators.

$= \dfrac{-43}{36} = -\dfrac{43}{36} = -1\dfrac{7}{36}$

To add or subtract rational numbers in decimal form, use the sign rules for adding integers.

| HOW TO 3 | Add: $47.034 + (-56.91)$ |

$\begin{array}{r} 56.910 \\ -\ 47.034 \\ \hline 9.876 \end{array}$

• The signs are different. Find the difference between the absolute values of the numbers.

$47.034 + (-56.91)$
$= -9.876$

• Attach the sign of the number with the greater absolute value.

HOW TO 4 Subtract: $-39.09 - 102.98$

$-39.09 - 102.98$
$= -39.09 + (-102.98)$

• Rewrite subtraction as addition of the opposite.

39.09
$+ 102.98$
142.07

• The signs of the addends are the same. Find the sum of the absolute values of the numbers.

$-39.09 - 102.98 = -142.07$

• Attach the sign of the addends.

EXAMPLE • 1

Subtract: $\frac{5}{16} - \frac{7}{40}$

Solution

$\frac{5}{16} - \frac{7}{40} = \frac{25}{80} - \frac{14}{80}$

• The LCM of 16 and 40 is **80**.

$= \frac{25}{80} + \frac{-14}{80}$

• Rewrite as addition of the opposite.

$= \frac{25 + (-14)}{80} = \frac{11}{80}$

YOU TRY IT • 1

Subtract: $\frac{5}{9} - \frac{11}{12}$

Your solution

EXAMPLE • 2

Simplify: $-\frac{3}{4} + \frac{1}{6} - \frac{5}{8}$

Solution

$-\frac{3}{4} + \frac{1}{6} - \frac{5}{8} = -\frac{18}{24} + \frac{4}{24} - \frac{15}{24}$

• The LCM of 4, 6, and 8 is **24**.

$= \frac{-18}{24} + \frac{4}{24} + \frac{-15}{24}$

$= \frac{-18 + 4 + (-15)}{24}$

$= \frac{-29}{24} = -\frac{29}{24} = -1\frac{5}{24}$

YOU TRY IT • 2

Simplify: $-\frac{7}{8} - \frac{5}{6} + \frac{2}{3}$

Your solution

EXAMPLE • 3

Subtract: $42.987 - 98.61$

Solution

$42.987 - 98.61 = 42.987 + (-98.61)$

98.610
$- 42.987$
55.623

$42.987 - 98.61 = -55.623$

YOU TRY IT • 3

Subtract: $16.127 - 67.91$

Your solution

Solutions on p. S24

EXAMPLE · 4

Simplify: $1.02 + (-3.6) + 9.24$

Solution

$1.02 + (-3.6) + 9.24 = -2.58 + 9.24$
$= 6.66$

YOU TRY IT · 4

Simplify: $2.7 + (-9.44) + 6.2$

Your solution

Solution on p. S24

OBJECTIVE B To multiply or divide rational numbers

The product of two rational numbers written as fractions is the product of the numerators over the product of the denominators. Use the sign rules for multiplying integers.

HOW TO · 5 Simplify: $-\frac{3}{8} \times \frac{12}{17}$

$-\frac{3}{8} \times \frac{12}{17} = -\left(\frac{3 \cdot 12}{8 \cdot 17}\right) = -\frac{9}{34}$ • The signs are different.
The product is negative.

To divide rational numbers written as fractions, invert the divisor and then multiply. Use the sign rules for dividing integers.

HOW TO · 6 Simplify: $-\frac{3}{10} \div \left(-\frac{18}{25}\right)$

$-\frac{3}{10} \div \left(-\frac{18}{25}\right) = \frac{3}{10} \times \frac{25}{18} = \frac{3 \cdot 25}{10 \cdot 18} = \frac{5}{12}$ • The signs are the same.
The quotient is positive.

To multiply or divide rational numbers written in decimal form, use the sign rules for integers.

HOW TO · 7 Simplify: $(-6.89) \times (-0.00035)$

$$
\begin{array}{r}
6.89 \\
\times \quad 0.00035 \\
\hline
3445 \\
2067 \\
\hline
0.0024115
\end{array}
$$

2 decimal places
5 decimal places

7 decimal places

• The signs are the same.
Multiply the absolute values.

$(-6.89) \times (-0.00035) = 0.0024115$ • The product is positive.

HOW TO 8 Divide $1.32 \div (-0.27)$. Round to the nearest tenth.

$$
\begin{array}{r}
4.88 \approx 4.9 \\
0.27.\overline{)1.32.00} \\
-1\,08 \\
\hline
24\,0 \\
-21\,6 \\
\hline
2\,40 \\
-2\,16 \\
\hline
24
\end{array}
$$

• Divide the absolute values. Move the decimal point two places in the divisor and then in the dividend. Place the decimal point in the quotient.

$1.32 \div (-0.27) \approx -4.9$

• The signs are different. The quotient is negative.

EXAMPLE 5

Multiply: $-\dfrac{7}{12} \times \dfrac{9}{14}$

Solution
The product is negative.

$-\dfrac{7}{12} \times \dfrac{9}{14} = -\left(\dfrac{7 \cdot 9}{12 \cdot 14}\right)$

$= -\dfrac{3}{8}$

YOU TRY IT 5

Multiply: $\left(-\dfrac{2}{3}\right)\left(-\dfrac{9}{10}\right)$

Your solution

EXAMPLE 6

Divide: $-\dfrac{3}{8} \div \left(-\dfrac{5}{12}\right)$

Solution
The quotient is positive.

$-\dfrac{3}{8} \div \left(-\dfrac{5}{12}\right) = \dfrac{3}{8} \times \dfrac{12}{5}$

$= \dfrac{3 \cdot 12}{8 \cdot 5}$

$= \dfrac{9}{10}$

YOU TRY IT 6

Divide: $-\dfrac{5}{8} \div \dfrac{5}{40}$

Your solution

EXAMPLE 7

Multiply: -4.29×8.2

Solution
The product is negative.

$$
\begin{array}{r}
4.29 \\
\times\ 8.2 \\
\hline
858 \\
3432\ \\
\hline
35.178
\end{array}
$$

$-4.29 \times 8.2 = -35.178$

YOU TRY IT 7

Multiply: -5.44×3.8

Your solution

Solutions on p. S24

EXAMPLE · 8

Multiply: $-3.2 \times (-0.4) \times 6.9$

Solution

$-3.2 \times (-0.4) \times 6.9$
$= 1.28 \times 6.9$
$= 8.832$

YOU TRY IT · 8

Multiply: $3.44 \times (-1.7) \times 0.6$

Your solution

EXAMPLE · 9

Divide: $-0.0792 \div (-0.42)$
Round to the nearest hundredth.

Solution

$$
\begin{array}{r}
0.188 \approx 0.19 \\
0.42{\overline{\smash{\big)}\,0.07{,}920}} \\
\underline{-4\,2} \\
3\,72 \\
\underline{-3\,36} \\
360 \\
\underline{-336} \\
24
\end{array}
$$

$-0.0792 \div (-0.42) \approx 0.19$

YOU TRY IT · 9

Divide: $-0.394 \div 1.7$
Round to the nearest hundredth.

Your solution

Solutions on pp. S24–S25

OBJECTIVE C   To solve application problems

EXAMPLE · 10

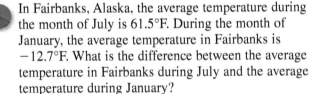

In Fairbanks, Alaska, the average temperature during the month of July is 61.5°F. During the month of January, the average temperature in Fairbanks is −12.7°F. What is the difference between the average temperature in Fairbanks during July and the average temperature during January?

Strategy

To find the difference, subtract the average temperature in January (−12.7°F) from the average temperature in July (61.5°F).

Solution

$61.5 - (-12.7) = 61.5 + 12.7 = 74.2$

The difference between the average temperature during July and the average temperature during January in Fairbanks is 74.2°F.

YOU TRY IT · 10

On January 10, 1911, in Rapid City, South Dakota, the temperature fell from 12.78°C at 7:00 A.M. to −13.33°C at 7:15 A.M. How many degrees did the temperature fall during the 15-minute period?

Your strategy

Your solution

Solution on p. S25

10.4 EXERCISES

OBJECTIVE A   To add or subtract rational numbers

For Exercises 1 to 43, simplify.

1. $\dfrac{5}{8} - \dfrac{5}{6}$

2. $\dfrac{1}{9} \quad \dfrac{5}{27}$

3. $-\dfrac{5}{12} - \dfrac{3}{8}$

4. $-\dfrac{5}{6} - \dfrac{5}{9}$

5. $-\dfrac{6}{13} + \dfrac{17}{26}$

6. $-\dfrac{7}{12} + \dfrac{5}{8}$

7. $-\dfrac{5}{8} - \left(-\dfrac{11}{12}\right)$

8. $-\dfrac{7}{12} - \left(-\dfrac{7}{8}\right)$

9. $\dfrac{5}{12} - \dfrac{11}{15}$

10. $\dfrac{2}{5} - \dfrac{14}{15}$

11. $-\dfrac{3}{4} - \dfrac{5}{8}$

12. $-\dfrac{2}{3} - \dfrac{5}{8}$

13. $-\dfrac{5}{2} - \left(-\dfrac{13}{4}\right)$

14. $-\dfrac{7}{3} - \left(-\dfrac{3}{2}\right)$

15. $-\dfrac{3}{8} - \dfrac{5}{12} - \dfrac{3}{16}$

16. $-\dfrac{5}{16} + \dfrac{3}{4} - \dfrac{7}{8}$

17. $\dfrac{1}{2} - \dfrac{3}{8} - \left(-\dfrac{1}{4}\right)$

18. $\dfrac{3}{4} - \left(-\dfrac{7}{12}\right) - \dfrac{7}{8}$

19. $\dfrac{1}{3} - \dfrac{1}{4} - \dfrac{1}{5}$

20. $\dfrac{5}{16} + \dfrac{1}{8} - \dfrac{1}{2}$

21. $\dfrac{1}{2} + \left(-\dfrac{3}{8}\right) + \dfrac{5}{12}$

22. $-\dfrac{3}{8} + \dfrac{3}{4} - \left(-\dfrac{3}{16}\right)$

23. $3.4 + (-6.8)$

24. $-4.9 + 3.27$

25. $-8.32 + (-0.57)$

26. $-3.5 + 7$

27. $-4.8 + (-3.2)$

28. $6.2 + (-4.29)$

29. $-4.6 + 3.92$

30. $7.2 + (-8.42)$

31. $-45.71 + (-135.8)$

32. $-35.274 + 12.47$

33. $4.2 + (-6.8) + 5.3$

34. $6.7 + 3.2 + (-10.5)$

35. $-4.5 + 3.2 + (-19.4)$

36. $2.09 - 6.72 - 5.4$

37. $-18.39 + 4.9 - 23.7$

38. $19 - (-3.72) - 82.75$

39. $-3.09 - 4.6 - 27.3$

40. $-3.89 + (-2.9) + 4.723 + 0.2$

41. $-4.02 + 6.809 - (-3.57) - (-0.419)$

42. $0.0153 + (-1.0294) + (-1.0726)$

43. $0.27 + (-3.5) - (-0.27) + (-5.44)$

 For Exercises 44 and 45, state whether the given sum or difference will be *positive* or *negative*.

44. A negative mixed number subtracted from a negative proper fraction

45. A positive improper fraction subtracted from a positive proper fraction

---

OBJECTIVE B   To multiply or divide rational numbers

For Exercises 46 to 87, simplify.

46. $\dfrac{1}{2} \times \left(-\dfrac{3}{4}\right)$

47. $-\dfrac{2}{9} \times \left(-\dfrac{3}{14}\right)$

48. $\left(-\dfrac{3}{8}\right)\left(-\dfrac{4}{15}\right)$

49. $\left(-\dfrac{3}{4}\right)\left(-\dfrac{8}{27}\right)$

50. $-\dfrac{1}{2} \times \dfrac{8}{9}$

51. $\dfrac{5}{12} \times \left(-\dfrac{8}{15}\right)$

52. $\left(-\dfrac{5}{12}\right)\left(\dfrac{42}{65}\right)$

53. $\left(\dfrac{3}{8}\right)\left(-\dfrac{15}{41}\right)$

54. $\left(-\dfrac{15}{8}\right)\left(-\dfrac{16}{3}\right)$

55. $\left(-\dfrac{5}{7}\right)\left(-\dfrac{14}{15}\right)$

56. $\dfrac{5}{8} \times \left(-\dfrac{7}{12}\right) \times \dfrac{16}{25}$

57. $\left(\dfrac{1}{2}\right)\left(-\dfrac{3}{4}\right)\left(-\dfrac{5}{8}\right)$

58. $\dfrac{1}{3} \div \left(-\dfrac{1}{2}\right)$

59. $-\dfrac{3}{8} \div \dfrac{7}{8}$

60. $\left(-\dfrac{3}{4}\right) \div \left(-\dfrac{7}{40}\right)$

61. $\dfrac{5}{6} \div \left(-\dfrac{3}{4}\right)$

62. $-\dfrac{5}{12} \div \dfrac{15}{32}$

63. $-\dfrac{5}{16} \div \left(-\dfrac{3}{8}\right)$

64. $\left(-\dfrac{3}{8}\right) \div \left(-\dfrac{5}{12}\right)$

65. $\left(-\dfrac{8}{19}\right) \div \dfrac{7}{38}$

66. $\left(-\dfrac{2}{3}\right) \div 4$

67. $-6 \div \dfrac{4}{9}$

68. $-6.7 \times (-4.2)$

69. $-8.9 \times (-3.5)$

70. -1.6×4.9

71. -14.3×7.9

72. $(-0.78)(-0.15)$

73. $(-1.21)(-0.03)$

74. $(-8.919) \div (-0.9)$

75. $-77.6 \div (-0.8)$

76. $59.01 \div (-0.7)$

77. $(-7.04) \div (-3.2)$

78. $(-84.66) \div 1.7$

79. $-3.312 \div (0.8)$

80. $1.003 \div (-0.59)$

81. $26.22 \div (-6.9)$

82. $(-19.08) \div 0.45$

83. $21.792 \div (-0.96)$

84. $(-38.665) \div (-9.5)$

85. $(-3.171) \div (-45.3)$ **86.** $27.738 \div (-60.3)$ **87.** $(-13.97) \div (-25.4)$

For Exercises 88 and 89, use the following information: When -3.54 is divided into a certain dividend, the result is a positive number less than 1. Determine whether each statement is *true* or *false*.

88. The dividend is a positive number.

89. The absolute value of the dividend is greater than 3.54.

OBJECTIVE C   **To solve application problems**

 90. Meteorology On January 23, 1916, the temperature in Browing, Montana, was 6.67°C. On January 24, 1916, the temperature in Browing was −48.9°C. Find the difference between the temperatures in Browing on these two days.

91. Meteorology On January 22, 1943, in Spearfish, South Dakota, the temperature fell from 12.22°C at 9 A.M. to −20°C at 9:27 A.M. How many degrees did the temperature fall during the 27-minute period?

 92. Temperature The date of the news clipping at the right is July 20, 2007. Find the difference between the record high and low temperatures for Slovakia.

In the News

Slovakia Hits Record High

Slovakia, which became an independent country in 1993 with the peaceful division of Czechoslovakia, hit a record high temperature today of 104.5°F. The record low temperature, set on February 11, 1929, was −41.8°F.

Source: wikipedia.org

 93. If the temperature begins at 4.8°C and ends up below 0°C, is the difference between the starting and ending temperatures less than or greater than 4.8?

 94. If the temperature rose 20.3°F during one day and ended up at a high temperature of 15.7°F, did the temperature begin above or below 0°F?

Slovakia

 95. Chemistry The boiling point of nitrogen is −195.8°C, and the melting point is −209.86°C. Find the difference between the boiling point and the melting point of nitrogen.

 96. Chemistry The boiling point of oxygen is −182.962°C. Oxygen's melting point is −218.4°C. What is the difference between the boiling point and the melting point of oxygen?

Investments The chart at the right shows the closing price of a share of stock on September 4, 2008, for each of five companies. Also shown is the change in the price from the previous day. To find the closing price on the previous day, subtract the change in price from the closing price on September 4. Use this chart for Exercises 97 and 98.

Company	Closing Price	Change in Price
Del Monte Foods Co.	8.18	−0.14
General Mills, Inc.	66.93	+0.21
Hershey Foods, Inc.	36.42	−0.41
Hormel Food Corp.	35.39	−0.07
Sara Lee Corp.	13.35	−0.07

97. **a.** Find the closing price on the previous day for General Mills.
   b. Find the closing price on the previous day for Hormel Foods.

98. **a.** Find the closing price on the previous day for Sara Lee.
   b. Find the closing price on the previous day for Hershey Foods.

Astronomy Stars are classified according to their **apparent magnitude,** or how bright they appear to be. The brighter a star appears, the lower the value of its apparent magnitude. A star's **absolute magnitude** is its apparent magnitude if all the stars were an equal distance from Earth. The **distance modulus** of a star is its apparent magnitude minus its absolute magnitude. The smaller the distance modulus, the closer the star is to Earth. Use the table at the right for Exercises 99 to 103.

Star	Apparent Magnitude	Absolute Magnitude
Sun	−26.8	4.83
Sirius	−1.47	1.41
Betelgeuse	0.41	−5.6
Vega	0.04	0.5
Polaris	1.99	−3.2

99. Arrange the stars in the table from brightest to least bright, as measured by absolute magnitude.

100. Find the distance modulus for the sun.

101. Find the distance modulus for Polaris.

102. Find the distance modulus for Sirius.

103. Which of the stars is farthest from Earth?

Applying the Concepts

104. Determine whether the statement is true or false.
   a. Every integer is a rational number.
   b. Every whole number is an integer.
   c. Every integer is a positive number.
   d. Every rational number is an integer.

105. **Number Problem** Find a rational number between $-\frac{3}{4}$ and $-\frac{2}{3}$.

106. **Number Problems**
   a. Find a rational number between 0.1 and 0.2.
   b. Find a rational number between 1 and 1.1.
   c. Find a rational number between 0 and 0.005.

107. Given any two different rational numbers, is it always possible to find a rational number between them? If so, explain how. If not, give an example of two different rational numbers for which there is no rational number between them.

SECTION

10.5

Scientific Notation and the Order of Operations Agreement

OBJECTIVE A

To write a number in scientific notation

 Point of Interest

The first woman mathematician for whom documented evidence exists is Hypatia (370–415). She lived in Alexandria, Egypt, and lectured at the Museum, the forerunner of our modern university. She made important contributions in mathematics, astronomy, and philosophy.

Scientific notation uses negative exponents. Therefore, we will discuss that topic before presenting scientific notation.

Look at the powers of 10 shown at the right. Note the pattern: The exponents are decreasing by 1, and each successive number on the right is one-tenth of the number above it. $(100,000 \div 10 = 10,000;$ $10,000 \div 10 = 1000;$ etc.)

$$10^5 = 100,000$$
$$10^4 = 10,000$$
$$10^3 = 1000$$
$$10^2 = 100$$
$$10^1 = 10$$

If we continue this pattern, the next exponent on 10 is $1 - 1 = 0$, and the number on the right side is $10 \div 10 = 1$.

$$10^0 = 1$$

The next exponent on 10 is $0 - 1 = -1$, and 10^{-1} is equal to $1 \div 10 = 0.1$.

$$10^{-1} = 0.1$$

The pattern is continued at the right. Note that a negative exponent does not indicate a negative number. Rather, each power of 10 with a negative exponent is equal to a number between 0 and 1. Also note that as the exponent on 10 decreases, so does the number it is equal to.

$$10^{-2} = 0.01$$
$$10^{-3} = 0.001$$
$$10^{-4} = 0.0001$$
$$10^{-5} = 0.00001$$
$$10^{-6} = 0.000001$$

Very large and very small numbers are encountered in the natural sciences. For example, the mass of an electron is 0.0000000000000000000000000000911 kg. Numbers such as this are difficult to read, so a more convenient system called **scientific notation** is used. In scientific notation, a number is expressed as the product of two factors, one a number between 1 and 10, and the other a power of 10.

To express a number in scientific notation, write it in the form $a \times 10^n$, where a is a number between 1 and 10 and n is an integer.

For numbers greater than 10, move the decimal point to the right of the first digit. The exponent n is positive and equal to the number of places the decimal point has been moved.

$$240,000 = 2.4 \times 10^5$$

$$93,000,000 = 9.3 \times 10^7$$

✓ **Take Note**

There are two steps in writing a number in scientific notation: (1) Determine the number between 1 and 10, and (2) determine the exponent on 10.

For numbers less than 1, move the decimal point to the right of the first nonzero digit. The exponent n is negative. The absolute value of the exponent is equal to the number of places the decimal point has been moved.

$0.0003 = 3 \times 10^{-4}$

$0.0000832 = 8.32 \times 10^{-5}$

Changing a number written in scientific notation to decimal notation also requires moving the decimal point.

When the exponent on 10 is positive, move the decimal point to the right the same number of places as the exponent.

$3.45 \times 10^9 = 3,450,000,000$

$2.3 \times 10^8 = 230,000,000$

When the exponent on 10 is negative, move the decimal point to the left the same number of places as the absolute value of the exponent.

$8.1 \times 10^{-3} = 0.0081$

$6.34 \times 10^{-6} = 0.00000634$

EXAMPLE • 1

Write 824,300,000,000 in scientific notation.

Solution

The number is greater than 10. Move the decimal point 11 places to the left. The exponent on 10 is 11.

$824,300,000,000 = 8.243 \times 10^{11}$

YOU TRY IT • 1

Write 0.000000961 in scientific notation.

Your solution

EXAMPLE • 2

Write 6.8×10^{-10} in decimal notation.

Solution

The exponent on 10 is negative. Move the decimal point 10 places to the left.

$6.8 \times 10^{-10} = 0.00000000068$

YOU TRY IT • 2

Write 7.329×10^6 in decimal notation.

Your solution

Solutions on p. S25

OBJECTIVE B   To use the Order of Operations Agreement to simplify expressions

The Order of Operations Agreement has been used throughout this book. In simplifying expressions with rational numbers, the same Order of Operations Agreement is used. This agreement is restated here.

The Order of Operations Agreement

Step 1 Do all operations inside parentheses.

Step 2 Simplify any number expressions containing exponents.

Step 3 Do multiplication and division as they occur from left to right.

Step 4 Rewrite subtraction as addition of the opposite. Then do additions as they occur from left to right.

Exponents may be confusing in expressions with signed numbers.

✓ Take Note

In $(-3)^2$, we are squaring -3; we multiply -3 times -3. In -3^2, we are finding the opposite of 3^2. The expression -3^2 is the same as $-(3^2)$.

$$(-3)^2 = (-3) \times (-3) = 9$$

$$-3^2 = -(3)^2 = -(3 \times 3) = -9$$

Note that -3 is squared only when the negative sign is *inside* the parentheses.

HOW TO 1 Simplify: $(-3)^2 - 2 \times (8 - 3) + (-5)$

$(-3)^2 - 2 \times \underbrace{(8 - 3)} + (-5)$

$\underbrace{(-3)^2} - 2 \times 5 + (-5)$

$9 - \underbrace{2 \times 5} + (-5)$

$9 - 10 + (-5)$

$\underbrace{9 + (-10)} + (-5)$

$\underbrace{(-1) + (-5)}$

-6

1. Perform operations inside parentheses.

2. Simplify expressions with exponents.

3. Do multiplications and divisions as they occur from left to right.

4. Rewrite subtraction as the addition of the opposite. Then add from left to right.

Integrating Technology

As shown above, the value of -3^2 is different from the value of $(-3)^2$. The keystrokes to evaluate each of these expressions on your calculator are different. To evaluate -3^2, enter

3 | **x²** | **+/−**

To evaluate $(-3)^2$, enter

3 | **+/−** | **x²**

HOW TO 2 Simplify: $\left(\frac{1}{4} - \frac{1}{2}\right)^2 \div \frac{3}{8}$

$\underbrace{\left(\frac{1}{4} - \frac{1}{2}\right)^2} \div \frac{3}{8}$

$\underbrace{\left(-\frac{1}{4}\right)^2} \div \frac{3}{8}$

$\underbrace{\frac{1}{16} \div \frac{3}{8}}$

$\frac{1}{16} \times \frac{8}{3}$

$\underbrace{\frac{1}{6}}$

1. Perform operations inside parentheses.

2. Simplify expressions with exponents.

3. Do multiplication and division as they occur from left to right.

EXAMPLE 3

Simplify: $8 - 4 \div (-2)$

Solution

$8 - 4 \div (-2)$
$= 8 - (-2)$ • **Do the division.**
$= 8 + 2$ • **Rewrite as**
$= 10$ **addition. Add.**

YOU TRY IT 3

Simplify: $9 - 9 \div (-3)$

Your solution

Solution on p. S25

EXAMPLE • 4

Simplify: $12 \div (-2)^2 + 5$

Solution

$12 \div (-2)^2 + 5$
$= 12 \div 4 + 5$ • Exponents
$= 3 + 5$ • Division
$= 8$ • Addition

YOU TRY IT • 4

Simplify: $8 \div 4 \cdot 4 - (-2)^2$

Your solution

EXAMPLE • 5

Simplify: $12 - (-10) \div (8 - 3)$

Solution

$12 - (-10) \div (8 - 3)$
$= 12 - (-10) \div 5$
$= 12 - (-2)$
$= 12 + 2$
$= 14$

YOU TRY IT • 5

Simplify: $8 - (-15) \div (2 - 7)$

Your solution

EXAMPLE • 6

Simplify: $(-3)^2 \times (5 - 7)^2 - (-9) \div 3$

Solution

$(-3)^2 \times (5 - 7)^2 - (-9) \div 3$
$= (-3)^2 \times (-2)^2 - (-9) \div 3$
$= 9 \times 4 - (-9) \div 3$
$= 36 - (-9) \div 3$
$= 36 - (-3)$
$= 36 + 3$
$= 39$

YOU TRY IT • 6

Simplify: $(-2)^2 \times (3 - 7)^2 - (-16) \div (-4)$

Your solution

EXAMPLE • 7

Simplify: $3 \div \left(\frac{1}{2} - \frac{1}{4}\right) - 3$

Solution

$3 \div \left(\frac{1}{2} - \frac{1}{4}\right) - 3$

$= 3 \div \frac{1}{4} - 3$

$= 3 \times \frac{4}{1} - 3$

$= 12 - 3$

$= 12 + (-3)$

$= 9$

YOU TRY IT • 7

Simplify: $7 \div \left(\frac{1}{7} - \frac{3}{14}\right) - 9$

Your solution

Solutions on p. S25

10.5 EXERCISES

OBJECTIVE A   To write a number in scientific notation

 1. In order to write a certain positive number in scientific notation, you move the decimal point to the right. Is the number greater than 1 or less than 1?

 2. In order to write a certain positive number in scientific notation, you move the decimal point to the left. Is the exponent on 10 greater than zero or less than zero?

For Exercises 3 to 14, write the number in scientific notation.

3. 2,370,000

4. 75,000

5. 0.00045

6. 0.000076

7. 309,000

8. 819,000,000

9. 0.000000601

10. 0.00000000096

11. 57,000,000,000

12. 934,800,000,000

13. 0.000000017

14. 0.0000009217

For Exercises 15 to 26, write the number in decimal notation.

15. 7.1×10^5

16. 2.3×10^7

17. 4.3×10^{-5}

18. 9.21×10^{-7}

19. 6.71×10^8

20. 5.75×10^9

21. 7.13×10^{-6}

22. 3.54×10^{-8}

23. 5×10^{12}

24. 1.0987×10^{11}

25. 8.01×10^{-3}

26. 4.0162×10^{-9}

 27. **Physics** Light travels 16,000,000,000 mi in 1 day. Write this number in scientific notation.

 28. **Earth Science** Write the mass of Earth, which is approximately 5,980,000,000,000,000,000,000,000 kg, in scientific notation.

29. **Wars** The graph at the right shows the monetary costs of four wars. Write the monetary cost of World War II in scientific notation.

Monetary Costs of Wars

Source: Congressional Research Service, using numbers from the Statistical Abstract of the United States

30. **Supernovas** See the news clipping at the right. SN2006gy is 2.27×10^{24} mi from Earth. Write this number in decimal notation.

In the News

NASA Discovers Giant Supernova

NASA announced the discovery of a star, SN2006gy, that is the brightest and biggest supernova even seen. Unlike typical supernovas that reach a peak brightness in a matter of days and then dim a few months later, SN2006gy took 10 weeks to reach full brightness and 8 months later remains as bright as a typical supernova at its peak.

Source: www.nasa.gov

31. **Spacecraft** Voyager 1 is the farthest man-made object from Earth. It was launched in 1977 with the mission of studying the boundaries of the solar system. Voyager 1 is now approximately 9,800,000,000 mi from Earth and remains operational. Write the number of miles from Earth to Voyager 1 in scientific notation.

32. **Air Travel** Use the news clipping at the right. Write the given number of passengers in scientific notation.

In the News

China Becomes Popular Destination

It is predicted that 83,000,000 passengers will pass through Beijing Capital Airport in 2015. Passenger numbers and cargo volume are skyrocketing in China.

Source: Lufthansa Magazine, May 2008

33. **Chemistry** The electric charge on an electron is 0.00000000000000000016 coulomb. Write this number in scientific notation.

34. **Physics** The length of an infrared light wave is approximately 0.0000037 m. Write this number in scientific notation.

In the News

Getting to Know the Universe

On September 10, 2008, scientists successfully fired the first beam of protons around the Large Hadron Collider, the world's largest particle collider. The scientists hope their research will lead to a greater understanding of the makeup of the universe.

Source: news.yahoo.com

35. **Computers** One unit used to measure the speed of a computer is the picosecond. One picosecond is 0.000000000001 of a second. Write this number in scientific notation.

36. **Protons** See the news clipping at the right. A proton is a subatomic particle that has a mass of 1.673×10^{-27} kg. Write this number in decimal notation.

OBJECTIVE B   To use the Order of Operations Agreement to simplify expressions

For Exercises 37 to 98, simplify.

37. $8 \div 4 + 2$

38. $3 - 12 \div 2$

39. $4 + (-7) + 3$

40. $-16 \div 2 + 8$

41. $4^2 - 4$

42. $6 - 2^2$

43. $2 \times (3 - 5) - 2$

44. $2 - (8 - 10) \div 2$

45. $4 - (-3)^2$

46. $(-2)^2 - 6$

47. $4 - (-3) - 5$

48. $6 + (-8) - (-3)$

49. $4 - (-2)^2 + (-3)$

50. $-3 + (-6)^2 - 1$

51. $3^2 - 4 \times 2$

52. $9 \div 3 - (-3)^2$

53. $3 \times (6 - 2) \div 6$

54. $4 \times (2 - 7) \div 5$

55. $2^2 - (-3)^2 + 2$

56. $3 \times (8 - 5) + 4$

57. $6 - 2 \times (1 - 5)$

58. $4 \times 2 \times (3 - 6)$

59. $(-2)^2 - (-3)^2 + 1$

60. $4^2 - 3^2 - 4$

61. $6 - (-3) \times (-3)^2$

62. $4 - (-5) \times (-2)^2$

63. $4 \times 2 - 3 \times 7$

64. $16 \div 2 - 9 \div 3$

65. $(-2)^2 - 5 \times 3 - 1$

66. $4 - 2 \times 7 - 3^2$

67. $7 \times 6 - 5 \times 6 + 3 \times 2 \quad 2 + 1$

68. $3 \times 2^2 + 5 \times (3 + 2) - 17$

69. $-4 \times 3 \times (-2) + 12 \times (3 - 4) + (-12)$

70. $3 \times 4^2 - 16 - 4 + 3 - (1 - 2)^2$

71. $-12 \times (6 - 8) + 1^2 \times 3^2 \times 2 - 6 \times 2$

72. $-3 \times (-2)^2 \times 4 \div 8 - (-12)$

73. $10 \times 9 - (8 + 7) \div 5 + 6 - 7 + 8$

74. $-27 - (-3)^2 - 2 - 7 + 6 \times 3$

75. $3^2 \times (4 - 7) \div 9 + 6 - 3 - 4 \times 2$

76. $16 - 4 \times 8 + 4^2 - (-18) \div (-9)$

77. $(-3)^2 \times (5 - 7)^2 - (-9) \div 3$

78. $-2 \times 4^2 - 3 \times (2 - 8) - 3$

79. $4 - 6(2 - 5)^3 \div (17 - 8)$

80. $5 + 7(3 - 8)^2 \div (-14 + 9)$

81. $(1.2)^2 - 4.1 \times 0.3$

82. $2.4 \times (-3) - 2.5$

83. $1.6 - (-1.6)^2$

84. $4.1 \times 8 \div (-4.1)$

85. $(4.1 - 3.9) - 0.7^2$

86. $1.8 \times (-2.3) - 2$

87. $(-0.4)^2 \times 1.5 - 2$

88. $(6.2 - 1.3) \times (-3)$

89. $4.2 - (-3.9) - 6$

90. $-\dfrac{1}{2} + \dfrac{3}{8} \div \left(-\dfrac{3}{4}\right)$

91. $\left(\dfrac{3}{4}\right)^2 - \dfrac{3}{8}$

92. $\left(\dfrac{1}{2}\right)^2 - \left(-\dfrac{1}{2}\right)^2$

93. $\dfrac{5}{16} - \dfrac{3}{8} + \dfrac{1}{2}$

94. $\dfrac{2}{7} \div \dfrac{5}{7} - \dfrac{3}{14}$

95. $\dfrac{1}{2} \times \dfrac{1}{4} \times \dfrac{1}{2} - \dfrac{3}{8}$

96. $\dfrac{2}{3} \times \dfrac{5}{8} \div \dfrac{2}{7}$

97. $\dfrac{1}{2} - \left(\dfrac{3}{4} - \dfrac{3}{8}\right) \div \dfrac{1}{3}$

98. $\dfrac{3}{8} \div \left(-\dfrac{1}{2}\right)^2 + 2$

99. Which expression is equivalent to $7 - (-2^2)$?
(i) $7 + 4$ (ii) 9^2 (iii) $7 - 4$ (iv) 7×4

100. Which expression is equivalent to $3 - 5 \times 4 - 7^2$?
(i) $-2 \times (-3)^2$ (ii) $3 - 20 + 49$
(iii) $3 - 20 - 49$ (iv) $-2 \times 4 - 49$

Applying the Concepts

101. Place the correct symbol, $<$ or $>$, between the two numbers.
a. 3.45×10^{-14} 3.45×10^{-15}
b. 5.23×10^{18} 5.23×10^{17}
c. 3.12×10^{12} 3.12×10^{11}

102. Astronomy Light travels 3×10^8 m in 1 s. How far does light travel in 1 year? (Astronomers refer to this distance as 1 light year.)

103. a. Evaluate $1^3 + 2^3 + 3^3 + 4^3$.
b. Evaluate $(-1)^3 + (-2)^3 + (-3)^3 + (-4)^3$.
c. Evaluate $1^3 + 2^3 + 3^3 + 4^3 + 5^3$.
d. On the basis of your answers to parts a, b, and c, evaluate $(-1)^3 + (-2)^3 + (-3)^3 + (-4)^3 + (-5)^3$.

104. Evaluate $2^{(3^2)}$ and $(2^3)^2$. Are the answers the same? If not, which is larger?

105. Abdul, Becky, Carl, and Diana were being questioned by their teacher. One of the students had left an apple on the teacher's desk, but the teacher did not know which one. Abdul said it was either Becky or Diana. Diana said it was neither Becky nor Carl. If both those statements are false, who left the apple on the teacher's desk? Explain how you arrived at your solution.

106. In your own words, explain how you know that a number is written in scientific notation.

107. a. Express the mass of the sun in kilograms using scientific notation.
b. Express the mass of a neutron in kilograms using scientific notation.

FOCUS ON PROBLEM SOLVING

Drawing Diagrams

How do you best remember something? Do you remember best what you hear? The word *aural* means "pertaining to the ear"; people with a strong aural memory remember best those things that they hear. The word *visual* means "pertaining to the sense of sight"; people with a strong visual memory remember best that which they see written down. Some people claim that their memory is in their writing hand—they remember something only if they write it down! The method by which you best remember something is probably also the method by which you can best learn something new.

In problem-solving situations, try to capitalize on your strengths. If you tend to understand the material better when you hear it spoken, read application problems aloud or have someone else read them to you. If writing helps you to organize ideas, rewrite application problems in your own words.

No matter what your main strength, visualizing a problem can be a valuable aid in problem solving. A drawing, sketch, diagram, or chart can be a useful tool in problem solving, just as calculators and computers are tools. A diagram can be helpful in gaining an understanding of the relationships inherent in a problem-solving situation. A sketch will help you to organize the given information and can lead to your being able to focus on the method by which the solution can be determined.

HOW TO 1 A tour bus drives 5 mi south, then 4 mi west, then 3 mi north, then 4 mi east. How far is the tour bus from the starting point?

Draw a diagram of the given information.

From the diagram, we can see that the solution can be determined by subtracting 3 from 5: $5 - 3 = 2$.

The bus is 2 mi from the starting point.

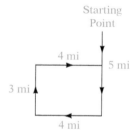

HOW TO 2 If you roll two ordinary six-sided dice and multiply the two numbers that appear on top, how many different possible products are there?

Make a chart of the possible products. In the chart below, repeated products are marked with an asterisk.

$1 \cdot 1 = 1$	$2 \cdot 1 = 2$ (*)	$3 \cdot 1 = 3$ (*)	$4 \cdot 1 = 4$ (*)	$5 \cdot 1 = 5$ (*)	$6 \cdot 1 = 6$ (*)
$1 \cdot 2 = 2$	$2 \cdot 2 = 4$ (*)	$3 \cdot 2 = 6$ (*)	$4 \cdot 2 = 8$ (*)	$5 \cdot 2 = 10$ (*)	$6 \cdot 2 = 12$ (*)
$1 \cdot 3 = 3$	$2 \cdot 3 = 6$ (*)	$3 \cdot 3 = 9$	$4 \cdot 3 = 12$ (*)	$5 \cdot 3 = 15$ (*)	$6 \cdot 3 = 18$ (*)
$1 \cdot 4 = 4$	$2 \cdot 4 = 8$	$3 \cdot 4 = 12$ (*)	$4 \cdot 4 = 16$	$5 \cdot 4 = 20$ (*)	$6 \cdot 4 = 24$ (*)
$1 \cdot 5 = 5$	$2 \cdot 5 = 10$	$3 \cdot 5 = 15$	$4 \cdot 5 = 20$	$5 \cdot 5 = 25$	$6 \cdot 5 = 30$ (*)
$1 \cdot 6 = 6$	$2 \cdot 6 = 12$	$3 \cdot 6 = 18$	$4 \cdot 6 = 24$	$5 \cdot 6 = 30$	$6 \cdot 6 = 36$

By counting the products that are not repeats, we can see that there are 18 different possible products.

Look at Sections 10.1 and 10.2. You will notice that number lines are used to help you visualize the integers, as an aid in ordering integers, to help you understand the concepts of opposite and absolute value, and to illustrate addition of integers. As you begin your work with integers, you may find that sketching a number line proves helpful in coming to understand a problem or in working through a calculation that involves integers.

PROJECTS AND GROUP ACTIVITIES

Deductive Reasoning

Suppose that during the last week of your math class, your instructor tells you that if you receive an A on the final exam, you will earn an A in the course. When the final exam grades are posted, you learn that you received an A on the final exam. You can then assume that you will earn an A in the course.

The process used to determine your grade in the math course is called deductive reasoning. **Deductive reasoning** involves drawing a conclusion that is based on given facts. The problems below require deductive reasoning.

1. Given that $\triangle\triangle\triangle = \Diamond\Diamond\Diamond\Diamond$ and $\Diamond\Diamond\Diamond\Diamond = \acute{O}\acute{O}$, then $\triangle\triangle\triangle\triangle\triangle\triangle =$ how many Ós?

2. Given that $\ddagger\ddagger = \bullet\bullet\bullet$ and $\bullet\bullet\bullet = \Lambda$, then $\ddagger\ddagger\ddagger\ddagger =$ how many Λs?

3. Given that $\acute{O}\acute{O}\acute{O} = \Omega\Omega$ and $\varnothing = \Omega\Omega$, then $\varnothing\varnothing =$ how many Ós?

4. Given that $\int\int\int\int\int = \partial\partial$ and $\partial\partial\partial\partial = ¥¥¥$, then $¥¥¥¥¥¥ =$ how many ∫s?

5. Given that $\hat{O}\hat{O}\hat{O}\hat{O}\hat{O} = \square\square\square$ and $\square\square\square\square\square\square = \S\S\S\S$, then $\S\S\S\S\S\S =$ how many Ôs?

6. Chris, Dana, Leslie, and Pat are neighbors. Each drives a different type of vehicle: an SUV, a sedan, a sports car, or a minivan. From the following statements, determine which type of vehicle each of the neighbors drives. It may be helpful to use the chart provided below.

    a. Although the vehicle owned by Chris has more mileage on it than does either the sedan or the sports car, it does not have the highest mileage of all four cars.

    b. Pat and the owner of the sports car live on one side of the street, and Leslie and the owner of the SUV live on the other side of the street.

    c. Leslie owns the vehicle with the most mileage on it.

√ Take Note

To use the chart to solve this problem, write an X in a box to indicate that a possibility has been eliminated. Write a √ to show that a match has been found. When a row or column has three X s, a √ is written in the remaining open box in that row or column of the chart.

	SUV	Sedan	Sports Car	Minivan
Chris				
Dana				
Leslie				
Pat				

7. Four neighbors, Anna, Kay, John, and Nicole, each plant a different vegetable (beans, cucumbers, squash, or tomatoes) in their garden. From the following statements, determine which vegetable each neighbor plants.

  a. Nicole's garden is bigger than the one that has tomatoes but smaller than the one that has cucumbers.

  b. Anna, who planted the largest garden, didn't plant the beans.

  c. The person who planted the beans has a garden the same size as Nicole's.

  d. Kay and the person who planted the tomatoes also have flower gardens.

8. The Ontkeans, Kedrovas, McIvers, and Levinsons are neighbors. Each of the four families specializes in a different national cuisine (Chinese, French, Italian, or Mexican). From the following statements, determine which cuisine each family specializes in.

  a. The Ontkeans invited the family that specializes in Chinese cuisine and the family that specializes in Mexican cuisine for dinner last night.

  b. The McIvers live between the family that specializes in Italian cuisine and the Ontkeans. The Levinsons live between the Kedrovas and the family that specializes in Chinese cuisine.

  c. The Kedrovas and the family that specializes in Italian cuisine both subscribe to the same culinary magazine.

CHAPTER 10

SUMMARY

KEY WORDS	EXAMPLES
Positive numbers are numbers greater than zero. *Negative numbers* are numbers less than zero. The *integers* are . . ., −4, −3, −2, −1, 0, 1, 2, 3, 4, ... *Positive integers* are to the right of zero on the number line. *Negative integers* are to the left of zero on the number line. [10.1A, p. 404]	9, 87, and 603 are positive numbers. They are also positive integers. −5, −41, and −729 are negative numbers. They are also negative integers.
Opposite numbers are two numbers that are the same distance from zero on the number line but on opposite sides of zero. [10.1B, p. 405]	8 is the opposite of −8. −2 is the opposite of 2.
The absolute value of a number is its distance from zero on the number line. The absolute value of a number is a positive number or zero. The symbol for absolute value is | |. [10.1B, p. 405]	$\|9\| = 9$ $\|-9\| = 9$ $-\|9\| = -9$
A *rational number* is a number that can be written in the form $\frac{a}{b}$, where a and b are integers and $b \neq 0$. [10.4A, p. 428]	$\frac{3}{7}, -\frac{5}{8}, 9, -2, 4\frac{1}{2}, 0.6,$ and $0.\overline{3}$ are rational numbers.

ESSENTIAL RULES AND PROCEDURES	EXAMPLES
Order Relations [10.1A, p. 404] A number that appears to the left of a given number on the number line is less than (<) the given number. A number that appears to the right of a given number on the number line is greater than (>) the given number.	$-6 > -12$ $-8 < 4$

To add numbers with the same sign, add the absolute values of the numbers. Then attach the sign of the addends. [10.2A, p. 411]	$6 + 4 = 10$ $-6 + (-4) = -10$
To add numbers with different signs, find the difference between the absolute values of the numbers. Then attach the sign of the addend with the greater absolute value. [10.2A, p. 411]	$-6 + 4 = -2$ $6 + (-4) = 2$
To subtract two numbers, add the opposite of the second number to the first number. [10.2B, p. 412]	$6 - 4 = 6 + (-4) = 2$ $6 - (-4) = 6 + 4 = 10$ $-6 - 4 = -6 + (-4) = -10$ $-6 - (-4) = -6 + 4 = -2$
To multiply numbers with the same sign, multiply the absolute values of the factors. The product is positive. [10.3A, p. 419]	$3 \cdot 5 = 15$ $-3(-5) = 15$
To multiply numbers with different signs, multiply the absolute values of the factors. The product is negative. [10.3A, p. 419]	$-3(5) = -15$ $3(-5) = -15$
To divide two numbers with the same sign, divide the absolute values of the numbers. The quotient is positive. [10.3B, p. 420]	$15 \div 3 = 5$ $(-15) \div (-3) = 5$
To divide two numbers with different signs, divide the absolute values of the numbers. The quotient is negative. [10.3B, p. 420]	$-15 \div 3 = -5$ $15 \div (-3) = -5$

Properties of Zero and One in Division   [10.3B, p. 421]
Zero divided by any number other than zero is zero.
Any number other than zero divided by itself is 1.
Any number divided by 1 is the number.
Division by zero is not defined.

$0 \div (-5) = 0$
$-5 \div (-5) = 1$
$-5 \div 1 = -5$
$-5 \div 0$ is undefined.

Scientific Notation   [10.5A, pp. 439–440]
To express a number in scientific notation, write it in the form
$a \times 10^n$, where a is a number between 1 and 10 and n is an integer.
If the number is greater than 10, the exponent on 10 will be positive.
If the number is less than 1, the exponent on 10 will be negative.

$367{,}000{,}000 = 3.67 \times 10^8$
$0.0000059 = 5.9 \times 10^{-6}$

To change a number written in scientific notation to decimal notation,
move the decimal point to the right if the exponent on 10 is positive
and to the left if the exponent on 10 is negative. Move the decimal
point the same number of places as the absolute value of the
exponent on 10.

$2.418 \times 10^7 = 24{,}180{,}000$
$9.06 \times 10^{-5} = 0.0000906$

The Order of Operations Agreement   [10.5B, p. 441]

Step 1 Do all operations inside parentheses.

Step 2 Simplify any numerical expressions containing exponents.

Step 3 Do multiplication and division as they occur from
left to right.

Step 4 Rewrite subtraction as addition of the opposite. Then do
additions as they occur from left to right.

$$(-4)^2 - 3(1 - 5) = (-4)^2 - 3(-4)$$
$$= 16 - 3(-4)$$
$$= 16 - (-12)$$
$$= 16 + 12$$
$$= 28$$

CHAPTER 10

CONCEPT REVIEW

Test your knowledge of the concepts presented in this chapter. Answer each question. Then check your answers against the ones provided in the Answer Section.

1. Find two numbers that are 6 units from 4 on the number line.

2. Find the absolute value of −6.

3. What is the rule for adding two integers?

4. What is the rule for subtracting two integers?

5. What operation is needed to find the change in temperature from −5°C to −14°C?

6. Show the result on the number line: 4 − 9.

7. If you multiply two nonzero numbers with different signs, what is the sign of the product?

8. If you divide two nonzero numbers with the same sign, what is the sign of the quotient?

9. What is the result when a number is divided by zero?

10. What is a terminating decimal?

11. What are the four steps in the Order of Operations Agreement?

12. How do you write the number 0.000754 in scientific notation?

CHAPTER 10

REVIEW EXERCISES

1. Find the opposite of 22.

2. Subtract: $-8 - (-2) - (-10) - 3$

3. Subtract: $\frac{5}{8} - \frac{5}{6}$

4. Simplify: $-0.33 + 1.98 - 1.44$

5. Multiply: $\left(-\frac{2}{3}\right)\left(\frac{6}{11}\right)\left(-\frac{22}{25}\right)$

6. Multiply: -0.08×16

7. Simplify: $12 - 6 \div 3$

8. Simplify: $\left(\frac{2}{3}\right)^2 - \frac{5}{6}$

9. Find the opposite of -4.

10. Place the correct symbol, $<$ or $>$, between the two numbers.
$0 \quad -3$

11. Evaluate $-|-6|$.

12. Divide: $-18 \div (-3)$

13. Add: $-\frac{3}{8} + \frac{5}{12} + \frac{2}{3}$

14. Multiply: $\frac{1}{3} \times \left(-\frac{3}{4}\right)$

15. Divide: $-\frac{7}{12} \div \left(-\frac{14}{39}\right)$

16. Simplify: $16 \div 4(8 - 2)$

17. Add: $-22 + 14 + (-18)$

18. Simplify: $3^2 - 9 + 2$

19. Write 0.0000397 in scientific notation.

20. Divide: $-1.464 \div 18.3$

21. Simplify: $-\frac{5}{12} + \frac{7}{9} - \frac{1}{3}$

22. Multiply: $\frac{6}{34} \times \frac{17}{40}$

23. Multiply: $1.2 \times (-0.035)$

24. Simplify: $-\frac{1}{2} + \frac{3}{8} \div \frac{9}{20}$

25. Evaluate $|-5|$.

26. Place the correct symbol, $<$ or $>$, between the two numbers.
$-2 \quad -40$

27. Find 2 times -13.

28. Simplify: $-0.4 \times 5 - (-3.33)$

29. Add: $\frac{5}{12} + \left(-\frac{2}{3}\right)$

30. Simplify: $-33.4 + 9.8 - (-16.2)$

31. Divide: $\left(-\frac{3}{8}\right) \div \left(-\frac{4}{5}\right)$

32. Write 2.4×10^5 in decimal notation.

33. **Temperatures** Find the temperature after a rise of $18°$ from $-22°$.

34. **Education** To discourage guessing on a multiple-choice exam, an instructor graded the test by giving 3 points for a correct answer, -1 point for an answer left blank, and -2 points for an incorrect answer. How many points did a student score who answered 38 questions correctly, answered 4 questions incorrectly, and left 8 questions blank?

35. **Chemistry** The boiling point of mercury is $356.58°C$. The melting point of mercury is $-38.87°C$. Find the difference between the boiling point and the melting point of mercury.

CHAPTER 10

TEST

1. Subtract: $-5 - (-8)$

2. Evaluate $-|-2|$.

3. Add: $-\dfrac{2}{5} + \dfrac{7}{15}$

4. Find the product of 0.032 and -1.9.

5. Place the correct symbol, $<$ or $>$, between the two numbers.
$-8 \quad -10$

6. Add: $1.22 + (-3.1)$

7. Simplify: $4 \times (4 - 7) \div (-2) - 4 \times 8$

8. Multiply: $-5(-6)(3)$

9. What is -1.004 decreased by 3.01?

10. Divide: $-72 \div 8$

11. Find the sum of -2, 3, and -8.

12. Add: $-\dfrac{3}{8} + \dfrac{2}{3}$

13. Write $87,600,000,000$ in scientific notation.

14. Find the product of -4 and 12.

15. Divide: $\dfrac{0}{-17}$

16. Subtract: $16 - 4 - (-5) - 7$

17. Find the quotient of $-\frac{2}{3}$ and $\frac{5}{6}$.

18. Place the correct symbol, $<$ or $>$, between the two numbers.
$$0 \quad -4$$

19. Add: $16 + (-10) + (-20)$

20. Simplify: $(-2)^2 - (-3)^2 \div (1 - 4)^2 \times 2 - 6$

21. Subtract: $-\frac{2}{5} - \left(-\frac{7}{10}\right)$

22. Write 9.601×10^{-8} in decimal notation.

23. Divide: $-15.64 \div (-4.6)$

24. Find the sum of $-\frac{1}{2}, \frac{1}{3},$ and $\frac{1}{4}$.

25. Multiply: $\frac{3}{8} \times \left(-\frac{5}{6}\right) \times \left(-\frac{4}{15}\right)$

26. Subtract: $2.113 - (-1.1)$

27. **Temperatures** Find the temperature after a rise of 11°C from −4°C.

28. **Chemistry** The melting point of radon is −71°C. The melting point of oxygen is three times the melting point of radon. Find the melting point of oxygen.

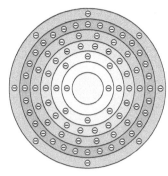

Radon

29. **Meteorology** On December 24, 1924, in Fairfield, Montana, the temperature fell from 17.22°C at noon to −29.4°C at midnight. How many degrees did the temperature fall in the 12-hour period?

30. **Meteorology** The daily low temperature readings for a 3-day period were as follows: −7°F, 9°F, −8°F. Find the average low temperature for the 3-day period.

CUMULATIVE REVIEW EXERCISES

1. Simplify: $16 - 4 \cdot (3 - 2)^2 \cdot 4$

2. Find the difference between $8\frac{1}{2}$ and $3\frac{4}{7}$.

3. Divide: $3\frac{7}{8} \div 1\frac{1}{2}$

4. Simplify: $\frac{3}{8} \div \left(\frac{3}{8} - \frac{1}{4}\right) \div \frac{7}{3}$

5. Subtract: $2.907 - 1.09761$

6. Solve the proportion $\frac{7}{12} = \frac{n}{32}$.
Round to the nearest hundredth.

7. 22 is 160% of what number?

8. Convert: $7 \text{ qt} = $ ____ gal ____ qt

9. Convert: $6692 \text{ ml} = $ ____ L

10. Convert 4.2 ft to meters. Round to the nearest hundredth. (1 m = 3.28 ft)

11. Find 32% of 180.

12. Convert $3\frac{2}{5}$ to a percent.

13. Add: $-8 + 5$

14. Add: $3\frac{1}{4} + \left(-6\frac{5}{8}\right)$

15. Subtract: $-6\frac{1}{8} - 4\frac{5}{12}$

16. Simplify: $-12 - (-7) - 3(-8)$

17. What is -3.2 times -1.09?

18. Multiply: $-6 \times 7 \times \left(\frac{3}{4}\right)$

19. Find the quotient of 42 and -6.

20. Divide: $-2\frac{1}{7} \div \left(-3\frac{3}{5}\right)$

21. Simplify: $3 \times (3 - 7) \div 6 - 2$

22. Simplify: $4 - (-2)^2 \div (1 - 2)^2 \times 3 + 4$

23. Carpentry A board $5\frac{2}{3}$ ft long is cut from a board 8 ft long. What is the length of the board remaining?

24. Banking Nimisha had a balance of $763.56 in her checkbook before writing checks for $135.88 and $47.81 and making a deposit of $223.44. Find her new checkbook balance.

25. Consumerism An 8-gigabyte video watch that regularly sells for $165 is on sale for $120. Find the percent decrease in price. Round to the nearest tenth of a percent.

26. Measurement A reception is planned for 80 guests. How many gallons of coffee should be prepared to provide 2 c of coffee for each guest?

27. Investments A stock selling for $82.625 per share paid a dividend of $1.50 per share before the dividend was increased by 12%. Find the dividend per share after the increase.

28. Family Night The circle graph at the right shows how often American households have a family night, during which they play a game as a family. Use this graph, and the fact that there are 114 million households in the United States, to answer the questions.
  a. How many households have a family night once a week?
  b. What fraction of U.S. households rarely or never have a family night?
  c. Is the number of households that have a family night only once a month more or less than three times the number of households that have a family night once a week?

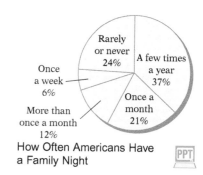

How Often Americans Have a Family Night

29. Elections A pre-election survey showed that 5 out of every 8 registered voters would cast ballots in a city election. At this rate, how many people would vote in a city of 960,000 registered voters?

30. Meteorology The daily high temperature readings for a 4-day period were recorded as follows: $-19°F$, $-7°F$, $1°F$, and $9°F$. Find the average high temperature for the 4-day period.

Introduction to Algebra

OBJECTIVES

SECTION 11.1
A To evaluate variable expressions
B To simplify variable expressions containing no parentheses
C To simplify variable expressions containing parentheses

SECTION 11.2
A To determine whether a given number is a solution of an equation
B To solve an equation of the form $x + a = b$
C To solve an equation of the form $ax = b$
D To solve application problems using formulas

SECTION 11.3
A To solve an equation of the form $ax + b = c$
B To solve application problems using formulas

SECTION 11.4
A To solve an equation of the form $ax + b = cx + d$
B To solve an equation containing parentheses

SECTION 11.5
A To translate a verbal expression into a mathematical expression given the variable
B To translate a verbal expression into a mathematical expression by assigning the variable

SECTION 11.6
A To translate a sentence into an equation and solve
B To solve application problems

Vito Palmisano/Getty Images

ARE YOU READY?

Take the Chapter 11 Prep Test to find out if you are ready to learn to:

- Simplify variable expressions
- Solve equations
- Translate a sentence into an equaiton and then solve the equation

PREP TEST

Do these exercises to prepare for Chapter 11.

For Exercises 1 to 9, simplify.

1. $2 - 9$

2. $-5(4)$

3. $-16 + 16$

4. $\dfrac{-7}{-7}$

5. $-\dfrac{3}{8}\left(-\dfrac{8}{3}\right)$

6. $\left(\dfrac{3}{5}\right)^3 \cdot \left(\dfrac{5}{9}\right)^2$

7. $\dfrac{2}{3} + \left(\dfrac{3}{4}\right)^2 \cdot \dfrac{2}{9}$

8. $-8 \div (-2)^2 + 6$

9. $4 + 5(2 - 7)^2 \div (-8 + 3)$

459

SECTION

11.1   Variable Expressions

OBJECTIVE A   To evaluate variable expressions

 Point of Interest

There are historical records indicating that mathematics has been studied for at least 4000 years. However, only in the last 400 years have variables been used. Prior to that, mathematics was written in words.

Often we discuss a quantity without knowing its exact value—for example, next year's inflation rate, the price of gasoline next summer, or the interest rate on a new-car loan next fall. In mathematics, a letter of the alphabet is used to stand for a quantity that is unknown or that can change, or *vary.* The letter is called a **variable.** An expression that contains one or more variables is called a **variable expression.**

A company's business manager has determined that the company will make a $10 profit on each DVD it sells. The manager wants to describe the company's total profit from the sale of DVDs. Because the number of DVDs that the company will sell is unknown, the manager lets the variable n stand for that number. Then the variable expression $10 \cdot n$, or simply $10n$, describes the company's profit from selling n DVDs.

The company's profit from selling n DVDs is $\$10 \cdot n = \$10n$.

If the company sells 12 DVDs, its profit is $\$10 \cdot 12 = \$120$.

If the company sells 75 DVDs, its profit is $\$10 \cdot 75 = \$750$.

Replacing the variable or variables in a variable expression and then simplifying the resulting numerical expression is called **evaluating a variable expression.**

HOW TO 1   Evaluate $3x^2 + xy - z$ when $x = -2$, $y = 3$, and $z = -4$.

$3x^2 + xy - z$

$3(-2)^2 + (-2)(3) - (-4)$ • Replace each variable in the expression with the number it stands for.

$= 3 \cdot 4 + (-2)(3) - (-4)$ • Use the Order of Operations Agreement to simplify the resulting numerical expression.

$= 12 + (-6) - (-4)$

$= 12 + (-6) + 4$

$= 6 + 4$

$= 10$

The value of the variable expression $3x^2 + xy - z$ when $x = -2$, $y = 3$, and $z = -4$ is 10.

EXAMPLE · 1

Evaluate $3x - 4y$ when $x = -2$ and $y = 3$.

Solution

$3x - 4y$

$3(-2) - 4(3) = -6 - 12$
$\qquad\qquad\quad = -6 + (-12) = -18$

YOU TRY IT · 1

Evaluate $6a - 5b$ when $a = -3$ and $b = 4$.

Your solution

EXAMPLE · 2

Evaluate $-x^2 - 6 \div y$ when $x = -3$ and $y = 2$.

Solution

$-x^2 - 6 \div y$

$-(-3)^2 - 6 \div 2 = -9 - 6 \div 2$
$\qquad\qquad\qquad\quad = -9 - 3$
$\qquad\qquad\qquad\quad = -9 + (-3) = -12$

YOU TRY IT · 2

Evaluate $-3s^2 - 12 \div t$ when $s = -2$ and $t = 4$.

Your solution

EXAMPLE · 3

Evaluate $-\dfrac{1}{2}y^2 - \dfrac{3}{4}z$ when $y = 2$ and $z = -4$.

Solution

$-\dfrac{1}{2}y^2 - \dfrac{3}{4}z$

$-\dfrac{1}{2}(2)^2 - \dfrac{3}{4}(-4) = -\dfrac{1}{2} \cdot 4 - \dfrac{3}{4}(-4)$
$\qquad\qquad\qquad\qquad = -2 - (-3)$
$\qquad\qquad\qquad\qquad = -2 + 3 = 1$

YOU TRY IT · 3

Evaluate $-\dfrac{2}{3}m + \dfrac{3}{4}n^3$ when $m = 6$ and $n = 2$.

Your solution

EXAMPLE · 4

Evaluate $-2ab + b^2 + a^2$ when $a = -\dfrac{3}{5}$ and $b = \dfrac{2}{5}$.

Solution

$-2ab + b^2 + a^2$

$-2\left(-\dfrac{3}{5}\right)\left(\dfrac{2}{5}\right) + \left(\dfrac{2}{5}\right)^2 + \left(-\dfrac{3}{5}\right)^2$

$= -2\left(-\dfrac{3}{5}\right)\left(\dfrac{2}{5}\right) + \left(\dfrac{4}{25}\right) + \left(\dfrac{9}{25}\right)$

$= \dfrac{12}{25} + \dfrac{4}{25} + \dfrac{9}{25} = \dfrac{25}{25} = 1$

YOU TRY IT · 4

Evaluate $-3yz - z^2 + y^2$ when $y = -\dfrac{2}{3}$ and $z = \dfrac{1}{3}$.

Your solution

Solutions on p. S25

OBJECTIVE B   To simplify variable expressions containing no parentheses

The **terms of a variable expression** are the addends of the expression. The variable expression at the right has four terms.

$$\overbrace{7x^2 \;+\; \underbrace{(-6xy)}_{} \;+\; x}^{\text{4 terms}} \;+\; \underbrace{(-8)}_{}$$

Variable terms   Constant term

Three of the terms are **variable terms:** $7x^2$, $-6xy$, and x.

One of the terms is a **constant term:** -8. A constant term has no variables.

Each variable term is composed of a **numerical coefficient** (the number part of a variable term) and a **variable part** (the variable or variables and their exponents). When the numerical coefficient is 1, the 1 is usually not written. $(1x = x)$

Like terms of a variable expression are the terms with the same variable part. (Because $y^2 = y \cdot y$, y^2 and y are not like terms.)

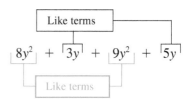

In variable expressions that contain constant terms, the constant terms are like terms.

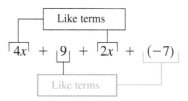

The Commutative and Associative Properties of Addition are used to simplify variable expressions. These properties can be stated in general form using variables.

Commutative Property of Addition

If a and b are two numbers, then $a + b = b + a$.

Associative Property of Addition

If a, b, and c are three numbers, then $a + (b + c) = (a + b) + c$.

The phrase **simplifying a variable expression** means *combining like terms* by adding their numerical coefficients. For example, to simplify $2y + 3y$, think

$$2y + 3y = (y + y) + (y + y + y) = 5y$$

HOW TO 2 Simplify: $8z - 5 + 2z$

$8z - 5 + 2z = 8z + 2z - 5$
$\qquad\qquad\quad = 10z - 5$

• Use the Commutative and Associative Properties of Addition to group like terms. Combine the like terms $8z + 2z$.

HOW TO 3 Simplify: $12a - 4b - 8a + 2b$

$$12a - 4b - 8a + 2b = 12a + (-4)b + (-8)a + 2b$$

• Change subtraction to addition of the opposite.

$$= 12a + (-8)a + (-4)b + 2b$$
$$= 4a + (-2)b$$

• Use the Commutative and Associative Properties of Addition to group like terms. Combine like terms.

$$= 4a - 2b$$

• Recall that $a + (-b) = a - b$.

HOW TO 4 Simplify: $6z^2 + 3 - z^2 - 7$

$$6z^2 + 3 - z^2 - 7 = 6z^2 + 3 + (-1)z^2 + (-7)$$

• Change subtraction to addition of the opposite.

$$= 6z^2 + (-1)z^2 + 3 + (-7)$$
$$= 5z^2 + (-4)$$
$$= 5z^2 - 4$$

• Use the Commutative and Associative Properties of Addition to group like terms. Combine like terms.

EXAMPLE 5

Simplify: $6xy - 8x + 5x - 9xy$

Solution

$6xy - 8x + 5x - 9xy$
$= 6xy + (-8)x + 5x + (-9)xy$
$= 6xy + (-9)xy + (-8)x + 5x$
$= (-3)xy + (-3)x$
$= -3xy - 3x$

YOU TRY IT 5

Simplify: $5a^2 - 6b^2 + 7a^2 - 9b^2$

Your solution

EXAMPLE 6

Simplify: $-4z^2 + 8 + 5z^2 - 3$

Solution

$-4z^2 + 8 + 5z^2 - 3$
$= -4z^2 + 8 + 5z^2 + (-3)$
$= -4z^2 + 5z^2 + 8 + (-3)$
$= z^2 + 5$

YOU TRY IT 6

Simplify: $-6x + 7 + 9x - 10$

Your solution

EXAMPLE 7

Simplify: $\frac{1}{4}m^2 - \frac{1}{2}n^2 + \frac{1}{2}m^2$

Solution

$$\frac{1}{4}m^2 - \frac{1}{2}n^2 + \frac{1}{2}m^2 = \frac{1}{4}m^2 + \left(-\frac{1}{2}\right)n^2 + \frac{1}{2}m^2$$

$$= \frac{1}{4}m^2 + \frac{1}{2}m^2 + \left(-\frac{1}{2}\right)n^2$$

$$= \frac{1}{4}m^2 + \frac{2}{4}m^2 + \left(-\frac{1}{2}\right)n^2$$

$$= \frac{3}{4}m^2 + \left(-\frac{1}{2}\right)n^2$$

$$= \frac{3}{4}m^2 - \frac{1}{2}n^2$$

YOU TRY IT 7

Simplify: $\frac{3}{8}w + \frac{1}{2} - \frac{1}{4}w - \frac{2}{3}$

Your solution

Solutions on pp. S25–S26

OBJECTIVE C   To simplify variable expressions containing parentheses

The Commutative and Associative Properties of Multiplication and the Distributive Property are used to simplify variable expressions that contain parentheses. These properties can be stated in general form using variables.

Commutative Property of Multiplication

If a and b are two numbers, then $a \cdot b = b \cdot a$.

Associative Property of Multiplication

If a, b, and c are three numbers, then $a \cdot (b \cdot c) = (a \cdot b) \cdot c$.

The Associative and Commutative Properties of Multiplication are used to simplify variable expressions such as the following.

HOW TO 5   Simplify: $-5(4x)$

$$-5(4x) = (-5 \cdot 4)x$$
$$= -20x$$

• Use the Associative Property of Multiplication.

HOW TO 6   Simplify: $(6y) \cdot 5$

$$(6y) \cdot 5 = 5 \cdot (6y)$$
$$= (5 \cdot 6)y = 30y$$

• Use the Commutative Property of Multiplication.
• Use the Associative Property of Multiplication.

The **Distributive Property** is used to remove parentheses from variable expressions that contain both multiplication and addition.

Distributive Property

If a, b, and c are three numbers, then $a(b + c) = ab + ac$.

HOW TO 7   Simplify: $4(z + 5)$

$$4(z + 5) = 4z + 4(5)$$
$$= 4z + 20$$

• The Distributive Property is used to rewrite the variable expression without parentheses.

HOW TO 8   Simplify: $-3(2x + 7)$

$$-3(2x + 7) = -3(2x) + (-3)(7)$$
$$= -6x + (-21)$$
$$= -6x - 21$$

• Use the Distributive Property.

• Recall that $a + (-b) = a - b$.

The Distributive Property can also be stated in terms of subtraction.

$$a(b - c) = ab - ac$$

HOW TO 9 Simplify: $8(2r - 3s)$

$$8(2r - 3s) = 8(2r) - 8(3s) \qquad \text{• Use the Distributive Property.}$$
$$= 16r - 24s$$

HOW TO 10 Simplify: $-5(2x - 4y)$

$$-5(2x - 4y) = (-5)(2x) - (-5)(4y) \qquad \text{• Use the Distributive Property.}$$
$$= -10x - (-20y)$$
$$= -10x + 20y \qquad \text{• Recall that } a - (-b) = a + b.$$

HOW TO 11 Simplify: $12 - 5(m + 2) + 2m$

$$12 - 5(m + 2) + 2m = 12 - 5m + (-5)(2) + 2m \qquad \text{• Use the Distributive}$$
$$= 12 - 5m + (-10) + 2m \qquad\qquad \text{Property to simplify}$$
the expression
$-5(m + 2)$.

$$= -5m + 2m + 12 + (-10) \qquad \text{• Use the Commutative and}$$
Associative Properties to
group like terms.

$$= -3m + 2 \qquad\qquad\qquad \text{• Combine like terms by}$$
adding their numerical
coefficients. Add constant
terms.

The answer $-3m + 2$ can also be written as $2 - 3m$. In this text, we will write answers with variable terms first, followed by the constant term.

EXAMPLE 8

Simplify: $4(x - 3)$

Solution

$$4(x - 3) = 4x - 4(3)$$
$$= 4x - 12$$

YOU TRY IT 8

Simplify: $5(a - 2)$

Your solution

EXAMPLE 9

Simplify: $5n - 3(2n - 4)$

Solution

$$5n - 3(2n - 4) = 5n - 3(2n) - (-3)(4)$$
$$= 5n - 6n - (-12)$$
$$= 5n - 6n + 12$$
$$= -n + 12$$

YOU TRY IT 9

Simplify: $8s - 2(3s - 5)$

Your solution

EXAMPLE 10

Simplify: $3(c - 2) + 2(c + 6)$

Solution

$$3(c - 2) + 2(c + 6) = 3c - 3(2) + 2c + 2(6)$$
$$= 3c - 6 + 2c + 12$$
$$= 3c + 2c - 6 + 12$$
$$= 5c + 6$$

YOU TRY IT 10

Simplify: $4(x - 3) - 2(x + 1)$

Your solution

Solutions on p. S26

11.1 EXERCISES

OBJECTIVE A   To evaluate variable expressions

For Exercises 1 to 34, evaluate the variable expression when $a = -3$, $b = 6$, and $c = -2$.

1. $5a - 3b$

2. $4c - 2b$

3. $2a + 3c$

4. $2c + 4a$

5. $-c^2$

6. $-a^2$

7. $b - a^2$

8. $b - c^2$

9. $ab - c^2$

10. $bc - a^2$

11. $2ab - c^2$

12. $3bc - a^2$

13. $a - (b \div a)$

14. $c - (b \div c)$

15. $2ac - (b \div a)$

16. $4ac \div (b \div a)$

17. $b^2 - c^2$

18. $b^2 - a^2$

19. $b^2 \div (ac)$

20. $3c^2 \div (ab)$

21. $c^2 - (b \div c)$

22. $a^2 - (b \div a)$

23. $a^2 + b^2 + c^2$

24. $a^2 - b^2 - c^2$

25. $ac + bc + ab$

26. $ac - bc - ab$

27. $a^2 + b^2 - ab$

28. $b^2 + c^2 - bc$

29. $2b - (3c + a^2)$

30. $\frac{2}{3}b + \left(\frac{1}{2}c - a\right)$

31. $\frac{1}{3}a + \left(\frac{1}{2}b - \frac{2}{3}a\right)$

32. $-\frac{2}{3}b - \left(\frac{1}{2}c + a\right)$

33. $\frac{1}{6}b + \frac{1}{3}(c + a)$

34. $\frac{1}{2}c + \left(\frac{1}{3}b - a\right)$

For Exercises 35 to 38, evaluate the variable expression when $a = -\frac{1}{2}$, $b = \frac{3}{4}$, and $c = \frac{1}{4}$.

35. $4a + (3b - c)$

36. $2b + (c - 3a)$

37. $2a - b^2 \div c$

38. $b \div (-c) + 2a$

For Exercises 39 to 42, evaluate the variable expression when $a = 3.48$, $b = -2.31$, and $c = -1.74$.

39. $a^2 - b^2$

40. $a^2 - b \cdot c$

41. $3ac - (a \div c)$

42. $2c + (b^2 - c)$

For Exercises 43 to 46, suppose a is positive and b is negative, with $|a|$ greater than $|b|$. Determine whether the value of the variable expression is positive or negative.

43. ab^2 **44.** $b^2 - a^2$ **45.** $a - (a + b)$ **46.** $a^2 + ab$

OBJECTIVE B   To simplify variable expressions containing no parentheses

For Exercises 47 to 50, name the terms of the variable expression. Then underline the constant term.

47. $2x^2 + 3x - 4$ **48.** $-4y^2 + 5$ **49.** $3a^2 - 4a + 8$ **50.** $7 - b$

For Exercises 51 to 54, name the variable terms of the expression. Then underline the coefficients of the variable terms.

51. $3x^2 - 4x + 9$ **52.** $-5a^2 + a - 4$ **53.** $y^2 + 6a - 1$ **54.** $8 - c$

 55. Which expressions are equivalent to $3 - 2a^2 + 8a^2 - 5$?

(i) $9a^2 - 5$ (ii) $6a^2 - 2$ (iii) $-2 + 6a^2$ (iv) $6a^2 - 8$

 56. Which expressions are equivalent to $6x^2 - x + 3x + x^2$?

(i) $6x^2 + 2x + x^2$ (ii) $2x + 7x^2$ (iii) $7x^2 + 2x$ (iv) $6x^2 + x^2 + 3x - x$

For Exercises 57 to 98, simplify.

57. $7z + 9z$ **58.** $6x + 5x$ **59.** $12m - 3m$

60. $5y - 12y$ **61.** $5at + 7at$ **62.** $12mn + 11mn$

63. $-4yt + 7yt$ **64.** $-12yt + 5yt$ **65.** $-3x - 12y$

66. $-12y - 7y$ **67.** $3t^2 - 5t^2$ **68.** $7t^2 + 8t^2$

69. $6c - 5 + 7c$ **70.** $7x - 5 + 3x$ **71.** $2t + 3t - 7t$

72. $9x^2 - 5 - 3x^2$ **73.** $7y^2 - 2 - 4y^2$ **74.** $3w - 7u + 4w$

75. $6w - 8u + 8w$ **76.** $4 - 6xy - 7xy$ **77.** $10 - 11xy - 12xy$

78. $7t^2 - 5t^2 - 4t^2$ **79.** $3v^2 - 6v^2 - 8v^2$ **80.** $5ab - 7a - 10ab$

81. $-10ab - 3a + 2ab$

82. $-4x^2 - x + 2x^2$

83. $-3y^2 - y + 7y^2$

84. $4x^2 - 8y - x^2 + y$

85. $2a - 3b^2 - 5a + b^2$

86. $8y - 4z - y + 2z$

87. $3x^2 - 7x + 4x^2 - x$

88. $5y^2 - y + 6y^2 - 5y$

89. $6s - t - 9s + 7t$

90. $5w - 2v - 9w + 5v$

91. $4m + 8n - 7m + 2n$

92. $z + 9y - 4z + 3y$

93. $-5ab + 7ac + 10ab - 3ac$

94. $-2x^2 - 3x - 11x^2 + 14x$

95. $\dfrac{4}{9}a^2 - \dfrac{1}{5}b^2 + \dfrac{2}{9}a^2 + \dfrac{4}{5}b^2$

96. $\dfrac{6}{7}x^2 + \dfrac{2}{5}x - \dfrac{3}{7}x^2 - \dfrac{4}{5}x$

97. $7.81m + 3.42n - 6.25m - 7.19n$

98. $8.34y^2 - 4.21y - 6.07y^2 - 5.39y$

OBJECTIVE C To simplify variable expressions containing parentheses

99. Which expressions are equivalent to $7a + 7b$?

(i) $9 - 2(a + b)$ (ii) $(9 - 2)(a + b)$ (iii) $3(a + b) + 4(a + b)$

100. Which expressions are equivalent to $5 + 3(m + 8)$?

(i) $3m + 29$ (ii) $5 + 3m + 24$ (iii) $5 + 3m + 8$

For Exercises 101 to 136, simplify.

101. $5(x + 4)$

102. $3(m + 6)$

103. $(y - 3)4$

104. $(z - 3)7$

105. $-2(a + 4)$

106. $-5(b + 3)$

107. $3(5x + 10)$

108. $2(4m - 7)$

109. $5(3c - 5)$

110. $-4(w - 3)$

111. $-3(y - 6)$

112. $3m + 4(m + z)$

113. $5x + 2(x + 7)$

114. $6z - 3(z + 4)$

115. $8y - 4(y + 2)$

116. $7w - 2(w - 3)$

117. $9x - 4(x - 6)$

118. $-5m + 3(m + 4)$

119. $-2y + 3(y - 2)$

120. $5m + 3(m + 4) - 6$

121. $4n + 2(n + 1) - 5$

122. $8z - 2(z - 3) + 8$

123. $9y - 3(y - 4) + 8$

124. $6 - 4(a + 4) + 6a$

125. $3x + 2(x + 2) + 5x$

126. $7x + 4(x + 1) + 3x$

127. $-7t + 2(t - 3) - t$

128. $-3y + 2(y - 4) - y$

129. $z - 2(1 - z) - 2z$

130. $2y - 3(2 - y) + 4y$

131. $3(y - 2) - 2(y - 6)$

132. $7(x + 2) + 3(x - 4)$

133. $2(t - 3) + 7(t + 3)$

134. $3(y - 4) - 2(y - 3)$

135. $3t - 6(t - 4) + 8t$

136. $5x + 3(x - 7) - 9x$

Applying the Concepts

137. The square and the rectangle at the right can be used to illustrate algebraic expressions. The illustration below represents the expression $2x + 1$.

    a. Using squares and rectangles in a similar manner, draw a figure that represents the expression $2 + 3x$.

    b. Draw a figure that represents the expression $5x$.

    c. Does the figure $2 + 3x$ equal the figure $5x$? Explain how this is related to combining like terms.

138.   **a.** Using squares and rectangles as in Exercise 137, draw a figure that represents the expression $2(2x + 3)$.

    b. Draw a figure that represents the expression $4x + 3$.

    c. Draw a figure that represents the expression $4x + 6$.

    d. Does the figure for $2(2x + 3)$ equal the figure for $4x + 3$? Does the figure for $2(2x + 3)$ equal the figure for $4x + 6$? Explain how your results are related to the Distributive Property.

SECTION

11.2 Introduction to Equations

OBJECTIVE A To determine whether a given number is a solution of an equation

 Point of Interest

Finding solutions of equations has been a principal aim of mathematics for thousands of years. However, the equals sign did not appear in any text until 1557.

An **equation** expresses the equality of two mathematical expressions. These expressions can be either numerical or variable expressions.

$$5 + 4 = 9$$
$$3x + 13 = x - 8$$
$$y^2 + 4 = 6y + 1$$
$$x = -3$$

Equations

In the equation at the right, if the variable is replaced by 4, the equation is true.

$$x + 3 = 7$$
$$4 + 3 = 7 \quad \text{A true equation}$$

If the variable is replaced by 6, the equation is false.

$$6 + 3 = 7 \quad \text{A false equation}$$

A **solution of an equation** is a number that, when substituted for the variable, results in a true equation. 4 is a solution of the equation $x + 3 = 7$. 6 is not a solution of the equation $x + 3 = 7$.

HOW TO 1 Is -2 a solution of the equation $-2x + 1 = 2x + 9$?

$$-2x + 1 = 2x + 9$$

$-2(-2) + 1$	$2(-2) + 9$	• Replace the variable by the given number.
$4 + 1$	$-4 + 9$	• Evaluate the numerical expressions.

$$5 = 5$$

• Compare the results. If the results are equal, the given number is a solution. If the results are not equal, the given number is not a solution.

Yes, -2 is a solution of the equation $-2x + 1 = 2x + 9$.

EXAMPLE 1

Is $\dfrac{1}{2}$ a solution of $2x(x + 2) = 3x + 1$?

Solution

$$2x(x + 2) = 3x + 1$$

$2\left(\dfrac{1}{2}\right)\left(\dfrac{1}{2} + 2\right)$	$3\left(\dfrac{1}{2}\right) + 1$
$2\left(\dfrac{1}{2}\right)\left(\dfrac{5}{2}\right)$	$\dfrac{3}{2} + 1$

$$\dfrac{5}{2} = \dfrac{5}{2}$$

Yes, $\dfrac{1}{2}$ is a solution.

YOU TRY IT 1

Is -2 a solution of $x(x + 3) = 4x + 6$?

Your solution

Solution on p. S26

EXAMPLE • 2

Is 5 a solution of $(x - 2)^2 = x^2 - 4x + 2$?

Solution

$$
\begin{array}{c|l}
(x - 2)^2 &= x^2 - 4x + 2 \\
\hline
(5 - 2)^2 & 5^2 - 4(5) + 2 \\
3^2 & 25 - 4(5) + 2 \\
9 & 25 - 20 + 2 \\
& 25 + (-20) + 2 \\
9 \neq 7 & (\neq \text{ means "is not} \\
& \text{equal to")}
\end{array}
$$

No, 5 is not a solution.

YOU TRY IT • 2

Is -3 a solution of $x^2 - x = 3x + 7$?

Your solution

Solution on p. S26

OBJECTIVE B   To solve an equation of the form $x + a = b$

A solution of an equation is a number that, when substituted for the variable, results in a true equation. The phrase **solving an equation** means finding a solution of the equation.

The simplest equation to solve is an equation of the form *variable = constant*. The constant is the solution of the equation.

If $x = 7$, then 7 is the solution of the equation because $7 = 7$ is a true equation.

In solving an equation of the form $x + a = b$, the goal is to simplify the given equation to one of the form *variable = constant*. The Addition Properties that follow are used to simplify equations to this form.

Addition Property of Zero

The sum of a term and zero is the term.

$$a + 0 = a \qquad 0 + a = a$$

Addition Property of Equations

If *a*, *b*, and *c* are algebraic expressions, then the equation $a = b$ has the same solutions as the equation $a + c = b + c$.

The Addition Property of Equations states that the same quantity can be added to each side of an equation without changing the solution of the equation.

The Addition Property of Equations is used to rewrite an equation in the form *variable = constant*. We remove a term from one side of the equation by adding the opposite of that term to each side of the equation.

HOW TO · 2   Solve: $x - 7 = -2$

$$x - 7 = -2$$

• The goal is to simplify the equation to one of the form *variable = constant*.

$$x - 7 + 7 = -2 + 7$$
$$x + 0 = 5$$
$$x = 5$$

• Add the opposite of the constant term -7 to each side of the equation. After we simplify and use the Addition Property of Zero, the equation will be in the form *variable = constant*.

The solution is 5.

 Take Note

Always check the solution to an equation.

Check: $x - 7 = -2$
$$\frac{5 - 7 \,|\, -2}{-2 = -2 \quad \text{True}}$$

Because subtraction is defined in terms of addition, the Addition Property of Equations allows the same number to be subtracted from each side of an equation.

Tips for Success

When we suggest that you check a solution, you should substitute the solution into the *original* equation. For instance,

$$x + 8 = 5$$
$$\frac{-3 + 8 \,|\, 5}{5 = 5}$$

The solution checks.

HOW TO · 3   Solve: $x + 8 = 5$

$$x + 8 = 5$$

• The goal is to simplify the equation to one of the form *variable = constant*.

$$x + 8 - 8 = 5 - 8$$
$$x + 0 = -3$$
$$x = -3$$

• Add the opposite of the constant term 8 to each side of the equation. This procedure is equivalent to subtracting 8 from each side of the equation.

The solution is -3. **You should check this solution.**

EXAMPLE · 3

Solve: $4 + m = -2$

Solution
$$4 + m = -2$$
$$4 - 4 + m = -2 - 4 \quad \text{• Subtract 4}$$
$$0 + m = -6 \qquad\qquad \text{from each}$$
$$m = -6 \qquad\qquad\quad \text{side.}$$

The solution is -6.

YOU TRY IT · 3

Solve: $-2 + y = -5$

Your solution

EXAMPLE · 4

Solve: $3 = y - 2$

Solution
$$3 = y - 2$$
$$3 + 2 = y - 2 + 2 \quad \text{• Add 2}$$
$$5 = y + 0 \qquad\qquad \text{to each side.}$$
$$5 = y$$

The solution is 5.

YOU TRY IT · 4

Solve: $7 = y + 8$

Your solution

EXAMPLE · 5

Solve: $\dfrac{2}{7} = \dfrac{5}{7} + t$

Solution
$$\frac{2}{7} = \frac{5}{7} + t$$
$$\frac{2}{7} - \frac{5}{7} = \frac{5}{7} - \frac{5}{7} + t \quad \text{• Subtract } \frac{5}{7} \text{ from}$$
$$\qquad\qquad\qquad\qquad \text{each side.}$$
$$-\frac{3}{7} = 0 + t$$
$$-\frac{3}{7} = t$$

The solution is $-\dfrac{3}{7}$.

YOU TRY IT · 5

Solve: $\dfrac{1}{5} = z + \dfrac{4}{5}$

Your solution

OBJECTIVE C   To solve an equation of the form $ax = b$

In solving an equation of the form $ax = b$, the goal is to simplify the given equation to one of the form *variable = constant*. The Multiplication Properties that follow are used to simplify equations to this form.

Multiplication Property of Reciprocals

The product of a nonzero term and its reciprocal equals 1.

Because $a = \frac{a}{1}$, the reciprocal of a is $\frac{1}{a}$.

$a\left(\dfrac{1}{a}\right) = 1 \qquad \dfrac{1}{a}(a) = 1$

The reciprocal of $\frac{a}{b}$ is $\frac{b}{a}$.

$\left(\dfrac{a}{b}\right)\left(\dfrac{b}{a}\right) = 1 \qquad \left(\dfrac{b}{a}\right)\left(\dfrac{a}{b}\right) = 1$

Multiplication Property of One

The product of a term and 1 is the term.

$a \cdot 1 = a \qquad 1 \cdot a = a$

Multiplication Property of Equations

If a, b, and c are algebraic expressions and $c \neq 0$, then the equation $a = b$ has the same solutions as the equation $ac = bc$.

The Multiplication Property of Equations states that each side of an equation can be multiplied by the same nonzero number without changing the solutions of the equation.

Recall that the goal of solving an equation is to rewrite the equation in the form *variable = constant*. The Multiplication Property of Equations is used to rewrite an equation in this form by multiplying each side of the equation by the reciprocal of the coefficient.

HOW TO 4   Solve: $\frac{2}{3}x = 8$

$$\frac{2}{3}x = 8$$

$$\left(\frac{3}{2}\right)\left(\frac{2}{3}\right)x = \left(\frac{3}{2}\right)8$$ • **Multiply** each side of the equation by $\frac{3}{2}$, the

$$1 \cdot x = 12$$ reciprocal of $\frac{2}{3}$. After simplifying, the

$$x = 12$$ equation will be in the form *variable = constant*.

Check: $\dfrac{2}{3}x = 8$

$$\dfrac{\dfrac{2}{3}(12)}{} \ \Big| \ 8$$

$$8 = 8$$

The solution is 12.

Because division is defined in terms of multiplication, the Multiplication Property of Equations allows each side of an equation to be divided by the same nonzero quantity.

HOW TO 5   Solve: $-4x = 24$

$-4x = 24$

• The goal is to rewrite the equation in the form *variable = constant*.

$$\frac{-4x}{-4} = \frac{24}{-4}$$

• Multiply each side of the equation by the reciprocal of -4. This is equivalent to dividing each side of the equation by -4. Then simplify.

$1x = -6$

$x = -6$

The solution is -6. You should check this solution.

Tips for Success

When we suggest that you check a solution, you should substitute the solution into the *original* equation. For instance,

$$\frac{-4x = 24}{-4(-6) \mid 24}$$
$$24 = 24$$

The solution checks.

In using the Multiplication Property of Equations, it is usually easier to multiply each side of the equation by the reciprocal of the coefficient when the coefficient is a fraction. Divide each side of the equation by the coefficient when the coefficient is an integer or a decimal.

EXAMPLE • 6

Solve: $-2x = 6$

Solution

$-2x = 6$

$$\frac{-2x}{-2} = \frac{6}{-2}$$ • **Divide** each side by -2.

$1x = -3$

$x = -3$

The solution is -3.

YOU TRY IT • 6

Solve: $4z = -20$

Your solution

EXAMPLE • 7

Solve: $-9 = \frac{3}{4}y$

Solution

$$-9 = \frac{3}{4}y$$

$$\left(\frac{4}{3}\right)(-9) = \left(\frac{4}{3}\right)\left(\frac{3}{4}y\right)$$ • **Multiply** each side by $\frac{4}{3}$.

$-12 = 1y$

$-12 = y$

The solution is -12.

YOU TRY IT • 7

Solve: $8 = \frac{2}{5}n$

Your solution

EXAMPLE • 8

Solve: $6z - 8z = -5$

Solution

$6z - 8z = -5$

$-2z = -5$ • **Combine** like terms.

$$\frac{-2z}{-2} = \frac{-5}{-2}$$ • **Divide** each side by -2.

$$1z = \frac{5}{2}$$

$$z = \frac{5}{2} = 2\frac{1}{2}$$

The solution is $2\frac{1}{2}$.

YOU TRY IT • 8

Solve: $\frac{2}{3}t - \frac{1}{3}t = -2$

Your solution

Solutions on p. S26

OBJECTIVE D To solve application problems using formulas

A **formula** is an equation that expresses a relationship among variables. Formulas are used in the examples below.

EXAMPLE 9

An accountant for a greeting card store found that the weekly profit for the store was $1700 and that the total amount spent during the week was $2400. Use the formula $P = R - C$, where P is the profit, R is the revenue, and C is the amount spent, to find the revenue for the week.

Strategy

To find the revenue for the week, replace the variables P and C in the formula by the given values, and solve for R.

Solution

$$P = R - C$$
$$1700 = R - 2400$$
$$1700 + 2400 = R - 2400 + 2400$$
$$4100 = R + 0$$
$$4100 = R$$

The revenue for the week was $4100.

YOU TRY IT 9

A clothing store's sale price for a college sweatshirt is $44. This is a discount of $16 off the regular price. Use the formula $S = R - D$, where S is the sale price, R is the regular price, and D is the discount, to find the regular price.

Your strategy

Your solution

EXAMPLE 10

A store manager uses the formula $S = R - d \cdot R$, where S is the sale price, R is the regular price, and d is the discount rate. During a clearance sale, all items are discounted 20%. Find the regular price of running shoes that are on sale for $120.

Strategy

To find the regular price of the running shoes, replace the variables S and d in the formula by the given values, and solve for R.

Solution

$$S = R - d \cdot R$$
$$120 = R - 0.20R$$
$$120 = 0.80R \qquad \bullet \; R - 0.20R = 1R - 0.20R$$
$$\frac{120}{0.80} = \frac{0.80R}{0.80}$$
$$150 = R$$

The regular price of the running shoes is $150.

YOU TRY IT 10

Find the monthly payment when the total amount paid on a loan is $6840 and the loan is paid off in 24 months. Use the formula $A = MN$, where A is the total amount paid on a loan, M is the monthly payment, and N is the number of monthly payments.

Your strategy

Your solution

Solutions on p. S26

11.2 EXERCISES

OBJECTIVE A   To determine whether a given number is a solution of an equation

1. Is -3 a solution of $2x + 9 = 3$?

2. Is -2 a solution of $5x + 7 = 12$?

3. Is 2 a solution of $4 - 2x = 8$?

4. Is 4 a solution of $5 - 2x = 4x$?

5. Is 3 a solution of $3x - 2 = x + 4$?

6. Is 2 a solution of $4x + 8 = 4 - 2x$?

7. Is 3 a solution of $x^2 - 5x + 1 = 10 - 5x$?

8. Is -5 a solution of $x^2 - 3x - 1 = 9 - 6x$?

9. Is -1 a solution of $2x(x - 1) = 3 - x$?

10. Is 2 a solution of $3x(x - 3) = x - 8$?

11. Is 2 a solution of $x(x - 2) = x^2 - 4$?

12. Is -4 a solution of $x(x + 4) = x^2 + 16$?

13. Is $-\dfrac{2}{3}$ a solution of $3x + 6 = 4$?

14. Is $\dfrac{1}{2}$ a solution of $2x - 7 = -3$?

15. Is $\dfrac{1}{4}$ a solution of $2x - 3 = 1 - 14x$?

16. Is 1.32 a solution of $x^2 - 3x = -0.8776 - x$?

17. Is -1.9 a solution of $x^2 - 3x = x + 3.8$?

18. Is 1.05 a solution of $x^2 + 3x = x(x + 3)$?

 For Exercises 19 and 20, determine whether the statement is *true* or *false*.

19. Any number that is a solution of the equation $5x + 1 = -9$ must be a negative number.

20. Any number that is a solution of the equation $3x + 6 = 3 - 4x$ must also be a solution of the equation $3(x + 2) = 3 + 4x - 8x$.

OBJECTIVE B   To solve an equation of the form $x + a = b$

For Exercises 21 to 48, solve.

21. $y - 6 = 16$

22. $z - 4 = 10$

23. $3 + n = 4$

24. $6 + x = 8$

25. $z + 7 = 2$

26. $w + 9 = 5$

27. $x - 3 = -7$

28. $m - 4 = -9$

29. $y + 6 = 6$ **30.** $t - 3 = -3$ **31.** $v - 7 = -4$ **32.** $x - 3 = -1$

33. $1 + x = 0$ **34.** $3 + y = 0$ **35.** $x - 10 = 5$ **36.** $y - 7 = 3$

37. $x + 4 = -7$ **38.** $t - 3 = -8$ **39.** $w + 5 = -5$ **40.** $z + 6 = -6$

41. $x + \dfrac{1}{2} = -\dfrac{1}{2}$ **42.** $x - \dfrac{5}{6} = -\dfrac{1}{6}$ **43.** $\dfrac{2}{5} + x = -\dfrac{3}{5}$ **44.** $\dfrac{7}{8} + y = -\dfrac{1}{8}$

45. $x + \dfrac{1}{2} = -\dfrac{1}{3}$ **46.** $y + \dfrac{3}{8} = \dfrac{1}{4}$ **47.** $t + \dfrac{1}{4} = -\dfrac{1}{2}$ **48.** $x + \dfrac{1}{3} = \dfrac{5}{12}$

For Exercises 49 to 52, use the given conditions for a and b to determine whether the value of x in the equation $x + a = b$ must be negative, must be positive, or could be either positive or negative.

49. a is positive and b is negative.   **50.** a is a positive proper fraction and b is greater than 1.

51. a is the opposite of b and b is positive.   **52.** a is negative and b is negative.

OBJECTIVE C   To solve an equation of the form $ax = b$

For Exercises 53 to 84, solve.

53. $3y = 12$ **54.** $5x = 30$ **55.** $5z = -20$ **56.** $3z = -27$

57. $-2x = 6$ **58.** $-4t = 20$ **59.** $-5x = -40$ **60.** $-2y = -28$

61. $40 = 8x$ **62.** $24 = 3y$ **63.** $-24 = 4x$ **64.** $-21 = 7y$

65. $\dfrac{x}{3} = 5$ **66.** $\dfrac{y}{2} = 10$ **67.** $\dfrac{n}{4} = -2$ **68.** $\dfrac{y}{7} = -3$

69. $-\dfrac{x}{4} = 1$

70. $-\dfrac{y}{3} = 5$

71. $\dfrac{2}{3}w = 4$

72. $\dfrac{5}{8}x = 10$

73. $\dfrac{3}{4}v = -3$

74. $\dfrac{2}{7}x = -12$

75. $-\dfrac{1}{3}x = -2$

76. $-\dfrac{1}{5}y = -3$

77. $-4 = -\dfrac{2}{3}z$

78. $-12 = -\dfrac{3}{8}y$

79. $\dfrac{2}{3}x = -\dfrac{2}{7}$

80. $\dfrac{3}{7}y = \dfrac{5}{6}$

81. $4x - 2x = 7$

82. $3a - 6a = 8$

83. $\dfrac{4}{5}m - \dfrac{1}{5}m = 9$

84. $\dfrac{1}{3}b - \dfrac{2}{3}b = -1$

For Exercises 85 to 87, determine whether the statement is *true* or *false*.

85. If a is positive and b is negative, then the value of x in the equation $ax = b$ must be negative.

86. If a is the opposite of b, then the value of x in the equation $ax = b$ must be -1.

87. If a is negative and b is negative, then the value of x in the equation $ax = b$ is negative.

OBJECTIVE D   To solve application problems using formulas

88. A store's cost for a pair of jeans is \$38. The store sells the jeans and makes a \$14 profit. Which equations can you use to find the selling price of the jeans? In each equation, S represents the selling price of the item.
(i) $S - 14 = 38$ (ii) $14 = 38 - S$
(iii) $38 - 14 = S$ (iv) $S - 38 = 14$

Fuel Efficiency In Exercises 89 to 92, use the formula $D = M \cdot G$, where D is the distance, M is the miles per gallon, and G is the number of gallons. Round to the nearest tenth.

89. Julio, a sales executive, averaged 28 mi/gal on a 621-mile trip. Find the number of gallons of gasoline used on the trip.

90. Over a 3-day weekend, you take a 592-mile trip. If you average 32 mi/gal on the trip, how many gallons of gasoline did you use?

91. The manufacturer of a hatchback estimates that the car can travel 560 mi on a 15-gallon tank of gas. Find the miles per gallon.

92. You estimate that your car can travel 410 mi on 12 gal of gasoline. Find the miles per gallon.

Investments In Exercises 93 to 96, use the formula $A = P + I$, where A is the value of the investment after 1 year, P is the original investment, and I is the increase in value of the investment.

93. The value of an investment in a high-tech company after 1 year was $17,700. The increase in value during the year was $2700. Find the amount of the original investment.

94. The value of an investment in a software company after 1 year was $26,440. The increase in value during the year was $2830. Find the amount of the original investment.

95. The original investment in a mutual fund was $8000. The value of the mutual fund after 1 year was $11,420. Find the increase in value of the investment.

96. The original investment in a money market fund was $7500. The value of the mutual fund after 1 year was $8690. Find the increase in value of the investment.

Markup In Exercises 97 and 98, use the formula $S = C + M$, where S is the selling price, C is the cost, and M is the markup.

97. A computer store sells a computer for $2240. The computer has a markup of $420. Find the cost of the computer.

98. A toy store buys stuffed animals for $23.50 and sells them for $39.80. Find the markup on each stuffed animal.

© James Leynse/Corbis

Markup In Exercises 99 and 100, use the formula $S = C + R \cdot C$, where S is the selling price, C is the cost, and R is the markup rate.

99. A store manager uses a markup rate of 24% on all children's furniture. Find the cost of a crib that sells for $232.50.

100. A music store uses a markup rate of 30%. Find the cost of a compact disc that sells for $18.85.

Applying the Concepts

101. Write out the steps for solving the equation $x - 3 = -5$. Identify each property of real numbers and each property of equations as you use it.

102. Write out the steps for solving the equation $\frac{3}{4}x = 6$. Identify each property of real numbers and each property of equations as you use it.

103. Write an equation of the form $x + a = b$ that has -4 as its solution.

SECTION

11.3 General Equations: Part I

OBJECTIVE A To solve an equation of the form $ax + b = c$

To solve an equation of the form $ax + b = c$, it is necessary to use both the Addition and the Multiplication Properties to simplify the equation to one of the form *variable = constant*.

 Point of Interest

Evariste Galois, despite being killed in a duel at the age of 21, made significant contributions to solving equations. In fact, there is a branch of mathematics called Galois theory that shows what kinds of equations can be solved and what kinds cannot.

✓ **Take Note**

$\dfrac{x}{4} = \dfrac{1}{4}x$

The reciprocal of $\frac{1}{4}$ is 4.

HOW TO 1 Solve: $\dfrac{x}{4} - 1 = 3$

$$\frac{x}{4} - 1 = 3$$

• The goal is to simplify the equation to one of the form *variable = constant*.

$$\frac{x}{4} - 1 + 1 = 3 + 1$$

$$\frac{x}{4} + 0 = 4$$

• Add the opposite of the constant term -1 to each side of the equation. Then simplify (Addition Properties).

$$\frac{x}{4} = 4$$

$$4 \cdot \frac{x}{4} = 4 \cdot 4$$

$$1x = 16$$

$$x = 16$$

• Multiply each side of the equation by the reciprocal of the numerical coefficient of the variable term. Then simplify (Multiplication Properties).

The solution is 16.

• Write the solution.

EXAMPLE 1

Solve: $3x + 7 = 2$

Solution

$$3x + 7 = 2$$
$$3x + 7 - 7 = 2 - 7 \qquad \text{• Subtract 7}$$
$$3x = -5 \qquad \text{from each side.}$$
$$\frac{3x}{3} = \frac{-5}{3} \qquad \text{• Divide each}$$
$$\text{side by 3.}$$
$$x = -\frac{5}{3} = -1\frac{2}{3}$$

The solution is $-1\frac{2}{3}$.

YOU TRY IT 1

Solve: $5x + 8 = 6$

Your solution

EXAMPLE 2

Solve: $5 - x = 6$

Solution

$$5 - x = 6$$
$$5 - 5 - x = 6 - 5 \qquad \text{• Subtract 5}$$
$$-x = 1 \qquad \text{from each side.}$$
$$(-1)(-x) = (-1) \cdot 1 \qquad \text{• Multiply each}$$
$$x = -1 \qquad \text{side by } -1.$$

The solution is -1.

YOU TRY IT 2

Solve: $7 - x = 3$

Your solution

Solutions on p. S26–S27

OBJECTIVE B

To solve application problems using formulas

Anders Celsius

The Fahrenheit temperature scale was devised by Daniel Gabriel Fahrenheit (1686—1736), a German physicist and maker of scientific instruments. He invented the mercury thermometer in 1714. On the Fahrenheit scale, the temperature at which water freezes is 32°F, and the temperature at which water boils is 212°F. *Note:* The small raised circle is the symbol for degrees, and the capital F is for Fahrenheit. The Fahrenheit scale is used only in the United States.

In the metric system, temperature is measured on the Celsius scale. The Celsius temperature scale was devised by Anders Celsius (1701–1744), a Swedish astronomer. On the Celsius scale, the temperature at which water freezes is 0°C, and the temperature at which water boils is 100°C. *Note:* The small raised circle is the symbol for degrees, and the capital C is for Celsius.

On both the Celsius scale and the Fahrenheit scale, temperatures below 0° are negative numbers.

The relationship between Celsius temperature and Fahrenheit temperature is given by the formula

$$F = 1.8C + 32$$

where F represents degrees Fahrenheit and C represents degrees Celsius.

HOW TO 2 Normal body temperature is 98.6°F. Convert this temperature to degrees Celsius.

$$F = 1.8C + 32$$
$$98.6 = 1.8C + 32 \qquad \text{• Substitute } 98.6 \text{ for } F.$$
$$98.6 - 32 = 1.8C + 32 - 32 \qquad \text{• Subtract } 32 \text{ from each side.}$$
$$66.6 = 1.8C \qquad \text{• Combine like terms on each side.}$$
$$\frac{66.6}{1.8} = \frac{1.8C}{1.8} \qquad \text{• Divide each side by } 1.8.$$
$$37 = C$$

Normal body temperature is 37°C.

Integrating Technology

You can check the solution to this equation using a calculator. Evaluate the right side of the equation after substituting 37 for C. Enter

1.8 **X** 37 **+** 32 **=**

The display reads 98.6, the given Fahrenheit temperature. The solution checks.

EXAMPLE • 3

Find the Celsius temperature when the Fahrenheit temperature is 212°. Use the formula $F = 1.8C + 32$, where F is the Fahrenheit temperature and C is the Celsius temperature.

Strategy

To find the Celsius temperature, replace the variable F in the formula by the given value and solve for C.

Solution

$$F = 1.8C + 32$$
$$212 = 1.8C + 32 \qquad \text{• Substitute } \textbf{212} \text{ for } F.$$
$$212 - 32 = 1.8C + 32 - 32 \qquad \text{• Subtract } \textbf{32} \text{ from each side.}$$

$$180 = 1.8C \qquad \text{• Combine like terms.}$$
$$\frac{180}{1.8} = \frac{1.8C}{1.8} \qquad \text{• Divide each side by } \textbf{1.8}.$$
$$100 = C$$

The Celsius temperature is 100°.

YOU TRY IT • 3

Find the Celsius temperature when the Fahrenheit temperature is −22°. Use the formula $F = 1.8C + 32$, where F is the Fahrenheit temperature and C is the Celsius temperature.

Your strategy

Your solution

EXAMPLE • 4

To find the total cost of production, an economist uses the formula $T = U \cdot N + F$, where T is the total cost, U is the cost per unit, N is the number of units made, and F is the fixed cost. Find the number of units made during a week in which the total cost was $8000, the cost per unit was $16, and the fixed costs were $2000.

Strategy

To find the number of units made, replace the variables T, U, and F in the formula by the given values and solve for N.

Solution

$$T = U \cdot N + F$$
$$8000 = 16 \cdot N + 2000$$
$$8000 - 2000 = 16 \cdot N + 2000 - 2000$$
$$6000 = 16 \cdot N$$
$$\frac{6000}{16} = \frac{16 \cdot N}{16}$$
$$375 = N$$

The number of units made was 375.

YOU TRY IT • 4

Find the cost per unit during a week in which the total cost was $4500, the number of units produced was 250, and the fixed costs were $1500. Use the formula $T = U \cdot N + F$, where T is the total cost, U is the cost per unit, N is the number of units made, and F is the fixed cost.

Your strategy

Your solution

Solutions on p. S27

11.3 EXERCISES

OBJECTIVE A   To solve an equation of the form $ax + b = c$

For Exercises 1 to 86, solve.

1. $3x + 5 = 14$

2. $5z + 6 = 31$

3. $2n - 3 = 7$

4. $4y - 4 = 20$

5. $5w + 8 = 3$

6. $3x + 10 = 1$

7. $3z - 4 = -16$

8. $6x - 1 = -13$

9. $5 + 2x = 7$

10. $12 + 7x = 33$

11. $6 - x = 3$

12. $4 - x = -2$

13. $3 - 4x = 11$

14. $2 - 3x = 11$

15. $5 - 4x = 17$

16. $8 - 6x = 14$

17. $3x + 6 = 0$

18. $5x - 20 = 0$

19. $-3x - 4 = -1$

20. $-7x - 22 = -1$

21. $12y - 30 = 6$

22. $9b - 7 = 2$

23. $3c + 7 = 4$

24. $8t + 13 = 5$

25. $-2x + 11 = -3$

26. $-4x + 15 = -1$

27. $14 - 5x = 4$

28. $7 - 3x = 4$

29. $-8x + 7 = -9$

30. $-7x + 13 = -8$

31. $9x + 13 = 13$

32. $-2x + 7 = 7$

33. $7d - 14 = 0$

34. $5z + 10 = 0$

35. $4n - 4 = -4$

36. $-13m - 1 = -1$

37. $3x + 5 = 7$

38. $4x + 6 = 9$

39. $6x - 1 = 16$

40. $12x - 3 = 7$

41. $2x - 3 = -8$ **42.** $5x - 3 = -12$ **43.** $-6x + 2 = -7$ **44.** $-3x + 9 = -1$

45. $-2x - 3 = -7$ **46.** $5x - 7 = -4$ **47.** $3x + 8 = 2$ **48.** $2x - 9 = 8$

49. $3w - 7 = 0$ **50.** $7b - 2 = 0$ **51.** $-2d + 9 = 12$ **52.** $-7c + 3 = 1$

53. $\frac{1}{2}x - 2 = 3$ **54.** $\frac{1}{3}x + 1 = 4$ **55.** $\frac{3}{5}w - 1 = 2$ **56.** $\frac{2}{5}w + 5 = 6$

57. $\frac{2}{9}t - 3 = 5$ **58.** $\frac{5}{9}t - 3 = 2$ **59.** $\frac{y}{3} - 6 = -8$ **60.** $\frac{y}{2} - 2 = 3$

61. $\frac{x}{3} - 2 = -5$ **62.** $\frac{x}{4} - 3 = 5$ **63.** $\frac{5}{8}v + 6 = 3$ **64.** $\frac{2}{3}v - 4 = 3$

65. $\frac{4}{7}z + 10 = 5$ **66.** $\frac{3}{8}v - 3 = 4$ **67.** $\frac{2}{9}x - 3 = 5$ **68.** $\frac{1}{2}x + 3 = -8$

69. $\frac{3}{4}x - 5 = -4$ **70.** $\frac{2}{3}x - 5 = -8$ **71.** $1.5x - 0.5 = 2.5$

72. $2.5w - 1.3 = 3.7$ **73.** $0.8t + 1.1 = 4.3$ **74.** $0.3v + 2.4 = 1.5$

75. $0.4x - 2.3 = 1.3$ **76.** $1.2t + 6.5 = 2.9$ **77.** $3.5y - 3.5 = 10.5$

78. $1.9x - 1.9 = -1.9$ **79.** $0.32x + 4.2 = 3.2$ **80.** $5x - 3x + 2 = 8$

81. $6m + 2m - 3 = 5$ **82.** $4a - 7a - 8 = 4$ **83.** $3y - 8y - 9 = 6$

84. $x - 4x + 5 = 11$ **85.** $-2y + y - 3 = 6$ **86.** $-4y - y - 8 = 12$

 For Exercises 87 to 89, suppose that a is positive and b is negative. Determine whether the value of x in the given equation *must be negative, must be positive,* or *could be either positive or negative.*

87. $bx - a = 12$ **88.** $bx + 5 = a$ **89.** $a - x = b$

OBJECTIVE B   **To solve application problems using formulas**

 90. In the formula $T = U \cdot N + F$, T is the total cost, U is the cost per unit, N is the number of units produced, and F is the fixed cost. When you know the number of units produced, the fixed cost, and the total cost for one week, which of the following is the first step in solving the formula for the cost per unit?
   (i)   Subtract the fixed cost from each side of the equation.
   (ii) Divide each side of the equation by the number of units produced.
   (iii) Subtract the number of units produced from each side of the equation.

Temperature Conversion In Exercises 91 and 92, use the relationship between Fahrenheit temperature and Celsius temperature, which is given by the formula $F = 1.8C + 32$, where F is the Fahrenheit temperature and C is the Celsius temperature.

91. Find the Celsius temperature when the Fahrenheit temperature is $-40°$.

92. Find the Celsius temperature when the Fahrenheit temperature is $72°$. Round to the nearest tenth of a degree.

Physics In Exercises 93 and 94, use the formula $V = V_0 + 32t$, where V is the final velocity of a falling object, V_0 is the starting velocity of the falling object, and t is the time for the object to fall.

93. Find the time required for an object to increase in velocity from 8 ft/s to 472 ft/s.

94. Find the time required for an object to increase in velocity from 16 ft/s to 128 ft/s.

Manufacturing In Exercises 95 and 96, use the formula $T = U \cdot N + F$, where T is the total cost, U is the cost per unit, N is the number of units made, and F is the fixed cost.

95. Find the number of units made during a week in which the total cost was $25,000, the cost per unit was $8, and the fixed costs were $5000.

96. Find the cost per unit during a week in which the total cost was $80,000, the total number of units produced was 500, and the fixed costs were $15,000.

Taxes In Exercises 97 and 98, use the formula $T = I \cdot R + B$, where T is the monthly tax, I is the monthly income, R is the income tax rate, and B is the base monthly tax.

97. The monthly tax that a clerk pays is $476. The clerk's monthly tax rate is 22%, and the base monthly tax is $80. Find the clerk's monthly income.

98. The monthly tax that Marcy, a teacher, pays is $770. Her monthly income is $3100, and the base monthly tax is $150. Find Marcy's income tax rate.

Compensation In Exercises 99 to 102, use the formula $M = S \cdot R + B$, where M is the monthly earnings, S is the total sales, R is the commission rate, and B is the base monthly salary.

99. A sales representative for an advertising firm earns a base monthly salary of $600 plus a 9% commission on total sales. Find the total sales during a month in which the representative earned $3480.

100. A sales executive earns a base monthly salary of $1000 plus a 5% commission on total sales. Find the total sales during a month in which the executive earned $2800.

101. Miguel earns a base monthly salary of $750. Find his commission rate during a month in which total sales were $42,000 and he earned $2640.

102. Tina earns a base monthly salary of $500. Find her commission rate during a month in which total sales were $42,500 and her earnings were $3560.

Applying the Concepts

103. Explain in your own words the steps you would take to solve the equation $\frac{2}{3}x - 4 = 10$. State the property of real numbers or the property of equations that is used at each step.

104. Make up an equation of the form $ax + b = c$ that has -3 as its solution.

105. Does the sentence "Solve $3x + 4(x - 3)$" make sense? Why or why not?

SECTION

11.4 General Equations: Part II

OBJECTIVE A To solve an equation of the form $ax + b = cx + d$

When a variable occurs on each side of an equation, the Addition Properties are used to rewrite the equation so that all variable terms are on one side of the equation and all constant terms are on the other side of the equation. Then the Multiplication Properties are used to simplify the equation to one of the form *variable* = *constant*.

> **HOW TO 1** Solve: $4x - 6 = 8 - 3x$
>
> $4x - 6 = 8 - 3x$
>
> • The goal is to write the equation in the form *variable* = *constant*.
>
> $4x + 3x - 6 = 8 - 3x + 3x$
> $7x - 6 = 8 + 0$
> $7x - 6 = 8$
>
> • Add $3x$ to each side of the equation. Then simplify (Addition Properties). Now only one variable term occurs in the equation.
>
> $7x - 6 + 6 = 8 + 6$
> $7x + 0 = 14$
> $7x = 14$
>
> • Add 6 to each side of the equation. Then simplify (Addition Properties). Now only one constant term occurs in the equation.
>
> $\dfrac{7x}{7} = \dfrac{14}{7}$
>
> $1x = 2$
> $x = 2$
>
> • Divide each side of the equation by the numerical coefficient of the variable term. Then simplify (Multiplication Properties).
>
> The solution is 2.
>
> • Write the solution.

Tips for Success

Always check the solution of an equation. For the equation at the right:

$$\begin{array}{c|c} 4x - 6 &= 8 - 3x \\ \hline 4(2) - 6 & 8 - 3(2) \\ 8 - 6 & 8 - 6 \\ 2 &= 2 \end{array}$$

The solution checks.

EXAMPLE 1

Solve: $\dfrac{2}{9}x - 3 = \dfrac{7}{9}x + 2$

Solution

$$\dfrac{2}{9}x - 3 = \dfrac{7}{9}x + 2$$

$$\dfrac{2}{9}x - \dfrac{7}{9}x - 3 = \dfrac{7}{9}x - \dfrac{7}{9}x + 2$$ • Subtract $\dfrac{7}{9}x$ from each side.

$$-\dfrac{5}{9}x - 3 = 2$$

$$-\dfrac{5}{9}x - 3 + 3 = 2 + 3$$ • Add 3 to each side.

$$-\dfrac{5}{9}x = 5$$

$$\left(-\dfrac{9}{5}\right)\left(-\dfrac{5}{9}\right)x = \left(-\dfrac{9}{5}\right)5$$ • Multiply each side by $\dfrac{9}{5}$.

$$x = -9$$

The solution is -9.

YOU TRY IT 1

Solve: $\dfrac{1}{5}x - 2 = \dfrac{2}{5}x + 4$

Your solution

Solution on p. S27

OBJECTIVE B   **To solve an equation containing parentheses**

When an equation contains parentheses, one of the steps involved in solving the equation requires use of the Distributive Property.

$$a(b + c) = ab + ac$$

The Distributive Property is used to rewrite a variable expression without parentheses.

HOW TO 2   Solve: $4(3 + x) - 2 = 2(x - 4)$

$4(3 + x) - 2 = 2(x - 4)$	• The goal is to write the equation in the form *variable = constant*.
$12 + 4x - 2 = 2x - 8$	• Use the Distributive Property to rewrite the equation without parentheses.
$10 + 4x = 2x - 8$	• Combine like terms.
$10 + 4x - 2x = 2x - 2x - 8$	• Use the Addition Property of Equations. Subtract $2x$ from each side of the equation.
$10 + 2x = -8$	
$10 - 10 + 2x = -8 - 10$	• Use the Addition Property of Equations. Subtract 10 from each side of the equation.
$2x = -18$	
$\dfrac{2x}{2} = \dfrac{-18}{2}$	• Use the Multiplication Property of Equations. Divide each side of the equation by the numerical coefficient of the variable term.
$x = -9$	

Check:

$$
\begin{array}{c|c}
4(3 + x) - 2 & = 2(x - 4) \\
\hline
4[3 + (-9)] - 2 & 2(-9 - 4) \\
4(-6) - 2 & 2(-13) \\
-24 - 2 & -26 \\
-26 & = -26
\end{array}
$$

• Check the solution.

A true equation

The solution is -9.   • Write the solution.

The solution to this last equation illustrates the steps involved in solving first-degree equations.

Steps in Solving General First-Degree Equations

1. Use the Distributive Property to remove parentheses.

2. Combine like terms on each side of the equation.

3. Use the Addition Property of Equations to rewrite the equation with only one variable term.

4. Use the Addition Property of Equations to rewrite the equation with only one constant term.

5. Use the Multiplication Property of Equations to rewrite the equation so that the coefficient of the variable term is 1.

PPT

EXAMPLE • 2

Solve: $3(x + 2) - x = 11$

Solution

$3(x + 2) - x = 11$

$3x + 6 - x = 11$ • Use the Distributive Property.

$2x + 6 = 11$ • Combine like terms on the left side.

$2x + 6 - 6 = 11 - 6$ • Use the Addition Property of Equations. **Subtract 6** from each side.

$2x = 5$ • Combine like terms on each side.

$\dfrac{2x}{2} = \dfrac{5}{2}$ • Use the Multiplication Property. **Divide** both sides **by 2**.

$x = 2\dfrac{1}{2}$ • The solution checks.

The solution is $2\dfrac{1}{2}$.

YOU TRY IT • 2

Solve: $4(x - 1) - x = 5$

Your solution

EXAMPLE • 3

Solve: $5x - 2(x - 3) = 6(x - 2)$

Solution

$5x - 2(x - 3) = 6(x - 2)$

$5x - 2x + 6 = 6x - 12$ • Distributive Property

$3x + 6 = 6x - 12$ • Combine like terms.

$3x - 6x + 6 = 6x - 6x - 12$ • **Subtract 6x** from each side.

$-3x + 6 = -12$ • Combine like terms.

$-3x + 6 - 6 = -12 - 6$ • **Subtract 6** from each side.

$-3x = -18$ • Combine like terms.

$\dfrac{-3x}{-3} = \dfrac{-18}{-3}$ • **Divide** both sides by **−3**.

$x = 6$ • The solution checks.

The solution is 6.

YOU TRY IT • 3

Solve: $2x - 7(3x + 1) = 5(5 - 3x)$

Your solution

Solutions on p. S27

11.4 EXERCISES

OBJECTIVE A To solve an equation of the form $ax + b = cx + d$

For Exercises 1 to 54, solve.

1. $6x + 3 = 2x + 5$

2. $7x + 1 = x + 19$

3. $3x + 3 = 2x + 2$

4. $6x + 3 = 3x + 6$

5. $5x + 4 = x - 12$

6. $3x - 12 = x - 8$

7. $7b - 2 = 3b - 6$

8. $2d - 9 = d - 8$

9. $9n - 4 = 5n - 20$

10. $8x - 7 = 5x + 8$

11. $2x + 1 = 16 - 3x$

12. $3x + 2 = -23 - 2x$

13. $5x - 2 = -10 - 3x$

14. $4x - 3 = 7 - x$

15. $2x + 7 = 4x + 3$

16. $7m - 6 = 10m - 15$

17. $c + 4 = 6c - 11$

18. $t - 6 = 4t - 21$

19. $3x - 7 = x - 7$

20. $2x + 6 = 7x + 6$

21. $3 - 4x = 5 - 3x$

22. $6 - 2x = 9 - x$

23. $7 + 3x = 9 + 5x$

24. $12 + 5x = 9 - 3x$

25. $5 + 2y = 7 + 5y$

26. $9 + z = 2 + 3z$

27. $8 - 5w = 4 - 6w$

28. $9 - 4x = 11 - 5x$

29. $6x + 1 = 3x + 2$

30. $7x + 5 = 4x + 7$

31. $5x + 8 = x + 5$

32. $9x + 1 = 3x - 4$

33. $2x - 3 = 6x - 4$

34. $4 - 3x = 4 - 5x$

35. $6 - 3x = 6 - 5x$

36. $2x + 7 = 4x - 3$

37. $6x - 2 = 2x - 9$

38. $4x - 7 = -3x + 2$

39. $6x - 3 = -5x + 8$

40. $7y - 5 = 3y + 9$

41. $-6t - 2 = -8t - 4$

42. $-7w + 2 = 3w - 8$

43. $-3 - 4x = 7 - 2x$

44. $-8 + 5x = 8 + 6x$

45. $3 - 7x = -2 + 5x$

46. $3x - 2 = 7 - 5x$

47. $5x + 8 = 4 - 2x$

48. $4 - 3x = 6x - 8$

49. $12z - 9 = 3z + 12$

50. $4c + 13 = -6c + 9$

51. $\frac{5}{7}m - 3 = \frac{2}{7}m + 6$

52. $\frac{4}{5}x - 1 = \frac{1}{5}x + 5$

53. $\frac{3}{7}x + 5 = \frac{5}{7}x - 1$

54. $\frac{3}{4}x + 2 = \frac{1}{4}x - 9$

 55. If a is a negative number, will solving the equation $-8x = a - 6x$ for x result in a positive solution or a negative solution?

 56. If a is a positive number, will solving the equation $9x - a = 13x$ for x result in a positive solution or a negative solution?

OBJECTIVE B   **To solve an equation containing parentheses**

 57. Which of the following equations is equivalent to $9 - 2(4y - 1) = 6$?
(i) $9 - 8y - 1 = 6$ (ii) $7(4y - 1) = 6$
(iii) $9 - 8y - 2 = 6$ (iv) $9 - 8y + 2 = 6$

 58. Which of the following equations is equivalent to $5x - 3(4x + 2) = 7(x + 3)$?
(i) $2x(4x + 2) = 7x + 21$ (ii) $5x - 12x - 6 = 7x + 21$
(iii) $5x - 12x - 6 = 7x + 3$ (iv) $5x(-12x - 6) = 7x + 21$

For Exercises 59 to 104, solve.

59. $6x + 2(x - 1) = 14$

60. $3x + 2(x + 4) = 13$

61. $-3 + 4(x + 3) = 5$

62. $8b - 3(b - 5) = 30$

63. $6 - 2(d + 4) = 6$

64. $5 - 3(n + 2) = 8$

65. $5 + 7(x + 3) = 20$

66. $6 - 3(x - 4) = 12$

67. $2x + 3(x - 5) = 10$

68. $3x - 4(x + 3) = 9$

69. $3(x - 4) + 2x = 3$

70. $4 + 3(x - 9) = -12$

71. $2x - 3(x - 4) = 12$

72. $4x - 2(x - 5) = 10$

73. $2x + 3(x + 4) = 7$

74. $3(x + 2) + 7 = 12$

75. $3(x - 2) + 5 = 5$

76. $4(x - 5) + 7 = 7$

77. $3y + 7(y - 2) = 5$

78. $-3z - 3(z - 3) = 3$

79. $4b - 2(b + 9) = 8$

80. $3x - 6(x - 3) = 9$

81. $3x + 5(x - 2) = 10$

82. $3x - 5(x - 1) = -5$

83. $3x + 4(x + 2) = 2(x + 9)$

84. $5x + 3(x + 4) = 4(x + 2)$

85. $2d - 3(d - 4) = 2(d + 6)$

86. $3t - 4(t - 1) = 3(t - 2)$

87. $7 - 2(x - 3) = 3(x - 1)$

88. $4 - 3(x + 2) = 2(x - 4)$

89. $6x - 2(x - 3) = 11(x - 2)$

90. $9x - 5(x - 3) = 5(x + 4)$

91. $6c - 3(c + 1) = 5(c + 2)$

92. $2w - 7(w - 2) = 3(w - 4)$

93. $7 - (x + 1) = 3(x + 3)$

94. $12 + 2(x - 9) = 3(x - 12)$

95. $2x - 3(x + 4) = 2(x - 5)$

96. $3x + 2(x - 7) = 7(x - 1)$

97. $x + 5(x - 4) = 3(x - 8) - 5$

98. $2x - 2(x - 1) = 3(x - 2) + 7$

99. $9b - 3(b - 4) = 13 + 2(b - 3)$

100. $3y - 4(y - 2) = 15 - 3(y - 2)$

101. $3(x - 4) + 3x = 7 - 2(x - 1)$

102. $2(x - 6) + 7x = 5 - 3(x - 2)$

103. $3.67x - 5.3(x - 1.932) = 6.9959$

104. $4.06x + 4.7(x + 3.22) = 1.775$

Applying the Concepts

105. If $2x - 2 = 4x + 6$, what is the value of $3x^2$?

106. If $3 + 2(4a - 3) = 4$ and $4 - 3(2 - 3b) = 11$, which is larger, *a* or *b?*

 107. The equation $x = x + 1$ has no solution, whereas the solution of the equation $2x + 3 = 3$ is zero. Is there a difference between no solution and a solution of zero? Explain.

SECTION

11.5 Translating Verbal Expressions into Mathematical Expressions

OBJECTIVE A

To translate a verbal expression into a mathematical expression given the variable

One of the major skills required in applied mathematics is to translate a verbal expression into a mathematical expression. Doing so requires recognizing the verbal phrases that translate into mathematical operations. Following is a partial list of the verbal phrases used to indicate the different mathematical operations.

Addition	more than	5 more than x	$x + 5$
	the sum of	the sum of w and 3	$w + 3$
	the total of	the total of 6 and z	$6 + z$
	increased by	x increased by 7	$x + 7$
Subtraction	less than	5 less than y	$y - 5$
	the difference between	the difference between w and 3	$w - 3$
	decreased by	8 decreased by a	$8 - a$
Multiplication	times	3 times c	$3c$
	the product of	the product of 4 and t	$4t$
	of	two-thirds of v	$\frac{2}{3}v$
	twice	twice d	$2d$
Division	divided by	n divided by 3	$\frac{n}{3}$
	the quotient of	the quotient of z and 4	$\frac{z}{4}$
	the ratio of	the ratio of s to 6	$\frac{s}{6}$

PPT

Translating phrases that contain the words *sum, difference, product,* and *quotient* can be difficult. In the examples at the right, note where the operation symbol is placed.

Note where we place the fraction bar when translating the word *ratio*.

the *sum* of x and y $\qquad$ $x + y$

the *difference* between x and y $\qquad$ $x - y$

the *product* of x and y $\qquad$ $x \cdot y$

the *quotient* of x and y $\qquad$ $\dfrac{x}{y}$

the *ratio* of x to y $\qquad$ $\dfrac{x}{y}$

HOW TO 1 Translate "the quotient of n and the sum of n and 6" into a mathematical expression.

the *quotient* of n and the *sum* of n and 6 $\qquad$ $\dfrac{n}{n + 6}$

EXAMPLE • 1

Translate "the sum of 5 and the product of 4 and n" into a mathematical expression.

Solution

$5 + 4n$

YOU TRY IT • 1

Translate "the difference between 8 and twice t" into a mathematical expression.

Your solution

EXAMPLE • 2

Translate "the product of 3 and the difference between z and 4" into a mathematical expression.

Solution

$3(z - 4)$

YOU TRY IT • 2

Translate "the quotient of 5 and the product of 7 and x" into a mathematical expression.

Your solution

Solutions on p. S27

OBJECTIVE B   To translate a verbal expression into a mathematical expression by assigning the variable

In most applications that involve translating phrases into mathematical expressions, the variable to be used is not given. To translate these phrases, we must assign a variable to the unknown quantity before writing the mathematical expression.

HOW TO • 2   Translate "the difference between seven and twice a number" into a mathematical expression.

The difference between seven and twice a number
• Identify the phrases that indicate the mathematical operations.

The unknown number: n
• Assign a variable to one of the unknown quantities.

Twice the number: $2n$
• Use the assigned variable to write an expression for any other unknown quantity.

$7 - 2n$
• Use the identified operations to write the mathematical expression.

EXAMPLE • 3

Translate "the total of a number and the square of the number" into a mathematical expression.

Solution

The *total* of a number and the *square* of the number

The unknown number: x
The square of the number: x^2

$x + x^2$

YOU TRY IT • 3

Translate "the product of a number and one-half of the number" into a mathematical expression.

Your solution

Solution on p. S27

11.5 EXERCISES

OBJECTIVE A   To translate a verbal expression into a mathematical expression given the variable

For Exercises 1 to 20, translate into a mathematical expression.

1. 9 less than y

2. w divided by 7

3. z increased by 3

4. the product of -2 and x

5. the sum of two-thirds of n and n

6. the difference between the square of r and r

7. the quotient of m and the difference between m and 3

8. v increased by twice v

9. the product of 9 and 4 more than x

10. the difference between n and the product of 5 and n

11. x decreased by the quotient of x and 2

12. the product of c and one-fourth of c

13. the quotient of 3 less than z and z

14. the product of y and the sum of y and 4

15. 2 times the sum of t and 6

16. the quotient of r and the difference between 8 and r

17. x divided by the total of 9 and x

18. the sum of z and the product of 6 and z

19. three times the sum of b and 6

20. the ratio of w to the sum of w and 8

 For Exercises 21 and 22, translate the mathematical expression into a verbal phrase. *Note:* Answers will vary.

21. a. $2x + 3$

b. $2(x + 3)$

22. a. $\dfrac{2x}{7}$

b. $\dfrac{2 + x}{7}$

OBJECTIVE B   To translate a verbal expression into a mathematical expression by assigning the variable

For Exercises 23 to 42, translate into a mathematical expression.

23. the square of a number

24. five less than some number

25. a number divided by twenty

26. the difference between a number and twelve

27. four times some number

28. the quotient of five and a number

29. three-fourths of a number

30. the sum of a number and seven

31. four increased by some number

32. the ratio of a number to nine

33. the difference between five times a number and the number

34. six less than the total of three and a number

35. the product of a number and two more than the number

36. the quotient of six and the sum of nine and a number

37. seven times the total of a number and eight

38. the difference between ten and the quotient of a number and two

39. the square of a number plus the product of three and the number

40. a number decreased by the product of five and the number

41. the sum of three more than a number and one-half of the number

42. eight more than twice the sum of a number and seven

For Exercises 43 and 44, determine whether the expression $4n^2 - 5$ is a correct translation of the given phrase.

43. five less than the square of the product of four and a number

44. the difference between five and the product of four and the square of a number

Applying the Concepts

45. In your own words, explain how variables are used.

46. **Chemistry** The chemical formula for water is H_2O. This formula means that there are two hydrogen atoms and one oxygen atom in each molecule of water. If x represents the number of atoms of oxygen in a glass of pure water, express the number of hydrogen atoms in the glass of water.

47. **Chemistry** The chemical formula for one molecule of glucose (sugar) is $C_6H_{12}O_6$, where C is carbon, H is hydrogen, and O is oxygen. If x represents the number of atoms of hydrogen in a sample of pure sugar, express the number of carbon atoms and the number of oxygen atoms in the sample in terms of x.

SECTION

11.6 Translating Sentences into Equations and Solving

OBJECTIVE A To translate a sentence into an equation and solve

An equation states that two mathematical expressions are equal. Therefore, to translate a sentence into an equation requires recognition of the words or phrases that mean "equals." Some of these phrases are

$$\left.\begin{array}{l} \text{equals} \\ \text{is} \\ \text{is equal to} \\ \text{amounts to} \\ \text{represents} \end{array}\right\} \text{ translate to } =$$

Once the sentence is translated into an equation, the equation can be simplified to one of the form *variable = constant* and the solution can be found.

HOW TO 1 Translate "three more than twice a number is seventeen" into an equation and solve.

The unknown number: n
• Assign a variable to the unknown quantity.

| Three more than twice a number | is | seventeen |

• Find two verbal expressions for the same value.

$$2n + 3 = 17$$
$$2n + 3 - 3 = 17 - 3$$
$$2n = 14$$
$$\frac{2n}{2} = \frac{14}{2}$$
$$n = 7$$

• Write a mathematical expression for each verbal expression. Write the equals sign. Solve the resulting equation.

The number is 7.

Point of Interest

Number puzzle problems similar to the one on this page have appeared in textbooks for hundreds of years. Here is one from a 1st-century Chinese textbook: "When a number is divided by 3, the remainder is 2; when it is divided by 5, the remainder is 3; when it is divided by 7, the remainder is 2. Find the number." There are actually infinitely many solutions to this problem. See whether you can find one of them.

EXAMPLE 1

Translate "a number decreased by six equals fifteen" into an equation and solve.

Solution
The unknown number: x

| A number decreased by six | equals | fifteen |

$$x - 6 = 15$$
$$x - 6 + 6 = 15 + 6$$
$$x = 21$$

The number is 21.

YOU TRY IT 1

Translate "a number increased by four equals twelve" into an equation and solve.

Your solution

Solution on p. S27

EXAMPLE • 2

The quotient of a number and six is five. Find the number.

Solution

The unknown number: z

The quotient of a number and six	is	five

$$\frac{z}{6} = 5$$

$$6 \cdot \frac{z}{6} = 6 \cdot 5$$

$$z = 30$$

The number is 30.

YOU TRY IT • 2

The product of two and a number is ten. Find the number.

Your solution

EXAMPLE • 3

Eight decreased by twice a number is four. Find the number.

Solution

The unknown number: t

Eight decreased by twice a number	is	four

$$8 - 2t = 4$$
$$8 - 8 - 2t = 4 - 8$$
$$-2t = -4$$
$$\frac{-2t}{-2} = \frac{-4}{-2}$$
$$t = 2$$

The number is 2.

YOU TRY IT • 3

The sum of three times a number and six equals four. Find the number.

Your solution

EXAMPLE • 4

Three less than the ratio of a number to seven is one. Find the number.

Solution

The unknown number: x

Three less than the ratio of a number to seven	is	one

$$\frac{x}{7} - 3 = 1$$

$$\frac{x}{7} - 3 + 3 = 1 + 3$$

$$\frac{x}{7} = 4$$

$$7 \cdot \frac{x}{7} = 7 \cdot 4$$

$$x = 28$$

The number is 28.

YOU TRY IT • 4

Three more than one-half of a number is nine. Find the number.

Your solution

Solutions on p. S28

OBJECTIVE B   To solve application problems

EXAMPLE 5

The cost of a portable DVD player with carrying case is $187. This amount is $38 more than the cost of the DVD player without the carrying case. Find the cost of the DVD player without the carrying case.

Strategy
To find the cost of the DVD player without the carrying case, write and solve an equation using C to represent the cost of the DVD player without the carrying case.

Solution

| $187 | is | $38 more than the cost of the DVD player without the carrying case |

$$187 = C + 38$$
$$187 - 38 = C + 38 - 38$$
$$149 = C$$

The cost of the DVD player without the carrying case is $149.

YOU TRY IT 5

The sale price of a baseball jersey is $38.95. This amount is $11 less than the regular price. Find the regular price.

Your strategy

Your solution

EXAMPLE 6

By purchasing a fleet of cars, a company receives a discount of $1972 on each car purchased. This amount is 8% of the regular price. Find the regular price.

Strategy
To find the regular price, write and solve an equation using P to represent the regular price of the car.

Solution

| $1972 | is | 8% of the regular price |

$$1972 = 0.08 \cdot P$$
$$\frac{1972}{0.08} = \frac{0.08P}{0.08}$$
$$24,650 = P$$

The regular price is $24,650.

YOU TRY IT 6

At a certain speed, the engine rpm (revolutions per minute) of a car in fourth gear is 2500. This is two-thirds of the rpm of the engine in third gear. Find the rpm of the engine when it is in third gear.

Your strategy

Your solution

Solutions on p. S28

EXAMPLE · 7

Ron Sierra charged $1775 for plumbing repairs in an office building. This charge included $180 for parts and $55 per hour for labor. Find the number of hours Ron worked in the office building.

Strategy
To find the number of hours worked, write and solve an equation using N to represent the number of hours worked.

Solution

$$1775 = 180 + 55N$$
$$1775 - 180 = 180 - 180 + 55N$$
$$1595 = 55N$$
$$\frac{1595}{55} = \frac{55N}{55}$$
$$29 = N$$

Ron worked 29 h.

YOU TRY IT · 7

The total cost to make a model Z100 television is $492. The cost includes $100 for materials plus $24.50 per hour for labor. How many hours of labor are required to make a model Z100 television?

Your strategy

Your solution

EXAMPLE · 8

The state income tax for Tim Fong last month was $256. This amount is $5 more than 8% of his monthly salary. Find Tim's monthly salary.

Strategy
To find Tim's monthly salary, write and solve an equation using S to represent his monthly salary.

Solution

| $256 | is | $5 more than 8% of the monthly salary |

$$256 = 0.08 \cdot S + 5$$
$$256 - 5 = 0.08S + 5 - 5$$
$$251 = 0.08S$$
$$\frac{251}{0.08} = \frac{0.08S}{0.08}$$
$$3137.50 = S$$

Tim's monthly salary is $3137.50.

YOU TRY IT · 8

Natalie Adams earned $2500 last month for temporary work. This amount was the sum of a base monthly salary of $800 and an 8% commission on total sales. Find Natalie's total sales for the month.

Your strategy

Your solution

Solutions on p. S28

11.6 EXERCISES

OBJECTIVE A   To translate a sentence into an equation and solve

For Exercises 1 to 26, write an equation and solve.

1. The sum of a number and seven is twelve. Find the number.

2. A number decreased by seven is five. Find the number.

3. The product of three and a number is eighteen. Find the number.

4. The quotient of a number and three is one. Find the number.

5. Five more than a number is three. Find the number.

6. A number divided by four is six. Find the number.

7. Six times a number is fourteen. Find the number.

8. Seven less than a number is three. Find the number.

9. Five-sixths of a number is fifteen. Find the number.

10. The total of twenty and a number is five. Find the number.

11. The sum of three times a number and four is eight. Find the number.

12. The sum of one-third of a number and seven is twelve. Find the number.

13. Seven less than one-fourth of a number is nine. Find the number.

14. The total of a number divided by four and nine is two. Find the number.

15. The ratio of a number to nine is fourteen. Find the number.

16. Five increased by the product of five and a number is equal to 30. Find the number.

17. Six less than the quotient of a number and four is equal to negative two. Find the number.

18. The product of a number plus three and two is eight. Find the number.

19. The difference between seven and twice a number is thirteen. Find the number.

20. Five more than the product of three and a number is eight. Find the number.

21. Nine decreased by the quotient of a number and two is five. Find the number.

22. The total of ten times a number and seven is twenty-seven. Find the number.

23. The sum of three-fifths of a number and eight is two. Find the number.

24. Five less than two-thirds of a number is three. Find the number.

25. The difference between a number divided by 4.18 and 7.92 is 12.52. Find the number.

26. The total of 5.68 times a number and 132.7 is the number minus 29.228. Find the number.

 For Exercises 27 and 28, determine whether the equation $5 - 7x = 9$ is a correct translation of the given sentence.

27. Five less than the product of seven and a number is equal to nine.

28. The difference between five and the product of seven and a number is nine.

OBJECTIVE B   To solve application problems

 29. A student writes the equation $314.7 = x + 260.9$ to represent the situation described in Exercise 31. What does x represent in this equation?

 30. A student writes the equation $184{,}040 = n - 21{,}660$ to represent the situation described in Exercise 32. What does n represent in this equation?

 31. The Military In 2007, the Army Reserve paid $314.7 million in cash bonuses to recruit and keep soldiers. This amount was $260.9 million more than the amount paid in cash bonuses in 2003. (*Source:* Army Reserve) Find the amount the Army Reserve paid in cash bonuses in 2003.

 32. Housing The median price of a house in 2001 was $184,040. This price was $21,660 less than the median price of a house in 2007. (*Source:* National Association of Realtors) Find the median price of a house in 2007.

 33. Bridges The length of the Akashi Kaikyo Bridge is 1991 m. This is 1505 m greater than the length of the Brooklyn Bridge. Find the length of the Brooklyn Bridge.

Akashi Kaikyo Bridge

34. Real Estate This year the value of a lakefront summer home is $525,000. This amount is twice the value of the home 6 years ago. What was its value 6 years ago?

35. Depreciation The value of a sport utility vehicle this year is $16,000, which is four-fifths of its value last year. Find the value of the vehicle last year.

World Energy Consumption Projected

World energy consumption in 2015 is projected to be 560 quadrillion Btu.

Source: EIA, System for the Analysis of Global Energy Markets

36. Energy Consumption See the news clipping at the right. The projected world energy consumption in 2015 is four-fifths the expected world energy consumption in 2030. What is the expected world energy consumption in 2030?

37. Sleep First- through fifth-graders get an average of 9.5 h of sleep daily. This is three-fourths the number of hours infants aged 3 months to 11 months sleep each day. (*Source:* National Sleep Foundation) How many hours do infants aged 3 months to 11 months sleep each day? Round to the nearest tenth.

38. Finances Each month the Manzanares family spends $1360 for their house payment and utilities, which amounts to one-fourth of the family's monthly income. Find the family's monthly income.

39. Consumerism The cost of a graphing calculator today is three-fourths of the cost of the calculator 5 years ago. The cost of the graphing calculator today is $72. Find the cost of the calculator 5 years ago.

40. Sports The average number of home runs per major league game today is 2.21. This represents 135% of the average number of home runs per game 40 years ago. (*Source:* Elias Sports Bureau) Find the average number of home runs per game 40 years ago. Round to the nearest hundredth.

41. Nutrition The nutrition label on a bag of Baked Tostitos tortilla chips lists the sodium content of one serving as 200 mg, which is 8% of the recommended daily allowance of sodium. What is the recommended daily allowance of sodium? Express the answer in grams.

42. Language Eighty percent of the people who speak Ojibwa, an ancient Native American language, are over the age of 60. Given that 8000 people who speak Ojibwa are over the age of 60, how many people speak Ojibwa? (*Source: Time*, May 26, 2008)

43. Diet Americans consume 7 billion hot dogs from Memorial Day through Labor Day. This is 35% of the hot dogs consumed annually in the United States. (*Source:* National Hot Dog & Sausage Council; American Meat Institute) How many hot dogs do Americans consume annually?

44. Compensation Sandy's monthly salary as a sales representative was $2580. This amount included her base monthly salary of $600 plus a 3% commission on total sales. Find her total sales for the month.

45. Conservation In Central America and Mexico, 1184 plants and animals are known to be at risk of extinction. This represents approximately 10.7% of all the species known to be at risk of extinction on Earth. (*Source:* World Conservation Union) Approximately how many plants and animals are known to be at risk of extinction in the world?

46. Insecticides Americans spend approximately $295 million a year on remedies for cockroaches. The table at the right shows the top U.S. cities for sales of roach insecticides. What percent of the total is spent in New York? Round to the nearest tenth of a percent.

City	Roach Insecticide Sales
Los Angeles	$16.8 million
New York	$9.8 million
Houston	$6.7 million

Source: IRI InfoScan for Combat

47. Contractors Budget Plumbing charged $445 for a water softener and installation. The charge included $310 for the water softener and $45 per hour for labor. How many hours of labor were required for the job?

48. Conservation A water flow restrictor has reduced the flow of water to 2 gal/min. This amount is 1 gal/min less than three-fifths the original flow rate. Find the original rate.

49. Vacation Days In Italy, workers take an average of 42 vacation days per year. This number is 3 more than three times the average number of vacation days workers take each year in the United States. (*Source:* World Tourism Organization) On average, how many vacation days do U.S. workers take per year?

50. Astronautics Four hundred seventeen men have flown into space. This number is 25 more than 8 times the number of women who have flown into space. (*Source:* Encyclopedia Astronautica) How many women have flown into space?

51. Compensation Assume that a sales executive receives a base monthly salary of $600 plus an 8.25% commission on total sales per month. Find the executive's total sales during a month in which she receives total compensation of $4109.55.

Applying the Concepts

52. A man's boyhood lasted $\frac{1}{6}$ of his life, he played football for the next $\frac{1}{8}$ of his life, and he married 5 years after quitting football. A daughter was born after he had been married $\frac{1}{12}$ of his life. The daughter lived $\frac{1}{2}$ as many years as her father. The man died 6 years after his daughter. How old was the man when he died? Use a number line to illustrate the time. Then write an equation and solve it.

FOCUS ON PROBLEM SOLVING

From Concrete to Abstract

As you progress in your study of algebra, you will find that the problems become less concrete and more abstract. Problems that are concrete provide information pertaining to a specific instance. Abstract problems are theoretical; they are stated without reference to a specific instance. Let's look at an example of an abstract problem.

How many cents are in d dollars?

How can you solve this problem? Are you able to solve the same problem if the information given is concrete?

How many cents are in 5 dollars?

You know that there are 100 cents in 1 dollar. To find the number of cents in 5 dollars, multiply 5 by 100.

$$100 \cdot 5 = 500 \qquad \text{There are 500 cents in 5 dollars.}$$

Use the same procedure to find the number of cents in d dollars: multiply d by 100.

$$100 \cdot d = 100d \qquad \text{There are } 100d \text{ cents in } d \text{ dollars.}$$

This problem might be taken a step further:

If one pen costs c cents, how many pens can be purchased with d dollars?

Consider the same problem using numbers in place of the variables.

If one pen costs 50 cents, how many pens can be purchased with 2 dollars?

To solve this problem, you need to calculate the number of cents in 2 dollars (multiply 2 by 100) and divide the result by the cost per pen (50 cents).

$$\frac{100 \cdot 2}{50} = \frac{200}{50} = 4 \qquad \begin{array}{l} \text{If one pen costs 50 cents,} \\ \text{4 pens can be purchased with 2 dollars.} \end{array}$$

Use the same procedure to solve the related abstract problem. Calculate the number of cents in d dollars (multiply d by 100) and divide the result by the cost per pen (c cents).

$$\frac{100 \cdot d}{c} = \frac{100d}{c} \qquad \begin{array}{l} \text{If one pen costs } c \text{ cents, } \frac{100d}{c} \text{ pens} \\ \text{can be purchased with } d \text{ dollars.} \end{array}$$

At the heart of the study of algebra is the use of variables. It is the variables in the problems above that make them abstract. But it is variables that enable us to generalize situations and state rules about mathematics.

Try the following problems.

1. How many nickels are in *d* dollars?

2. How many copies can you make on a coin-operated copy machine if you have only *d* dollars and each copy costs *c* cents?

3. If you travel *m* miles on 1 gal of gasoline, how far can you travel on *g* gallons of gasoline?

4. If you walk a mile in *x* minutes, how far can you walk in *h* hours?

5. If one photocopy costs *n* nickels, how many photocopies can you make for *q* quarters?

PROJECTS AND GROUP ACTIVITIES

Averages   We often discuss temperature in terms of average high or average low temperature. Temperatures collected over a period of time are analyzed to determine, for example, the average high temperature for a given month in your city or state. The following activity is planned to help you better understand the concept of "average."

1. Choose two cities in the United States. We will refer to them as City X and City Y. Over an 8-day period, record the daily high temperature each day in each city.

2. Determine the average high temperature for City X for the 8-day period. (Add the eight numbers, and then divide the sum by 8.) Do not round your answer.

3. Subtract the average high temperature for City X from each of the eight daily high temperatures for City X. You should have a list of eight numbers; the list should include positive numbers, negative numbers, and possibly zero.

Buffalo, NY

4. Find the sum of the list of eight differences recorded in Step 3.

5. Repeat Steps 2 through 4 for City Y.

6. Compare the two sums found in Steps 4 and 5 for City X and City Y.

Phoenix, AZ

7. If you were to conduct this activity again, what would you expect the outcome to be? Use the results to explain what an average high temperature is. In your own words, explain what "average" means.

CHAPTER 11

SUMMARY

KEY WORDS	EXAMPLES
A *variable* is a letter of the alphabet used to stand for a quantity that is unknown or that can change. An expression that contains one or more variables is a *variable expression*. Replacing the variable or variables in a variable expression and then simplifying the resulting numerical expression is called *evaluating the variable expression*. [11.1A, p.460]	Evaluate $5x^3 + 2y - 6$ when $x = -1$ and $y = 4$. $5x^3 + 2y - 6$ $5(-1)^3 + 2(4) - 6 = 5(-1) + 2(4) - 6$ $= -5 + 8 - 6$ $= 3 - 6$ $= 3 + (-6) = -3$
The *terms of a variable expression* are the addends of the expression. A *variable term* consists of a *numerical coefficient* and a *variable part*. A *constant term* has no variable part. [11.1B, pp. 461–462]	The variable expression $-3x^2 + 2x - 5$ has three terms: $-3x^2$, $2x$, and -5. $-3x^2$ and $2x$ are variable terms. -5 is a constant term. For the term $-3x^2$, the coefficient is -3 and the variable part is x^2.
Like terms of a variable expression have the same variable part. Constant terms are considered like terms. [11.1B, p. 462]	$-6a^3b^2$ and $4a^3b^2$ are like terms.
An *equation* expresses the equality of two mathematical expressions. [11.2A, p. 470]	$5x + 6 = 7x - 3$ $y = 4x - 10$ $3a^2 - 6a + 4 = 0$
A *solution of an equation* is a number that, when substituted for the variable, results in a true equation. [11.2A, p. 470]	6 is a solution of $x - 4 = 2$ because $6 - 4 = 2$ is a true equation.
Solving an equation means finding a solution of the equation. The goal is to rewrite the equation in the form *variable = constant*. [11.2B, p. 471]	$x = 5$ is in the form *variable = constant*. The solution of the equation $x = 5$ is the constant 5 because $5 = 5$ is a true equation.
A *formula* is an equation that expresses a relationship among variables. [11.2D, p. 475]	The relationship between Celsius temperature and Fahrenheit temperature is given by the formula $F = 1.8C + 32$, where F represents degrees Fahrenheit and C represents degrees Celsius.
Some of the words and phrases that translate to *equals* are *is, is equal to, amounts to,* and *represents*. [11.6A, p. 498]	"Eight plus a number is ten" translates to $8 + x = 10$.

ESSENTIAL RULES AND PROCEDURES	EXAMPLES
Commutative Property of Addition [11.1B, p. 462] $a + b = b + a$	$-9 + 5 = 5 + (-9)$

Associative Property of Addition   [11.1B, p. 462]
$(a + b) + c = a + (b + c)$

$(-6 + 4) + 2 = -6 + (4 + 2)$

Commutative Property of Multiplication   [11.1C, p. 464]
$a \cdot b = b \cdot a$

$-5(10) = 10(-5)$

Associative Property of Multiplication   [11.1C, p. 464]
$(a \cdot b) \cdot c = a \cdot (b \cdot c)$

$(-3 \cdot 4) \cdot 6 = -3 \cdot (4 \cdot 6)$

Distributive Property   [11.1C, p. 464]
$a(b + c) = ab + ac$
$a(b - c) = ab - ac$

$2(x + 7) = 2(x) + 2(7) = 2x + 14$
$5(4x - 3) = 5(4x) - 5(3) = 20x - 15$

Addition Property of Zero   [11.2B, p. 471]
The sum of a term and zero is the term.
$a + 0 = a$ or $0 + a = a$

$-16 + 0 = -16$

Addition Property of Equations   [11.2B, p. 471]
If a, b, and c are algebraic expressions, then the equations $a = b$
and $a + c = b + c$ have the same solutions.

The same number or variable expression can be added to each side
of an equation without changing the solution of the equation.

$$x + 7 = 20$$
$$x + 7 + (-7) = 20 + (-7)$$
$$x = 13$$

Multiplication Property of Reciprocals   [11.2C, p. 473]
The product of a nonzero term and its reciprocal equals 1.

$$8 \cdot \frac{1}{8} = 1$$

Multiplication Property of One   [11.2C, p. 473]
The product of a term and 1 is the term.

$-7(1) = -7$

Multiplication Property of Equations   [11.2C, p. 473]
If a, b, and c are algebraic expressions and $c \neq 0$, then the
equation $a = b$ has the same solutions as the equation $ac = bc$.

Each side of an equation can be multiplied by the same nonzero
number without changing the solution of the equation.

$$\frac{3}{4}x = 24$$
$$\frac{4}{3} \cdot \frac{3}{4}x = \frac{4}{3} \cdot 24$$
$$x = 32$$

Steps in Solving General First-Degree Equations   [11.4B, p. 488]

1. Use the Distributive Property to remove parentheses.

2. Combine like terms on each side of the equation.

3. Use the Addition Property of Equations to rewrite the equation with
   only one variable term.

4. Use the Addition Property of Equations to rewrite the equation with
   only one constant term.

5. Use the Multiplication Property of Equations to rewrite the equation
   so that the coefficient of the variable term is 1.

$$8 - 4(2x + 3) = 2(1 - x)$$
$$8 - 8x - 12 = 2 - 2x$$
$$-8x - 4 = 2 - 2x$$
$$-8x + 2x - 4 = 2 - 2x + 2x$$
$$-6x - 4 = 2$$
$$-6x - 4 + 4 = 2 + 4$$
$$-6x = 6$$
$$\frac{-6x}{-6} = \frac{6}{-6}$$
$$x = -1$$

CHAPTER 11

CONCEPT REVIEW

Test your knowledge of the concepts presented in this chapter. Answer each question.
Then check your answers against the ones provided in the Answer Section.

1. How do you evaluate a variable expression?

2. How do you add like terms?

3. How do you simplify a variable expression containing parentheses?

4. What is a solution of an equation?

5. How do you check the solution to an equation?

6. What is the Multiplication Property of Equations?

7. What is the Addition Property of Equations?

8. What properties are applied to solve the equation $5x - 4 = 26$?

9. After you substitute into a formula, how do you solve it?

10. How do you isolate the variable in the equation $5x - 3 = 10 - 8x$?

11. Name some mathematical terms that translate into addition.

12. Name some mathematical terms that translate into "equals."

CHAPTER 11

REVIEW EXERCISES

1. Simplify: $-2(a - b)$

2. Is -2 a solution of the equation $3x - 2 = -8$?

3. Solve: $x - 3 = -7$

4. Solve: $-2x + 5 = -9$

5. Evaluate $a^2 - 3b$ when $a = 2$ and $b = -3$.

6. Solve: $-3x = 27$

7. Solve: $\frac{2}{3}x + 3 = -9$

8. Simplify: $3x - 2(3x - 2)$

9. Solve: $6x - 9 = -3x + 36$

10. Solve: $x + 3 = -2$

11. Is 5 a solution of the equation $3x - 5 = -10$?

12. Evaluate $a^2 - (b \div c)$ when $a = -2$, $b = 8$, and $c = -4$.

13. Solve: $3(x - 2) + 2 = 11$

14. Solve: $35 - 3x = 5$

15. Simplify: $6bc - 7bc + 2bc - 5bc$

16. Solve: $7 - 3x = 2 - 5x$

17. Solve: $-\frac{3}{8}x = -\frac{15}{32}$

18. Simplify: $\frac{1}{2}x^2 - \frac{1}{3}x^2 + \frac{1}{5}x^2 + 2x^2$

19. Solve: $5x - 3(1 - 2x) = 4(2x - 1)$

20. Solve: $\frac{5}{6}x - 4 = 5$

21. **Fuel Efficiency** A tourist drove a rental car 621 mi on 27 gal of gas. Find the number of miles per gallon of gas. Use the formula $D = M \cdot G$, where D is distance, M is miles per gallon, and G is the number of gallons.

22. **Temperature Conversion** Find the Celsius temperature when the Fahrenheit temperature is 100°. Use the formula $F = 1.8C + 32$, where F is the Fahrenheit temperature and C is the Celsius temperature. Round to the nearest tenth.

23. Translate "the total of n and the quotient of n and five" into a mathematical expression.

24. Translate "the sum of five more than a number and one-third of the number" into a mathematical expression.

25. The difference between nine and twice a number is five. Find the number.

26. The product of five and a number is fifty. Find the number.

27. **Discount** An MP3 video player is now on sale for $228. This is 60% of the regular price. Find the regular price of the MP3 video player.

28. **Agriculture** A farmer harvested 28,336 bushels of corn. This amount represents a 12% increase over last year's crop. How many bushels of corn did the farmer harvest last year?

CHAPTER 11

TEST

1. Solve: $\frac{x}{5} - 12 = 7$

2. Solve: $x - 12 = 14$

3. Simplify: $3y - 2x - 7y - 9x$

4. Solve: $8 - 3x = 2x - 8$

5. Solve: $3x - 12 = -18$

6. Evaluate $c^2 - (2a + b^2)$ when $a = 3$, $b = -6$, and $c = -2$.

7. Is 3 a solution of the equation $x^2 + 3x - 7 = 3x - 2$?

8. Simplify: $9 - 8ab - 6ab$

9. Solve: $-5x = 14$

10. Simplify: $3y + 5(y - 3) + 8$

11. Solve: $3x - 4(x - 2) = 8$

12. Solve: $5 = 3 - 4x$

13. Evaluate $\frac{x^2}{y} - \frac{y^2}{x}$ for $x = 3$ and $y = -2$.

14. Solve: $\frac{5}{8}x = -10$

15. Solve: $y - 4y + 3 = 12$

16. Solve: $2x + 4(x - 3) = 5x - 1$

17. **Finance** A loan of $6600 is to be paid in 48 equal monthly installments. Find the monthly payment. Use the formula $L = P \cdot N$, where L is the loan amount, P is the monthly payment, and N is the number of months.

18. **Manufacturing** A clock manufacturer's fixed costs per month are $5000. The unit cost for each clock is $15. Find the number of clocks made during a month in which the total cost was $65,000. Use the formula $T = U \cdot N + F$, where T is the total cost, U is the cost per unit, N is the number of units made, and F is the fixed costs.

19. **Physics** Find the time required for a falling object to increase in velocity from 24 ft/s to 392 ft/s. Use the formula $V = V_0 + 32t$, where V is the final velocity of a falling object, V_0 is the starting velocity of the falling object, and t is the time for the object to fall.

20. Translate "the sum of x and one-third of x" into a mathematical expression.

21. Translate "five times the sum of a number and three" into a mathematical expression.

22. Translate "three less than two times a number is seven" into an equation and solve.

23. The total of five and three times a number is the number minus two. Find the number.

24. **Compensation** Eduardo Santos earned $3600 last month. This salary is the sum of a base monthly salary of $1200 and a 6% commission on total sales. Find Eduardo's total sales for the month.

25. **Consumerism** Your mechanic charges you $338 for performing a 30,000-mile checkup on your car. This charge includes $152 for parts and $62 per hour for labor. Find the number of hours the mechanic worked on your car.

CUMULATIVE REVIEW EXERCISES

1. Simplify: $6^2 - (18 - 6) \div 4 + 8$

2. Subtract: $3\frac{1}{6} - 1\frac{7}{15}$

3. Simplify: $\left(\frac{3}{8} - \frac{1}{4}\right) \div \frac{3}{4} + \frac{4}{9}$

4. Multiply: 9.67×0.0049

5. Write "$182 earned in 20 hours" as a unit rate.

6. Solve the proportion $\frac{2}{3} = \frac{n}{40}$. Round to the nearest hundredth.

7. Write $5\frac{1}{3}\%$ as a fraction.

8. What percent of 30 is 42?

9. 8 is 125% of what number?

10. Multiply: 3 ft 9 in. $\times$ 5

11. Convert $1\frac{3}{8}$ lb to ounces.

12. Convert 282 mg to grams.

13. Add: $-2 + 5 + (-8) + 4$

14. Find -6 less than 13.

15. Simplify: $(-2)^2 - (-8) \div (3 - 5)^2$

16. Evaluate $3ab - 2ac$ when $a = -2$, $b = 6$, and $c = -3$.

17. Simplify: $3z - 2x + 5z - 8x$

18. Simplify: $6y - 3(y - 5) + 8$

19. Solve: $2x - 5 = -7$

20. Solve: $7x - 3(x - 5) = -10$

21. Solve: $-\frac{2}{3}x = 5$

22. Solve: $\frac{x}{3} - 5 = -12$

23. Education In a mathematics class of 34 students, 6 received an A grade. Find the percent of students in the mathematics class who received an A grade. Round to the nearest tenth of a percent.

24. Markup The manager of a pottery store uses a markup rate of 40%. Find the price of a piece of pottery that cost the store $28.50.

25. Discount A laptop computer regularly priced at $450 is on sale for $369.
a. What is the discount?
b. What is the discount rate?

26. Simple Interest A toy store borrowed $80,000 at a simple interest rate of 11% for 4 months. What is the simple interest due on the loan? Round to the nearest cent.

27. Fuel Prices See the news clipping at the right. If an airline ticket from Manchester, New Hampshire, to Chicago, Illinois, cost $375, what dollar amount of that cost went to pay for fuel?

28. Probability A tetrahedral die is one with four triangular sides numbered 1 to 4. If two tetrahedral dice are rolled, what is the probability that the sum of the upward faces is 7?

29. Compensation Sunah Yee, a sales executive, receives a base salary of $800 plus an 8% commission on total sales. Find the total sales during a month in which Sunah earned $3400. Use the formula $M = S \cdot R + B$, where M is the monthly earnings, S is the total sales, R is the commission rate, and B is the base monthly salary.

30. Three less than eight times a number is three more than five times the number. Find the number.

In the News

Oil Prices Hit All-Time High

On Friday, July 11, 2008, the price of oil hit an all-time high of $147 a barrel, which sent the cost of jet fuel to $175 a barrel. For airline passengers, this meant that, on average, 40% of the cost of an airline ticket went to pay for fuel.

Source: United Hemispheres, September 2008

AP Images

Geometry

OBJECTIVES

SECTION 12.1
A To define and describe lines and angles
B To define and describe geometric figures
C To solve problems involving angles formed by intersecting lines

SECTION 12.2
A To find the perimeter of plane geometric figures
B To find the perimeter of composite geometric figures
C To solve application problems

SECTION 12.3
A To find the area of geometric figures
B To find the area of composite geometric figures
C To solve application problems

SECTION 12.4
A To find the volume of geometric solids
B To find the volume of composite geometric solids
C To solve application problems

SECTION 12.5
A To find the square root of a number
B To find the unknown side of a right triangle using the Pythagorean Theorem
C To solve application problems

SECTION 12.6
A To solve similar and congruent triangles
B To solve application problems

ARE YOU READY?

Take the Chapter 12 Prep Test to find out if you are ready to learn to:

- Solve problems involving the angles formed by intersecting lines
- Find the perimeter and area of geometric figures
- Find the volume of geometric solids
- Use the Pythagorean Theorem
- Solve similar and congruent triangles

PREP TEST

Do these exercises to prepare for Chapter 12.

1. Solve: $x + 47 = 90$

2. Solve: $32 + 97 + x = 180$

3. Simplify: $2(18) + 2(10)$

4. Evaluate abc when $a = 2$, $b = 3.14$, and $c = 9$.

5. Evaluate xyz^3 when $x = \frac{4}{3}$, $y = 3.14$, and $z = 3$.

6. Solve: $\frac{5}{12} = \frac{6}{x}$

12.1 Angles, Lines, and Geometric Figures

OBJECTIVE A   To define and describe lines and angles

Point of Interest

Geometry is one of the oldest branches of mathematics. Around 350 B.C., the Greek mathematician Euclid wrote the *Elements,* which contained all of the known concepts of geometry. Euclid's contribution was to unify various concepts into a single deductive system that was based on a set of axioms.

The word *geometry* comes from the Greek words for "earth" (*geo*) and "measure" (*metron*). The original purpose of geometry was to measure land. Today geometry is used in many sciences, such as physics, chemistry, and geology, and in applied fields such as mechanical drawing and astronomy. Geometric form is used in art and design.

Two basic geometric concepts are plane and space.

A **plane** is a flat surface, such as a table-top or a blackboard. Figures that can lie totally in a plane are called **plane figures.**

Space extends in all directions. Objects in space, such as trees, ice cubes, and doors, are called **solids.**

A **line** extends indefinitely in two directions in a plane. A line has no width.

Tips for Success

A great many new vocabulary words are introduced in this chapter. There are eight new terms on this page alone: plane, plane figures, space, solids, line, line segment, parallel lines, and intersecting lines. All of these terms are in **bold type**. The bold type indicates that these are concepts you must know to learn the material. Be sure to study each new term as it is presented.

A **line segment** is part of a line and has two endpoints. The line segment AB is shown in the figure.

The length of a line segment is the distance between the endpoints of the line segment. The length of a line segment may be expressed as the sum of two or more shorter line segments, as shown. For this example, $AB = 5$, $BC = 3$, and $AC = AB + BC = 5 + 3 = 8$.

> **HOW TO 1**   Given that $AB = 22$ and $AC = 31$, find the length of BC.
>
> $AC = AB + BC$
> $31 = 22 + BC$ • Substitute 22 for AB and 31 for AC,
> $31 - 22 = 22 - 22 + BC$ and solve for BC.
> $9 = BC$

Lines in a plane can be parallel or intersecting. **Parallel lines** never meet; the distance between them is always the same. **Intersecting lines** cross at a point in the plane.

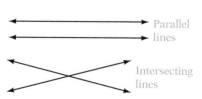

The symbol ∥ means "is parallel to." In the accompanying figure, $AB \parallel CD$ and $p \parallel q$. Note that line p contains line segment AB and that line q contains line segment CD. Parallel lines contain parallel line segments.

A **ray** starts at a point and extends indefinitely in one direction.

An **angle** is formed when two rays start from the same point. Rays r_1 and r_2 start from point B. The common endpoint is called the **vertex** of the angle.

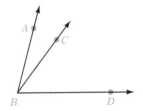
If A and C are points on rays r_1 and r_2 above, respectively, then the angle is called $\angle ABC$, $\angle CBA$, or $\angle B$, where $\angle$ is the symbol for angle. Note that an angle is named by giving three points, with the vertex as the second point listed, or by giving the point at the vertex.

An angle can also be named by writing a variable between the rays close to the vertex. In the figure, $\angle x = \angle QRS = \angle SRQ$ and $\angle y = \angle SRT = \angle TRS$. Note that in this figure, more than two rays meet at the vertex. In this case, the vertex cannot be used to name the angle.

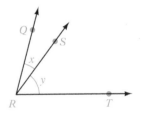

A unit in which angles are measured is the **degree.** The symbol for degree is °. One complete revolution is 360° (360 degrees).

One-quarter of a revolution is 90°. A 90° angle is called a **right angle.** The symbol ∟ represents a right angle.

Perpendicular lines are intersecting lines that form right angles.

The symbol ⊥ means "is perpendicular to." In the accompanying figure, $AB \perp CD$ and $p \perp q$. Note that line p contains line segment AB and line q contains line segment CD. Perpendicular lines contain perpendicular line segments.

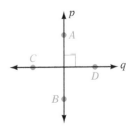

Complementary angles are two angles whose sum is 90°.

$\angle A + \angle B = 70° + 20° = 90°$
$\angle A$ and $\angle B$ are complementary angles.

One-half of a revolution is 180°. A 180° angle is called a **straight angle**. $\angle AOB$ in the figure is a straight angle.

Supplementary angles are two angles whose sum is 180°.

$\angle C + \angle D = 130° + 50° = 180°$
$\angle C$ and $\angle D$ are supplementary angles.

An **acute angle** is an angle whose measure is between 0° and 90°. $\angle D$ in the figure above is an acute angle. An **obtuse angle** is an angle whose measure is between 90° and 180°. $\angle C$ in the figure above is an obtuse angle.

In the accompanying figure, $\angle DAC = 45°$ and $\angle CAB = 55°$.

$\angle DAB = \angle DAC + \angle CAB$
$\quad\quad = 45° + 55° = 100°$

EXAMPLE • 1

Given that $MN = 15$, $NO = 18$, and $MP = 48$, find the length of OP.

Solution

$MP = MN + NO + OP$
$48 = 15 + 18 + OP$
$48 = 33 + OP$
$48 - 33 = 33 - 33 + OP$
$15 = OP$

YOU TRY IT • 1

Given that $QR = 24$, $ST = 17$, and $QT = 62$, find the length of RS.

Your solution

EXAMPLE • 2

Find the complement of a 32° angle.

Solution

Let x represent the complement of 32°.

$x + 32° = 90°$ • The sum of
$x + 32° - 32° = 90° - 32°$ complementary
$x = 58°$ angles is 90°.

58° is the complement of 32°.

YOU TRY IT • 2

Find the supplement of a 32° angle.

Your solution

Solutions on p. S29

EXAMPLE · 3

Find the measure of ∠x.

Solution

$$\angle x + 47° = 90°$$
$$\angle x + 47° - 47° = 90° - 47°$$
$$\angle x = 43°$$

YOU TRY IT · 3

Find the measure of ∠a.

Your solution

Solution on p. S29

OBJECTIVE B

To define and describe geometric figures

✓ Take Note

Any side of a triangle can be considered the base, but generally the **base of a triangle** refers to the side that the triangle rests on. The **height of a triangle** is a line segment drawn perpendicular to the base from the opposite vertex.

A **triangle** is a closed, three-sided plane figure. Figure *ABC* is a triangle. *AB* is called the **base.** The line *CD*, perpendicular to the base, is called the **height.**

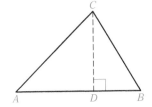

> **The Angles of a Triangle**
>
> The sum of the three angles of a triangle is 180°.
>
> $\angle A + \angle B + \angle C = 180°$

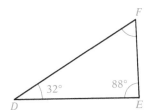

HOW TO · 2 In triangle *DEF*, ∠D = 32° and ∠E = 88°. Find the measure of ∠F.

$$\angle D + \angle E + \angle F = 180°$$
$$32° + 88° + \angle F = 180°$$
$$120° + \angle F = 180°$$
$$120° - 120° + \angle F = 180° - 120°$$
$$\angle F = 60°$$

• The sum of the three angles of a triangle is 180°.
• ∠D = 32° and ∠E = 88°
• Solve for ∠F.

A **right triangle** contains one right angle. The side opposite the right angle is called the **hypotenuse.** The **legs of a right triangle** are its other two sides. In a right triangle, the two acute angles are complementary.

$$\angle A + \angle B = 90°$$

HOW TO · 3 In the right triangle at the left, ∠A = 30°. Find the measure of ∠B.

$$\angle A + \angle B = 90°$$
$$30° + \angle B = 90°$$
$$30° - 30° + \angle B = 90° - 30°$$
$$\angle B = 60°$$

• The two acute angles are complementary.
• ∠A = 30°
• Solve for ∠B.

A **quadrilateral** is a closed, four-sided plane figure. Three quadrilaterals with special characteristics are described here.

A **parallelogram** has opposite sides parallel and equal. The distance *AE* between the parallel sides is called the **height.**

Parallelogram

A **rectangle** is a parallelogram that has four right angles.

A **square** is a rectangle that has four equal sides.

Rectangle

Square

A **circle** is a plane figure in which all points are the same distance from point *O*, which is called the **center** of the circle.

The **diameter of a circle** (*d*) is a line segment through the center of the circle with endpoints on the circle. *AB* is a diameter of the circle shown.

The **radius of a circle** (*r*) is a line segment from the center to a point on the circle. *OC* is a radius of the circle.

$$d = 2r \quad \text{or} \quad r = \frac{1}{2}d$$

Circle

HOW TO 4 The line segment *AB* is a diameter of the circle shown. Find the radius of the circle.

The radius is one-half the diameter. Therefore,

$$r = \frac{1}{2}d$$

$$= \frac{1}{2}(8 \text{ in.}) \qquad \bullet \; d = 8 \text{ in.}$$

$$= 4 \text{ in.}$$

A **geometric solid** is a figure in space, or space figure. Four common space figures are the rectangular solid, cube, sphere, and cylinder.

A **rectangular solid** is a solid in which all six faces are rectangles.

Rectangular solid

A **cube** is a rectangular solid in which all six faces are squares.

Cube

A **sphere** is a solid in which all points on the surface are the same distance from point *O*, which is called the **center** of the sphere.

The **diameter of a sphere** is a line segment going through the center with endpoints on the sphere. *AB* is a diameter of the sphere shown.

Sphere

The **radius of a sphere** is a line segment from the center to a point on the sphere. *OC* is a radius of the sphere.

$$d = 2r \quad \text{or} \quad r = \frac{1}{2}d$$

> **HOW TO 5** The radius of the sphere shown at the right is 5 cm. Find the diameter of the sphere.

$d = 2r$ • **The diameter equals twice the radius.**

$\quad = 2(5 \text{ cm})$ • $r = 5$ **cm**

$\quad = 10 \text{ cm}$

The diameter is 10 cm.

The most common **cylinder** is one in which the bases are circles and are perpendicular to the side.

Height

Base

Cylinder

EXAMPLE • 4

One angle in a right triangle is equal to 50°. Find the measures of the other angles.

Solution

In a right triangle, one angle measures 90° and the two acute angles are complementary.

$$\angle A + \angle B = 90°$$
$$\angle A + 50° = 90°$$
$$\angle A + 50° - 50° = 90° - 50°$$
$$\angle A = 40°$$

The other angles measure 90° and 40°.

YOU TRY IT • 4

A right triangle has one angle equal to 7°. Find the measures of the other angles.

Your solution

EXAMPLE • 5

Two angles of a triangle measure 42° and 103°. Find the measure of the third angle.

Solution

The sum of the three angles of a triangle is 180°.

$$\angle A + \angle B + \angle C = 180°$$
$$\angle A + 42° + 103° = 180°$$
$$\angle A + 145° = 180°$$
$$\angle A + 145° - 145° = 180° - 145°$$
$$\angle A = 35°$$

The measure of the third angle is 35°.

YOU TRY IT • 5

Two angles of a triangle measure 62° and 45°. Find the measure of the other angle.

Your solution

EXAMPLE • 6

A circle has a radius of 8 cm. Find the diameter.

Solution

$d = 2r$

$\quad = 2 \cdot 8 \text{ cm} = 16 \text{ cm}$

The diameter is 16 cm.

YOU TRY IT • 6

A circle has a diameter of 8 in. Find the radius.

Your solution

Solutions on p. S29

OBJECTIVE C   **To solve problems involving angles formed by intersecting lines**

Four angles are formed by the intersection of two lines. If the two lines are perpendicular, then each of the four angles is a right angle. If the two lines are not perpendicular, then two of the angles formed are acute angles and two of the angles are obtuse angles. The two acute angles are always opposite each other, and the two obtuse angles are always opposite each other.

In the figure, $\angle w$ and $\angle y$ are acute angles. $\angle x$ and $\angle z$ are obtuse angles. Two angles that are on opposite sides of the intersection of two lines are called **vertical angles.** Vertical angles have the same measure. $\angle w$ and $\angle y$ are vertical angles. $\angle x$ and $\angle z$ are vertical angles.

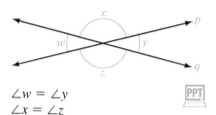

$$\angle w = \angle y$$
$$\angle x = \angle z$$

Two angles that share a common side are called **adjacent angles.** In the previous figure, $\angle x$ and $\angle y$ are adjacent angles, as are $\angle y$ and $\angle z$, $\angle z$ and $\angle w$, and $\angle w$ and $\angle x$. Adjacent angles of intersecting lines are supplementary angles.

$$\angle x + \angle y = 180°$$
$$\angle y + \angle z = 180°$$
$$\angle z + \angle w = 180°$$
$$\angle w + \angle x = 180°$$

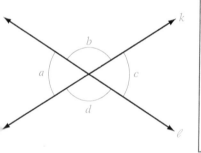

HOW TO 6   In the figure at the left, $\angle c = 65°$. Find the measures of angles a, b, and d.

$\angle a = 65°$

• $\angle a = \angle c$ because $\angle c$ and $\angle a$ are vertical angles.

$$\angle b + \angle c = 180°$$

• $\angle c$ is supplementary to $\angle b$ because $\angle c$ and $\angle b$ are adjacent angles.

$$\angle b + 65° = 180°$$
$$\angle b + 65° - 65° = 180° - 65°$$
$$\angle b = 115°$$

• $\angle c = 65°$

$\angle d = 115°$

• $\angle d = \angle b$ because $\angle b$ and $\angle d$ are vertical angles.

A line intersecting two other lines at two different points is called a **transversal.**

If the lines cut by a transversal are parallel lines and the transversal is perpendicular to the parallel lines, then all eight angles formed are right angles.

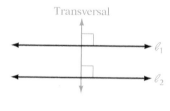

If the lines cut by a transversal are parallel lines and the transversal is not perpendicular to the parallel lines, then all four acute angles have the same measure and all four obtuse angles have the same measure. For the figure at the right,

$$\angle a = \angle c = \angle w = \angle y \quad \text{and} \quad \angle b = \angle d = \angle x = \angle z$$

Transversal

Alternate interior angles are two nonadjacent angles that are on opposite sides of the transversal and between the parallel lines. For the figure at the left, $\angle c$ and $\angle w$ are alternate interior angles. $\angle d$ and $\angle x$ are alternate interior angles. Alternate interior angles have the same measure.

Alternate exterior angles are two nonadjacent angles that are on opposite sides of the transversal and outside the parallel lines. For the figure at the left, $\angle a$ and $\angle y$ are alternate exterior angles. $\angle b$ and $\angle z$ are alternate exterior angles. Alternate exterior angles have the same measure.

Corresponding angles are two angles that are on the same side of the transversal and are both acute angles or are both obtuse angles. For the figure at the top left, the following pairs of angles are corresponding angles: $\angle a$ and $\angle w$, $\angle d$ and $\angle z$, $\angle b$ and $\angle x$, $\angle c$ and $\angle y$. Corresponding angles have the same measure.

HOW TO 7 In the figure at the left, $\ell_1 \| \ell_2$ and $\angle c = 58°$. Find the measures of $\angle f$, $\angle h$, and $\angle g$.

$\angle f = 58°$ • $\angle f = \angle c$ because $\angle f$ and $\angle c$ are alternate interior angles.

$\angle h = 58°$ • $\angle h = \angle c$ because $\angle c$ and $\angle h$ are corresponding angles.

$\angle g + \angle h = 180°$ • $\angle g$ is supplementary to $\angle h$.
$\angle g + 58° = 180°$ • $\angle h = 58°$
$\angle g = 122°$ • Subtract 58° from each side.

EXAMPLE 7

In the figure, $\angle a = 75°$. Find $\angle b$.

Solution

$\angle a + \angle b = 180°$ • $\angle a$ and $\angle b$ are supplementary.

$75° + \angle b = 180°$ • $\angle a = 75°$
$\angle b = 105°$ • Subtract 75° from each side.

YOU TRY IT 7

In the figure, $\angle a = 125°$. Find $\angle b$.

Your solution

EXAMPLE 8

In the figure, $\ell_1 \| \ell_2$ and $\angle a = 70°$. Find $\angle b$.

Solution

$\angle c = \angle a = 70°$ • Corresponding angles are equal.

$\angle b + \angle c = 180°$ • $\angle b$ and $\angle c$ are supplementary.
$\angle b + 70° = 180°$ • $\angle c = 70°$
$\angle b = 110°$ • Subtract 70° from each side.

YOU TRY IT 8

In the figure, $\ell_1 \| \ell_2$ and $\angle a = 120°$. Find $\angle b$.

Your solution

Solutions on p. S29

12.1 EXERCISES

OBJECTIVE A   To define and describe lines and angles

1. The measure of an acute angle is between ___ and ___.

2. The measure of an obtuse angle is between ___ and ___.

3. How many degrees are in a straight angle?

4. Two lines that intersect at right angles are _____ lines.

5. In the figure, $EF = 20$ and $FG = 10$. Find the length of EG.

6. In the figure, $EF = 18$ and $FG = 6$. Find the length of EG.

7. In the figure, it is given that $QR = 7$ and $QS = 28$. Find the length of RS.

8. In the figure, it is given that $QR = 15$ and $QS = 45$. Find the length of RS.

9. In the figure, it is given that $AB = 12$, $CD = 9$, and $AD = 35$. Find the length of BC.

10. In the figure, it is given that $AB = 21$, $BC = 14$, and $AD = 54$. Find the length of CD.

11. Find the complement of a 31° angle.

12. Find the complement of a 62° angle.

13. Find the supplement of a 72° angle.

14. Find the supplement of a 162° angle.

15. Find the complement of a 13° angle.

16. Find the complement of an 88° angle.

17. Find the supplement of a 127° angle.

18. Find the supplement of a 7° angle.

 For Exercises 19 to 22, determine whether the described angle is an acute angle, is a right angle, is an obtuse angle, or does not exist.

19. The complement of an acute angle

20. The supplement of a right angle

21. The supplement of an acute angle

22. The complement of an obtuse angle

In Exercises 23 and 24, find the measure of angle *AOB*.

23.

24.

For Exercises 25 to 28, find the measure of angle *a*.

25.

26.

27.

28.

29. In the figure, it is given that ∠*LOM* = 53° and ∠*LON* = 139°. Find the measure of ∠*MON*.

30. In the figure, it is given that ∠*MON* = 38° and ∠*LON* = 85°. Find the measure of ∠*LOM*.

OBJECTIVE B   To define and describe geometric figures

31. What is the sum of the three angles of a triangle?

32. Name the side opposite the right angle in a right triangle.

33. Name the rectangle with four equal sides.

34. Name the solid in which all points are the same distance from the center.

35. Name the plane figure in which all points are the same distance from the center.

36. Name the solid in which the bases are circular and perpendicular to the side.

37. A triangle has a 13° angle and a 65° angle. Find the measure of the other angle.

38. A triangle has a 105° angle and a 32° angle. Find the measure of the other angle.

39. A right triangle has a 45° angle. Find the measures of the other two angles.

40. A right triangle has a 62° angle. Find the measures of the other two angles.

41. A triangle has a 62° angle and a 104° angle. Find the measure of the other angle.

42. A triangle has a 30° angle and a 45° angle. Find the measure of the other angle.

43. A right triangle has a 25° angle. Find the measures of the other two angles.

44. Two angles of a triangle are 42° and 105°. Find the measure of the other angle.

45. Find the radius of a circle with a diameter of 16 in.

46. Find the radius of a circle with a diameter of 9 ft.

47. Find the diameter of a circle with a radius of $2\frac{1}{3}$ ft.

48. Find the diameter of a circle with a radius of 24 cm.

49. The radius of a sphere is 3.5 cm. Find the diameter.

50. The radius of a sphere is $1\frac{1}{2}$ ft. Find the diameter.

51. The diameter of a sphere is 4 ft 8 in. Find the radius.

52. The diameter of a sphere is 1.2 m. Find the radius.

 For Exercises 53 to 55, determine whether the statement is *true* or *false*.

53. A triangle can have two obtuse angles.

54. The legs of a right triangle are perpendicular.

55. If the sum of two angles of a triangle is less than 90°, then the third angle is an obtuse angle.

OBJECTIVE C   To solve problems involving angles formed by intersecting lines

For Exercises 56 to 59, find the measures of angles *a* and *b*.

56.

57.

58.

59.

For Exercises 60 to 67, $\ell_1 \| \ell_2$. Find the measures of angles a and b.

60.

61.

62.

63.

64.

65.

66.

67.

 For Exercises 68 to 70, use the diagram at the right. Determine whether the given statement is *true* or *false*.

68. $\angle a$ and $\angle b$ have the same measure even if ℓ_1 and ℓ_2 are not parallel.

69. $\angle c$ and $\angle d$ are supplementary even if ℓ_1 and ℓ_2 are not parallel.

70. If $\angle a$ is greater than $\angle c$, then ℓ_1 and ℓ_2 are not parallel.

Applying the Concepts

 71. If AB and CD intersect at point O, and $\angle AOC = \angle BOC$, explain why AB is perpendicular to CD.

SECTION

12.2   Plane Geometric Figures

OBJECTIVE A   To find the perimeter of plane geometric figures

A **polygon** is a closed figure determined by three or more line segments that lie in a plane. The **sides of a polygon** are the line segments that form the polygon. The figures below are examples of polygons.

    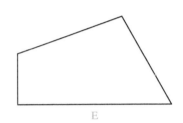

A   B   C   D   E

A **regular polygon** is one in which each side has the same length and each angle has the same measure. The polygons in Figures A, C, and D above are regular polygons.

Point of Interest

Although a polygon is defined in terms of its *sides* (see the definition above), the word actually comes from the Latin word *polygonum*, which means "having many *angles*." This is certainly the case for a polygon.

The name of a polygon is based on the number of its sides. The table below lists the names of polygons that have from 3 to 10 sides.

Number of Sides	Name of the Polygon
3	Triangle
4	Quadrilateral
5	Pentagon
6	Hexagon
7	Heptagon
8	Octagon
9	Nonagon
10	Decagon

The Pentagon in Arlington, Virginia

Triangles and quadrilaterals are two of the most common types of polygons. Triangles are distinguished by the number of equal sides and also by the measures of their angles.

An **isosceles triangle** has two sides of equal length. The angles opposite the equal sides are of equal measure.
$AC = BC$
$\angle A = \angle B$

The three sides of an **equilateral triangle** are of equal length. The three angles are of equal measure.
$AB = BC = AC$
$\angle A = \angle B = \angle C$

A **scalene triangle** has no two sides of equal length. No two angles are of equal measure.

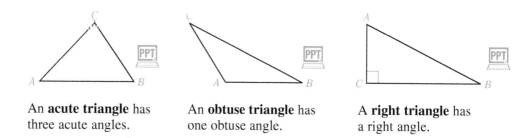

An **acute triangle** has three acute angles.

An **obtuse triangle** has one obtuse angle.

A **right triangle** has a right angle.

Quadrilaterals are also distinguished by their sides and angles, as shown below. Note that a rectangle, a square, and a rhombus are different forms of a parallelogram.

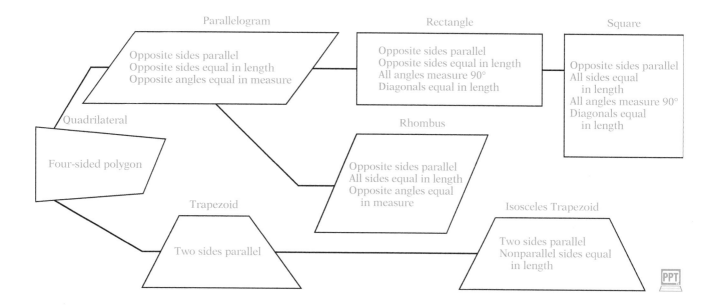

Parallelogram
Opposite sides parallel
Opposite sides equal in length
Opposite angles equal in measure

Rectangle
Opposite sides parallel
Opposite sides equal in length
All angles measure 90°
Diagonals equal in length

Square
Opposite sides parallel
All sides equal in length
All angles measure 90°
Diagonals equal in length

Quadrilateral
Four-sided polygon

Rhombus
Opposite sides parallel
All sides equal in length
Opposite angles equal in measure

Trapezoid
Two sides parallel

Isosceles Trapezoid
Two sides parallel
Nonparallel sides equal in length

The **perimeter** of a plane geometric figure is a measure of the distance around the figure. Perimeter is used in buying fencing for a lawn or determining how much baseboard is needed for a room.

The perimeter of a triangle is the sum of the lengths of the three sides.

Perimeter of a Triangle
$P = a + b + c$

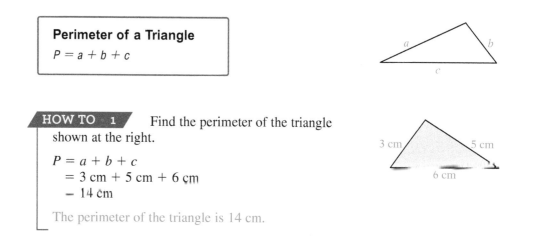

HOW TO 1 Find the perimeter of the triangle shown at the right.

$P = a + b + c$
$= 3 \text{ cm} + 5 \text{ cm} + 6 \text{ cm}$
$= 14 \text{ cm}$

The perimeter of the triangle is 14 cm.

The perimeter of a quadrilateral is the sum of the lengths of the four sides. The perimeter of a square is the sum of the four equal sides.

Perimeter of a Square

$P = 4s$

HOW TO 2 Find the perimeter of the square shown at the right.

$P = 4s$
$\quad = 4(3 \text{ ft})$ • $s = 3$ ft
$\quad = 12$ ft

The perimeter of the square is 12 ft.

A rectangle is a quadrilateral with opposite sides of equal length. The length of a rectangle refers to the longer side, and the width refers to the length of the shorter side.

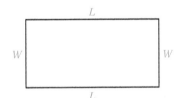

Perimeter of a Rectangle

$P = 2L + 2W$

HOW TO 3 Find the perimeter of the rectangle shown at the right.

$P = 2L + 2W$
$\quad = 2(6 \text{ m}) + 2(3 \text{ m})$ • $L = 6$ m; $W = 3$ m
$\quad = 12 \text{ m} + 6 \text{ m}$
$\quad = 18$ m

The perimeter of the rectangle is 18 m.

The distance around a circle is called the **circumference.** The circumference of a circle is equal to the product of π (pi) and the **diameter.**

Circumference of a Circle

$C = \pi d$

or

$C = 2\pi r$ • Because diameter $= 2r$

The formula for circumference uses the number π (pi). The value of π can be approximated by a decimal or a fraction.

$$\pi \approx 3.14 \qquad \pi \approx \frac{22}{7}$$

The π key on a calculator gives a closer approximation of π than 3.14.

HOW TO 4 Find the circumference of the circle shown at the right.

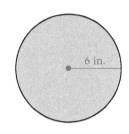

$C = 2\pi r$

$\approx 2 \cdot 3.14 \cdot 6$ in.   • $r = 6$ in.

$= 37.68$ in.

The circumference of the circle is approximately 37.68 in.

EXAMPLE 1

Find the perimeter of a rectangle with a width of $\frac{2}{3}$ ft and a length of 2 ft.

Solution

$P = 2L + 2W$

$= 2(2 \text{ ft}) + 2\left(\frac{2}{3}\text{ ft}\right)$ • $L = 2$ ft, $W = \frac{2}{3}$ ft

$= 4 \text{ ft} + \frac{4}{3} \text{ ft}$

$= 5\frac{1}{3}$ ft

The perimeter of the rectangle is $5\frac{1}{3}$ ft.

YOU TRY IT 1

Find the perimeter of a rectangle with a length of 2 m and a width of 0.85 m.

Your solution

EXAMPLE 2

Find the perimeter of a triangle with sides of lengths 5 in., 7 in., and 8 in.

Solution

$P = a + b + c$

$= 5 \text{ in.} + 7 \text{ in.} + 8 \text{ in.}$

$= 20$ in.

The perimeter of the triangle is 20 in.

YOU TRY IT 2

Find the perimeter of a triangle with sides of lengths 12 cm, 15 cm, and 18 cm.

Your solution

EXAMPLE 3

Find the circumference of a circle with a radius of 18 cm. Use 3.14 for π.

Solution

18 cm

$C = 2\pi r$

$\approx 2 \cdot 3.14 \cdot 18$ cm

$= 113.04$ cm

The circumference is approximately 113.04 cm.

YOU TRY IT 3

Find the circumference of a circle with a diameter of 6 in. Use 3.14 for π.

Your solution

Solutions on p. S29

OBJECTIVE B　　　To find the perimeter of composite geometric figures

A **composite geometric figure** is a figure made from two or more geometric figures. The following composite is made from part of a rectangle and part of a circle:

$$\text{Perimeter of the composite figure} = 3 \text{ sides of a rectangle} + \frac{1}{2} \text{ the circumference of a circle}$$

$$\text{Perimeter of the composite figure} = 2L + W + \frac{1}{2}\pi d$$

The perimeter of the composite figure below is found by adding the measures of twice the length plus the width plus one-half the circumference of the circle.

$$P = 2L + W + \frac{1}{2}\pi d$$
$$\approx 2(12 \text{ m}) + 4 \text{ m} + \frac{1}{2}(3.14)(4 \text{ m})$$
$$= 34.28 \text{ m}$$

• $L = 12$ m, $W = 4$ m, $d = 4$ m. *Note:* The diameter of the circle is equal to the width of the rectangle.

The perimeter is approximately 34.28 m.

EXAMPLE • 4

Find the perimeter of the composite figure. Use $\frac{22}{7}$ for π.

Solution

Perimeter of composite figure	=	sum of lengths of the 4 sides	+	$\frac{1}{2}$ the circumference of the circle
P	=	$4s$	+	$\frac{1}{2}\pi d$

$$\approx 4(5 \text{ cm}) + \frac{1}{2}\left(\frac{22}{7}\right)(7 \text{ cm})$$
$$= 20 \text{ cm} + 11 \text{ cm} = 31 \text{ cm}$$

The perimeter is approximately 31 cm.

YOU TRY IT • 4

Find the perimeter of the composite figure. Use 3.14 for π.

Your solution

OBJECTIVE C   To solve application problems

EXAMPLE • 5

The dimensions of a triangular sail are 18 ft, 11 ft, and 15 ft. What is the perimeter of the sail?

Strategy

To find the perimeter, use the formula for the perimeter of a triangle.

Solution
$P = a + b + c$
   $= 18 \text{ ft} + 11 \text{ ft} + 15 \text{ ft}$
   $= 44 \text{ ft}$

The perimeter of the sail is 44 ft.

YOU TRY IT • 5

What is the perimeter of a standard piece of computer paper that measures $8\frac{1}{2}$ in. by 11 in.?

Your strategy

Your solution

EXAMPLE • 6

If fencing costs $6.75 per foot, how much will it cost to fence a rectangular lot that is 108 ft wide and 240 ft long?

Strategy

To find the cost of the fence:
• Find the perimeter of the lot.
• Multiply the perimeter by the per-foot cost of fencing.

Solution
$P = 2L + 2W$
   $= 2(240 \text{ ft}) + 2(108 \text{ ft})$
   $= 480 \text{ ft} + 216 \text{ ft}$
   $= 696 \text{ ft}$

Cost $= 696 \times 6.75 = 4698$

The cost is $4698.

YOU TRY IT • 6

Metal stripping is being installed around a workbench that is 0.74 m wide and 3 m long. At $4.49 per meter, find the cost of the metal stripping. Round to the nearest cent.

Your strategy

Your solution

Solutions on p. S29

12.2 EXERCISES

OBJECTIVE A   To find the perimeter of plane geometric figures

For Exercises 1 to 8, find the perimeter or circumference of the given figure. Use 3.14 for π.

1.

2.

3.

4.

5.

6.

7.

8.

9. Find the perimeter of a triangle with sides of lengths 2 ft 4 in., 3 ft, and 4 ft 6 in.

10. Find the perimeter of a rectangle with a length of 2 m and a width of 0.8 m.

11. Find the circumference of a circle with a radius of 8 cm. Use 3.14 for π.

12. Find the circumference of a circle with a diameter of 14 in. Use $\frac{22}{7}$ for π.

13. Find the perimeter of a square in which each side is equal to 60 m.

14. Find the perimeter of a triangle in which each side is $1\frac{2}{3}$ ft.

15. Find the perimeter of a five-sided figure with sides of 22 cm, 47 cm, 29 cm, 42 cm, and 17 cm.

16. Find the perimeter of a rectangular farm that is $\frac{1}{2}$ mi wide and $\frac{3}{4}$ mi long.

17. The length of a side of a square is equal to the diameter of a circle. Which is greater, the perimeter of the square or the circumference of the circle?

18. The length of a rectangle is equal to the diameter of a circle, and the width of the rectangle is equal to the radius of the same circle. Which is greater, the perimeter of the rectangle or the circumference of the circle?

OBJECTIVE B   To find the perimeter of composite geometric figures

For Exercises 19 to 26, find the perimeter. Use 3.14 for π.

19.

20.

21.

22.

23.

24.

25.

26.

For Exercises 27 and 28, determine whether the perimeter of the given composite figure is *less than, equal to,* or *greater than* the perimeter of the figure shown at the right. The figure at the right is made up of a square and one-half a circle of diameter *d*.

27. A figure formed by an equilateral triangle and one-half a circle of diameter *d*

28. A figure formed by a square and one-half a circle of diameter *d*

OBJECTIVE C   To solve application problems

29. **Landscaping** How many feet of fencing should be purchased for a rectangular garden that is 18 ft long and 12 ft wide?

30. **Interior Design** Wall-to-wall carpeting is installed in a room that is 12 ft long and 10 ft wide. The edges of the carpet are nailed to the floor. Along how many feet must the carpet be nailed down?

31. **Quilting** How many feet of binding are required to bind the edge of a rectangular quilt that measures 3.5 ft by 8.5 ft?

32. **Carpentry** Find the length of molding needed to put around a circular table that is 3.8 ft in diameter. Use 3.14 for π.

33. **Race Tracks** The first circular dog race track opened in 1919 in Emeryville, California. The radius of the circular track was 157.64 ft. Find the circumference of the track. Use 3.14 for π. Round to the nearest whole number.

34. **Landscaping** The rectangular lot shown in the figure at the right is being fenced. The fencing along the road is to cost $6.20 per foot. The rest of the fencing costs $5.85 per foot. Find the total cost to fence the lot.

35. **Sewing** Bias binding is to be sewed around the edge of a rectangular quilt measuring 72 in. by 45 in. Each package of bias binding costs $5.50 and contains 15 ft of binding. How many packages of bias binding are needed for the quilt?

36. **Travel** A bicycle tire has a diameter of 24 in. How many feet does the bicycle travel when the wheel makes 5 revolutions? Use 3.14 for π.

37. **Travel** A tricycle tire has a diameter of 12 in. How many feet does the tricycle travel when the wheel makes 8 revolutions? Use 3.14 for π.

Architecture For Exercises 38 and 39, use the floor plan of a roller rink shown at the right.

 38. Use estimation to determine whether the perimeter is less than 70 m or greater than 70 m.

39. Calculate the perimeter of the roller rink. Use 3.14 for π.

40. **Home Improvement** A rain gutter is being installed on a home that has the dimensions shown in the figure at the right. At a cost of $11.30 per meter, how much will it cost to install the rain gutter?

41. **Home Improvement** Find the length of weather stripping installed around the arched door shown in the figure at the right. Use 3.14 for π.

 42. **Astronomy** The distance from Earth to the sun is 93,000,000 mi. Approximate the distance Earth travels in making one revolution about the sun. Use 3.14 for π.

Applying the Concepts

43. **a.** If the diameter of a circle is doubled, how many times larger is the resulting circumference?
b. If the radius of a circle is doubled, how many times larger is the resulting circumference?

44. **Geometry** In the pattern at the right, the length of one side of a square is 1 unit. Find the perimeter of the eighth figure in the pattern.

45. **Geometry** An equilateral triangle is placed inside an equilateral triangle as shown at the right. Now three more equilateral triangles are placed inside the unshaded equilateral triangles. The process is repeated again. Determine the perimeter of all the shaded triangles in Figure C.

Figure A   Figure B   Figure C

 46. A forest ranger must determine the diameter of a redwood tree. Explain how the ranger could do this without cutting down the tree.

SECTION

12.3   Area

OBJECTIVE A   **To find the area of geometric figures**

Area is a measure of the amount of surface in a region. Area can be used to describe the size of a rug, a parking lot, a farm, or a national park. Area is measured in square units.

A square that measures 1 in. on each side has an area of 1 square inch, which is written 1 in².

A square that measures 1 cm on each side has an area of 1 square centimeter, which is written 1 cm².

Larger areas can be measured in square feet (ft²), square meters (m²), square miles (mi²), acres (43,560 ft²), or any other square unit.

The area of a geometric figure is the number of squares that are necessary to cover the figure. In the figures below, two rectangles have been drawn and covered with squares. In the figure on the left, 12 squares, each of area 1 cm², were used to cover the rectangle. The area of the rectangle is 12 cm². In the figure on the right, 6 squares, each of area 1 in², were used to cover the rectangle. The area of the rectangle is 6 in².

The area of the rectangle is 12 cm².

The area of the rectangle is 6 in².

Note from the above figures that the area of a rectangle can be found by multiplying the length of the rectangle by its width.

> **Area of a Rectangle**
> $A = LW$

HOW TO 1   Find the area of the rectangle shown at the right.

$A = LW$
$\quad = (8 \text{ ft})(5 \text{ ft})$ • $L = 8$ ft, $W = 5$ ft
$\quad = 40 \text{ ft}^2$

The area of the rectangle is 40 ft².

5 ft
8 ft

A square is a rectangle in which all sides are the same length. Therefore, both the length and the width can be represented by a side. Remember that $s \cdot s = s^2$.

Area of a Square

$A = s^2$

HOW TO 2 Find the area of the square shown at the right.

$A = s^2$

$\quad = (14 \text{ cm})^2 \qquad \bullet \; s = 14 \text{ cm}$

$\quad = 196 \text{ cm}^2$

The area of the square is 196 cm².

The area of a circle is equal to the product of π and the square of the radius.

Area of a Circle

$A = \pi r^2$

Radius

HOW TO 3 Find the area of the circle shown at the right.

8 in.

$A = \pi r^2$

$\quad = \pi (8 \text{ in.})^2 = 64\pi \text{ in}^2$

$\quad \approx 64 \cdot 3.14 \text{ in}^2 = 200.96 \text{ in}^2$

The area is exactly 64π in².
The area is approximately 200.96 in².

In the figure below, *AB* is the base of the triangle, and *CD*, which is perpendicular to the base, is the height. The area of a triangle is one-half the product of the base and the height.

Integrating Technology

To calculate the area of the triangle shown at the right, you can enter

20 **X** 5 **÷** 2 **=**

or

.5 **X** 20 **X** 5 **=**

Area of a Triangle

$A = \frac{1}{2}bh$

HOW TO 4 Find the area of the triangle shown below.

5 m

20 m

$A = \frac{1}{2}bh$

$\quad = \frac{1}{2}(20 \text{ m})(5 \text{ m}) \qquad \bullet \; b = 20 \text{ m}, h = 5 \text{ m}$

$\quad - 50 \text{ m}^2$

The area of the triangle is 50 m².

EXAMPLE 1

Find the area of a circle with a diameter of 9 cm. Use 3.14 for π.

Solution

$r = \dfrac{1}{2}d = \dfrac{1}{2}(9 \text{ cm}) = 4.5 \text{ cm}$

$A = \pi r^2$

$\approx 3.14(4.5 \text{ cm})^2 = 63.585 \text{ cm}^2$

The area is approximately 63.585 cm².

Find the area of a triangle with a base of 24 in. and a height of 14 in.

Your solution

Solution on p. S30

OBJECTIVE B   **To find the area of composite geometric figures**

The area of the composite figure shown below is found by calculating the area of the rectangle and then subtracting the area of the triangle.

$A = LW - \dfrac{1}{2}bh$

$= (20 \text{ in.})(8 \text{ in.}) - \dfrac{1}{2}(20 \text{ in.})(3 \text{ in.}) = 160 \text{ in}^2 - 30 \text{ in}^2 = 130 \text{ in}^2$

EXAMPLE 2

Find the area of the shaded portion of the figure. Use 3.14 for π.

Solution

$\begin{array}{ccc} \text{Area of} \\ \text{shaded} & = & \text{area of} & - & \text{area of} \\ \text{portion} & & \text{square} & & \text{circle} \end{array}$

$A = \quad s^2 \quad - \quad \pi r^2$

$= (8 \text{ m})^2 \quad - \quad \pi(4 \text{ m})^2$

$\approx 64 \text{ m}^2 \quad - \quad 3.14(16 \text{ m}^2)$

$= 64 \text{ m}^2 \quad - \quad 50.24 \text{ m}^2 = 13.76 \text{ m}^2$

The area is approximately 13.76 m².

Find the area of the composite figure.

Your solution

Solution on p. S30

OBJECTIVE C   To solve application problems

EXAMPLE · 3

A walkway 2 m wide is built along the front and along both sides of a building, as shown in the figure. Find the area of the walkway.

New carpet is installed in a room measuring 9 ft by 12 ft. Find the area of the room in square yards. $(9 \text{ ft}^2 = 1 \text{ yd}^2)$

Strategy

To find the area of the walkway, add the area of the front section (54 m · 2 m) and the area of the two side sections (each 35 m · 2 m).

Your strategy

Solution

$$\underset{\text{walkway}}{\text{Area of}} = \underset{\text{section}}{\underset{\text{front}}{\underbrace{\text{area of}}}} + \underset{\text{section)}}{\underset{\text{one side}}{\underbrace{2(\text{area of}}}}$$

$A = (54 \text{ m})(2 \text{ m}) + 2(35 \text{ m})(2 \text{ m})$
$= 108 \text{ m}^2 \quad + 140 \text{ m}^2$
$= 248 \text{ m}^2$

The area of the walkway is 248 m².

Your solution

Solution on p. S30

12.3 EXERCISES

OBJECTIVE A   To find the area of geometric figures

For Exercises 1 to 8, find the area of the given figure. Use 3.14 for π.

1.

6 ft

24 ft

2.

8 in.

18 in.

3.
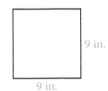
9 in.

9 in.

4.
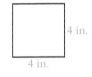
4 in.

4 in.

5.

4 ft

6.

3 cm

7.

4 in.

10 in.

8.

6 m

7 m

9. Find the area of a right triangle with a base of 3 cm and a height of 1.42 cm.

10. Find the area of a triangle with a base of 3 ft and a height of $\frac{2}{3}$ ft.

11. Find the area of a square with a side of 4 ft.

12. Find the area of a square with a side of 10 cm.

13. Find the area of a rectangle with a length of 43 in. and a width of 19 in.

14. Find the area of a rectangle with a length of 82 cm and a width of 20 cm.

15. Find the area of a circle with a radius of 7 in. Use $\frac{22}{7}$ for π.

16. Find the area of a circle with a diameter of 40 cm. Use 3.14 for π.

 For Exercises 17 and 18, determine whether the area of the first figure is *less than, equal to,* or *greater than* the area of the second figure.

17.

y

y

x

x

18.

2x

x

2x

OBJECTIVE B   To find the area of composite geometric figures

For Exercises 19 to 26, find the area. Use 3.14 for π.

19.

20.

21.

22.

23.

24.

25.

26.

 27. Determine whether the area of Figure 1 at the right is *less than, equal to,* or *greater than* the area of Figure 2.

Figure 1

Figure 2

OBJECTIVE C   To solve application problems

28. **Interior Design** See the news clipping at the right. What would be the cost of carpeting the entire living space if the cost of the carpet were $36 per square yard?

29. **Sports** Artificial turf is being used to cover a playing field. The field is rectangular with a length of 100 yd and a width of 75 yd. How much artificial turf must be purchased to cover the field?

In the News

Billion-Dollar Home Built in Mumbai

The world's first billion-dollar home is a 27-story skyscraper in downtown Mumbai, India (formerly known as Bombay). It is 550 ft high with 400,000 square feet of living space.

Source: Forbes.com

30. Telescopes The telescope lens of the Hale telescope at Mount Palomar, California, has a diameter of 200 in. Find the area of the lens. Use 3.14 for π.

31. Agriculture An irrigation system waters a circular field that has a 50-foot radius. Find the area watered by the irrigation system. Use 3.14 for π.

32. Interior Design A fabric wall hanging is to fill a space that measures 5 m by 3.5 m. Allowing for 0.1 m of the fabric to be folded back along each edge, how much fabric must be purchased for the wall hanging?

In the News

Animal Sanctuary Established

The government of the Republic of Congo in Africa has set aside a vast expanse of land in the Sankuru Province to be used as a nature reserve. It will be a sanctuary for elephants; 11 species of primates, including the bonobos; and the Okapi, a short-necked relative of the giraffe, which is on the endangered list.

Source: www.time.com

33. Home Improvement You plan to stain the wooden deck attached to your house. The deck measures 10 ft by 8 ft. A quart of stain will cost $11.87 and will cover 50 ft^2. How many quarts of stain should you buy?

34. Conservation See the news clipping at the right. The nature reserve in Sankuru is about the size of Massachusetts. Consider Massachusetts a rectangle with a length of 150 mi and a width of 70 mi. Use these dimensions to approximate the area of the reserve in the Congo.

Interior Design A carpet is to be installed in one room and a hallway, as shown in the diagram at the right. For Exercises 35 to 38, state whether the given expression can be used to calculate the area of the carpet in square meters.

35. 4.5(6.8) + 10.8(1)

36. 4.5(10.8) − 3.5(4)

37. 10.8(1) + 3.5(6.8)

38. 4.5(6.8) + 1(4)

39. Interior Design Use the diagram for Exercises 35 to 38. At a cost of $28.50 per square meter, how much will it cost to carpet the area?

40. Landscaping Find the area of a concrete driveway with the measurements shown in the figure.

41. Interior Design You want to tile your kitchen floor. The floor measures 12 ft by 9 ft. How many tiles, each a square with side $1\frac{1}{2}$ ft, should you purchase for the job?

42. Interior Design You are wallpapering two walls of a child's room. One wall measures 9 ft by 8 ft, and the other measures 11 ft by 8 ft. The wallpaper costs $34.50 per roll, and each roll of the wallpaper will cover 40 ft^2. What is the cost to wallpaper the two walls?

43. Construction Find the area of the 2-meter boundary around the swimming pool shown in the figure.

44. Parks An urban renewal project involves reseeding a park that is in the shape of a square, 60 ft on each side. Each bag of grass seed costs $5.75 and will seed 1200 ft². How much money should be budgeted for buying grass seed for the park?

45. Architecture The roller rink shown in the figure at the right is to be covered with hardwood floor.
    a. Without doing the calculations, indicate whether the area of the rink is more than 8000 ft² or less than 8000 ft².
    b. Calculate how much hardwood floor is needed to cover the roller rink. Use 3.14 for π.

46. Parks Find the total area of the national park with the dimensions shown in the figure. Use 3.14 for π.

47. Interior Design Find the cost of plastering the walls of a room 22 ft wide, 25 ft 6 in. long, and 8 ft high. Subtract 120 ft² for windows and doors. The cost is $3 per square foot.

48. Lake Tahoe One way to measure the area of an irregular figure, such as a lake, is to divide the area into trapezoids that have the same height. Then measure the length of each base, calculate the area of each trapezoid, and add the areas. The figure at the right gives approximate dimensions for Lake Tahoe, which straddles the California and Nevada borders. Approximate the area of Lake Tahoe using the given trapezoids. Round to the nearest tenth. *Note:* The formula for the area A of a trapezoid is $A = \frac{1}{2}h(b_1 + b_2)$, where h is the height of the trapezoid and b_1 and b_2 are the lengths of the bases.

Applying the Concepts

49.   **a.** If both the length and the width of a rectangle are doubled, how many times larger is the area of the resulting rectangle?
    b. If the radius of a circle is doubled, what happens to the area?
    c. If the diameter of a circle is doubled, what happens to the area?

50. The circles at the right are identical. Is the area in the circles to the left of the line equal to, less than, or greater than the area in the circles to the right of the line? Explain your answer.

51. Determine whether the statement is always true, sometimes true, or never true.
    a. If two triangles have the same perimeter, then they have the same area.
    b. If two rectangles have the same area, then they have the same perimeter.
    c. If two squares have the same area, then the sides of the squares have the same length.

SECTION

12.4 Volume

OBJECTIVE A To find the volume of geometric solids

Volume is a measure of the amount of space inside a closed surface, or figure in space. Volume can be used to describe the amount of heating gas used for cooking, the amount of concrete delivered for the foundation of a house, or the amount of water in storage for a city's water supply.

A cube that is 1 ft on each side has a volume of 1 cubic foot, which is written 1 ft³.

A cube that measures 1 cm on each side has a volume of 1 cubic centimeter, which is written 1 cm³.

The volume of a solid is the number of cubes that are necessary to fill the solid exactly. The volume of the rectangular solid at the right is 24 cm³ because it will hold exactly 24 cubes, each 1 cm on a side. Note that the volume can be found by multiplying the length times the width times the height.

Volume of a Rectangular Solid

$V = LWH$

HOW TO · 1 Find the volume of a rectangular solid with a length of 9 in., a width of 3 in., and a height of 4 in.

$V = LWH$

$= (9 \text{ in.})(3 \text{ in.})(4 \text{ in.})$ • $L = 9$ in., $W = 3$ in.,

$= 108 \text{ in}^3$ $H = 4$ in.

The volume of the rectangular solid is 108 in³.

The length, width, and height of a cube have the same measure. The volume of a cube is found by multiplying the length of a side of the cube times itself three times (side cubed).

Volume of a Cube

$V = s^3$

HOW TO 2 Find the volume of the cube shown at the right.

$V = s^3$

$\quad = (3 \text{ ft})^3 \qquad \bullet \ s = 3 \text{ ft}$

$\quad = 27 \text{ ft}^3$

The volume of the cube is 27 ft³.

The volume of a sphere is found by multiplying four-thirds times pi (π) times the radius cubed.

Volume of a Sphere

$V = \dfrac{4}{3}\pi r^3$

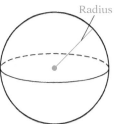

HOW TO 3 Find the volume of the sphere shown below. Use 3.14 for π. Round to the nearest hundredth.

$V = \dfrac{4}{3}\pi r^3$

$\quad \approx \dfrac{4}{3}(3.14)(2 \text{ in.})^3 \qquad \bullet \ r = 2 \text{ in.}$

$\quad = \dfrac{4}{3}(3.14)(8 \text{ in}^3)$

$\quad \approx 33.49 \text{ in}^3$

The volume is approximately 33.49 in³.

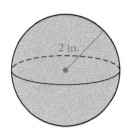

The volume of a cylinder is found by multiplying the area of the base of the cylinder (a circle) times the height.

Volume of a Cylinder

$V = \pi r^2 h$

> **HOW TO • 4**   Find the volume of the cylinder shown below. Use 3.14 for π.
>
> $V = \pi r^2 h$
>
> $\approx 3.14(3 \text{ cm})^2(8 \text{ cm})$ • $r = 3$ cm, $h = 8$ cm
>
> $= 3.14(9 \text{ cm}^2)(8 \text{ cm})$
>
> $= 226.08 \text{ cm}^3$
>
> The volume of the cylinder is approximately 226.08 cm³.

8 cm

Radius = 3 cm

EXAMPLE • 1

Find the volume of a rectangular solid with a length of 3 ft, a width of 1.5 ft, and a height of 2 ft.

Solution

$V = LWH$

$ = (3 \text{ ft})(1.5 \text{ ft})(2 \text{ ft})$

$ = 9 \text{ ft}^3$

The volume is 9 ft³.

YOU TRY IT • 1

Find the volume of a rectangular solid with a length of 8 cm, a width of 3.5 cm, and a height of 4 cm.

Your solution

EXAMPLE • 2

Find the volume of a cube that has a side measuring 2.5 in.

Solution

$V = s^3$

$ = (2.5 \text{ in.})^3$

$ = 15.625 \text{ in}^3$

The volume is 15.625 in³.

YOU TRY IT • 2

Find the volume of a cube with a side of length 5 cm.

Your solution

EXAMPLE • 3

Find the volume of a cylinder with a radius of 12 cm and a height of 65 cm. Use 3.14 for π.

Solution

$V = \pi r^2 h$

$ \approx 3.14(12 \text{ cm})^2(65 \text{ cm})$

$ = 3.14(144 \text{ cm}^2)(65 \text{ cm})$

$ = 29{,}390.4 \text{ cm}^3$

The volume is approximately 29,390.4 cm³.

YOU TRY IT • 3

Find the volume of a cylinder with a diameter of 14 in. and a height of 15 in. Use $\frac{22}{7}$ for π.

Your solution

Solutions on p. S30

EXAMPLE · 4

Find the volume of a sphere with a diameter of 12 in.
Use 3.14 for π.

Solution

$r = \dfrac{1}{2}d = \dfrac{1}{2}(12 \text{ in.}) = 6 \text{ in.}$ • Find the radius.

$V = \dfrac{4}{3}\pi r^3$ • Use the formula
for the volume
of a sphere.

$\approx \dfrac{4}{3}(3.14)(6 \text{ in.})^3$

$= \dfrac{4}{3}(3.14)(216 \text{ in}^3)$

$= 904.32 \text{ in}^3$

The volume is approximately 904.32 in³.

YOU TRY IT · 4

Find the volume of a sphere with a radius of 3 m.
Use 3.14 for π.

Your solution

Solution on p. S30

OBJECTIVE B   To find the volume of composite geometric solids

A **composite geometric solid** is a solid made from two or more geometric solids. The solid shown is made from a cylinder and one-half of a sphere.

Volume of the composite solid   = volume of the cylinder $+ \dfrac{1}{2}$ the volume of the sphere

HOW TO · 5   Find the volume of the composite solid shown above if the radius of the base of the cylinder is 3 in. and the height of the cylinder is 10 in. Use 3.14 for π.

The volume equals the volume of a cylinder plus one-half the volume of a sphere. The radius of the sphere equals the radius of the base of the cylinder.

$V = \pi r^2 h + \dfrac{1}{2}\left(\dfrac{4}{3}\pi r^3\right)$

$\approx 3.14(3 \text{ in.})^2(10 \text{ in.}) + \dfrac{1}{2}\left(\dfrac{4}{3}\right)(3.14)(3 \text{ in.})^3$

$= 3.14(9 \text{ in}^2)(10 \text{ in.}) + \dfrac{1}{2}\left(\dfrac{4}{3}\right)(3.14)(27 \text{ in}^3)$

$= 282.6 \text{ in}^3 + 56.52 \text{ in}^3$

$= 339.12 \text{ in}^3$

The volume is approximately 339.12 in³.

EXAMPLE · 5

Find the volume of the solid in the figure.
Use 3.14 for π.

Solution

| Volume of the solid | = | volume of rectangular solid | + volume of cylinder |

$V = LWH + \pi r^2 h$
$\approx (8 \text{ cm})(8 \text{ cm})(2 \text{ cm}) + 3.14(1 \text{ cm})^2(2 \text{ cm})$
$= 128 \text{ cm}^3 + 6.28 \text{ cm}^3$
$= 134.28 \text{ cm}^3$

The volume is approximately 134.28 cm³.

YOU TRY IT · 5

Find the volume of the solid in the figure.
Use 3.14 for π.

Your solution

EXAMPLE · 6

Find the volume of the solid in the figure.
Use 3.14 for π.

Solution

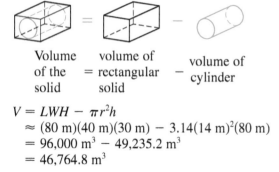

| Volume of the solid | = | volume of rectangular solid | − volume of cylinder |

$V = LWH - \pi r^2 h$
$\approx (80 \text{ m})(40 \text{ m})(30 \text{ m}) - 3.14(14 \text{ m})^2(80 \text{ m})$
$= 96,000 \text{ m}^3 - 49,235.2 \text{ m}^3$
$= 46,764.8 \text{ m}^3$

The volume is approximately 46,764.8 m³.

YOU TRY IT · 6

Find the volume of the solid in the figure.
Use 3.14 for π.

Your solution

Solutions on p. S30

OBJECTIVE C   To solve application problems

EXAMPLE • 7

An aquarium is 28 in. long, 14 in. wide, and 16 in. high. Find the volume of the aquarium.

Strategy
To find the volume of the aquarium, use the formula for the volume of a rectangular solid.

Solution
$V = LWH$
$\quad = (28 \text{ in.})(14 \text{ in.})(16 \text{ in.})$
$\quad = 6272 \text{ in}^3$

The volume of the aquarium is 6272 in³.

YOU TRY IT • 7

Find the volume of a freezer that is 7 ft long, 3 ft high, and 2.5 ft wide.

Your strategy

Your solution

EXAMPLE • 8

Find the volume of the bushing shown in the figure below. Use 3.14 for π.

Strategy
To find the volume of the bushing, subtract the volume of the half-cylinder from the volume of the rectangular solid.

Solution

Volume of the bushing	=	volume of rectangular solid	−	$\frac{1}{2}$ the volume of cylinder

$V = LWH - \frac{1}{2}\pi r^2 h$

$\quad \approx (8 \text{ cm})(4 \text{ cm})(4 \text{ cm}) - \frac{1}{2}(3.14)(1 \text{ cm})^2(8 \text{ cm})$

$\quad = 128 \text{ cm}^3 - 12.56 \text{ cm}^3$

$\quad = 115.44 \text{ cm}^3$

The volume of the bushing is approximately 115.44 cm³.

YOU TRY IT • 8

Find the volume of the channel iron shown in the figure below.

Your strategy

Your solution

Solutions on p. S30

12.4 EXERCISES

OBJECTIVE A   To find the volume of geometric solids

For Exercises 1 to 8, find the volume. Round to the nearest hundredth. Use 3.14 for π.

1.

3 cm
12 cm
4 cm

2.

5 ft
6 ft
8 ft

3.

8 in.
8 in.
8 in.

4.

12 m
12 in
12 m

5.

8 in.

6.

7 in.

7.

12 cm
2 cm

8.

8 ft
5 ft

For Exercises 9 to 16, find the volume.

9. Find the volume, in cubic meters, of a rectangular solid with a length of 2 m, a width of 80 cm, and a height of 4 m.

10. Find the volume of a cylinder with a radius of 7 cm and a height of 14 cm. Use $\frac{22}{7}$ for π.

11. Find the volume of a sphere with an 11-millimeter radius. Use 3.14 for π. Round to the nearest hundredth.

12. Find the volume of a cube with a side of length 2.14 m. Round to the nearest tenth.

13. Find the volume of a cylinder with a diameter of 12 ft and a height of 30 ft. Use 3.14 for π.

14. Find the volume of a sphere with a 6-foot diameter. Use 3.14 for π.

15. Find the volume of a cube with a side of length $3\frac{1}{2}$ ft.

16. Find the volume, in cubic meters, of a rectangular solid with a length of 1.15 m, a width of 60 cm, and a height of 25 cm.

 17. The length of a side of a cube is equal to the radius of a sphere. Which solid has the greater volume?

 18. A sphere and a cylinder have the same radius. The height of the cylinder is equal to the radius of its base. Which solid has the greater volume?

OBJECTIVE B   **To find the volume of composite geometric solids**

For Exercises 19 to 24, find the volume. Use 3.14 for π.

19.

2 in.
1 in.
9 in.
6 in.

20.

6 ft
12 ft

21.

40 cm
80 cm
2 m
1.2 m

22.

1.5 m
1.5 m
50 cm
2 m
2 m

23.

2 in.
6 in.
2 in.
4 in.

24.

9 cm
24 cm
18 cm

 For Exercises 25 and 26, use the solid shown in Exercise 24. If the solid is changed as described, will its volume increase or decrease?

25. The outer cylinder is changed to a rectangular solid with a square base. The height and width of the outer solid remain the same.

26. The inner cylinder is changed to a rectangular solid with a square base. The height and width of the inner solid remain the same.

OBJECTIVE C   To solve application problems

For Exercises 27 to 40, solve. Use 3.14 for π.

27. Fish Hatchery A rectangular tank at a fish hatchery is 9 m long, 3 m wide, and 1.5 m deep. Find the volume of the water in the tank when the tank is full.

28. Rocketry A fuel tank in a booster rocket is a cylinder 10 ft in diameter and 52 ft high. Find the volume of the fuel tank.

29. Ballooning A hot air balloon is in the shape of a sphere. Find the volume of a hot air balloon that is 32 ft in diameter. Round to the nearest hundredth.

30. Petroleum An oil tank, which is in the shape of a cylinder, is 4 m high and has a diameter of 6 m. The oil tank is two-thirds full. Find the number of cubic meters of oil in the tank. Round to the nearest hundredth.

31. Agriculture A silo, which is in the shape of a cylinder, is 16 ft in diameter and has a height of 30 ft. The silo is three-fourths full. Find the volume of the portion of the silo that is not being used for storage.

32. The Panama Canal The Gatun Lock of the Panama Canal is 1000 ft long, 110 ft wide, and 60 ft deep. Find the volume of the lock in cubic feet.

Panama Canal

33. The Panama Canal When the lock is full, the water in the Pedro Miguel Lock near the Pacific Ocean side of the Panama Canal fills a rectangular solid of dimensions 1000 ft long, 110 ft wide, and 43 ft deep. There are 7.48 gal of water in each cubic foot. How many gallons of water are in the lock?

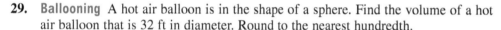

34. Guacamole Consumption See the news clipping at the right. What is the volume of the guacamole in cubic feet?

35. Guacamole Consumption See the news clipping at the right. Assuming that each person eats 1 c of guacamole, how many people could be fed from the covered football field? (1 ft³ = 59.84 pt)

In the News

Super Bowl Win for Guacamole

Guacamole has become the dish of choice at Super Bowl parties. If all the guacamole eaten during the Super Bowl were piled onto a football field—that's a football field which, including endzones, is 360 ft long and 160 ft wide—it would cover the field to a depth of 3 ft!

Source: www.azcentral.com

36. Architecturo An architect is designing the heating system for an auditorium and needs to know the volume of the structure. Find the volume of the auditorium with the measurements shown in the figure.

32 ft
94 ft
125 ft

37. Aquariums How many gallons of water will fill an aquarium that is 12 in. wide, 18 in. long, and 16 in. high? Round to the nearest tenth. $(1 \text{ gal} = 231 \text{ in}^3)$

38. Aquariums How many gallons of water will fill a fish tank that is 12 in. long, 8 in. wide, and 9 in. high? Round to the nearest tenth. $(1 \text{ gal} = 231 \text{ in}^3)$

8 in.
3 in.
12 in.
4 in.

39. Metal Works Find the volume of the bushing shown at the right.

40. Petroleum A truck carrying an oil tank is shown in the figure at the right.
   a. Without doing the calculations, determine whether the volume of the oil tank is more than 240 ft^3 or less than 240 ft^3.
   b. If the tank is half full, how many cubic feet of oil is the truck carrying? Round to the nearest hundredth.

30 ft
8 ft

 Construction For Exercises 41 to 44, use the diagram at the right showing the concrete floor of a building. State whether the given expression can be used to calculate the volume of the concrete floor in cubic feet.

50 ft
25 ft
6 in.

41. $(25)(50)(6) + (0.5)(3.14)(25^2)(6)$ **42.** $(25)(50)(0.5) + (0.5)(3.14)(25^2)(0.5)$

43. $0.5[(25)(50) + (0.5)(3.14)(25^2)]$ **44.** $(25)(50)(0.5) + (0.5)(3.14)(50^2)(0.5)$

45. Construction Use the diagram for Exercises 41 to 44. At a cost of $10 per cubic foot, find the cost of having the floor poured.

Applying the Concepts

 For Exercises 46 to 49, explain how you could cut through a cube so that the face of the resulting solid is the given geometric figure.

46. A square   **47.** An equilateral triangle

48. A trapezoid   **49.** A hexagon

 50. Suppose a cylinder is cut into 16 equal pieces, which are then arranged as shown at the right. The figure resembles a rectangular solid. What variable expressions could be used to represent the length, width, and height of the rectangular solid? Explain how the formula for the volume of a cylinder is derived from this approach.

SECTION

12.5 The Pythagorean Theorem

OBJECTIVE A **To find the square root of a number**

The area of a square is 36 in². What is the length of one side?

Area of the square = (side)²

$$36 = \text{side} \cdot \text{side}$$

What number multiplied times itself equals 36?

$$36 = 6 \cdot 6$$

The side of the square is 6 in.

> Area =
> 36 in²

The **square root** of a number is one of two identical factors of that number. The square root symbol is $\sqrt{}$.

The square root of 36 is **6**.

$$\sqrt{36} = 6$$

A **perfect square** is the product of a whole number times itself.

Point of Interest

The square root of a number that is not a perfect square is an irrational number. There is evidence of irrational numbers as early as 500 B.C. These numbers were not very well understood, and they were given the name *numerus surdus*. This phrase comes from the Latin word *surdus*, which means "deaf" or "mute." Thus irrational numbers were "inaudible numbers."

1, 4, 9, 16, 25, and 36 are perfect squares.

The square root of a perfect square is a whole number.

$1 \cdot 1 = 1$	$\sqrt{1} = 1$
$2 \cdot 2 = 4$	$\sqrt{4} = 2$
$3 \cdot 3 = 9$	$\sqrt{9} = 3$
$4 \cdot 4 = 16$	$\sqrt{16} = 4$
$5 \cdot 5 = 25$	$\sqrt{25} = 5$
$6 \cdot 6 = 36$	$\sqrt{36} = 6$

If a number is not a perfect square, its square root can only be approximated. The approximate square roots of numbers can be found using a calculator. For example:

Number	Square Root
33	$\sqrt{33} \approx 5.745$
34	$\sqrt{34} \approx 5.831$
35	$\sqrt{35} \approx 5.916$

EXAMPLE • 1

a. Find the square roots of the perfect squares 49 and 81.
b. Find the square roots of 27 and 108. Round to the nearest thousandth.

Solution

a. $\sqrt{49} = 7$ $\sqrt{81} = 9$
b. $\sqrt{27} \approx 5.196$ $\sqrt{108} \approx 10.392$

YOU TRY IT • 1

a. Find the square roots of the perfect squares 16 and 169.
b. Find the square roots of 32 and 162. Round to the nearest thousandth.

Your solution

Solution on p. S30

OBJECTIVE D

To find the unknown side of a right triangle using the Pythagorean Theorem

Point of Interest
The first known proof of the Pythagorean Theorem is in a Chinese textbook that dates from 150 B.C. The book is called *Nine Chapters on the Mathematical Art.* The diagram below is from that book and was used in the proof of the theorem.

The Greek mathematician Pythagoras is generally credited with the discovery that the square of the hypotenuse of a right triangle is equal to the sum of the squares of the two legs. This is called the **Pythagorean Theorem.** However, the Babylonians used this theorem more than 1000 years before Pythagoras lived.

Square of the hypotenuse	equals	sum of the squares of the two legs
5^2	$=$	$3^2 + 4^2$
25	$=$	$9 + 16$
25	$=$	25

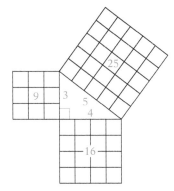

If the length of one side of a right triangle is unknown, one of the following formulas can be used to find it.

If the hypotenuse is unknown, use

$$\text{Hypotenuse} = \sqrt{(\text{leg})^2 + (\text{leg})^2}$$
$$= \sqrt{(3)^2 + (4)^2}$$
$$= \sqrt{9 + 16}$$
$$= \sqrt{25} = 5$$

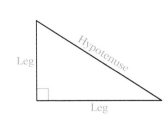

If the length of a leg is unknown, use

$$\text{Leg} = \sqrt{(\text{hypotenuse})^2 - (\text{leg})^2}$$
$$= \sqrt{(5)^2 - (4)^2}$$
$$= \sqrt{25 - 16}$$
$$= \sqrt{9} = 3$$

EXAMPLE • 2

Find the hypotenuse of the triangle in the figure. Round to the nearest thousandth.

Solution

$$\text{Hypotenuse} = \sqrt{(\text{leg})^2 + (\text{leg})^2}$$
$$= \sqrt{8^2 + 4^2}$$
$$= \sqrt{64 + 16}$$
$$= \sqrt{80} \approx 8.944$$

The hypotenuse is approximately 8.944 in.

YOU TRY IT • 2

Find the hypotenuse of the triangle in the figure. Round to the nearest thousandth.

Your solution

Solution on p. S30

EXAMPLE · 3

Find the length of the leg of
the triangle in the figure.
Round to the nearest thousandth.

12 cm

9 cm

YOU TRY IT · 3

Find the length of the
leg of the triangle in
the figure. Round to the
nearest thousandth.

5 ft

12 ft

Solution

$\text{Leg} = \sqrt{(\text{hypotenuse})^2 - (\text{leg})^2}$

$= \sqrt{12^2 - 9^2}$

$= \sqrt{144 - 81}$

$= \sqrt{63} \approx 7.937$

The length of the leg is approximately 7.937 cm.

Your solution

Solution on p. S31

OBJECTIVE C   **To solve application problems**

EXAMPLE · 4

A 25-foot ladder
is placed against
a building at a
point 21 ft from
the ground, as
shown in the
figure. Find the
distance from
the base of the
building to the
base of the ladder. Round to the nearest thousandth.

25 ft

21 ft

YOU TRY IT · 4

Find the distance between the centers of the holes in
the metal plate in the figure. Round to the nearest
thousandth.

3 cm

8 cm

Strategy

To find the distance from the base of the building to
the base of the ladder, use the Pythagorean Theorem.
The hypotenuse is the length of the ladder (25 ft).
One leg is the distance along the building from the
ground to the top of the ladder (21 ft). The distance
from the base of the building to the base of the ladder
is the unknown leg.

Your strategy

Solution

$\text{Leg} = \sqrt{(\text{hypotenuse})^2 - (\text{leg})^2}$

$= \sqrt{25^2 - 21^2}$

$= \sqrt{625 - 441}$

$= \sqrt{184} \approx 13.565$

The distance is approximately 13.565 ft.

Your solution

Solution on p. S31

12.5 EXERCISES

OBJECTIVE A   To find the square root of a number

For Exercises 1 to 8, find the square root. Round to the nearest thousandth.

1. 7

2. 34

3. 42

4. 64

5. 165

6. 144

7. 189

8. 130

 9. True or false? If a number is between 100 and 400, then its square root is between 10 and 20.

 10. True or false? There are no perfect squares between 50 and 60.

OBJECTIVE B   To find the unknown side of a right triangle using the Pythagorean Theorem

For Exercises 11 to 28, find the unknown side of the triangle. Round to the nearest thousandth.

11.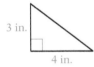
3 in.
4 in.

12.
5 in.
12 in.

13.
5 cm
7 cm

14.
7 cm
9 cm

15.
15 ft
10 ft

16.
20 ft
18 ft

17.
4 cm   6 cm

18.
9 m   12 m

19.
9 yd
9 yd

20.
20 cm
10 cm

21.
12 ft
6 ft

22.
16 cm
8 cm

23.
15 cm
15 cm

24.
6 in.   6 in.

25.
8 m   4 m

26.

8.6 cm

4.3 cm

27.
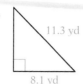
11.3 yd

8.1 yd

28.

13.9 ft

8.2 ft

29. Describe a triangle for which the expression $\sqrt{50^2 - 40^2}$ could be used to find the length of one side of the triangle.

OBJECTIVE C   To solve application problems

30. Ramps Find the length of the ramp used to roll barrels up to the loading dock, which is 3.5 ft high. Round to the nearest hundredth.

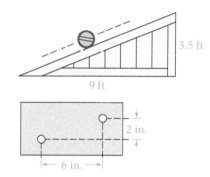
3.5 ft

9 ft

31. Metal Works Find the distance between the centers of the holes in the metal plate in the figure at the right. Round to the nearest hundredth.

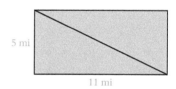
2 in.

6 in.

32. Travel If you travel 18 mi east and then 12 mi north, how far are you from your starting point? Round to the nearest tenth.

33. Travel If you travel 12 mi west and 16 mi south, how far are you from your starting point?

5 mi

34. Geometry The diagonal of a rectangle is a line drawn from one vertex to the opposite vertex. Find the length of the diagonal in the rectangle shown at the right. Round to the nearest tenth.

11 mi

35. Geometry A diagonal of a rectangle is a line drawn from one vertex to the opposite vertex. (See Exercise 34.) Find the length of a diagonal of a rectangle that has a length of 8 m and a width of 3.5 m. Round to the nearest tenth.

36. Home Maintenance A ladder 8 m long is placed against a building in preparation for washing the windows. How high on the building does the ladder reach when the bottom of the ladder is 3 m from the base of the building? Round to the nearest tenth.

8 m

3 m

For Exercises 37 and 38, use the following information. A ladder c feet long leans against the side of a building with its bottom a feet from the building. The ladder reaches a height of b feet. Refer to the following lengths: (i) 15 ft (ii) 20 ft (iii) 30 ft

37. If $c = 18$ ft, which of the given lengths is possible as a value for b?

38. If $b = 18$ ft, which of the given lengths are possible as values for c?

39. Metal Works Four holes are drilled in the circular plate in the figure at the right. The centers of the holes are 3 in. from the center of the plate. Find the distance between the centers of adjacent holes. Round to the nearest thousandth.

40. **Metal Works** Find the distance between the centers of the holes in the metal plate shown in the diagram at the right. Round to the nearest tenth.

41. **Parks** An L-shaped sidewalk from a parking lot to a memorial is shown in the figure at the right. The distance directly across the grass to the memorial is 650 ft. The distance to the corner is 600 ft. Find the distance from the corner to the memorial.

42. **Geometry** Find the perimeter of a right triangle with legs that measure 5 cm and 9 cm. Round to the nearest tenth.

43. **Geometry** Find the perimeter of a right triangle with legs that measure 6 in. and 10 in. Round to the nearest tenth.

44. **Landscaping** A vinyl fence is built around the plot shown in the figure at the right. At $12.90 per meter, how much did it cost to fence the plot? (*Hint:* Use the Pythagorean Theorem to find the unknown length.)

45. **Plumbing** Find the offset distance, *d*, of the length of pipe shown in the diagram at the right. The total length of the pipe is 62 in.

46. **Home Maintenance** You need to clean the gutters of your home. The gutters are 24 ft above the ground. For safety, the distance a ladder reaches up a wall should be four times the distance from the bottom of the ladder to the base of the side of the house. Therefore, the ladder must be 6 ft from the base of the house. Will a 25-foot ladder be long enough to reach the gutters? Explain how you determined your answer.

Applying the Concepts

47. Can the Pythagorean Theorem be used to find the length of side *c* of the triangle at the right? If so, determine *c*. If not, explain why the theorem cannot be used.

48. **a.** What is a Pythagorean triple?
   b. Provide at least three examples of Pythagorean triples.

SECTION

12.6 Similar and Congruent Triangles

OBJECTIVE A   To solve similar and congruent triangles

Similar objects have the same shape but not necessarily the same size. A baseball is similar to a basketball. A model airplane is similar to an actual airplane.

Similar objects have corresponding parts; for example, the propellers on the model airplane correspond to the propellers on the actual airplane. The relationship between the sizes of each of the corresponding parts can be written as a ratio, and all such ratios will be the same. If the propellers on the model plane are $\frac{1}{50}$ the size of the propellers on the actual plane, then the model wing is $\frac{1}{50}$ the size of the actual wing, the model fuselage is $\frac{1}{50}$ the size of the actual fuselage, and so on.

The two triangles ABC and DEF shown are similar. The ratios of corresponding sides are equal.

$$\frac{AB}{DE} = \frac{2}{6} = \frac{1}{3}, \frac{BC}{EF} = \frac{3}{9} = \frac{1}{3}, \text{ and } \frac{AC}{DF} = \frac{4}{12} = \frac{1}{3}$$

The ratio of corresponding sides $= \frac{1}{3}$.

Because the ratios of corresponding sides are equal, three proportions can be formed:

$$\frac{AB}{DE} = \frac{BC}{EF}, \frac{AB}{DE} = \frac{AC}{DF}, \text{ and } \frac{BC}{EF} = \frac{AC}{DF}$$

The ratio of heights equals the ratio of corresponding sides, as shown in the figure.

Ratio of corresponding sides $= \frac{1.5}{6} = \frac{1}{4}$

Ratio of heights $= \frac{2}{8} = \frac{1}{4}$

Congruent objects have the same shape *and* the same size.

The two triangles shown are congruent. They have exactly the same size.

For triangles, *congruent* means that the corresponding sides *and* angles of the triangle are equal (this contrasts with similar triangles, in which corresponding angles, but not necessarily corresponding sides, are equal).

Here are two major rules that can be used to determine whether two triangles are congruent.

Side-Side-Side (SSS) Rule

Two triangles are congruent if three sides of one triangle equal the corresponding sides of the second triangle.

In the two triangles at the right, $AB = DE$, $AC = DF$, and $BC = EF$. The corresponding sides of triangles ABC and DEF are equal. The triangles are congruent by the SSS rule.

 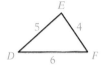

Side-Angle-Side (SAS) Rule

Two triangles are congruent if two sides and the included angle of one triangle equal the corresponding sides and included angle of the second triangle.

In the two triangles at the right, $AB = EF$, $AC = DE$, and $\angle CAB = \angle DEF$. The triangles are congruent by the SAS rule.

HOW TO 1 Determine whether the two triangles in the figure at the right are congruent.

Because $AC = DF$, $AB = FE$, and $BC = DE$, all three sides of one triangle equal the corresponding sides of the second triangle. The triangles are congruent by the SSS rule.

EXAMPLE • 1

Find the ratio of corresponding sides for the similar triangles ABC and DEF in the figure.

Solution

$$\frac{7 \text{ m}}{12 \text{ m}} = \frac{7}{12}$$

YOU TRY IT • 1

Find the ratio of corresponding sides for the similar triangles ABC and DEF in the figure.

Your solution

Solution on p. S31

EXAMPLE · 2

Triangles *ABC* and *DEF* in the figure are similar. Find *x*, the length of side *EF*.

Solution

$$\frac{AB}{DE} = \frac{BC}{EF}$$ • The ratios of corresponding sides of similar triangles are equal.

$$\frac{8 \text{ m}}{12 \text{ m}} = \frac{6 \text{ m}}{x}$$

$$8x = 12 \cdot 6 \text{ m}$$

$$8x = 72 \text{ m}$$

$$\frac{8x}{8} = \frac{72 \text{ m}}{8}$$

$$x = 9 \text{ m}$$

Side *EF* is 9 m.

YOU TRY IT · 2

Triangles *ABC* and *DEF* in the figure are similar. Find *x*, the length of side *DF*.

Your solution

EXAMPLE · 3

Determine whether triangle *ABC* in the figure is congruent to triangle *FDE*.

Solution

Because *AB* = *DF*, *AC* = *EF*, and angle *BAC* = angle *DFE*, the triangles are congruent by the SAS rule.

YOU TRY IT · 3

Determine whether triangle *ABC* in the figure is congruent to triangle *DEF*.

Your solution

EXAMPLE · 4

Triangles *ABC* and *DEF* in the figure are similar. Find *h*, the height of triangle *DEF*.

Solution

$$\frac{8 \text{ cm}}{12 \text{ cm}} = \frac{4 \text{ cm}}{h}$$ • The ratios of corresponding sides of similar triangles equal the ratio of corresponding heights:

$$8h = 12 \cdot 4 \text{ cm}$$

$$8h = 48 \text{ cm}$$

$$\frac{8h}{8} = \frac{48 \text{ cm}}{8}$$ $$\frac{AB}{DE} = \frac{CH}{FG}.$$

$$h = 6 \text{ cm}$$

The height of *DEF* is 6 cm.

YOU TRY IT · 4

Triangles *ABC* and *DEF* in the figure are similar. Find *h*, the height of triangle *DEF*.

Your solution

Solutions on p. S31

OBJECTIVE B   To solve application problems

| EXAMPLE · 5 | YOU TRY IT · 5 |

Triangles *ABC* and *DEF* in the figure are similar. Find the area of triangle *DEF*.

Triangles *ABC* and *DEF* in the figure are similar right triangles. Find the perimeter of triangle *ABC*.

Strategy

To find the area of triangle *DEF*:

- Solve a proportion to find the height of triangle *DEF*. Let h = the height.
- Use the formula $A = \frac{1}{2}bh$.

Your strategy

Solution

$$\frac{AB}{DE} = \frac{\text{height of triangle } ABC}{\text{height of triangle } DEF}$$

$$\frac{4 \text{ cm}}{12 \text{ cm}} = \frac{3 \text{ cm}}{h}$$

$$4h = 12 \cdot 3 \text{ cm}$$

$$4h = 36 \text{ cm}$$

$$\frac{4h}{4} = \frac{36 \text{ cm}}{4}$$

$$h = 9 \text{ cm}$$

$$A = \frac{1}{2}bh$$

$$= \frac{1}{2}(12 \text{ cm})(9 \text{ cm})$$

$$= 54 \text{ cm}^2$$

The area is 54 cm².

Your solution

Solution on p. S31

12.6 EXERCISES

OBJECTIVE A **To solve similar and congruent triangles**

Find the ratio of corresponding sides for the similar triangles in Exercises 1 to 4.

1.

2.

3.

4.

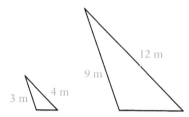

Determine whether the two triangles in Exercises 5 to 8 are congruent.

5.

6.

7.

8.

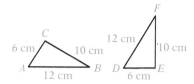

Triangles *ABC* and *DEF* in Exercises 9 to 12 are similar. Find the indicated distance. Round to the nearest tenth.

9. Find side *DE*.

10. Find side *DE*.

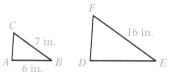

11. Find the height of triangle *DEF*.

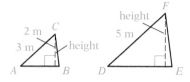

12. Find the height of triangle *ABC*.

 13. True or false? If the ratio of the corresponding sides of two similar triangles is 1 to 1, then the two triangles are congruent.

OBJECTIVE B   To solve application problems

Measurement The sun's rays, objects on Earth, and the shadows cast by those objects form similar triangles. For Exercises 14 and 15, find the height of the building.

14.

15.

In Exercises 16 to 19, triangles *ABC* and *DEF* are similar.

16. Find the perimeter of triangle *ABC*.

17. Find the perimeter of triangle *DEF.*

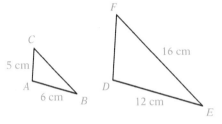

18. Find the area of triangle *ABC*.

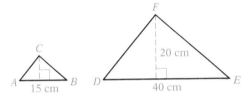

19. Find the area of triangle *DEF.*

20. Determine whether the statement is always true, sometimes true, or never true.
a. If two angles of one triangle are equal to two angles of a second triangle, then the triangles are similar triangles.
b. Two isosceles triangles are similar triangles.
c. Two equilateral triangles are similar triangles.
d. The ratio of the perimeters of two similar triangles is the same as the ratio of corresponding sides of the two triangles.

Applying the Concepts

21. Are all squares similar? Are all rectangles similar? Explain. Use a drawing in your explanation.

22. Figure *ABC* is a right triangle and *DE* is parallel to *AB*. What is the perimeter of the trapezoid *ABED?*

23. Explain how, by using only a yardstick, you could determine the approximate height of a tree without climbing it.

FOCUS ON PROBLEM SOLVING

Trial and Error   Some problems in mathematics are solved by using **trial and error.** The trial-and-error method of arriving at a solution to a problem involves performing repeated tests or experiments until a satisfactory conclusion is reached.

Many of the Applying the Concepts exercises in this text require a trial-and-error method of solution. For example, Exercises 46 to 49 in Section 12.4 ask you to explain how you could cut through a cube so that the face of the resulting solid is a square, an equilateral triangle, a trapezoid, and a hexagon.

There is no formula to apply to these problems; there is no computation to perform. These problems require that you picture a cube and the results after it is cut through at different places on its surface and at different angles. For Exercise 46, cutting perpendicular to the top and bottom of the cube and parallel to two of its sides will result in a square. The other shapes may prove more difficult.

When solving problems of this type, keep an open mind. Sometimes when using the trial-and-error method, we are hampered by our narrowness of vision; we cannot expand our thinking to include other possibilities. Then, when we see someone else's solution, it appears so obvious to us! For example, for the Applying the Concepts question above, it is necessary to conceive of cutting through the cube at places other than the top surface; we need to be open to the idea of beginning the cut at one of the corner points of the cube.

One topic of the Projects and Group Activities in this chapter is symmetry. Here again, the trial-and-error method is used to determine the lines of symmetry inherent in an object. For example, in determining lines of symmetry for a square, begin by drawing a square. The horizontal line of symmetry and the vertical line of symmetry may be immediately obvious to you.

But there are two others. Do you see that a line drawn through opposite corners of the square is also a line of symmetry?

Many of the questions in this text that require an answer of "always true, sometimes true, or never true" are best solved by the trial-and-error method. For example, consider the following statement, which is presented in Section 12.3.

If two rectangles have the same area, then they have the same perimeter.

Try some numbers. Each of two rectangles, one measuring 6 units by 2 units and another measuring 4 units by 3 units, has an area of 12 square units, but the perimeter of the first is 16 units and the perimeter of the second is 14 units, so the answer "always true" has been eliminated. We still need to determine whether there is a case when the statement is true. After experimenting with a lot of numbers, you may come to realize that we are trying to determine whether it is possible for two different pairs of factors of a number to have the same sum. Is it?

Don't be afraid to make many experiments, and remember that *errors,* or tests that "don't work," are a part of the trial-and-*error* process.

PROJECTS AND GROUP ACTIVITIES

Investigating Perimeter   The perimeter of the square at the right is 4 units.

If two squares are joined along one of the sides, the perimeter is 6 units. Note that it does not matter which sides are joined; the perimeter is still 6 units.

If three squares are joined, the perimeter of the resulting figure is 8 units for each possible placement of the squares.

Four squares can be joined in five different ways as shown. There are two possible perimeters: 10 units for A, B, C, and D, and 8 for E.

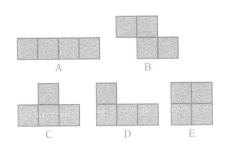

1. If five squares are joined, what is the maximum perimeter possible?

2. If five squares are joined, what is the minimum perimeter possible?

3. If six squares are joined, what is the maximum perimeter possible?

4. If six squares are joined, what is the minimum perimeter possible?

Symmetry   Look at the letter A printed at the left. If the letter were folded along line ℓ, the two sides of the letter would match exactly. This letter has **symmetry** with respect to line ℓ. Line ℓ is called the axis of **symmetry.**

Now consider the letter H printed below at the left. Both lines ℓ_1 and ℓ_2 are axes of symmetry for this letter; the letter could be folded along either line and the two sides would match exactly.

1. Does the letter A have more than one axis of symmetry?

2. Find axes of symmetry for other capital letters of the alphabet.

3. Which lowercase letters have one axis of symmetry?

4. Do any of the lowercase letters have more than one axis of symmetry?

5. Find the number of axes of symmetry for each of the plane geometric figures presented in this chapter.

6. There are other types of symmetry. Look up the meaning of *point symmetry* and *rotational symmetry*. Which plane geometric figures provide examples of these types of symmetry?

7. Find examples of symmetry in nature, art, and architecture.

CHAPTER 12

SUMMARY

KEY WORDS	EXAMPLES
A *line* extends indefinitely in two directions. A *line segment* is part of a line and has two endpoints. The length of a line segment is the distance between the endpoints of the line segment. [12.1A, p. 518]	
Parallel lines never meet; the distance between them is always the same. The symbol ‖ means "is parallel to." *Intersecting lines* cross at a point in the plane. *Perpendicular lines* are intersecting lines that form right angles. The symbol ⊥ means "is perpendicular to." [12.1A, pp. 518–519]	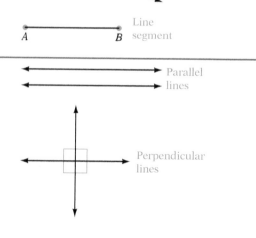

A *ray* starts at a point and extends indefinitely in one direction. An *angle* is formed when two rays start from the same point. The common point is called the *vertex* of the angle. An angle is measured in *degrees*. A 90° angle is a *right angle*. A 180° angle is a *straight angle*. *Complementary angles* are two angles whose measures have the sum 90°. *Supplementary angles* are two angles whose measures have the sum 180°. An *acute angle* is an angle whose measure is between 0° and 90°. An *obtuse angle* is an angle whose measure is between 90° and 180°.   [12.1A, pp. 519–520]

Two angles that are on opposite sides of the intersection of two lines are *vertical angles*. Vertical angles have the same measure. Two angles that share a common side are *adjacent angles*. Adjacent angles of intersecting lines are supplementary angles.   [12.1C, p. 524]

Angles *w* and *y* are vertical angles.
Angles *x* and *y* are adjacent angles.

A line that intersects two other lines at two different points is a *transversal*. If the lines cut by a transversal are parallel lines, equal angles are formed: *alternate interior angles, alternate exterior angles,* and *corresponding angles.*   [12.1C, pp. 524–525]

Parallel lines l_1 and l_2 are cut by transversal *t*. All four acute angles have the same measure. All four obtuse angles have the same measure.

A *quadrilateral* is a four-sided polygon. A *parallelogram*, a *rectangle*, and a *square* are quadrilaterals.   [12.1B, pp. 521–522]

A *polygon* is a closed figure determined by three or more line segments. The line segments that form the polygon are its *sides*. A *regular polygon* is one in which each side has the same length and each angle has the same measure. Polygons are classified by the number of sides.   [12.2A, p. 530]

Number of Sides	Name of the Polygon
3	Triangle
4	Quadrilateral
5	Pentagon
6	Hexagon
7	Heptagon
8	Octagon
9	Nonagon
10	Decagon

A *triangle* is a closed, three-sided plane figure.   [12.1B, p. 521]

An *isosceles triangle* has two sides of equal length. The three sides of an *equilateral triangle* are of equal length. A *scalene triangle* has no two sides of equal length. An *acute triangle* has three acute angles. An *obtuse triangle* has one obtuse angle.   [12.2A, pp. 530–531]

A *right triangle* contains a right angle. The side opposite the right angle is called the *hypotenuse*. The other two sides are called legs.   [12.1B, p. 521]

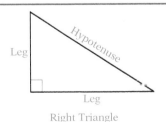

A *circle* is a plane figure in which all points are the same distance from the center of the circle. A *diameter* of a circle is a line segment across the circle through the center. A *radius* of a circle is a line segment from the center of the circle to a point on the circle. [12.1B, p. 522]

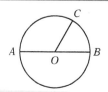

AB is a diameter of the circle.
OC is a radius.

Geometric solids are figures in space. Four common space figures are the rectangular solid, cube, sphere, and cylinder. A *rectangular solid* is a solid in which all six faces are rectangles. A *cube* is a rectangular solid in which all six faces are squares. A *sphere* is a solid in which all points on the sphere are the same distance from the center of the sphere. The most common *cylinder* is one in which the bases are circles and are perpendicular to the side. [12.1B, pp. 522–523]

Rectangular Solid

The *square root* of a number is one of two identical factors of that number. The symbol for square root is $\sqrt{}$. A *perfect square* is the product of a whole number times itself. The square root of a perfect square is a whole number. [12.5A, p. 558]

$1^2 = 1, 2^2 = 4, 3^2 = 9, 4^2 = 16,$
$5^2 = 25, \ldots\ldots 1, 4, 9, 16, 25, \ldots$ are
perfect squares.
$\sqrt{1} = 1, \sqrt{4} = 2, \sqrt{9} = 3, \sqrt{16} = 4, \ldots$

Similar triangles have the same shape but not necessarily the same size. The ratios of corresponding sides are equal. The ratio of heights is equal to the ratio of corresponding sides. *Congruent triangles* have the same shape and the same size. [12.6A, p. 564]

Triangles *ABC* and *DEF* are similar triangles. The ratio of corresponding sides is $\frac{1}{2}$.

ESSENTIAL RULES AND PROCEDURES	EXAMPLES

Triangles [12.1B, p. 521]
Sum of three angles = 180°

Two angles of a triangle measure 32° and 48°. Find the measure of the third angle.

$\angle A + \angle B + \angle C = 180°$
$\angle A + 32° + 48° = 180°$
$\angle A + 80° = 180°$
$\angle A + 80° - 80° = 180° - 80°$
$\angle A = 100°$

The measure of the third angle is 100°.

Formulas for Perimeter (the distance around a figure)
[12.2A, pp. 531–532]
Triangle: $P = a + b + c$
Square: $P = 4s$
Rectangle: $P = 2L + 2W$
Circumference of a circle: $C = \pi d$ or $C = 2\pi r$

The length of a rectangle is 8 m. The width is 5.5 m. Find the perimeter of the rectangle.
$P = 2L + 2W$
$P = 2(8 \text{ m}) + 2(5.5 \text{ m})$
$P = 16 \text{ m} + 11 \text{ m}$
$P = 27 \text{ m}$
The perimeter is 27 m.

Formulas for Area (the amount of surface in a region)
[12.3A, pp. 540–541]
Triangle: $A = \frac{1}{2}bh$
Square: $A = s^2$
Rectangle: $A = LW$
Circle: $A = \pi r^2$

Find the area of a circle with a radius of 4 cm. Use 3.14 for π.
$A = \pi r^2$
$A \approx 3.14(4 \text{ cm})^2$
$A \approx 50.24 \text{ cm}^2$
The area is approximately 50.24 cm².

Formulas for Volume (the amount of space inside a figure in space)
[12.4A, pp. 548–549]
Rectangular solid: $V = LWH$
Cube: $V = s^3$
Sphere: $V = \frac{4}{3}\pi r^3$
Cylinder: $V = \pi r^2 h$

Find the volume of a cube that measures 3 in. on a side.
$V = s^3$
$V = 3^3$
$V = 27$
The volume is 27 in³.

Pythagorean Theorem [12.5B, p. 559]
The square of the hypotenuse of a right triangle is equal to the sum of the squares of the two legs.
If the length of one side of a triangle is unknown, one of the following formulas can be used to find it.
If the hypotenuse is unknown, use
$$\text{Hypotenuse} = \sqrt{(\text{leg})^2 + (\text{leg})^2}$$
If the length of a leg is unknown, use
$$\text{Leg} = \sqrt{(\text{hypotenuse})^2 - (\text{leg})^2}$$

Two legs of a right triangle measure 6 ft and 8 ft. Find the hypotenuse of the right triangle.
$$\begin{aligned}
\text{Hypotenuse} &= \sqrt{(\text{leg})^2 + (\text{leg})^2} \\
&= \sqrt{6^2 + 8^2} \\
&= \sqrt{36 + 64} \\
&= \sqrt{100} \\
&= 10
\end{aligned}$$
The length of the hypotenuse is 10 ft.

Side-Side-Side (SSS) Rule
Two triangles are congruent if three sides of one triangle equal the corresponding sides of the second triangle.

Side-Angle-Side (SAS) Rule
Two triangles are congruent if two sides and the included angle of one triangle equal the corresponding sides and included angle of the second triangle. [12.6A, p. 565]

Triangles *ABC* and *EFD* are congruent by the SAS rule.

CHAPTER 12

CONCEPT REVIEW

Test your knowledge of the concepts presented in this chapter. Answer each question. Then check your answers against the ones provided in the Answer Section.

1. What are perpendicular lines?

2. When are two angles complementary?

3. How are the angles in a triangle related?

4. If you know the diameter of a circle, how can you find the radius?

5. What is the formula for the perimeter of a rectangle?

6. How do you find the circumference of a circle?

7. What is the formula for the area of a triangle?

8. What three dimensions are needed to find the volume of a rectangular solid?

9. When is a number a perfect square?

10. How do you identify the hypotenuse of a right triangle?

11. How can you use a proportion to solve for a missing length in similar triangles?

12. What is the side-angle-side rule?

CHAPTER 12

REVIEW EXERCISES

1. The diameter of a sphere is 1.5 m. Find the radius of the sphere.

2. Find the circumference of a circle with a radius of 5 cm. Use 3.14 for π.

3. Find the perimeter of the rectangle in the figure below.

4. Given $AB = 15$, $CD = 6$, and $AD = 24$, find the length of BC.

5. Find the volume of the rectangular solid shown below.

6. Find the unknown side of the triangle in the figure below.

7. Find the supplement of a 105° angle.

8. Find the square root of 15. Round to the nearest thousandth.

9. Triangles *ABC* and *DEF* are similar. Find the height of triangle *DEF*.

10. Find the area of the circle shown below. Use 3.14 for π.

11. In the figure below, $\ell_1 \parallel \ell_2$.
  a. Find the measure of angle *b*.
  b. Find the measure of angle *a*.

12. Find the area of the rectangle shown below.

13. Find the volume of the composite figure shown below.

14. Find the area of the composite figure shown below. Use 3.14 for π.

15. Find the volume of a sphere with a diameter of 8 ft. Use 3.14 for π. Round to the nearest tenth.

16. Triangles *ABC* and *DEF* are similar. Find the area of triangle *DEF*.

17. Find the perimeter of the composite figure shown below. Use 3.14 for π.

10 in.

16 in.   16 in.

18. In the figure below, $\ell_1 \parallel \ell_2$.
  a. Find the measure of angle *b*.
  b. Find the measure of angle *a*.

19. Home Maintenance How high on a building will a 17-foot ladder reach when the bottom of the ladder is 8 ft from the building?

20. A right triangle has a 32° angle. Find the measures of the other two angles.

28 in.

21. Travel A bicycle tire has a diameter of 28 in. How many feet does the bicycle travel when the wheel makes 10 revolutions? Use 3.14 for π. Round to the nearest tenth of a foot.

22. Aquariums See the news clipping at the right. The glass on Window on Washington Waters is a rectangular solid with a length of 40 ft, a width of 20 ft, and a thickness of 12.5 in.
  a. Inside and out, what is the area of glass that must be cleaned?

  b. What volume, in cubic inches, does the pane of glass in this exhibit fill?

23. Agriculture A silo, which is in the shape of a cylinder, is 9 ft in diameter and has a height of 18 ft. Find the volume of the silo. Use 3.14 for π.

24. Find the area of a right triangle with a base of 8 m and a height of 2.75 m.

25. Travel If you travel 20 mi west and then 21 mi south, how far are you from your starting point?

CHAPTER 12

TEST

1. Find the volume of a cylinder with a height of 6 m and a radius of 3 m. Use 3.14 for π.

2. Find the perimeter of a rectangle that has a length of 2 m and a width of 1.4 m.

3. Find the volume of the composite figure. Use 3.14 for π.

4. Triangles *ABC* and *FED* are congruent right triangles. Find the length of *FE*.

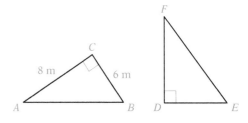

5. Find the complement of a 32° angle.

6. Find the area of a circle that has a diameter of 2 m. Use $\frac{22}{7}$ for π.

7. In the figure below, lines ℓ_1 and ℓ_2 are parallel. Angle *x* measures 30°. Find the measure of angle *y*.

8. Find the perimeter of the composite figure. Use 3.14 for π.

9. Find the square root of 189. Round to the nearest thousandth.

10. Find the unknown side of the triangle shown below. Round to the nearest thousandth.

11. Find the area of the composite figure.

$1\frac{1}{2}$ ft

3 ft

$4\frac{1}{2}$ ft

12. In the figure below, lines ℓ_1 and ℓ_2 are parallel. Angle x measures 45°. Find the measures of angles a and b.

13. Triangles *ABC* and *DEF* are similar. Find side *BC*.

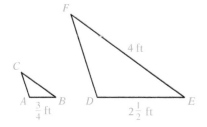

$\frac{3}{4}$ ft

4 ft

$2\frac{1}{2}$ ft

14. A right triangle has a 40° angle. Find the measures of the other two angles.

15. **Measurement** Use similar triangles to find the width of the canal shown in the figure at the right.

12 ft

5 ft

60 ft

Canal

16. **Consumerism** How much more pizza is contained in a pizza with radius 10 in. than in one with radius 8 in.? Use 3.14 for π.

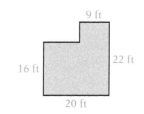

9 ft

16 ft

22 ft

20 ft

17. **Interior Design** A carpet is to be placed as shown in the diagram at the right. At $26.80 per square yard, how much will it cost to carpet the area? Round to the nearest cent. ($9 \text{ ft}^2 = 1 \text{ yd}^2$)

18. **Construction** Find the length of the rafter needed for the roof shown in the figure.

2 ft

5 ft

24 ft

19. **Forestry** Find the cross-sectional area of a redwood tree that is 11 ft 6 in. in diameter. Use 3.14 for π. Round to the nearest hundredth.

20. **Alcatraz** Inmate cells at Alcatraz were 9 ft long and 5 ft wide. The height of a cell was 7 ft.
a. Find the area of the floor of a cell at Alcatraz.
b. Find the volume of a cell at Alcatraz.

CUMULATIVE REVIEW EXERCISES

1. Find the GCF of 96 and 144.

2. Add: $3\frac{5}{12} + 2\frac{9}{16} + 1\frac{7}{8}$

3. Find the quotient of $4\frac{1}{3}$ and $6\frac{2}{9}$.

4. Simplify: $\left(\frac{2}{3}\right)^2 \div \left(\frac{1}{3} + \frac{1}{2}\right) - \frac{2}{5}$

5. Simplify: $-\frac{2}{3} - \left(-\frac{5}{8}\right)$

6. Write "$348.80 earned in 20 hours" as a unit rate.

7. Solve the proportion $\frac{3}{8} = \frac{n}{100}$.

8. Write $37\frac{1}{2}\%$ as a fraction.

9. Evaluate $a^2 - (b^2 - c)$ when $a = 2$, $b = -2$, and $c = -4$.

10. 30.94 is 36.4% of what number?

11. Solve: $\frac{x}{3} + 3 = 1$

12. Solve: $2(x - 3) + 2 = 5x - 8$

13. Convert 32.5 km to meters.

14. Subtract: $32 \text{ m} - 42 \text{ cm}$

15. Solve: $\frac{2}{3}x = -10$

16. Solve: $2x - 4(x - 3) = 8$

17. **Finance** You bought a car for $26,488 and made a down payment of $1000. You paid the balance in 36 equal monthly installments. Find the monthly payment.

18. **Taxes** The sales tax on a color printer costing $175 is $6.75. At the same rate, find the sales tax on a home theater system costing $1220.

19. **Compensation** A heavy-equipment operator receives an hourly wage of $32.12 an hour after receiving a 10% wage increase. Find the operator's hourly wage before the increase.

20. **Discount** An after-Christmas sale has a discount rate of 55%. Find the sale price of a PDA that had a regular price of $240.

21. **Investments** An IRA pays 7% annual interest, compounded daily. What would be the value of an investment of $25,000 after 20 years? Use the table in the Appendix.

22. **Shipping** A square tile measuring 4 in. by 4 in. weighs 6 oz. Find the weight, in pounds, of a package of 144 such tiles.

23. **Metal Works** Twenty-five rivets are used to fasten two steel plates together. The plates are 5.4 m long, and the rivets are equally spaced with a rivet at each end. Find the distance, in centimeters, between the rivets.

24. **Integer Problems** The total of four times a number and two is negative six. Find the number.

25. The lines ℓ_1 and ℓ_2 in the figure below are parallel.
    a. Find angle a.
    b. Find angle b.

26. Find the perimeter of the composite figure. Use 3.14 for π.

27. Find the area of the composite figure.

28. Find the volume of the composite figure. Use 3.14 for π.

29. Find the unknown side of the triangle shown in the figure below. Round to the nearest hundredth.

30. Triangles *ABC* and *FED* below are similar. Find the perimeter of *FED*.

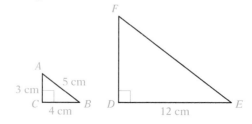

FINAL EXAM

1. Subtract: $100{,}914 - 97{,}655$

2. Find 34,821 divided by 657.

3. Find 90,001 decreased by 29,796.

4. Simplify: $3^2 \cdot (5 - 3)^2 \div 3 + 4$

5. Find the LCM of 9, 12, and 16.

6. Add: $\frac{3}{8} + \frac{5}{6} + \frac{1}{5}$

7. Subtract: $7\frac{5}{12} - 3\frac{13}{16}$

8. Find the product of $3\frac{5}{8}$ and $1\frac{5}{7}$.

9. Divide: $1\frac{2}{3} \div 3\frac{3}{4}$

10. Simplify: $\left(\frac{2}{3}\right)^3 \cdot \left(\frac{3}{4}\right)^2$

11. Simplify: $\left(\frac{2}{3}\right)^2 \div \left(\frac{3}{4} + \frac{1}{3}\right) - \frac{1}{3}$

12. Add: 4.972 28.6 1.88 $+\ 128.725$

13. Multiply: 2.97 $\times\ 0.0094$

14. Divide: $0.062\overline{)0.0426}$ Round to the nearest hundredth.

15. Convert 0.45 to a fraction in simplest form.

16. Write "323.4 miles on 13.2 gallons of gas" as a unit rate.

17. Solve the proportion $\frac{12}{35} = \frac{n}{160}$. Round to the nearest tenth.

18. Write $22\frac{1}{2}\%$ as a fraction.

19. Write 1.35 as a percent.

20. Write $\frac{5}{4}$ as a percent.

21. Find 120% of 30.

22. 12 is what percent of 9?

23. 42 is 60% of what number?

24. Convert $1\frac{2}{3}$ ft to inches.

25. Subtract: 3 ft 2 in. − 1 ft 10 in.

26. Convert 40 oz to pounds.

27. Find the sum of 3 lb 12 oz and 2 lb 10 oz.

28. Convert 18 pt to gallons.

29. Divide: $3\overline{)5\text{ gal 1 qt}}$

30. Convert 2.48 m to centimeters.

31. Convert 4 m 62 cm to meters.

32. Convert 1 kg 614 g to kilograms.

33. Convert 2 L 67 ml to milliliters.

34. Convert 55 mi to kilometers. Round to the nearest hundredth. (1.61 km ≈ 1 mi)

35. Consumerism How much does it cost to run a 2400-watt air conditioner for 6 h at 8¢ per kilowatt-hour? Round to the nearest cent.

36. Write 0.0000000679 in scientific notation.

37. Find the perimeter of a rectangle with a length of 1.2 m and a width of 0.75 m.

38. Find the area of a rectangle with a length of 9 in. and a width of 5 in.

39. Find the volume of a box with a length of 20 cm, a width of 12 cm, and a height of 5 cm.

40. Add: $-2 + 8 + (-10)$

41. Subtract: $-30 - (-15)$

42. Multiply: $2\frac{1}{2} \times \left(-\frac{1}{5}\right)$

43. Find the quotient of $-1\frac{3}{8}$ and $5\frac{1}{2}$.

44. Simplify: $(-4)^2 \div (1 - 3)^2 - (-2)$

45. Simplify: $2x - 3(x - 4) + 5$

46. Solve: $\frac{2}{3}x = -12$

47. Solve: $3x - 5 = 10$

48. Solve: $8 - 3x = x + 4$

49. Banking You have $872.48 in your checking account. You write checks for $321.88 and $34.23 and then make a deposit of $443.56. Find your new checking account balance.

50. Elections On the basis of a pre-election survey, it is estimated that 5 out of 8 eligible voters will vote in an election. How many people will vote in an election with 102,000 eligible voters?

51. Investments This month a company is paying its stockholders a dividend of $1.60 per share. This is 80% of what the dividend per share was 1 year ago. What was the dividend per share 1 year ago?

52. Compensation A sales executive received commissions of $4320, $3572, $2864, and $4420 during a 4-month period. Find the mean monthly income from commissions for the 4 months.

53. Simple Interest A contractor borrows $120,000 for 9 months at an annual interest rate of 8%. What is the simple interest due on the loan?

54. Probability If two dice are tossed, what is the probability that the sum of the dots on the upward faces is divisible by 3?

55. Wars The top four highest death counts, by country, in World War II are shown in the circle graph. What percent of the total death count in all four countries is the death count of China? Round to the nearest tenth of a percent.

Top Four Highest Death Counts in World War II (in thousands)

Source: U.S. Department of Defense

56. Discounts A pair of Bose headphones that regularly sells for $314.00 is on sale for $226.08. What is the discount rate?

57. Shipping A square tile measuring 8 in. by 8 in. weighs 9 oz. Find the weight, in pounds, of a box containing 144 tiles.

58. Find the perimeter of the composite figure. Use 3.14 for π.

8 in.   8 in.

8 in.

59. Find the area of the composite figure. Use 3.14 for π.

|←————— 10 cm ————→|

2 cm

60. Integer Problems Five less than the quotient of a number and two is equal to three. Find the number.

Appendix

Table of Geometric Formulas

Pythagorean Theorem

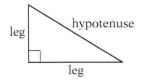

$\text{Hypotenuse} = \sqrt{(\text{leg})^2 + (\text{leg})^2}$

$\text{Leg} = \sqrt{(\text{hypotenuse})^2 - (\text{leg})^2}$

Perimeter and Area of a Triangle

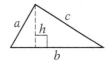

$P = a + b + c$

$A = \dfrac{1}{2}bh$

Perimeter and Area of a Rectangle

$P = 2L + 2W$

$A = LW$

Perimeter and Area of a Square

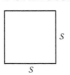

$P = 4s$

$A = s^2$

Circumference and Area of a Circle

$C = 2\pi r \text{ or } C = \pi d$

$A = \pi r^2$

Area of a Trapezoid

$A = \dfrac{1}{2}h(b_1 + b_2)$

Volume of a Rectangular Solid

$V = LWH$

Volume of a Cube

$V = s^3$

Volume of a Sphere

$V = \dfrac{4}{3}\pi r^2$

Volume of a Right Circular Cylinder

$V = \pi r^2 h$

Compound Interest Table

Compounded Annually							
	4%	5%	6%	7%	8%	9%	10%
1 year	1.04000	1.05000	1.06000	1.07000	1.08000	1.09000	1.10000
5 years	1.21665	1.27628	1.33823	1.40255	1.46933	1.53862	1.61051
10 years	1.48024	1.62890	1.79085	1.96715	2.15893	2.36736	2.59374
15 years	1.80094	2.07893	2.39656	2.75903	3.17217	3.64248	4.17725
20 years	2.19112	2.65330	3.20714	3.86968	4.66095	5.60441	6.72750

Compounded Semiannually							
	4%	5%	6%	7%	8%	9%	10%
1 year	1.04040	1.05062	1.06090	1.07123	1.08160	1.09203	1.10250
5 years	1.21899	1.28008	1.34392	1.41060	1.48024	1.55297	1.62890
10 years	1.48595	1.63862	1.80611	1.98979	2.19112	2.41171	2.65330
15 years	1.81136	2.09757	2.42726	2.80679	3.24340	3.74531	4.32194
20 years	2.20804	2.68506	3.26204	3.95926	4.80102	5.81634	7.03999

Compounded Quarterly							
	4%	5%	6%	7%	8%	9%	10%
1 year	1.04060	1.05094	1.06136	1.07186	1.08243	1.09308	1.10381
5 years	1.22019	1.28204	1.34686	1.41478	1.48595	1.56051	1.63862
10 years	1.48886	1.64362	1.81402	2.00160	2.20804	2.43519	2.68506
15 years	1.81670	2.10718	2.44322	2.83182	3.28103	3.80013	4.39979
20 years	2.21672	2.70148	3.29066	4.00639	4.87544	5.93015	7.20957

Compounded Monthly							
	4%	5%	6%	7%	8%	9%	10%
1 year	1.04074	1.051162	1.061678	1.072290	1.083000	1.093807	1.104713
5 years	1.220997	1.283359	1.348850	1.417625	1.489846	1.565681	1.645309
10 years	1.490833	1.647009	1.819397	2.009661	2.219640	2.451357	2.707041
15 years	1.820302	2.113704	2.454094	2.848947	3.306921	3.838043	4.453920
20 years	2.222582	2.712640	3.310204	4.038739	4.926803	6.009152	7.328074

Compounded Daily							
	4%	5%	6%	7%	8%	9%	10%
1 year	1.04080	1.05127	1.06183	1.07250	1.08328	1.09416	1.10516
5 years	1.22139	1.28400	1.34983	1.41902	1.49176	1.56823	1.64861
10 years	1.49179	1.64866	1.82203	2.01362	2.22535	2.45933	2.71791
15 years	1.82206	2.11689	2.45942	2.85736	3.31968	3.85678	4.48077
20 years	2.22544	2.71810	3.31979	4.05466	4.95217	6.04830	7.38703

To use this table:
1. Locate the section that gives the desired compounding period.
2. Locate the interest rate in the top row of that section.
3. Locate the number of years in the left-hand column of that section.
4. Locate the number where the interest-rate column and the number-of-years row meet. This is the compound interest factor.

Example An investment yields an annual interest rate of 10% compounded quarterly for 5 years.
The compounding period is "compounded quarterly."
The interest rate is 10%.
The number of years is 5.
The number where the row and column meet is 1.63862. This is the compound interest factor.

Compound Interest Table

Compounded Annually

	11%	12%	13%	14%	15%	16%	17%
1 year	1.11000	1.12000	1.13000	1.14000	1.15000	1.16000	1.17000
5 years	1.68506	1.76234	1.84244	1.92542	2.01136	2.10034	2.19245
10 years	2.83942	3.10585	3.39457	3.70722	4.04556	4.41144	4.80683
15 years	4.78459	5.47357	6.25427	7.13794	8.13706	9.26552	10.53872
20 years	8.06239	9.64629	11.52309	13.74349	16.36654	19.46076	23.10560

Compounded Semiannually

	11%	12%	13%	14%	15%	16%	17%
1 year	1.11303	1.12360	1.13423	1.14490	1.15563	1.16640	1.17723
5 years	1.70814	1.79085	1.87714	1.96715	2.06103	2.15893	2.26098
10 years	2.91776	3.20714	3.52365	3.86968	4.24785	4.66096	5.11205
15 years	4.98395	5.74349	6.61437	7.61226	8.75496	10.06266	11.55825
20 years	8.51331	10.28572	12.41607	14.97446	18.04424	21.72452	26.13302

Compounded Quarterly

	11%	12%	13%	14%	15%	16%	17%
1 year	1.11462	1.12551	1.13648	1.14752	1.15865	1.16986	1.18115
5 years	1.72043	1.80611	1.89584	1.98979	2.08815	2.19112	2.29891
10 years	2.95987	3.26204	3.59420	3.95926	4.36038	4.80102	5.28497
15 years	5.09225	5.89160	6.81402	7.87809	9.10513	10.51963	12.14965
20 years	8.76085	10.64089	12.91828	15.67574	19.01290	23.04980	27.93091

Compounded Monthly

	11%	12%	13%	14%	15%	16%	17%
1 year	1.115719	1.126825	1.138032	1.149342	1.160755	1.172271	1.183892
5 years	1.728916	1.816697	1.908857	2.005610	2.107181	2.213807	2.325733
10 years	2.989150	3.300387	3.643733	4.022471	4.440213	4.900941	5.409036
15 years	5.167988	5.995802	6.955364	8.067507	9.356334	10.849737	12.579975
20 years	8.935015	10.892554	13.276792	16.180270	19.715494	24.019222	29.257669

Compounded Daily

	11%	12%	13%	14%	15%	16%	17%
1 year	1.11626	1.12747	1.13880	1.15024	1.16180	1.17347	1.18526
5 years	1.73311	1.82194	1.91532	2.01348	2.11667	2.22515	2.33918
10 years	3.00367	3.31946	3.66845	4.05411	4.48031	4.95130	5.47178
15 years	5.20569	6.04786	7.02625	8.16288	9.48335	11.01738	12.79950
20 years	9.02203	11.01883	13.45751	16.43582	20.07316	24.51534	29.94039

Monthly Payment Table

	4%	5%	6%	7%	8%	9%
1 year	0.0851499	0.0856075	0.0860664	0.0865267	0.0869884	0.0874515
2 years	0.0434249	0.0438714	0.0443206	0.0447726	0.0452273	0.0456847
3 years	0.0295240	0.0299709	0.0304219	0.0308771	0.0313364	0.0317997
4 years	0.0225791	0.0230293	0.0234850	0.0239462	0.0244129	0.0248850
5 years	0.0184165	0.0188712	0.0193328	0.0198012	0.0202764	0.0207584
15 years	0.0073969	0.0079079	0.0084386	0.0089883	0.0095565	0.0101427
20 years	0.0060598	0.0065996	0.0071643	0.0077530	0.0083644	0.0089973
25 years	0.0052784	0.0058459	0.0064430	0.0070678	0.0077182	0.0083920
30 years	0.0047742	0.0053682	0.0059955	0.0066530	0.0073376	0.0080462

	10%	11%	12%	13%	
1 year	0.0879159	0.0883817	0.0888488	0.0893173	
2 years	0.0461449	0.0466078	0.0470735	0.0475418	
3 years	0.0322672	0.0327387	0.0332143	0.0336940	
4 years	0.0253626	0.0258455	0.0263338	0.0268275	
5 years	0.0212470	0.0217424	0.0222445	0.0227531	
15 years	0.0107461	0.0113660	0.0120017	0.0126524	
20 years	0.0096502	0.0103219	0.0110109	0.0117158	
25 years	0.0090870	0.0098011	0.0105322	0.0112784	
30 years	0.0087757	0.0095232	0.0102861	0.0110620	

To use this table:
1. Locate the desired interest rate in the top row.
2. Locate the number of years in the left-hand column.
3. Locate the number where the interest-rate column and the number-of-years row meet. This is the monthly payment factor.

Example A home has a 30-year mortgage at an annual interest rate of 12%.
The interest rate is 12%.
The number of years is 30.
The number where the row and column meet is 0.0102861. This is the monthly payment factor.

Table of Measurements

Prefixes in the Metric System of Measurement

kilo-	1000	deci-	0.1
hecto-	100	centi-	0.01
deca-	10	milli-	0.001

Measurement Abbreviations
U.S. Customary System

Length

in.	inches
ft	feet
yd	yards
mi	miles

Capacity

oz	fluid ounces
c	cups
qt	quarts
gal	gallons

Weight

oz	ounces
lb	pounds

Area

in^2	square inches
ft^2	square feet
yd^2	square yards
mi^2	square miles

Rate

ft/s	feet per second
mi/h	miles per hour

Time

h	hours
min	minutes
s	seconds

Metric System

Length

mm	millimeters
cm	centimeters
m	meters
km	kilometers

Capacity

ml	milliliters
cl	centiliters
L	liters
kl	kiloliters

Weight/Mass

mg	milligrams
cg	centigrams
g	grams
kg	kilograms

Area

cm^2	square centimeters
m^2	square meters
km^2	square kilometers

Rate

m/s	meters per second
km/s	kilometers per second
km/h	kilometers per hour

Time

h	hours
min	minutes
s	seconds

Table of Properties

Properties of Real Numbers

Commutative Property of Addition
If a and b are real numbers, then $a + b = b + a$.

Commutative Property of Multiplication
If a and b are real numbers, then $a \cdot b = b \cdot a$.

Associative Property of Addition
If a, b, and c are real numbers, then
$(a + b) + c = a + (b + c)$.

Associative Property of Multiplication
If a, b, and c are real numbers, then
$(a \cdot b) \cdot c = a \cdot (b \cdot c)$.

Addition Property of Zero
If a is a real number, then $a + 0 = 0 + a = a$.

Multiplication Property of Zero
If a is a real number, then $a \cdot 0 = 0 \cdot a = 0$.

Multiplication Property of One
If a is a real number, then $a \cdot 1 = 1 \cdot a = a$.

Inverse Property of Addition
If a is a real number, then $a + (-a) = (-a) + a = 0$.

Inverse Property of Multiplication
If a is a real number and $a \neq 0$, then
$a \cdot \dfrac{1}{a} = \dfrac{1}{a} \cdot a = 1$.

Distributive Property
If a, b, and c are real numbers, then
$a(b + c) = ab + ac$.

Properties of Zero and One in Division

Any number divided by 1 is the number.
Division by zero is not allowed.

Any number other than zero divided by itself is 1.
Zero divided by a number other than zero is zero.

Solutions to "You Try It"

SOLUTIONS TO CHAPTER 1 "YOU TRY IT"

SECTION 1.1

You Try It 1

$$0\ 1\ 2\ 3\ 4\ 5\ 6\ 7\ 8\ 9\ 10\ 11\ 12\ 13\ 14$$

You Try It 2   **a.** $45 > 29$ **b.** $27 > 0$

You Try It 3   Thirty-six million four hundred sixty-two thousand seventy-five

You Try It 4   452,007

You Try It 5 $60,000 + 8000 + 200 + 80 + 1$

You Try It 6 $100,000 + 9000 + 200 + 7$

You Try It 7

Given place value

368,492

$8 > 5$

368,492 rounded to the nearest ten-thousand is 370,000.

You Try It 8

Given place value

3962

$6 > 5$

3962 rounded to the nearest hundred is 4000.

SECTION 1.2

You Try It 1

$$\begin{array}{r} 1\ 1 \\ 347 \\ +12,453 \\ \hline 12,800 \end{array}$$

• $7 + 3 = 10$
Write the 0 in the ones column. Carry the 1 to the tens column.
$1 + 4 + 5 = 10$
Write the 0 in the tens column. Carry the 1 to the hundreds column.
$1 + 3 + 4 = 8$

347 increased by 12,453 is 12,800.

You Try It 2

$$\begin{array}{r} 2 \\ 95 \\ 88 \\ + 67 \\ \hline 250 \end{array}$$

• $5 + 8 + 7 = 20$
Write the 0 in the ones column.
Carry the 2 to the tens column.

You Try It 3

$$\begin{array}{r} 1\ 1\ 2\ 1 \\ 392 \\ 4,079 \\ 89,035 \\ + 4,992 \\ \hline 98,498 \end{array}$$

You Try It 4

Strategy   To find the total square footage of Wal-Mart stores in the United States, read the table to find the square footage of discount stores, Supercenters, Sam's Clubs, and neighborhood markets. Then add the four numbers.

Solution

$$\begin{array}{r} 105 \\ 457 \\ 78 \\ + 5 \\ \hline 645 \end{array}$$

The total square footage of Wal-Mart stores in the United States is 645 million square feet.

SECTION 1.3

You Try It 1

$$\begin{array}{r} 8925 \\ - 6413 \\ \hline 2512 \end{array}$$
   Check:
$$\begin{array}{r} 6413 \\ + 2512 \\ \hline 8925 \end{array}$$

You Try It 2

$$\begin{array}{r} 17,504 \\ - 9,302 \\ \hline 8,202 \end{array}$$
   Check:
$$\begin{array}{r} 9,302 \\ + 8,202 \\ \hline 17,504 \end{array}$$

You Try It 3

$$\begin{array}{r} 2\ 14\ 7\ 11 \\ 3\,4\,8\,1 \\ - \quad 8\,6\,5 \\ \hline 2\,6\,1\,6 \end{array}$$
   Check:
$$\begin{array}{r} 865 \\ + 2616 \\ \hline 3481 \end{array}$$

You Try It 4

$$\begin{array}{r} 15 \\ 4\ \,8\ 12 \\ 5\,4,8\,6\,2 \\ -\,1\,4,1\,8\,5 \\ \hline 4\,0,0\,7\,7 \end{array}$$
   Check:
$$\begin{array}{r} 14,485 \\ + 40,077 \\ \hline 54,562 \end{array}$$

You Try It 5

$$\begin{array}{r} {}^{3}\ \ {}^{10} \\ 6\,4,\cancel{0}\,0\,3 \\ -\,5\,4,9\,3\,6 \end{array}$$

- There are two zeros in the minuend. Borrow 1 thousand from the thousands column and write 10 in the hundreds column.

$$\begin{array}{r} \quad\ {}^{9} \\ {}^{3}\ {}^{10}\ {}^{10} \\ 6\,4,\cancel{0}\,\cancel{0}\,3 \\ -\,5\,4,9\,3\,6 \end{array}$$

- Borrow 1 hundred from the hundreds column and write 10 in the tens column.

$$\begin{array}{r} {}^{13}\ \ {}^{9}\ \ {}^{9} \\ {}^{5}\ {}^{3}\ {}^{10}\ {}^{10}\ {}^{13} \\ \cancel{6}\,\cancel{4},\cancel{0}\,\cancel{0}\,\cancel{3} \\ -\,5\,4,9\,3\,6 \\ \hline 9,0\,6\,7 \end{array}$$

- Borrow 1 ten from the tens column and add 10 to the 3 in the ones column.

Check:
$$\begin{array}{r} 54,936 \\ +\ 9,067 \\ \hline 64,003 \end{array}$$

You Try It 6

Strategy　To find the difference, subtract the number of personnel on active duty in the Air Force in 1945 (2,282,259) from the number of personnel on active duty in the Navy in 1945 (3,380,817).

Solution
$$\begin{array}{r} 3,380,817 \\ -\ 2,282,259 \\ \hline 1,098,558 \end{array}$$

The difference was 1,098,558 personnel.

You Try It 7

Strategy　To find your take-home pay:

- Add to find the total of the deductions (127 + 18 + 35).
- Subtract the total of the deductions from your total salary (638).

Solution
$$\begin{array}{r} 127 \\ 18 \\ +\ 35 \\ \hline 180 \end{array} \qquad \begin{array}{r} 638 \\ -\ 180 \\ \hline 458 \end{array}$$

180 in deductions

Your take-home pay is $458.

SECTION 1.4

You Try It 1

$$\begin{array}{r} {}^{3\,5} \\ 648 \\ \times\quad 7 \\ \hline 4536 \end{array}$$

- $7 \times 8 = 56$
  Write the 6 in the ones column.
  Carry the 5 to the tens column.
  $7 \times 4 = 28,\ 28 + 5 = 33$
  $7 \times 6 = 42,\ 42 + 3 = 45$

You Try It 2

$$\begin{array}{r} 756 \\ \times\ 305 \\ \hline 3780 \\ 22680 \\ \hline 230,580 \end{array}$$

- $5 \times 756 = 3780$
  Write a zero in the tens column for 0×756.
  $3 \times 756 = 2268$

You Try It 3

Strategy　To find the number of cars the dealer will receive in 12 months, multiply the number of months (12) by the number of cars received each month (37).

Solution
$$\begin{array}{r} 37 \\ \times\ 12 \\ \hline 74 \\ 37 \\ \hline 444 \end{array}$$

The dealer will receive 444 cars in 12 months.

You Try It 4

Strategy　To find the total cost of the order:

- Find the cost of the sports jackets by multiplying the number of jackets (25) by the cost for each jacket (23).
- Add the product to the cost for the suits (4800).

Solution
$$\begin{array}{r} 23 \\ \times\ 25 \\ \hline 115 \\ 46 \\ \hline 575 \end{array} \qquad \begin{array}{r} 4800 \\ +\ 575 \\ \hline 5375 \end{array}$$

575 cost for jackets

The total cost of the order is $5375.

SECTION 1.5

You Try It 1

$$9\overline{)63}\ \ \ ^{7}$$

Check: $7 \times 9 = 63$

You Try It 2

$$\begin{array}{r} 453 \\ 9\overline{)\,4077} \\ -36 \\ \hline 47 \\ -45 \\ \hline 27 \\ -27 \\ \hline 0 \end{array}$$

Check: $453 \times 9 = 4077$

You Try It 3

$$
\begin{array}{r}
705 \\
9\overline{)6345} \\
-63 \\
\hline
04 \\
-\ 0 \\
\hline
45 \\
-45 \\
\hline
0
\end{array}
$$

• Think $9\overline{)4}$. Place 0 in quotient.
• Subtract 0×9.
• Bring down the 5.

Check: $705 \times 9 = 6345$

You Try It 4

$$
\begin{array}{r}
870 \text{ r5} \\
6\overline{)5225} \\
-48 \\
\hline
42 \\
-42 \\
\hline
05 \\
-\ 0 \\
\hline
5
\end{array}
$$

• Think $6\overline{)5}$. Place 0 in quotient.
• Subtract 0×6.

Check: $(870 \times 6) + 5 =$
$5220 + 5 = 5225$

You Try It 5

$$
\begin{array}{r}
3{,}058 \text{ r3} \\
7\overline{)21{,}409} \\
-21 \\
\hline
04 \\
-\ 0 \\
\hline
40 \\
-35 \\
\hline
59 \\
-56 \\
\hline
3
\end{array}
$$

• Think $7\overline{)4}$. Place 0 in quotient.
• Subtract 0×7.

Check: $(3058 \times 7) + 3 =$
$21{,}406 + 3 = 21{,}409$

You Try It 6

$$
\begin{array}{r}
109 \\
42\overline{)4578} \\
-42 \\
\hline
37 \\
-\ 0 \\
\hline
378 \\
-\ 378 \\
\hline
0
\end{array}
$$

• Think $42\overline{)37}$. Place 0 in quotient.
• Subtract 0×42.

Check: $(109 \times 42) = 4578$

You Try It 7

$$
\begin{array}{r}
470 \text{ r29} \\
39\overline{)18{,}359} \\
-15\ 6 \\
\hline
2\ 75 \\
-2\ 73 \\
\hline
29 \\
-\ 0 \\
\hline
29
\end{array}
$$

• Think $3\overline{)18}$. 6×39 is too large. Try 5. 5×39 is too large. Try 4.

Check: $(470 \times 39) + 29 =$
$18{,}330 + 29 = 18{,}359$

You Try It 8

$$
\begin{array}{r}
62 \text{ r111} \\
534\overline{)33{,}219} \\
-32\ 04 \\
\hline
1\ 179 \\
-1\ 068 \\
\hline
111
\end{array}
$$

Check: $(62 \times 534) + 111 =$
$33{,}108 + 111 = 33{,}219$

You Try It 9

$$
\begin{array}{r}
421 \text{ r33} \\
515\overline{)216{,}848} \\
-206\ 0 \\
\hline
10\ 84 \\
-10\ 30 \\
\hline
548 \\
-515 \\
\hline
33
\end{array}
$$

Check: $(421 \times 515) + 33 =$
$216{,}815 + 33 = 216{,}848$

You Try It 10

Strategy To find the number of tires that can be stored on each shelf, divide the number of tires (270) by the number of shelves (15).

Solution

$$
\begin{array}{r}
18 \\
15\overline{)270} \\
-15 \\
\hline
120 \\
-120 \\
\hline
0
\end{array}
$$

Each shelf can store 18 tires.

You Try It 11

Strategy To find the number of cases produced in 8 hours:

• Find the number of cases produced in 1 hour by dividing the number of cans produced (12,600) by the number of cans to a case (24).
• Multiply the number of cases produced in 1 hour by 8.

Solution

$$
\begin{array}{r}
525 \\
24\overline{)12{,}600} \\
-12\ 0 \\
\hline
60 \\
-48 \\
\hline
120 \\
-120 \\
\hline
0
\end{array}
$$

cases produced in 1 hour

$$
\begin{array}{r}
525 \\
\times\quad 8 \\
\hline
4200
\end{array}
$$

In 8 hours, 4200 cases are produced.

SECTION 1.6

You Try It 1 $2 \cdot 2 \cdot 2 \cdot 2 \cdot 3 \cdot 3 \cdot 3 = 2^4 \cdot 3^3$

You Try It 2 $10 \cdot 10 \cdot 10 \cdot 10 \cdot 10 \cdot 10 \cdot 10 = 10^7$

You Try It 3 $2^3 \cdot 5^2 = (2 \cdot 2 \cdot 2) \cdot (5 \cdot 5) = 8 \cdot 25$
$$= 200$$

You Try It 4
$$
\begin{aligned}
5 \cdot (8 - 4)^2 \div 4 - 2 \quad &\bullet \text{ Parentheses}\\
= 5 \cdot 4^2 \div 4 - 2 \quad &\bullet \text{ Exponents}\\
= 5 \cdot 16 \div 4 - 2 \quad &\bullet \text{ Multiplication}\\
= 80 \div 4 - 2 \quad &\quad\ \text{and division}\\
= 20 - 2 \quad &\bullet \text{ Subtraction}\\
= 18
\end{aligned}
$$

SECTION 1.7

You Try It 1
$40 \div 1 = 40$
$40 \div 2 = 20$
$40 \div 3 \qquad$ Will not divide evenly
$40 \div 4 = 10$
$40 \div 5 = 8$
$40 \div 6 \qquad$ Will not divide evenly
$40 \div 7 \qquad$ Will not divide evenly
$40 \div 8 = 5$

1, 2, 4, 5, 8, 10, 20, and 40 are factors of 40.

You Try It 2

44	
2	22
2	11
11	1

• $44 \div 2 = 22$
• $22 \div 2 = 11$
• $11 \div 11 = 1$

$44 = 2 \cdot 2 \cdot 11$

You Try It 3

177	
3	59
59	1

• Try only 2, 3, 4, 7, and 11 because $11^2 > 59$.

$177 = 3 \cdot 59$

SOLUTIONS TO CHAPTER 2 "YOU TRY IT"

SECTION 2.1

You Try It 1

	2	3	5
12 =	(2 · 2)	3	
27 =		(3 · 3 · 3)	
50 =	2		(5 · 5)

The LCM $= 2 \cdot 2 \cdot 3 \cdot 3 \cdot 3 \cdot 5 \cdot 5 =$
2700

You Try It 2

	2	3	5
36 =	(2 · 2)	3 · 3	
60 =	2 · 2	(3)	5
72 =	2 · 2 · 2	3 · 3	

The GCF $= 2 \cdot 2 \cdot 3 = 12$.

You Try It 3

	2	3	5	11
11 =				11
24 =	2 · 2 · 2	3		
30 =	2	3	5	

Because no numbers are circled, the GCF = 1.

SECTION 2.2

You Try It 1 $4\dfrac{1}{4}$

You Try It 2 $\dfrac{17}{4}$

You Try It 3
$$
\begin{array}{r}
4 \\
5\overline{)22} \\
-20 \\
\hline
2
\end{array}
\qquad \frac{22}{5} = 4\frac{2}{5}
$$

You Try It 4
$$
\begin{array}{r}
4 \\
7\overline{)28} \\
-28 \\
\hline
0
\end{array}
\qquad \frac{28}{7} = 4
$$

You Try It 5 $14\dfrac{5}{8} = \dfrac{112 + 5}{8} = \dfrac{117}{8}$

SECTION 2.3

You Try It 1 $45 \div 5 = 9 \qquad \dfrac{3}{5} = \dfrac{3 \cdot 9}{5 \cdot 9} = \dfrac{27}{45}$

$\dfrac{27}{45}$ is equivalent to $\dfrac{3}{5}$.

You Try It 2 Write 6 as $\dfrac{6}{1}$.

$18 \div 1 = 18 \quad 6 = \dfrac{6 \cdot 18}{1 \cdot 18} = \dfrac{108}{18}$

$\dfrac{108}{18}$ is equivalent to 6.

You Try It 3 $\dfrac{16}{24} = \dfrac{\cancel{2} \cdot \cancel{2} \cdot \cancel{2} \cdot 2}{\cancel{2} \cdot \cancel{2} \cdot \cancel{2} \cdot 3} = \dfrac{2}{3}$

You Try It 4 $\dfrac{8}{56} = \dfrac{\cancel{2} \cdot \cancel{2} \cdot \cancel{2}}{\cancel{2} \cdot \cancel{2} \cdot \cancel{2} \cdot 7} = \dfrac{1}{7}$

You Try It 5 $\dfrac{15}{32} = \dfrac{3 \cdot 5}{2 \cdot 2 \cdot 2 \cdot 2 \cdot 2} = \dfrac{15}{32}$

You Try It 6 $\dfrac{48}{36} = \dfrac{\cancel{2} \cdot \cancel{2} \cdot 2 \cdot 2 \cdot \cancel{3}}{\cancel{2} \cdot \cancel{2} \cdot \cancel{3} \cdot 3} = \dfrac{4}{3} = 1\dfrac{1}{3}$

SECTION 2.4

You Try It 1

$$\dfrac{3}{8}$$
$$+\dfrac{7}{8}$$

• The denominators are the same. Add the numerators. Place the sum over the common denominator.

$$\dfrac{10}{8} = \dfrac{5}{4} = 1\dfrac{1}{4}$$

You Try It 2

$$\dfrac{5}{12} = \dfrac{20}{48}$$
$$+\dfrac{9}{16} = \dfrac{27}{48}$$
$$\dfrac{47}{48}$$

• The LCM of 12 and 16 is 48.

You Try It 3

$$\dfrac{7}{8} = \dfrac{105}{120}$$
$$+\dfrac{11}{15} = \dfrac{88}{120}$$
$$\dfrac{193}{120} = 1\dfrac{73}{120}$$

• The LCM of 8 and 15 is 120.

You Try It 4

$$\dfrac{3}{4} = \dfrac{30}{40}$$
$$\dfrac{4}{5} = \dfrac{32}{40}$$
$$+\dfrac{5}{8} = \dfrac{25}{40}$$
$$\dfrac{87}{40} = 2\dfrac{7}{40}$$

• The LCM of 4, 5, and 8 is 40.

You Try It 5

$$7 + \dfrac{6}{11} = 7\dfrac{6}{11}$$

You Try It 6

$$29$$
$$+17\dfrac{5}{12}$$
$$46\dfrac{5}{12}$$

You Try It 7

$$7\dfrac{4}{5} = 7\dfrac{24}{30}$$
$$6\dfrac{7}{10} = 6\dfrac{21}{30}$$
$$+13\dfrac{11}{15} = 13\dfrac{22}{30}$$
$$26\dfrac{67}{30} = 28\dfrac{7}{30}$$

• LCM = 30

You Try It 8

$$9\dfrac{3}{8} = 9\dfrac{45}{120}$$
$$17\dfrac{7}{12} = 17\dfrac{70}{120}$$
$$+10\dfrac{14}{15} = 10\dfrac{112}{120}$$
$$36\dfrac{227}{120} = 37\dfrac{107}{120}$$

• LCM = 120

You Try It 9

Strategy To find the total time spent on the activities, add the three times $\left(4\dfrac{1}{2}, 3\dfrac{3}{4}, 1\dfrac{1}{3}\right)$.

Solution

$$4\dfrac{1}{2} = 4\dfrac{6}{12}$$
$$3\dfrac{3}{4} = 3\dfrac{9}{12}$$
$$+1\dfrac{1}{3} = 1\dfrac{4}{12}$$
$$8\dfrac{19}{12} = 9\dfrac{7}{12}$$

The total time spent on the three activities was $9\dfrac{7}{12}$ hours.

You Try It 10

Strategy To find the overtime pay:
- Find the total number of overtime ours $\left(1\dfrac{2}{3} + 3\dfrac{1}{3} + 2\right)$.
- Multiply the total number of hours by the overtime hourly wage (36).

Solution

$$1\dfrac{2}{3}$$
$$3\dfrac{1}{3}$$
$$+2$$

$$\begin{array}{r} 36 \\ \times\ 7 \\ \hline 252 \end{array}$$

$$6\dfrac{3}{3} = 7 \text{ hours}$$

Jeff earned $252 in overtime pay.

SECTION 2.5

You Try It 1

$$\dfrac{16}{27}$$
$$-\dfrac{7}{27}$$
$$\dfrac{9}{27} = \dfrac{1}{3}$$

• The denominators are the same. Subtract the numerators. Place the difference over the common denominator.

You Try It 2

$$\dfrac{13}{18} = \dfrac{52}{72}$$
$$-\dfrac{7}{24} = \dfrac{21}{72}$$
$$\dfrac{31}{72}$$

• LCM = 72

You Try It 3

$$17\frac{5}{9} = 17\frac{20}{36} \qquad \text{• LCM} = 36$$
$$-11\frac{5}{12} = 11\frac{15}{36}$$
$$\rule{3cm}{0.4pt}$$
$$6\frac{5}{36}$$

You Try It 4

$$8 = 7\frac{13}{13} \qquad \text{• LCM} = 13$$
$$-2\frac{4}{13} = 2\frac{4}{13}$$
$$\rule{3cm}{0.4pt}$$
$$5\frac{9}{13}$$

You Try It 5

$$21\frac{7}{9} = 21\frac{28}{36} = 20\frac{64}{36} \quad \text{• LCM} = 36$$
$$-7\frac{11}{12} = 7\frac{33}{36} = 7\frac{33}{36}$$
$$\rule{5cm}{0.4pt}$$
$$13\frac{31}{36}$$

You Try It 6

Strategy   To find the time remaining before the plane lands, subtract the number of hours already in the air $\left(2\frac{3}{4}\right)$ from the total time of the trip $\left(5\frac{1}{2}\right)$.

Solution

$$5\frac{1}{2} = 5\frac{2}{4} = 4\frac{6}{4}$$
$$-2\frac{3}{4} = 2\frac{3}{4} = 2\frac{3}{4}$$
$$\rule{4cm}{0.4pt}$$
$$2\frac{3}{4} \text{ hours}$$

The plane will land in $2\frac{3}{4}$ hours.

You Try It 7

Strategy   To find the amount of weight to be lost during the third month:

- Find the total weight loss during the first two months $\left(7\frac{1}{2} + 5\frac{3}{4}\right)$.

- Subtract the total weight loss from the goal (24 pounds).

Solution

$$7\frac{1}{2} = 7\frac{2}{4}$$
$$+5\frac{3}{4} = 5\frac{3}{4}$$
$$\rule{3cm}{0.4pt}$$
$$12\frac{5}{4} = 13\frac{1}{4} \text{ pounds lost}$$

$$24 = 23\frac{4}{4}$$
$$-13\frac{1}{4} = 13\frac{1}{4}$$
$$\rule{3cm}{0.4pt}$$
$$10\frac{3}{4} \text{ pounds}$$

The patient must lose $10\frac{3}{4}$ pounds to achieve the goal.

SECTION 2.6

You Try It 1

$$\frac{4}{21} \times \frac{7}{44} = \frac{4 \cdot 7}{21 \cdot 44}$$
$$= \frac{\overset{1}{\cancel{2}} \cdot 2 \cdot \overset{1}{\cancel{7}}}{3 \cdot \underset{1}{\cancel{7}} \cdot \underset{1}{\cancel{2}} \cdot 2 \cdot 11} = \frac{1}{33}$$

You Try It 2

$$\frac{2}{21} \times \frac{10}{33} = \frac{2 \cdot 10}{21 \cdot 33}$$
$$= \frac{2 \cdot 2 \cdot 5}{3 \cdot 7 \cdot 3 \cdot 11} = \frac{20}{693}$$

You Try It 3

$$\frac{16}{5} \times \frac{15}{24} = \frac{16 \cdot 15}{5 \cdot 24}$$
$$= \frac{\overset{1}{\cancel{2}} \cdot \overset{1}{\cancel{2}} \cdot \overset{1}{\cancel{2}} \cdot 2 \cdot \overset{1}{\cancel{3}} \cdot \overset{1}{\cancel{5}}}{\underset{1}{\cancel{5}} \cdot \underset{1}{\cancel{2}} \cdot \underset{1}{\cancel{2}} \cdot \underset{1}{\cancel{2}} \cdot \underset{1}{\cancel{3}}} = \frac{2}{1} = 2$$

You Try It 4

$$5\frac{2}{5} \times \frac{5}{9} = \frac{27}{5} \times \frac{5}{9} = \frac{27 \cdot 5}{5 \cdot 9}$$
$$= \frac{\overset{1}{\cancel{3}} \cdot \overset{1}{\cancel{3}} \cdot 3 \cdot \overset{1}{\cancel{5}}}{\underset{1}{\cancel{5}} \cdot \underset{1}{\cancel{3}} \cdot \underset{1}{\cancel{3}}} = \frac{3}{1} = 3$$

You Try It 5

$$3\frac{2}{5} \times 6\frac{1}{4} = \frac{17}{5} \times \frac{25}{4} = \frac{17 \cdot 25}{5 \cdot 4}$$
$$= \frac{17 \cdot \overset{1}{\cancel{5}} \cdot 5}{\underset{1}{\cancel{5}} \cdot 2 \cdot 2} = \frac{85}{4} = 21\frac{1}{4}$$

You Try It 6

$$3\frac{2}{7} \times 6 = \frac{23}{7} \times \frac{6}{1} = \frac{23 \cdot 6}{7 \cdot 1}$$
$$= \frac{23 \cdot 2 \cdot 3}{7 \cdot 1} = \frac{138}{7} = 19\frac{5}{7}$$

You Try It 7

Strategy   To find the value of the house today, multiply the old value of the house (170,000) by $2\frac{1}{2}$.

Solution

$$170,000 \times 2\frac{1}{2} = \frac{170,000}{1} \times \frac{5}{2}$$
$$= \frac{170,000 \cdot 5}{1 \cdot 2}$$
$$= 425,000$$

The value of the house today is $425,000.

You Try It 8

Strategy   To find the cost of the air compressor:

- Multiply to find the value of the drying chamber $\left(\dfrac{4}{5} \times 160{,}000\right)$.

- Subtract the value of the drying chamber from the total value of the two items (160,000).

Solution $\dfrac{4}{5} \times \dfrac{160{,}000}{1} = \dfrac{640{,}000}{5}$

$= 128{,}000$ • Value of the drying chamber

$160{,}000 - 128{,}000 = 32{,}000$

The cost of the air compressor was $32,000.

SECTION 2.7

You Try It 1 $\dfrac{3}{7} \div \dfrac{2}{3} = \dfrac{3}{7} \times \dfrac{3}{2} = \dfrac{3 \cdot 3}{7 \cdot 2} = \dfrac{9}{14}$

You Try It 2 $\dfrac{3}{4} \div \dfrac{9}{10} = \dfrac{3}{4} \times \dfrac{10}{9}$

$= \dfrac{3 \cdot 10}{4 \cdot 9} = \dfrac{\overset{1}{\cancel{3}} \cdot \overset{1}{\cancel{2}} \cdot 5}{\cancel{2} \cdot 2 \cdot \cancel{3} \cdot 3} = \dfrac{5}{6}$

You Try It 3 $\dfrac{5}{7} \div 6 = \dfrac{5}{7} \div \dfrac{6}{1}$ • $6 = \dfrac{6}{1}$. The reciprocal of $\dfrac{6}{1}$ is $\dfrac{1}{6}$.

$= \dfrac{5}{7} \times \dfrac{1}{6} = \dfrac{5 \cdot 1}{7 \cdot 6}$

$= \dfrac{5}{7 \cdot 2 \cdot 3} = \dfrac{5}{42}$

You Try It 4 $12\dfrac{3}{5} \div 7 = \dfrac{63}{5} \div \dfrac{7}{1} = \dfrac{63}{5} \times \dfrac{1}{7}$

$= \dfrac{63 \cdot 1}{5 \cdot 7} = \dfrac{3 \cdot 3 \cdot \overset{1}{\cancel{7}}}{5 \cdot \cancel{7}} = \dfrac{9}{5} = 1\dfrac{4}{5}$

You Try It 5 $3\dfrac{2}{3} \div 2\dfrac{2}{5} = \dfrac{11}{3} \div \dfrac{12}{5}$

$= \dfrac{11}{3} \times \dfrac{5}{12} = \dfrac{11 \cdot 5}{3 \cdot 12}$

$= \dfrac{11 \cdot 5}{3 \cdot 2 \cdot 2 \cdot 3} = \dfrac{55}{36} = 1\dfrac{19}{36}$

You Try it 6 $2\dfrac{5}{6} \div 8\dfrac{1}{2} = \dfrac{17}{6} \div \dfrac{17}{2}$

$= \dfrac{17}{6} \times \dfrac{2}{17} = \dfrac{17 \cdot 2}{6 \cdot 17}$

$= \dfrac{\overset{1}{\cancel{17}} \cdot \overset{1}{\cancel{2}}}{\underset{1}{\cancel{2}} \cdot 3 \cdot \underset{1}{\cancel{17}}} = \dfrac{1}{3}$

You Try It 7 $6\dfrac{2}{5} \div 4 = \dfrac{32}{5} \div \dfrac{4}{1}$

$= \dfrac{32}{5} \times \dfrac{1}{4} = \dfrac{32 \cdot 1}{5 \cdot 4}$

$= \dfrac{2 \cdot 2 \cdot 2 \cdot \overset{1}{\cancel{2}} \cdot \overset{1}{\cancel{2}}}{5 \cdot \underset{1}{\cancel{2}} \cdot \underset{1}{\cancel{2}}} = \dfrac{8}{5} = 1\dfrac{3}{5}$

You Try It 8

Strategy   To find the number of products, divide the number of minutes in 1 hour (60) by the time to assemble one product $\left(7\dfrac{1}{2}\right)$.

Solution $60 \div 7\dfrac{1}{2} = \dfrac{60}{1} \div \dfrac{15}{2} = \dfrac{60}{1} \cdot \dfrac{2}{15}$

$= \dfrac{60 \cdot 2}{1 \cdot 15} = 8$

The factory worker can assemble 8 products in 1 hour.

You Try It 9

Strategy   To find the length of the remaining piece:

- Divide the total length of the board (16) by the length of each shelf $\left(3\dfrac{1}{3}\right)$. This will give you the number of shelves cut, with a certain fraction of a shelf left over.
- Multiply the fractional part of the result in step 1 by the length of one shelf to determine the length of the remaining piece.

Solution $16 \div 3\dfrac{1}{3} = \dfrac{16}{1} \div \dfrac{10}{3}$

$= \dfrac{16}{1} \times \dfrac{3}{10} = \dfrac{16 \cdot 3}{1 \cdot 10}$

$= \dfrac{\overset{1}{\cancel{2}} \cdot 2 \cdot 2 \cdot 2 \cdot 3}{\underset{1}{\cancel{2}} \cdot 5} = \dfrac{24}{5}$

$= 4\dfrac{4}{5}$

There are 4 pieces that are each $3\dfrac{1}{3}$ feet long. There is 1 piece that is $\dfrac{4}{5}$ of $3\dfrac{1}{3}$ feet long.

$\dfrac{4}{5} \times 3\dfrac{1}{3} = \dfrac{4}{5} \times \dfrac{10}{3}$

$= \dfrac{4 \cdot 10}{5 \cdot 3} = \dfrac{2 \cdot 2 \cdot 2 \cdot \overset{1}{\cancel{5}}}{\underset{1}{\cancel{5}} \cdot 3}$

$= \dfrac{8}{3} = 2\dfrac{2}{3}$

The length of the piece remaining is $2\dfrac{2}{3}$ feet.

SECTION 2.8

You Try It 1 $\dfrac{9}{14} = \dfrac{27}{42}$ $\dfrac{13}{21} = \dfrac{26}{42}$ $\dfrac{9}{14} > \dfrac{13}{21}$

You Try It 2 $\left(\dfrac{7}{11}\right)^2 \cdot \left(\dfrac{2}{7}\right) = \left(\dfrac{7}{11} \cdot \dfrac{7}{11}\right) \cdot \left(\dfrac{2}{7}\right)$

$$= \dfrac{\overset{1}{\cancel{7}} \cdot 7 \cdot 2}{11 \cdot 11 \cdot \underset{1}{\cancel{7}}} = \dfrac{14}{121}$$

You Try It 3 $\left(\dfrac{1}{13}\right)^2 \cdot \left(\dfrac{1}{4} + \dfrac{1}{6}\right) \div \dfrac{5}{13} =$

$$\left(\dfrac{1}{13}\right)^2 \cdot \left(\dfrac{5}{12}\right) \div \dfrac{5}{13} =$$

$$\left(\dfrac{1}{169}\right) \cdot \left(\dfrac{5}{12}\right) \div \dfrac{5}{13} =$$

$$\left(\dfrac{1 \cdot 5}{13 \cdot 13 \cdot 12}\right) \div \dfrac{5}{13} =$$

$$\left(\dfrac{1 \cdot 5}{13 \cdot 13 \cdot 12}\right) \times \dfrac{13}{5} =$$

$$\dfrac{1 \cdot \overset{1}{\cancel{5}} \cdot \overset{1}{\cancel{13}}}{\underset{1}{\cancel{13}} \cdot 13 \cdot 12 \cdot \underset{1}{\cancel{5}}} = \dfrac{1}{156}$$

SOLUTIONS TO CHAPTER 3 "YOU TRY IT"

SECTION 3.1

You Try It 1   The digit 4 is in the thousandths place.

You Try It 2 $\dfrac{501}{1000} = 0.501$
(five hundred one thousandths)

You Try It 3 $0.67 = \dfrac{67}{100}$ (sixty-seven hundredths)

You Try It 4   Fifty-five and six thousand eighty-three ten-thousandths

You Try It 5   806.00491   • 1 is in the hundred-thousandths place.

You Try It 6   ┌─── Given place value
3.675849
       └── 4 < 5

3.675849 rounded to the nearest ten-thousandth is 3.6758.

You Try It 7   ┌─── Given place value
48.907
      └── 0 < 5

48.907 rounded to the nearest tenth is 48.9.

You Try It 8   ┌─── Given place value
31.8652
      └── 8 > 5

31.8652 rounded to the nearest whole number is 32.

You Try It 9   2.65 rounded to the nearest whole number is 3.

To the nearest inch, the average annual precipitation in Yuma is 3 inches.

SECTION 3.2

You Try It 1

$$\begin{array}{r} \overset{1\,2}{4.62} \\ 27.9 \\ +\ \ 0.62054 \\ \hline 33.14054 \end{array}$$

• Place the decimal points on a vertical line.

You Try It 2

$$\begin{array}{r} \overset{1}{6.05} \\ 12. \\ +\ \ 0.374 \\ \hline 18.424 \end{array}$$

You Try It 3

Strategy   To determine the number, add the numbers of hearing-impaired Americans ages 45 to 54, 55 to 64, 65 to 74, and 75 and over.

Solution

$$\begin{array}{r} 4.48 \\ 4.31 \\ 5.41 \\ +3.80 \\ \hline 18.00 \end{array}$$

18 million Americans ages 45 and older are hearing-impaired.

You Try It 4

Strategy   To find the total income, add the four commissions (985.80, 791.46, 829.75, and 635.42) to the salary (875).

Solution $875 + 985.80 + 791.46 + 829.75$
$+ 635.42 = 4117.43$

Anita's total income was $4117.43.

SECTION 3.3

You Try It 1

$$\begin{array}{r} \overset{11\ \ 9}{\overset{6\ \ \cancel{7}\ \cancel{10}13}{7\,2.0\,3\,9}} \\ -\ \ \ \ 8.4\,7 \\ \hline 6\,3.5\,6\,9 \end{array}$$

Check:
$$\begin{array}{r} \overset{1\ \ 11}{8.47} \\ +\ 63.569 \\ \hline 72.039 \end{array}$$

You Try It 2

$$\begin{array}{r} {\scriptstyle 14\ 9} \\ {\scriptstyle 2\ \ 4\ \cancel{1}0\ 10} \\ \cancel{3}\ \cancel{3}.\cancel{0}\ \cancel{0} \\ -\quad 9.6\ 7 \\ \hline 2\ 5.3\ 3 \end{array}$$

Check:
$$\begin{array}{r} {\scriptstyle 1\ 1\ \ 1} \\ 9.67 \\ +\ 25.33 \\ \hline 35.00 \end{array}$$

You Try It 3

$$\begin{array}{r} {\scriptstyle 16} \\ {\scriptstyle 2\ \ \cancel{8}\ 9\ 9\ 10} \\ \cancel{3}.\cancel{7}\ \cancel{0}\ \cancel{0}\ \cancel{0} \\ -\ 1.9\ 7\ 1\ 5 \\ \hline 1.7\ 2\ 8\ 5 \end{array}$$

Check:
$$\begin{array}{r} {\scriptstyle 1\ 1\ 1\ 1} \\ 1.9715 \\ +\ 1.7285 \\ \hline 3.7000 \end{array}$$

You Try It 4

Strategy To find the amount of change, subtract the amount paid (6.85) from 10.00.

Solution
$$\begin{array}{r} 10.00 \\ -\ 6.85 \\ \hline 3.15 \end{array}$$

Your change was $3.15.

You Try It 5

Strategy To find the new balance:

- Add to find the total of the three checks (1025.60 + 79.85 + 162.47).
- Subtract the total from the previous balance (2472.69).

Solution
$$\begin{array}{r} 1025.60 \\ 79.85 \\ +\ 162.47 \\ \hline 1267.92 \end{array}$$
$$\begin{array}{r} 2472.69 \\ -\ 1267.92 \\ \hline 1204.77 \end{array}$$

The new balance is $1204.77.

SECTION 3.4

You Try It 1
$$\begin{array}{r} 870 \\ \times\quad 4.6 \\ \hline 522\ 0 \\ 3480 \\ \hline 4002.0 \end{array}$$
• 1 decimal place

• 1 decimal place

You Try It 2
$$\begin{array}{r} 0.000086 \\ \times\quad 0.057 \\ \hline 602 \\ 430 \\ \hline 0.000004902 \end{array}$$
• 6 decimal places
• 3 decimal places

• 9 decimal places

You Try It 3
$$\begin{array}{r} 4.68 \\ \times\ 6.03 \\ \hline 1404 \\ 28\ 080 \\ \hline 28.2204 \end{array}$$
• 2 decimal places
• 2 decimal places

• 4 decimal places

You Try It 4 $6.9 \times 1000 = 6900$

You Try It 5 $4.0273 \times 10^2 = 402.73$

You Try It 6

Strategy To find the total bill:

- Find the number of gallons of water used by multiplying the number of gallons used per day (5000) by the number of days (62).
- Find the cost of water by multiplying the cost per 1000 gallons (1.39) by the number of 1000-gallon units used.
- Add the cost of the water to the meter fee (133.70).

Solution Number of gallons = 5000(62) = 310,000

$$\text{Cost of water} = \frac{310{,}000}{1000} \times 1.39 = 430.90$$

Total cost = 430.90 + 133.70 = 564.60

The total bill is $564.60.

You Try It 7

Strategy To find the cost of running the freezer for 210 hours, multiply the hourly cost (0.035) by the number of hours the freezer has run (210).

Solution
$$\begin{array}{r} 0.035 \\ \times\quad 210 \\ \hline 350 \\ 7\ 0 \\ \hline 7.350 \end{array}$$

The cost of running the freezer for 210 hours is $7.35.

You Try It 8

Strategy To find the total cost of the electronic drum kit:

- Multiply the monthly payment (37.18) by the number of months (18).
- Add that product to the down payment (175.00).

Solution
$$\begin{array}{r} 37.18 \\ \times\quad 18 \\ \hline 29744 \\ 3718 \\ \hline 669.24 \end{array}$$
$$\begin{array}{r} 175.00 \\ +\ 669.24 \\ \hline 844.24 \end{array}$$

The total cost of the electronic drum kit is $844.24.

SECTION 3.5

You Try It 1
$$\begin{array}{r} 2.7 \\ 0.052\overline{)0.140.4} \\ -104 \\ \hline 36\ 4 \\ -36\ 4 \\ \hline 0 \end{array}$$

• Move the decimal point 3 places to the right in the divisor and the dividend. Write the decimal point in the quotient directly over the decimal point in the dividend.

You Try It 2

$$\begin{array}{r} 0.4873 \approx 0.487 \\ 76\overline{)37.0420} \\ -30\ 4\ \ \ \ \ \\ \overline{6\ 64\ \ \ \ } \\ -6\ 08\ \ \ \ \\ \overline{562\ \ } \\ -532\ \ \\ \overline{300} \\ -228 \\ \overline{72} \end{array}$$

• Write the decimal point in the quotient directly over the decimal point in the dividend.

You Try It 3

$$\begin{array}{r} 72.73 \approx 72.7 \\ 5.09.\overline{)370.20.00} \\ -356\ 3 \\ \overline{13\ 90} \\ 10\ 18 \\ \overline{3\ 720} \\ -3\ 563 \\ \overline{1570} \\ -1527 \\ \overline{43} \end{array}$$

You Try It 4 $309.21 \div 10,000 = 0.030921$

You Try It 5 $42.93 \div 10^4 = 0.004293$

You Try It 6

Strategy   To find how many times greater the average hourly earnings were, divide the 1998 average hourly earnings (12.88) by the 1978 average hourly earnings (5.70).

Solution $12.88 \div 5.70 \approx 2.3$
The average hourly earnings in 1998 were about 2.3 times greater than in 1978.

You Try It 7

Strategy   To find the average number of people watching TV:

• Add the numbers of people watching each day of the week.
• Divide the total number of people watching by 7.

Solution $91.9 + 89.8 + 90.6 + 93.9 + 78.0$
$ + 77.1 + 87.7 = 609$

$$\frac{609}{7} = 87$$

An average of 87 million people watch television per day.

SECTION 3.6

You Try It 1

$$\begin{array}{r} 0.56 \approx 0.6 \\ 16\overline{)9.00} \end{array}$$

You Try It 2 $4\frac{1}{6} = \frac{25}{6}$

$$\begin{array}{r} 4.166 \approx 4.17 \\ 6\overline{)25.000} \end{array}$$

You Try It 3 $0.56 = \dfrac{56}{100} = \dfrac{14}{25}$

$5.35 = 5\dfrac{35}{100} = 5\dfrac{7}{20}$

You Try It 4 $0.12\dfrac{7}{8} = \dfrac{12\frac{7}{8}}{100} = 12\dfrac{7}{8} \div 100$

$$= \dfrac{103}{8} \times \dfrac{1}{100} = \dfrac{103}{800}$$

You Try It 5 $\dfrac{5}{8} = 0.625$

$0.630 > 0.625$

$0.63 > \dfrac{5}{8}$

• Convert the fraction $\frac{5}{8}$ to a decimal.
• Compare the two decimals.
• Convert 0.625 back to a fraction.

SOLUTIONS TO CHAPTER 4 "YOU TRY IT"

SECTION 4.1

You Try It 1 $\dfrac{20 \text{ pounds}}{24 \text{ pounds}} = \dfrac{20}{24} = \dfrac{5}{6}$

20 pounds:24 pounds = 20:24 = 5:6

20 pounds to 24 pounds = 20 to 24
$\phantom{20 \text{ pounds to 24 pounds}} = 5 \text{ to } 6$

You Try It 2 $\dfrac{64 \text{ miles}}{8 \text{ miles}} = \dfrac{64}{8} = \dfrac{8}{1}$

64 miles:8 miles = 64:8 = 8:1

64 miles to 8 miles = 64 to 8 = 8 to 1

You Try It 3

Strategy   To find the ratio, write the ratio of board feet of cedar (12,000) to board feet of ash (18,000) in simplest form.

Solution $\dfrac{12,000}{18,000} = \dfrac{2}{3}$
The ratio is $\frac{2}{3}$.

You Try It 4

Strategy   To find the ratio, write the ratio of the amount spent on radio advertising (450,000) to the amount spent on radio and television advertising (450,000 + 600,000) in simplest form.

Solution $\dfrac{\$450,000}{\$450,000 + \$600,000} = \dfrac{450,000}{1,050,000} = \dfrac{3}{7}$
The ratio is $\frac{3}{7}$.

SECTION 4.2

You Try It 1
$$\frac{15 \text{ pounds}}{12 \text{ trees}} = \frac{5 \text{ pounds}}{4 \text{ trees}}$$

You Try It 2
$$\frac{260 \text{ miles}}{8 \text{ hours}}$$

$$\begin{array}{r} 32.5 \\ 8)\overline{260.0} \end{array}$$

32.5 miles/hour

You Try It 3

Strategy   To find Erik's profit per ounce:
- Find the total profit by subtracting the cost ($1625) from the selling price ($1720).
- Divide the total profit by the number of ounces (5).

Solution $1720 - 1625 = 95$
$95 \div 5 = 19$

The profit was $19/ounce.

SECTION 4.3

You Try It 1
$$\frac{6}{10} \diagup\!\!\!\diagdown \frac{9}{15} \quad \begin{array}{l} 10 \times 9 = 90 \\ 6 \times 15 = 90 \end{array}$$

The cross products are equal. The proportion is true.

You Try It 2
$$\frac{32}{6} \diagup\!\!\!\diagdown \frac{90}{8} \quad \begin{array}{l} 6 \times 90 = 540 \\ 32 \times 8 = 256 \end{array}$$

The cross products are not equal. The proportion is not true.

You Try It 3
$$\frac{n}{14} = \frac{3}{7}$$
• Find the cross products. Then solve for n.
$$n \times 7 = 14 \times 3$$
$$n \times 7 = 42$$
$$n = 42 \div 7$$
$$n = 6$$

Check: $\dfrac{6}{14} \diagup\!\!\!\diagdown \dfrac{3}{7} \quad \begin{array}{l} 14 \times 3 = 42 \\ 6 \times 7 = 42 \end{array}$

You Try It 4
$$\frac{5}{7} = \frac{n}{20}$$
• Find the cross products. Then solve for n.
$$5 \times 20 = 7 \times n$$
$$100 = 7 \times n$$
$$100 \div 7 = n$$
$$14.3 \approx n$$

You Try It 5
$$\frac{15}{20} = \frac{12}{n}$$
• Find the cross products. Then solve for n.
$$15 \times n = 20 \times 12$$
$$15 \times n = 240$$
$$n = 240 \div 15$$
$$n = 16$$

Check: $\dfrac{15}{20} \diagup\!\!\!\diagdown \dfrac{12}{16} \quad \begin{array}{l} 20 \times 12 = 240 \\ 15 \times 16 = 240 \end{array}$

You Try It 6
$$7.5 = n$$
$$12 \times 4 = n \times 7$$
$$48 = n \times 7$$
$$48 \div 7 = n$$
$$6.86 \approx n$$

You Try It 7
$$\frac{n}{12} = \frac{4}{1}$$
$$n \times 1 = 12 \times 4$$
$$n \times 1 = 48$$
$$n = 48 \div 1$$
$$n = 48$$

Check: $\dfrac{48}{12} \diagup\!\!\!\diagdown \dfrac{4}{1} \quad \begin{array}{l} 12 \times 4 = 48 \\ 48 \times 1 = 48 \end{array}$

You Try It 8

Strategy   To find the number of tablespoons of fertilizer needed, write and solve a proportion using n to represent the number of tablespoons of fertilizer.

Solution
$$\frac{3 \text{ tablespoons}}{4 \text{ gallons}} = \frac{n \text{ tablespoons}}{10 \text{ gallons}}$$
• The unit "tablespoons" is in the numerator. The unit "gallons" is in the denominator.
$$3 \times 10 = 4 \times n$$
$$30 = 4 \times n$$
$$30 \div 4 = n$$
$$7.5 = n$$

For 10 gallons of water, 7.5 tablespoons of fertilizer are required.

You Try It 9

Strategy   To find the number of jars that can be packed in 15 boxes, write and solve a proportion using n to represent the number of jars.

Solution
$$\frac{24 \text{ jars}}{6 \text{ boxes}} = \frac{n \text{ jars}}{15 \text{ boxes}}$$
$$24 \times 15 = 6 \times n$$
$$360 = 6 \times n$$
$$360 \div 6 = n$$
$$60 = n$$

60 jars can be packed in 15 boxes.

SOLUTIONS TO CHAPTER 5 "YOU TRY IT"

SECTION 5.1

You Try It 1
a. $125\% = 125 \times \dfrac{1}{100} = \dfrac{125}{100} = 1\dfrac{1}{4}$

b. $125\% = 125 \times 0.01 = 1.25$

You Try It 2 $33\frac{1}{3}\% = 33\frac{1}{3} \times \frac{1}{100}$

$= \frac{100}{3} \times \frac{1}{100}$

$= \frac{100}{300} = \frac{1}{3}$

You Try It 3 $0.25\% = 0.25 \times 0.01 = 0.0025$

You Try It 4 $0.048 = 0.048 \times 100\% = 4.8\%$

$3.67 = 3.67 \times 100\% = 367\%$

$0.62\frac{1}{2} = 0.62\frac{1}{2} \times 100\%$

$= 62\frac{1}{2}\%$

You Try It 5 $\frac{5}{6} = \frac{5}{6} \times 100\% = \frac{500}{6}\% = 83\frac{1}{3}\%$

You Try It 6 $1\frac{4}{9} = \frac{13}{9} = \frac{13}{9} \times 100\%$

$= \frac{1300}{9}\% \approx 144.4\%$

SECTION 5.2

You Try It 1   Percent × base = amount

$0.063 \times 150 = n$

$9.45 = n$

You Try It 2   Percent × base = amount

$\frac{1}{6} \times 66 = n$ $\cdot \; 16\frac{2}{3}\% = \frac{1}{6}$

$11 = n$

You Try It 3

Strategy   To determine the amount that came from corporations, write and solve the basic percent equation using n to represent the amount. The percent is 4%. The base is $212 billion.

Solution   Percent × base = amount

$4\% \times 212 = n$

$0.04 \times 212 = n$

$8.48 = n$

Corporations gave $8.48 billion to charities.

You Try It 4

Strategy   To find the new hourly wage:

- Find the amount of the raise. Write and solve the basic percent equation using n to represent the amount of the raise (amount). The percent is 8%. The base is $33.50.
- Add the amount of the raise to the old wage (33.50).

Solution $8\% \times 33.50 = n$ $\quad\quad 33.50$

$0.08 \times 33.50 = n$ $\quad + \; 2.68$

$2.68 = n$ $\quad\quad\quad 36.18$

The new hourly wage is $36.18.

SECTION 5.3

You Try It 1   Percent × base = amount

$n \times 32 = 16$

$n = 16 \div 32$

$n = 0.50$

$n = 50\%$

You Try It 2   Percent × base = amount

$n \times 15 = 48$

$n = 48 \div 15$

$n = 3.2$

$n = 320\%$

You Try It 3   Percent × base = amount

$n \times 45 = 30$

$n = 30 \div 45$

$n = \frac{2}{3} = 66\frac{2}{3}\%$

You Try It 4

Strategy   To find what percent of the income the income tax is, write and solve the basic percent equation using n to represent the percent. The base is $33,500, and the amount is $5025.

Solution $n \times 33,500 = 5025$

$n = 5025 \div 33,500$

$n = 0.15 = 15\%$

The income tax is 15% of the income.

You Try It 5

Strategy   To find the percent who were women:

- Subtract to find the number of enlisted personnel who were women (518,921 − 512,370).
- Write and solve the basic percent equation using n to represent the percent. The base is 518,921, and the amount is the number of enlisted personnel who were women.

Solution $518,921 - 512,370 = 6551$

$n \times 518,921 = 6551$

$n = 6551 \div 518,921$

$n \approx 0.013 = 1.3\%$

In 1950, 1.3% of the enlisted personnel in the U.S. Army were women.

SECTION 5.4

You Try It 1

$$\text{Percent} \times \text{base} = \text{amount}$$
$$0.86 \times n = 215$$
$$n = 215 \div 0.86$$
$$n = 250$$

You Try It 2

$$\text{Percent} \times \text{base} = \text{amount}$$
$$0.025 \times n = 15$$
$$n = 15 \div 0.025$$
$$n = 600$$

You Try It 3

$$\text{Percent} \times \text{base} = \text{amount}$$
$$\frac{1}{6} \times n = 5 \qquad \cdot 16\frac{2}{3}\% = \frac{1}{6}$$
$$n = 5 \div \frac{1}{6}$$
$$n = 30$$

You Try It 4

Strategy   To find the original value of the car, write and solve the basic percent equation using n to represent the original value (base). The percent is 42%, and the amount is $10,458.

Solution
$$42\% \times n = 10{,}458$$
$$0.42 \times n = 10{,}458$$
$$n = 10{,}458 \div 0.42$$
$$n = 24{,}900$$

The original value of the car was $24,900.

You Try It 5

Strategy   To find the difference between the original price and the sale price:

- Find the original price. Write and solve the basic percent equation using n to represent the original price (base). The percent is 80%, and the amount is $89.60.
- Subtract the sale price (89.60) from the original price.

Solution
$$80\% \times n = 89.60$$
$$0.80 \times n = 89.60$$
$$n = 89.60 \div 0.80$$
$$n = 112.00 \quad \text{(original price)}$$

$$112.00 - 89.60 = 22.40$$

The difference between the original price and the sale price is $22.40.

SECTION 5.5

You Try It 1

$$\frac{26}{100} = \frac{22}{n}$$
$$26 \times n = 100 \times 22$$
$$26 \times n = 2200$$
$$n = 2200 \div 26$$
$$n \approx 84.62$$

You Try It 2

$$\frac{16}{100} = \frac{n}{132}$$
$$16 \times 132 = 100 \times n$$
$$2112 = 100 \times n$$
$$2112 \div 100 = n$$
$$21.12 = n$$

You Try It 3

Strategy   To find the number of days it snowed, write and solve a proportion using n to represent the number of days it snowed (amount). The percent is 64%, and the base is 150.

Solution
$$\frac{64}{100} = \frac{n}{150}$$
$$64 \times 150 = 100 \times n$$
$$9600 = 100 \times n$$
$$9600 \div 100 = n$$
$$96 = n$$

It snowed 96 days.

You Try It 4

Strategy   To find the percent of alarms that were not false alarms:

- Subtract to find the number of alarms that were not false alarms $(200 - 24)$.
- Write and solve a proportion using n to represent the percent of alarms that were not false. The base is 200, and the amount is the number of alarms that were not false.

Solution
$$200 - 24 = 176$$

$$\frac{n}{100} = \frac{176}{200}$$
$$n \times 200 = 100 \times 176$$
$$n \times 200 = 17{,}600$$
$$n = 17{,}600 \div 200$$
$$n = 88$$

88% of the alarms were not false alarms.

SOLUTIONS TO CHAPTER 6 "YOU TRY IT"

SECTION 6.1

You Try It 1

Strategy   To find the unit cost, divide the total cost by the number of units.

Solution   **a.** $7.67 \div 8 = 0.95875$
$.959 per battery
b. $2.29 \div 15 \approx 0.153$
$.153 per ounce

You Try It 2

Strategy To find the more economical purchase, compare the unit costs.

Solution
$8.70 \div 6 = 1.45$
$6.96 \div 4 = 1.74$
$\$1.45 < \$1.74$

The more economical purchase is 6 cans for $8.70.

You Try It 3

Strategy To find the total cost, multiply the unit cost (9.96) by the number of units (7).

Solution

Unit cost	×	number of units	=	total cost
9.96	×	7	=	69.72

The total cost is $69.72.

SECTION 6.2

You Try It 1

Strategy To find the percent increase:

- Find the amount of the increase.
- Solve the basic percent equation for *percent*.

Solution

New value	−	original value	=	amount of increase
3.83	−	3.46	=	0.37

Percent × base = amount
$n \times 3.46 = 0.37$
$n = 0.37 \div 3.46$
$n \approx 0.11 = 11\%$

The percent increase was 11%.

You Try It 2

Strategy To find the new hourly wage:

- Solve the basic percent equation for *amount*.
- Add the amount of the increase to the original wage.

Solution
Percent × base = amount
$0.14 \times 12.50 = n$
$1.75 = n$

$12.50 + 1.75 = 14.25$

The new hourly wage is $14.25.

You Try It 3

Strategy To find the markup, solve the basic percent equation for *amount*.

Solution
Percent × base = amount

Markup rate	×	cost	=	markup
0.20	×	32	=	n
				6.4 = n

The markup is $6.40.

You Try It 4

Strategy To find the selling price:

- Find the markup by solving the basic percent equation for *amount*.
- Add the markup to the cost.

Solution
Percent × base = amount

Markup rate	×	cost	=	markup
0.55	×	72	=	n
				39.60 = n

Cost	+	markup	=	selling price
72	+	39.60	=	111.60

The selling price is $111.60.

You Try It 5

Strategy To find the percent decrease:

- Find the amount of the decrease.
- Solve the basic percent equation for *percent*.

Solution

Original value	−	new value	=	amount of decrease
261,000	−	215,000	=	46,000

Percent × base = amount
$n \times 261,000 = 46,000$
$n = 46,000 \div 261,000$
$n \approx 0.176$

The percent decrease is 17.6%.

You Try It 6

Strategy To find the visibility:

- Find the amount of decrease by solving the basic percent equation for *amount*.
- Subtract the amount of decrease from the original visibility.

Solution
Percent × base = amount
$0.40 \times 5 = n$
$2 = n$

$5 - 2 = 3$

The visibility was 3 miles.

You Try It 7

Strategy   To find the discount rate:

- Find the discount.
- Solve the basic percent equation for *percent*.

Solution

Regular price	−	sale price	=	discount

$$12.50 \quad - \quad 10.99 \quad = \quad 1.51$$

Percent	×	base	=	amount

Discount rate	×	regular price	=	discount

$$n \quad \times \quad 12.50 \quad = \quad 1.51$$
$$n = 1.51 \div 12.50$$
$$n = 0.1208$$

The discount rate is 12.1%.

You Try It 8

Strategy   To find the sale price:

- Find the discount by solving the basic percent equation for *amount*.
- Subtract to find the sale price.

Solution

Percent	×	base	=	amount

Discount rate	×	regular price	=	discount

$$0.15 \quad \times \quad 225 \quad = \quad n$$
$$33.75 = n$$

Regular price	−	discount	=	sale price

$$225 \quad - \quad 33.75 \quad = \quad 191.25$$

The sale price is $191.25.

SECTION 6.3

You Try It 1

Strategy   To find the simple interest due, multiply the principal (15,000) times the annual interest rate (8% = 0.08) times the time in years $\left(18 \text{ months} = \dfrac{18}{12} \text{ years} = 1.5 \text{ years}\right)$.

Solution

Principal	×	annual interest rate	×	time (in years)	=	interest

$$15,000 \quad \times \quad 0.08 \quad \times \quad 1.5 \quad = \quad 1800$$

The interest due is $1800.

You Try It 2

Strategy   To find the maturity value:

- Use the simple interest formula to find the simple interest due.
- Find the maturity value by adding the principal and the interest.

Solution

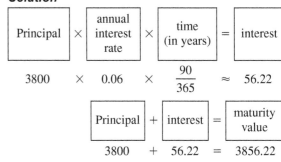

Principal	×	annual interest rate	×	time (in years)	=	interest

$$3800 \quad \times \quad 0.06 \quad \times \quad \dfrac{90}{365} \quad \approx \quad 56.22$$

Principal	+	interest	=	maturity value

$$3800 \quad + \quad 56.22 \quad = \quad 3856.22$$

The maturity value is $3856.22.

You Try It 3

Strategy   To find the monthly payment:

- Find the maturity value by adding the principal and the interest.
- Divide the maturity value by the length of the loan in months (12).

Solution   Principal + interest = maturity value

$$1900 \quad + \quad 152 \quad = \quad 2052$$

Maturity value ÷ length of the loan = payment

$$2052 \quad \div \quad 12 \quad = \quad 171$$

The monthly payment is $171.

You Try It 4

Strategy   To find the finance charge, multiply the principal, or unpaid balance (1250), times the monthly interest rate (1.6%) times the number of months (1).

Solution

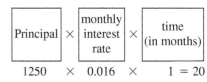

Principal	×	monthly interest rate	×	time (in months)

$$1250 \quad \times \quad 0.016 \quad \times \quad 1 = 20$$

The finance charge is $20.

You Try It 5

Strategy   To find the interest earned:

- Find the new principal by multiplying the original principal (1000) by the factor found in the Compound Interest Table (3.29066).
- Subtract the original principal from the new principal.

(Continued)

(Continued)

Solution $1000 \times 3.29066 = 3290.66$

The new principal is \$3290.66.

$3290.66 - 1000 = 2290.66$

The interest earned is \$2290.66.

SECTION 6.4

You Try It 1

Strategy To find the mortgage:

- Find the down payment by solving the basic percent equation for *amount*.
- Subtract the down payment from the purchase price.

Solution

Percent $\times$ base $=$ amount

$$\boxed{\text{Percent}} \times \boxed{\begin{array}{c}\text{purchase}\\\text{price}\end{array}} = \boxed{\begin{array}{c}\text{down}\\\text{payment}\end{array}}$$

$\quad 0.25 \quad \times \ 1{,}500{,}000 \ = \quad n$

$\qquad\qquad 375{,}000 = n$

$$\boxed{\begin{array}{c}\text{Purchase}\\\text{price}\end{array}} - \boxed{\begin{array}{c}\text{down}\\\text{payment}\end{array}} = \boxed{\text{mortgage}}$$

$1{,}500{,}000 \ - \ 375{,}000 \ = 1{,}125{,}000$

The mortgage is \$1,125,000.

You Try It 2

Strategy To find the loan origination fee, solve the basic percent equation for *amount*.

Solution

Percent $\times$ base $=$ amount

$$\boxed{\text{Points}} \times \boxed{\text{mortgage}} = \boxed{\text{fee}}$$

$\quad 0.045 \quad \times \quad 180{,}000 \quad = \quad n$

$\qquad\qquad\qquad 8100 = n$

The loan origination fee was \$8100.

You Try It 3

Strategy To find the monthly mortgage payment:

- Subtract the down payment from the purchase price to find the mortgage.
- Multiply the mortgage by the factor found in the Monthly Payment Table.

Solution

$$\boxed{\begin{array}{c}\text{Purchase}\\\text{price}\end{array}} - \boxed{\begin{array}{c}\text{down}\\\text{payment}\end{array}} = \boxed{\text{mortgage}}$$

$175{,}000 \ - \ 17{,}500 \ = 157{,}500$

$157{,}500 \quad \times 0.0089973 \approx 1417.08$

$\qquad\qquad\qquad\uparrow$

$\qquad\qquad$ From the table

The monthly mortgage payment is \$1417.08.

You Try It 4

Strategy To find the interest:

- Multiply the mortgage by the factor found in the Monthly Payment Table to find the monthly mortgage payment.
- Subtract the principal from the monthly mortgage payment.

Solution $625{,}000 \times 0.0070678 \approx 4417.38$

$\qquad\qquad\qquad\quad \uparrow \qquad\qquad\quad \uparrow$

$\qquad\qquad\qquad$ From the Monthly mortgage

$\qquad\qquad\qquad\quad$ table $\qquad\qquad$ payment

$$\boxed{\begin{array}{c}\text{Monthly}\\\text{mortgage}\\\text{payment}\end{array}} - \boxed{\text{principal}} = \boxed{\text{interest}}$$

$\quad 4417.38 \quad - \quad 2516.08 \ = 1901.30$

The interest on the mortgage is \$1901.30.

You Try It 5

Strategy To find the monthly payment:

- Divide the annual property tax by 12 to find the monthly property tax.
- Add the monthly property tax to the monthly mortgage payment.

Solution $3000 \div 12 = 250$

The monthly property tax is \$250.

$815.20 + 250 = 1065.20$

The total monthly payment is \$1065.20.

SECTION 6.5

You Try It 1

Strategy To find the amount financed:

- Find the down payment by solving the basic percent equation for *amount*.
- Subtract the down payment from the purchase price.

Solution

Percent $\times$ base $=$ amount

$$\boxed{\text{Percent}} \times \boxed{\begin{array}{c}\text{purchase}\\\text{price}\end{array}} = \boxed{\begin{array}{c}\text{down}\\\text{payment}\end{array}}$$

$\quad 0.20 \quad \times \quad 19{,}200 \quad = \quad n$

$\qquad\qquad\qquad 3840 = n$

The down payment is \$3840.

$19{,}200 - 3840 = 15{,}360$

The amount financed is \$15,360.

You Try It 2

Strategy To find the license fee, solve the basic percent equation for *amount*.

Solution

Percent × base = amount

| Percent | × | purchase price | = | license fee |

0.015 × 27,350 = n

410.25 = n

The license fee is $410.25.

You Try It 3

Strategy To find the cost, multiply the cost per mile (0.41) by the number of miles driven (23,000).

Solution 23,000 × 0.41 = 9430

The cost is $9430.

You Try It 4

Strategy To find the cost per mile for car insurance, divide the cost for insurance (360) by the number of miles driven (15,000).

Solution 360 ÷ 15,000 = 0.024

The cost per mile for insurance is $.024.

You Try It 5

Strategy To find the monthly payment:
- Subtract the down payment from the purchase price to find the amount financed.
- Multiply the amount financed by the factor found in the Monthly Payment Table.

Solution 25,900 − 6475 = 19,425

19,425 × 0.0244129 ≈ 474.22

The monthly payment is $474.22.

SECTION 6.6

You Try It 1

Strategy To find the worker's earnings:
- Find the worker's overtime wage by multiplying the hourly wage by 2.
- Multiply the number of overtime hours worked by the overtime wage.

Solution 28.50 × 2 = 57

The hourly wage for overtime is $57.

57 × 8 = 456

The construction worker earns $456.

You Try It 2

Strategy To find the salary per month, divide the annual salary by the number of months in a year (12).

Solution 70,980 ÷ 12 = 5915

The contractor's monthly salary is $5915.

You Try It 3

Strategy To find the total earnings:
- Find the sales over $50,000.
- Multiply the commission rate by sales over $50,000.
- Add the commission to the annual salary.

Solution 175,000 − 50,000 = 125,000

Sales over $50,000 totaled $125,000.

125,000 × 0.095 = 11,875

Earnings from commissions totaled $11,875.

37,000 + 11,875 = 48,875

The insurance agent earned $48,875.

SECTION 6.7

You Try It 1

Strategy To find the current balance:
- Subtract the amount of the check from the old balance.
- Add the amount of each deposit.

Solution

302.46
− 20.59   check
281.87
176.86   first deposit
+ 94.73 second deposit
553.46

The current checking account balance is $553.46.

You Try It 2

Current checkbook balance: 623.41
Check: 237   + 78.73
702.14
Interest: + 2.11
704.25
Deposit: −523.84
180.41

Closing bank balance from bank statement: $180.41

Checkbook balance: $180.41

The bank statement and the checkbook balance.

SOLUTIONS TO CHAPTER 7 "YOU TRY IT"

SECTION 7.1

You Try It 1

    Strategy   To find what percent of the total the number of cellular phones purchased in March represents:

- Read the pictograph to determine the number of cellular phones purchased for each month.
- Find the total cellular phone purchases for the 4-month period.
- Solve the basic percent equation for percent (*n*). Amount = 3000; the base is the total sales for the 4-month period.

    Solution

$$
\begin{array}{r}
4{,}500 \\
3{,}500 \\
3{,}000 \\
\underline{1{,}500} \\
12{,}500
\end{array}
$$

Percent × base = amount

$$n \times 12{,}500 = 3000$$
$$n = 3000 \div 12{,}500$$
$$n = 0.24$$

The number of cellular phones purchased in March represents 24% of the total number of cellular phones purchased.

You Try It 2

    Strategy   To find the ratio of the annual cost of fuel to the annual cost of maintenance:

- Locate the annual fuel cost and the annual maintenance cost in the circle graph.
- Write the ratio of the annual fuel cost to the annual maintenance cost in simplest form.

    Solution   Annual fuel cost: $700
Annual maintenance cost: $500

$$\frac{700}{500} = \frac{7}{5}$$

The ratio is $\frac{7}{5}$.

You Try It 3

    Strategy   To find the amount paid for medical/dental insurance:

- Locate the percent of the distribution that is medical/dental insurance.
- Solve the basic percent equation for amount.

    Solution   The percent of the distribution that is medical/dental insurance: 6%

Percent × base = amount
$$0.06 \times 2900 = n$$
$$174 = n$$

The amount paid for medical/dental insurance is $174.

SECTION 7.2

You Try It 1

    Strategy   To write the ratio:

- Read the graph to find the lung capacity of an inactive female and of an athletic female.
- Write the ratio in simplest form.

    Solution   Lung capacity of an inactive female: 25

Lung capacity of an athletic female: 55

$$\frac{25}{55} = \frac{5}{11}$$

The ratio is $\frac{5}{11}$.

You Try It 2

    Strategy   To determine between which two years the net income of Math Associates increased the most:

- Read the line graph to determine the net income of Math Associates for each of the years shown.
- Subtract to find the difference between each two years.
- Find the greatest difference.

    Solution   2000: $1 million
2001: $3 million
2002: $5 million
2003: $8 million
2004: $12 million

Between 2000 and 2001: $3 - 1 = 2$
Between 2001 and 2002: $5 - 3 = 2$
Between 2002 and 2003: $8 - 5 = 3$
Between 2003 and 2004: $12 - 8 = 4$

$$4 > 3 > 2$$

The net income of Math Associates increased the most between 2003 and 2004.

SECTION 7.3

You Try It 1

Strategy To find the number of employees:
- Read the histogram to find the number of employees whose hourly wage is between $10 and $12 and the number whose hourly wage is between $12 and $14.
- Add the two numbers.

Solution Number whose wage is between $10 and $12: 7
Number whose wage is between $12 and $14: 15

$7 + 15 = 22$

22 employees earn between $10 and $14.

You Try It 2

Strategy To write the ratio:
- Read the graph to find the number of states with a per capita income between $36,000 and $40,000 and the number with a per capita income between $44,000 and $48,000.
- Write the ratio in simplest form.

Solution Number of states with a per capita income between $36,000 and $40,000: 7
Number of states with a per capita income between $44,000 and $48,000: 1

The ratio is $\dfrac{7}{1}$.

SECTION 7.4

You Try It 1

Strategy To find the mean amount spent by the 12 customers:
- Find the sum of the numbers.
- Divide the sum by the number of customers (12).

Solution $11.01 + 10.75 + 12.09 + 15.88 + 13.50 + 12.29 + 10.69 + 9.36 + 11.66 + 15.25 + 10.09 + 12.72 = 145.29$

$$\bar{x} = \frac{145.29}{12} \approx 12.11$$

The mean amount spent by the 12 customers was $12.11.

You Try It 2

Strategy To find the median weight loss:
- Arrange the weight losses from least to greatest.
- Because there is an even number of values, the median is the mean of the middle two numbers.

Solution 10, 14, 16, 16, 22, 27, 29, 31, 31, 40

$$\text{Median} = \frac{22 + 27}{2} = \frac{49}{2} = 24.5$$

The median weight loss was 24.5 pounds.

You Try It 3

Strategy To draw the box-and-whiskers plot:
- Find the median, Q_1, and Q_3.
- Use the least value, Q_1, the median, Q_3, and the greatest value to draw the box-and-whiskers plot.

Solution

a.

b. Answers about the spread of the data will vary. For example, in You Try It 3, the values in the interquartile range are all very close to the median. They are not so close to the median in Example 3. The whiskers are long with respect to the box in You Try It 3, whereas they are short with respect to the box in Example 3. This shows that the data values outside the interquartile range are closer together in Example 3 than in You Try It 3.

SECTION 7.5

You Try It 1

Strategy To find the probability:
- List the outcomes of the experiment in a systematic way. We will use a table.
- Use the table to count the number of possible outcomes of the experiment.

(Continued)

(Continued)

- Count the number of outcomes of the experiment that are favorable to the event of two true questions and one false question.
- Use the probability formula.

Solution

Question 1	Question 2	Question 3
T	T	T
T	T	F
T	F	T
T	F	F
F	T	T
F	T	F
F	F	T
F	F	F

There are 8 possible outcomes:

S = {TTT, TTF, TFT, TFF, FTT, FTF, FFT, FFF}

There are 3 outcomes favorable to the event:

{TTF, TFT, FTT}

Probability of an event

$$= \frac{\text{number of favorable outcomes}}{\text{number of possible outcomes}} = \frac{3}{8}$$

The probability of two true questions and one false question is $\frac{3}{8}$.

SOLUTIONS TO CHAPTER 8 "YOU TRY IT"

SECTION 8.1

You Try It 1 $14 \text{ ft} = 14 \cancel{\text{ft}} \times \dfrac{1 \text{ yd}}{3 \cancel{\text{ft}}} = \dfrac{14 \text{ yd}}{3} = 4\dfrac{2}{3} \text{ yd}$

You Try It 2 $9240 \text{ ft} = 9240 \cancel{\text{ft}} \times \dfrac{1 \text{ mi}}{5280 \cancel{\text{ft}}}$

$$= \dfrac{9240 \text{ mi}}{5280} = 1\dfrac{3}{4} \text{ mi}$$

You Try It 3

```
        3 ft 6 in.
   12) 42              • 12 in. = 1 ft
      -36
        6
```
42 in. = 3 ft 6 in.

You Try It 4

```
        4 yd 2 ft
    3) 14              • 3 ft = 1 yd
      -12
        2
```
14 ft = 4 yd 2 ft

You Try It 5

```
   3 ft   5 in.
 + 4 ft   9 in.
   7 ft  14 in.        • 14 in. = 1 ft 2 in.
```
7 ft 14 in. = 8 ft 2 in.

You Try It 6

```
   3 ft  14 in.
   4 ft   2 in.        • Borrow 1 ft (12 in.) from
 - 1 ft   8 in.          4 ft and add it to 2 in.
   2 ft   6 in.
```

You Try It 7

```
   4 yd 1 ft
 ×      8
  32 yd 8 ft           • 8 ft = 2 yd 2 ft
```

32 yd 8 ft = 34 yd 2 ft

You Try It 8

```
          3 yd    2 ft
    2) 7 yd    1 ft
      -6 yd
       1 yd = 3 ft
              4 ft
             -4 ft
              0
```

You Try It 9 $6\dfrac{1}{4} \text{ ft} = 6\dfrac{3}{12} \text{ ft} = 5\dfrac{15}{12} \text{ ft}$

$-3\dfrac{2}{3} \text{ ft} = 3\dfrac{8}{12} \text{ ft} = 3\dfrac{8}{12} \text{ ft}$

$2\dfrac{7}{12} \text{ ft}$

You Try It 10

Strategy   To find the width of the storage room:

- Multiply the number of tiles (8) by the width of each tile (9 in.).
- Divide the result by the number of inches in 1 foot (12) to find the width in feet.

Solution   9 in. × 8 = 72 in.

72 ÷ 12 = 6

The width is 6 ft.

You Try It 11

Strategy   To find the length of each piece, divide the total length (9 ft 8 in.) by the number of pieces (4).

Solution

```
        2 ft    5 in.
    4) 9 ft    8 in.
      -8 ft
       1 ft = 12 in.
              20 in.
             -20 in.
              0 in.
```

Each piece is 2 ft 5 in. long.

SECTION 8.2

You Try It 1 $3 \text{ lb} = 3 \cancel{\text{lb}} \times \dfrac{16 \text{ oz}}{1 \cancel{\text{lb}}} = 48 \text{ oz}$

You Try It 2 $4200 \text{ lb} = 4200 \cancel{\text{lb}} \times \dfrac{1 \text{ ton}}{2000 \cancel{\text{lb}}}$

$$= \dfrac{4200 \text{ tons}}{2000} = 2\dfrac{1}{10} \text{ tons}$$

You Try It 3

$$\begin{array}{r} {}^{6\text{ lb}}{}^{17\text{ oz}} \\ \cancel{7\text{ lb}}\cancel{1\text{ oz}} \\ -\ 3\text{ lb}\ \ 4\text{ oz} \\ \hline 3\text{ lb}\ 13\text{ oz} \end{array}$$

• Borrow 1 lb (16 oz) from 7 lb and add it to 1 oz.

You Try It 4

$$\begin{array}{r} 3\text{ lb}\ 6\text{ oz} \\ \times4 \\ \hline 12\text{ lb}\ 24\text{ oz} = 13\text{ lb}\ 8\text{ oz} \end{array}$$

You Try It 5

Strategy   To find the weight of 12 bars of soap:

• Multiply the number of bars (12) by the weight of each bar (7 oz).
• Convert the number of ounces to pounds.

Solution

$$\begin{array}{r} 12 \\ \times\ 7\text{ oz} \\ \hline \end{array}$$

$$84\text{ oz} = 84\ \cancel{\text{oz}} \times \frac{1\text{ lb}}{16\ \cancel{\text{oz}}} = 5\frac{1}{4}\text{ lb}$$

The 12 bars of soap weigh $5\frac{1}{4}$ lb.

SECTION 8.3

You Try It 1

$$18\text{ pt} = 18\ \cancel{\text{pt}} \times \frac{1\ \cancel{\text{qt}}}{2\ \cancel{\text{pt}}} \times \frac{1\text{ gal}}{4\ \cancel{\text{qt}}}$$

$$= \frac{18\text{ gal}}{8} = 2\frac{1}{4}\text{ gal}$$

You Try It 2

$$\begin{array}{r} 1\text{ gal}\ \ 2\text{ qt} \\ 3)\ \overline{\ 4\text{ gal}\ \ 2\text{ qt}} \\ -3\text{ gal} \\ \hline 1\text{ gal} = 4\text{ qt} \\ 6\text{ qt} \\ -6\text{ qt} \\ \hline 0 \end{array}$$

You Try It 3

Strategy   To find the number of gallons of water needed:

• Find the number of quarts required by multiplying the number of quarts one student needs per day (5) by the number of students (5) by the number of days (3).
• Convert the number of quarts to gallons.

Solution $5 \times 5 \times 3 = 75$ qt

$$75\text{ qt} = 75\ \cancel{\text{qt}} \cdot \frac{1\text{ gal}}{4\ \cancel{\text{qt}}} = \frac{75\text{ gal}}{4} = 18\frac{3}{4}\text{ gal}$$

The students should take $18\frac{3}{4}$ gal of water.

SECTION 8.4

You Try It 1

$$18{,}000\text{ s} = 18{,}000\ \cancel{\text{s}} \times \frac{1\ \cancel{\text{min}}}{60\ \cancel{\text{s}}} \times \frac{1\text{ h}}{60\ \cancel{\text{min}}}$$

$$= \frac{18{,}000\text{ h}}{3600} = 5\text{ h}$$

SECTION 8.5

You Try It 1

$$4.5\text{ Btu} = 4.5\ \cancel{\text{Btu}} \times \frac{778\text{ ft}\cdot\text{lb}}{1\ \cancel{\text{Btu}}}$$

$$= 3501\text{ ft}\cdot\text{lb}$$

You Try It 2   Energy $= 800\text{ lb} \times 16\text{ ft} = 12{,}800\text{ ft}\cdot\text{lb}$

You Try It 3 $56{,}000\text{ Btu} =$

$$56{,}000\ \cancel{\text{Btu}} \times \frac{778\text{ ft}\cdot\text{lb}}{1\ \cancel{\text{Btu}}} =$$

$$43{,}568{,}000\text{ ft}\cdot\text{lb}$$

You Try It 4   Power $= \dfrac{90\text{ ft} \times 1200\text{ lb}}{24\text{ s}}$

$$= 4500\ \frac{\text{ft}\cdot\text{lb}}{\text{s}}$$

You Try It 5 $\dfrac{3300}{550} = 6\text{ hp}$

SOLUTIONS TO CHAPTER 9 "YOU TRY IT"

SECTION 9.1

You Try It 1 $3.07\text{ m} = 307\text{ cm}$

You Try It 2 $750\text{ m} = 0.750\text{ km}$

$$\begin{aligned} 3\text{ km } 750\text{ m} &= 3\text{ km} + 0.750\text{ km} \\ &= 3.750\text{ km} \end{aligned}$$

You Try It 3

Strategy   To find the cost of the shelves:

• Multiply the length of the bookcase (175 cm) by the number of shelves (4).
• Convert centimeters to meters.
• Multiply the number of meters by the cost per meter ($15.75).

Solution

$$\begin{array}{r} 175\text{ cm} \\ \times4 \\ \hline 700\text{ cm} \end{array}$$

$$700\text{ cm} = 7\text{ m}$$

$$\begin{array}{r} 15.75 \\ \times7 \\ \hline 110.25 \end{array}$$

The cost of the shelves is $110.25.

SECTION 9.2

You Try It 1 42.3 mg = 0.0423 g

You Try It 2 54 mg = 0.054 g

$$3 \text{ g } 54 \text{ mg} = 3 \text{ g} + 0.054 \text{ g}$$
$$= 3.054 \text{ g}$$

You Try It 3

Strategy To find how much fertilizer is required:
- Convert 300 g to kilograms.
- Multiply the number of kilograms by the number of trees (400).

Solution 300 g = 0.3 kg

$$
\begin{array}{r}
400 \\
\times \ 0.3 \text{ kg} \\
\hline
120.0 \text{ kg}
\end{array}
$$

To fertilize the trees, 120 kg of fertilizer are required.

SECTION 9.3

You Try It 1 2 kl = 2000 L

$$2 \text{ kl } 167 \text{ L} = 2000 \text{ L} + 167 \text{ L}$$
$$= 2167 \text{ L}$$

You Try It 2 325 cm³ = 325 ml = 0.325 L

You Try It 3

Strategy To find the profit:
- Convert 5 L to milliliters.
- Find the number of jars by dividing the number of milliliters by the number of milliliters in each jar (125).
- Multiply the number of jars by the cost per jar ($.85).
- Find the total cost by adding the cost of the jars to the cost for the moisturizer ($299.50).
- Find the income by multiplying the number of jars by the selling price per jar ($29.95).
- Subtract the total cost from the income.

Solution 5 L = 5000 ml

5000 ÷ 125 = 40   This is the number of jars.

$$
\begin{array}{r}
0.85 \\
\times \quad 40 \\
\hline
34.00
\end{array}
$$
This is the cost of the jars.

$$
\begin{array}{r}
299.50 \\
+ \quad 34.00 \\
\hline
333.50
\end{array}
$$
This is the total cost.

$$
\begin{array}{r}
29.95 \\
\times \quad 40 \\
\hline
1198.00
\end{array}
$$
This is the income from sales.

$$
\begin{array}{r}
1198.00 \\
- \quad 333.50 \\
\hline
864.50
\end{array}
$$

The profit on the 5 L of moisturizer is $864.50.

SECTION 9.4

You Try It 1

Strategy To find the number of Calories burned off, multiply the number of hours spent doing housework $\left(4\frac{1}{2}\right)$ by the Calories used per hour (240).

Solution $4\frac{1}{2} \times 240 = \dfrac{9}{2} \times 240 = 1080$

Doing $4\frac{1}{2}$ h of housework burns off 1080 Calories.

You Try It 2

Strategy To find the number of kilowatt-hours used:
- Find the number of watt-hours used.
- Convert watt-hours to kilowatt-hours.

Solution 150 × 200 = 30,000

30,000 Wh = 30 kWh

30 kWh of energy are used.

You Try It 3

Strategy To find the cost:
- Convert 20 min to hours.
- Multiply to find the total number of hours the oven is used.
- Multiply the number of hours used by the number of watts to find the watt-hours.
- Convert watt-hours to kilowatt-hours.
- Multiply the number of kilowatt-hours by the cost per kilowatt-hour.

Solution $20 \text{ min} = 20 \text{ min} \times \dfrac{1 \text{ h}}{60 \text{ min}}$

$= \dfrac{20}{60} \text{ h} = \dfrac{1}{3} \text{ h}$

$\dfrac{1}{3} \text{ h} \times 30 = 10 \text{ h}$

$10 \text{ h} \times 500 \text{ W} = 5000 \text{ Wh}$

$5000 \text{ Wh} = 5 \text{ kWh}$

$5 \times 13.7\cancel{c} = 68.5\cancel{c}$

The cost is 68.5¢.

SECTION 9.5

You Try It 1

$\dfrac{60 \text{ ft}}{\text{s}} \approx \dfrac{60 \text{ ft}}{\text{s}} \times \dfrac{1 \text{ m}}{3.28 \text{ ft}}$

$= \dfrac{60 \text{ m}}{3.28 \text{ s}} \approx 18.29 \text{ m/s}$

$60 \text{ ft/s} \approx 18.29 \text{ m/s}$

• The conversion rate is $\dfrac{1 \text{ m}}{3.28 \text{ ft}}$ with m in the numerator and ft in the denominator.

You Try It 2

$\dfrac{\$3.69}{\text{gal}} \approx \dfrac{\$3.69}{\text{gal}} \times \dfrac{1 \text{ gal}}{3.79 \text{ L}}$

$= \dfrac{\$3.69}{3.79 \text{ L}} \approx \dfrac{\$.97}{\text{L}}$

$\$3.69/\text{gal} \approx \$.97/\text{L}$

You Try It 3

$45 \text{ cm} = \dfrac{45 \text{ cm}}{1} \times \dfrac{1 \text{ in.}}{2.54 \text{ cm}}$

$= \dfrac{45 \text{ in.}}{2.54} \approx 17.72 \text{ in.}$

$45 \text{ cm} \approx 17.72 \text{ in.}$

You Try It 4

$\dfrac{75 \text{ km}}{\text{h}} \approx \dfrac{75 \text{ km}}{\text{h}} \times \dfrac{1 \text{ mi}}{1.61 \text{ km}}$

$= \dfrac{75 \text{ mi}}{1.61 \text{ h}} \approx 46.58 \text{ mi/h}$

$75 \text{ km/h} \approx 46.58 \text{ mi/h}$

You Try It 5

$\dfrac{\$1.75}{\text{L}} \approx \dfrac{\$1.75}{\text{L}} \times \dfrac{3.79 \text{ L}}{1 \text{ gal}}$

$= \dfrac{\$6.6325}{1 \text{ gal}} \approx \$6.63/\text{gal}$

$\$1.75/\text{L} \approx \$6.63/\text{gal}$

SOLUTIONS TO CHAPTER 10 "YOU TRY IT"

SECTION 10.1

You Try It 1 -232 ft

You Try It 2

You Try It 3 $-12 < -8$ • -12 is to the left of -8 on the number line.

You Try It 4 $|-7| = 7; |21| = 21$

You Try It 5 $|2| = 2; |-9| = 9$

You Try It 6 $-|-12| = -12$

SECTION 10.2

You Try It 1 $-154 + (-37) = -191$ • The signs of the addends are the same.

You Try It 2 $-5 + (-2) + 9 + (-3)$
$= -7 + 9 + (-3)$
$= 2 + (-3)$
$= -1$

You Try It 3 $-8 - 14$
$= -8 + (-14)$ • Rewrite "−" as "+"; the opposite of 14 is -14.
$= -22$

You Try It 4 $3 - (-15)$
$= 3 + 15$ • Rewrite "−" as "+"; the opposite of -15 is 15.
$= 18$

You Try It 5 $4 - (-3) - 12 - (-7) - 20$
$= 4 + 3 + (-12) + 7 + (-20)$
$= 7 + (-12) + 7 + (-20)$
$= -5 + 7 + (-20)$
$= 2 + (-20)$
$= -18$

You Try It 6

Strategy   To find the temperature, add the increase (12°C) to the previous temperature (−10°C).

Solution $-10 + 12 = 2$

After an increase of 12°C, the temperature is 2°C.

SECTION 10.3

You Try It 1 $(-3)(-5) = 15$ • The signs are the same. The product is positive.

You Try It 2 The signs are different. The product is negative.

$-38 \cdot 51 = -1938$

You Try It 3 $-7(-8)(9)(-2) = 56(9)(-2)$
$= 504(-2)$
$= -1008$

You Try It 4 $(-135) \div (-9) = 15$ • The signs are the same. The quotient is positive.

You Try It 5 $84 \div (-6) = -14$ • The signs are different. The quotient is negative.

You Try It 6 $-72 \div 4 = -18$

You Try It 7 Division by zero is undefined.
$-39 \div 0$ is undefined.

You Try It 8

Strategy To find the melting point of argon, multiply the melting point of mercury $(-38°C)$ by 5.

Solution $5(-38) = -190$

The melting point of argon is $-190°C$.

You Try It 9

Strategy To find the average daily low temperature:

• Add the seven temperature readings.
• Divide by 7.

Solution

$-6 + (-7) + 1 + 0 + (-5) + (-10) + (-1)$
$= -13 + 1 + 0 + (-5) + (-10) + (-1)$
$= -12 + 0 + (-5) + (-10) + (-1)$
$= -12 + (-5) + (-10) + (-1)$
$= -17 + (-10) + (-1)$
$= -27 + (-1)$
$= -28$

$-28 \div 7 = -4$

The average daily low temperature was $-4°F$.

SECTION 10.4

You Try It 1 The LCM of 9 and 12 is 36.

$\dfrac{5}{9} - \dfrac{11}{12} = \dfrac{20}{36} - \dfrac{33}{36} = \dfrac{20}{36} + \dfrac{-33}{36}$

$= \dfrac{20 + (-33)}{36} = \dfrac{-13}{36}$

$= -\dfrac{13}{36}$

You Try It 2 The LCM of 8, 6, and 3 is 24.

$-\dfrac{7}{8} - \dfrac{5}{6} + \dfrac{2}{3}$

$= -\dfrac{21}{24} - \dfrac{20}{24} + \dfrac{16}{24}$

$= \dfrac{-21}{24} + \dfrac{-20}{24} + \dfrac{16}{24}$

$= \dfrac{-21 + (-20) + 16}{24} = \dfrac{225}{24}$

$= -\dfrac{25}{24} = -1\dfrac{1}{24}$

You Try It 3 $16.127 - 67.91 = 16.127 + (-67.91)$

$\begin{array}{r} 67.910 \\ -16.127 \\ \hline 51.783 \end{array}$

$16.127 - 67.91 = -51.783$

You Try It 4 $2.7 + (-9.44) + 6.2$
$= --6.74 + 6.2$
$= -0.54$

You Try It 5 The product is positive.

$\left(-\dfrac{2}{3}\right)\left(-\dfrac{9}{10}\right) = \dfrac{2 \cdot 9}{3 \cdot 10}$

$= \dfrac{3}{5}$

You Try It 6 The quotient is negative.

$-\dfrac{5}{8} \div \dfrac{5}{40} = -\left(\dfrac{5}{8} \div \dfrac{5}{40}\right)$

$= -\left(\dfrac{5}{8} \times \dfrac{40}{5}\right)$

$= -\left(\dfrac{5 \cdot 40}{8 \cdot 5}\right)$

$= -5$

You Try It 7 The product is negative.

$\begin{array}{r} 5.44 \\ \times 3.8 \\ \hline 4352 \\ 1632 \\ \hline 20.672 \end{array}$

$-5.44 \times 3.8 = -20.672$

You Try It 8 $3.44 \times (-1.7) \times 0.6$
$= (-5.848) \times 0.6$
$= -3.5088$

You Try It 9

$$\begin{array}{r} 0.231 \\ 1{,}7{\overline{)0{,}3{,}940}} \\ -3\ 4 \\ \hline 54 \\ -51 \\ \hline 30 \\ -17 \\ \hline 13 \end{array}$$

$$-0.394 \div 1.7 \approx -0.23$$

You Try It 10

Strategy To find how many degrees the temperature fell, subtract the lower temperature ($-13.33°C$) from the higher temperature ($12.78°C$).

Solution $12.78 - (-13.33) = 12.78 + 13.33$
$$= 26.11$$

The temperature fell $26.11°C$ in the 15-minute period.

SECTION 10.5

You Try It 1 The number is less than 1. Move the decimal point 7 places to the right. The exponent on 10 is -7.

$$0.000000961 = 9.61 \times 10^{-7}$$

You Try It 2 The exponent on 10 is positive. Move the decimal point 6 places to the right.

$$7.329 \times 10^6 = 7{,}329{,}000$$

You Try It 3
$$9 - 9 \div (-3)$$
$$= 9 - (-3) \qquad \bullet \text{ Do the division.}$$
$$= 9 + 3 \qquad \bullet \text{ Rewrite as}$$
$$= 12 \qquad\qquad \text{addition. Add.}$$

You Try It 4
$$8 \div 4 \cdot 4 - (-2)^2$$
$$= 8 \div 4 \cdot 4 - 4 \qquad \bullet \text{ Exponents}$$
$$= 2 \cdot 4 - 4 \qquad\quad \bullet \text{ Division}$$
$$= 8 - 4 \qquad\qquad \bullet \text{ Multiplication}$$
$$= 8 + (-4) \qquad\;\; \bullet \text{ Subtraction}$$
$$= 4$$

You Try It 5
$$8 - (-15) \div (2 - 7)$$
$$= 8 - (-15) \div (-5)$$
$$= 8 - 3$$
$$= 8 + (-3)$$
$$= 5$$

You Try It 6
$$(-2)^2 \times (3 - 7)^2 - (-16) \div (-4)$$
$$= (-2)^2 \times (-4)^2 - (-16) \div (-4)$$
$$= 4 \times 16 - (-16) \div (-4)$$
$$= 64 - (-16) \div (-4)$$
$$= 64 - 4$$
$$= 64 + (-4)$$
$$= 60$$

You Try It 7 $7 \div \left(\dfrac{1}{7} - \dfrac{3}{14}\right) - 9$

$$= 7 \div \left(-\dfrac{1}{14}\right) - 9$$
$$= 7(-14) - 9$$
$$= -98 - 9$$
$$= -98 + (-9)$$
$$= -107$$

SOLUTIONS TO CHAPTER 11 "YOU TRY IT"

SECTION 11.1

You Try It 1 $6a - 5b$
$$6(-3) - 5(4) = -18 - 20$$
$$= -18 + (-20)$$
$$= -38$$

You Try It 2 $-3s^2 - 12 \div t$
$$-3(-2)^2 - 12 \div 4 = -3(4) - 12 \div 4$$
$$= -12 - 12 \div 4$$
$$= -12 - 3$$
$$= -12 + (-3)$$
$$= -15$$

You Try It 3 $-\dfrac{2}{3}m + \dfrac{3}{4}n^3$

$$-\dfrac{2}{3}(6) + \dfrac{3}{4}(2)^3 = -\dfrac{2}{3}(6) + \dfrac{3}{4}(8)$$
$$= -4 + 6$$
$$= 2$$

You Try It 4 $-3yz - z^2 + y^2$

$$-3\left(-\dfrac{2}{3}\right)\left(\dfrac{1}{3}\right) - \left(\dfrac{1}{3}\right)^2 + \left(-\dfrac{2}{3}\right)^2$$
$$= -3\left(-\dfrac{2}{3}\right)\left(\dfrac{1}{3}\right) - \dfrac{1}{9} + \dfrac{4}{9}$$
$$= \dfrac{2}{3} - \dfrac{1}{9} + \dfrac{4}{9} = \dfrac{6}{9} - \dfrac{1}{9} + \dfrac{4}{9} = \dfrac{9}{9}$$
$$= 1$$

You Try It 5 $5a^2 - 6b^2 + 7a^2 - 9b^2$
$$= 5a^2 + (-6)b^2 + 7a^2 + (-9)b^2$$
$$= 5a^2 + 7a^2 + (-6)b^2 + (-9)b^2$$
$$= 12a^2 + (-15)b^2$$
$$= 12a^2 - 15b^2$$

You Try It 6 $-6x + 7 + 9x - 10$
$$= (-6)x + 7 + 9x + (-10)$$
$$= (-6)x + 9x + 7 + (-10)$$
$$= 3x + (-3)$$
$$= 3x - 3$$

You Try It 7　　$\frac{3}{8}w + \frac{1}{2} - \frac{1}{4}w - \frac{2}{3}$

$$= \frac{3}{8}w + \frac{1}{2} + \left(-\frac{1}{4}\right)w + \left(-\frac{2}{3}\right)$$

$$= \frac{3}{8}w + \left(-\frac{1}{4}\right)w + \frac{1}{2} + \left(-\frac{2}{3}\right)$$

$$= \frac{3}{8}w + \left(-\frac{2}{8}\right)w + \frac{3}{6} + \left(-\frac{4}{6}\right)$$

$$= \frac{1}{8}w + \left(-\frac{1}{6}\right)$$

$$= \frac{1}{8}w - \frac{1}{6}$$

You Try It 8　　$5(a - 2) = 5a - 5(2)$
$$= 5a - 10$$

You Try It 9　　$8s - 2(3s - 5)$
$$= 8s - 2(3s) - (-2)(5)$$
$$= 8s - 6s + 10$$
$$= 2s + 10$$

You Try It 10　　$4(x - 3) - 2(x + 1)$
$$= 4x - 4(3) - 2x - 2(1)$$
$$= 4x - 12 - 2x - 2$$
$$= 4x - 2x - 12 - 2$$
$$= 2x - 14$$

SECTION 11.2

You Try It 1

$$\begin{array}{c|c} \multicolumn{2}{c}{x(x + 3) = 4x + 6} \\ \hline (-2)(-2 + 3) & 4(-2) + 6 \\ (-2)(1) & (-8) + 6 \\ \multicolumn{2}{c}{-2 = -2} \end{array}$$

Yes, -2 is a solution.

You Try It 2

$$\begin{array}{c|c} \multicolumn{2}{c}{x^2 - x = 3x + 7} \\ \hline (-3)^2 - (-3) & 3(-3) + 7 \\ 9 + 3 & -9 + 7 \\ \multicolumn{2}{c}{12 \neq -2} \end{array}$$

No, -3 is not a solution.

You Try It 3

$$-2 + y = -5$$
$$-2 + 2 + y = -5 + 2 \qquad \text{• Add 2 to}$$
$$0 + y = -3 \qquad\qquad\quad \text{each side.}$$
$$y = -3$$

The solution is -3.

You Try It 4

$$7 = y + 8$$
$$7 - 8 = y + 8 - 8 \qquad \text{• Subtract}$$
$$-1 = y + 0 \qquad\qquad\quad \text{8 from}$$
$$-1 = y \qquad\qquad\qquad\;\; \text{each side.}$$

The solution is -1.

You Try It 5

$$\frac{1}{5} = z + \frac{4}{5}$$
$$\frac{1}{5} - \frac{4}{5} = z + \frac{4}{5} - \frac{4}{5} \qquad \text{• Subtract}$$
$$-\frac{3}{5} = z + 0 \qquad\qquad\qquad \frac{4}{5} \text{ from}$$
$$-\frac{3}{5} = z \qquad\qquad\qquad\quad \text{each side.}$$

The solution is $-\frac{3}{5}$.

You Try It 6　　$4z = -20$

$$\frac{4z}{4} = \frac{-20}{4} \qquad \text{• Divide each side by 4.}$$
$$1z = -5$$
$$z = -5$$

The solution is -5.

You Try It 7　　$8 = \frac{2}{5}n$

$$\left(\frac{5}{2}\right)(8) = \left(\frac{5}{2}\right)\frac{2}{5}n \qquad \text{• Multiply each}$$
$$20 = 1n \qquad\qquad\qquad\;\; \text{side by } \frac{5}{2}.$$
$$20 = n$$

The solution is 20.

You Try It 8　　$\frac{2}{3}t - \frac{1}{3}t = -2$　　• Combine like terms.

$$\frac{1}{3}t = -2$$

$$\left(\frac{3}{1}\right)\frac{1}{3}t = \left(\frac{3}{1}\right)(-2) \qquad \text{• Multiply each}$$
$$1t = -6 \qquad\qquad\qquad\;\; \text{side by 3.}$$
$$t = -6$$

The solution is -6.

You Try It 9

Strategy　To find the regular price, replace the variables S and D in the formula by the given values and solve for R.

Solution
$$S = R - D$$
$$44 = R - 16$$
$$44 + 16 = R - 16 + 16$$
$$60 = R$$

The regular price is $60.

You Try It 10

Strategy　To find the monthly payment, replace the variables A and N in the formula by the given values and solve for M.

Solution
$$A = MN$$
$$6840 = M \cdot 24$$
$$6840 = 24M$$
$$\frac{6840}{24} = \frac{24M}{24}$$
$$285 = M$$

The monthly payment is $285.

SECTION 11.3

You Try It 1
$$5x + 8 = 6$$
$$5x + 8 - 8 = 6 - 8 \qquad \text{• Subtract 8 from}$$
$$5x = -2 \qquad\qquad\quad \text{each side.}$$
$$\frac{5x}{5} = \frac{-2}{5} \qquad\qquad \text{• Divide each}$$
$$x = -\frac{2}{5} \qquad\qquad\quad \text{side by 5.}$$

The solution is $-\frac{2}{5}$.

You Try It 2

$$7 - x = 3$$
$$7 - 7 - x = 3 - 7 \quad \bullet \text{ Subtract 7}$$
$$-x = -4 \qquad\qquad \text{from each side.}$$
$$(-1)(-x) = (-1)(-4) \quad \bullet \text{ Multiply each}$$
$$x = 4 \qquad\qquad\quad \text{side by } -1.$$

The solution is 4.

You Try It 3

Strategy To find the Celsius temperature, replace the variable F in the formula by the given value and solve for C.

Solution

$$F = 1.8C + 32$$
$$-22 = 1.8C + 32 \qquad\qquad \bullet \text{ Substitute } -22 \text{ for } F.$$
$$-22 - 32 = 1.8C + 32 - 32 \quad \bullet \text{ Subtract 32 from}$$
$$\qquad\qquad\qquad\qquad\qquad \text{each side.}$$
$$-54 = 1.8C \qquad\qquad\qquad \bullet \text{ Combine like terms.}$$
$$\frac{-54}{1.8} = \frac{1.8C}{1.8} \qquad\qquad\quad \bullet \text{ Divide each side}$$
$$\qquad\qquad\qquad\qquad\qquad \text{by 1.8.}$$
$$-30 = C$$

The temperature is $-30°C$.

You Try It 4

Strategy To find the cost per unit, replace the variables T, N, and F in the formula by the given values and solve for U.

Solution

$$T = U \cdot N + F$$
$$4500 = U \cdot 250 + 1500$$
$$4500 = 250U + 1500$$
$$4500 - 1500 = 250U + 1500 - 1500$$
$$3000 = 250U$$
$$\frac{3000}{250} = \frac{250U}{250}$$
$$12 = U$$

The cost per unit is $12.

SECTION 11.4

You Try It 1

$$\tfrac{1}{5}x - 2 = \tfrac{2}{5}x + 4$$
$$\tfrac{1}{5}x - \tfrac{2}{5}x - 2 = \tfrac{2}{5}x - \tfrac{2}{5}x + 4 \quad \bullet \text{ Subtract } \tfrac{2}{5}x \text{ from}$$
$$\qquad\qquad\qquad\qquad\qquad \text{each side.}$$
$$-\tfrac{1}{5}x - 2 = 4$$
$$-\tfrac{1}{5}x - 2 + 2 = 4 + 2 \quad \bullet \text{ Add 2 to each side.}$$
$$-\tfrac{1}{5}x = 6$$
$$(-5)\left(-\tfrac{1}{5}x\right) = (-5)6 \quad \bullet \text{ Multiply each side}$$
$$\qquad\qquad\qquad\qquad \text{by } -5.$$
$$x = -30$$

The solution is -30.

You Try It 2

$$4(x - 1) - x = 5$$
$$4x - 4 - x = 5$$
$$3x - 4 = 5$$
$$3x - 4 + 4 = 5 + 4$$
$$3x = 9$$
$$\frac{3x}{3} = \frac{9}{3}$$
$$x = 3$$

The solution is 3.

You Try It 3

$$2x - 7(3x + 1) = 5(5 - 3x)$$
$$2x - 21x - 7 = 25 - 15x \qquad\qquad\quad \bullet \text{ Distributive Property}$$
$$-19x - 7 = 25 - 15x \qquad\qquad\qquad \bullet \text{ Combine like terms.}$$
$$-19x + 15x - 7 = 25 - 15x + 15x \quad \bullet \text{ Add } 15x \text{ to both sides.}$$
$$-4x - 7 = 25 \qquad\qquad\qquad\qquad \bullet \text{ Combine like terms.}$$
$$-4x - 7 + 7 = 25 + 7 \qquad\qquad\quad \bullet \text{ Add 7 to both sides.}$$
$$-4x = 32 \qquad\qquad\qquad\qquad\quad \bullet \text{ Combine like terms.}$$
$$\frac{-4x}{-4} = \frac{32}{-4} \qquad\qquad\qquad\qquad \bullet \text{ Divide each side by } -4.$$
$$x = -8$$

The solution is -8.

SECTION 11.5

You Try It 1 $8 - 2t$

You Try It 2 $\dfrac{5}{7x}$

You Try It 3   The *product* of a number and *one-half* of the number

The unknown number: x

One-half the number: $\tfrac{1}{2}x$

$$(x)\left(\tfrac{1}{2}x\right)$$

SECTION 11.6

You Try It 1   The unknown number: x

A number increased by four	equals	twelve

$$x + 4 = 12$$
$$x + 4 - 4 = 12 - 4$$
$$x = 8$$

The number is 8.

You Try It 2   The unknown number: x

$$2x = 10$$
$$\frac{2x}{2} = \frac{10}{2}$$
$$x = 5$$

The number is 5.

You Try It 3   The unknown number: x

$$3x + 6 = 4$$
$$3x + 6 - 6 = 4 - 6$$
$$3x = -2$$
$$\frac{3x}{3} = \frac{-2}{3}$$
$$x = -\frac{2}{3}$$

The number is $-\frac{2}{3}$.

You Try It 4   The unknown number: x

$$\frac{1}{2}x + 3 = 9$$
$$\frac{1}{2}x + 3 - 3 = 9 - 3$$
$$\frac{1}{2}x = 6$$
$$2 \cdot \frac{1}{2}x = 2 \cdot 6$$
$$x = 12$$

The number is 12.

You Try It 5

Strategy   To find the regular price, write and solve an equation using R to represent the regular price of the baseball jersey.

Solution

| $38.95 | is | $11 less than the regular price |

$$38.95 = R - 11$$
$$38.95 + 11 = R - 11 + 11$$
$$49.95 = R$$

The regular price of the baseball jersey is $49.95.

You Try It 6

Strategy   To find the rpm of the engine when it is in third gear, write and solve an equation using R to represent the rpm of the engine in third gear.

Solution

$$2500 = \frac{2}{3}R$$
$$\frac{3}{2}(2500) = \left(\frac{3}{2}\right)\frac{2}{3}R$$
$$3750 = R$$

The rpm of the engine in third gear is 3750 rpm.

You Try It 7

Strategy   To find the number of hours of labor, write and solve an equation using H to represent the number of hours of labor required.

Solution

| $492 | includes | $100 for materials plus $24.50 per hour of labor |

$$492 = 100 + 24.50H$$
$$492 - 100 = 100 - 100 + 24.50H$$
$$392 = 24.50H$$
$$\frac{392}{24.50} = \frac{24.50H}{24.50}$$
$$16 = H$$

16 h of labor are required.

You Try It 8

Strategy   To find the total sales, write and solve an equation using S to represent the total sales.

Solution

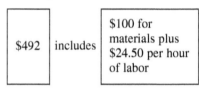

$$2500 = 800 + 0.08S$$
$$2500 - 800 = 800 - 800 + 0.08S$$
$$1700 = 0.08S$$
$$\frac{1700}{0.08} = \frac{0.08S}{0.08}$$
$$21{,}250 = S$$

Natalie's total sales for the month were $21,250.

SOLUTIONS TO CHAPTER 12 "YOU TRY IT"

SECTION 12.1

You Try It 1

$$QT = QR + RS + ST$$
$$62 = 24 + RS + 17$$
$$62 = 41 + RS$$
$$62 - 41 = 41 - 41 + RS$$
$$21 = RS$$

You Try It 2

Let x represent the supplement of a $32°$ angle. The sum of supplementary angles is $180°$.

$$x + 32° = 180°$$
$$x + 32° - 32° = 180° - 32°$$
$$x = 148°$$

$148°$ is the supplement of $32°$.

You Try It 3

$$\angle a + 68° = 118°$$
$$\angle a + 68° - 68° = 118° - 68°$$
$$\angle a = 50°$$

You Try It 4

In a right triangle, one angle measures $90°$ and the two acute angles are complementary.

$$\angle A + \angle B = 90°$$
$$\angle A + 7° = 90°$$
$$\angle A + 7° - 7° = 90° - 7°$$
$$\angle A = 83°$$

The other angles measure $90°$ and $83°$.

You Try It 5

The sum of the three angles of a triangle is $180°$.

$$\angle A + \angle B + \angle C = 180°$$
$$\angle A + 62° + 45° = 180°$$
$$\angle A + 107° = 180°$$
$$\angle A + 107° - 107° = 180° - 107°$$
$$\angle A = 73°$$

The measure of the other angle is $73°$.

You Try It 6

$$r = \frac{1}{2}d$$

$$r = \frac{1}{2}(8 \text{ in.}) = 4 \text{ in.}$$

The radius is 4 in.

You Try It 7

Angles a and b are supplementary angles.

$$\angle a + \angle b = 180°$$
$$125° + \angle b = 180°$$
$$125° - 125° + \angle b = 180° - 125°$$
$$\angle b = 55°$$

You Try It 8

$\angle c$ and $\angle a$ are corresponding angles. Corresponding angles are equal.

$$\angle c = \angle a = 120°$$

$\angle b$ and $\angle c$ are supplementary angles.

$$\angle b + \angle c = 180°$$
$$\angle b + 120° = 180°$$
$$\angle b + 120° - 120° = 180° - 120°$$
$$\angle b = 60°$$

SECTION 12.2

You Try It 1

$$P = 2L + 2W$$
$$= 2(2 \text{ m}) + 2(0.85 \text{ m})$$
$$= 4 \text{ m} + 1.7 \text{ m}$$
$$= 5.7 \text{ m}$$

The perimeter of the rectangle is 5.7 m.

You Try It 2

$$P = a + b + c$$
$$= 12 \text{ cm} + 15 \text{ cm} + 18 \text{ cm}$$
$$= 45 \text{ cm}$$

The perimeter of the triangle is 45 cm.

You Try It 3

$$C = \pi d$$
$$\approx 3.14 \cdot 6 \text{ in.}$$
$$= 18.84 \text{ in.}$$

The circumference is approximately 18.84 in.

You Try It 4

$$\begin{matrix} \text{Perimeter} & \text{two} & \text{the} \\ \text{of} & \text{lengths} & \text{circumference} \\ \text{composite} = \text{of a} & + \text{of a} \\ \text{figure} & \text{rectangle} & \text{circle} \end{matrix}$$

$$P = 2L + \pi d$$
$$\approx 2(8 \text{ in.}) + 3.14(3 \text{ in.})$$
$$= 16 \text{ in.} + 9.42 \text{ in.}$$
$$= 25.42 \text{ in.}$$

The perimeter is approximately 25.42 in.

You Try It 5

Strategy To find the perimeter, use the formula for the perimeter of a rectangle.

Solution
$$P = 2L + 2W$$
$$= 2(11 \text{ in.}) + 2\left(8\frac{1}{2} \text{ in.}\right)$$
$$= 22 \text{ in.} + 17 \text{ in.}$$
$$= 39 \text{ in.}$$

The perimeter of the computer paper is 39 in.

You Try It 6

Strategy To find the cost:

- Find the perimeter of the workbench.
- Multiply the perimeter by the per-meter cost of the stripping.

Solution
$$P = 2L + 2W$$
$$= 2(3 \text{ m}) + 2(0.74 \text{ m})$$
$$= 6 \text{ m} + 1.48 \text{ m}$$
$$= 7.48 \text{ m}$$

$$\$4.49 \times 7.48 = \$33.5852$$

The cost is $33.59.

SECTION 12.3

You Try It 1 $A = \frac{1}{2}bh = \frac{1}{2}(24 \text{ in.})(14 \text{ in.}) = 168 \text{ in}^2$
The area is 168 in².

You Try It 2

$A = $ area of rectangle $-$ area of triangle

$A = LW - \frac{1}{2}bh$

$= (10 \text{ in.} \times 6 \text{ in.}) - \left(\frac{1}{2} \times 6 \text{ in.} \times 4 \text{ in.}\right)$

$= 60 \text{ in}^2 - 12 \text{ in}^2$

$= 48 \text{ in}^2$

The area is 48 in².

You Try It 3

Strategy   To find the area of the room:
- Find the area in square feet.
- Convert to square yards.

Solution $A = LW$
$= 12 \text{ ft} \cdot 9 \text{ ft}$
$= 108 \text{ ft}^2$

$108 \text{ ft}^2 \times \frac{1 \text{ yd}^2}{9 \text{ ft}^2} = \frac{108}{9} \text{ yd}^2$
$= 12 \text{ yd}^2$

The area of the room is 12 yd².

SECTION 12.4

You Try It 1 $V = LWH$
$= (8 \text{ cm})(3.5 \text{ cm})(4 \text{ cm})$
$= 112 \text{ cm}^3$

The volume is 112 cm³.

You Try It 2 $V = s^3$
$= (5 \text{ cm})^3$
$= 125 \text{ cm}^3$

The volume is 125 cm³.

You Try It 3 $r = \frac{1}{2}d = \frac{1}{2}(14 \text{ in.}) = 7 \text{ in.}$

$V = \pi r^2 h$

$\approx \frac{22}{7}(7 \text{ in.})^2(15 \text{ in.})$

$= 2310 \text{ in}^3$

The volume is approximately 2310 in³.

You Try It 4 $V = \frac{4}{3}\pi r^3$

$\approx \frac{4}{3}(3.14)(3 \text{ m})^3$

$= 113.04 \text{ m}^3$

The volume is approximately 113.04 m³.

You Try It 5

$V = $ volume of rectangular solid $+$ volume of cylinder

$V = LWH + \pi r^2 h$

$\approx (1.5 \text{ m})(0.4 \text{ m})(0.4 \text{ m}) + 3.14(0.8 \text{ m})^2(0.2 \text{ m})$

$= 0.24 \text{ m}^3 + 0.40192 \text{ m}^3$

$= 0.64192 \text{ m}^3$

The volume is approximately 0.64192 m³.

You Try It 6

$V = $ volume of rectangular solid $+ \frac{1}{2}$ volume of cylinder

$V = LWH + \frac{1}{2}\pi r^2 h$

$\approx (24 \text{ in.})(6 \text{ in.})(4 \text{ in.}) + \frac{1}{2}(3.14)(3 \text{ in.})^2(24 \text{ in.})$

$= 576 \text{ in}^3 + 339.12 \text{ in}^3$

$= 915.12 \text{ in}^3$

The volume is approximately 915.12 in³.

You Try It 7

Strategy   To find the volume of the freezer, use the formula for the volume of a rectangular solid.

Solution $V = LWH$
$= (7 \text{ ft})(2.5 \text{ ft})(3 \text{ ft})$
$= 52.5 \text{ ft}^3$

The volume of the freezer is 52.5 ft³.

You Try It 8

Strategy   To find the volume of the channel iron, subtract the volume of the cut-out rectangular solid from the volume of the large rectangular solid.

Solution

$V = LWH - LWH$
$= (10 \text{ ft})(0.5 \text{ ft})(0.8 \text{ ft}) - (10 \text{ ft})(0.3 \text{ ft})(0.2 \text{ ft})$
$= 4 \text{ ft}^3 - 0.6 \text{ ft}^3$
$= 3.4 \text{ ft}^3$

The volume of the channel iron is 3.4 ft³.

SECTION 12.5

You Try It 1   a. $\sqrt{16} = 4$
$\sqrt{169} = 13$

b. $\sqrt{32} \approx 5.657$
$\sqrt{162} \approx 12.728$

You Try It 2   Hypotenuse $= \sqrt{(\text{leg})^2 + (\text{leg})^2}$
$= \sqrt{8^2 + 11^2}$
$= \sqrt{64 + 121}$
$= \sqrt{185}$
≈ 13.601

The hypotenuse is approximately 13.601 in.

You Try It 3

$$\text{Leg} = \sqrt{(\text{hypotenuse})^2 - (\text{leg})^2}$$
$$= \sqrt{12^2 - 5^2}$$
$$= \sqrt{144 - 25}$$
$$= \sqrt{119}$$
$$\approx 10.909$$

The length of the leg is approximately 10.909 ft.

You Try It 4

Strategy To find the distance between the holes, use the Pythagorean Theorem. The hypotenuse is the distance between the holes. The length of each leg is given (3 cm and 8 cm).

Solution
$$\text{Hypotenuse} = \sqrt{(\text{leg})^2 + (\text{leg})^2}$$
$$= \sqrt{3^2 + 8^2}$$
$$= \sqrt{9 + 64}$$
$$= \sqrt{73}$$
$$\approx 8.544$$

The distance is approximately 8.544 cm.

SECTION 12.6

You Try It 1 $\dfrac{4 \cancel{\text{ cm}}}{7 \cancel{\text{ cm}}} = \dfrac{4}{7}$

You Try It 2

$$\frac{AB}{DE} = \frac{AC}{DF}$$
$$\frac{7 \cancel{\text{ cm}}}{14 \cancel{\text{ cm}}} = \frac{3 \text{ cm}}{x}$$
$$7x = 14 \cdot 3 \text{ cm}$$
$$7x = 42 \text{ cm}$$
$$\frac{7x}{7} = \frac{42 \text{ cm}}{7}$$
$$x = 6 \text{ cm}$$

Side DF is 6 cm.

You Try It 3 $AC = DF$, $\angle ACB = \angle DFE$, but $BC \neq EF$ because $\angle CAB \neq \angle FDE$. Therefore, the triangles are not congruent.

You Try It 4

$$\frac{AC}{DF} = \frac{\text{height } CH}{\text{height } FG}$$
$$\frac{10 \cancel{\text{ m}}}{15 \cancel{\text{ m}}} = \frac{7 \text{ m}}{h}$$
$$10h = 15 \cdot 7 \text{ m}$$
$$10h = 105 \text{ m}$$
$$\frac{10h}{10} = \frac{105 \text{ m}}{10}$$
$$h = 10.5 \text{ m}$$

The height FG is 10.5 m.

You Try It 5

Strategy To find the perimeter of triangle ABC:

- Solve a proportion to find the lengths of sides BC and AC.
- Use the formula $P = a + b + c$.

Solution

$$\frac{BC}{EF} = \frac{AB}{DE}$$
$$\frac{BC}{10 \text{ in.}} = \frac{4 \cancel{\text{ in.}}}{8 \cancel{\text{ in.}}}$$
$$8(BC) = 10 \text{ in.}(4)$$
$$8(BC) = 40 \text{ in.}$$
$$\frac{8(BC)}{8} = \frac{40 \text{ in.}}{8}$$
$$BC = 5 \text{ in.}$$

$$\frac{AC}{DF} = \frac{AB}{DE}$$
$$\frac{AC}{6 \text{ in.}} = \frac{4 \cancel{\text{ in.}}}{8 \cancel{\text{ in.}}}$$
$$8(AC) = 6 \text{ in.}(4)$$
$$8(AC) = 24 \text{ in.}$$
$$\frac{8(AC)}{8} = \frac{24 \text{ in.}}{8}$$
$$AC = 3 \text{ in.}$$

$$\text{Perimeter} = 4 \text{ in.} + 5 \text{ in.} + 3 \text{ in.}$$
$$= 12 \text{ in.}$$

The perimeter of triangle ABC is 12 in.

Answers to Selected Exercises

ANSWERS TO CHAPTER 1 SELECTED EXERCISES

PREP TEST
1. 8   **2.** 1 2 3 4 5 6 7 8 9 10   **3.** a and D; b and E; c and A; d and B; e and F; f and C

SECTION 1.1
1.   **3.**   **5.** $37 < 49$ **7.** $101 > 87$
9. $2701 > 2071$ **11.** $107 > 0$ **13.** Yes   **15.** Millions   **17.** Hundred-thousands   **19.** Three thousand seven hundred ninety   **21.** Fifty-eight thousand four hundred seventy-three   **23.** Four hundred ninety-eight thousand five hundred twelve   **25.** Six million eight hundred forty-two thousand seven hundred fifteen   **27.** 357   **29.** 63,780
31. 7,024,709   **33.** $5000 + 200 + 80 + 7$ **35.** $50,000 + 8000 + 900 + 40 + 3$ **37.** $200,000 + 500 + 80 + 3$
39. $400,000 + 3000 + 700 + 5$ **41.** No   **43.** 850   **45.** 4000   **47.** 53,000   **49.** 630,000   **51.** 250,000
53. 72,000,000   **55.** No. Round 3846 to the nearest hundred.

SECTION 1.2
1. 28   **3.** 125   **5.** 102   **7.** 154   **9.** 1489   **11.** 828   **13.** 1584   **15.** 1219   **17.** 102,317   **19.** 79,326
21. 1804   **23.** 1579   **25.** 19,740   **27.** 7420   **29.** 120,570   **31.** 207,453   **33.** 24,218   **35.** 11,974
37. 9323   **39.** 77,139   **41.** 14,383   **43.** 9473   **45.** 33,247   **47.** 5058   **49.** 1992   **51.** 68,263
53. Cal.: 17,754   **55.** Cal.: 2872   **57.** Cal.: 101,712   **59.** Cal.: 158,763   **61.** Cal.: 261,595   **63.** Cal.: 946,718
    Est.: 17,700   Est.: 2900   Est.: 101,000   Est.: 158,000   Est.: 260,000   Est.: 940,000
65. Commutative Property of Addition   **67.** There were 118,295 multiple births during the year.   **69.** The estimated income from the four *Star Wars* movies was $1,500,000,000.   **71. a.** The income from the two movies with the lowest box-office returns is $599,300,000.   **b.** Yes, this income exceeds the income from the 1977 *Star Wars* production.   **73. a.** During the three days, 1285 miles will be driven.   **b.** At the end of the trip, the odometer will read 69,977 miles.   **75.** The total length of the trail is 740 miles.   **77.** 11 different sums   **79.** No. $0 + 0 = 0$ **81.** 10 numbers

SECTION 1.3
1. 4   **3.** 4   **5.** 10   **7.** 4   **9.** 9   **11.** 22   **13.** 60   **15.** 66   **17.** 31   **19.** 901   **21.** 791   **23.** 1125
25. 3131   **27.** 47   **29.** 925   **31.** 4561   **33.** 3205   **35.** 1222   **37.** 53   **39.** 29   **41.** 8   **43.** 37   **45.** 58
47. 574   **49.** 337   **51.** 1423   **53.** 754   **55.** 2179   **57.** 6489   **59.** 889   **61.** 71,129   **63.** 698   **65.** 29,405
67. 49,624   **69.** 628   **71.** 6532   **73.** 4286   **75.** 4042   **77.** 5209   **79.** 10,378   **81.** (ii) and (iii)   **83.** 11,239
85. 8482   **87.** 625   **89.** 76,725   **91.** 23   **93.** 4648   **95.** Cal.: 29,837   **97.** Cal.: 36,668   **99.** Cal.: 101,998
                                        Est.: 30,000   Est.: 40,000   Est.: 100,000
101. a. The honey bee has 91 more smell genes than the mosquito.   **b.** The mosquito has eight more taste genes than the fruit fly.   **c.** The honey bee has the best sense of smell.   **d.** The honey bee has the worst sense of taste.   **103.** The difference between the maximum eruption heights is 15 feet.   **105.** 202,345 more women than men earned a bachelor's degree.   **107. a.** The smallest expected increase occurs from 2010 to 2012.   **b.** The greatest expected increase occurs from 2018 to 2020.   **109.** Your new credit card balance is $360.

SECTION 1.4

1. 6×2 or $6 \cdot 2$ **3.** 4×7 or $4 \cdot 7$ **5.** 12   **7.** 35   **9.** 25   **11.** 0   **13.** 72   **15.** 198   **17.** 335
19. 2492   **21.** 5463   **23.** 4200   **25.** 6327   **27.** 1896   **29.** 5056   **31.** 1685   **33.** 46,963   **35.** 59,976
37. 19,120   **39.** 19,790   **41.** 140   **43.** 22,456   **45.** 18,630   **47.** 336   **49.** 910   **51.** 63,063   **53.** 33,520
55. 380,834   **57.** 541,164   **59.** 400,995   **61.** 105,315   **63.** 428,770   **65.** 260,000   **67.** 344,463   **69.** 41,808
71. 189,500   **73.** 401,880   **75.** 1,052,763   **77.** 4,198,388   **79.** For example, 5 and 20   **81.** 198,423
83. 18,834   **85.** 260,178   **87.** Cal.: 440,076   **89.** Cal.: 6,491,166   **91.** Cal.: 18,728,744   **93.** Cal.: 57,691,192
Est.: 450,000   Est.: 6,300,000   Est.: 18,000,000   Est.: 54,000,000
95. The car could travel 516 miles on 12 gallons of gas.   **97.** The perimeter is 64 miles.   **99. a.** eHarmony can take credit for 630 marriages a week.   **b.** eHarmony can take credit for 32,850 marriages a year.   **101.** The estimated cost of the electricians' labor is $5100.   **103.** The total cost is $2138.   **105.** There are 12 accidental deaths each hour; 288 deaths each day; and 105,120 deaths each year.

SECTION 1.5

1. 2   **3.** 6   **5.** 7   **7.** 16   **9.** 210   **11.** 44   **13.** 703   **15.** 910   **17.** 21,560   **19.** 3580   **21.** 1075
23. 1   **25.** 47   **27.** 23   **29.** 3 r1   **31.** 9 r7   **33.** 16 r1   **35.** 10 r4   **37.** 90 r3   **39.** 120 r5   **41.** 309 r3
43. 1160 r4   **45.** 708 r2   **47.** 3825 r1   **49.** 9044 r2   **51.** 11,430   **53.** 510   **55.** False   **57.** 1 r38
59. 1 r26   **61.** 21 r21   **63.** 30 r22   **65.** 5 r40   **67.** 9 r17   **69.** 200 r21   **71.** 303 r1   **73.** 67 r13
75. 176 r13   **77.** 1086 r7   **79.** 403   **81.** 12 r456   **83.** 4 r160   **85.** 160 r27   **87.** 1669 r14   **89.** 7950
91. Cal.: 5129   **93.** Cal.: 21,968   **95.** Cal.: 24,596   **97.** Cal.: 2836   **99.** Cal.: 3024   **101.** Cal.: 32,036
Est.: 5000   Est.: 20,000   Est.: 22,500   Est.: 3000   Est.: 3000   Est.: 30,000
103. The average monthly claim for theft is $25,000.   **105.** The average number of hours worked by employees in Britain is 35 hours.   **107.** 380 pennies are in circulation for each person.   **109.** On average, each household will receive 175 pieces of mail.   **111.** (i) and (iii)   **113.** The total of the deductions is $350.   **115.** 49,500,000 more cases of eggs were sold by retail stores.   **117.** The average monthly expense for housing is $976.   **119.** A major's annual pay is $75,024.   **121.** The total amount paid is $11,860.

SECTION 1.6

1. 2^3 **3.** $6^3 \cdot 7^4$ **5.** $2^3 \cdot 3^3$ **7.** $5 \cdot 7^5$ **9.** $3^3 \cdot 6^4$ **11.** $3^3 \cdot 5 \cdot 9^3$ **13.** 8   **15.** 400   **17.** 900   **19.** 972
21. 120   **23.** 360   **25.** 0   **27.** 90,000   **29.** 540   **31.** 4050   **33.** 11,025   **35.** 25,920   **37.** 4,320,000   **39.** 5
41. 4   **43.** 23   **45.** 5   **47.** 10   **49.** 7   **51.** 8   **53.** 6   **55.** 52   **57.** 26   **59.** 52   **61.** 42   **63.** 8   **65.** 16
67. 6   **69.** 8   **71.** 3   **73.** 4   **75.** 13   **77.** 0   **79.** $8 - 2 \cdot (3 + 1)$

SECTION 1.7

1. 1, 2, 4   **3.** 1, 2, 5, 10   **5.** 1, 7   **7.** 1, 3, 9   **9.** 1, 13   **11.** 1, 2, 3, 6, 9, 18   **13.** 1, 2, 4, 7, 8, 14, 28, 56
15. 1, 3, 5, 9, 15, 45   **17.** 1, 29   **19.** 1, 2, 11, 22   **21.** 1, 2, 4, 13, 26, 52   **23.** 1, 2, 41, 82   **25.** 1, 3, 19, 57
27. 1, 2, 3, 4, 6, 8, 12, 16, 24, 48   **29.** 1, 5, 19, 95   **31.** 1, 2, 3, 6, 9, 18, 27, 54   **33.** 1, 2, 3, 6, 11, 22, 33, 66
35. 1, 2, 4, 5, 8, 10, 16, 20, 40, 80   **37.** 1, 2, 3, 4, 6, 8, 12, 16, 24, 32, 48, 96   **39.** 1, 2, 3, 5, 6, 9, 10, 15, 18, 30, 45, 90
41. True   **43.** $2 \cdot 3$ **45.** Prime   **47.** $2 \cdot 2 \cdot 2 \cdot 3$ **49.** $3 \cdot 3 \cdot 3$ **51.** $2 \cdot 2 \cdot 3 \cdot 3$ **53.** Prime   **55.** $2 \cdot 3 \cdot 3 \cdot 5$
57. $5 \cdot 23$ **59.** $2 \cdot 3 \cdot 3$ **61.** $2 \cdot 2 \cdot 7$ **63.** Prime   **65.** $2 \cdot 31$ **67.** $2 \cdot 11$ **69.** Prime   **71.** $2 \cdot 3 \cdot 11$
73. $2 \cdot 37$ **75.** Prime   **77.** $5 \cdot 11$ **79.** $2 \cdot 2 \cdot 2 \cdot 3 \cdot 5$ **81.** $2 \cdot 2 \cdot 2 \cdot 2 \cdot 2 \cdot 5$ **83.** $2 \cdot 2 \cdot 2 \cdot 3 \cdot 3 \cdot 3$
85. $5 \cdot 5 \cdot 5 \cdot 5$ **87.** False

CHAPTER 1 CONCEPT REVIEW*

1. The symbol $<$ means "is less than." A number that appears to the left of a given number on the number line is less than ($<$) the given number. For example, $4 < 9$. The symbol $>$ means "is greater than." A number that appears to the right of a given number on the number line is greater than ($>$) the given number. For example, $5 > 2$.   [1.1A]

*Note: The numbers in brackets following the answers in the Concept Review are a reference to the objective that corresponds to that problem. For example, the reference [1.2A] stands for Section 1.2, Objective A. This notation will be used for all Prep Tests, Concept Reviews, Chapter Reviews, Chapter Tests, and Cumulative Reviews throughout the text.

2. To round a four-digit whole number to the nearest hundred, look at the digit in the tens place. If the digit in the tens place is less than 5, that digit and the digit in the ones place are replaced by zeros. If the digit in the tens place is greater than or equal to 5, increase the digit in the hundreds place by 1 and replace the digits in the tens place and the ones place by zeros.   [1.1D]

3. The Commutative Property of Addition states that two numbers can be added in either order; the sum is the same. For example, $3 + 5 = 5 + 3$. The Associative Property of Addition states that changing the grouping of three or more addends does not change their sum. For example, $3 + (4 + 5) = (3 + 4) + 5$. Note that in the Commutative Property of Addition, the order in which the numbers appear changes, while in the Associative Property of Addition, the order in which the numbers appear does not change.   [1.2A]

4. To estimate the sum of two numbers, round each number to the same place value. Then add the numbers. For example, to estimate the sum of 562,397 and 41,086, round the numbers to 560,000 and 40,000. Then add $560,000 + 40,000 = 600,000$.   [1.2A]

5. It is necessary to borrow when performing subtraction if, in any place value, the lower digit is larger than the upper digit.   [1.3B]

6. The Multiplication Property of Zero states that the product of a number and zero is zero. For example, $8 \times 0 = 0$. The Multiplication Property of One states that the product of a number and one is the number. For example, $8 \times 1 = 8$.   [1.4A]

7. To multiply a whole number by 100, write two zeros to the right of the number. For example, $64 \times 100 = 6400$.   [1.4B]

8. To estimate the product of two numbers, round each number so that it contains only one nonzero digit. Then multiply. For example, to estimate the product of 87 and 43, round the two numbers to 90 and 40; then multiply $90 \times 40 = 3600$.   [1.4B]

9. $0 \div 9 = 0$. Zero divided by any whole number except zero is zero. $9 \div 0$ is undefined. Division by zero is not allowed.   [1.5A]

10. To check the answer to a division problem that has a remainder, multiply the quotient by the divisor. Add the remainder to the product. The result should be the dividend. For example, $16 \div 5 = 3$ r1. Check: $(3 \times 5) + 1 = 16$, the dividend.   [1.5B]

11. The steps in the Order of Operations Agreement are:
 1. Do all operations inside parentheses.
 2. Simplify any number expressions containing exponents.
 3. Do multiplication and division as they occur from left to right.
 4. Do addition and subtraction as they occur from left to right.   [1.6B]

12. A number is a factor of another number if it divides that number evenly (there is no remainder). For example, 7 is a factor of 21 because $21 \div 7 = 3$, with a remainder of 0.   [1.7A]

13. Three is a factor of a number if the sum of the digits of the number is divisible by 3. For the number 285, $2 + 8 + 5 = 15$, which is divisible by 3. Thus 285 is divisible by 3.   [1.7A]

CHAPTER 1 REVIEW EXERCISES

1. 600   [1.6A]   **2.** $10,000 + 300 + 20 + 7$ [1.1C]   **3.** 1, 2, 3, 6, 9, 18   [1.7A]   **4.** 12,493   [1.2A]   **5.** 1749   [1.3B]

6. 2135   [1.5A]   **7.** $101 > 87$ [1.1A]   **8.** $5^2 \cdot 7^5$ [1.6A]   **9.** 619,833   [1.4B]   **10.** 5409   [1.3B]   **11.** 1081   [1.2A]

12. 2   [1.6B]   **13.** 45,700   [1.1D]   **14.** Two hundred seventy-six thousand fifty-seven   [1.1B]   **15.** 1306 r59   [1.5C]

16. 2,011,044   [1.1B]   **17.** 488 r2   [1.5B]   **18.** 17   [1.6B]   **19.** 32   [1.6B]   **20.** $2 \cdot 2 \cdot 2 \cdot 3 \cdot 3$ [1.7B]

21. 2133   [1.3A]   **22.** 22,761   [1.4B]   **23.** The total pay for last week's work is $768.   [1.4C]   **24.** He drove 27 miles per gallon of gasoline.   [1.5D]   **25.** Each monthly car payment is $560.   [1.5D]   **26.** The total income from commissions is $2567.   [1.2B]   **27.** The total amount deposited is $301. The new checking account balance is $817.   [1.2B]   **28.** The total of the car payments is $2952.   [1.4C]   **29.** More males were involved in college sports in 2005 than in 1972.   [1.1A]   **30.** The difference between the numbers of male and female athletes in 1972 was 140,407 students.   [1.3C]   **31.** The number of female athletes increased by 175,515 students from 1972 to 2005.   [1.3C]   **32.** 296,928 more students were involved in athletics in 2005 than in 1972.   [1.3C]

CHAPTER 1 TEST

1. 432   [1.6A, Example 3]   **2.** Two hundred seven thousand sixty-eight   [1.1B, Example 3]   **3.** 9333   [1.3B, Example 3]

4. 1, 2, 4, 5, 10, 20   [1.7A, Example 1]   **5.** 6,854,144   [1.4B, HOW TO 3]   **6.** 9   [1.6B, Example 4]

7. $900,000 + 6000 + 300 + 70 + 8$ [1.1C, Example 6]   **8.** 75,000   [1.1D, Example 8]   **9.** 1121 r27   [1.5C, Example 8]

10. $3^3 \cdot 7^2$ [1.6A, Example 1]   **11.** 54,915   [1.2A, Example 1]   **12.** $2 \cdot 2 \cdot 3 \cdot 7$ [1.7B, Example 2]   **13.** 4 [1.6B, Example 4]   **14.** 726,104   [1.4A, Example 1]   **15.** 1,204,006   [1.1B, Example 4]   **16.** 8710 r2   [1.5B, Example 5]

17. $21 > 19$ [1.1A, Example 2]   **18.** 703   [1.5A, Example 3]   **19.** 96,798   [1.2A, Example 3]   **20.** 19,922 [1.3B, Example 4]   **21.** The difference in projected total enrollment between 2016 and 2013 is 1,908,000 students.   [1.3C, Example 6]

22. The average projected enrollment in each of the grades 9 through 12 in 2016 is 4,171,000 students.   [1.5D, HOW TO 3]

23. 3000 boxes were needed to pack the lemons.   [1.5D, Example 10]   **24.** The investor receives $2844 over the 12-month period. [1.4C, Example 3]   **25. a.** 855 miles were driven during the 3 days.   **b.** The odometer reading at the end of the 3 days is 48,481 miles.   [1.2B, Example 4]

ANSWERS TO CHAPTER 2 SELECTED EXERCISES

PREP TEST

1. 20 [1.4A] **2.** 120 [1.4A] **3.** 9 [1.4A] **4.** 10 [1.2A] **5.** 7 [1.3A] **6.** 2 r3 [1.5C]
7. 1, 2, 3, 4, 6, 12 [1.7A] **8.** 59 [1.6B] **9.** 7 [1.3A] **10.** $44 < 48$ [1.1A]

SECTION 2.1

1. 40 **3.** 24 **5.** 30 **7.** 12 **9.** 24 **11.** 60 **13.** 56 **15.** 9 **17.** 32 **19.** 36 **21.** 660 **23.** 9384
25. 24 **27.** 30 **29.** 24 **31.** 576 **33.** 1680 **35.** True **37.** 1 **39.** 3 **41.** 5 **43.** 25 **45.** 1 **47.** 4
49. 4 **51.** 6 **53.** 4 **55.** 1 **57.** 7 **59.** 5 **61.** 8 **63.** 1 **65.** 25 **67.** 7 **69.** 8 **71.** True
73. They will have another day off together in 12 days.

SECTION 2.2

1. Improper fraction **3.** Proper fraction **5.** $\frac{3}{4}$ **7.** $\frac{7}{8}$ **9.** $1\frac{1}{2}$ **11.** $2\frac{5}{8}$ **13.** $3\frac{3}{5}$ **15.** $\frac{5}{4}$ **17.** $\frac{8}{3}$ **19.** $\frac{28}{8}$

21. **23.** **25.** False **27.** $5\frac{1}{3}$ **29.** 2 **31.** $3\frac{1}{4}$ **33.** $14\frac{1}{2}$ **35.** 17 **37.** $1\frac{7}{9}$

39. $1\frac{4}{5}$ **41.** 23 **43.** $1\frac{15}{16}$ **45.** $6\frac{1}{3}$ **47.** 5 **49.** 1 **51.** $\frac{14}{3}$ **53.** $\frac{26}{3}$ **55.** $\frac{59}{8}$ **57.** $\frac{25}{4}$ **59.** $\frac{121}{8}$ **61.** $\frac{41}{12}$
63. $\frac{34}{9}$ **65.** $\frac{38}{3}$ **67.** $\frac{38}{7}$ **69.** $\frac{63}{5}$ **71.** $\frac{41}{9}$ **73.** $\frac{117}{14}$

SECTION 2.3

1. $\frac{5}{10}$ **3.** $\frac{9}{48}$ **5.** $\frac{12}{32}$ **7.** $\frac{9}{51}$ **9.** $\frac{12}{16}$ **11.** $\frac{27}{9}$ **13.** $\frac{20}{60}$ **15.** $\frac{44}{60}$ **17.** $\frac{12}{18}$ **19.** $\frac{35}{49}$ **21.** $\frac{10}{18}$ **23.** $\frac{21}{3}$
25. $\frac{35}{45}$ **27.** $\frac{60}{64}$ **29.** $\frac{21}{98}$ **31.** $\frac{30}{48}$ **33.** $\frac{15}{42}$ **35.** $\frac{102}{144}$ **37.** $\frac{1}{3}$ **39.** $\frac{1}{2}$ **41.** $\frac{1}{6}$ **43.** $1\frac{1}{9}$ **45.** 0 **47.** $\frac{9}{22}$
49. 3 **51.** $\frac{4}{21}$ **53.** $\frac{12}{35}$ **55.** $\frac{7}{11}$ **57.** $1\frac{1}{3}$ **59.** $\frac{3}{5}$ **61.** $\frac{1}{11}$ **63.** 4 **65.** $\frac{1}{3}$ **67.** $\frac{3}{5}$ **69.** $2\frac{1}{4}$ **71.** $\frac{1}{5}$
73. Answers will vary. For example, $\frac{4}{6}, \frac{6}{9}, \frac{8}{12}, \frac{10}{15}, \frac{12}{8}$. **75. a.** $\frac{4}{25}$ **b.** $\frac{4}{25}$

SECTION 2.4

1. $\frac{3}{7}$ **3.** 1 **5.** $1\frac{4}{11}$ **7.** $3\frac{2}{5}$ **9.** $2\frac{4}{5}$ **11.** $2\frac{1}{4}$ **13.** $1\frac{3}{8}$ **15.** $1\frac{7}{15}$ **17.** $1\frac{5}{12}$ **19.** A whole number other than 1
21. The number 1 **23.** $1\frac{1}{6}$ **25.** $\frac{13}{14}$ **27.** $\frac{53}{60}$ **29.** $1\frac{1}{56}$ **31.** $\frac{23}{60}$ **33.** $1\frac{17}{18}$ **35.** $1\frac{11}{48}$ **37.** $1\frac{9}{20}$ **39.** $2\frac{17}{120}$
41. $2\frac{5}{72}$ **43.** $\frac{39}{40}$ **45.** $1\frac{19}{24}$ **47.** (ii) **49.** $10\frac{1}{12}$ **51.** $9\frac{2}{7}$ **53.** $9\frac{47}{48}$ **55.** $8\frac{3}{13}$ **57.** $16\frac{29}{120}$ **59.** $24\frac{29}{40}$
61. $33\frac{7}{24}$ **63.** $10\frac{5}{36}$ **65.** $10\frac{5}{12}$ **67.** $14\frac{73}{90}$ **69.** $10\frac{13}{48}$ **71.** $9\frac{5}{24}$ **73.** $14\frac{1}{18}$ **75.** $11\frac{11}{12}$ **77.** No
79. The length of the shaft is $8\frac{9}{16}$ inches. **81.** The sum represents the height of the table. **83.** The total length of the course
is $10\frac{1}{2}$ miles. **85.** The wall is $6\frac{5}{8}$ inches thick. **87.** The minimum length of the bolt needed is $1\frac{7}{16}$ inches.

SECTION 2.5

1. $\frac{2}{17}$ **3.** $\frac{1}{3}$ **5.** $\frac{1}{10}$ **7.** $\frac{5}{13}$ **9.** $\frac{1}{3}$ **11.** $\frac{4}{7}$ **13.** $\frac{1}{4}$ **15.** Yes **17.** $\frac{1}{2}$ **19.** $\frac{19}{56}$ **21.** $\frac{1}{2}$ **23.** $\frac{11}{60}$ **25.** $\frac{1}{32}$
27. $\frac{19}{60}$ **29.** $\frac{5}{72}$ **31.** $\frac{11}{60}$ **33.** $\frac{29}{60}$ **35.** (i) **37.** $5\frac{1}{5}$ **39.** $4\frac{7}{8}$ **41.** $\frac{16}{21}$ **43.** $5\frac{1}{2}$ **45.** $5\frac{4}{7}$ **47.** $7\frac{5}{24}$ **49.** $1\frac{2}{5}$
51. $15\frac{11}{20}$ **53.** $4\frac{37}{45}$ **55.** No **57.** The missing dimension is $9\frac{1}{2}$ inches. **59.** The difference between Meyfarth's distance and
Coachman's distance was $9\frac{3}{8}$ inches. The difference between Kostadinova's distance and Meyfarth's distance was $5\frac{1}{4}$ inches.
61. a. The hikers plan to travel $17\frac{17}{24}$ miles the first two days. **b.** There will be $9\frac{19}{24}$ miles left to travel on the third day.
63. The difference represents how much farther the hikers plan to travel on the second day than on the first day. **65. a.** Yes
b. The wrestler needs to lose $3\frac{1}{4}$ pounds to reach the desired weight. **67.** $\frac{11}{15}$ of the electrician's income is not spent for housing.
69. $6\frac{1}{8}$

SECTION 2.6

1. $\frac{7}{12}$ **3.** $\frac{7}{48}$ **5.** $\frac{1}{48}$ **7.** $\frac{11}{14}$ **9.** 6 **11.** $\frac{5}{12}$ **13.** 6 **15.** $\frac{2}{3}$ **17.** $\frac{3}{16}$ **19.** $\frac{3}{80}$ **21.** 10 **23.** $\frac{1}{15}$ **25.** $\frac{2}{3}$

27. $\frac{7}{26}$ **29.** 4 **31.** $\frac{100}{357}$ **33.** Answers will vary. For example, $\frac{3}{4}$ and $\frac{4}{3}$. **35.** $1\frac{1}{3}$ **37.** $2\frac{1}{2}$ **39.** $\frac{9}{34}$ **41.** 10 **43.** $16\frac{2}{3}$

45. 1 **47.** $\frac{1}{2}$ **49.** 30 **51.** 42 **53.** $12\frac{2}{3}$ **55.** $1\frac{4}{5}$ **57.** $1\frac{2}{3}$ **59.** $1\frac{2}{3}$ **61.** 0 **63.** $27\frac{2}{3}$ **65.** $17\frac{85}{128}$ **67.** $\frac{2}{5}$

69. $8\frac{1}{16}$ **71.** 8 **73.** 9 **75.** $\frac{5}{8}$ **77.** $3\frac{1}{40}$ **79.** Less than **81.** The cost is $11. **83.** The length is $3\frac{1}{12}$ feet.

85. The area is $27\frac{9}{16}$ square feet. **87.** Each year $5\frac{1}{2}$ billion bushels of corn are turned into ethanol. **89.** The weight is

$54\frac{19}{36}$ pounds. **91.** The total cost of the material is $363. **93.** $\frac{1}{2}$ **95.** A

SECTION 2.7

1. $\frac{5}{6}$ **3.** 1 **5.** 0 **7.** $\frac{1}{2}$ **9.** $\frac{1}{6}$ **11.** $\frac{7}{10}$ **13.** 2 **15.** 2 **17.** $\frac{1}{6}$ **19.** 6 **21.** $\frac{1}{15}$ **23.** 2 **25.** $2\frac{1}{2}$

27. 3 **29.** $1\frac{1}{6}$ **31.** $3\frac{1}{3}$ **33.** True **35.** 6 **37.** $\frac{1}{2}$ **39.** $\frac{1}{30}$ **41.** $1\frac{4}{5}$ **43.** 13 **45.** 3 **47.** $\frac{1}{5}$ **49.** $\frac{11}{28}$

51. 120 **53.** $\frac{11}{40}$ **55.** $\frac{33}{40}$ **57.** $4\frac{4}{9}$ **59.** $\frac{13}{32}$ **61.** $10\frac{2}{3}$ **63.** $\frac{12}{53}$ **65.** $4\frac{62}{191}$ **67.** 68 **69.** $8\frac{2}{7}$ **71.** $3\frac{13}{49}$

73. 4 **75.** $1\frac{3}{5}$ **77.** $\frac{9}{34}$ **79.** False **81.** Less than **83.** There are 12 servings in 16 ounces of cereal. **85.** Each acre

costs $24,000. **87.** The nut will make 12 turns in moving $1\frac{7}{8}$ inches. **89. a.** The total weight of the fat and bone is $1\frac{5}{12}$ pounds.

b. The chef can cut 28 servings from the roast. **91.** The distance between each post is $2\frac{3}{4}$ inches. **93.** $\frac{31}{50}$ of the money borrowed

on home equity loans is spent on debt consolidation and home improvement. **95.** $\frac{1}{6}$ of the puzzle is left to complete.

97. The difference was $\frac{3}{32}$ inch. **99.** The dimensions of the board when it is closed are 14 inches by 7 inches by $1\frac{3}{4}$ inches.

101. The average teenage boy drinks 7 more cans of soda per week than the average teenage girl. **103. a.** $\frac{2}{3}$ **b.** $2\frac{5}{8}$

SECTION 2.8

1. $\frac{11}{40} < \frac{19}{40}$ **3.** $\frac{2}{3} < \frac{5}{7}$ **5.** $\frac{5}{8} > \frac{7}{12}$ **7.** $\frac{7}{9} < \frac{11}{12}$ **9.** $\frac{13}{14} > \frac{19}{21}$ **11.** $\frac{7}{24} < \frac{11}{30}$ **13.** $\frac{4}{5}$ **15.** $\frac{25}{144}$ **17.** $\frac{2}{9}$ **19.** $\frac{3}{125}$

21. $\frac{4}{45}$ **23.** $\frac{16}{1225}$ **25.** $\frac{4}{49}$ **27.** $\frac{9}{125}$ **29.** $\frac{27}{88}$ **31.** $\frac{5}{6}$ **33.** $1\frac{5}{12}$ **35.** $\frac{7}{48}$ **37.** $\frac{29}{36}$ **39.** $\frac{55}{72}$ **41.** $\frac{35}{54}$ **43.** 2

45. $\frac{9}{19}$ **47.** $\frac{7}{32}$ **49.** $\frac{64}{75}$ **51. a.** More people choose a fast-food restaurant on the basis of its location. **b.** Location was

the criterion cited by the most people.

CHAPTER 2 CONCEPT REVIEW*

1. To find the LCM of 75, 30, and 50, find the prime factorization of each number and write the factorization of each number in a table. Circle the greatest product in each column. The LCM is the product of the circled numbers.

	2	3	5
75 =		③	⑤·5
30 =	②	3	5
50 =	2		5·5

LCM = 2 · 3 · 5 · 5 = 150   [2.1A]

*Note: The numbers in brackets following the answers in the Concept Review are a reference to the objective that corresponds to that problem. For example, the reference [1.2A] stands for Section 1.2, Objective A. This notation will be used for all Prep Tests, Concept Reviews, Chapter Reviews, Chapter Tests, and Cumulative Reviews throughout the text.

2. To find the GCF of 42, 14, and 21, find the prime factorization of each number and write the factorization of each number in a table. Circle the least product in each column that does not have a blank. The GCF is the product of the circled numbers.

	2	3	7
42 =	2	3	⑦
14 =	2		7
21 =		3	7

GCF = 7   [2.1B]

3. To write an improper fraction as a mixed number, divide the numerator by the denominator. The quotient without the remainder is the whole number part of the mixed number. To write the fractional part of the mixed number, write the remainder over the divisor. [2.2B]

4. A fraction is in simplest form when the numerator and denominator have no common factors other than 1. For example, $\frac{8}{12}$ is not in simplest form because 8 and 12 have a common factor of 4. $\frac{5}{7}$ is in simplest form because 5 and 7 have no common factors other than 1.   [2.3B]

5. When adding fractions, you have to convert to equivalent fractions with a common denominator. One way to explain this is that you can combine like things, but you cannot combine unlike things. You can combine 3 apples and 4 apples and get 7 apples. You cannot combine 4 apples and 3 oranges and get a sum consisting of just one item. In adding whole numbers, you add like things: ones, tens, hundreds, and so on. In adding fractions, you can combine 2 *ninths* and 5 *ninths* and get 7 *ninths*, but you cannot add 2 *ninths* and 3 *fifths*.   [2.4B]

6. To add mixed numbers, add the fractional parts and then add the whole number parts. Then reduce the sum to simplest form. [2.4C]

7. To subtract mixed numbers, the first step is to subtract the fractional parts. If we are subtracting a mixed number from a whole number, there is no fractional part in the whole number from which to subtract the fractional part of the mixed number. Therefore, we must borrow a 1 from the whole number and write 1 as an equivalent fraction with a denominator equal to the denominator of the fraction in the mixed number. Then we can subtract the fractional parts and then subtract the whole numbers.   [2.5C]

8. When multiplying two fractions, it is better to eliminate the common factors before multiplying the remaining factors in the numerator and denominator so that (1) we don't end up with very large products and (2) we don't have the added step of simplifying the resulting fraction.   [2.6A]

9. Let's look at an example, $\frac{1}{2} \times \frac{1}{3} = \frac{1}{6}$. The fractions $\frac{1}{2}$ and $\frac{1}{3}$ are less than 1. The product, $\frac{1}{6}$, is less than $\frac{1}{2}$ and less than $\frac{1}{3}$. Therefore, the product is less than the smaller number.   [2.6A]

10. Reciprocals are used to rewrite division problems as related multiplication problems. Since "divided by" means the same thing as "times the reciprocal of," we can change the division sign to a multiplication sign and change the divisor to its reciprocal. For example, $9 \div 3 = 9 \times \frac{1}{3}$.   [2.7A]

11. When a fraction is divided by a whole number, we write the whole number as a fraction before dividing so that we can easily determine the reciprocal of the whole number.   [2.7B]

12. When comparing fractions, we must first look at the denominators. If they are not the same, we must rewrite the fractions as equivalent fractions with a common denominator. If the denominators are the same, we must look at the numerators. The fraction that has the smaller numerator is the smaller fraction.   [2.8A]

13. We must follow the Order of Operations Agreement in simplifying the expression $\left(\frac{5}{6}\right)^2 - \left(\frac{3}{4} - \frac{2}{3}\right) \div \frac{1}{2}$. Therefore, we must

first simplify the expression $\frac{3}{4} - \frac{2}{3}$ inside the parentheses: $\frac{3}{4} - \frac{2}{3} = \frac{1}{12}$. Then simplify the exponential expression: $\left(\frac{5}{6}\right)^2 = \frac{25}{36}$.

Then perform the division: $\frac{1}{12} \div \frac{1}{2} = \frac{1}{6}$. Then perform the subtraction: $\frac{25}{36} - \frac{1}{6} = \frac{19}{36}$.   [2.8C]

CHAPTER 2 REVIEW EXERCISES

1. $\frac{2}{3}$ [2.3B]   2. $\frac{5}{16}$ [2.8B]   3. $\frac{13}{4}$ [2.2A]   4. $1\frac{13}{18}$ [2.4B]   5. $\frac{11}{18} < \frac{17}{24}$ [2.8A]   6. $14\frac{19}{42}$ [2.5C]   7. $\frac{5}{36}$ [2.8C]

8. $9\frac{1}{24}$ [2.6B]   9. 2   [2.7B]   10. $\frac{25}{48}$ [2.5B]   11. $3\frac{1}{3}$ [2.7B]   12. 4   [2.1B]   13. $\frac{24}{36}$ [2.3A]   14. $\frac{3}{4}$ [2.7A]

15. $\frac{32}{44}$ [2.3A]   16. $16\frac{1}{2}$ [2.6B]   17. 36   [2.1A]   18. $\frac{4}{11}$ [2.3B]   19. $1\frac{1}{8}$ [2.4A]   20. $10\frac{1}{8}$ [2.5C]

21. $18\frac{13}{54}$ [2.4C]   22. 5   [2.1B]   23. $3\frac{2}{5}$ [2.2B]   24. $\frac{1}{15}$ [2.8C]   25. $5\frac{7}{8}$ [2.4C]   26. 54   [2.1A]

27. $\frac{1}{3}$ [2.5A]   28. $\frac{19}{7}$ [2.2B]   29. 2   [2.7A]   30. $\frac{1}{15}$ [2.6A]   31. $\frac{1}{8}$ [2.6A]   32. $1\frac{7}{8}$ [2.2A]

33. The total rainfall for the 3 months was $21\frac{7}{24}$ inches. [2.4D]   **34.** The cost per acre was $36,000. [2.7C]   **35.** The second checkpoint is $4\frac{3}{4}$ miles from the finish line. [2.5D]   **36.** The car can travel 243 miles. [2.6C]

CHAPTER 2 TEST

1. $\frac{4}{9}$ [2.6A, Example 1]   **2.** 8 [2.1B, Example 2]   **3.** $1\frac{3}{7}$ [2.7A, Example 2]   **4.** $\frac{7}{24}$ [2.8C, You Try It 3]

5. $\frac{49}{5}$ [2.2B, Example 5]   **6.** 8 [2.6B, Example 5]   **7.** $\frac{5}{8}$ [2.3B, Example 3]   **8.** $\frac{3}{8} < \frac{5}{12}$ [2.8A, Example 1]

9. $\frac{5}{6}$ [2.8C, Example 3]   **10.** 120 [2.1A, Example 1]   **11.** $\frac{1}{4}$ [2.5A, Example 1]   **12.** $3\frac{3}{5}$ [2.2B, Example 3]

13. $2\frac{2}{19}$ [2.7B, Example 4]   **14.** $\frac{45}{72}$ [2.3A, Example 1]   **15.** $1\frac{61}{90}$ [2.4B, Example 4]   **16.** $13\frac{81}{88}$ [2.5C, Example 5]

17. $\frac{7}{48}$ [2.5B, Example 2]   **18.** $\frac{1}{6}$ [2.8B, You Try It 2]   **19.** $1\frac{11}{12}$ [2.4A, Example 1]   **20.** $22\frac{4}{15}$ [2.4C, Example 7]

21. $\frac{11}{4}$ [2.2A, Example 2]   **22.** The electrician earns $840. [2.6C, Example 7]   **23.** 11 lots were available for sale. [2.7C, Example 8]   **24.** The actual length of wall a is $12\frac{1}{2}$ feet. The actual length of wall b is 18 feet. The actual length of wall c is $15\frac{3}{4}$ feet. [2.7C, Example 8]   **25.** The total rainfall for the 3-month period was $21\frac{11}{24}$ inches. [2.4D, Example 9]

CUMULATIVE REVIEW EXERCISES

1. 290,000 [1.1D]   **2.** 291,278 [1.3B]   **3.** 73,154 [1.4B]   **4.** 540 r12 [1.5C]   **5.** 1 [1.6B]   **6.** $2 \cdot 2 \cdot 11$ [1.7B]

7. 210 [2.1A]   **8.** 20 [2.1B]   **9.** $\frac{23}{3}$ [2.2B]   **10.** $6\frac{1}{4}$ [2.2B]   **11.** $\frac{15}{48}$ [2.3A]   **12.** $\frac{2}{5}$ [2.3B]   **13.** $1\frac{7}{48}$ [2.4B]

14. $14\frac{11}{48}$ [2.4C]   **15.** $\frac{13}{24}$ [2.5B]   **16.** $1\frac{7}{9}$ [2.5C]   **17.** $\frac{7}{20}$ [2.6A]   **18.** $7\frac{1}{2}$ [2.6B]   **19.** $1\frac{1}{20}$ [2.7A]

20. $2\frac{5}{8}$ [2.7B]   **21.** $\frac{1}{9}$ [2.8B]   **22.** $5\frac{5}{24}$ [2.8C]   **23.** The amount in the checking account at the end of the week was $862. [1.3C]   **24.** The total income from the tickets was $1410. [1.4C]   **25.** The total weight is $12\frac{1}{24}$ pounds. [2.4D]

26. The length of the remaining piece is $4\frac{17}{24}$ feet. [2.5D]   **27.** The car travels 225 miles on $8\frac{1}{3}$ gallons of gas. [2.6C]

28. 25 parcels can be sold from the remaining land. [2.7C]

ANSWERS TO CHAPTER 3 SELECTED EXERCISES

PREP TEST

1. $\frac{3}{10}$ [2.2A]   **2.** 36,900 [1.1D]   **3.** Four thousand seven hundred ninety-one [1.1B]   **4.** 6842 [1.1B]   **5.** 9394 [1.2A]

6. 1638 [1.3B]   **7.** 76,804 [1.4B]   **8.** 278 r18 [1.5C]

SECTION 3.1

1. Thousandths   **3.** Ten-thousandths   **5.** Hundredths   **7.** 0.3   **9.** 0.21   **11.** 0.461   **13.** $\frac{1}{10}$ **15.** $\frac{47}{100}$ **17.** $\frac{289}{1000}$

19. Thirty-seven hundredths   **21.** Nine and four tenths   **23.** Fifty-three ten-thousandths   **25.** Forty-five thousandths

27. Twenty-six and four hundredths   **29.** 3.0806   **31.** 407.03   **33.** 246.024   **35.** 73.02684   **37.** 6.2   **39.** 21.0

41. 18.41   **43.** 72.50   **45.** 936.291   **47.** 47   **49.** 7015   **51.** 2.97527   **53.** An entrant who completes the Boston Marathon runs 26.2 miles.   **55.** For example, 0.2701   **57.** For example, **a.** 0.15   **b.** 1.05   **c.** 0.001

SECTION 3.2

1. 150.1065   **3.** 95.8446   **5.** 69.644   **7.** 92.883   **9.** 113.205   **11.** 0.69   **13.** 16.305   **15.** 110.7666

17. 104.4959   **19.** Cal.: 234.192   **21.** Cal.: 781.943   **23.** You   **25.** The length of the shaft is 4.35 feet.
Est.: 234   Est.: 782

27. The perimeter is 18.5 meters.   **29.** The total number of people who watched the three news programs is 26.3 million.

31. No, a 4-foot rope cannot be wrapped around the box.   **33.** Three possible answers are bread, butter, and mayonnaise; raisin bran, butter, and bread; and lunch meat, milk, and toothpaste.

SECTION 3.3

1. 5.627 **3.** 113.6427 **5.** 6.7098 **7.** 215.697 **9.** 53.8776 **11.** 72.7091 **13.** 0.3142 **15.** 1.023 **17.** 261.166 **19.** 655.32 **21.** 342.9268 **23.** 8.628 **25.** 7.01 − 2.325 **27.** 19.35 − 8.967 **29.** Cal.: 2.74506 **31.** Cal.: 7.14925
Est.: 3 Est.: 7

33. The missing dimension is 2.59 feet. **35.** The difference in the average number of tickets sold is 320,000 tickets.

37. 33.5 million more people watched Super Bowl XLII than watched the Super Bowl post-game show. **39. a.** 0.1

b. 0.01 **c.** 0.001

SECTION 3.4

1. 0.36 **3.** 0.25 **5.** 6.93 **7.** 1.84 **9.** 0.74 **11.** 39.5 **13.** 2.72 **15.** 0.603 **17.** 13.50 **19.** 4.316 **21.** 0.1323 **23.** 0.03568 **25.** 0.0784 **27.** 0.076 **29.** 34.48 **31.** 580.5 **33.** 20.148 **35.** 0.04255 **37.** 0.17686 **39.** 0.19803 **41.** 0.0006608 **43.** 0.536335 **45.** 0.429 **47.** 2.116 **49.** 0.476 **51.** 1.022 **53.** 37.96 **55.** 2.318 **57.** 3.2 **59.** 6.5 **61.** 6285.6 **63.** 3200 **65.** 35,700 **67.** 6.3 **69.** 3.9 **71.** 49,000 **73.** 6.7 **75.** 0.012075 **77.** 0.0117796 **79.** 0.31004 **81.** 0.082845 **83.** 5.175 **85.** Cal.: 91.2 **87.** Cal.: 1.0472 **89.** Cal.: 3.897 **91.** Cal.: 11.2406 **93.** Cal.: 0.371096 **95.** Cal.: 31.8528
Est.: 90 Est.: 0.8 Est.: 4.5 Est.: 12 Est.: 0.32 Est.: 30

97. A U.S. homeowner's average annual cost of electricity is $1147.92. **99.** The amount received is $14.06. **101.** You will pay $2.40 in taxes. **103.** The area is 23.625 square feet. **105. a.** The nurse's overtime pay is $785.25. **b.** The nurse's total income is $2181.25. **107.** It would cost the company $406.25. **109.** The expression represents the cost of mailing 4 express mail packages, each weighing $\frac{1}{2}$ pound or less, and 9 express mail packages, each weighing between $\frac{1}{2}$ pound and 1 pound, from the post office to the addressee. **111.** The added cost is $3,200,000. **113. a.** The total cost for grade 1 is $56.32.

b. The total cost for grade 2 is $74.04. **c.** The total cost for grade 3 is $409.56. **d.** The total cost is $539.92.

115. $1\frac{3}{10} \times 2\frac{31}{100} = \frac{13}{10} \times \frac{231}{100} = \frac{3003}{1000} = 3\frac{3}{1000} = 3.003$

SECTION 3.5

1. 0.82 **3.** 4.8 **5.** 89 **7.** 60 **9.** 84.3 **11.** 32.3 **13.** 5.06 **15.** 1.3 **17.** 0.11 **19.** 3.8 **21.** 6.3 **23.** 0.6 **25.** 2.5 **27.** 1.1 **29.** 130.6 **31.** 0.81 **33.** 0.09 **35.** 40.70 **37.** 0.46 **39.** 0.019 **41.** 0.087 **43.** 0.360 **45.** 0.103 **47.** 0.009 **49.** 1 **51.** 3 **53.** 1 **55.** 57 **57.** 0.407 **59.** 4.267 **61.** 0.01037 **63.** 0.008295 **65.** 0.82537 **67.** 0.032 **69.** 0.23627 **71.** 0.000053 **73.** 0.0018932 **75.** 18.42 **77.** 16.07 **79.** 0.0135 **81.** 0.023678 **83.** 0.112 **85.** Cal.: 11.1632 **87.** Cal.: 884.0909 **89.** Cal.: 1.8269 **91.** Cal.: 58.8095 **93.** Cal.: 72.3053
Est.: 10 Est.: 1000 Est.: 1.5 Est.: 50 Est.: 100

95. a. Use division to find the cost. **b.** Use multiplication to find the cost. **97.** 6.23 yards are gained per carry.

99. The trucker must drive 35 miles. **101.** Three complete shelves can be cut from a 12-foot board. **103.** The dividend is $1.72 per share. **105.** The car travels 25.5 miles on 1 gallon of gasoline. **107.** You will use 0.405 barrel of oil.

109. 2.57 million more women than men were attending institutions of higher learning. **111.** The Army's advertising budget was 4.2 times greater than the Navy's advertising budget. **113.** The population of this segment is expected to be 2.1 times greater in 2030 than in 2000. **117.** ÷ **119.** − **121.** + **123.** 5.217 **125.** 0.025

SECTION 3.6

1. 0.625 **3.** 0.667 **5.** 0.167 **7.** 0.417 **9.** 1.750 **11.** 1.500 **13.** 4.000 **15.** 0.003 **17.** 7.080 **19.** 37.500 **21.** 0.160 **23.** 8.400 **25.** Less than 1 **27.** Greater than 1 **29.** $\frac{4}{5}$ **31.** $\frac{8}{25}$ **33.** $\frac{1}{8}$ **35.** $1\frac{1}{4}$ **37.** $16\frac{9}{10}$ **39.** $8\frac{2}{5}$ **41.** $8\frac{437}{1000}$ **43.** $2\frac{1}{4}$ **45.** $\frac{23}{150}$ **47.** $\frac{703}{800}$ **49.** $7\frac{19}{50}$ **51.** $\frac{57}{100}$ **53.** $\frac{2}{3}$ **55.** 0.15 < 0.5 **57.** 6.65 > 6.56 **59.** 2.504 > 2.054 **61.** $\frac{3}{8} > 0.365$ **63.** $\frac{2}{3} > 0.65$ **65.** $\frac{5}{9} > 0.55$ **67.** $0.62 > \frac{7}{15}$ **69.** $0.161 > \frac{1}{7}$ **71.** 0.86 > 0.855 **73.** 1.005 > 0.5 **75.** 0.172 < 17.2 **77.** Cars 2 and 5 would fail the emissions test.

CHAPTER 3 CONCEPT REVIEW*

1. To round a decimal to the nearest tenth, look at the digit in the hundredths place. If the digit in the hundredths place is less than 5, that digit and all digits to the right are dropped. If the digit in the hundredths place is greater than or equal to 5, increase the digit in the tenths place by 1 and drop all digits to its right.   [3.1B]

2. The decimal 0.37 is read 37 hundredths. To write the decimal as a fraction, put 37 in the numerator and 100 in the denominator: $\frac{37}{100}$.   [3.1A]

3. The fraction $\frac{173}{10,000}$ is read 173 ten-thousandths. To write the fraction as a decimal, insert one 0 after the decimal point so that the 3 is in the ten-thousandths place: 0.0173.   [3.1A]

4. When adding decimals of different place values, write the numbers so that the decimal points are on a vertical line.   [3.2A]

5. Write the decimal point in the product of two decimals so that the number of decimal places in the product is the sum of the numbers of decimal places in the factors.   [3.4A]

6. To estimate the product of two decimals, round each number so that it contains one nonzero digit. Then multiply. For example, to estimate the product of 0.068 and 0.0052, round the two numbers to 0.07 and 0.005; then multiply $0.07 \times 0.005 = 0.00035$.   [3.4A]

7. When dividing decimals, move the decimal point in the divisor to the right to make the divisor a whole number. Move the decimal point in the dividend the same number of places to the right. Place the decimal point in the quotient directly over the decimal point in the dividend, and then divide as with whole numbers.   [3.5A]

8. First convert the fraction to a decimal: The fraction $\frac{5}{8}$ is equal to 0.625. Now compare the decimals: $0.63 > 0.625$. In the inequality $0.63 > 0.625$, replace the decimal 0.625 with the fraction $\frac{5}{8}$: $0.63 > \frac{5}{8}$. The answer is that the decimal 0.63 is greater than the fraction $\frac{5}{8}$.   [3.6C]

9. When dividing 0.763 by 0.6, the decimal points will be moved one place to the right: $7.63 \div 6$. The decimal 7.63 has digits in the tenths and hundredths places. We need to write a zero in the thousandths place in order to determine the digit in the thousandths place of the quotient so that we can then round the quotient to the nearest hundredth.   [3.5A]

10. To subtract a decimal from a whole number that has no decimal point, write a decimal point in the whole number to the right of the ones place. Then write as many zeros to the right of that decimal point as there are places in the decimal being subtracted from the whole number. For example, the subtraction $5 - 3.578$ would be written $5.000 - 3.578$.   [3.3A]

CHAPTER 3 REVIEW EXERCISES

1. 54.5 [3.5A]   2. 833.958 [3.2A]   3. $0.055 < 0.1$ [3.6C]   4. Twenty-two and ninety-two ten-thousandths [3.1A]

5. 0.05678 [3.1B]   6. 2.33 [3.6A]   7. $\frac{3}{8}$ [3.6B]   8. 36.714 [3.2A]   9. 34.025 [3.1A]   10. $\frac{5}{8} > 0.62$ [3.6C]

11. 0.778 [3.6A]   12. $\frac{33}{50}$ [3.6B]   13. 22.8635 [3.3A]   14. 7.94 [3.1B]   15. 8.932 [3.4A]   16. Three hundred

forty-two and thirty-seven hundredths [3.1A]   17. 3.06753 [3.1A]   18. 25.7446 [3.4A]   19. 6.594 [3.5A]

20. 4.8785 [3.3A]   21. The new balance in your account is $661.51. [3.3B]   22. The difference between the amount United expects to pay per gallon of fuel and the amount Southwest expects to pay is $.96. [3.3B]   23. Northwest's cost per gallon of fuel is $3.34. Northwest's cost per gallon is more than United's cost per gallon. [3.5B; 3.6C]   24. The number who drove is 6.4 times greater than the number who flew. [3.5B]   25. During a 5-day school week, 9.5 million gallons of milk are served. [3.4B]

CHAPTER 3 TEST

1. $0.66 < 0.666$ [3.6C, Example 5]   2. 4.087 [3.3A, Example 1]   3. Forty-five and three hundred two ten-thousandths [3.1A, Example 4]   4. 0.692 [3.6A, You Try It 1]   5. $\frac{33}{40}$ [3.6B, Example 3]   6. 0.0740 [3.1B, Example 6]

7. 1.583 [3.5A, Example 3]   8. 27.76626 [3.3A, Example 2]   9. 7.095 [3.1B, Example 6]   10. 232 [3.5A, Example 1]

11. 458.581 [3.2A, Example 2]   12. The missing dimension is 1.37 inches. [3.3B, Example 4]   13. 0.00548 [3.4A, Example 2]   14. 255.957 [3.2A, Example 1]   15. 209.07086 [3.1A, Example 4]   16. Each payment is $395.40. [3.5B, Example 7]   17. Your total income is $3087.14. [3.2B, You Try It 4]   18. The cost of the call is $4.63. [3.4B, Example 8]

*Note: The numbers in brackets following the answers in the Concept Review are a reference to the objective that corresponds to that problem. For example, the reference [1.2A] stands for Section 1.2, Objective A. This notation will be used for all Prep Tests, Concept Reviews, Chapter Reviews, Chapter Tests, and Cumulative Reviews throughout the text.

19. The yearly average computer use by a 10th-grade student is 348.4 hours. [3.4B, Example 7] **20.** On average, a 2nd-grade student uses a computer 36.4 hours more per year than a 5th-grade student. [3.4B, Example 8]

CUMULATIVE REVIEW EXERCISES

1. 235 r17 [1.5C] **2.** 128 [1.6A] **3.** 3 [1.6B] **4.** 72 [2.1A] **5.** $4\frac{2}{5}$ [2.2B] **6.** $\frac{37}{8}$ [2.2B] **7.** $\frac{25}{60}$ [2.3A]

8. $1\frac{17}{48}$ [2.4B] **9.** $8\frac{35}{36}$ [2.4C] **10.** $5\frac{23}{36}$ [2.5C] **11.** $\frac{1}{12}$ [2.6A] **12.** $9\frac{1}{8}$ [2.6B] **13.** $1\frac{2}{9}$ [2.7A] **14.** $\frac{19}{20}$ [2.7B]

15. $\frac{3}{16}$ [2.8B] **16.** $2\frac{5}{18}$ [2.8C] **17.** Sixty-five and three hundred nine ten-thousandths [3.1A] **18.** 504.6991 [3.2A]

19. 21.0764 [3.3A] **20.** 55.26066 [3.4A] **21.** 2.154 [3.5A] **22.** 0.733 [3.6A] **23.** $\frac{1}{6}$ [3.6B]

24. $\frac{8}{9} < 0.98$ [3.6C] **25.** Sweden mandates 14 more vacation days than Germany. [1.3C] **26.** The patient must lose

$7\frac{3}{4}$ pounds the third month to achieve the goal. [2.5D] **27.** Your checking account balance is $617.38. [3.3B]

28. The resulting thickness is 1.395 inches. [3.3B] **29.** You paid $6008.80 in income tax last year. [3.4B]

30. The amount of each payment is $46.37. [3.5B]

ANSWERS TO CHAPTER 4 SELECTED EXERCISES

PREP TEST

1. $\frac{4}{5}$ [2.3B] **2.** $\frac{1}{2}$ [2.3B] **3.** 24.8 [3.6A] **4.** 4 × 33 [1.4A] **5.** 4 [1.5A]

SECTION 4.1

1. $\frac{1}{5}$ 1:5 1 to 5 **3.** $\frac{2}{1}$ 2:1 2 to 1 **5.** $\frac{3}{8}$ 3:8 3 to 8 **7.** $\frac{1}{1}$ 1:1 1 to 1 **9.** $\frac{7}{10}$ 7:10 7 to 10 **11.** $\frac{1}{2}$ 1:2 1 to 2

13. $\frac{2}{1}$ 2:1 2 to 1 **15.** $\frac{5}{2}$ 5:2 5 to 2 **17.** $\frac{5}{7}$ 5:7 5 to 7 **19.** days **21.** The ratio is $\frac{1}{3}$. **23.** The ratio is $\frac{3}{8}$.

25. The ratio is $\frac{1}{25,000}$. **27.** The ratio is $\frac{1}{5}$. **29.** The ratio is $\frac{1}{56}$. **31.** The ratio is 24 to 59.

SECTION 4.2

1. $\frac{3\text{ pounds}}{4\text{ people}}$ **3.** $\frac{\$20}{3\text{ boards}}$ **5.** $\frac{20\text{ miles}}{1\text{ gallon}}$ **7.** $\frac{8\text{ gallons}}{1\text{ hour}}$ **9. a.** Dollars **b.** Seconds **11.** 1 **13.** 2.5 feet/second

15. $975/week **17.** 110 trees/acre **19.** $18.84/hour **21.** 35.6 miles/gallon **23.** The rate is 7.4 miles per dollar.

25. An average of 179.86 bushels of corn were harvested from each acre. **27.** The cost was $2.72 per disk. **29. a.** Australia has the least population density. **b.** There are 807 more people per square mile in India than in the United States.

31. 1.0179 × 2500 represents the value of 2500 American dollars in Canadian dollars.

SECTION 4.3

1. True **3.** Not true **5.** Not true **7.** True **9.** True **11.** True **13.** True **15.** Not true **17.** True

19. Yes **21.** Yes **23.** 3 **25.** 105 **27.** 2 **29.** 60 **31.** 2.22 **33.** 6.67 **35.** 21.33 **37.** 16.25

39. 2.44 **41.** 47.89 **43.** A 0.5-ounce serving contains 50 calories. **45.** The car can travel 329 miles. **47.** 12.5 gallons of water are required. **49.** The distance is 16 miles. **51.** 1.25 ounces are required. **53.** 160,000 people would vote.

55. The monthly payment is $176.75. **57.** 750 defective circuit boards can be expected in a run of 25,000. **59.** A bowling ball would weigh 2.67 pounds on the moon. **61.** The dividend would be $1071.

CHAPTER 4 CONCEPT REVIEW*

1. If the units in a comparison are different, then the comparison is a rate. For example, the comparison "50 miles in 2 hours" is a rate. [4.2A]

*Note: The numbers in brackets following the answers in the Concept Review are a reference to the objective that corresponds to that problem. For example, the reference [1.2A] stands for Section 1.2, Objective A. This notation will be used for all Prep Tests, Concept Reviews, Chapter Reviews, Chapter Tests, and Cumulative Reviews throughout the text.

2. To find a unit rate, divide the number in the numerator of the rate by the number in the denominator of the rate.   [4.2B]

3. To write the ratio $\frac{6}{7}$ using a colon, write the two numbers 6 and 7 separated by a colon: $6:7$.   [4.1A]

4 To write the ratio $12:15$ in simplest form, divide both numbers by the GCF of 3: $\frac{12}{3} : \frac{15}{3} = 4:5$.   [4.1A]

5. To write the rate $\frac{342 \text{ miles}}{9.5 \text{ gallons}}$ as a unit rate, divide the number in the numerator by the number in the denominator: $342 \div 9.5 = 36$. $\frac{342 \text{ miles}}{9.5 \text{ gallons}}$ is the rate. 36 miles/gallon is the unit rate.   [4.2B]

6. A proportion is true if the fractions are equal when written in lowest terms. Another way to describe a true proportion is to say that in a true proportion, the cross products are equal.   [4.3A]

7. When one of the numbers in a proportion is unknown, we can solve the proportion to find the unknown number. We do this by setting the cross products equal to each other and then solving for the unknown number.   [4.3B]

8. When setting up a proportion, keep the same units in the numerator and the same units in the denominator.   [4.3C]

9. To check the solution of a proportion, replace the unknown number in the proportion with the solution. Then find the cross products. If the cross products are equal, the solution is correct. If the cross products are not equal, the solution is not correct.   [4.3B]

10. To write the ratio $19:6$ as a fraction, write the first number as the numerator of the fraction and the second number as the denominator: $\frac{19}{6}$.   [4.1A]

CHAPTER 4 REVIEW EXERCISES

1. True [4.3A]   **2.** $\frac{2}{5}$ $2:5$ 2 to 5 [4.1A]   **3.** 62.5 miles/hour [4.2B]   **4.** True [4.3A]   **5.** 68 [4.3B]

6. $12.50/hour [4.2B]   **7.** $1.75/pound [4.2B]   **8.** $\frac{2}{7}$ $2:7$ 2 to 7 [4.1A]   **9.** 36 [4.3B]   **10.** 19.44 [4.3B]

11. $\frac{2}{5}$ $2:5$ 2 to 5 [4.1A]   **12.** Not true [4.3A]   **13.** $\frac{\$35}{4 \text{ hours}}$ [4.2A]   **14.** 27.2 miles/gallon [4.2B]

15. $\frac{1}{1}$ $1:1$ 1 to 1 [4.1A]   **16.** True [4.3A]   **17.** 65.45 [4.3B]   **18.** $\frac{100 \text{ miles}}{3 \text{ hours}}$ [4.2A]   **19.** The ratio is $\frac{2}{5}$. [4.1B]

20. The property tax is $6400. [4.3C]   **21.** The ratio is $\frac{3}{8}$. [4.1B]   **22.** The cost per phone is $37.50. [4.2C]

23. 1344 blocks would be needed. [4.3C]   **24.** The ratio is $\frac{5}{2}$. [4.1B]   **25.** The turkey costs $.93/pound. [4.2C]

26. The average was 56.8 miles/hour. [4.2C]   **27.** The cost is $493.50. [4.3C]   **28.** The cost is $44.75/share. [4.2C]

29. 22.5 pounds of fertilizer will be used. [4.3C]   **30.** The ratio is $\frac{1}{2}$. [4.1B]

CHAPTER 4 TEST

1. $3836.40/month [4.2B, Example 2]   **2.** $\frac{1}{6}$ $1:6$ 1 to 6 [4.1A, You Try It 1]   **3.** $\frac{9 \text{ supports}}{4 \text{ feet}}$ [4.2A, Example 1]

4. Not true [4.3A, Example 2]   **5.** $\frac{3}{2}$ $3:2$ 3 to 2 [4.1A, Example 2]   **6.** 144 [4.3B, Example 3]

7. 30.5 miles/gallon [4.2B, Example 2]   **8.** $\frac{1}{3}$ $1:3$ 1 to 3 [4.1A, Example 2]   **9.** True [4.3A, Example 1]

10. 40.5 [4.3B, Example 3]   **11.** $\frac{\$27}{2 \text{ boards}}$ [4.2A, Example 1]   **12.** $\frac{3}{5}$ $3:5$ 3 to 5 [4.1A, You Try It 1]

13. The dividend is $625. [4.3C, Example 8]   **14.** The ratio is $\frac{1}{12}$. [4.3C, Example 8]   **15.** The plane's speed is

538 miles/hour. [4.2C, How To 1]   **16.** The college student's body contains 132 pounds of water. [4.3C, Example 8]

17. The cost of the lumber is $1.73/foot. [4.2C, How To 1]   **18.** The amount of medication required is 0.875 ounce.

[4.3C, Example 8]   **19.** The ratio is $\frac{4}{5}$. [4.1B, Example 4]   **20.** 36 defective hard drives are expected to be found in the

production of 1200 hard drives. [4.3C, Example 8]

CUMULATIVE REVIEW EXERCISES

1. 9158 [1.3B]   **2.** $2^4 \cdot 3^3$ [1.6A]   **3.** 3 [1.6B]   **4.** $2 \cdot 2 \cdot 2 \cdot 2 \cdot 2 \cdot 5$ [1.7B]   **5.** 36 [2.1A]   **6.** 14 [2.1B]

7. $\frac{5}{8}$ [2.3B]   **8.** $8\frac{3}{10}$ [2.4C]   **9.** $5\frac{11}{18}$ [2.5C]   **10.** $2\frac{5}{6}$ [2.6B]   **11.** $4\frac{2}{3}$ [2.7B]   **12.** $\frac{23}{30}$ [2.8C]

13. Four and seven hundred nine ten-thousandths [3.1A]   **14.** 2.10 [3.1B]   **15.** 1.990 [3.5A]   **16.** $\frac{1}{15}$ [3.6B]

17. $\frac{1}{8}$ [4.1A] **18.** $\frac{29¢}{2 \text{ pencils}}$ [4.2A] **19.** 33.4 miles/gallon [4.2B] **20.** 4.25 [4.3B] **21.** The car's speed is 57.2 miles/hour. [4.2C] **22.** 36 [4.3B] **23.** Your new balance is $744. [1.3C] **24.** The monthly payment is $570. [1.5D] **25.** 105 pages remain to be read. [2.6C] **26.** The cost per acre was $36,000. [2.7C] **27.** The change was $35.24. [3.3B] **28.** Your monthly salary is $3468.25. [3.5B] **29.** 25 inches will erode in 50 months. [4.3C] **30.** 1.6 ounces are required. [4.3C]

ANSWERS TO CHAPTER 5 SELECTED EXERCISES

PREP TEST

1. $\frac{19}{100}$ [2.6B] **2.** 0.23 [3.4A] **3.** 47 [3.4A] **4.** 2850 [3.4A] **5.** 4000 [3.5A] **6.** 32 [2.7B] **7.** 62.5 [3.6A] **8.** $66\frac{2}{3}$ [2.2B] **9.** 1.75 [3.5A]

SECTION 5.1

1. $\frac{1}{4}$, 0.25 **3.** $1\frac{3}{10}$, 1.30 **5.** 1, 1.00 **7.** $\frac{73}{100}$, 0.73 **9.** $3\frac{83}{100}$, 3.83 **11.** $\frac{7}{10}$, 0.70 **13.** $\frac{22}{25}$, 0.88 **15.** $\frac{8}{25}$, 0.32 **17.** $\frac{2}{3}$ **19.** $\frac{5}{6}$ **21.** $\frac{1}{9}$ **23.** $\frac{5}{11}$ **25.** $\frac{3}{70}$ **27.** $\frac{1}{15}$ **29.** 0.065 **31.** 0.123 **33.** 0.0055 **35.** 0.0825 **37.** 0.0505 **39.** 0.02 **41.** Greater than **43.** 73% **45.** 1% **47.** 294% **49.** 0.6% **51.** 310.6% **53.** 70% **55.** 37% **57.** 40% **59.** 12.5% **61.** 150% **63.** 166.7% **65.** 87.5% **67.** 48% **69.** $33\frac{1}{3}\%$ **71.** $166\frac{2}{3}\%$ **73.** $87\frac{1}{2}\%$ **75.** Less than **77.** 6% of those surveyed named something other than corn on the cob, cole slaw, corn bread, or fries. **79.** This represents $\frac{1}{2}$ off the regular price.

SECTION 5.2

1. 8 **3.** 10.8 **5.** 0.075 **7.** 80 **9.** 51.895 **11.** 7.5 **13.** 13 **15.** 3.75 **17.** 20 **19.** 5% of 95 **21.** 79% of 16 **23.** Less than **25.** The number of people in the United States aged 18 to 24 without life insurance is less than 44 million. **27.** 58,747 new student pilots are flying single-engine planes this year. **29.** The piece contains 29.25 grams of gold, 8.75 grams of silver, and 12 grams of copper. **31.** 77 million returns were filed electronically. **33.** 49.59 billion of the email messages sent per day are not spam. **35.** 6232 respondents did not answer yes to the question.

SECTION 5.3

1. 32% **3.** $16\frac{2}{3}\%$ **5.** 200% **7.** 37.5% **9.** 18% **11.** 0.25% **13.** 20% **15.** 400% **17.** 2.5% **19.** 37.5% **21.** 0.25% **23.** False **25.** 70% of couples disagree about financial matters. **27.** Approximately 25.4% of the vegetables were wasted. **29.** 29.8% of Americans with diabetes have not been diagnosed. **31.** 98.5% of the slabs did meet safety requirements. **33.** 26.7% of the total is spent on veterinary care.

SECTION 5.4

1. 75 **3.** 50 **5.** 100 **7.** 85 **9.** 1200 **11.** 19.2 **13.** 7.5 **15.** 32 **17.** 200 **19.** 9 **21.** 504 **23.** Less than **25.** There were 15.8 million travelers who allowed their children to miss school to go along on a trip. **27.** 22,366 runners started the Boston Marathon in 2008. **29.** The cargo ship's daily fuel bill is $8000. **31. a.** 3000 boards were tested. **b.** 2976 of the boards tested were not defective. **33.** The recommended daily amount of thiamin for an adult is 1.5 milligrams.

SECTION 5.5

1. 65 **3.** 25% **5.** 75 **7.** 12.5% **9.** 400 **11.** 19.5 **13.** 14.8% **15.** 62.62 **17.** 15 **19. a.** (ii) and (iii) **b.** (i) and (iv) **21.** The drug will be effective for 4.8 hours. **23. a.** $175 million is generated annually from sales of Thin Mints. **b.** $63 million is generated annually from sales of Trefoil shortbread cookies. **25.** The U.S. total turkey production was 7 billion pounds. **27.** 57.7% of baby boomers have some college experience but have not earned a degree. **29.** 46.8% of the deaths were due to traffic accidents.

CHAPTER 5 CONCEPT REVIEW*

1. To write 197% as a fraction, remove the percent sign and multiply by $\frac{1}{100}$: $197 \times \frac{1}{100} = \frac{197}{100}$.   [5.1A]

2. To write 6.7% as a decimal, remove the percent sign and multiply by 0.01: $6.7 \times 0.01 = 0.067$.   [5.1A]

3. To write $\frac{9}{5}$ as a percent, multiply by 100%: $\frac{9}{5} \times 100\% = 180\%$.   [5.1B]

4. To write 56.3 as a percent, multiply by 100% : $56.3 \times 100\% = 5630\%$.   [5.1B]

5. The basic percent equation is Percent $\times$ base = amount.   [5.2A]

6. To find what percent of 40 is 30, use the basic percent equation: $n \times 40 = 30$. To solve for n, we *divide* 30 by 40: $30 \div 40 = 0.75 = 75\%$.   [5.3A]

7. To find 11.7% of 532, use the basic percent equation and write the percent as a decimal: $0.117 \times 532 = n$. To solve for n, we *multiply* 0.117 by 532: $0.117 \times 532 = 62.244$.   [5.2A]

8. To answer the question "36 is 240% of what number?", use the basic percent equation and write the percent as a decimal: $2.4 \times n = 36$. To solve for n, we *divide* 36 by 2.4: $36 \div 2.4 = 15$.   [5.4A]

9. To use the proportion method to solve a percent problem, identify the percent, the amount, and the base. Then use the proportion $\frac{\text{percent}}{100} = \frac{\text{amount}}{\text{base}}$. Substitute the known values into this proportion and solve for the unknown.   [5.5A]

10. To answer the question "What percent of 1400 is 763?" by using the proportion method, first identify the base (1400) and the amount (763). The percent is unknown. Substitute 1400 for the base and 763 for the amount in the proportion $\frac{\text{percent}}{100} = \frac{\text{amount}}{\text{base}}$. Then solve the proportion $\frac{n}{100} = \frac{763}{1400}$ for n. $n = 54.5$, so 54.5% of 1400 is 763.   [5.5A]

CHAPTER 5 REVIEW EXERCISES

1. 60   [5.2A]   **2.** 20%   [5.3A]   **3.** 175%   [5.1B]   **4.** 75   [5.4A]   **5.** $\frac{3}{25}$ [5.1A]   **6.** 19.36   [5.2A]

7. 150%   [5.3A]   **8.** 504   [5.4A]   **9.** 0.42   [5.1A]   **10.** 5.4   [5.2A]   **11.** 157.5   [5.4A]   **12.** 0.076   [5.1A]

13. 77.5   [5.2A]   **14.** $\frac{1}{6}$ [5.1A]   **15.** 160%   [5.5A]   **16.** 75   [5.5A]   **17.** 38%   [5.1B]   **18.** 10.9   [5.4A]

19. 7.3%   [5.3A]   **20.** 613.3%   [5.3A]   **21.** The student answered 85% of the questions correctly.   [5.5B]

22. The company spent $4500 for newspaper advertising.   [5.2B]   **23.** 31.7% of the cost is for electricity.   [5.3B]

24. The total cost of the camcorder was $1041.25.   [5.2B]   **25.** Approximately 78.6% of the women wore sunscreen often.   [5.3B]

26. The world's population in 2000 was approximately 6,100,000,000 people.   [5.4B]   **27.** The cost of the computer 4 years ago was $3000.   [5.5B]   **28.** The total cranberry crop in that year was 572 million pounds.   [5.3B/5.5B]

CHAPTER 5 TEST

1. 0.973   [5.1A, Example 3]   **2.** $\frac{5}{6}$ [5.1A, Example 2]   **3.** 30%   [5.1B, Example 4]   **4.** 163%   [5.1B, Example 4]

5. 150%   [5.1B, HOW TO 1]   **6.** $66\frac{2}{3}\%$ [5.1B, Example 5]   **7.** 50.05   [5.2A, Example 1]   **8.** 61.36   [5.2A, Example 2]

9. 76% of 13   [5.2A, Example 1]   **10.** 212% of 12   [5.2A, Example 1]   **11.** The amount spent for advertising is $45,000. [5.2B, Example 3]   **12.** 1170 pounds of vegetables were not spoiled.   [5.2B, Example 4]   **13.** 14.7% of the daily recommended amount of potassium is provided.   [5.3B, Example 4]   **14.** 9.1% of the daily recommended number of calories is provided. [5.3B, Example 4]   **15.** The number of temporary employees is 16% of the number of permanent employees.   [5.3B, Example 4] **16.** The student answered approximately 91.3% of the questions correctly.   [5.3B, Example 5]   **17.** 80   [5.4A, Example 2] **18.** 28.3   [5.4A, Example 2]   **19.** 32,000 PDAs were tested.   [5.4B, Example 4]   **20.** The increase was 60% of the original price.   [5.3B, Example 5]   **21.** 143.0   [5.5A, Example 1]   **22.** 1000%   [5.5A, Example 1]   **23.** The dollar increase is $1.74.   [5.5B, Example 3]   **24.** The population now is 220% of the population 10 years ago.   [5.5B, Example 4] **25.** The value of the car is $25,000.   [5.5B, Example 3]

Note: The numbers in brackets following the answers in the Concept Review are a reference to the objective that corresponds to that problem. For example, the reference [1.2A] stands for Section 1.2, Objective A. This notation will be used for all Prep Tests, Concept Reviews, Chapter Reviews, Chapter Tests, and Cumulative Reviews throughout the text.

CUMULATIVE REVIEW EXERCISES

1. 4 [1.6B] **2.** 240 [2.1A] **3.** $10\frac{11}{24}$ [2.4C] **4.** $12\frac{41}{48}$ [2.5C] **5.** $12\frac{4}{7}$ [2.6B] **6.** $\frac{7}{24}$ [2.7B] **7.** $\frac{1}{3}$ [2.8B]

8. $\frac{13}{36}$ [2.8C] **9.** 3.08 [3.1B] **10.** 1.1196 [3.3A] **11.** 34.2813 [3.5A] **12.** 3.625 [3.6A] **13.** $1\frac{3}{4}$ [3.6B]

14. $\frac{3}{8} < 0.87$ [3.6C] **15.** 53.3 [4.3B] **16.** $19.20/hour [4.2B] **17.** $\frac{11}{60}$ [5.1A] **18.** $83\frac{1}{3}\%$ [5.1B]

19. 19.56 [5.2A/5.5A] **20.** $133\frac{1}{3}\%$ [5.3A/5.5A] **21.** 9.92 [5.4A/5.5A] **22.** 342.9% [5.3A/5.5A]

23. Sergio's take-home pay is $592. [2.6C] **24.** Each monthly payment is $292.50. [3.5B] **25.** 420 gallons were used during the month. [3.5B] **26.** The real estate tax is $10,000. [4.3C] **27.** 22,577 hotels in the United States are located along highways. [5.2B/5.5B] **28.** 45% of the people did not favor the candidate. [5.3B/5.5B] **29.** The approximate average number of hours spent watching TV in a week is 61.3 hours. [5.2B/5.5B] **30.** 18% of the children tested had levels of lead that exceeded federal standards. [5.3B/5.5B]

ANSWERS TO CHAPTER 6 SELECTED EXERCISES

PREP TEST
1. 0.75 [3.5A] **2.** 52.05 [3.4A] **3.** 504.51 [3.3A] **4.** 9750 [3.4A] **5.** 45 [3.4A] **6.** 1417.24 [3.2A]
7. 3.33 [3.5A] **8.** 0.605 [3.5A] **9.** $0.379 < 0.397$ [3.6C]

SECTION 6.1
1. The unit cost is $.055 per ounce. **3.** The unit cost is $.374 per ounce. **5.** The unit cost is $.080 per tablet.
7. The unit cost is $6.975 per clamp. **9.** The unit cost is $.199 per ounce. **11.** Divide the price of one pint by 2.
13. The Kraft mayonnaise is the more economical purchase. **15.** The Cortexx shampoo is the more economical purchse.
17. The Ultra Mr. Clean is the more economical purchase. **19.** The Bertolli olive oil is the more economical purchase.
21. The Wagner's vanilla extract is the more economical purchase. **23.** Tea A is the more economical purchase.
25. The total cost is $73.50. **27.** The total cost is $16.88. **29.** The total cost is $3.89. **31.** The total cost is $26.57.

SECTION 6.2
1. The percent increase is 182.9%. **3.** The percent increase is 117.6%. **5.** The percent increase is 500%.
7. The percent increase is 111.2%. **9.** Yes **11.** Use equation (3) and then equation (2). **13.** The markup is $17.10.
15. The markup rate is 60%. **17.** The selling price is $304.50. **19.** The selling price is $74. **21.** The percent decrease is 40%. **23. a.** The percent decrease in the population of Detroit is 13.7%. **b.** The percent decrease in the population of Philadelphia is 7.7%. **c.** The percent decrease in the population of Chicago is 1.8%. **25.** The car loses $8460 in value.
27. a. The amount of the decrease was $35.20. **b.** The new average monthly gasoline bill is $140.80. **29.** The percent decrease is 31.9%. **31.** Use equation (3) and then (2). **33.** The discount rate is $33\frac{1}{3}\%$. **35.** The discount is $80. **37.** The discount rate is 30%. **39. a.** The discount is $.25 per pound. **b.** The sale price is $1.00 per pound. **41.** The discount rate is 20%.

SECTION 6.3
1. a. $10,000 **b.** $850 **c.** 4.25% **d.** 2 years **3.** The simple interest owed is $960. **5.** The simple interest due is $3375. **7.** The simple interest due is $1320. **9.** The simple interest due is $84.76. **11.** The maturity value is $5120.
13. The total amount due on the loan is $12,875. **15.** The maturity value is $14,543.70. **17.** The monthly payment is $6187.50.
19. a. The interest charged is $1080. **b.** The monthly payment is $545. **21.** The monthly payment is $3039.02.
23. a. Student A's principal is equal to student B's principal. **b.** Student A's maturity value is greater than student B's maturity value. **c.** Student A's monthly payment is less than student B's monthly payment. **25.** The finance charge is $6.85.
27. You owe the company $11.94. **29.** The difference between the finance charges is $3.44. **31.** The finance charges for the first and second months are the same. No, you will not be able to pay off the balance. **33.** The value of the investment after 20 years is $12,380.43. **35.** The value of the investment after 5 years is $28,352.50. **37. a.** The value of the investment after 10 years will be $6040.86. **b.** The amount of interest earned will be $3040.86. **39.** The amount of interest earned is $505.94.

SECTION 6.4

1. The mortgage is $172,450.   **3.** The down payment is $212,500.   **5.** The loan origination fee is $3750.   **7.** The mortgage is $315,000.   **9.** The mortgage is $189,000.   **11.** (iii)   **13.** The monthly mortgage payment is $644.79.   **15.** Yes, the lawyer can afford the monthly mortgage payment.   **17.** The monthly property tax is $166.   **19. a.** The monthly mortgage payment is $898.16.   **b.** The interest payment is $505.69.   **21.** The total monthly payment for the mortgage and property tax is $842.40.   **23.** The monthly mortgage payment is $2295.29.   **25.** The couple can save $63,408 in interest.

SECTION 6.5

1. No, Amanda does not have enough money for the down payment.   **3.** The sales tax is $1192.50.   **5.** The license fee is $650.   **7. a.** The sales tax is $1120.   **b.** The total cost of the sales tax and the license fee is $1395.   **9.** The amount financed is $12,150.   **11.** The amount financed is $36,000.   **13.** The expression represents the total cost of buying the car.   **15.** Use multiplication to find the cost.   **17.** The monthly car payment is $531.43.   **19.** The cost is $252.   **21.** Your cost per mile for gasoline was $.16.   **23. a.** The amount financed is $12,000.   **b.** The monthly car payment is $359.65.   **25.** The monthly payment is $810.23.   **27.** The amount of interest paid is $4665.

SECTION 6.6

1. Lewis earns $460.   **3.** The real estate agent's commission is $3930.   **5.** The stockbroker's commission is $84.   **7.** The teacher's monthly salary is $3244.   **9.** Carlos earned a commission of $540.   **11.** Steven receives $920 for installing the carpet.   **13.** The chemist's hourly wage is $125.   **15. a.** Gil's hourly wage for working overtime is $21.56.   **b.** Gil earns $344.96 for working 16 hours of overtime.   **17. a.** The increase in hourly pay is $1.23.   **b.** The clerk's hourly wage for working the night shift is $9.43.   **19.** Nicole's earnings were $475.   **21.** The starting salary for an accountant in the previous year was $40,312.   **23.** The starting salary for a computer science major in the previous year was $50,364.

SECTION 6.7

1. Your current checking account balance is $486.32.   **3.** The nutritionist's current balance is $825.27.   **5.** The current checking account balance is $3000.82.   **7.** Yes, there is enough money in the carpenter's account to purchase the refrigerator.   **9.** Yes, there is enough money in the account to make the two purchases.   **11.** The account's ending balance might be less than its starting balance on that day.   **13.** The bank statement and the checkbook balance.   **15.** The bank statement and the checkbook balance.   **17.** The ending balance on the monthly bank statement is greater than the ending balance on the check register.

CHAPTER 6 CONCEPT REVIEW*

1. To find the unit cost, divide the total cost by the number of units. The cost is $2.96. The number of units is 4. $2.96 \div 4 = 0.74$, so the unit cost is $.74 per can.   [6.1A]

2. To find the total cost, multiply the unit cost by the number of units purchased. The unit cost is $.85. The number of units purchased is 3.4 pounds. $0.85 \times 3.4 = 2.89$, so the total cost is $2.89.   [6.1C]

3. To find the selling price when you know the cost and the markup, add the cost and the markup.   [6.2B]

4. To find the markup when you know the markup rate, multiply the markup rate by the cost.   [6.2B]

5. If you know the percent decrease, you can find the amount of decrease by multiplying the percent decrease by the original value.   [6.2C]

6. If you know the regular price and the sale price, you can find the discount by subtracting the sale price from the regular price.   [6.2D]

7. To find the discount rate, first subtract the sale price from the regular price to find the discount. Then divide the discount by the regular price.   [6.2D]

8. To find simple interest, multiply the principal by the annual interest rate by the time (in years).   [6.3A]

9. To find the maturity value for a simple interest loan, add the principal and the interest.   [6.3A]

10. Principal is the original amount deposited in an account or the original amount borrowed for a loan.   [6.3A]

11. If you know the maturity value of an 18-month loan, you can find the monthly payment by dividing the maturity value by 18 (the length of the loan in months).   [6.3A]

Note: The numbers in brackets following the answers in the Concept Review are a reference to the objective that corresponds to that problem. For example, the reference [1.2A] stands for Section 1.2, Objective A. This notation will be used for all Prep Tests, Concept Reviews, Chapter Reviews, Chapter Tests, and Cumulative Reviews throughout the text.

12. Compound interest is computed not only on the original principal but also on interest already earned.　[6.3C]

13. For a fixed-rate mortgage, the monthly payment remains the same throughout the life of the loan.　[6.4B]

14. The following expenses are involved in owning a car: car insurance, gas, oil, general maintenance, and the monthly car payment.　[6.5B]

15. To balance a checkbook:

　1. In the checkbook register, put a check mark by each check paid by the bank and by each deposit recorded by the bank.

　2. Add to the current checkbook balance all checks that have been written but have not yet been paid by the bank and any interest paid on the account.

　3. Subtract any service charges and any deposits not yet recorded by the bank. This is the checkbook balance.

　4. Compare the balance with the bank balance listed on the bank statement. If the two numbers are equal, the bank statement and the checkbook "balance."　[6.7B]

CHAPTER 6 REVIEW EXERCISES

1. The unit cost is $.195 per ounce or 19.5¢ per ounce.　[6.1A]　　**2.** The cost is $.279 or 27.9¢ per mile.　[6.5B]
3. The percent increase is 30.4%.　[6.2A]　　**4.** The markup is $72.　[6.2B]　　**5.** The simple interest due is $3000.　[6.3A]
6. The value of the investment after 10 years is $45,550.75.　[6.3C]　　**7.** The percent increase is 15%.　[6.2A]　　**8.** The total
monthly payment for the mortgage and property tax is $1138.90.　[6.4B]　　**9.** The monthly payment is $518.02.　[6.5B]
10. The value of the investment will be $53,593.　[6.3C]　　**11.** The down payment is $29,250.　[6.4A]　　**12.** The total cost of
the sales tax and license fee is $2096.25.　[6.5A]　　**13.** The selling price is $2079.　[6.2B]　　**14.** The interest paid is $157.33.　[6.5B]
15. The commission was $3240.　[6.6A]　　**16.** The sale price is $141.　[6.2D]　　**17.** The current checkbook balance
is $943.68.　[6.7A]　　**18.** The maturity value is $31,200.　[6.3A]　　**19.** The origination fee is $1875.　[6.4A]　　**20.** The more
economical purchase is 33 ounces for $6.99.　[6.1B]　　**21.** The monthly mortgage payment is $2131.62.　[6.4B]　　**22.** The total
income was $655.20.　[6.6A]　　**23.** The donut shop's checkbook balance is $8866.58.　[6.7A]　　**24.** The monthly payment is
$14,093.75.　[6.3A]　　**25.** The finance charge is $7.20.　[6.3B]

CHAPTER 6 TEST

1. The cost per foot is $6.92.　[6.1A, Example 1]　　**2.** The more economical purchase is 3 pounds for $7.49.　[6.1B, Example 2]
3. The total cost is $14.53.　[6.1C, Example 3]　　**4.** The percent increase in the cost of the exercise bicycle is 20%.
[6.2A, Example 1]　　**5.** The selling price of the blu-ray disc player is $441.　[6.2B, Example 4]　　**6.** The percent increase is 8.9%.
[6.2A, Example 1]　　**7.** The percent decrease is 20%.　[6.2C, HOW TO 3]　　**8.** The sale price of the corner hutch is $209.30.
[6.2D, Example 8]　　**9.** The discount rate is 40%.　[6.2D, Example 7]　　**10.** The simple interest due is $2000.　[6.3A, You Try It 1]
11. The maturity value is $26,725.　[6.3A, Example 2]　　**12.** The finance charge is $4.50.　[6.3B, Example 4]　　**13.** The amount
of interest earned in 10 years was $24,420.60.　[6.3C, You Try It 5]　　**14.** The loan origination fee is $3350.　[6.4A, Example 2]
15. The monthly mortgage payment is $1713.44.　[6.4B, HOW TO 2]　　**16.** The amout financed is $19,000.　[6.5A, Example 1]
17. The monthly truck payment is $686.22.　[6.5B, Example 5]　　**18.** Shaney earnes $1596.　[6.6A, Example 1]
19. The current checkbook balance is $6612.25.　[6.7A, Example 1]　　**20.** The bank statement and the checkbook balance.
[6.7B, Example 2]

CUMULATIVE REVIEW EXERCISES

1. 13　[1.6B]　　**2.** $8\frac{13}{24}$　[2.4C]　　**3.** $2\frac{37}{48}$　[2.5C]　　**4.** 9　[2.6B]　　**5.** 2　[2.7B]　　**6.** 5　[2.8C]　　**7.** 52.2　[3.5A]

8. 1.417　[3.6A]　　**9.** $51.25/hour　[4.2B]　　**10.** 10.94　[4.3B]　　**11.** 62.5%　[5.1B]　　**12.** 27.3　[5.2A]
13. 0.182　[5.1A]　　**14.** 42%　[5.3A]　　**15.** 250　[5.4A]　　**16.** 154.76　[5.4A/5.5A]　　**17.** The total rainfall is
$13\frac{11}{12}$ inches.　[2.4D]　　**18.** The amount paid in taxes is $970.　[2.6C]　　**19.** The ratio is $\frac{3}{5}$.　[4.1B]　　**20.** 33.4 miles are
driven per gallon.　[4.2C]　　**21.** The unit cost is $1.10 per pound.　[4.2C]　　**22.** The dividend is $280.　[4.3C]　　**23.** The sale
price is $720.　[6.2D]　　**24.** The selling price of the grinding rail is $119.　[6.2B]　　**25.** The percent increase in Sook Kim's
salary is 8%.　[6.2A]　　**26.** The simple interest due is $2700.　[6.3A]　　**27.** The monthly car payment is $791.81.　[6.5B]
28. The family's new checking account balance is $2243.77.　[6.7A]　　**29.** The cost per mile is $.33.　[6.5B]　　**30.** The monthly
mortgage payment is $1232.26.　[6.4B]

ANSWERS TO CHAPTER 7 SELECTED EXERCISES

PREP TEST

1. 48.0% was bill-related mail.   [5.3B]   **2. a.** The greatest cost increase is between 2009 and 2010.
b. Between those years, there was an increase of $5318.   [1.3C]   **3.** The ratio is $\frac{5}{3}$.   [4.1B]   **4.** $\frac{3}{50}$ of the women in the military are in the Marine Corps.   [5.1A]

SECTION 7.1

1. The gross revenue is $1 billion.   **3.** The percent is 35%.   **5.** 50 more people agreed that humanity should explore planets than agreed that space exploration impacts daily life.   **7.** 150 children said they hid their vegetables under a napkin.
11. Sample answers: Sergey Brin and Larry Page have the same net worth. Bill Gates's net worth is twice Larry Ellison's net worth.
13. The number of units required in humanities is less than twice the number of units required in science.   **15. a.** The complaint mentioned most often was people talking.   **b.** The complaint mentioned least often was uncomfortable seats.   **17.** The ratio is $\frac{9}{11}$.
19. Americans spent $1,085,000,000 on TV game machines.   **21.** $\frac{2}{25}$ of the total money spent was spent on accessories.
23. The age group 35 to 44 represents the largest segment of the homeless population.   **25.** 37% of the homeless population is under age 35.   **27.** The total land area is 57,240,000 square miles.   **29.** The land area of Asia is 30.0% of the total land area.
31. Approximately 13,500,000 people living in the United States are of Asian ethnic origin.   **33.** On average, there would be 64,500 people of black ethnic origin in a random sample of 500,000.

SECTION 7.2

1. The maximum height is 150 feet.   **3.** The ratio is 4 to 3.   **5.** The difference is 12 million passenger cars.   **7.** The maximum salary of police officers in the suburbs is $16,000 higher than the maximum salary of police officers in the city.   **9.** The greatest difference in salaries is in Philadelphia.   **11.** (iii)   **13.** The average snowfall during January is 20 inches.   **15.** The total average snowfall during March and April is 25 inches.   **17.** 4500 megawatts of wind power capacity was produced in 2002.
19. The wind power capacity increased most between 2006 and 2007.   **21.** Women ages 75 and over have the lowest recommended number of Calories.   **23.** True

25.

	Wind Power Capacity (in megawatts)	
Year	Texas	California
2000	200	1600
2001	1100	1700
2002	1100	1800
2003	1300	2000
2004	1300	2100
2005	2000	2200
2006	2700	2400
2007	4300	2400

26.

Year	Greater capacity	Difference (megawatts)
2000	California	1400
2001	California	600
2002	California	700
2003	California	700
2004	California	800
2005	California	200
2006	Texas	300
2007	Texas	1900

The wind power capacity of Texas exceeded that of California in 2006 and 2007.

SECTION 7.3

1. 13 account balances were between $1500 and $2000.   **3.** 22% of the account balances were between $2000 and $2500.
5. 410 cars are between 6 and 12 years old.   **7.** 230 cars are more than 12 years old.   **9.** The class intervals 9–12 and 12–15 have the greatest difference in class frequency.   **11.** 18 adults spend between 3 and 4 hours at the mall.
13. Approximately 22,000 runners finished with times between $2\frac{1}{2}$ hours and 6 hours.   **15.** Yes   **17.** 10.8% of the people purchased between 20 and 30 tickets each month.   **19.** 170,000 students scored between 1200 and 1400 on the exam.
21. 530,000 students scored below 1000.

SECTION 7.4

1. a. Median   **b.** Mean   **c.** Mode   **d.** Median   **e.** Mode   **f.** Mean   **3.** The mean number of seats filled is 381.5625 seats. The median number of seats filled is 394.5 seats. Since each number occurs only once, there is no mode.   **5.** The mean cost is $85.615. The median cost is $85.855.   **7.** The mean monthly rate is $403.625. The median monthly rate is $404.50. **9.** The mean life expectancy is 73.4 years. The median life expectancy is 74 years.   **13. a.** 25%   **b.** 75%   **c.** 75%   **d.** 25% **15.** Lowest is $46,596. Highest is $82,879. Q_1 = $56,067. Q_3 = $66,507. Median = $61,036. Range = $36,283. Interquartile range= $10,440.   **17. a.** There were 40 adults who had cholesterol levels above 217.   **b.** There were 60 adults who had cholesterol levels below 254.   **c.** There are 20 cholesterol levels in each quartile.   **d.** 25% of the adults had cholesterol levels not more than 198.   **19. a.** Range = 5.6 million metric tons. Q_1 = 0.59 million metric tons. Q_3 = 1.52 million metric tons. Interquartile range = 0.93 million metric tons.   **b.**

0.45   1.52   6.05
0.59 1.035   **c.** 6.05

21. a. No, the difference in the means is not greater than 1 inch.   **b.** The difference in medians is 0.3 inch.

c.

1.8   3.0   5.95   7.8
2.15

0.5   1.55 2.7   4.55   6.4

23. Answers will vary. For example, 55, 55, 55, 55, 55, or 50, 55, 55, 55, 60   **27.** Answers will vary. For example, 20, 21, 22, 24, 26, 27, 29, 31, 31, 32, 32, 33, 33, 36, 37, 37, 39, 40, 41, 43, 45, 46, 50, 54, 57

SECTION 7.5

1. {(HHHH), (HHHT), (HHTT), (HHTH), (HTTT), (HTHH), (HTTH), (HTHT), (TTTT), (TTTH), (TTHH), (THHH), (TTHT), (THHT), (THTT), (THTH)}   **3.** {(1, 1), (1, 2), (1, 3), (1, 4), (2, 1), (2, 2), (2, 3), (2, 4), (3, 1), (3, 2), (3, 3), (3, 4), (4, 1), (4, 2), (4, 3), (4, 4)}   **5. a.** {1, 2, 3, 4, 5, 6, 7, 8}   **b.** {1, 2, 3}   **7. a.** The probability that the sum is 5 is $\frac{1}{9}$.   **b.** The probability that the sum is 15 is 0.   **c.** The probability that the sum is less than 15 is 1.   **d.** The probability that the sum is 2 is $\frac{1}{36}$. **9. a.** The probability that the number is divisible by 4 is $\frac{1}{4}$.   **b.** The probability that the number is a multiple of 3 is $\frac{1}{3}$. **11.** The probability of throwing a sum of 5 is greater.   **13. a.** The probability is $\frac{4}{11}$ that the letter I is drawn.   **b.** The probability of choosing an S is greater.   **15.** The fraction $\frac{1}{11}$ represents the probability that the card has the letter M on it.   **17.** Drawing a spade has the greater probability.   **19.** The empirical probability that a person prefers a cash discount is 0.39. **21.** The probability is $\frac{185}{377}$ that a customer rated the service as satisfactory or excellent.

CHAPTER 7 CONCEPT REVIEW*

1. A sector of a circle is one of the "pieces of the pie" into which a circle graph is divided.   [7.1B]

2. A pictograph uses symbols to represent numerical information.   [7.1A]

3. A jagged portion of the vertical axis on a bar graph is used to indicate that the vertical scale is missing numbers from 0 to the lowest number shown on the vertical axis.   [7.2A]

4. In a broken-line graph, points are connected by line segments to show data trends. If a line segment goes up from left to right, it indicates an increase in the quantity represented on the vertical axis during that time period. If a line segment goes down from left to right, it indicates a decrease in the quantity represented on the vertical axis during that time period.   [7.2B]

5. Class frequency in a histogram is the number of occurrences of data in each class interval.   [7.3A]

6. A class interval in a histogram is a range of numbers that corresponds to the width of each bar.   [7.3A]

7. A class midpoint is the center of a class interval in a frequency polygon.   [7.3B]

*Note: The numbers in brackets following the answers in the Concept Review are a reference to the objective that corresponds to that problem. For example, the reference [1.2A] stands for Section 1.2, Objective A. This notation will be used for all Prep Tests, Concept Reviews, Chapter Reviews, Chapter Tests, and Cumulative Reviews throughout the text.

8. The formula for the mean is $\bar{x} = \dfrac{\text{sum of the data values}}{\text{number of data values}}$ [7.4A]

9. To find the median, the data must be arranged in order from least to greatest in order to determine the "middle" number, or the number which separates the data so that half of the numbers are less than the median and half of the numbers are greater than the median.   [7.4A]

10. If a set of numbers has no number occurring more than once, then the data have no mode.   [7.4A]

11. A box-and-whiskers plot shows the least number in the data; the first quartile, Q_1; the median; the third quartile, Q_3; and the greatest number in the data.   [7.4B]

12. To find the first quartile, find the median of the data values that lie below the median.   [7.4B]

13. The empirical probability formula states that the empirical probability of an event is the ratio of the number of observations of the event to the total number of observations.   [7.5A]

14. The theoretical probability formula states that the theoretical probability of an event is a fraction with the number of favorable outcomes of the experiment in the numerator and the total number of possible outcomes in the denominator.   [7.5A]

CHAPTER 7 REVIEW EXERCISES

1. The agencies spent $349 million on maintaining websites.   [7.1B]   **2.** The ratio is $\dfrac{9}{8}$.   [7.1B]   **3.** NASA spent 8.9% of the total amount of money.   [7.1B]   **4.** Texas had the larger population.   [7.2B]   **5.** The population of California was 12.5 million people more than the population of Texas.   [7.2B]   **6.** The Texas population increased the least from 1925 to 1950.   [7.2B]   **7.** There were 54 games in which the Knicks scored fewer than 100 points.   [7.3B]   **8.** The ratio is $\dfrac{31}{8}$.   [7.3B]   **9.** The percent is 11.3%.   [7.3B]   **10.** From the pictograph, O'Hare airport had 10 million more passengers than the Denver airport.   [7.1A]   **11.** The ratio is $4:3$.   [7.1A]   **12.** The difference was 50 days.   [7.2A]   **13.** The percent is 50%.   [7.2A]   **14.** The Southeast had the lowest number of days of full operation. This region had 30 days of full operation.   [7.2A]   **15.** The probability of one tail and three heads is $\dfrac{1}{4}$.   [7.5A]   **16.** There were 15 people who slept 8 hours or more.   [7.3A]   **17.** The percent is 28.3%.   [7.3A]   **18. a.** The mean heart rate is 91.6 heartbeats per minute. The median heart rate is 93.5 heartbeats per minute. The mode is 96 heartbeats per minute.   [7.4A]   **b.** The range is 36 heartbeats per minute. The interquartile range is 15 heartbeats per minute.   [7.4B]

CHAPTER 7 TEST

1. 19 students spent between $45 and $75 each week.   [7.3B, You Try It 2]   **2.** The ratio is $\dfrac{2}{3}$.   [7.3B, You Try It 2]   **3.** The percent is 45%. [7.3B, Example 2]   **4.** 36 people were surveyed for the Gallup poll.   [7.1A, Example 1]   **5.** The ratio is $\dfrac{5}{2}$.   [7.1A, Example 1]   **6.** The percent is 58.3%.   [7.1A, You Try It 1]   **7.** During 1995 and 1996, the numbers of fatalities were the same.   [7.2A, Example 1]   **8.** There were 32 fatal accidents from 1991 to 1999.   [7.2A, Example 1]   **9.** There were 4 more fatalities from 1995 to 1998.   [7.2A, Example 1]   **10.** There were 355 more films rated R.   [7.1B, HOW TO 2]   **11.** There were 16 times more films rated PG-13.   [7.1B, HOW TO 2]   **12.** The percent of films rated G was 5.6%.   [7.1B, Example 3]   **13.** There are 24 states that have a median income between $40,000 and $60,000.   [7.3A, Example 1]   **14.** The percent is 72%.   [7.3A, Example 1]   **15.** The percent is 18%.   [7.3A, Example 1]   **16.** The probability is $\dfrac{3}{10}$ that the ball chosen is red.   [7.5A, HOW TO 2]   **17.** The student enrollment increased the least during the 1990s.   [7.2B, You Try It 2]   **18.** The increase in enrollment was 11 million students.   [7.2B, Example 2]   **19. a.** The mean time is 2.53 days.   [7.4A, Example 1]   **b.** The median time is 2.55 days.   [7.4A, Example 2]   **c.**

[7.4B, Example 3]

CUMULATIVE REVIEW EXERCISES

1. 540   [1.6A]   **2.** 14   [1.6B]   **3.** 120   [2.1A]   **4.** $\dfrac{5}{12}$ [2.3B]   **5.** $12\dfrac{3}{40}$ [2.4C]   **6.** $4\dfrac{17}{24}$ [2.5C]

7. 2   [2.6B]   **8.** $\dfrac{64}{85}$ [2.7B]   **9.** $8\dfrac{1}{4}$ [2.8C]   **10.** 209.305   [3.1A]   **11.** 2.82348   [3.4A]   **12.** 16.67   [3.6A]

13. 26.4 miles/gallon   [4.2B]   **14.** 3.2   [4.3B]   **15.** 80%   [5.1B]   **16.** 80   [5.4A]   **17.** 16.34   [5.2A]

18. 40%   [5.3A]   **19.** The salesperson's income for the week was $650   [6.6A]   **20.** The cost is $407.50.   [4.3C]

21. The interest due is $3750.   [6.3A]   **22.** The markup rate is 55%.   [6.2B]   **23.** The amount budgeted for food is $855.   [7.1B]

24. The difference in the number answered correctly is 12 problems.   [7.2B]   **25.** The mean high temperature is 69.6°F.   [7.4A]

26. The probability is $\dfrac{5}{36}$ that the sum of the dots on the upward faces is 8.   [7.5A]

ANSWERS TO CHAPTER 8 SELECTED EXERCISES

PREP TEST

1. 702 [1.2A] **2.** 58 [1.3B] **3.** 4 [2.6B] **4.** 10 [2.6B] **5.** 25 [2.6B] **6.** 46 [2.6B] **7.** 238 [1.5A]
8. 1.5 [3.5A]

SECTION 8.1

1. Greater than **3.** Greater than **5.** 108 in. **7.** $5\frac{1}{3}$ ft **9.** $13\frac{1}{2}$ ft **11.** $1\frac{1}{2}$ yd **13.** 180 in. **15.** 7920 ft

17. 5280 **19.** 1 mi 1120 ft **21.** 13 ft 5 in. **23.** $11\frac{1}{6}$ ft **25.** 5 yd 2 ft **27.** $14\frac{2}{3}$ ft **29.** $4\frac{1}{6}$ in.

31. The length of material used was 96 ft. **33.** The total length of the shaft is 3 ft 1 in. **35.** The missing dimension is $1\frac{1}{2}$ in.

37. The board must be 13 ft 4 in. long. **39.** You purchased 384 in. of baseboard. **41.** The total number of feet of material needed is 75 ft. **43.** True

SECTION 8.2

1. Less than **3.** Greater than **5.** $2\frac{1}{4}$ lb **7.** 112 oz **9.** $4\frac{1}{2}$ tons **11.** 2500 lb **13.** $5\frac{5}{8}$ lb **15.** 42 oz

17. $2\frac{1}{2}$ tons **19.** 2000 **21.** 4 tons 1000 lb **23.** 8 lb 3 oz **25.** 3 lb 13 oz **27.** $3\frac{13}{24}$ lb **29.** 33 lb

31. 1 lb 12 oz **33.** The total weight will be greater than 25 lb. **35.** The total weight of the rods is $31\frac{1}{4}$ lb.

37. The total weight of the textbooks is 675 lb. **39.** The weight of the case of soft drinks is 9 lb. **41.** Each container holds 1 lb 5 oz of shampoo. **43.** One million tons of plastic bottles are not recycled each year. **45.** The cost of mailing the manuscript is $11.90.

SECTION 8.3

1. Greater than **3.** Less than **5.** 6 c **7.** 20 fl oz **9.** $2\frac{1}{2}$ pt **11.** 6 qt **13.** $2\frac{1}{2}$ gal **15.** 28 qt

17. $4\frac{1}{4}$ qt **19.** 4 **21.** 3 gal 2 qt **23.** 8 gal 1 qt **25.** 5 gal 1 qt **27.** 1 gal 3 qt **29.** 1 c 7 fl oz **31.** $2\frac{3}{4}$ gal

33. $17\frac{1}{2}$ pt **35.** $\frac{7}{8}$ gal **37.** $7\frac{1}{2}$ gal of coffee should be prepared. **39.** The more economical purchase is $1.59 for 1 qt.

41. On average, an American drinks 37.7 c of bottled water per month. **43.** The farmer used $8\frac{3}{4}$ gal of oil.

45. Orlando made $890 profit on one 50-gallon container of oil. **47.** The product represents the number of cups of lemonade in the punch.

SECTION 8.4

1. Greater than **3.** Greater than **5.** 84 days **7.** $4\frac{3}{4}$ days **9.** 465 min **11.** $12\frac{1}{2}$ min **13.** $4\frac{1}{4}$ h **15.** 20,700 s

17. $4\frac{3}{4}$ days **19.** 9000 min **21.** $3\frac{1}{2}$ weeks **23.** 924 h **25.** 2 weeks **27.** 30,240 min **29.** No

SECTION 8.5

1. 19,450 ft · lb **3.** 19,450,000 ft · lb **5.** 1500 ft · lb **7.** 29,700 ft · lb **9.** 30,000 ft · lb **11.** 25,500 ft · lb

13. 35,010,000 ft · lb **15.** 9,336,000 ft · lb **17.** Less than **19.** 2 hp **21.** 8 hp **23.** $4950\frac{ft \cdot lb}{s}$ **25.** $3850\frac{ft \cdot lb}{s}$

27. $500\frac{ft \cdot lb}{s}$ **29.** $4800\frac{ft \cdot lb}{s}$ **31.** $720\frac{ft \cdot lb}{s}$ **33.** 30 hp

CHAPTER 8 CONCEPT REVIEW*

1. To convert from feet to inches, use multiplication. [8.1A]

2. To convert 5 ft 7 in. to all inches, multiply 5 by 12 and add the product to 7. [8.1B]

Note: The numbers in brackets following the answers in the Concept Review are a reference to the objective that corresponds to that problem. For example, the reference [1.2A] stands for Section 1.2, Objective A. This notation will be used for all Prep Tests, Concept Reviews, Chapter Reviews, Chapter Tests, and Cumulative Reviews throughout the text.

3. To divide 7 ft 8 in. by 2, first divide 7 ft by 2. The result is 3 ft with 1 ft remaining. Convert the 1 ft remaining to 12 in., and add the 12 in. to 8 in.: 8 in. + 12 in. = 20 in. Divide 20 in. by 2. The result is 10 in. Therefore, 7 ft 8 in. divided by 2 is 3 ft 10 in. [8.1B]

4. To convert 5240 lb to tons, use the conversion rate $\dfrac{1 \text{ ton}}{2000 \text{ lb}}$.   [8.2A]

5. To multiply 6 lb 9 oz by 3, first multiply 9 oz by 3. The result is 27 oz. Multiply 6 lb by 3. The result is 18 lb. 18 lb 27 oz = 19 lb 11 oz. Therefore, 6 lb 9 oz times 3 equals 19 lb 11 oz.   [8.2B]

6. Five measures of capacity are fluid ounces, cups, pints, quarts, and gallons.   [8.3A]

7. To convert 7 gal to quarts, use the conversion rate $\dfrac{4 \text{ qt}}{1 \text{ gal}}$.   [8.3A]

8. To convert 374 min to hours, use the conversion rate $\dfrac{1 \text{ h}}{60 \text{ min}}$.   [8.4A]

9. A foot-pound of energy is the amount of energy needed to lift 1 lb a distance of 1 ft.   [8.5A]

10. To find the power needed to raise 200 lb a distance of 24 ft in 12 s, use multiplication and division.

Power $= \dfrac{24 \text{ ft} \times 200 \text{ lb}}{12 \text{ s}} = 400\, \dfrac{\text{ft} \cdot \text{lb}}{\text{s}}$ [8.5B]

CHAPTER 8 REVIEW EXERCISES

1. 48 in.   [8.1A]   2. 2 ft 6 in.   [8.1B]   3. 1600 ft · lb   [8.5A]   4. 40 fl oz   [8.3A]   5. $4\frac{2}{3}$ yd   [8.1A]

6. $1\frac{1}{5}$ tons   [8.2A]   7. 2 lb 7 oz   [8.2B]   8. 54 oz   [8.2A]   9. 9 ft 3 in.   [8.1B]   10. 1 ton 1000 lb   [8.2B]

11. 7 c 2 fl oz   [8.3B]   12. 1 yd 2 ft   [8.1B]   13. 3 qt   [8.3A]   14. $6\frac{1}{4}$ h   [8.4A]   15. $1375\dfrac{\text{ft} \cdot \text{lb}}{\text{s}}$ [8.5B]

16. 44 lb   [8.2B]   17. 38,900 ft · lb   [8.5A]   18. 7 hp   [8.5B]   19. The length of the remaining piece is 3 ft 6 in.

[8.1C]   20. The cost of mailing the book is $10.15.   [8.2C]   21. There are $13\frac{1}{2}$ qt in a case.   [8.3C]

22. 16 gal of milk were sold that day.   [8.3C]   23. The furnace releases 27,230,000 ft · lb of energy.   [8.5A]

24. The power is $480\dfrac{\text{ft} \cdot \text{lb}}{\text{s}}$.   [8.5B]

CHAPTER 8 TEST

1. 30 in.   [8.1A, Example 2]   2. 2 ft 5 in.   [8.1B, Example 6]   3. Each piece is $1\frac{1}{3}$ ft long.   [8.1C, You Try It 11]

4. The wall is 48 ft long.   [8.1C, You Try It 10]   5. 46 oz   [8.2A, You Try It 1]   6. 2 lb 8 oz   [8.2B, Example 4]

7. 17 lb 1 oz   [8.2B, Example 3]   8. 1 lb 11 oz   [8.2B, Example 4]   9. The total weight of the workbooks is 750 lb.

[8.2C, You Try It 5]   10. The amount the class received for recycling was $28.13.   [8.2C, You Try It 5]

11. $3\frac{1}{4}$ gal   [8.3A, How To 1]   12. 28 pt   [8.3A, How To 2]   13. $12\frac{1}{4}$ gal   [8.3B, You Try It 2]   14. 8 gal 1 qt

[8.3B, Example 2]   15. $4\frac{1}{2}$ weeks   [8.4A, How To 2]   16. 4680 min   [8.4A, How To 1]   17. There are 60 c in a case.

[8.3C, Example 3]   18. Nick's profit is $686.   [8.3C, Example 3]   19. 3750 ft · lb   [8.5A, You Try It 2]

20. The furnace releases 31,120,000 ft · lb of energy.   [8.5A, Example 3]   21. $160\dfrac{\text{ft} \cdot \text{lb}}{\text{s}}$ [8.5B, Example 4]

22. 4 hp   [8.5B, Example 5]

CUMULATIVE REVIEW EXERCISES

1. 180   [2.1A]   2. $5\frac{3}{8}$ [2.2B]   3. $3\frac{7}{24}$ [2.5C]   4. 2   [2.7B]   5. $4\frac{3}{8}$ [2.8C]   6. 2.10   [3.1B]

7. 0.038808   [3.4A]   8. 8.8   [4.3B]   9. 1.25   [5.2A]   10. 42.86   [5.4A]   11. The unit cost is $5.15/lb.   [6.1A]

12. $8\frac{11}{15}$ in.   [8.1B]   13. 1 lb 8 oz   [8.2D]   14. 31 lb 8 oz   [8.2B]   15. $2\frac{1}{2}$ qt   [8.3B]   16. 1 lb 12 oz   [8.2B]

17. The dividend would be $280.   [4.3C]   18. Anna's balance is $642.79.   [6.7A]   19. The executive's monthly income is $4100.   [6.6A]   20. The truck driver must drive 35 mi.   [3.5B]   21. The percent is 18%.   [7.3A]

22. The selling price of a DMB television is $308.   [6.2B]   23. The interest paid on the loan is $8000.   [6.3A]

24. Each student received $2533.   [8.2C]   **25.** The cost of mailing the books is $20.16.   [8.2C]   **26.** The better buy is 36 oz for $2.98.   [6.1B]   **27.** The probability is $\frac{1}{9}$ that the sum of the dots on the upward faces is 9.   [7.5A]

28. 3200 ft · lb   [8.5A]   **29.** $400\dfrac{\text{ft} \cdot \text{lb}}{\text{s}}$ [8.5B]

ANSWERS TO CHAPTER 9 SELECTED EXERCISES

PREP TEST

1. 37,320   [3.4A]   **2.** 659,000   [3.4A]   **3.** 0.04107   [3.5A]   **4.** 28.496   [3.5A]   **5.** 5.125   [3.3A]
6. 5.96   [3.2A]   **7.** 0.13   [3.4A]   **8.** 56.35   [2.6A, 3.4A]   **9.** 0.5   [2.6B, 3.5A]   **10.** 675   [2.6B]

SECTION 9.1

1. 420 mm   **3.** 8.1 cm   **5.** 6.804 km   **7.** 2109 m   **9.** 4.32 m   **11.** 88 cm   **13.** 7.038 km   **15.** 3500 m
17. 2.60 m   **19.** 168.5 cm   **21.** 148 mm   **23.** 62.07 m   **25.** 31.9 cm   **27.** 8.075 km   **29.** m   **31.** Three shelves can be cut from the board. There is no length of board remaining.   **33.** The walk was 4.4 km long.   **35.** The total length of the shaft is 181 cm.   **37.** The average number of meters adopted by a group is 2254 m.   **39.** It takes 500 s for light to travel from the sun to Earth.   **41.** Light travels 25,920,000,000 km in 1 day.

SECTION 9.2

1. 0.420 kg   **3.** 0.127 g   **5.** 4200 g   **7.** 450 mg   **9.** 1.856 kg   **11.** 4.057 g   **13.** 1370 g   **15.** 45.6 mg
17. 18.000 kg   **19.** 3.922 kg   **21.** 7.891 g   **23.** 4.063 kg   **25.** mg   **27.** g   **29.** The patient should take 4 tablets per day.   **31.** There are 3.288 g of cholesterol in 12 eggs.   **33.** The cost of the three packages of meat is $14.62.
35. Two of the power-generating knee braces weigh 3200 g.   **37.** The profit is $117.50   **39.** The total weight is 12,645 kg.

SECTION 9.3

1. 4.2 L   **3.** 3420 ml   **5.** 423 cm³   **7.** 642 ml   **9.** 0.042 L   **11.** 435 cm³   **13.** 4620 L   **15.** 1.423 kl
17. 1267 cm³   **19.** 3.042 L   **21.** 3.004 kl   **23.** 8.200 L   **25.** L   **27.** L   **29.** There are 16 servings in the container.
31. 4000 patients can be immunized.   **33.** One carafe of Pure Premium contains 12.5 servings.   **35.** The 12 one-liter bottles are the better buy.   **37.** The profit is $271.80.   **39.** The distributor's profit is $76,832.   **41.** 2.72 L; 2720 ml; 2 L 720 ml

SECTION 9.4

1. You can eliminate 3300 Calories from your diet.   **3. a.** There are 90 Calories in $1\frac{1}{2}$ servings.   **b.** There are 30 fat Calories in 6 slices of bread.   **5.** 2025 Calories would be needed.   **7.** You burn 10,125 Calories.   **9.** You would have to hike for 2.6 h.   **11.** 1250 Wh are used.   **13.** The fax machine used 0.567 kWh.   **15.** The cost of running the air conditioner is $2.11.   **17. a.** 400 lumens is less than half the output of the Soft White Bulb.   **b.** The Energy Saver Bulb costs $.42 less to operate.   **19.** (iii)

SECTION 9.5

1. $\dfrac{1 \text{ qt}}{4 \text{ c}} \times \dfrac{1 \text{ L}}{1.06 \text{ qt}}$ **3.** 65.91 kg   **5.** 0.47 L   **7.** 54.20 L   **9.** 8.89 m   **11.** 48.3 km/h   **13.** $1.30/kg

15. $8.70/L   **17.** 260.64 km   **19.** 328 ft   **21.** 1.58 gal   **23.** 4920 ft   **25.** 6.33 gal   **27.** 49.69 mi/h
29. $3.85/gal   **31.** 4.62 lb   **33.** Gary will lose 1 lb.   **35.** Cliff Gullet's speed was 239.13 mph.

CHAPTER 9 CONCEPT REVIEW*

1. To convert from millimeters to meters, move the decimal point three places to the left.   [9.1A]

2. To convert 51 m 2 cm to meters, first convert 2 cm to meters by moving the decimal point two places to the left: 2 cm = 0.02 m. Add 0.02 m to 51 m: 51 m + 0.02 m = 51.02 m.   [9.1A]

3. On the surface of Earth, mass and weight are the same.   [9.2A]

4. To convert 4 kg 7 g to kilograms, first convert 7 g to kilograms by moving the decimal point three places to the left: 7 g = 0.007 kg. Add 0.007 kg to 4 kg: 4 kg + 0.007 kg = 4.007 kg.   [9.2B]

5. One liter is defined as the capacity of a box that is 10 cm long on each side.   [9.3A]

6. To convert 4 L 27 ml to liters, first convert 27 ml to liters by moving the decimal point three places the left: 27 ml = 0.027 L. Add 0.027 L to 4 L: 4 L + 0.027 L = 4.027 L.   [9.3A]

*Note: The numbers in brackets following the answers in the Concept Review are a reference to the objective that corresponds to that problem. For example, the reference [1.2A] stands for Section 1.2, Objective A. This notation will be used for all Prep Tests, Concept Reviews, Chapter Reviews, Chapter Tests, and Cumulative Reviews throughout the text.

7. To find the cost per week to run a microwave oven rated at 650 W that is used for 30 min per day, at 8.8 cents per kilowatt-hour, use multiplication and conversion. The microwave oven is used $\frac{1}{2} \times 7 = \frac{7}{2}$ hours per week. The microwave uses $\frac{7}{2} \times 650 = 2275$ Wh, or 2.275 kWh, of energy. The cost is $2.275 \times 8.8 = 20.02$ cents per week.   [9.4A]

8. To find the Calories from fat for a daily intake of 1800 Calories, multiply 1800 times 30%: $1800 \times 0.30 = 540$ Calories.   [9.4A]

9. To convert 100 ft to meters, use the conversion rate $\frac{1 \text{ m}}{3.28 \text{ ft}}$.   [9.5A]

10. To convert the price of gas from \$3.19/gal to a cost per liter, use the conversion rate $\frac{1 \text{ gal}}{3.79 \text{ L}}$.   [9.5A]

CHAPTER 9 REVIEW EXERCISES

1. 1250 m   [9.1A]   **2.** 450 mg   [9.2A]   **3.** 5.6 ml   [9.3A]   **4.** 1090 yd   [9.5B]   **5.** 7.9 cm   [9.1A]
6. 5.34 m   [9.1A]   **7.** 0.990 kg   [9.2A]   **8.** 2.550 L   [9.3A]   **9.** 4.870 km   [9.1A]   **10.** 3.7 mm   [9.1A]
11. 6.829 g   [9.2A]   **12.** 1200 cm³   [9.3A]   **13.** 4050 g   [9.2A]   **14.** 870 cm   [9.1A]   **15.** 192 cm³   [9.3A]
16. 0.356 g   [9.2A]   **17.** 3.72 m   [9.1A]   **18.** 8300 L   [9.3A]   **19.** 2.089 L   [9.3A]   **20.** 5.410 L   [9.3A]
21. 3.792 kl   [9.3A]   **22.** 468 ml   [9.3A]   **23.** There are 37.2 m of wire left on the roll.   [9.1B]   **24.** The total cost of the chicken is \$12.72.   [9.2B]   **25.** \$9.68/kg   [9.5A]   **26.** The amount of coffee that should be prepared is 50 L.   [9.3B]
27. You can eliminate 2700 Calories.   [9.4A]   **28.** The cost of running the TV set is \$3.42.   [9.4A]
29. The backpack weighs 4.18 lb.   [9.5B]   **30.** 8.75 h of cycling are needed.   [9.4A]   **31.** The profit was \$52.80.   [9.3B]
32. The color TV used 1.120 kWh of electricity.   [9.4A]   **33.** The amount of fertilizer needed is 125 kg.   [9.2B]

CHAPTER 9 TEST

1. 2960 m   [9.1A, Example 1]   **2.** 378 mg   [9.2A, Example 1]   **3.** 46 ml   [9.3A, Example 2]   **4.** 919 ml   [9.3A, You Try It 2]   **5.** 4.26 cm   [9.1A, Example 1]   **6.** 7.96 m   [9.1A, Example 2]   **7.** 0.847 kg   [9.2A, Example 1]
8. 3.920 L   [9.3A, Example 2]   **9.** 5.885 km   [9.1A, Example 1]   **10.** 15 mm   [9.1A, Example 1]
11. 3.089 g   [9.2A, Example 2]   **12.** 1600 cm³   [9.3A, Example 2]   **13.** 3290 g   [9.2A, Example 1]
14. 420 cm   [9.1A, Example 1]   **15.** 96 cm³   [9.3A, Example 2]   **16.** 1.375 g   [9.2A, Example 1]
17. 4.02 m   [9.1A, Example 1]   **18.** 8920 L   [9.3A, You Try It 1]   **19.** A 140-pound sedentary person should consume 2100 Calories per day to maintain that weight.   [9.4A, Example 1]   **20.** 3.15 kWh of energy are used during the week to operate the television.   [9.4A, Example 2]   **21.** The total length of the rafters is 114 m.   [9.1B, Example 3]   **22.** The weight of the box is 36 kg.   [9.2B, Example 3]   **23.** The amount of vaccine needed is 5.2 L.   [9.3B, Example 3]
24. 56.4 km/h   [9.5A, Example 1]   **25.** The distance between the rivets is 17.5 cm.   [9.1B, Example 3]   **26.** The cost to fertilize the trees is \$660.   [9.2B, Example 3]   **27.** The total cost is \$24.00.   [9.4A, Example 3]   **28.** The assistant should order 11 L of acid.   [9.3B, Example 3]   **29.** The measure of the large hill is 393.6 ft.   [9.5B, Example 3]
30. 4.8 in. is approximately 12.2 cm.   [9.5A, HOW TO 1]

CUMULATIVE REVIEW EXERCISES

1. 6   [1.6B]   **2.** $12\frac{13}{36}$ [2.4C]   **3.** $\frac{29}{36}$ [2.5C]   **4.** $3\frac{1}{14}$ [2.7B]   **5.** 1   [2.8B]   **6.** 2.0702   [3.3A]
7. 31.3   [4.3B]   **8.** 175%   [5.1B]   **9.** 145   [5.4A]   **10.** 2.25 gal   [8.3A]   **11.** 8.75 m   [9.1A]
12. 3.420 km   [9.1A]   **13.** 5050 g   [9.2A]   **14.** 3.672 g   [9.2A]   **15.** 6000 ml   [9.3A]   **16.** 2400 L   [9.3A]
17. \$3933 is left after the rent is paid.   [2.6C]   **18.** The business paid \$7207.20 in state income tax.   [3.4B]   **19.** The property tax is \$5500.   [4.3C]   **20.** The car buyer will receive a rebate of \$2820.   [5.2B]   **21.** The percent is 6.5%.   [5.3B]
22. Your mean grade is 76.   [7.4A]   **23.** Karla's salary next year will be \$25,200.   [5.2B]   **24.** The discount rate is 22%.   [6.2D]   **25.** The length of the wall is 36 ft.   [8.1C]   **26.** The tank travels 0.56 mi on 1 gal of fuel.   [3.5B]   **27.** The profit was \$616.00.   [8.3C]   **28.** 24 L of chlorine are used.   [9.3B]   **29.** The total cost of operating the hair dryer is \$2.43.   [9.4A]
30. 96.6 km/h   [9.5A]

ANSWERS TO CHAPTER 10 SELECTED EXERCISES

PREP TEST

1. 54 > 45   [1.1A]   **2.** 4 units [1.3A]   **3.** 15,847   [1.2A]   **4.** 3779   [1.3B]   **5.** 26,432   [1.4B]   **6.** 6   [2.3B]

7. $1\frac{4}{15}$ [2.4B]   **8.** $\frac{7}{16}$ [2.5D]   **9.** 11.050   [3.2A]   **10.** 3.781   [3.3A]   **11.** $\frac{2}{5}$ [2.6A]   **12.** $\frac{5}{9}$ [2.7A]

13. 9.4   [3.4A]   **14.** 0.4   [3.5A]   **15.** 31   [1.6B]

SECTION 10.1

1. −120 ft **3.** −324 dollars **5.** **7.**
9. 1 **11.** −1 **13.** 3 **15. a.** A is −4. **b.** C is −2. **17. a.** A is −7. **b.** D is −4. **19.** −2 > −5 **21.** −16 < 1
23. 3 > −7 **25.** −11 < −8 **27.** 35 > 28 **29.** −42 < 27 **31.** 21 > −34 **33.** −27 > −39 **35.** −87 < 63
37. 86 > −79 **39.** −62 > −84 **41.** −131 < 101 **43.** −7, −2, 0, 3 **45.** −5, −3, 1, 4 **47.** −4, 0, 5, 9
49. −10, −7, −5, 4, 12 **51.** −11, −7, −2, 5, 10 **53.** Always true **55.** Sometimes true **57.** −16 **59.** 3 **61.** −45
63. 59 **65.** 88 **67.** 4 **69.** 9 **71.** 11 **73.** 12 **75.** 2 **77.** 6 **79.** 5 **81.** 1 **83.** −5 **85.** 16 **87.** 12
89. −29 **91.** −14 **93.** 15 **95.** −33 **97.** 32 **99.** −42 **101.** 61 **103.** −52 **105.** $|-12| > |8|$
107. $|6| < |13|$ **109.** $|-1| < |-17|$ **111.** $|17| = |-17|$ **113.** −9, −|6|, −4, |−7| **115.** −9, −|−7|, |4|, 5
117. −|10|, −|−8|, −3, |5| **119.** Positive integers **121.** Negative integers **123. a.** 8 and −2 are 5 units from 3.
b. 2 and −4 are 3 units from −1. **125.** −12 min and counting is closer to blastoff. **127.** The loss was greater during the first quarter.

SECTION 10.2

1. −14, −364 **3.** −2 **5.** 20 **7.** −11 **9.** −9 **11.** −3 **13.** 1 **15.** −5 **17.** −30 **19.** 9 **21.** 1
23. −10 **25.** −28 **27.** −41 **29.** −392 **31.** −20 **33.** −23 **35.** −2 **37.** −6 **39.** −6 **41.** Always true
43. Sometimes true **45.** Negative six minus positive four **47.** Positive six minus negative four **49.** 9 + 5 **51.** 1 + (−8)
53. 8 **55.** −7 **57.** −9 **59.** 36 **61.** −3 **63.** 18 **65.** −9 **67.** 11 **69.** 0 **71.** 11 **73.** 2 **75.** −138
77. 86 **79.** −337 **81.** 4 **83.** −12 **85.** −12 **87.** 3 **89.** Never true **91.** Sometimes true
93. The difference between the temperatures is 11°C. **95.** The temperature is −1°C. **97.** The new temperature is above 0°C.
99. Nick's score was −15 points after his opponent shot the moon. **101.** The change in the price of the stock is −9 dollars.
103. The difference between the temperatures is 115°F. **105.** The difference in elevation is 6051 m. **107.** The difference is greatest in Asia. **109.** The difference is 82°C. **111.** The largest difference that can be obtained is 22. The smallest positive difference is 2. **113.** The possible pairs of integers are −7 and −1, −6 and −2, −5 and −3, and −4 and −4.

SECTION 10.3

1. Subtraction **3.** Multiplication **5.** 42 **7.** −24 **9.** 6 **11.** 18 **13.** −20 **15.** −16 **17.** 25 **19.** 0
21. −72 **23.** −102 **25.** 140 **27.** −228 **29.** −320 **31.** −156 **33.** −70 **35.** 162 **37.** 120 **39.** 36
41. 192 **43.** −108 **45.** −2100 **47.** 20 **49.** −48 **51.** 140 **53.** Negative **55.** Zero **57.** 3(−12) = −36
59. −5(11) = −55 **61.** −2 **63.** 8 **65.** 0 **67.** −9 **69.** −9 **71.** 9 **73.** −24 **75.** −12 **77.** 31 **79.** 17
81. 15 **83.** −13 **85.** −18 **87.** 19 **89.** 13 **91.** −19 **93.** 17 **95.** 26 **97.** 23 **99.** 25 **101.** −34
103. 11 **105.** −13 **107.** 13 **109.** 12 **111.** −14 **113.** −14 **115.** −6 **117.** −4 **119.** Never true
121. Always true **123.** The average high temperature was −4°F. **125.** False **127.** The average score was −2.
129. The average score is −10. **131.** The wind chill factor is −45°F. **133. a.** 25 **b.** −17 **135. a.** True **b.** True

SECTION 10.4

1. $-\frac{5}{24}$ **3.** $-\frac{19}{24}$ **5.** $\frac{5}{26}$ **7.** $\frac{7}{24}$ **9.** $-\frac{19}{60}$ **11.** $-1\frac{3}{8}$ **13.** $\frac{3}{4}$ **15.** $-\frac{47}{48}$ **17.** $\frac{3}{8}$ **19.** $-\frac{7}{60}$ **21.** $\frac{13}{24}$ **23.** −3.4
25. −8.89 **27.** −8.0 **29.** −0.68 **31.** −181.51 **33.** 2.7 **35.** −20.7 **37.** −37.19 **39.** −34.99 **41.** 6.778
43. −8.4 **45.** Negative **47.** $\frac{1}{21}$ **49.** $\frac{2}{9}$ **51.** $-\frac{2}{9}$ **53.** $-\frac{45}{328}$ **55.** $\frac{2}{3}$ **57.** $\frac{15}{64}$ **59.** $-\frac{3}{7}$ **61.** $-1\frac{1}{9}$ **63.** $\frac{5}{6}$
65. $-2\frac{2}{7}$ **67.** $-13\frac{1}{2}$ **69.** 31.15 **71.** −112.97 **73.** 0.0363 **75.** 97 **77.** 2.2 **79.** −4.14 **81.** −3.8
83. −22.70 **85.** 0.07 **87.** 0.55 **89.** False **91.** The temperature fell 32.22°C in 27 min. **93.** The difference is greater than 4.8. **95.** The difference is 14.06°C. **97. a.** The closing price for General Mills was $66.72. **b.** The closing price for Hormel Foods was $35.46. **99.** From brightest to least bright, the stars are Betelgeuse, Polaris, Vega, Sirius, Sun.
101. The distance modulus for Polaris is 5.19. **103.** Betelgeuse is farthest from Earth. **105.** Answers will vary. $-\frac{17}{24}$ is one example.

SECTION 10.5

1. Less than 1   **3.** 2.37×10^6 **5.** 4.5×10^{-4} **7.** 3.09×10^5 **9.** 6.01×10^{-7} **11.** 5.7×10^{10} **13.** 1.7×10^{-8}
15. 710,000   **17.** 0.000043   **19.** 671,000,000   **21.** 0.00000713   **23.** 5,000,000,000,000   **25.** 0.00801
27. 1.6×10^{10} mi   **29.** $\$3.1 \times 10^{12}$ **31.** 9.8×10^9 **33.** 1.6×10^{-19} coulomb   **35.** 1×10^{-12} **37.** 4
39. 0   **41.** 12   **43.** −6   **45.** −5   **47.** 2   **49.** −3   **51.** 1   **53.** 2   **55.** −3   **57.** 14   **59.** −4   **61.** 33
63. −13   **65.** −12   **67.** 17   **69.** 0   **71.** 30   **73.** 94   **75.** −8   **77.** 39   **79.** 22   **81.** 0.21   **83.** −0.96
85. −0.29   **87.** −1.76   **89.** 2.1   **91.** $\frac{3}{16}$ **93.** $\frac{7}{16}$ **95.** $-\frac{5}{16}$ **97.** $-\frac{5}{8}$ **99.** (i)   **101. a.** $3.45 \times 10^{-14} > 3.45 \times 10^{-15}$
b. $5.23 \times 10^{18} > 5.23 \times 10^{17}$ **c.** $3.12 \times 10^{12} > 3.12 \times 10^{11}$ **103. a.** 100   **b.** −100   **c.** 225   **d.** −225
107. a. 1.99×10^{30} kg   **b.** 1.67×10^{-27} kg

CHAPTER 10 CONCEPT REVIEWS*

1. To find two numbers that are 6 units from 4, move 6 units to the right of 4 and 6 units to the left of 4. The two numbers are 10 and −2.   [10.1B]

2. The absolute value of −6 is the distance from 0 to −6 on the number line. The absolute value of −6 is 6.   [10.1B]

3. Rule for adding two integers:

To add numbers with the same sign, add the absolute values of the numbers. Then attach the sign of the addends.

To add numbers with different signs, find the difference between the absolute values of the numbers. Then attach the sign of the addend with the greater absolute value.   [10.2A]

4. The rule for subtracting two integers is to add the opposite of the second number to the first number.   [10.2B]

5. To find the change in temperature from −5°C to −14°C, use subtraction.   [10.2C]

6. $4 − 9 = 4 + (−9)$. To show this addition on the number line, start at 4 on the number line. Draw an arrow 9 units long pointing to the left, with its tail at 4. The head of the arrow is at −5. $4 + (−9) = −5$.   [10.2A]

7. If you multiply two nonzero numbers with different signs, the product is negative.   [10.3A]

8. If you divide two nonzero numbers with the same sign, the quotient is positive.   [10.3B]

9. When a number is divided by zero, the result is undefined because division by zero is undefined.   [10.3B]

10. A terminating decimal is one that ends. For example, 0.75 is a terminating decimal.   [10.4A]

11. The steps in the Order of Operations Agreement are:

1. Do all operations inside parentheses.

2. Simplify any number expressions containing exponents.

3. Do multiplication and division as they occur from left to right.

4. Rewrite subtraction as addition of the opposite. Then do additions as they occur from left to right.   [10.5B]

12. To write the number 0.000754 in scientific notation, move the decimal point to the right of the 7. Because we move the decimal point 4 places to the right, the exponent on 10 is −4. 0.000754 written in scientific notation is 7.54×10^{-4}.   [10.5A]

CHAPTER 10 REVIEW EXERCISES

1. −22 [10.1B]   **2.** 1 [10.2B]   **3.** $-\frac{5}{24}$ [10.4A]   **4.** 0.21 [10.4A]   **5.** $\frac{8}{25}$ [10.4B]   **6.** −1.28 [10.4B]

7. 10 [10.5B]   **8.** $-\frac{7}{18}$ [10.5B]   **9.** 4 [10.1B]   **10.** $0 > −3$ [10.1A]   **11.** −6 [10.1B]   **12.** 6 [10.3B]

13. $\frac{17}{24}$ [10.4A]   **14.** $-\frac{1}{4}$ [10.4B]   **15.** $1\frac{5}{8}$ [10.4B]   **16.** 24 [10.5B]   **17.** −26 [10.2A]   **18.** 2 [10.5B]

19. 3.97×10^{-5} [10.5A]   **20.** −0.08 [10.4B]   **21.** $\frac{1}{36}$ [10.4A]   **22.** $\frac{3}{40}$ [10.4B]   **23.** −0.042 [10.4B]

24. $\frac{1}{3}$ [10.5B]   **25.** 5 [10.1B]   **26.** $−2 > −40$ [10.1A]   **27.** −26 [10.3A]   **28.** 1.33 [10.5B]

29. $-\frac{1}{4}$ [10.4A]   **30.** −7.4 [10.4A]   **31.** $\frac{15}{32}$ [10.4B]   **32.** 240,000 [10.5A]   **33.** The temperature is −4°. [10.2C]

34. The student's score was 98. [10.3C]   **35.** The difference between the boiling and melting points is 395.45°C. [10.4C]

Note: The numbers in brackets following the answers in the Concept Review are a reference to the objective that corresponds to that problem. For example, the reference [1.2A] stands for Section 1.2, Objective A. This notation will be used for all Prep Tests, Concept Reviews, Chapter Reviews, Chapter Tests, and Cumulative Reviews throughout the text.

CHAPTER 10 TEST

1. 3   [10.2B, Example 4]   **2.** −2   [10.1B, Example 6]   **3.** $\frac{1}{15}$ [10.4A, HOW TO 1]   **4.** −0.0608   [10.4B, Example 7]

5. −8 > −10   [10.1A, Example 3]   **6.** −1.88   [10.4A, HOW TO 3]   **7.** −26   [10.5B, Example 6]

8. 90   [10.3A, Example 3]   **9.** −4.014   [10.4A, Example 3]   **10.** −9   [10.3B, Example 5]   **11.** −7   [10.2A, Example 2]

12. $\frac{7}{24}$ [10.4A, HOW TO 1]   **13.** 8.76×10^{10} [10.5A, Example 1]   **14.** −48   [10.3A, Example 2]

15. 0   [10.3B, Example 7]   **16.** 10   [10.2B, Example 5]   **17.** $-\frac{4}{5}$ [10.4B, You Try It 6]   **18.** 0 > −4   [10.1A, Example 3]

19. −14   [10.2A, Example 2]   **20.** −4   [10.5B, Example 6]   **21.** $\frac{3}{10}$ [10.4A, HOW TO 2]   **22.** 0.00000009601

[10.5A, Example 2]   **23.** 3.4   [10.4B, Example 9]   **24.** $\frac{1}{12}$ [10.4A, Example 2]   **25.** $\frac{1}{12}$ [10.4B, Example 5]

26. 3.213   [10.4A, HOW TO 4]   **27.** The temperature is 7°C.   [10.2C, Example 6]   **28.** The melting point of oxygen is

−213°C.   [10.3C, You Try It 8]   **29.** The temperature fell 46.62°C.   [10.4C, Example 10]   **30.** The average low temperature

was −2°F.   [10.3C, Example 9]

CUMULATIVE REVIEW EXERCISES

1. 0   [1.6B]   **2.** $4\frac{13}{14}$ [2.5C]   **3.** $2\frac{7}{12}$ [2.7B]   **4.** $1\frac{2}{7}$ [2.8C]   **5.** 1.80939   [3.3A]   **6.** 18.67   [4.3B]
7. 13.75   [5.4A]   **8.** 1 gal 3 qt   [8.3A]   **9.** 6.692 L   [9.3A]   **10.** 1.28 m   [9.5A]   **11.** 57.6   [5.2A]
12. 340%   [5.1B]   **13.** −3   [10.2A]   **14.** $-3\frac{3}{8}$ [10.4A]   **15.** $-10\frac{13}{24}$ [10.4A]   **16.** 19   [10.5B]
17. 3.488   [10.4B]   **18.** $31\frac{1}{2}$ [10.4B]   **19.** −7   [10.3B]   **20.** $\frac{25}{42}$ [10.4B]   **21.** −4   [10.5B]   **22.** −4   [10.5B]
23. The length remaining is $2\frac{1}{3}$ ft.   [2.5D]   **24.** Nimisha's new balance is $803.31.   [6.7A]   **25.** The percent decrease
is 27.3%.   [6.2C]   **26.** The amount of coffee that should be prepared is 10 gal.   [8.3C]   **27.** The dividend per share after
the increase was $1.68.   [6.2A]   **28. a.** 6.84 million households have a family night once a week.   **b.** $\frac{6}{25}$ of U.S.
households rarely or never have a family night.   **c.** The number of households that have a family night only once a month is
more than three times the number that have a family night once a week.   [7.1B]   **29.** 600,000 people would vote.   [4.3C]
30. The average high temperature was −4°F.   [10.3C]

ANSWERS TO CHAPTER 11 SELECTED EXERCISES

PREP TEST
1. −7   [10.2B]   **2.** −20   [10.3A]   **3.** 0   [10.2A]   **4.** 1   [10.3B]   **5.** 1   [10.4B]   **6.** $\frac{1}{15}$ [2.8B]   **7.** $\frac{19}{24}$ [2.8C]
8. 4   [10.5B]   **9.** −21   [10.5B]

SECTION 11.1
1. −33   **3.** −12   **5.** −4   **7.** −3   **9.** −22   **11.** −40   **13.** −1   **15.** 14   **17.** 32   **19.** 6   **21.** 7   **23.** 49
25. −24   **27.** 63   **29.** 9   **31.** 4   **33.** $-\frac{2}{3}$ **35.** 0   **37.** $-3\frac{1}{4}$ **39.** 6.7743   **41.** −16.1656   **43.** Positive
45. Positive   **47.** $2x^2, 3x, -4$ **49.** $3a^2, -4a, 8$ **51.** $3x^2, -4x$ **53.** $1y^2, 6a$ **55.** (ii), (iii)   **57.** 16z   **59.** 9m
61. 12at   **63.** 3yt   **65.** Unlike terms   **67.** $-2t^2$ **69.** $13c - 5$ **71.** −2t   **73.** $3y^2 - 2$ **75.** $14w - 8u$
77. $-23xy + 10$ **79.** $-11v^2$ **81.** $-8ab - 3a$ **83.** $4y^2 - y$ **85.** $-3a - 2b^2$ **87.** $7x^2 - 8x$ **89.** $-3s + 6t$
91. $-3m + 10n$ **93.** $5ab + 4ac$ **95.** $\frac{2}{3}a^2 + \frac{3}{5}b^2$ **97.** $1.56m - 3.77n$ **99.** (ii), (iii)   **101.** $5x + 20$ **103.** $4y - 12$
105. $-2a - 8$ **107.** $15x + 30$ **109.** $15c - 25$ **111.** $-3y + 18$ **113.** $7x + 14$ **115.** $4y - 8$ **117.** $5x + 24$
119. $y - 6$ **121.** $6n - 3$ **123.** $6y + 20$ **125.** $10x + 4$ **127.** $-6t - 6$ **129.** $z - 2$ **131.** $y + 6$ **133.** $9t + 15$
135. $5t + 24$ **137. a.** | 1 | 1 | x | x | x | **b.** | x | x | x | x | x | **c.** No

SECTION 11.2

1. Yes   **3.** No   **5.** Yes   **7.** Yes   **9.** Yes   **11.** Yes   **13.** Yes   **15.** Yes   **17.** No   **19.** True   **21.** 22   **23.** 1

25. -5 **27.** -4 **29.** 0   **31.** 3   **33.** -1 **35.** 15   **37.** -11 **39.** -10 **41.** -1 **43.** -1 **45.** $-\frac{5}{6}$

47. $-\frac{3}{4}$ **49.** The value of x must be negative.   **51.** The value of x must be positive.   **53.** 4   **55.** -4 **57.** -3 **59.** 8

61. 5   **63.** -6 **65.** 15   **67.** -8 **69.** -4 **71.** 6   **73.** -4 **75.** 6   **77.** 6   **79.** $-\frac{3}{7}$ **81.** $3\frac{1}{2}$ **83.** 15

85. True   **87.** False   **89.** Julio used 22.2 gal of gasoline on the trip.   **91.** The hatchback gets 37.3 mi/gal.   **93.** The amount of the original investment was \$15,000.   **95.** The value of the investment increased by \$3420.   **97.** The computer costs \$1820.   **99.** The crib costs \$187.50.   **103.** Answers will vary. For example, $x + 5 = 1$.

SECTION 11.3

1. 3   **3.** 5   **5.** -1 **7.** -4 **9.** 1   **11.** 3   **13.** -2 **15.** -3 **17.** -2 **19.** -1 **21.** 3   **23.** -1

25. 7   **27.** 2   **29.** 2   **31.** 0   **33.** 2   **35.** 0   **37.** $\frac{2}{3}$ **39.** $2\frac{5}{6}$ **41.** $-2\frac{1}{2}$ **43.** $1\frac{1}{2}$ **45.** 2   **47.** -2

49. $2\frac{1}{3}$ **51.** $-1\frac{1}{2}$ **53.** 10   **55.** 5   **57.** 36   **59.** -6 **61.** -9 **63.** $-4\frac{4}{5}$ **65.** $-8\frac{3}{4}$ **67.** 36   **69.** $1\frac{1}{3}$

71. 2   **73.** 4   **75.** 9   **77.** 4   **79.** -3.125 **81.** 1   **83.** -3 **85.** -9 **87.** Must be negative

89. Must be positive   **91.** The temperature is $-40°C$.   **93.** The time is 14.5 s.   **95.** 2500 units were made.

97. The clerk's monthly income is \$1800.   **99.** The total sales were \$32,000.   **101.** Miguel's commission rate was 4.5%.

SECTION 11.4

1. $\frac{1}{2}$ **3.** -1 **5.** -4 **7.** -1 **9.** -4 **11.** 3   **13.** -1 **15.** 2   **17.** 3   **19.** 0   **21.** -2 **23.** -1

25. $-\frac{2}{3}$ **27.** -4 **29.** $\frac{1}{3}$ **31.** $-\frac{3}{4}$ **33.** $\frac{1}{4}$ **35.** 0   **37.** $-1\frac{3}{4}$ **39.** 1   **41.** -1 **43.** -5 **45.** $\frac{5}{12}$

47. $-\frac{4}{7}$ **49.** $2\frac{1}{3}$ **51.** 21   **53.** 21   **55.** Positive   **57.** (iv)   **59.** 2   **61.** -1 **63.** -4 **65.** $-\frac{6}{7}$ **67.** 5

69. 3   **71.** 0   **73.** -1 **75.** 2   **77.** $1\frac{9}{10}$ **79.** 13   **81.** $2\frac{1}{2}$ **83.** 2   **85.** 0   **87.** $3\frac{1}{5}$ **89.** 4   **91.** $-6\frac{1}{2}$

93. $-\frac{3}{4}$ **95.** $-\frac{2}{3}$ **97.** -3 **99.** $-1\frac{1}{4}$ **101.** $2\frac{5}{8}$ **103.** 1.99   **105.** 48

SECTION 11.5

1. $y - 9$ **3.** $z + 3$ **5.** $\frac{2}{3}n + n$ **7.** $\frac{m}{m-3}$ **9.** $9(x+4)$ **11.** $x - \frac{x}{2}$ **13.** $\frac{z-3}{z}$ **15.** $2(t+6)$ **17.** $\frac{x}{9+x}$

19. $3(b+6)$ **21. a.** 3 more than twice x **b.** Twice the sum of x and 3   **23.** x^2 **25.** $\frac{x}{20}$ **27.** $4x$ **29.** $\frac{3}{4}x$ **31.** $4 + x$

33. $5x - x$ **35.** $x(x+2)$ **37.** $7(x+8)$ **39.** $x^2 + 3x$ **41.** $(x+3) + \frac{1}{2}x$ **43.** No   **47.** carbon: $\frac{1}{2}x$; oxygen: $\frac{1}{2}x$

SECTION 11.6

1. $x + 7 = 12$; 5   **3.** $3x = 18$; 6   **5.** $x + 5 = 3$; -2 **7.** $6x = 14$; $2\frac{1}{3}$ **9.** $\frac{5}{6}x = 15$; 18   **11.** $3x + 4 = 8$; $1\frac{1}{3}$

13. $\frac{1}{4}x - 7 = 9$; 64   **15.** $\frac{x}{9} = 14$; 126   **17.** $\frac{x}{4} - 6 = -2$; 16   **19.** $7 - 2x = 13$; -3 **21.** $9 - \frac{x}{2} = 5$; 8

23. $\frac{3}{5}x + 8 = 2$;   10   **25.** $\frac{x}{4.18} - 7.92 = 12.52$; 85.4392   **27.** No   **29.** x represents the amount the Army Reserve paid

in cash bonuses in 2003.   **31.** The Army Reserve paid \$53.8 million in cash bonuses in 2003.   **33.** The length of the Brooklyn Bridge is 486 m.   **35.** The value of the SUV last year was \$20,000.   **37.** Infants aged 3 months to 11 months sleep 12.7 h each day.   **39.** Five years ago the calculator cost \$96.   **41.** The recommended daily allowance of sodium is 2.5 g.

43. Americans consume 20 billion hot dogs annually.   **45.** About 11,065 plants and animals are known to be at risk of extinction in the world.   **47.** It took 3 h to install the water softener.   **49.** On average, U.S. workers take 13 vacation days per year.

51. The total sales for the month were \$42,540.

CHAPTER 11 CONCEPT REVIEWS*

1. To evaluate a variable expression, replace the variable or variables with numbers and then simplify the resulting numerical expression. [11.1A]

2. To add like terms, add the numerical coefficients. The variable part stays the same. For example, $6x + 7x = 13x$. [11.1B]

3. To simplify a variable expression containing parentheses, use the Distributive Property to remove the parentheses from the variable expression. Then combine like terms. [11.1C]

4. A solution of an equation is a number that, when substituted for the variable, results in a true equation. 6 is a solution of $x + 3 = 9$ because $6 + 3 = 9$. [11.2A]

5. To check the solution to an equation, substitute the solution into the original equation. Evaluate the resulting numerical expressions. Compare the results. If the results are the same, the given number is a solution. If the results are not the same, the given number is not a solution. [11.2B]

6. The Multiplication Property of Equations states that both sides of an equation can be multiplied by the same nonzero number without changing the solution of the equation. [11.2C]

7. The Addition Property of Equations states that the same number or variable expression can be added to each side of an equation without changing the solution of the equation. [11.2B]

8. To solve the equation $5x - 4 = 26$, apply both the Addition Property of Equations and the Multiplication Property of Equations. [11.3A]

9. After you substitute a number or numbers into a formula, solve the formula by using the Distributive Property if there are parentheses to remove. Then use the Addition and Multiplication Properties of Equations to rewrite the equation in the form *variable* = *constant* or *constant* = *variable*. [11.3B]

10. To isolate the variable in the equation $5x - 3 = 10 - 8x$, first add $8x$ to each side of the equation so that there is only one variable term. The result is $13x - 3 = 10$. Then add 3 to both sides of the equation so that there is only one constant term. The result is $13x = 13$. Next divide each side of the equation by the coefficient of x, 13. The result is $x = 1$. The variable is alone on the left side of the equation. The constant on the right side of the equation is the solution. The solution is 1. [11.4A]

11. Some mathematical terms that translate into addition are *more than, the sum of, the total of,* and *increased by.* [11.5A]

12. Some mathematical terms that translate into "equals" are *is, is equal to, amounts to,* and *represents.* [11.6A]

CHAPTER 11 REVIEW EXERCISES

1. $-2a + 2b$ [11.1C]   2. Yes [11.2A]   3. -4 [11.2B]   4. 7 [11.3A]   5. 13 [11.1A]   6. -9 [11.2C]

7. -18 [11.3A]   8. $-3x + 4$ [11.1C]   9. 5 [11.4A]   10. -5 [11.2B]   11. No [11.2A]   12. 6 [11.1A]

13. 5 [11.4B]   14. 10 [11.3A]   15. $-4bc$ [11.1B]   16. $-2\frac{1}{2}$ [11.4A]   17. $1\frac{1}{4}$ [11.2C]   18. $\frac{71}{30}x^2$ [11.1B]

19. $-\frac{1}{3}$ [11.4B]   20. $10\frac{4}{5}$ [11.3A]   21. The car averaged 23 mi/gal. [11.2D]   22. The temperature is 37.8°C. [11.3B]

23. $n + \frac{n}{5}$ [11.5A]   24. $(n + 5) + \frac{1}{3}n$ [11.5B]   25. The number is 2. [11.6A]   26. The number is 10. [11.6A]

27. The regular price of the MP3 video player is $380. [11.6B]   28. Last year's crop was 25,300 bushels. [11.6B]

CHAPTER 11 TEST

1. 95 [11.3A, HOW TO 1]   2. 26 [11.2B, HOW TO 2]   3. $-11x - 4y$ [11.1B, HOW TO 3]   4. $3\frac{1}{5}$ [11.4A, HOW TO 1]

5. -2 [11.3A, Example 1]   6. -38 [11.1A, HOW TO 1]   7. No [11.2A, Example 2]   8. $-14ab + 9$ [11.1B, HOW TO 2]   9. $-2\frac{4}{5}$ [11.2C, Example 6]   10. $8y - 7$ [11.1C, Example 9]   11. 0 [11.4B, Example 2]

12. $-\frac{1}{2}$ [11.3A, Example 2]   13. $-5\frac{5}{6}$ [11.1A, Example 3]   14. -16 [11.2C, Example 7]   15. -3 [11.3A, Example 2]

16. 11 [11.4B, Example 3]   17. The monthly payment is $137.50. [11.2D, You Try It 10]   18. 4000 clocks were made during the month. [11.3B, Example 4]   19. The time required is 11.5 s. [11.3B, Example 3]   20. $x + \frac{1}{3}x$ [11.5A, Example 1]

Note: The numbers in brackets following the answers in the Concept Review are a reference to the objective that corresponds to that problem. For example, the reference [1.2A] stands for Section 1.2, Objective A. This notation will be used for all Prep Tests, Concept Reviews, Chapter Reviews, Chapter Tests, and Cumulative Reviews throughout the text.

21. $5(x + 3)$ [11.5B, HOW TO 2]   **22.** $2x - 3 = 7; 5$ [11.6A, HOW TO 1]   **23.** The number is $-3\frac{1}{2}$.   [11.6A, HOW TO 1]

24. Eduardo's total sales for the month were $40,000.   [11.6B, You Try It 8]   **25.** The mechanic worked for 3 h.   [11.6B, Example 7]

CUMULATIVE REVIEW EXERCISES

1. 41   [1.6B]   **2.** $1\frac{7}{10}$ [2.5C]   **3.** $\frac{11}{18}$ [2.8C]   **4.** 0.047383   [3.4A]   **5.** $9.10/h   [4.2B]   **6.** 26.67   [4.3B]

7. $\frac{4}{75}$ [5.1A]   **8.** 140%   [5.3A]   **9.** 6.4   [5.4A]   **10.** 18 ft 9 in.   [8.1B]   **11.** 22 oz   [8.2A]   **12.** 0.282 g   [9.2A]

13. -1 [10.2A]   **14.** 19   [10.2B]   **15.** 6   [10.5B]   **16.** -48 [11.1A]   **17.** $-10x + 8z$ [11.1B]

18. $3y + 23$ [11.1C]   **19.** -1 [11.3A]   **20.** $-6\frac{1}{4}$ [11.4B]   **21.** $-7\frac{1}{2}$ [11.2C]   **22.** -21 [11.3A]

23. The percent is 17.6%.   [5.3B]   **24.** The price of the piece of pottery is $39.90.   [6.2B]   **25. a.** The discount is $81.
b. The discount rate is 18%.   [6.2D]   **26.** The simple interest due on the loan is $2933.33.   [6.3A]   **27.** The cost for fuel
was $150.   [5.2B]   **28.** The probability is $\frac{1}{8}$ that the sum of the upward faces on the two dice is 7.   [7.5A]   **29.** The total
sales were $32,500.   [11.3B]   **30.** The number is 2.   [11.6A]

ANSWERS TO CHAPTER 12 SELECTED EXERCISES

PREP TEST
1. 43   [11.2B]   **2.** 51   [11.2B]   **3.** 56   [1.6B]   **4.** 56.52   [11.1A]   **5.** 113.04   [11.1A]   **6.** 14.4   [4.3B]

SECTION 12.1
1. 0°; 90°   **3.** 180°   **5.** 30   **7.** 21   **9.** 14   **11.** 59°   **13.** 108°   **15.** 77°   **17.** 53°   **19.** Acute angle
21. Obtuse angle   **23.** 77°   **25.** 118°   **27.** 133°   **29.** 86°   **31.** 180°   **33.** Square   **35.** Circle   **37.** 102°
39. 90° and 45°   **41.** 14°   **43.** 90° and 65°   **45.** 8 in.   **47.** $4\frac{2}{3}$ ft   **49.** 7 cm   **51.** 2 ft 4 in.   **53.** False   **55.** True
57. $\angle a = 131°, \angle b = 49°$ **59.** $\angle a = 131°, \angle b = 49°$ **61.** $\angle a = 44°, \angle b = 44°$ **63.** $\angle a = 55°, \angle b = 125°$
65. $\angle a = 105°, \angle b = 75°$ **67.** $\angle a = 62°, \angle b = 118°$ **69.** True

SECTION 12.2
1. 56 in.   **3.** 20 ft   **5.** 92 cm   **7.** 47.1 cm   **9.** 9 ft 10 in.   **11.** 50.24 cm   **13.** 240 m   **15.** 157 cm
17. Perimeter of the square   **19.** 121 cm   **21.** 50.56 m   **23.** 3.57 ft   **25.** 139.3 m   **27.** Less than   **29.** The amount
of fencing needed is 60 ft.   **31.** The amount of binding needed is 24 ft.   **33.** The circumferene of the track was 990 ft.
35. Two packages of bias binding are needed.   **37.** The tricycle travels 25.12 ft.   **39.** The perimeter of the roller rink is 81.4 m.
41. The length of the weather stripping is 20.71 ft.   **43. a.** The circumference is two times larger.   **b.** The circumference is two
times larger.   **45.** The total perimeter is $14\frac{1}{4}$ cm.

SECTION 12.3
1. 144 ft²   **3.** 81 in²   **5.** 50.24 ft²   **7.** 20 in²   **9.** 2.13 cm²   **11.** 16 ft²   **13.** 817 in²   **15.** 154 in²
17. Equal to   **19.** 26 cm²   **21.** 2220 cm²   **23.** 150.72 in²   **25.** 8.851323 ft²   **27.** Equal to   **29.** 7500 yd² of artificial
turf must be purchased.   **31.** The area watered by the irrigation system is approximately 7850 ft².   **33.** You should buy 2 qt
of stain.   **35.** No, the expression cannot be used to calculate the area of the carpet.   **37.** Yes, the expression can be used to
calculate the area of the carpet.   **39.** It will cost $986.10 to carpet the area.   **41.** You should purchase 48 tiles.
43. The area is 68 m².   **45. a.** The area of the rink is more than 8000 ft².   **b.** 19,024 ft² of hardwood floor is needed to
cover the rink.   **47.** The cost is $1920.   **49. a.** If the length and width are doubled, the area is increased 4 times.
b. If the radius is doubled, the area is quadrupled.   **c.** If the diameter is doubled, the area is quadrupled.
51. a. Sometimes true   **b.** Sometimes true   **c.** Always true

SECTION 12.4

1. 144 cm³   **3.** 512 in³   **5.** 2143.57 in³   **7.** 150.72 cm³   **9.** 6.4 m³   **11.** 5572.45 mm³   **13.** 3391.2 ft³

15. $42\frac{7}{8}$ ft³   **17.** Sphere   **19.** 82.26 in³   **21.** 1.6688 m³   **23.** 69.08 in³   **25.** Increase   **27.** The volume of the

water in the tank is 40.5 m³.   **29.** The volume is approximately 17,148.59 ft³.   **31.** The volume of the portion of the silo

not being used for storage is 1507.2 ft³.   **33.** The lock contains 35,380,400 gal of water.   **35.** 20,680,704 people could be fed.

37. The tank will hold 15.0 gal.   **39.** The volume of the bushing is approximately 212.64 in³.   **41.** No, the expression cannot

be used to calculate the volume of the concrete floor.   **43.** Yes, the expression can be used to calculate the volume of the

concrete floor.   **45.** The cost of having the floor poured is $11,156.25.

SECTION 12.5

1. 2.646   **3.** 6.481   **5.** 12.845   **7.** 13.748   **9.** True   **11.** 5 in.   **13.** 8.602 cm   **15.** 11.180 ft   **17.** 4.472 cm

19. 12.728 yd   **21.** 10.392 ft   **23.** 21.213 cm   **25.** 8.944 m   **27.** 7.879 yd   **29.** A right triangle with hypotenuse of

length 50 units and a leg of length 40 units   **31.** The distance is 6.32 in.   **33.** You are 20 mi from your starting point.

35. The length of the diagonal is 8.7 m.   **37.** (i)   **39.** The distance is 4.243 in.   **41.** The distance is 250 ft.

43. The perimeter is 27.7 in.   **45.** The offset distance is $3\frac{3}{4}$ in.

SECTION 12.6

1. $\frac{1}{2}$ **3.** $\frac{3}{4}$ **5.** Congruent   **7.** Congruent   **9.** 7.2 cm   **11.** 3.3 m   **13.** True   **15.** The height of the building is 20.8 ft.

17. The perimeter of triangle *DEF* is 38 cm.   **19.** The area of triangle *DEF* is 49 m².

CHAPTER 12 CONCEPT REVIEW*

1. Perpendicular lines are intersecting lines that form right angles.   [12.1A]

2. Two angles are complementary when the sum of their measures is 90°.   [12.1A]

3. The sum of the measures of the three angles in a triangle is 180°.   [12.1B]

4. To find the radius of a circle when you know the diameter, multiply the diameter by $\frac{1}{2}$.   [12.1B]

5. The formula for the perimeter of a rectangle is $P = 2L + 2W$, where P is the perimeter, L is the length, and W is the width.   [12.2A]

6. To find the circumference of a circle, multiply π times the diameter or multiply 2π times the radius.   [12.2A]

7. The formula for the area of a triangle is $A = \frac{1}{2}bh$, where A is the area, b is the base, and h is the height of the triangle.   [12.3A]

8. To find the volume of a rectangular solid, you need to know the length, the width, and the height.   [12.4A]

9. A perfect square is the product of a whole number times itself. Therefore, a number is a perfect square if it is the square of a whole number.   [12.5A]

10. The hypotenuse of a right triangle is the side of the triangle that is opposite the right angle.   [12.5B]

11. In similar triangles, the ratios of corresponding sides are equal. Therefore, we can write proportions setting the ratios of corresponding sides equal to each other. If one side of a triangle is unknown, we can write a proportion using two ratios and then solve for the unknown side.   [12.6A]

12. The side-angle-side rule states that two triangles are congruent if two sides and the included angle of one triangle equal the corresponding sides and included angle of the second triangle.   [12.6A]

CHAPTER 12 REVIEW EXERCISES

1. 0.75 m   [12.1B]   **2.** 31.4 cm   [12.2A]   **3.** 26 ft   [12.2A]   **4.** 3   [12.1A]   **5.** 200 ft³   [12.4A]

6. 26 cm   [12.5B]   **7.** 75°   [12.1A]   **8.** 3.873   [12.5A]   **9.** 16 cm   [12.6A]   **10.** 63.585 cm²   [12.3A]

11. a. 45°   **b.** 135°   [12.1C]   **12.** 55 m²   [12.3A]   **13.** 240 in³   [12.4B]   **14.** 57.12 in²   [12.3B]

Note: The numbers in brackets following the answers in the Concept Review are a reference to the objective that corresponds to that problem. For example, the reference [1.2A] stands for Section 1.2, Objective A. This notation will be used for all Prep Tests, Concept Reviews, Chapter Reviews, Chapter Tests, and Cumulative Reviews throughout the text.

15. 267.9 ft³ [12.4A]   **16.** 64.8 m² [12.6B]   **17.** 47.7 in. [12.2B]   **18. a.** 80° **b.** 100° [12.1C]
19. The ladder will reach 15 ft up the building. [12.5C]   **20.** The other angles of the triangle are 90° and 58°. [12.1B]
21. The bicycle travels approximately 73.3 ft in 10 revolutions. [12.2C]   **22. a.** The area of the glass that must be cleaned
is 1600 ft². [12.3C] **b.** The pane of glass in the exhibit fills 1,440,000 in³. [12.4C]   **23.** The volume of the silo is approximately
1144.53 ft³. [12.4C]   **24.** The area is 11 m². [12.3A]   **25.** The distance from the starting point is 29 mi. [12.5C]

CHAPTER 12 TEST

1. 169.56 m³ [12.4A, Example 3]   **2.** 6.8 m [12.2A, Example 1]   **3.** 1406.72 cm³ [12.4B, Example 6]   **4.** 10 m

[12.6A, You Try It 5]   **5.** 58° [12.1A, Example 2]   **6.** $3\frac{1}{7}$ m² [12.3A, Example 1]   **7.** 150° [12.1C, Example 8]

8. 15.85 ft [12.2B, Example 4]   **9.** 13.748 [12.5A, Example 1]   **10.** 9.747 ft [12.5B, Example 3]   **11.** $10\frac{1}{8}$ ft²

[12.3B, You Try It 2]   **12.** $\angle a = 45°$; $\angle b = 135°$ [12.1C, Example 8]   **13.** $1\frac{1}{5}$ ft [12.6A, Example 2]   **14.** 90° and 50°

[12.1B, Example 4]   **15.** The width of the canal is 25 ft. [12.6B, Example 5]   **16.** The amount of extra pizza is 113.04 in².
[12.3C, Example 3]   **17.** It will cost $1113.69 to carpet the area. [12.3C, You Try It 3]   **18.** The length of the rafter is 15 ft.
[12.5C, Example 4]   **19.** The area is 103.82 ft².   [12.3C, Example 3]   **20. a.** The area of the floor of a cell is 45 ft².
[12.3C, Example 3] **b.** The volume of a cell is 315 ft³. [12.4C, Example 7]

CUMULATIVE REVIEW EXERCISES

1. 48 [2.1B]   **2.** $7\frac{41}{48}$ [2.4C]   **3.** $\frac{39}{56}$ [2.7B]   **4.** $\frac{2}{15}$ [2.8C]   **5.** $-\frac{1}{24}$ [10.4A]   **6.** $17.44/h [4.2B]
7. 37.5 [4.3B]   **8.** $\frac{3}{8}$ [5.1A]   **9.** −4 [11.1A]   **10.** 85 [5.4A]   **11.** −6 [11.3A]   **12.** $\frac{4}{3}$ [11.4B]
13. 32,500 m [9.1A]   **14.** 31.58 m [9.1A]   **15.** −15 [11.2C]   **16.** 2 [11.4B]   **17.** The monthly payment
is $708. [1.5D]   **18.** The sales tax on the home theater system is $47.06. [4.3C]   **19.** The original wage was $29.20. [5.4B]
20. The sale price of the PDA is $108. [6.2D]   **21.** The value of the investment after 20 years would be $101,366.50. [6.3C]
22. The weight of the package is 54 lb. [8.2C]   **23.** The distance between the rivets is 22.5 cm. [9.1B]   **24.** The number
is −2. [11.6A]   **25. a.** 74° **b.** 106° [12.1C]   **26.** 29.42 cm [12.2B]   **27.** 50 in² [12.3B]   **28.** 92.86 in³ [12.4B]
29. 10.63 ft [12.5B]   **30.** 36 cm [12.6B]

FINAL EXAM

1. 3259 [1.3B]   **2.** 53 [1.5C]   **3.** 60,205 [1.3B]   **4.** 16 [1.6B]   **5.** 144 [2.1A]   **6.** $1\frac{49}{120}$ [2.4B]
7. $3\frac{29}{48}$ [2.5C]   **8.** $6\frac{3}{14}$ [2.6B]   **9.** $\frac{4}{9}$ [2.7B]   **10.** $\frac{1}{6}$ [2.8B]   **11.** $\frac{1}{13}$ [2.8C]   **12.** 164.177 [3.2A]
13. 0.027918 [3.4A]   **14.** 0.69 [3.5A]   **15.** $\frac{9}{20}$ [3.6B]   **16.** 24.5 mi/gal [4.2B]   **17.** 54.9 [4.3B]
18. $\frac{9}{40}$ [5.1A]   **19.** 135% [5.1B]   **20.** 125% [5.1B]   **21.** 36 [5.2A]   **22.** $133\frac{1}{3}\%$ [5.3A]   **23.** 70 [5.4A]
24. 20 in. [8.1A]   **25.** 1 ft 4 in. [8.1B]   **26.** 2.5 lb [8.2A]   **27.** 6 lb 6 oz [8.2B]   **28.** 2.25 gal [8.3A]
29. 1 gal 3 qt [8.3B]   **30.** 248 cm [9.1A]   **31.** 4.62 m [9.1A]   **32.** 1.614 kg [9.2A]   **33.** 2067 ml [9.3A]
34. 88.55 km [9.5A]   **35.** The cost is $1.15. [9.4A]   **36.** 6.79×10^{-8} [10.5A]   **37.** 3.9 m [12.2A]
38. 45 in² [12.3A]   **39.** 1200 cm³ [12.4A]   **40.** −4 [10.2A]   **41.** −15 [10.2B]   **42.** $-\frac{1}{2}$ [10.4B]
43. $-\frac{1}{4}$ [10.4B]   **44.** 6 [10.5B]   **45.** $-x + 17$ [11.1C]   **46.** −18 [11.2C]   **47.** 5 [11.3A]   **48.** 1 [11.4A]
49. Your new balance is $959.93. [6.7A]   **50.** 63,750 people will vote. [4.3C]   **51.** One year ago the dividend
per share was $2.00. [5.4B]   **52.** The average monthly income is $3794. [7.4A]   **53.** The simple interest due is
$7200. [6.3A]   **54.** The probability is $\frac{1}{3}$. [7.5A]   **55.** The death count of China is 6.7% of the death count of the four
countries. [7.1B]   **56.** The discount rate for the Bose headphones is 28%. [6.2D]   **57.** The weight of the box is 81 lb.
[8.2C]   **58.** The perimeter is approximately 28.56 in. [12.2B]   **59.** The area is approximately 16.86 cm². [12.3B]
60. The number is 16. [11.6A]

Glossary

absolute value of a number The distance between zero and the number on the number line. [10.1]

acute angle An angle whose measure is between 0° and 90°. [12.1]

acute triangle A triangle that has three acute angles. [12.2]

addend In addition, one of the numbers added. [1.2]

addition The process of finding the total of two numbers. [1.2]

Addition Property of Zero Zero added to a number does not change the number. [1.2]

adjacent angles Two angles that share a common side. [12.1]

alternate exterior angles Two non-adjacent angles that are on opposite sides of the transversal and outside the parallel lines. [12.1]

alternate interior angles Two nonadjacent angles that are on opposite sides of the transversal and between the parallel lines. [12.1]

angle An angle is formed when two rays start at the same point; it is measured in degrees. [12.1]

approximation An estimated value obtained by rounding an exact value. [1.1]

area A measure of the amount of surface in a region. [12.3]

Associative Property of Addition Numbers to be added can be grouped (with parentheses, for example) in any order; the sum will be the same. [1.2]

Associative Property of Multiplication Numbers to be multiplied can be grouped (with parentheses, for example) in any order; the product will be the same. [1.4]

average The sum of all the numbers divided by the number of those numbers. [1.5]

average value The sum of all values divided by the number of those values; also known as the mean value. [7.4]

balancing a checkbook Determining whether the checking account balance is accurate. [6.7]

bank statement A document showing all the transactions in a bank account during the month. [6.7]

bar graph A graph that represents data by the height of the bars. [7.2]

base of a triangle The side that the triangle rests on. [12.1]

basic percent equation Percent times base equals amount. [5.2]

borrowing In subtraction, taking a unit from the next larger place value in the minuend and adding it to the number in the given place value in order to make that number larger than the number to be subtracted from it. [1.3]

box-and-whiskers plot A graph that shows the smallest value in a set of numbers, the first quartile, the median, the third quartile, and the greatest value. [7.4]

British thermal unit A unit of energy. 1 British thermal unit = 778 foot-pounds. [8.5]

broken-line graph A graph that represents data by the position of the lines and shows trends and comparisons. [7.2]

Calorie A unit of energy in the metric system. [9.4]

capacity A measure of liquid substances. [8.3]

carrying In addition, transferring a number to another column. [1.2]

center of a circle The point from which all points on the circle are equidistant. [12.1]

center of a sphere The point from which all points on the surface of the sphere are equidistant. [12.1]

centi- The metric system prefix that means one-hundredth. [9.1]

check A printed form that, when filled out and signed, instructs a bank to pay a specified sum of money to the person named on it. [6.7]

checking account A bank account that enables you to withdraw money or make payments to other people, using checks. [6.7]

circle A plane figure in which all points are the same distance from point O, which is called the center of the circle. [12.1]

circle graph A graph that represents data by the size of the sectors. [7.1]

circumference The distance around a circle. [12.2]

class frequency The number of occurrences of data in a class interval on a histogram; represented by the height of each bar. [7.3]

class interval Range of numbers represented by the width of a bar on a histogram. [7.3]

class midpoint The center of a class interval in a frequency polygon. [7.3]

commission That part of the pay earned by a salesperson that is calculated as a percent of the salesperson's sales. [6.6]

common factor A number that is a factor of two or more numbers is a common factor of those numbers. [2.1]

common multiple A number that is a multiple of two or more numbers is a common multiple of those numbers. [2.1]

Commutative Property of Addition Two numbers can be added in either order; the sum will be the same. [1.2]

Commutative Property of Multiplication Two numbers can be multiplied in either order; the product will be the same. [1.4]

complementary angles Two angles whose sum is 90°. [12.1]

composite geometric figure A figure made from two or more geometric figures. [12.2]

composite geometric solid A solid made from two or more geometric solids. [12.4]

composite number A number that has whole-number factors besides 1 and itself. For instance, 18 is a composite number. [1.7]

compound interest Interest computed not only on the original principal but also on interest already earned. [6.3]

congruent objects Objects that have the same shape and the same size. [12.6]

congruent triangles Triangles that have the same shape and the same size. [12.6]

constant term A term that has no variables. [11.1]

conversion rate A relationship used to change one unit of measurement to another. [8.1]

corresponding angles Two angles that are on the same side of the transversal and are both acute angles or are both obtuse angles. [12.1]

cost The price that a business pays for a product. [6.2]

cross product In a proportion, the product of the numerator on the left side of the proportion times the denominator on the right, and the product of the denominator on the left side of the proportion times the numerator on the right. [4.3]

cube A rectangular solid in which all six faces are squares. [12.1]

cubic centimeter A unit of capacity equal to 1 milliliter. [9.3]

cup A U.S. Customary measure of capacity. 2 cups = 1 pint. [8.3]

cylinder A geometric solid in which the bases are circles and are perpendicular to the height. [12.1]

data Numerical information. [7.1]

day A unit of time. 24 hours = 1 day. [8.4]

decimal A number written in decimal notation. [3.1]

decimal notation Notation in which a number consists of a whole-number part, a decimal point, and a decimal part. [3.1]

decimal part In decimal notation, that part of the number that appears to the right of the decimal point. [3.1]

decimal point In decimal notation, the point that separates the whole-number part from the decimal part. [3.1]

degree Unit used to measure angles; one complete revolution is 360°. [12.1]

denominator The part of a fraction that appears below the fraction bar. [2.2]

deposit slip A form for depositing money in a checking account. [6.7]

diameter of a circle A line segment with endpoints on the circle and going through the center. [12.1]

diameter of a sphere A line segment with endpoints on the sphere and going through the center. [12.1]

difference In subtraction, the result of subtracting two numbers. [1.3]

discount The difference between the regular price and the sale price. [6.2]

discount rate The percent of a product's regular price that is represented by the discount. [6.2]

dividend In division, the number into which the divisor is divided to yield the quotient. [1.5]

division The process of finding the quotient of two numbers. [1.5]

divisor In division, the number that is divided into the dividend to yield the quotient. [1.5]

double-bar graph A graph used to display data for purposes of comparison. [7.2]

down payment The percent of a home's purchase price that the bank, when issuing a mortgage, requires the borrower to provide. [6.4]

empirical probability The ratio of the number of observations of an event to the total number of observations. [7.5]

energy The ability to do work. [8.5]

equation A statement of the equality of two mathematical expressions. [11.2]

equilateral triangle A triangle that has three sides of equal length; the three angles are also of equal measure. [12.2]

equivalent fractions Equal fractions with different denominators. [2.3]

evaluating a variable expression Replacing the variable or variables with numbers and then simplifying the resulting numerical expression. [11.1]

event One or more outcomes of an experiment. [7.5]

expanded form The number 46,208 can be written in expanded form as $40{,}000 + 6000 + 200 + 0 + 8$. [1.1]

experiment Any activity that has an observable outcome. [7.5]

exponent In exponential notation, the raised number that indicates how many times the number to which it is attached is taken as a factor. [1.6]

exponential notation The expression of a number to some power, indicated by an exponent. [1.6]

factors In multiplication, the numbers that are multiplied. [1.4]

factors of a number The whole-number factors of a number divide that number evenly. [1.7]

favorable outcomes The outcomes of an experiment that satisfy the requirements of a particular event. [7.5]

finance charges Interest charges on purchases made with a credit card. [6.3]

first quartile In a set of numbers, the number below which one-quarter of the data lie. [7.4]

fixed-rate mortgage A mortgage in which the monthly payment remains the same for the life of the loan. [6.4]

fluid ounce A U.S. Customary measure of capacity. 8 fluid ounces = 1 cup. [8.3]

foot A U.S. Customary unit of length. 3 feet = 1 yard. [8.1]

foot-pound A U.S. Customary unit of energy. One foot-pound is the amount of energy required to lift 1 pound a distance of 1 foot. [8.5]

foot-pounds per second A U.S. Customary unit of power. [8.5]

formula An equation that expresses a relationship among variables. [11.2]

fraction The notation used to represent the number of equal parts of a whole. [2.2]

fraction bar The bar that separates the numerator of a fraction from the denominator. [2.2]

frequency polygon A graph that displays information similarly to a histogram. A dot is placed above the center of each class interval at a height corresponding to that class's frequency. [7.3]

gallon A U.S. Customary measure of capacity. 1 gallon = 4 quarts. [8.3]

geometric solid A figure in space. [12.1]

gram The basic unit of mass in the metric system. [9.2]

graph A display that provides a pictorial representation of data. [7.1]

graph of a whole number A heavy dot placed directly above that number on the number line. [1.1]

greater than A number that appears to the right of a given number on the number line is greater than the given number. [1.1]

greatest common factor (GCF) The largest common factor of two or more numbers. [2.1]

height of a parallelogram The distance between parallel sides. [12.1]

height of a triangle A line segment perpendicular to the base from the opposite vertex. [12.1]

histogram A bar graph in which the width of each bar corresponds to a range of numbers called a class interval. [7.3]

horsepower The U.S. Customary unit of power. 1 horsepower = 550 foot-pounds per second. [8.5]

hourly wage Pay calculated on the basis of a certain amount for each hour worked. [6.6]

hypotenuse The side opposite the right angle in a right triangle. [12.1]

improper fraction A fraction greater than or equal to 1. [2.2]

inch A U.S. Customary unit of length. 12 inches = 1 foot. [8.1]

integers The numbers . . . , −3, −2, −1, 0, 1, 2, 3, ... [10.1]

interest Money paid for the privilege of using someone else's money. [6.3]

interest rate The percent used to determine the amount of interest. [6.3]

interquartile range The difference between the third quartile and the first quartile. [7.4]

intersecting lines Lines that cross at a point in the plane. [12.1]

inverting a fraction Interchanging the numerator and denominator. [2.7]

isosceles triangle A triangle that has two sides of equal length; the angles opposite the equal sides are of equal measure. [12.2]

kilo- The metric system prefix that means one thousand. [9.1]

kilowatt-hour A unit of electrical energy in the metric system equal to 1000 watt-hours. [9.4]

least common denominator (LCD) The least common multiple of denominators. [2.4]

least common multiple (LCM) The smallest common multiple of two or more numbers. [2.1]

legs of a right triangle The two shortest sides of a right triangle. [12.1]

length A measure of distance. [8.1]

less than A number that appears to the left of a given number on the number line is less than the given number. [1.1]

license fees Fees charged for authorization to operate a vehicle. [6.5]

like terms Terms of a variable expression that have the same variable part. [11.1]

line A line extends indefinitely in two directions in a plane; it has no width. [12.1]

line segment Part of a line; it has two endpoints. [12.1]

liter The basic unit of capacity in the metric system. [9.3]

loan origination fee The fee a bank charges for processing mortgage papers. [6.4]

markup The difference between selling price and cost. [6.2]

markup rate The percent of a product's cost that is represented by the markup. [6.2]

mass The amount of material in an object. On the surface of Earth, mass is the same as weight. [9.2]

maturity value of a loan The principal of a loan plus the interest owed on it. [6.3]

mean The sum of all values divided by the number of those values; also known as the average value. [7.4]

measurement A measurement has both a number and a unit. Examples include 7 feet, 4 ounces, and 0.5 gallon. [8.1]

median The value that separates a list of values in such a way that there is the same number of values below the median as above it. [7.4]

meter The basic unit of length in the metric system. [9.1]

metric system A system of measurement based on the decimal system. [9.1]

mile A U.S. Customary unit of length. 5280 feet = 1 mile. [8.1]

milli- The metric system prefix that means one-thousandth. [9.1]

minuend In subtraction, the number from which another number (the subtrahend) is subtracted. [1.3]

minute A unit of time. 60 minutes = 1 hour. [8.4]

mixed number A number greater than 1 that has a whole-number part and a fractional part. [2.2]

mode In a set of numbers, the value that occurs most frequently. [7.4]

monthly mortgage payment One of 12 payments due each year to the lender of money to buy real estate. [6.4]

mortgage The amount borrowed to buy real estate. [6.4]

multiples of a number The products of that number and the numbers 1, 2, 3, ... [2.1]

multiplication The process of finding the product of two numbers. [1.4]

Multiplication Property of One The product of a number and 1 is the number. [1.4]

Multiplication Property of Zero The product of a number and zero is zero. [1.4]

negative integers The numbers . . . , −5, −4, −3, −2, −1. [10.1]

natural numbers The numbers 1, 2, 3, 4, 5, . . . ; also called the positive integers. [10.1]

negative numbers Numbers less than zero. [10.1]

number line A line on which a number can be graphed. [1.1]

numerator The part of a fraction that appears above the fraction bar. [2.2]

numerical coefficient The number part of a variable term. When the numerical coefficient is 1 or −1, the 1 is usually not written. [11.1]

obtuse angle An angle whose measure is between 90° and 180°. [12.1]

obtuse triangle A triangle that has one obtuse angle. [12.2]

opposite numbers Two numbers that are the same distance from zero on the number line, but on opposite sides. [10.1]

Order of Operations Agreement A set of rules that tells us in what order to perform the operations that occur in a numerical expression. [1.6]

ounce A U.S. Customary unit of weight. 16 ounces = 1 pound. [8.2]

parallel lines Lines that never meet; the distance between them is always the same. [12.1]

parallelogram A quadrilateral that has opposite sides equal and parallel. [12.1]

percent Parts per hundred. [5.1]

percent decrease A decrease of a quantity, expressed as a percent of its original value. [6.2]

percent increase An increase of a quantity, expressed as a percent of its original value. [6.2]

perfect square The product of a whole number and itself. [12.5]

perimeter The distance around a plane figure. [12.2]

period In a number written in standard form, each group of digits separated from other digits by a comma or commas. [1.1]

perpendicular lines Intersecting lines that form right angles. [12.1]

pictograph A graph that uses symbols to represent information. [7.1]

pint A U.S. Customary measure of capacity. 2 pints = 1 quart. [8.3]

place value The position of each digit in a number written in standard form determines that digit's place value. [1.1]

place-value chart A chart that indicates the place value of every digit in a number. [1.1]

plane A flat surface. [12.1]

plane figures Figures that lie totally in a plane. [12.1]

points A term banks use to mean percent of a mortgage; used to express the loan origination fee. [6.4]

polygon A closed figure determined by three or more line segments that lie in a plane. [12.2]

positive integers The numbers 1, 2, 3, 4, 5, . . . ; also called the natural numbers. [10.1]

positive numbers Numbers greater than zero. [10.1]

pound A U.S. Customary unit of weight. 1 pound = 16 ounces. [8.2]

power The rate at which work is done or energy is released. [8.5]

prime factorization The expression of a number as the product of its prime factors. [1.7]

prime number A number whose only whole-number factors are 1 and itself. For instance, 13 is a prime number. [1.7]

principal The amount of money originally deposited or borrowed. [6.3]

probability A number from 0 to 1 that tells us how likely it is that a certain outcome of an experiment will happen. [7.5]

product In multiplication, the result of multiplying two numbers. [1.4]

proper fraction A fraction less than 1. [2.2]

property tax A tax based on the value of real estate. [6.4]

proportion An expression of the equality of two ratios or rates. [4.3]

Pythagorean Theorem The square of the hypotenuse of a right triangle is equal to the sum of the squares of the two legs. [12.5]

quadrilateral A four-sided closed figure. [12.1]

quart A U.S. Customary measure of capacity. 4 quarts = 1 gallon. [8.3]

quotient In division, the result of dividing the divisor into the dividend. [1.5]

radius of a circle A line segment going from the center to a point on the circle. [12.1]

radius of a sphere A line segment going from the center to a point on the sphere. [12.1]

range In a set of numbers, the difference between the largest and smallest values. [7.4]

rate A comparison of two quantities that have different units. [4.2]

ratio A comparison of two quantities that have the same units. [4.1]

rational number A number that can be written as the ratio of two integers, where the denominator is not zero. [10.4]

ray A ray starts at a point and extends indefinitely in one direction. [12.1]

reciprocal of a fraction The fraction with the numerator and denominator interchanged. [2.7]

rectangle A parallelogram that has four right angles. [12.1]

rectangular solid A solid in which all six faces are rectangles. [12.1]

regular polygon A polygon in which each side has the same length and each angle has the same measure. [12.2]

remainder In division, the quantity left over when it is not possible to separate objects or numbers into a whole number of equal groups. [1.5]

repeating decimal A decimal in which a block of one or more digits repeats forever. [10.4]

right angle A 90° angle. [12.1]

right triangle A triangle that contains one right angle. [12.1]

rounding Giving an approximate value of an exact number. [1.1]

salary Pay based on a weekly, biweekly, monthly, or annual time schedule. [6.6]

sale price The reduced price. [6.2]

sales tax A tax levied by a state or municipality on purchases. [6.5]

sample space All the possible outcomes of an experiment. [7.5]

scalene triangle A triangle that has no sides of equal length; no two of its angles are of equal measure. [12.2]

scientific notation Notation in which a number is expressed as a product of two factors, one a number between 1 and 10 and the other a power of 10. [10.5]

second A unit of time. 60 seconds = 1 minute. [8.4]

sector of a circle One of the "pieces of the pie" in a circle graph. [7.1]

selling price The price for which a business sells a product to a customer. [6.2]

service charge An amount of money charged by a bank for handling a transaction. [6.7]

sides of a polygon The line segments that form the polygon. [12.2]

similar objects Objects that have the same shape but not necessarily the same size. [12.6]

similar triangles Triangles that have the same shape but not necessarily the same size. [12.6]

simple interest Interest computed on the original principal. [6.3]

simplest form of a fraction A fraction is in simplest form when there are no common factors in the numerator and denominator. [2.3]

simplest form of a rate A rate is in simplest form when the numbers that make up the rate have no common factor. [4.2]

simplest form of a ratio A ratio is in simplest form when the two numbers do not have a common factor. [4.1]

simplifying a variable expression Combining like terms by adding their numerical coefficients. [11.1]

solids Objects in space. [12.1]

solution of an equation A number that, when substituted for the variable, results in a true equation. [11.2]

solving an equation Finding a solution of the equation. [11.2]

sphere A solid in which all points are the same distance from point O, which is called the center of the sphere. [12.1]

square A rectangle that has four equal sides. [12.1]

square root A square root of a number is one of two identical factors of that number. [12.5]

standard form A whole number is in standard form when it is written using the digits 0, 1, 2, . . . , 9. An example is 46,208. [1.1]

statistics The branch of mathematics concerned with data, or numerical information. [7.1]

straight angle A 180° angle. [12.1]

subtraction The process of finding the difference between two numbers. [1.3]

subtrahend In subtraction, the number that is subtracted from another number (the minuend). [1.3]

sum In addition, the total of the numbers added. [1.2]

supplementary angles Two angles whose sum is 180°. [12.1]

terminating decimal A decimal that has a finite number of digits after the decimal point, which means that it comes to an end and does not go on forever. [10.4]

terms of a variable expression The addends of the expression. [11.1]

theoretical probability A fraction with the number of favorable outcomes of an experiment in the numerator and the total number of possible outcomes of the experiment in the denominator. [7.5]

third quartile In a set of numbers, the number above which one-quarter of the data lie. [7.4]

ton A U.S. Customary unit of weight. 1 ton = 2000 pounds. [8.2]

total cost The unit cost multiplied by the number of units purchased. [6.1]

transversal A line intersecting two other lines at two different points. [12.1]

triangle A three-sided closed figure. [12.1]

true proportion A proportion in which the fractions are equal. [4.3]

unit cost The cost of one item. [6.1]

unit rate A rate in which the number in the denominator is 1. [4.2]

variable A letter used to stand for a quantity that is unknown or that can change. [11.1]

variable expression An expression that contains one or more variables. [11.1]

variable part In a variable term, the variable or variables and their exponents. [11.1]

variable term A term composed of a numerical coefficient and a variable part. [11.1]

vertex The common endpoint of two rays that form an angle. [12.1]

vertical angles Two angles that are on opposite sides of the intersection of two lines. [12.1]

volume A measure of the amount of space inside a closed surface. [12.4]

watt-hour A unit of electrical energy in the metric system. [9.4]

week A unit of time. 7 days = 1 week. [8.4]

weight A measure of how strongly Earth is pulling on an object. [8.2]

whole numbers The whole numbers are 0, 1, 2, 3, ... [1.1]

whole-number part In decimal notation, that part of the number that appears to the left of the decimal point. [3.1]

yard A U.S. Customary unit of length. 36 inches = 1 yard. [8.1]

Index

Index of Applications